MATHEMATICS RESEARCH DEVELOPMENTS

UNDERSTANDING BANACH SPACES

MATHEMATICS RESEARCH DEVELOPMENTS

Additional books and e-books in this series can be found on Nova's website under the Series tab.

MATHEMATICS RESEARCH DEVELOPMENTS

UNDERSTANDING BANACH SPACES

DANIEL GONZÁLEZ SÁNCHEZ
EDITOR

Copyright © 2020 by Nova Science Publishers, Inc.

All rights reserved. No part of this book may be reproduced, stored in a retrieval system or transmitted in any form or by any means: electronic, electrostatic, magnetic, tape, mechanical photocopying, recording or otherwise without the written permission of the Publisher.

We have partnered with Copyright Clearance Center to make it easy for you to obtain permissions to reuse content from this publication. Simply navigate to this publication's page on Nova's website and locate the "Get Permission" button below the title description. This button is linked directly to the title's permission page on copyright.com. Alternatively, you can visit copyright.com and search by title, ISBN, or ISSN.

For further questions about using the service on copyright.com, please contact:
Copyright Clearance Center
Phone: +1-(978) 750-8400   Fax: +1-(978) 750-4470   E-mail: info@copyright.com.

NOTICE TO THE READER

The Publisher has taken reasonable care in the preparation of this book, but makes no expressed or implied warranty of any kind and assumes no responsibility for any errors or omissions. No liability is assumed for incidental or consequential damages in connection with or arising out of information contained in this book. The Publisher shall not be liable for any special, consequential, or exemplary damages resulting, in whole or in part, from the readers' use of, or reliance upon, this material. Any parts of this book based on government reports are so indicated and copyright is claimed for those parts to the extent applicable to compilations of such works.

Independent verification should be sought for any data, advice or recommendations contained in this book. In addition, no responsibility is assumed by the Publisher for any injury and/or damage to persons or property arising from any methods, products, instructions, ideas or otherwise contained in this publication.

This publication is designed to provide accurate and authoritative information with regard to the subject matter covered herein. It is sold with the clear understanding that the Publisher is not engaged in rendering legal or any other professional services. If legal or any other expert assistance is required, the services of a competent person should be sought. FROM A DECLARATION OF PARTICIPANTS JOINTLY ADOPTED BY A COMMITTEE OF THE AMERICAN BAR ASSOCIATION AND A COMMITTEE OF PUBLISHERS.

Additional color graphics may be available in the e-book version of this book.

Library of Congress Cataloging-in-Publication Data

Names: González Sánchez, Daniel, Escuela de Ciencias Físicas y Matemáticas,
   Universidad de Las Américas, Quito, Ecuador, editor.
Title: Understanding Banach Spaces
Description: New York: Nova Science Publishers, [2019] | Series: Mathematics Research Developments | Includes bibliographical references and index.
Identifiers: LCCN 2019954662 (print) | ISBN 9781536167450 (hardcover) |
   ISBN 9781536167467 (adobe pdf)

Published by Nova Science Publishers, Inc. † New York

CONTENTS

Preface ... vii

Chapter 1 Right Complex Caputo Fractional Inequalities ... 1
George A. Anastassiou

Chapter 2 Mixed Complex Fractional Inequalities ... 13
George A. Anastassiou

Chapter 3 Advanced Complex Fractional Ostrowski Inequalities ... 23
George A. Anastassiou

Chapter 4 Improved Qualitative Analysis for Newton-Like Methods with R-Order of Convergence at Least Three in Banach Spaces ... 35
Ioannis K. Argyros and Santhosh George

Chapter 5 Developments on the Convergence Region of Newton-Like Methods with Generalized Inverses in Banach Spaces ... 47
Ioannis K. Argyros and Santhosh George

Chapter 6 Modified Newton-Type Compositions for Solving Equations in Banach Spaces ... 57
Ioannis K. Argyros and Santhosh George

Chapter 7 Ball Convergence for Optimal Derivative Free Methods ... 71
Ioannis K. Argyros and Ramandeep Behl

Chapter 8 Ball Convergence for a Derivative Free Method with Memory ... 83
Ioannis K. Argyros and Ramandeep Behl

Chapter 9 Weaker Convergence Conditions of an Iterative Method for Nonlinear Ill-Posed Equations ... 97
Ioannis K. Argyros and Santhosh George

Chapter 10 Ball Convergence Theorem for a Fifth-Order Method in Banach Spaces ... 115
Ioannis K. Argyros and Santhosh George

Chapter 11	Extended Convergence of King-Werner-Like Methods without Derivatives *Ioannis K. Argyros and Santhosh George*	**125**
Chapter 12	On an Eighth-Order Steffensen-Type Solver Free of Derivatives *Ioannis K. Argyros and Santhosh George*	**137**
Chapter 13	Local Convergence of Osada's Method for Finding Zeros with Multiplicity *Ioannis K. Argyros and Santhosh George*	**147**
Chapter 14	Expanding the Applicability of an Eighth-Order Method in Banach Space under Weak Conditions *Ioannis K. Argyros, Ramandeep Behl, Daniel González Sánchez and S. S. Motsa*	**153**
Chapter 15	Local Convergence for Three-Step Eighth-Order Method under Weak Conditions *Ioannis K. Argyros, Ramandeep Behl and Daniel González Sánchez*	**165**
Chapter 16	Ball Convergence for a Three-Step One Parameter Efficient Method in Banach Space under Generalized Conditions *Ioannis K. Argyros, Ramandeep Behl and Daniel González Sánchez*	**177**
Chapter 17	On the Convergence of Newton-Moser Method from Data at One Point *José M. Gutiérrez and Miguel A. Hernández-Verón*	**189**
Chapter 18	Approximating Inverse Operators by a Fourth-Order Iterative Method *J. A. Ezquerro, Miguel A. Hernández-Verón and J. L. Varona*	**199**
Chapter 19	Initial Value Problems in Clifford Analysis Using Associated Spaces *David Armendáriz and Johan Ceballos and Antonio Di Teodoro*	**209**
Chapter 20	Computational Approach of Initial Value Problems in Clifford Analysis Using Associated Spaces *David Armendáriz, Johan Ceballos and Antonio Di Teodoro*	**231**
Chapter 21	Asymmetric Convexity and Smoothness: Quantitative Lyapunov Theorems, Lozanovskii Factorisation, Walsh Chaos and Convex Duality *Sergey S. Ajiev*	**245**
Chapter 22	Copies of Sequence Spaces and Basic Properties of Anisotropic Function Spaces *Sergey S. Ajiev*	**313**

Chapter 23	Interpolation and Bounded Extension of Hölder-Lipschitz Mappings between Function, Non-Commutative and Other Banach Spaces *Sergey S. Ajiev*	**369**
Chapter 24	Differential Equations with a Small Parameter in a Banach Space *Vasiliy Kachalov*	**415**
Chapter 25	Role of Hanson-Antczak--Type *V*-Invex Functions in Sufficient Efficiency Conditions for Semiinfinite Multiobjective Fractional Programming *Ram U. Verma*	**429**
About the Editor		**457**
Index		**459**
Related Nova Publications		**465**

PREFACE

This book is developed with the study of several properties of Banach spaces applied to diverse problems in functional and numerical analysis. Many problems in science, engineering and other disciplines can be expressed in form of equations, inequalities or systems of equations using mathematical modelling. In particular, a large number of these problems can be solved using these spaces.

The work includes twenty-five contributions grouped into several blocks, depending of the topics, using definitions, properties, algorithms and applications of specific cases in the use of Banach spaces.

Three chapters start the book establishing several important right complex Caputo type fractional inequalities of the following kinds: Ostrowski's, Poincare's, Sobolev's, Opial's and Hilbert-Pachpatte's. Besides, we present some important mixed generalized fractional complex analytic inequalities of Polya, Ostrowski and Poincare.

The problem of finding zeros of a nonlinear function or the solutions of a nonlinear equation is the main objective of the following fifteen chapters, the main block of the book. Although some of these equations are solved analytically, usually this is not the case and numerical approximations of the solutions of the equation must be sought, as a consequence of finding exact roots is difficult. For this, we usually appeal to iterative techniques that approximate the solutions of the nonlinear equations. Clearly, Newton's method stands out among the iterative techniques, since it is the most studied and used method in practice because of its simplicity, easy implementation and efficiency. The study about convergence matter of iterative procedures is usually based on two types: semilocal and local convergence analysis. The semilocal convergence matter is, based on the information around an initial point, to give conditions ensuring the convergence of the iterative procedure; while the local one is, based on the information around a solution, to find estimates of the radii of convergence balls. This method converges if the initial guess is close enough to the solution. In particular, the practice for finding a solution of equations is essentially connected to variants of Newton's method. Iterative methods of higher convergence order require the evaluation of the second Fréchet-derivative, which is very expensive in general. A lot of works change the conditions about the involved operators to ensure the convergence of the method and improve the domains of convergence and error bounds. Then, in these chapters we give several methods with different convergence orders and techniques analyzing the mentioned convergence types, radii of convergence, computational costs and other properties.

In the next two chapters, nineteenth and twentieth, we propose and solve an initial value problem where the initial data belong to an associated space in a Clifford algebra. The method of associated spaces allows to construct and guaranteed the existence of a fixed point from the contractivity in a Banach space. We present the advantage of using a matrix representation for the basis of a Clifford algebra, the result of the product rule for the derivatives of two Clifford valued function and the calculation of the conditions over the coefficients in order to construct the associated space.

The limits of the applicability of abstract methods of modern functional analysis in PDE, real analysis, optimization and other mathematical disciplines are governed by understanding the basic properties of particular Banach spaces employed. Proceeding from abstract to numerous specific Banach spaces and starting with the choice of geometrically friendly equivalent norms, the chapter twenty first lays the foundation for our quasi-Euclidean approach to understanding the local geometric properties of the majority of specific parameterised families of Banach spaces in use. Being also helpful in non-local geometry of Banach spaces, these results have found applications in the approximation of uniformly continuous mappings between Banach spaces by Hölder ones, isometric and bounded extension and interpolation of the latter and the Hölder classification of Banach spaces. Quantitative and occasionally sharp study of the uniform convexity and smoothness of convex functionals on abstract and specific Banach spaces in terms of our asymmetric and double moduli, their comparison to power functions and related convexity and smoothness classes of functionals and spaces, including a reverse triangle and multi-homogeneous inequalities, explicit Hölder continuity and quantitative monotonicity of the subgradient mapping, sharp counterparts of the Fenchel-Young-Lindenstrauss duality formulae, cosine and Babylonian/Pythagorean type theorems and explicit estimates for various groups of parameterised classes of specific spaces, leads to occasionally sharp immediate quantitative applications.

Employing both harmonic/real and functional analysis tools, we study in the twenty second chapter the existence of isomorphic, almost isometric and complemented copies of a number of basic sequence spaces and their finite-dimensional subspaces in various classes of spaces of functions defined on an open subset of an Euclidean space and extensively employed in various branches of mathematics. The spaces under consideration include the classes of anisotropic Besov, Lizorkin-Triebel and Sobolev spaces whose norms are defined in terms of averaged differences, local approximations by polynomials and wavelet decompositions with underlying mixed Lebesgue, Lorentz, or general ideal norms. Our main functional analysis tools are three theorems due to James, Krivine, Maurey and Pisier, lattice convexity and concavity, upper and lower estimates and interplay between them leading to the Rademacher type and cotype sets of the spaces under consideration, among which are simple examples of spaces with semi-closed type or cotype sets complementing the classical constructions based on the Tzirel'son space. The presence of the properties UMD, stability, separability and superreflexivity is also studied with applications to semi-embeddings.

In the twenty third chapter we expose the mostly precise relation between the geometric properties of pairs of abstract and concrete Banach and, occasionally, metric spaces and such isomorphic properties as the approximation, real interpolation and bounded extension of the Hölder-Lipschitz mappings between them, their subspaces and quotients in a quantitative manner, sharpening the uniform upper bounds for the extension operators even in the known setting of the pairs of Lebesgue spaces. The pairs of spaces are drawn from ten parameterised groups of Banach spaces including auxiliary classes, noncommutative spaces and various types of anisotropic Besov and Lizorkin-Triebel spaces of functions on open subsets with underlying mixed Lebesgue or Lorentz (quasi)norms, their duals and sums covering the majority of such spaces considered in analysis and PDE. The geometric properties, quantified in terms of asymmetric and abstract uniform convexity and smoothness, our generalised counterpart of Pisier's martingale inequality and the presence of certain subspaces, lead to the identification of the Markov type and cotype sets with explicit estimates of the related constants, allowing an explicit quantitative and mostly sharp adaptation of K. Ball's extension scheme.

Many studies have been dedicated to the study of differential equations with a small parameter. In the penultimate chapter, the twenty-fourth, regardless of whether a small parameter in the equation is introduced in a regular or singular way, as a rule asymptotic solutions are constructed. We find the conditions for the usual convergence of series in the powers of a small parameter, representing solutions of both linear and nonlinear equations in Banach spaces. In the case of singularly perturbed problems, the author of the regularization method S. A. Lomov called such solutions pseudo-analytical. When the problem is regularly perturbed, in some cases the analytical dependence of the solution on the parameter is guaranteed by the Poincare decomposition theorem.

Finally, in the last chapter, the twenty fifth, we discuss numerous sets of global parametric sufficient efficiency conditions under various Hanson-Antczak-type generalized assumptions for a semiinfinite multiobjective fractional programming problem. A semiinfinite multiobjective fractional programming problem with a finite number of variables and infinitely many constraints is called a semiinfinite programming problem. Problems of this nature have been utilized for the modelling and analysis of a wide range of theoretical as well as concrete, real-world, practical problems. More specifically, semiinfinite programming concepts and techniques have found relevance and applications in approximation theory, statistics, game theory, engineering design, boundary value problems, defect minimization for operator equations, geometry, random graphs, graphs related to Newton flows, wavelet analysis, reliability testing, environmental protection planning, decision making under uncertainty, semidefinite programming, geometric programming, disjunctive programming, optimal control problems, robotics and continuum mechanics, among other business and industry. Furthermore, utilizing two partitioning schemes, we establish several sets of generalized parametric sufficient efficiency results each of which is in fact a family of such results whose members can easily be identified by appropriate choices of certain sets and functions, which can be applied to various aspects of semiinfinite programming, including, but not limited to, optimality conditions, duality relations, and numerical algorithms, anticipatory systems and gene-environment networks, the analysis and implementation of Vasicek-type interest rate models, an interesting gemstone cutting problem. Plus, we observe the obtained results can be applied as a major research based on previous works on general approximation-solvability of nonlinear equations involving regular operators.

This book is aimed at researchers from various disciplines whose field of interest is related to functional and numerical analysis.

In: Understanding Banach Spaces
Editor: Daniel González Sánchez

ISBN: 978-1-53616-745-0
© 2020 Nova Science Publishers, Inc.

Chapter 1

RIGHT COMPLEX CAPUTO FRACTIONAL INEQUALITIES

George A. Anastassiou[*]
Department of Mathematical Sciences
University of Memphis, Memphis, TN, US

Abstract

Here we establish several important right complex Caputo type fractional inequalities of the following kinds: Ostrowski's, Poincare's, Sobolev's, Opial's and Hilbert–Pachpatte's.

Keywords: complex inequalities, fractional inequalities, right Caputo fractional derivative

AMS Subject Classification: 26D10, 26D15, 30A10

1. INTRODUCTION

Here we follow [1].

Suppose γ is a smooth path parametrized by $z(t)$, $t \in [a, b]$ and f is a complex function which is continuous on γ. Put $z(a) = u$ and $z(b) = w$ with $u, w \in \mathbb{C}$. We define the integral of f on $\gamma_{u,w} = \gamma$ as

$$\int_\gamma f(z)\,dz = \int_{\gamma_{u,w}} f(z)\,dz := \int_a^b f(z(t))\,z'(t)\,dt.$$

We observe that the actual choice of parametrization of γ does not matter.

This definition immediately extends to paths that are piecewise smooth. Suppose γ is parametrized by $z(t)$, $t \in [a, b]$, which is differentiable on the intervals $[a, c]$ and $[c, b]$, then assuming that f is continuous on γ we define

$$\int_{\gamma_{u,w}} f(z)\,dz := \int_{\gamma_{u,v}} f(z)\,dz + \int_{\gamma_{v,w}} f(z)\,dz,$$

---
[*]Corresponding Author's E-mail: ganastss@memphis.edu.

where $v := z(c)$. This can be extended for a finite number of intervals.

We also define the integral with respect to arc–length

$$\int_{\gamma_{u,w}} f(z)\, |dz| := \int_a^b f(z(t))\, |z'(t)|\, dt$$

and the length of the curve γ is then

$$l(\gamma) = \int_{\gamma_{u,w}} |dz| := \int_a^b |z'(t)|\, dt,$$

where $|\cdot|$ is the complex absolute value.

Let f and g be holomorphic in G, and open domain and suppose $\gamma \subset G$ is a piecewise smooth path from $z(a) = u$ to $z(b) = w$. Then we have the integration by parts formula

$$\int_{\gamma_{u,w}} f(z) g'(z)\, dz = f(w) g(w) - f(u) g(u) - \int_{\gamma_{u,w}} f'(z) g(z)\, dz. \tag{1}$$

We recall also the triangle inequality for the complex integral, namely

$$\left| \int_\gamma f(z)\, dz \right| \leq \int_\gamma |f(z)|\, |dz| \leq \|f\|_{\gamma,\infty}\, l(\gamma), \tag{2}$$

where $\|f\|_{\gamma,\infty} := \sup_{z \in \gamma} |f(z)|$. We also define the p–norm with $p \geq 1$ by

$$\|f\|_{\gamma,p} := \left(\int_\gamma |f(z)|^p\, |dz| \right)^{\frac{1}{p}}.$$

For $p = 1$ we have

$$\|f\|_{\gamma,1} := \int_\gamma |f(z)|\, |dz|.$$

If $p, q > 1$ with $\frac{1}{p} + \frac{1}{q} = 1$, then by Hölder's inequality we have

$$\|f\|_{\gamma,1} \leq [l(\gamma)]^{\frac{1}{q}} \|f\|_{\gamma,p}. \tag{3}$$

We are inspired by the following extensions of Stekloff and Almansi inequalities to the complex integral:

Theorem 1. *([1]) Let f be analytic in G, a domain of complex numbers and suppose $\gamma \subset G$ is a smooth path parametrized by $z(t)$, $t \in [a, b]$ from $z(a) = u$ to $z(b) = w$ and $z'(t) \neq 0$ for $t \in (a, b)$.*

(i) If $\int_\gamma f(z)\, |dz| = 0$, then

$$\int_\gamma |f(z)|^2\, |dz| \leq \frac{1}{\pi^2} l^2(\gamma) \int_\gamma |f'(z)|^2\, |dz|. \tag{4}$$

(ii) In addition, if $f(u) = f(w) = 0$, then

$$\int_\gamma |f(z)|^2\, |dz| \leq \frac{1}{4\pi^2} l^2(\gamma) \int_\gamma |f'(z)|^2\, |dz|. \tag{5}$$

We are also inspired by complex Ostrowski type results:

Theorem 2. *([2]) Let $f : D \subseteq \mathbb{C} \to \mathbb{C}$ be an analytic function on the convex domain D with $z, x, y \in D$ and $\lambda \in \mathbb{C}$. Suppose $\gamma \subset D$ is a smooth path parametrized by $z(t)$, $t \in [a, b]$ with $z(a) = u$ and $z(b) = w$ where $u, w \in D$. Then*

$$\int_\gamma f(z)\,dz = [(1-\lambda)f(x) + \lambda f(y)](w-u) + \tag{6}$$

$$(1-\lambda)\sum_{k=1}^n \frac{1}{(k+1)!} f^{(k)}(x)\left[(w-x)^{k+1} + (-1)^k(x-u)^{k+1}\right] +$$

$$\lambda \sum_{k=1}^n \frac{1}{(k+1)!} f^{(k)}(y)\left[(w-y)^{k+1} + (-1)^k(y-u)^{k+1}\right] + T_{n,\lambda}(\gamma, x, y),$$

where the remainder $T_{n,\lambda}(\gamma, x, y)$ is given by

$$T_{n,\lambda}(\gamma, x, y) :=$$

$$\frac{1}{n!}\left[(1-\lambda)\int_\gamma (z-x)^{n+1}\left(\int_0^1 f^{(n+1)}[(1-s)x + sz](1-s)^n\,ds\right)dz\right.$$

$$\left.+(-1)^{n+1}\lambda \int_\gamma (y-z)^{n+1}\left(\int_0^1 f^{(n+1)}[(1-s)z + sy]s^n\,ds\right)dz\right] = \tag{7}$$

$$\frac{1}{n!}\left[(1-\lambda)\int_0^1 (1-s)^n\left(\int_\gamma (z-x)^{n+1} f^{(n+1)}[(1-s)x + sz]\,dz\right)ds\right.$$

$$\left.+(-1)^{n+1}\lambda \int_0^1 s^n\left(\int_\gamma (y-z)^{n+1} f^{(n+1)}[(1-s)z + sy]\,dz\right)ds\right].$$

Estimations of the above remainder follow:

Theorem 3. *([2]) Let $f : D \subseteq \mathbb{C} \to \mathbb{C}$ be an analytic function on the convex domain D with $x, y \in D$ and $\lambda \in \mathbb{C}$. Suppose $\gamma \subset D$ is a smooth path parametrized by $z(t)$, $t \in [a, b]$ with $z(a) = u$ and $z(b) = w$ where $u, w \in D$. Then we have the representation (6) and the remainder $T_{n,\lambda}(\gamma, x, y)$ satisfies the inequalities*

$$|T_{n,\lambda}(\gamma, x, y)| \le$$

$$\frac{1}{n!}\left[|1-\lambda|\left|\int_\gamma |z-x|^{n+1}\left(\int_0^1 \left|f^{(n+1)}[(1-s)x + sz]\right|(1-s)^n\,ds\right)|dz|\right|\right.$$

$$\left.+|\lambda|\left|\int_\gamma |y-z|^{n+1}\left(\int_0^1 \left|f^{(n+1)}[(1-s)z + sy]\right|s^n\,ds\right)|dz|\right|\right] \tag{8}$$

$$\leq \frac{1}{n!} |1-\lambda| \begin{cases} \frac{1}{n+1} \int_\gamma |z-x|^{n+1} \left(\max_{s\in[0,1]} \left|f^{(n+1)}\left[(1-s)x+sz\right]\right|\right) |dz| \\ \frac{1}{(qn+1)^{\frac{1}{p}}} \int_\gamma |z-x|^{n+1} \left(\int_0^1 \left|f^{(n+1)}\left[(1-s)x+sz\right]\right|^p ds\right)^{\frac{1}{p}} |dz| \\ \text{where } p,q>1 \text{ and } \frac{1}{p}+\frac{1}{q}=1 \\ \int_\gamma |z-x|^{n+1} \left(\int_0^1 \left|f^{(n+1)}\left[(1-s)x+sz\right]\right| ds\right) |dz| \end{cases}$$

$$+\frac{1}{n!} |\lambda| \begin{cases} \frac{1}{n+1} \int_\gamma |y-z|^{n+1} \left(\max_{s\in[0,1]} \left|f^{(n+1)}\left[(1-s)z+sy\right]\right|\right) |dz| \\ \frac{1}{(qn+1)^{\frac{1}{p}}} \int_\gamma |y-z|^{n+1} \left(\int_0^1 \left|f^{(n+1)}\left[(1-s)z+sy\right]\right|^p ds\right)^{\frac{1}{p}} |dz| \\ \text{where } p,q>1 \text{ and } \frac{1}{p}+\frac{1}{q}=1 \\ \int_\gamma |y-z|^{n+1} \left(\int_0^1 \left|f^{(n+1)}\left[(1-s)z+sy\right]\right| ds\right) |dz|, \end{cases}$$

and

$$|T_{n,\lambda}(\gamma,x,y)| \leq$$

$$\frac{1}{n!}\left[|1-\lambda|\int_0^1 (1-s)^n \left(\int_\gamma |z-x|^{n+1} \left|f^{(n+1)}\left[(1-s)x+sz\right]\right| |dz|\right) ds + \right.$$

$$\left. |\lambda|\int_0^1 s^n \left(\int_\gamma |y-z|^{n+1} \left|f^{(n+1)}\left[(1-s)z+sy\right]\right| |dz|\right) ds\right] \leq$$

$$\frac{1}{n!} |1-\lambda| \begin{cases} \int_\gamma |z-x|^{n+1} |dz| \int_0^1 (1-s)^n \left(\max_{z\in\gamma} \left|f^{(n+1)}\left[(1-s)x+sz\right]\right|\right) ds \\ \left(\int_\gamma |z-x|^{(n+1)q} |dz|\right)^{\frac{1}{q}} \int_0^1 (1-s)^n \left(\int_\gamma \left|f^{(n+1)}\left[(1-s)x+sz\right]\right|^p |dz|\right)^{\frac{1}{p}} ds \\ \text{where } p,q>1 \text{ and } \frac{1}{p}+\frac{1}{q}=1 \\ \max_{z\in\gamma}\left(|z-x|^{n+1}\right) \int_0^1 (1-s)^n \left(\int_\gamma \left|f^{(n+1)}\left[(1-s)x+sz\right]\right| ds\right) |dz| \end{cases} \quad (9)$$

$$+\frac{1}{n!} |\lambda| \begin{cases} \int_\gamma |z-y|^{n+1} |dz| \int_0^1 s^n \left(\max_{z\in\gamma} \left|f^{(n+1)}\left[(1-s)z+sy\right]\right|\right) ds \\ \left(\int_\gamma |z-y|^{(n+1)q} |dz|\right)^{\frac{1}{q}} \int_0^1 s^n \left(\int_\gamma \left|f^{(n+1)}\left[(1-s)z+sy\right]\right|^p |dz|\right)^{\frac{1}{p}} ds \\ \text{where } p,q>1 \text{ and } \frac{1}{p}+\frac{1}{q}=1 \\ \max_{z\in\gamma}\left(|z-y|^{n+1}\right) \int_0^1 s^n \left(\int_\gamma \left|f^{(n+1)}\left[(1-s)z+sy\right]\right| |dz|\right) ds. \end{cases}$$

In this work we utilize on $\mathbb{C}$ the results of [3] which are for general Banach space valued functions.

Mainly we give different cases of the right fractional $\mathbb{C}$–Ostrowski type inequality and we continue with the right fractional: $\mathbb{C}$–Poincaré like and Sobolev like inequalities.

We present an Opial type right $\mathbb{C}$–fractional inequality, and we finish with the Hilbert–Pachpatte right $\mathbb{C}$–fractional inequalities.

2. BACKGROUND

In this section all integrals are of Bochner type.
   We need:

Definition 4. *(see [4]) A definition of the Hausdorff measure h_α goes as follows: if (T, d) is a metric space, $A \subseteq T$ and $\delta > 0$, let $\Lambda(A, \delta)$ be the set of all arbitrary collections $(C)_i$ of subsets of T, such that $A \subseteq \cup_i C_i$ and $\operatorname{diam}(C_i) \leq \delta$ (diam =diameter) for every i. Now, for every $\alpha > 0$ define*

$$h_\alpha^\delta(A) := \inf \left\{ \sum (\operatorname{diam} C_i)^\alpha \,|\, (C_i)_i \in \Lambda(A, \delta) \right\}. \tag{10}$$

Then there exists $\lim_{\delta \to 0} h_\alpha^\delta(A) = \sup_{\delta > 0} h_\alpha^\delta(A)$, and $h_\alpha(A) := \lim_{\delta \to 0} h_\alpha^\delta(A)$ gives an outer measure on the power set $\mathcal{P}(T)$, which is countably additive on the σ–field of all Borel subsets of T. If $T = \mathbb{R}^n$, then the Hausdorff measure h_n, restricted to the σ–field of the Borel subsets of $\mathbb{R}^n$, equals the Lebesgue measure on $\mathbb{R}^n$ up to a constant multiple. In particular, $h_1(C) = \mu(C)$ for every Borel set $C \subseteq \mathbb{R}$, where μ is the Lebesgue measure.

Definition 5. *([3]) Let $[a, b] \subset \mathbb{R}$, $(X, \|\cdot\|)$ be a Banach space, $\alpha > 0$, $m := \lceil \alpha \rceil$, ($\lceil \cdot \rceil$ the ceiling of the number). We assume that $f^{(m)} \in L_1([a, b], X)$, where $f : [a, b] \to X$. We call the Caputo–Bochner right fractional derivative of order α:*

$$\left(D_{b-}^\alpha f\right)(x) := \frac{(-1)^m}{\Gamma(m - \alpha)} \int_x^b (J - x)^{m-\alpha-1} f^{(m)}(J) \, dJ, \quad \forall\, x \in [a, b], \tag{11}$$

where $f^{(m)}$ is the ordinary X–valued derivative, defined similarly to the numerical one.
   We observe that $D_{b-}^m f(x) = (-1)^m f^{(m)}(x)$, for $m \in \mathbb{N}$, and $D_{b-}^0 f(x) = f(x)$.

   By [3] $\left(D_{b-}^\alpha f\right)(x)$ exists almost everywhere on $[a, b]$ and $\left(D_{b-}^\alpha f\right) \in L_1([a, b], X)$.
   If $\|f^{(m)}\|_{L_\infty([a,b],X)} < \infty$, and $\alpha \notin \mathbb{N}$, then by [3], $D_{b-}^\alpha f \in C([a, b], X)$, hence $\|D_{b-}^\alpha f\| \in C([a, b])$.
   We need the right–fractional Taylor's formula:

Theorem 6. *([3]) Let $[a, b] \subset \mathbb{R}$, X be a Banach space, $\alpha > 0$, $m = \lceil \alpha \rceil$, $f \in C^{m-1}([a, b], X)$. Set*

$$F_x(t) := \sum_{i=0}^{m-1} \frac{(x - t)^i}{i!} f^{(i)}(t), \quad \forall\, t \in [x, b], \tag{12}$$

where $x \in [a, b]$.
   Assume that $f^{(m)}$ exists outside a μ–null Borel set $B_x \subseteq [x, b]$, such that

$$h_1(F_x(B_x)) = 0, \quad \forall\, x \in [a, b]. \tag{13}$$

We also assume that $f^{(m)} \in L_1([a, b], X)$. Then

$$f(x) = \sum_{i=0}^{m-1} \frac{(x - b)^i}{i!} f^{(i)}(b) + \frac{1}{\Gamma(\alpha)} \int_x^b (z - x)^{\alpha-1} \left(D_{b-}^\alpha f\right)(z) \, dz, \tag{14}$$

$\forall\, x \in [a, b]$.

Next we mention Ostrowski type inequalities at right fractional level for Banach valued functions. See also [5].

Theorem 7. *([3]) Let $\alpha > 0$, $m = \lceil \alpha \rceil$. Here all as in Theorem 6. Assume $f^{(k)}(b) = 0$, $k = 1, \ldots, m-1$, and $D_{b-}^{\alpha} f \in L_{\infty}([a,b], X)$. Then*

$$\left\| \frac{1}{b-a} \int_a^b f(x)\, dx - f(b) \right\| \leq \frac{\|D_{b-}^{\alpha} f\|_{L_{\infty}([a,b],X)}}{\Gamma(\alpha+2)} (b-a)^{\alpha}. \tag{15}$$

We also give:

Theorem 8. *([3]) Let $\alpha \geq 1$, $m = \lceil \alpha \rceil$. Here all as in Theorem 6. Assume that $f^{(k)}(b) = 0$, $k = 1, \ldots, m-1$, and $D_{b-}^{\alpha} f \in L_1([a,b], X)$. Then*

$$\left\| \frac{1}{b-a} \int_a^b f(x)\, dx - f(b) \right\| \leq \frac{\|D_{b-}^{\alpha} f\|_{L_1([a,b],X)}}{\Gamma(\alpha+1)} (b-a)^{\alpha-1}. \tag{16}$$

We mention also an L_p abstract Ostrowski type inequality:

Theorem 9. *([3]) Let $p, q > 1 : \frac{1}{p} + \frac{1}{q} = 1$, $\alpha > \frac{1}{q}$, $m = \lceil \alpha \rceil$. Here all as in Theorem 6. Assume that $f^{(k)}(b) = 0$, $k = 1, \ldots, m-1$, and $D_{b-}^{\alpha} f \in L_q([a,b], X)$. Then*

$$\left\| \frac{1}{b-a} \int_a^b f(x)\, dx - f(b) \right\| \leq \frac{\|D_{b-}^{\alpha} f\|_{L_q([a,b],X)}}{\Gamma(\alpha)(p(\alpha-1)+1)^{\frac{1}{p}} \left(\alpha + \frac{1}{p}\right)} (b-a)^{\alpha - \frac{1}{q}}. \tag{17}$$

It follows:

Corollary 10. *([3]) Let $\alpha > \frac{1}{2}$, $m = \lceil \alpha \rceil$. All as in Theorem 6. Assume $f^{(k)}(b) = 0$, $k = 1, \ldots, m-1$, $D_{b-}^{\alpha} f \in L_2([a,b], X)$. Then*

$$\left\| \frac{1}{b-a} \int_a^b f(x)\, dx - f(b) \right\| \leq \frac{\|D_{b-}^{\alpha} f\|_{L_2([a,b],X)}}{\Gamma(\alpha)\left(\sqrt{2\alpha-1}\right)\left(\alpha+\frac{1}{2}\right)} (b-a)^{\alpha - \frac{1}{2}}. \tag{18}$$

We continue with a Poincaré like right fractional inequality:

Theorem 11. *([3]) Let $p, q > 1 : \frac{1}{p} + \frac{1}{q} = 1$, and $\alpha > \frac{1}{q}$, $m = \lceil \alpha \rceil$. Here all as in Theorem 6. Assume that $f^{(k)}(b) = 0$, $k = 0, 1, \ldots, m-1$, and $D_{b-}^{\alpha} f \in L_q([a,b], X)$, where X is a Banach space. Then*

$$\|f\|_{L_q([a,b],X)} \leq \frac{(b-a)^{\alpha} \|D_{b-}^{\alpha} f\|_{L_q([a,b],X)}}{\Gamma(\alpha)(p(\alpha-1)+1)^{\frac{1}{p}} (q\alpha)^{\frac{1}{q}}}. \tag{19}$$

Next follows a right Sobolev like fractional inequality:

Theorem 12. *([3]) All as in the last Theorem 11. Let $r > 0$. Then*

$$\|f\|_{L_r([a,b],X)} \leq \frac{(b-a)^{\alpha - \frac{1}{q} + \frac{1}{r}} \|D_{b-}^{\alpha} f\|_{L_q([a,b],X)}}{\Gamma(\alpha)(p(\alpha-1)+1)^{\frac{1}{p}} \left(r\left(\alpha - \frac{1}{q}\right) + 1\right)^{\frac{1}{r}}}. \tag{20}$$

We also mention the following Opial type right fractional inequality:

Theorem 13. ([3]) Let $p, q > 1 : \frac{1}{p} + \frac{1}{q} = 1$, and $\alpha > \frac{1}{q}$, $m := \lceil \alpha \rceil$. Let $[a, b] \subset \mathbb{R}$, X a Banach space, and $f \in C^{m-1}([a, b], X)$. Set

$$F_x(t) := \sum_{i=0}^{m-1} \frac{(x-t)^i}{i!} f^{(i)}(t), \ \forall\, t \in [x, b], \text{ where } x \in [a, b]. \tag{21}$$

Assume that $f^{(m)}$ exists outside a μ–null Borel set $B_x \subseteq [x, b]$, such that

$$h_1(F_x(B_x)) = 0, \ \forall\, x \in [a, b]. \tag{22}$$

We assume that $f^{(m)} \in L_\infty([a, b], X)$. Assume also that $f^{(k)}(b) = 0$, $k = 0, 1, \ldots, m-1$. Then

$$\int_x^b \|f(w)\| \, \|(D_{b-}^\alpha f)(w)\| \, dw \leq$$

$$\frac{(b-x)^{\alpha-1+\frac{2}{p}}}{2^{\frac{1}{q}} \Gamma(\alpha) \left((p(\alpha-1)+1)(p(\alpha-1)+2)\right)^{\frac{1}{p}}} \left(\int_x^b \|(D_{b-}^\alpha f)(z)\|^q \, dz\right)^{\frac{2}{q}}, \tag{23}$$

$\forall\, x \in [a, b]$.

Next we describe an abstract Hilbert–Pachpatte right fractional inequality:

Theorem 14. ([3]) Let $p, q > 1 : \frac{1}{p} + \frac{1}{q} = 1$, and $\alpha_1 > \frac{1}{q}$, $\alpha_2 > \frac{1}{p}$, $m_i := \lceil \alpha_i \rceil$, $i = 1, 2$. Here $[a_i, b_i] \subset \mathbb{R}$, $i = 1, 2$; X is a Banach space. Let $f_i \in C^{m_i-1}([a_i, b_i], X)$, $i = 1, 2$. Set

$$F_{x_i}(t_i) := \sum_{j_i=0}^{m_i-1} \frac{(x_i-t_i)^{j_i}}{j_i!} f_i^{(j_i)}(t_i), \tag{24}$$

$\forall\, t_i \in [x_i, b_i]$, where $x_i \in [a_i, b_i]$; $i = 1, 2$. Assume that $f_i^{(m_i)}$ exists outside a μ–null Borel set $B_{x_i} \subseteq [x_i, b_i]$, such that

$$h_1(F_{x_i}(B_{x_i})) = 0, \ \forall\, x_i \in [a_i, b_i]; \ i = 1, 2. \tag{25}$$

We also assume that $f_i^{(m_i)} \in L_1([a_i, b_i], X)$, and

$$f_i^{(k_i)}(b_i) = 0, \ k_i = 0, 1, \ldots, m_i - 1; \ i = 1, 2, \tag{26}$$

and

$$\left(D_{b_1-}^{\alpha_1} f_1\right) \in L_q([a_1, b_1], X), \ \left(D_{b_2-}^{\alpha_2} f_2\right) \in L_p([a_2, b_2], X). \tag{27}$$

Then

$$\int_{a_1}^{b_1} \int_{a_2}^{b_2} \frac{\|f_1(x_1)\| \, \|f_2(x_2)\| \, dx_1 dx_2}{\left(\frac{(b_1-x_1)^{p(\alpha_1-1)+1}}{p(p(\alpha_1-1)+1)} + \frac{(b_2-x_2)^{q(\alpha_2-1)+1}}{q(q(\alpha_2-1)+1)}\right)} \leq$$

$$\frac{(b_1-a_1)(b_2-a_2)}{\Gamma(\alpha_1)\Gamma(\alpha_2)} \left\|D_{b_1-}^{\alpha_1} f_1\right\|_{L_q([a_1,b_1],X)} \left\|D_{b_2-}^{\alpha_2} f_2\right\|_{L_p([a_2,b_2],X)}. \tag{28}$$

3. MAIN RESULTS

We need a special case of Definition 5 over $\mathbb{C}$.

Definition 15. Let $[a, b] \subset \mathbb{R}$, $\nu > 0$; $n := \lceil \nu \rceil \in \mathbb{N}$, $\lceil \cdot \rceil$ is the ceiling of the number and $f \in C^n([a, b], \mathbb{C})$. We call Caputo–Complex right fractional derivative of order ν:

$$\left(D_{b-}^{\nu} f\right)(x) := \frac{(-1)^n}{\Gamma(n-\nu)} \int_x^b (\lambda - x)^{n-\nu-1} f^{(n)}(\lambda) d\lambda, \quad \forall \, x \in [a, b], \tag{29}$$

where the derivatives $f', \ldots, f^{(n)}$ are defined as the numerical derivative.

If $\nu \in \mathbb{N}$, we set $D_{b-}^{\nu} f := (-1)^{\nu} f^{(\nu)}$ the ordinary $\mathbb{C}$–valued derivative and also $D_{b-}^0 f := f$.

Notice here (by [3]) that $D_{b-}^{\nu} f \in C([a, b], \mathbb{C})$.

We give the following right–fractional $\mathbb{C}$–Taylor's formula:

Theorem 16. Let $h \in C^n([a, b], \mathbb{C})$, $n = \lceil \nu \rceil$, $\nu \geq 0$. Then

$$h(t) = \sum_{i=0}^{n-1} \frac{(t-b)^i}{i!} h^{(i)}(b) + \frac{1}{\Gamma(\nu)} \int_t^b (\lambda - t)^{\nu-1} \left(D_{b-}^{\nu} h\right)(\lambda) d\lambda, \tag{30}$$

$\forall \, t \in [a, b]$, in particular when $h(t) := f(z(t)) z'(t) \in C^n([a, b], \mathbb{C})$, where $f(z)$, $z(t)$, $t \in [a, b]$ are as in 1. Introduction, it holds,

$$f(z(t)) z'(t) = \sum_{i=0}^{n-1} \frac{(t-b)^i}{i!} \left(f(z(b)) z'(b)\right)^{(i)} +$$

$$\frac{1}{\Gamma(\nu)} \int_t^b (\lambda - t)^{\nu-1} \left(D_{b-}^{\nu} f(z(\cdot)) z'(\cdot)\right)(\lambda) d\lambda, \tag{31}$$

$\forall \, t \in [a, b]$.

Proof. By Theorem 6. □

It follows a right fractional $\mathbb{C}$–Ostroswski type inequality:

Theorem 17. Let $n \in \mathbb{N}$ and $h \in C^n([a, b], \mathbb{C})$, where $[a, b] \subset \mathbb{R}$, and let $\nu > 0 : n = \lceil \nu \rceil$. Assume that $h^{(i)}(b) = 0$, $i = 1, \ldots, n-1$. Then

$$\left| \frac{1}{b-a} \int_a^b h(t) dt - h(b) \right| \leq \frac{\|D_{b-}^{\nu} h\|_{\infty, [a,b]}}{\Gamma(\nu + 2)} (b-a)^{\nu}, \tag{32}$$

in particular when $h(t) := f(z(t)) z'(t) \in C^n([a, b], \mathbb{C})$, where $f(z)$, $z(t)$, $t \in [a, b]$ are as in 1. Introduction, and $(f(z(t)) z'(t))^{(i)}|_{t=b} = 0$, $i = 1, \ldots, n-1$, we get

$$\left| \frac{1}{b-a} \int_{\gamma_{u,w}} f(z) dz - f(w) z'(b) \right| = \left| \frac{1}{b-a} \int_a^b f(z(t)) z'(t) dt - f(z(b)) z'(b) \right|$$

$$\leq \frac{\|D_{b-}^{\nu} f(z(t)) z'(t)\|_{\infty, [a,b]}}{\Gamma(\nu + 2)} (b-a)^{\nu}. \tag{33}$$

Proof. By Theorem 7. □

The corresponding $\mathbb{C}$–Ostrowski type L_p inequality follows:

Theorem 18. Let $p, q > 1 : \frac{1}{p} + \frac{1}{q} = 1$, and $\nu > \frac{1}{q}$, $n = \lceil \nu \rceil$. Here $h \in C^n([a,b], \mathbb{C})$. Assume that $h^{(i)}(b) = 0$, $i = 1, \ldots, n-1$. Then

$$\left| \frac{1}{b-a} \int_a^b h(t) \, dt - h(b) \right| \leq \frac{\|D_{b-}^\nu h\|_{L_q([a,b],\mathbb{C})}}{\Gamma(\nu) (p(\nu-1)+1)^{\frac{1}{p}} \left(\nu + \frac{1}{p}\right)} (b-a)^{\nu - \frac{1}{q}}, \qquad (34)$$

in particular when $h(t) := f(z(t)) z'(t) \in C^n([a,b], \mathbb{C})$, where $f(z), z(t), t \in [a,b]$ are as in 1. Introduction, and $(f(z(t)) z'(t))^{(i)}|_{t=b} = 0$, $i = 1, \ldots, n-1$, we get:

$$\left| \frac{1}{b-a} \int_{\gamma_{u,w}} f(z) \, dz - f(w) z'(b) \right| = \left| \frac{1}{b-a} \int_a^b f(z(t)) z'(t) \, dt - f(z(b)) z'(b) \right|$$

$$\leq \frac{\|D_{b-}^\nu (f(z(t)) z'(t))\|_{L_q([a,b],\mathbb{C})}}{\Gamma(\nu)(p(\nu-1)+1)^{\frac{1}{p}} \left(\nu + \frac{1}{p}\right)} (b-a)^{\nu - \frac{1}{q}}. \qquad (35)$$

Proof. By Theorem 9. □

It follows:

Corollary 19. (to Theorem 18, case of $p = q = 2$). We have that

$$\left| \frac{1}{b-a} \int_{\gamma_{u,w}} f(z) \, dz - f(w) z'(b) \right| \leq \frac{\|D_{b-}^\nu (f(z(t)) z'(t))\|_{L_2([a,b],\mathbb{C})}}{\Gamma(\nu) \sqrt{2\nu - 1} \left(\nu + \frac{1}{2}\right)} (b-a)^{\nu - \frac{1}{2}}. \qquad (36)$$

We continue with an L_1 fractional $\mathbb{C}$–Ostrowski type inequality:

Theorem 20. Let $\nu \geq 1$, $n = \lceil \nu \rceil$. Assume that $h(t) := f(z(t)) z'(t) \in C^n([a,b], \mathbb{C})$, where $f(z), z(t), t \in [a,b]$ are as in 1. Introduction, and such that $h^{(i)}(b) = 0$, $i = 1, \ldots, n-1$. Then

$$\left| \frac{1}{b-a} \int_{\gamma_{u,w}} f(z) \, dz - f(w) z'(b) \right| \leq \frac{\|D_{b-}^\nu (f(z(t)) z'(t))\|_{L_1([a,b],\mathbb{C})}}{\Gamma(\nu + 1)} (b-a)^{\nu - 1}. \qquad (37)$$

Proof. By Theorem 8. □

It follows a Poincaré like $\mathbb{C}$–fractional inequality:

Theorem 21. Let $p, q > 1 : \frac{1}{p} + \frac{1}{q} = 1$, and $\nu > \frac{1}{q}$, $n = \lceil \nu \rceil$. Let $h \in C^n([a,b], \mathbb{C})$. Assume that $h^{(i)}(b) = 0$, $i = 1, \ldots, n-1$. Then

$$\|h\|_{L_q([a,b],\mathbb{C})} \leq \frac{(b-a)^\nu \|D_{b-}^\nu h\|_{L_q([a,b],\mathbb{C})}}{\Gamma(\nu)(p(\nu-1)+1)^{\frac{1}{p}} (q\nu)^{\frac{1}{q}}}, \qquad (38)$$

in particular when $h(t) := f(z(t)) z'(t) \in C^n([a,b], \mathbb{C})$, where $f(z), z(t), t \in [a,b]$ are as in 1. Introduction, and $(f(z(t)) z'(t))^{(i)}|_{t=b} = 0$, $i = 1, \ldots, n-1$, we get:

$$\|f(z(t)) z'(t)\|_{L_q([a,b],\mathbb{C})} \le$$

$$\frac{(b-a)^\nu}{\Gamma(\nu)(p(\nu-1)+1)^{\frac{1}{p}}(q\nu)^{\frac{1}{q}}} \|D_{b-}^\nu (f(z(t)) z'(t))\|_{L_q([a,b],\mathbb{C})}. \tag{39}$$

Proof. By Theorem 11. $\square$

The corresponding Sobolev like inequality follows:

Theorem 22. *All as in Theorem 21. Let $r > 0$. Then*

$$\|f(z(t)) z'(t)\|_{L_r([a,b],\mathbb{C})} \le$$

$$\frac{(b-a)^{\nu - \frac{1}{q} + \frac{1}{r}}}{\Gamma(\nu)(p(\nu-1)+1)^{\frac{1}{p}} \left(r\left(\nu - \frac{1}{q}\right) + 1\right)^{\frac{1}{r}}} \|D_{b-}^\nu (f(z(t)) z'(t))\|_{L_q([a,b],\mathbb{C})}. \tag{40}$$

Proof. By Theorem 12. $\square$

We continue with an Opial type $\mathbb{C}$–fractional inequality:

Theorem 23. *Let $p, q > 1 : \frac{1}{p} + \frac{1}{q} = 1$, and $\nu > \frac{1}{q}$, $n := \lceil \nu \rceil$, $h \in C^n([a,b], \mathbb{C})$. Assume $h^{(k)}(b) = 0$, $k = 0, 1, \ldots, n-1$. Then*

$$\int_x^b |h(t)| \left|(D_{b-}^\nu h)(t)\right| dt \le$$

$$\frac{(b-x)^{\nu - 1 + \frac{2}{p}}}{2^{\frac{1}{q}} \Gamma(\nu) ((p(\nu-1)+1)(p(\nu-1)+2))^{\frac{1}{p}}} \left(\int_x^b |(D_{b-}^\nu h)(t)|^q dt\right)^{\frac{2}{q}}, \tag{41}$$

$\forall x \in [a,b]$, in particular when $h(t) := f(z(t)) z'(t) \in C^n([a,b], \mathbb{C})$, where $f(z), z(t), t \in [a,b]$ are as in 1. Introduction, and $(f(z(t)) z'(t))^{(i)}|_{t=b} = 0$, $i = 1, \ldots, n-1$, we get:

$$\int_x^b |f(z(t))| \left|(D_{b-}^\nu (f(z(t)) z'(t)))\right| |z'(t)| dt \le$$

$$\frac{(b-x)^{\nu - 1 + \frac{2}{p}}}{2^{\frac{1}{q}} \Gamma(\nu) ((p(\nu-1)+1)(p(\nu-1)+2))^{\frac{1}{p}}} \left(\int_x^b |D_{b-}^\nu (f(z(t)) z'(t))|^q dt\right)^{\frac{2}{q}}, \tag{42}$$

$\forall x \in [a,b]$.

Proof. By Theorem 13. $\square$

We finish with Hilbert–Pachpatte left $\mathbb{C}$–fractional inequalities:

Theorem 24. Let $p, q > 1 : \frac{1}{p} + \frac{1}{q} = 1$, and $\nu_1 > \frac{1}{q}$, $\nu_2 > \frac{1}{p}$, $n_i := \lceil \nu_i \rceil$, $i = 1, 2$. Let $h_i \in C^{n_i}([a_i, b_i], \mathbb{C})$, $i = 1, 2$. Assume $h_i^{(k_i)}(b_i) = 0$, $k_i = 0, 1, \ldots, n_i - 1$; $i = 1, 2$. Then

$$\int_{a_1}^{b_1} \int_{a_2}^{b_2} \frac{|h_1(t_1)| |h_2(t_2)| dt_1 dt_2}{\left(\frac{(b_1-t_1)^{p(\nu_1-1)+1}}{p(p(\nu_1-1)+1)} + \frac{(b_2-t_2)^{q(\nu_2-1)+1}}{q(q(\nu_2-1)+1)} \right)} \leq$$

$$\frac{(b_1 - a_1)(b_2 - a_2)}{\Gamma(\nu_1) \Gamma(\nu_2)} \left\| D_{b_1-}^{\nu_1} h_1 \right\|_{L_q([a_1,b_1], \mathbb{C})} \left\| D_{b_2-}^{\nu_2} h_2 \right\|_{L_p([a_2,b_2], \mathbb{C})}, \quad (43)$$

in particular when $h_1(t_1) := f_1(z_1(t_1)) z_1'(t_1)$ and $h_2(t_2) := f_2(z_2(t_2)) z_2'(t_2)$ as in 1. Introduction, with $h_i^{(k_i)}(b_i) = 0$, $k_i = 0, 1, \ldots, n_i - 1$; $i = 1, 2$, we get:

$$\int_{a_1}^{b_1} \int_{a_2}^{b_2} \frac{\left| f_1(z_1(t_1)) z_1^{prime}(t_1) \right| |f_2(z_2(t_2)) z_2'(t_2)| dt_1 dt_2}{\left(\frac{(b_1-t_1)^{p(\nu_1-1)+1}}{p(p(\nu_1-1)+1)} + \frac{(b_2-t_2)^{q(\nu_2-1)+1}}{q(q(\nu_2-1)+1)} \right)} \leq \frac{(b_1 - a_1)(b_2 - a_2)}{\Gamma(\nu_1) \Gamma(\nu_2)} \cdot$$

$$\left\| D_{b_1-}^{\nu_1} \left(f_1(z_1(t_1)) z_1'(t_1) \right) \right\|_{L_q([a_1,b_1], \mathbb{C})} \left\| D_{b_2-}^{\nu_2} \left(f_2(z_2(t_2)) z_2'(t_2) \right) \right\|_{L_p([a_2,b_2], \mathbb{C})}. \quad (44)$$

Proof. By Theorem 14. □

REFERENCES

[1] Dragomir, S. S. 2019. "Extensions of Stekloff and Almansi inequalities to the complex integral". *RGMIA Res. Rep. Coll.* Art. 10(10): 10 pp., rgmia.org/v22.php.

[2] Dragomir, S. S. 2019. "Two points Ostrowski type rules for approximating the integral of analytic complex functions on paths from convex domains". *RGMIA Res. Rep. Coll.* Art. 14(22): 16 pp., rgmia.org/v22.php.

[3] Anastassiou, G. 2017. "Strong right fractional calculus for Banach space valued functions". *Revista Proyecciones* 36(1): 149–186.

[4] Volintiru, C. 2000. "A proof of the fundamental theorem of Calculus using Hausdorff measures". *Real Analysis Exchange* 26(1): 381–390.

[5] Anastassiou, G. 2011. *Advances on fractional inequalities*. New York: Springer.

In: Understanding Banach Spaces
Editor: Daniel González Sánchez
ISBN: 978-1-53616-745-0
© 2020 Nova Science Publishers, Inc.

Chapter 2

MIXED COMPLEX FRACTIONAL INEQUALITIES

*George A. Anastassiou**
Department of Mathematical Sciences
University of Memphis, Memphis, TN, US

Abstract

Here we present some important mixed generalized fractional complex analytic inequalities of the following kinds: Polya's, Ostrowski's and Poincaré's.

Keywords: complex inequalities, fractional inequalities, generalized fractional derivative

AMS Subject Classification: 26D10, 26D15, 30A10

1. INTRODUCTION

Here we follow [1].

Suppose γ is a smooth path parametrized by $z(t)$, $t \in [a, b]$ and f is a complex function which is continuous on γ. Put $z(a) = u$ and $z(b) = w$ with $u, w \in \mathbb{C}$. We define the integral of f on $\gamma_{u,w} = \gamma$ as

$$\int_\gamma f(z)\,dz = \int_{\gamma_{u,w}} f(z)\,dz := \int_a^b f(z(t))\, z'(t)\, dt.$$

We observe that the actual choice of parametrization of γ does not matter.

This definition immediately extends to paths that are piecewise smooth. Suppose γ is parametrized by $z(t)$, $t \in [a, b]$, which is differentiable on the intervals $[a, c]$ and $[c, b]$, then assuming that f is continuous on γ we define

$$\int_{\gamma_{u,w}} f(z)\,dz := \int_{\gamma_{u,v}} f(z)\,dz + \int_{\gamma_{v,w}} f(z)\,dz,$$

where $v := z(c)$. This can be extended for a finite number of intervals.

---
*Corresponding Author's E-mail: ganastss@memphis.edu.

We also define the integral with respect to arc–length

$$\int_{\gamma_{u,w}} f(z)\,|dz| := \int_a^b f(z(t))\,|z'(t)|\,dt$$

and the length of the curve γ is then

$$l(\gamma) = \int_{\gamma_{u,w}} |dz| := \int_a^b |z'(t)|\,dt.$$

Let f and g be holomorphic in G, and open domain and suppose $\gamma \subset G$ is a piecewise smooth path from $z(a) = u$ to $z(b) = w$. Then we have the integration by parts formula

$$\int_{\gamma_{u,w}} f(z)\,g'(z)\,dz = f(w)g(w) - f(u)g(u) - \int_{\gamma_{u,w}} f'(z)g(z)\,dz. \quad (1)$$

We recall also the triangle inequality for the complex integral, namely

$$\left| \int_\gamma f(z)\,dz \right| \le \int_\gamma |f(z)|\,|dz| \le \|f\|_{\gamma,\infty}\, l(\gamma), \quad (2)$$

where $\|f\|_{\gamma,\infty} := \sup_{z \in \gamma} |f(z)|$.

We also define the p–norm with $p \ge 1$ by

$$\|f\|_{\gamma,p} := \left(\int_\gamma |f(z)|^p\,|dz| \right)^{\frac{1}{p}}.$$

For $p = 1$ we have

$$\|f\|_{\gamma,1} := \int_\gamma |f(z)|\,|dz|.$$

If $p,q > 1$ with $\frac{1}{p} + \frac{1}{q} = 1$, then by Hölder's inequality we have

$$\|f\|_{\gamma,1} \le [l(\gamma)]^{\frac{1}{q}} \|f\|_{\gamma,p}.$$

Motivations to our work follow:

We mention the following Wirtinger type inequality for complex functions:

Theorem 1. *([1]) Let f be analytic in G, a domain of complex numbers and suppose $\gamma \subset G$ is a smooth path parametrized by $z(t)$, $t \in [a,b]$ from $z(a) = u$ to $z(b) = w$ and $z'(t) \ne 0$ for $t \in (a,b)$.*

(i) If $f(u) = f(w) = 0$, then

$$\int_\gamma |f(z)|^2\,|dz| \le \frac{1}{\pi^2} l^2(\gamma) \int_\gamma |f'(z)|^2\,|dz|. \quad (3)$$

The equality holds in (3) iff

$$f(v) = K \sin\left[\frac{\pi l(\gamma_{u,v})}{l(\gamma)} \right], \quad K \in \mathbb{C}, \quad (4)$$

where $v = z(t)$, $t \in [a,b]$ and $l(\gamma_{u,v}) = \int_a^t |z'(s)| \, ds$.

(ii) If $f(u) = 0$, then

$$\int_\gamma |f(z)|^2 \, |dz| \leq \frac{4}{\pi^2} l^2(\gamma) \int_\gamma |f'(z)|^2 \, |dz|. \tag{5}$$

The equality holds in (5) if only if

$$f(v) = K \sin\left[\frac{\pi l(\gamma_{u,v})}{2 l(\gamma)}\right], \quad K \in \mathbb{C}, \tag{6}$$

where $v = z(t)$, $t \in [a, b]$.

We mention some complex trapezoid type inequalities:

Proposition 2. ([1]) *Let g be analytic in G, a domain of complex numbers and suppose $\gamma \subset G$ is a smooth path parametrized by $z(t)$, $t \in [a, b]$ from $z(a) = u$ to $z(b) = w$, $w \neq u$ and $z'(t) \neq 0$ for $t \in (a, b)$. Then*

$$\left| \frac{1}{w-u} \int_\gamma g(z) \, dz - \frac{g(u) + g(w)}{2} \right| \leq$$

$$\frac{1}{\pi} \frac{l(\gamma)}{|w-u|} \left(\frac{1}{l(\gamma)} \int_\gamma \left| g'(z) - \frac{g(w) - g(u)}{w-u} \right|^2 |dz| \right)^{\frac{1}{2}}. \tag{7}$$

Proposition 3. ([1]) *Let g be analytic in G, a domain of complex numbers and suppose $\gamma \subset G$ is a smooth path parametrized by $z(t)$, $t \in [a, b]$ from $z(a) = u$ to $z(b) = w$, $w \neq u$ and $z'(t) \neq 0$ for $t \in (a, b)$. If $u + w - z \in G$ for $z \in \gamma$, then*

$$\left| \frac{1}{w-u} \int_\gamma \widehat{g(z)} \, dz - \frac{g(u) + g(w)}{2} \right| \leq$$

$$\frac{1}{2\pi} \frac{l(\gamma)}{|w-u|} \left(\frac{1}{l(\gamma)} \int_\gamma |g'(z) - g'(u+w-z)|^2 |dz| \right)^{\frac{1}{2}}, \tag{8}$$

where $\widehat{g(z)} := \frac{g(z) + g(u+w-z)}{2}$, $z \in \gamma$.

In this chapter we utilize on $\mathbb{C}$ the results of [2] which are for general Banach space valued functions.

We give mixed fractional: $\mathbb{C}$–Polya type integral inequality and $\mathbb{C}$–Ostrowski type integral inequality. We finish with right and left fractional $\mathbb{C}$–Poincaré like inequalities.

2. BACKGROUND

Here $C([a,b], X)$ stands for the space of continuous functions from $[a,b]$ into X, where $(X, \|\cdot\|)$ is a Banach space.

All integrals here are of Bochner type ([3]). By [4], we have that: if $f \in C([a,b], X)$, then $f \in L_\infty([a,b], X)$ and $f \in L_1([a,b], X)$. Derivatives for vector valued functions are defined according to [5], p. 83, similar to numerical ones.

We need:

Definition 4. *([2]) Let $f \in C([a,b], X)$, where X is a Banach space. Let $\nu > 0$, we define the right Riemann–Liouville fractional Bochner integral operator*

$$\left(J_{b-}^{\nu} f\right)(x) := \frac{1}{\Gamma(\nu)} \int_{x}^{b} (z-x)^{\nu-1} f(z) \, dz, \quad \forall \, x \in [a,b], \tag{9}$$

where Γ is the gamma function.

In [6], we have proved that

$$\left(J_{b-}^{\nu} f\right) \in C([a,b], X).$$

Furthermore in [6], we have proved that

$$J_{b-}^{\nu} J_{b-}^{\mu} f = J_{b-}^{\nu+\mu} f = J_{b-}^{\mu} J_{b-}^{\nu} f,$$

for any $\mu, \nu > 0$; any $f \in C([a,b], X)$.

We need:

Definition 5. *([2]) Let $\nu > 0$, $n := \lceil \nu \rceil$, where $\lceil \cdot \rceil$ is the integral part, $\alpha = \nu - n$, $0 < \alpha < 1$, $\nu \notin \mathbb{N}$. Define the subspace of functions*

$$C_{b-}^{\nu}([a,b], X) := \left\{ f \in C^{n}([a,b], X) : J_{b-}^{1-\alpha} f^{(n)} \in C^{1}([a,b], X) \right\}. \tag{10}$$

Define the Banach space valued right generalized ν–fractional derivative of f over $[a,b]$ as

$$D_{b-}^{\nu} f := (-1)^{n-1} \left(J_{b-}^{1-\alpha} f^{(n)} \right)'. \tag{11}$$

Notice that

$$J_{b-}^{1-\alpha} f^{(n)}(x) = \frac{1}{\Gamma(1-\alpha)} \int_{x}^{b} (z-x)^{-\alpha} f^{(n)}(z) \, dz \tag{12}$$

exists for $f \in C_{b-}^{\nu}([a,b], X)$, and

$$\left(D_{b-}^{\nu} f\right)(x) = \frac{(-1)^{n-1}}{\Gamma(1-\alpha)} \frac{d}{dx} \int_{x}^{b} (z-x)^{-\alpha} f^{(n)}(z) \, dz. \tag{13}$$

I. e.

$$\left(D_{b-}^{\nu} f\right)(x) = \frac{(-1)^{n-1}}{\Gamma(n-\nu+1)} \frac{d}{dx} \int_{x}^{b} (z-x)^{n-\nu} f^{(n)}(z) \, dz. \tag{14}$$

If $\nu \in \mathbb{N}$, then $\alpha = 0$, $n = \nu$, and

$$\left(D_{b-}^{\nu} f\right)(x) = \left(D_{b-}^{n} f\right)(x) = (-1)^{n} f^{(n)}(x). \tag{15}$$

Notice that $D_{b-}^{\nu} f \in C([a,b], X)$.

We mention the following right fractional Taylor's formula.

Theorem 6. ([2]) Let $f \in C_{b-}^{\nu}([a,b], X)$, $\nu > 0$, $n := [\nu]$. Then
(1) If $\nu \geq 1$, we get

$$f(x) = \sum_{k=0}^{n-1} \frac{f^{(k)}(b)}{k!}(x-b)^k + \left(J_{b-}^{\nu} D_{b-}^{\nu} f\right)(x), \quad \forall\, x \in [a,b]. \tag{16}$$

(2) If $0 < \nu < 1$, we get

$$f(x) = J_{b-}^{\nu} D_{b-}^{\nu} f(x), \quad \forall\, x \in [a,b].$$

We have that

$$J_{b-}^{\nu} D_{b-}^{\nu} f(x) = \frac{1}{\Gamma(\nu)} \int_x^b (z-x)^{\nu-1} \left(D_{b-}^{\nu} f\right)(z)\, dz, \quad \forall\, x \in [a,b].$$

Definition 7. ([2]) Let $f \in C([a,b], X)$. Let $\nu > 0$, we define the left Riemann–Liouville fractional Bochner integral operator

$$(J_a^{\nu} f)(x) := \frac{1}{\Gamma(\nu)} \int_a^x (x-z)^{\nu-1} f(z)\, dz, \quad \forall\, x \in [a,b]. \tag{17}$$

In [4], we have proved that

$$(J_a^{\nu} f) \in C([a,b], X).$$

Furthermore in [4], we have proved that

$$J_a^{\nu} J_a^{\mu} f = J_a^{\nu+\mu} f = J_a^{\mu} J_a^{\nu} f, \tag{18}$$

$\forall\, \mu, \nu > 0$, $\forall\, f \in C([a,b], X)$.

Definition 8. ([2]) Let $\nu > 0$, $n := [\nu]$, $\alpha = \nu - n$, $0 < \alpha < 1$, $\nu \notin \mathbb{N}$. Define the subspace of functions

$$C_a^{\nu}([a,b], X) := \left\{ f \in C^n([a,b], X) : J_a^{1-\alpha} f^{(n)} \in C^1([a,b], X) \right\}. \tag{19}$$

Define the Banach space valued left generalized ν–fractional derivative of f over $[a,b]$ as

$$(D_a^{\nu} f) := \left(J_a^{1-\alpha} f^{(n)}\right)'. \tag{20}$$

Notice that

$$J_a^{1-\alpha} f^{(n)}(x) = \frac{1}{\Gamma(1-\alpha)} \int_a^x (x-z)^{-\alpha} f^{(n)}(z)\, dz \tag{21}$$

exists for $f \in C_a^{\nu}([a,b], X)$, and

$$(D_a^{\nu} f)(x) = \frac{1}{\Gamma(1-\alpha)} \frac{d}{dx} \int_a^x (x-z)^{-\alpha} f^{(n)}(z)\, dz. \tag{22}$$

I. e.

$$(D_a^{\nu} f)(x) = \frac{1}{\Gamma(n-\nu+1)} \frac{d}{dx} \int_a^x (x-z)^{n-\nu} f^{(n)}(z)\, dz. \tag{23}$$

If $\nu \in \mathbb{N}$, then $\alpha = 0$, $n = \nu$, and

$$(D_a^{\nu} f)(x) = (D_a^n f)(x) = f^{(n)}(x). \tag{24}$$

Notice that $D_a^{\nu} f \in C([a,b], X)$.

We mention the following left fractional Taylor's formula.

Theorem 9. *([2]) Let $f \in C_a^\nu([a,b], X)$, $\nu > 0$, $n := \lceil \nu \rceil$. Then*
(1) If $\nu \geq 1$, we get

$$f(x) = \sum_{k=0}^{n-1} \frac{f^{(k)}(a)}{k!} (x-a)^k + (J_a^\nu D_a^\nu f)(x), \quad \forall x \in [a,b]. \tag{25}$$

(2) If $0 < \nu < 1$, we get

$$f(x) = J_a^\nu D_a^\nu f(x), \quad \forall x \in [a,b]. \tag{26}$$

We have that

$$J_a^\nu D_a^\nu f(x) = \frac{1}{\Gamma(\nu)} \int_a^x (x-z)^{\nu-1} (D_a^\nu f)(z) \, dz, \quad \forall x \in [a,b]. \tag{27}$$

We mention the following fractional Polya type integral inequality without any boundary conditions, see also [7], p. 4.

Theorem 10. *([2]) Let $0 < \nu < 1$, $f \in C([a,b], X)$. Assume that $f \in C_a^\nu\left(\left[a, \frac{a+b}{2}\right], X\right)$ and $f \in C_{b-}^\nu\left(\left[\frac{a+b}{2}, b\right], X\right)$. Set*

$$M(f) = \max\left\{ \|\|D_a^\nu f\|\|_{\infty, \left[a, \frac{a+b}{2}\right]}, \|\|D_{b-}^\nu f\|\|_{\infty, \left[\frac{a+b}{2}, b\right]} \right\}. \tag{28}$$

Then

$$\left\| \int_a^b f(x) \, dx \right\| \leq \int_a^b \|f(x)\| \, dx \leq M(f) \frac{(b-a)^{\nu+1}}{\Gamma(\nu+2) 2^\nu}. \tag{29}$$

Inequality (29) is sharp, namely it is attained by

$$f_*(x) = \begin{cases} (x-a)^\nu \, \vec{i}, & x \in \left[a, \frac{a+b}{2}\right], \\ (b-x)^\nu \, \vec{i}, & x \in \left[\frac{a+b}{2}, b\right] \end{cases}, \quad 0 < \nu < 1, \tag{30}$$

$\vec{i} \in X : \|\vec{i}\| = 1$.

Clearly here non zero constant vector function f are excluded.

We also mention the following fractional Ostrowski type inequality, see also [7], pp. 379-381.

Theorem 11. *([2]) Let $\nu \geq 1$, $n = \lceil \nu \rceil$, $f \in C([a,b], X)$, $x_0 \in [a,b]$. Assume that $f|_{[a,x_0]} \in C_{x_0-}^\nu([a, x_0], X)$, $f|_{[x_0,b]} \in C_{x_0}^\nu([x_0, b], X)$, and $f^{(i)}(x_0) = 0$, for $i = 1, \ldots, n-1$, which is void when $1 \leq \nu < 2$. Then*

$$\left\| \frac{1}{b-a} \int_a^b f(x) \, dx - f(x_0) \right\| \leq \frac{1}{(b-a)\Gamma(\nu+2)} \cdot$$

$$\left\{ \|\|D_{x_0-}^\nu f\|\|_{\infty, [a, x_0]} (x_0 - a)^{\nu+1} + \|\|D_{x_0}^\nu f\|\|_{\infty, [x_0, b]} (b - x_0)^{\nu+1} \right\} \leq$$

$$\frac{1}{(b-a)\Gamma(\nu+2)} \max\left(\||D^{\nu}_{x_0-}f|\|_{\infty,[a,x_0]}, \||D^{\nu}_{x_0}f|\|_{\infty,[x_0,b]}\right) \cdot$$
$$\left[(b-x_0)^{\nu+1} + (x_0-a)^{\nu+1}\right] \leq \tag{31}$$
$$\max\left(\||D^{\nu}_{x_0-}f|\|_{\infty,[a,x_0]}, \||D^{\nu}_{x_0}f|\|_{\infty,[x_0,b]}\right) \frac{(b-a)^{\nu}}{\Gamma(\nu+2)}.$$

We continue with a right fractional Poincaré like inequality:

Theorem 12. ([2]) Let $p, q > 1 : \frac{1}{p} + \frac{1}{q} = 1$, $\alpha > \frac{1}{q}$, $m = \lceil \alpha \rceil$. Let $f \in C^{\alpha}_{b-}([a,b], X)$. Assume that $f^{(k)}(b) = 0$, $k = 0, 1, \ldots, m-1$, when $\alpha \geq 1$. Then

$$\|f\|_{L_q([a,b],X)} \leq \frac{(b-a)^{\alpha} \|D^{\alpha}_{b-}f\|_{L_q([a,b],X)}}{\Gamma(\alpha)(p(\alpha-1)+1)^{\frac{1}{p}}(q\alpha)^{\frac{1}{q}}}. \tag{32}$$

We finally mention a Poincaré like left fractional inequality:

Theorem 13. ([2]) Let $p, q > 1 : \frac{1}{p} + \frac{1}{q} = 1$, and $\nu > \frac{1}{q}$, $n = \lceil \nu \rceil$. Let $f \in C^{\nu}_a([a,b], X)$. Assume that $f^{(k)}(a) = 0$, $k = 0, 1, \ldots, n-1$, if $\nu \geq 1$. Then

$$\|f\|_{L_q([a,b],X)} \leq \frac{(b-a)^{\nu}}{\Gamma(\nu)(p(\nu-1)+1)^{\frac{1}{p}}(q\nu)^{\frac{1}{q}}} \|D^{\nu}_a f\|_{L_q([a,b],X)}. \tag{33}$$

All this background next is applied for $X = \mathbb{C}$, the complex numbers with $\|\cdot\| = |\cdot|$ the absolute value.

3. MAIN RESULTS

From now on here $f(z)$ and $z(t)$, $t \in [a,b]$, are as in section 1. Introduction.

We give a fractional $\mathbb{C}$–Polya inequality:

Theorem 14. Let $0 < \nu < 1$, $h \in C([a,b], \mathbb{C})$. Assume that $h \in C^{\nu}_a\left([a, \frac{a+b}{2}], \mathbb{C}\right)$ and $h \in C^{\nu}_{b-}\left([\frac{a+b}{2}, b], \mathbb{C}\right)$. Set

$$M(h) = \max\left\{\||D^{\nu}_a h|\|_{\infty,[a,\frac{a+b}{2}]}, \||D^{\nu}_{b-} h|\|_{\infty,[\frac{a+b}{2},b]}\right\}. \tag{34}$$

Then

$$\left|\int_a^b h(x)\,dx\right| \leq \int_a^b |h(x)|\,dx \leq M(h) \frac{(b-a)^{\nu+1}}{\Gamma(\nu+2)2^{\nu}}. \tag{35}$$

Inequality (35) is sharp, namely it is attained by

$$h_*(x) = \begin{cases} (x-a)^{\nu}\widetilde{c}, & x \in [a, \frac{a+b}{2}], \\ (b-x)^{\nu}\widetilde{c}, & x \in [\frac{a+b}{2}, b] \end{cases}, \quad 0 < \nu < 1, \tag{36}$$

where $\widetilde{c} \in \mathbb{C} : |\widetilde{c}| = 1$.

Clearly here non–zero constant functions h are excluded.

Proof. By Theorem 10 for $X = \mathbb{C}$. $\square$

Next we apply Theorem 14 for $h(t) = f(z(t)) z'(t)$, $t \in [a, b]$, to derive the following complex fractional Polya inequality:

Theorem 15. *Let $0 < \nu < 1$, $f(z(\cdot)) z'(\cdot) \in C([a, b], \mathbb{C})$. Assume that $f(z(\cdot)) z'(\cdot) \in C_a^\nu\left(\left[a, \frac{a+b}{2}\right], \mathbb{C}\right)$ and $f(z(\cdot)) z'(\cdot) \in C_{b-}^\nu\left(\left[\lceil\frac{a+b}{2}\rceil, b\right], \mathbb{C}\right)$. Set*

$$M\left(f(z(\cdot)) z'(\cdot)\right) =$$

$$\max\left\{\||D_a^\nu(f(z(\cdot)) z'(\cdot))|\|_{\infty,[a,\frac{a+b}{2}]}, \||D_{b-}^\nu(f(z(\cdot)) z'(\cdot))|\|_{\infty,[\frac{a+b}{2},b]}\right\}. \quad (37)$$

Then

$$\left|\int_\gamma f(z) \, dz\right| = \left|\int_{\gamma_{u,w}} f(z) \, dz\right| = \left|\int_a^b f(z(t)) z'(t) \, dt\right|$$

$$\leq \int_a^b |f(z(t))| |z'(t)| \, dt = \int_{\gamma_{u,w}} |f(z)| |dz| \leq M\left(f(z(\cdot)) z'(\cdot)\right) \frac{(b-a)^{\nu+1}}{\Gamma(\nu+2) 2^\nu}. \quad (38)$$

Proof. By Theorem 14. $\square$

Note: No boundary conditions are needed in Theorems 14, 15.

We continue with a fractional $\mathbb{C}$–Ostrowski type inequality:

Theorem 16. *Let $\nu \geq 1$, $n = \lceil\nu\rceil$, $h \in C([a, b], \mathbb{C})$, $x_0 \in [a, b]$. Assume that $h|_{[a,x_0]} \in C_{x_0-}^\nu([a, x_0], \mathbb{C})$, $h|_{[x_0,b]} \in C_{x_0}^\nu([x_0, b], \mathbb{C})$, and $h^{(i)}(x_0) = 0$, for $i = 1, \ldots, n-1$, which is void when $1 \leq \nu < 2$. Then*

$$\left|\frac{1}{b-a}\int_a^b h(x) \, dx - h(x_0)\right| \leq \frac{1}{(b-a)\Gamma(\nu+2)} \cdot$$

$$\left\{\||D_{x_0-}^\nu h|\|_{\infty,[a,x_0]} (x_0 - a)^{\nu+1} + \||D_{x_0}^\nu h|\|_{\infty,[x_0,b]} (b - x_0)^{\nu+1}\right\} \leq$$

$$\frac{1}{(b-a)\Gamma(\nu+2)} \cdot$$

$$\max\left(\||D_{x_0-}^\nu h|\|_{\infty,[a,x_0]}, \||D_{x_0}^\nu h|\|_{\infty,[x_0,b]}\right) \left[(b-x_0)^{\nu+1} + (x_0 - a)^{\nu+1}\right] \leq \quad (39)$$

$$\max\left(\||D_{x_0-}^\nu h|\|_{\infty,[a,x_0]}, \||D_{x_0}^\nu h|\|_{\infty,[x_0,b]}\right) \frac{(b-a)^\nu}{\Gamma(\nu+2)}.$$

Proof. By Theorem 11. $\square$

Next we apply Theorem 16 for $h(t) = f(z(t)) z'(t)$, $t \in [a, b]$, to derive the following complex fractional Ostrowski type inequality:

Theorem 17. Let $\nu \geq 1$, $n = \lceil\nu\rceil$, $f(z(\cdot))z'(\cdot) \in C([a,b],\mathbb{C})$, $c \in [a,b]$; $v := z(c)$. Assume that $f(z(\cdot))z'(\cdot)|_{[a,c]} \in C^\nu_{c-}([a,c],\mathbb{C})$, $f(z(\cdot))z'(\cdot)|_{[c,b]} \in C^\nu_c([c,b],\mathbb{C})$, and $(f(z(\cdot))z'(\cdot))^{(i)}(c) = 0$, for $i = 1, \ldots, n-1$, which is void when $1 \leq \nu < 2$. Then

$$\left|\frac{1}{b-a}\int_a^b f(z(t))z'(t)\,dt - f(z(c))z'(c)\right| = \left|\frac{1}{b-a}\int_{\gamma_{u,w}} f(z)\,dz - f(v)z'(c)\right|$$

$$= \left|\frac{1}{b-a}\int_\gamma f(z)\,dz - f(v)z'(c)\right| \leq \frac{1}{(b-a)\Gamma(\nu+2)} \cdot$$

$$\left\{\left\|\left|D^\nu_{c-}(f(z(\cdot))z'(\cdot))\right|\right\|_{\infty,[a,c]}(c-a)^{\nu+1} + \left\|\left|D^\nu_c(f(z(\cdot))z'(\cdot))\right|\right\|_{\infty,[c,b]}(b-c)^{\nu+1}\right\} \leq$$

$$\frac{1}{(b-a)\Gamma(\nu+2)}\max\left(\left\|\left|D^\nu_{c-}(f(z(\cdot))z'(\cdot))\right|\right\|_{\infty,[a,c]}, \left\|\left|D^\nu_c(f(z(\cdot))z'(\cdot))\right|\right\|_{\infty,[c,b]}\right)$$

$$\times\left[(b-c)^{\nu+1} + (c-a)^{\nu+1}\right] \leq \quad (40)$$

$$\max\left(\left\|\left|D^\nu_{c-}(f(z(\cdot))z'(\cdot))\right|\right\|_{\infty,[a,c]}, \left\|\left|D^\nu_c(f(z(\cdot))z'(\cdot))\right|\right\|_{\infty,[c,b]}\right)\frac{(b-a)^\nu}{\Gamma(\nu+2)}.$$

Proof. By Theorem 16. □

Next comes a right fractional $\mathbb{C}$–Poincaré like inequality:

Theorem 18. Let $p, q > 1 : \frac{1}{p} + \frac{1}{q} = 1$, $\alpha > \frac{1}{q}$, $m = \lceil\alpha\rceil$. Let $f(z(\cdot))z'(\cdot) \in C^\alpha_{b-}([a,b],\mathbb{C})$. Assume that $(f(z(\cdot))z'(\cdot))^{(k)}(b) = 0$, $k = 0, 1, \ldots, m-1$, when $\alpha \geq 1$. Then

$$\|f(z(\cdot))z'(\cdot)\|_{L_q([a,b],\mathbb{C})} \leq \frac{(b-a)^\alpha \left\|D^\alpha_{b-}(f(z(\cdot))z'(\cdot))\right\|_{L_q([a,b],\mathbb{C})}}{\Gamma(\alpha)(p(\alpha-1)+1)^{\frac{1}{p}}(q\alpha)^{\frac{1}{q}}}. \quad (41)$$

Proof. By Theorem 12. □

We finish with a left fractional $\mathbb{C}$–Poincaré like inequality:

Theorem 19. Let $p, q > 1 : \frac{1}{p} + \frac{1}{q} = 1$, and $\nu > \frac{1}{q}$, $n = \lceil\nu\rceil$. Let $f(z(\cdot))z'(\cdot) \in C^\nu_a([a,b],\mathbb{C})$. Assume that $(f(z(\cdot))z'(\cdot))^{(k)}(a) = 0$, $k = 0, 1, \ldots, n-1$, if $\nu \geq 1$. Then

$$\|f(z(\cdot))z'(\cdot)\|_{L_q([a,b],\mathbb{C})} \leq \frac{(b-a)^\nu}{\Gamma(\nu)(p(\nu-1)+1)^{\frac{1}{p}}(q\nu)^{\frac{1}{q}}}\left\|D^\nu_a(f(z(\cdot))z'(\cdot))\right\|_{L_q([a,b],\mathbb{C})}. \quad (42)$$

Proof. By Theorem 13. □

REFERENCES

[1] Dragomir, S. S. 2019. "An extension of Wirtinger's inequality to the complex integral". *RGMIA Res. Rep. Coll.* 22, Art. 8, 9 pp. rgmia.org/v22.php.

[2] Anastassiou, G. A. 2017. "Strong mixed and generalized fractional calculus for Banach space valued functions". *Mat. Vesnik* 69(3): 176–191.

[3] Mikusinski, J. 1978. *The Bochner integral*. New York: Academic Press.

[4] Anastassiou, G. A. 2017. "A strong Fractional Calculus Theory for Banach space valued functions". *Nonlinear Functional Analysis and Applications* 22(3): 495–524.

[5] Shilov, G. E. 1996. *Elementary Functional Analysis*. New York: Dover Publications, Inc.

[6] Anastassiou, G. A. 2017. "Strong Right Fractional Calculus for Banach space valued functions". *Revista Proyecciones* 36(1): 149–186.

[7] Anastassiou, G. A. 2016. *Intelligent Comparisons: Analytic Inequalities*. New York, Heidelberg: Springer.

Chapter 3

ADVANCED COMPLEX FRACTIONAL OSTROWSKI INEQUALITIES

George A. Anastassiou[*]
Department of Mathematical Sciences
University of Memphis, Memphis, TN, US

Abstract

Here we present very general and advanced fractional complex analytic inequalities of the Ostrowski type.

Keywords: complex inequalities, fractional inequalities, Ostrowski inequalities

AMS Subject Classification: 26D10, 26D15, 30A10

1. Introduction

Here we follow [1].

Suppose γ is a smooth path parametrized by $z(t)$, $t \in [a, b]$ and f is a complex function which is continuous on γ. Put $z(a) = u$ and $z(b) = w$ with $u, w \in \mathbb{C}$. We define the integral of f on $\gamma_{u,w} = \gamma$ as

$$\int_\gamma f(z)\,dz = \int_{\gamma_{u,w}} f(z)\,dz := \int_a^b f(z(t))\,z'(t)\,dt. \qquad (1)$$

We observe that the actual choice of parametrization of γ does not matter.

This definition immediately extends to paths that are piecewise smooth. Suppose γ is parametrized by $z(t)$, $t \in [a, b]$, which is differentiable on the intervals $[a, c]$ and $[c, b]$, then assuming that f is continuous on γ we define

$$\int_{\gamma_{u,w}} f(z)\,dz := \int_{\gamma_{u,v}} f(z)\,dz + \int_{\gamma_{v,w}} f(z)\,dz, \qquad (2)$$

---

[*]Corresponding Author's E-mail: ganastss@memphis.edu.

where $v := z(c)$. This can be extended for a finite number of intervals.

We also define the integral with respect to arc–length

$$\int_{\gamma_{u,w}} f(z) |dz| := \int_a^b f(z(t)) |z'(t)| dt \tag{3}$$

and the length of the curve γ is then

$$l(\gamma) = \int_{\gamma_{u,w}} |dz| := \int_a^b |z'(t)| dt. \tag{4}$$

Let f and g be holomorphic in G, and open domain and suppose $\gamma \subset G$ is a piecewise smooth path from $z(a) = u$ to $z(b) = w$. Then we have the integration by parts formula

$$\int_{\gamma_{u,w}} f(z) g'(z) dz = f(w) g(w) - f(u) g(u) - \int_{\gamma_{u,w}} f'(z) g(z) dz. \tag{5}$$

We recall also the triangle inequality for the complex integral, namely

$$\left| \int_\gamma f(z) dz \right| \le \int_\gamma |f(z)| |dz| \le \|f\|_{\gamma,\infty} l(\gamma), \tag{6}$$

where $\|f\|_{\gamma,\infty} := \sup_{z \in \gamma} |f(z)|$.

We also define the p–norm with $p \ge 1$ by

$$\|f\|_{\gamma,p} := \left(\int_\gamma |f(z)|^p |dz| \right)^{\frac{1}{p}}.$$

For $p = 1$ we have

$$\|f\|_{\gamma,1} := \int_\gamma |f(z)| |dz|.$$

If $p, q > 1$ with $\frac{1}{p} + \frac{1}{q} = 1$, then by Hölder's inequality we have

$$\|f\|_{\gamma,1} \le [l(\gamma)]^{\frac{1}{q}} \|f\|_{\gamma,p}. \tag{7}$$

A motivation to our work follows: These are two complex Opial type inequalities.

Theorem 1. *([1]) Let f be analytic in G, a domain of complex numbers and suppose $\gamma \subset G$ is a smooth path parametrized by $z(t)$, $t \in [a, b]$ from $z(a) = u$ to $z(b) = w$ and $z'(t) \ne 0$ for $t \in (a, b)$.*

(i) If $f(u) = 0$ or $f(w) = 0$, then

$$\int_\gamma |f(z) f'(z)| |dz| \le \left(\int_\gamma l(\gamma_{u,z}) |f'(z)|^2 |dz| \right)^{\frac{1}{2}} \left(\int_\gamma l(\gamma_{z,w}) |f'(z)|^2 |dz| \right)^{\frac{1}{2}} \tag{8}$$

$$\le \frac{1}{2} l(\gamma_{u,w}) \int_\gamma |f'(z)|^2 |dz|.$$

(ii) If $f(u) = f(w) = 0$, then

$$\int_\gamma |f(z) f'(z)| |dz| \leq$$

$$\frac{1}{2} \left[\int_\gamma (l(\gamma_{u,w}) - |l(\gamma_{u,z}) - l(\gamma_{z,w})|) |f'(z)|^2 |dz| \right]^{\frac{1}{2}} \cdot \quad (9)$$

$$\left[\int_\gamma |l(\gamma_{u,z}) - l(\gamma_{z,w})| |f'(z)|^2 |dz| \right]^{\frac{1}{2}}$$

$$\leq \frac{1}{4} l(\gamma_{u,w}) \int_\gamma |f'(z)|^2 |dz|.$$

In this chapter we utilize on $\mathbb{C}$ the results of [2] related to Ostrowski type inequalities for general Banach space valued functions. So we produce here advanced and general complex Ostrowski type inequalities.

2. BACKGROUND

Here we follow [2].

We need:

Definition 2. ([2]) Let $[a,b] \subset \mathbb{R}$, $(X, \|\cdot\|)$ a Banach space, $g \in C^1([a,b])$ and increasing, $f \in C([a,b], X)$, $\nu > 0$. We define the left Riemann–Liouville generalized fractional Bochner integral operator

$$\left(J_{a;g}^\nu f \right)(x) := \frac{1}{\Gamma(\nu)} \int_a^x (g(x) - g(z))^{\nu-1} g'(z) f(z) dz, \quad (10)$$

$\forall\, x \in [a,b]$, where Γ is the gamma function.

The last integral is of Bochner type ([3]). Since $f \in C([a,b], X)$, then $f \in L_\infty([a,b], X)$. By [2] we get that $\left(J_{a;g}^\nu f \right) \in C([a,b], X)$. Above we set $J_{a;g}^0 f := f$ and see that $\left(J_{a;g}^\nu f \right)(a) = 0$.

We mention:

Theorem 3. ([2]) Let all as in Definition 2. Let $m, n > 0$ and $f \in C([a,b], X)$. Then

$$J_{a;g}^m J_{a;g}^n f = J_{a;g}^{m+n} f = J_{a;g}^n J_{a;g}^m f. \quad (11)$$

We need:

Definition 4. ([2]) Let $[a,b] \subset \mathbb{R}$, $(X, \|\cdot\|)$ a Banach space, $g \in C^1([a,b])$ and increasing, $f \in C([a,b], X)$, $\nu > 0$.

We define the right Riemann–Liouville generalized fractional Bochner integral operator

$$\left(J_{b-;g}^\nu f \right)(x) := \frac{1}{\Gamma(\nu)} \int_x^b (g(z) - g(x))^{\nu-1} g'(z) f(z) dz, \quad (12)$$

$\forall\, x \in [a,b]$, where Γ is the gamma function.

The last integral is of Bochner type. Since $f \in C([a,b], X)$, then $f \in L_\infty([a,b], X)$. By ([2]) we get that $\left(J_{b-;g}^\nu f\right) \in C([a,b], X)$. Above we set $J_{b-;g}^0 f := f$ and see that $\left(J_{b-;g}^\nu f\right)(b) = 0$.

We mention:

Theorem 5. *([2]) Let all as in Definition 4. Let $\alpha, \beta > 0$ and $f \in C([a,b], X)$. Then*

$$\left(J_{b-;g}^\alpha J_{b-;g}^\beta f\right)(x) = \left(J_{b-;g}^{\alpha+\beta} f\right)(x) = \left(J_{b-;g}^\beta J_{b-;g}^\alpha f\right)(x), \tag{13}$$

$\forall\, x \in [a,b]$.

We need:

Definition 6. *([2]) Let $\alpha > 0$, $\lceil \alpha \rceil = n$, $\lceil \cdot \rceil$ the ceiling of the number. Let $f \in C^n([a,b], X)$, where $[a,b] \subset \mathbb{R}$, and $(X, \|\cdot\|)$ is a Banach space. Let $g \in C^1([a,b])$, strictly increasing, such that $g^{-1} \in C^n([g(a), g(b)])$. We define the left generalized g–fractional derivative X–valued of f of order α as follows:*

$$\left(D_{a+;g}^\alpha f\right)(x) := \frac{1}{\Gamma(n-\alpha)} \int_a^x (g(x) - g(t))^{n-\alpha-1} g'(t) \left(f \circ g^{-1}\right)^{(n)}(g(t))\, dt, \tag{14}$$

$\forall\, x \in [a,b]$. The last integral is of Bochner type.

Derivatives for vector valued functions are defined according to [4], p. 83, similar to numerical ones.

If $\alpha \notin \mathbb{N}$, by [2], we have that $\left(D_{a+;g}^\alpha f\right) \in C([a,b], X)$.

We see that

$$\left(J_{a;g}^{n-\alpha}\left(\left(f \circ g^{-1}\right)^{(n)} \circ g\right)\right)(x) = \left(D_{a+;g}^\alpha f\right)(x), \quad \forall\, x \in [a,b]. \tag{15}$$

We set

$$D_{a+;g}^n f(x) := \left(\left(f \circ g^{-1}\right)^n \circ g\right)(x) \in C([a,b], X), \quad n \in \mathbb{N}, \tag{16}$$

$$D_{a+;g}^0 f(x) = f(x), \quad \forall\, x \in [a,b].$$

When $g = id$, then

$$D_{a+;g}^\alpha f = D_{a+;id}^\alpha f = D_{*a}^\alpha f, \tag{17}$$

the usual left X–valued Caputo fractional derivative, see [5].

We need

Definition 7. *([2]) Let $\alpha > 0$, $\lceil \alpha \rceil = n$, $\lceil \cdot \rceil$ the ceiling of the number. Let $f \in C^n([a,b], X)$, where $[a,b] \subset \mathbb{R}$, and $(X, \|\cdot\|)$ is a Banach space. Let $g \in C^1([a,b])$, strictly increasing, such that $g^{-1} \in C^n([g(a), g(b)])$. We define the right generalized g–fractional derivative X–valued of f of order α as follows:*

$$\left(D_{b-;g}^\alpha f\right)(x) := \frac{(-1)^n}{\Gamma(n-\alpha)} \int_x^b (g(t) - g(x))^{n-\alpha-1} g'(t) \left(f \circ g^{-1}\right)^{(n)}(g(t))\, dt, \tag{18}$$

$\forall\, x \in [a,b]$. The last integral is of Bochner type.

If $\alpha \notin \mathbb{N}$, by [2], we have that $\left(D_{b-;g}^{\alpha} f\right) \in C\left([a,b], X\right)$.

We see that
$$J_{b-;g}^{n-\alpha}\left((-1)^{n} \left(f \circ g^{-1}\right)^{(n)} \circ g\right)(x) = \left(D_{b-;g}^{\alpha} f\right)(x), \quad a \le x \le b. \tag{19}$$

We set
$$D_{b-;g}^{n} f(x) := (-1)^{n}\left(\left(f \circ g^{-1}\right)^{n} \circ g\right)(x) \in C\left([a,b], X\right), \quad n \in \mathbb{N}, \tag{20}$$
$$D_{b-;g}^{0} f(x) := f(x), \quad \forall\, x \in [a,b].$$

When $g = id$, then
$$D_{b-;g}^{\alpha} f(x) = D_{b-;id}^{\alpha} f(x) = D_{b-}^{\alpha} f, \tag{21}$$

the usual right X-valued Caputo fractional derivative, see [6], [7].

We mention the following general left fractional Taylor's formula:

Theorem 8. *([2]) Let $\alpha > 0$, $n = \lceil \alpha \rceil$, and $f \in C^{n}\left([a,b], X\right)$, where $[a,b] \subset \mathbb{R}$ and $(X, \|\cdot\|)$ is a Banach space. Let $g \in C^{1}\left([a,b]\right)$, strictly increasing, such that $g^{-1} \in C^{n}\left([g(a), g(b)]\right)$, $a \le x \le b$. Then*

$$f(x) = f(a) + \sum_{i=1}^{n-1} \frac{(g(x) - g(a))^{i}}{i!} \left(f \circ g^{-1}\right)^{(i)}(g(a)) +$$
$$\frac{1}{\Gamma(\alpha)} \int_{a}^{x} (g(x) - g(t))^{\alpha-1} g'(t) \left(D_{a+;g}^{\alpha} f\right)(t)\, dt =$$
$$f(a) + \sum_{i=1}^{n-1} \frac{(g(x) - g(a))^{i}}{i!} \left(f \circ g^{-1}\right)^{(i)}(g(a)) + \tag{22}$$
$$\frac{1}{\Gamma(\alpha)} \int_{g(a)}^{g(x)} (g(x) - z)^{\alpha-1} \left(\left(D_{a+;g}^{\alpha} f\right) \circ g^{-1}\right)(z)\, dz.$$

We also mention the following general right fractional Taylor's formula:

Theorem 9. *([2]) Let $\alpha > 0$, $n = \lceil \alpha \rceil$, and $f \in C^{n}\left([a,b], X\right)$, where $[a,b] \subset \mathbb{R}$ and $(X, \|\cdot\|)$ is a Banach space. Let $g \in C^{1}\left([a,b]\right)$, strictly increasing, such that $g^{-1} \in C^{n}\left([g(a), g(b)]\right)$, $a \le x \le b$. Then*

$$f(x) = f(b) + \sum_{i=1}^{n-1} \frac{(g(x) - g(b))^{i}}{i!} \left(f \circ g^{-1}\right)^{(i)}(g(b)) +$$
$$\frac{1}{\Gamma(\alpha)} \int_{x}^{b} (g(t) - g(x))^{\alpha-1} g'(t) \left(D_{b-;g}^{\alpha} f\right)(t)\, dt =$$
$$f(b) + \sum_{i=1}^{n-1} \frac{(g(x) - g(b))^{i}}{i!} \left(f \circ g^{-1}\right)^{(i)}(g(b)) + \tag{23}$$
$$\frac{1}{\Gamma(\alpha)} \int_{g(x)}^{g(b)} (z - g(x))^{\alpha-1} \left(\left(D_{b-;g}^{\alpha} f\right) \circ g^{-1}\right)(z)\, dz.$$

From Theorem 8 when $0 < \alpha \le 1$, we get that

$$\left(I_{a+;g}^{\alpha} D_{a+;g}^{\alpha} f\right)(x) = f(x) - f(a) =$$

$$\frac{1}{\Gamma(\alpha)} \int_{a}^{x} (g(x) - g(t))^{\alpha-1} g'(t) \left(D_{a+;g}^{\alpha} f\right)(t) dt = \qquad (24)$$

$$\frac{1}{\Gamma(\alpha)} \int_{g(a)}^{g(x)} (g(x) - z)^{\alpha-1} \left(\left(D_{a+;g}^{\alpha} f\right) \circ g^{-1}\right)(z) dz,$$

and by Theorem 9 when $0 < \alpha \le 1$ we get

$$\left(I_{b-;g}^{\alpha} D_{b-;g}^{\alpha} f\right)(x) = f(x) - f(b) =$$

$$\frac{1}{\Gamma(\alpha)} \int_{x}^{b} (g(t) - g(x))^{\alpha-1} g'(t) \left(D_{b-;g}^{\alpha} f\right)(t) dt = \qquad (25)$$

$$\frac{1}{\Gamma(\alpha)} \int_{g(x)}^{g(b)} (z - g(x))^{\alpha-1} \left(\left(D_{b-;g}^{\alpha} f\right) \circ g^{-1}\right)(z) dz,$$

all $a \le x \le b$.

Above we considered $f \in C^1([a,b], X)$, $g \in C^1([a,b])$, strictly increasing, such that $g^{-1} \in C^1([g(a), g(b)])$.

Denote by

$$D_{a+;g}^{n\alpha} := D_{a+;g}^{\alpha} D_{a+;g}^{\alpha} \cdots D_{a+;g}^{\alpha} \quad (n \text{ times}), n \in \mathbb{N}. \qquad (26)$$

Also denote by

$$I_{a+;g}^{n\alpha} := I_{a+;g}^{\alpha} I_{a+;g}^{\alpha} \cdots I_{a+;g}^{\alpha} \quad (n \text{ times}), \qquad (27)$$

and remind

$$\left(I_{a+;g}^{\alpha} f\right)(x) = \frac{1}{\Gamma(\alpha)} \int_{a}^{x} (g(x) - g(t))^{\alpha-1} g'(t) f(t) dt, \ x \ge a. \qquad (28)$$

By convention $I_{a+;g}^{0} = D_{a+;g}^{0} = I$ (identity operator).

We mention the following g–left generalized modified X–valued Taylor's formula.

Theorem 10. *([2]) Let $0 < \alpha \le 1$, $n \in \mathbb{N}$, $f \in C^1([a,b], X)$, $g \in C^1([a,b])$, strictly increasing, such that $g^{-1} \in C^1([g(a), g(b)])$. Let $F_k := D_{a+;g}^{k\alpha} f$, $k = 1, \ldots, n$, that fulfill $F_k \in C^1([a,b], X)$, and $F_{n+1} \in C([a,b], X)$.*

Then

$$f(x) = \sum_{i=0}^{n} \frac{(g(x) - g(a))^{i\alpha}}{\Gamma(i\alpha + 1)} \left(D_{a+;g}^{i\alpha} f\right)(a) +$$

$$\frac{1}{\Gamma((n+1)\alpha)} \int_{a}^{x} (g(x) - g(t))^{(n+1)\alpha-1} g'(t) \left(D_{a+;g}^{(n+1)\alpha} f\right)(t) dt, \qquad (29)$$

$\forall\, x \in [a, b]$.

Denote by
$$D_{b-;g}^{n\alpha} := D_{b-;g}^{\alpha} D_{b-;g}^{\alpha} \cdots D_{b-;g}^{\alpha} \quad (n \text{ times}), n \in \mathbb{N}. \tag{30}$$

Also denote by
$$I_{b-;g}^{n\alpha} := I_{b-;g}^{\alpha} I_{b-;g}^{\alpha} \cdots I_{b-;g}^{\alpha} \quad (n \text{ times}), \tag{31}$$

and remind
$$\left(I_{b-;g}^{\alpha} f\right)(x) = \frac{1}{\Gamma(\alpha)} \int_x^b (g(t) - g(x))^{\alpha-1} g'(t) f(t) \, dt, \quad x \leq b. \tag{32}$$

We also mention the following g–right generalized modified X–valued Taylor's formula.

Theorem 11. *([2]) Let $f \in C^1([a,b], X)$, $g \in C^1([a,b])$, strictly increasing, such that $g^{-1} \in C^1([g(a), g(b)])$. Suppose that $F_k := D_{b-;g}^{k\alpha} f$, $k = 1, \ldots, n$, fulfill $F_k \in C^1([a,b], X)$, and $F_{n+1} \in C([a,b], X)$, where $0 < \alpha \leq 1$, $n \in \mathbb{N}$.*
Then
$$f(x) = \sum_{i=0}^n \frac{(g(b) - g(x))^{i\alpha}}{\Gamma(i\alpha + 1)} \left(D_{b-;g}^{i\alpha} f\right)(b) +$$
$$\frac{1}{\Gamma((n+1)\alpha)} \int_x^b (g(t) - g(x))^{(n+1)\alpha - 1} g'(t) \left(D_{b-;g}^{(n+1)\alpha} f\right)(t) \, dt, \tag{33}$$

$\forall\, x \in [a,b]$.

Next we refer to a related generalized fractional Ostrowski type inequality:

Theorem 12. *([2]) Let $g \in C^1([a,b])$ and strictly increasing, such that $g^{-1} \in C^1([g(a), g(b)])$, and $0 < \alpha < 1$, $n \in \mathbb{N}$, $f \in C^1([a,b], X)$, where $(X, \|\cdot\|)$ is a Banach space. Let $x_0 \in [a,b]$ be fixed. Assume that $F_k^{x_0} := D_{x_0-;g}^{k\alpha} f$, for $k = 1, \ldots, n$, fulfill $F_k^{x_0} \in C^1([a,b], X)$ and $F_{n+1}^{x_0} \in C([a, x_0], X)$ and $\left(D_{x_0-;g}^{i\alpha} f\right)(x_0) = 0$, $i = 1, \ldots, n$.*
Similarly, we assume that $G_k^{x_0} := D_{x_0+;g}^{k\alpha} f$, for $k = 1, \ldots, n$, fulfill $G_k^{x_0} \in C^1([x_0, b], X)$ and $G_{n+1}^{x_0} \in C([x_0, b], X)$ and $\left(D_{x_0+;g}^{i\alpha} f\right)(x_0) = 0$, $i = 1, \ldots, n$.
Then
$$\left\| \frac{1}{b-a} \int_a^b f(x) \, dx - f(x_0) \right\| \leq \frac{1}{(b-a)\Gamma((n+1)\alpha + 1)} \cdot$$
$$\left\{ (g(b) - g(x_0))^{(n+1)\alpha} (b - x_0) \left\| D_{x_0+;g}^{(n+1)\alpha} f \right\|_{\infty, [x_0, b]} + \right.$$
$$\left. (g(x_0) - g(a))^{(n+1)\alpha} (x_0 - a) \left\| D_{x_0-;g}^{(n+1)\alpha} f \right\|_{\infty, [a, x_0]} \right\}. \tag{34}$$

We mention:

Remark 13. *Some examples for g follow:*
$$g(x) = x, \quad x \in [a,b], \tag{35}$$
$$g(x) = e^x, \quad x \in [a,b] \subset \mathbb{R},$$

also

$$g(x) = \sin x,$$
$$g(x) = \tan x, \text{ when } x \in [a,b] := \left[-\frac{\pi}{2} + \varepsilon, \frac{\pi}{2} - \varepsilon\right], \varepsilon > 0 \text{ small}, \quad (36)$$

and

$$g(x) = \cos x, \text{ when } x \in [a,b] := [\pi + \varepsilon, 2\pi - \varepsilon], \varepsilon > 0 \text{ small}. \quad (37)$$

Above all g's are strictly increasing, $g \in C^1([a,b])$, and $g^{-1} \in C^n([g(a), g(b)])$, for any $n \in \mathbb{N}$.

Applications od Theorem 12 follow:
We give the following exponential Ostrowski type fractional inequality:

Theorem 14. *([2]) Let $0 < \alpha < 1$, $n \in \mathbb{N}$, $f \in C^1([a,b], X)$, where $(X, \|\cdot\|)$ is a Banach space, $x_0 \in [a,b]$. Assume that $F_k^{x_0} := D_{x_0-;e^t}^{k\alpha} f$, for $k = 1, \ldots, n$, fulfill $F_k^{x_0} \in C^1([a, x_0], X)$ and $F_{n+1}^{x_0} \in C([a, x_0], X)$ and $\left(D_{x_0-;e^t}^{i\alpha} f\right)(x_0) = 0$, $i = 1, \ldots, n$.*

Similarly, we assume that $G_k^{x_0} := D_{x_0+;e^t}^{k\alpha} f$, for $k = 1, \ldots, n$, fulfill $G_k^{x_0} \in C^1([x_0, b], X)$ and $G_{n+1}^{x_0} \in C([x_0, b], X)$ and $\left(D_{x_0+;e^t}^{i\alpha} f\right)(x_0) = 0$, $i = 1, \ldots, n$.

Then

$$\left\|\frac{1}{b-a}\int_a^b f(x)\,dx - f(x_0)\right\| \leq \frac{1}{(b-a)\Gamma((n+1)\alpha+1)}. \quad (38)$$

$$\left\{\left(e^b - e^{x_0}\right)^{(n+1)\alpha}(b - x_0)\left\|D_{x_0+;e^t}^{(n+1)\alpha} f\right\|_{\infty,[x_0,b]} + \right.$$
$$\left. \left(e^{x_0} - e^a\right)^{(n+1)\alpha}(x_0 - a)\left\|D_{x_0-;e^t}^{(n+1)\alpha} f\right\|_{\infty,[a,x_0]}\right\}.$$

We finish this section with the following trigonometric Ostrowski type fractional inequality:

Theorem 15. *([2]) Let $0 < \alpha < 1$, $n \in \mathbb{N}$, $f \in C^1([\pi + \varepsilon, 2\pi - \varepsilon], X)$, $\varepsilon > 0$ small, where $(X, \|\cdot\|)$ is a Banach space, $x_0 \in [\pi + \varepsilon, 2\pi - \varepsilon]$. Assume that $F_k^{x_0} := D_{x_0-;\cos}^{k\alpha} f$, for $k = 1, \ldots, n$, fulfill $F_k^{x_0} \in C^1([\pi + \varepsilon, x_0], X)$ and $F_{n+1}^{x_0} \in C([\pi + \varepsilon, x_0], X)$ and $\left(D_{x_0-;\cos}^{i\alpha} f\right)(x_0) = 0$, $i = 1, \ldots, n$.*

Similarly, we assume that $G_k^{x_0} := D_{x_0+;\cos}^{k\alpha} f$, for $k = 1, \ldots, n$, fulfill $G_k^{x_0} \in C^1([x_0, 2\pi - \varepsilon], X)$ and $G_{n+1}^{x_0} \in C([x_0, 2\pi - \varepsilon], X)$ and $\left(D_{x_0+;\cos}^{i\alpha} f\right)(x_0) = 0$, $i = 1, \ldots, n$.

Then

$$\left\|\frac{1}{\pi - 2\varepsilon}\int_{\pi+\varepsilon}^{2\pi-\varepsilon} f(x)\,dx - f(x_0)\right\| \leq \frac{1}{(\pi - 2\varepsilon)\Gamma((n+1)\alpha+1)}.$$

$$\left\{(\cos(2\pi - \varepsilon) - \cos x_0)^{(n+1)\alpha}(2\pi - \varepsilon - x_0)\left\|D_{x_0+;\cos}^{(n+1)\alpha} f\right\|_{\infty,[x_0, 2\pi-\varepsilon]} + \right.$$
$$\left. (\cos x_0 - \cos(\pi + \varepsilon))^{(n+1)\alpha}(x_0 - \pi - \varepsilon)\left\|D_{x_0-;\cos}^{(n+1)\alpha} f\right\|_{\infty,[\pi+\varepsilon, x_0]}\right\}. \quad (39)$$

Important results of this background: Theorems 12, 14, 15 next are applied for $X = \mathbb{C}$, the Banach space of complex numbers with $\|\cdot\| = |\cdot|$, the absolute value.

3. Main Results

We start with some history of the topic of Ostrowski type inequalities:

In 1938, A. Ostrowski [8], proved the following inequality concerning the distance between the integral mean $\frac{1}{b-a}\int_a^b f(t)\,dt$ and the value $f(x)$, $x \in [a,b]$.

Theorem 16. *(Ostrowski, 1938 [8]) Let $f : [a,b] \to \mathbb{R}$ be continuous on $[a,b]$ and differentiable on (a,b) such that $f' : (a,b) \to \mathbb{R}$ is bounded on (a,b), i.e., $\|f'\|_\infty := \sup_{t\in(a,b)} |f'(t)| < \infty$. Then*

$$\left| f(x) - \frac{1}{b-a} \int_a^b f(t)\,dt \right| \leq \left[\frac{1}{4} + \left(\frac{x - \frac{a+b}{2}}{b-a} \right)^2 \right] \|f'\|_\infty (b-a), \qquad (40)$$

for all $x \in [a,b]$ and the constant $\frac{1}{4}$ is the best possible.

We present the following advanced generalized fractional $\mathbb{C}$–Ostrowski type inequalities:

Theorem 17. *Let $g \in C^1([a,b])$ and strictly increasing, such that $g^{-1} \in C^1([g(a), g(b)])$, and $0 < \alpha < 1$, $n \in \mathbb{N}$, $h \in C^1([a,b], \mathbb{C})$. Let $x_0 \in [a,b]$ be fixed. Assume that $F_k^{x_0} := D_{x_0-;g}^{k\alpha} h$, for $k = 1, \ldots, n$, fulfill $F_k^{x_0} \in C^1([a,b], \mathbb{C})$ and $F_{n+1}^{x_0} \in C([a, x_0], \mathbb{C})$ and $\left(D_{x_0-;g}^{i\alpha} h\right)(x_0) = 0$, $i = 1, \ldots, n$.*

Similarly, we assume that $G_k^{x_0} := D_{x_0+;g}^{k\alpha} h$, for $k = 1, \ldots, n$, fulfill $G_k^{x_0} \in C^1([x_0, b], \mathbb{C})$ and $G_{n+1}^{x_0} \in C([x_0, b], \mathbb{C})$ and $\left(D_{x_0+;g}^{i\alpha} h\right)(x_0) = 0$, $i = 1, \ldots, n$.

Then

$$\left| \frac{1}{b-a} \int_a^b h(x)\,dx - h(x_0) \right| \leq \frac{1}{(b-a)\Gamma((n+1)\alpha + 1)} \cdot \qquad (41)$$

$$\left\{ (g(b) - g(x_0))^{(n+1)\alpha} (b - x_0) \left\| D_{x_0+;g}^{(n+1)\alpha} h \right\|_{\infty, [x_0, b]} + \right.$$

$$\left. (g(x_0) - g(a))^{(n+1)\alpha} (x_0 - a) \left\| D_{x_0-;g}^{(n+1)\alpha} h \right\|_{\infty, [a, x_0]} \right\}.$$

Proof. By Theorem 12. $\square$

Theorem 18. *Let $0 < \alpha < 1$, $n \in \mathbb{N}$, $h \in C^1([a,b], \mathbb{C})$, $x_0 \in [a,b]$. Assume that $F_k^{x_0} := D_{x_0-;e^t}^{k\alpha} h$, for $k = 1, \ldots, n$, fulfill $F_k^{x_0} \in C^1([a, x_0], \mathbb{C})$ and $F_{n+1}^{x_0} \in C([a, x_0], \mathbb{C})$ and $\left(D_{x_0-;e^t}^{i\alpha} h\right)(x_0) = 0$, $i = 1, \ldots, n$.*

Similarly, we assume that $G_k^{x_0} := D_{x_0+;e^t}^{k\alpha} h$, for $k = 1, \ldots, n$, fulfill $G_k^{x_0} \in C^1([x_0, b], \mathbb{C})$ and $G_{n+1}^{x_0} \in C([x_0, b], \mathbb{C})$ and $\left(D_{x_0+;e^t}^{i\alpha} h\right)(x_0) = 0$, $i = 1, \ldots, n$.

Then

$$\left| \frac{1}{b-a} \int_a^b h(x)\,dx - h(x_0) \right| \leq \frac{1}{(b-a)\Gamma((n+1)\alpha + 1)} \cdot \qquad (42)$$

$$\left\{ \left(e^b - e^{x_0}\right)^{(n+1)\alpha} (b - x_0) \left\| D_{x_0+;e^t}^{(n+1)\alpha} h \right\|_{\infty, [x_0, b]} + \right.$$

$$(e^{x_0} - e^a)^{(n+1)\alpha} (x_0 - a) \left\| D_{x_0-;e^t}^{(n+1)\alpha} h \right\|_{\infty,[a,x_0]} \Big\}.$$

Proof. By Theorem 14. □

Theorem 19. *Let* $0 < \alpha < 1$, $n \in \mathbb{N}$, $h \in C^1([\pi + \varepsilon, 2\pi - \varepsilon], \mathbb{C})$, $\varepsilon > 0$ *small*, $x_0 \in [\pi + \varepsilon, 2\pi - \varepsilon]$. *Assume that* $F_k^{x_0} := D_{x_0-;\cos}^{k\alpha} h$, *for* $k = 1, \ldots, n$, *fulfill* $F_k^{x_0} \in C^1([\pi + \varepsilon, x_0], \mathbb{C})$ *and* $F_{n+1}^{x_0} \in C([\pi + \varepsilon, x_0], \mathbb{C})$ *and* $\left(D_{x_0-;\cos}^{i\alpha} h \right)(x_0) = 0$, $i = 1, \ldots, n$.

Similarly, we assume that $G_k^{x_0} := D_{x_0+;\cos}^{k\alpha} h$, *for* $k = 1, \ldots, n$, *fulfill* $G_k^{x_0} \in C^1([x_0, 2\pi - \varepsilon], \mathbb{C})$ *and* $G_{n+1}^{x_0} \in C([x_0, 2\pi - \varepsilon], \mathbb{C})$ *and* $\left(D_{x_0+;\cos}^{i\alpha} h \right)(x_0) = 0$, $i = 1, \ldots, n$.
Then

$$\left| \frac{1}{\pi - 2\varepsilon} \int_{\pi+\varepsilon}^{2\pi-\varepsilon} h(x)\, dx - h(x_0) \right| \leq \frac{1}{(\pi - 2\varepsilon)\Gamma((n+1)\alpha + 1)}. \quad (43)$$

$$\left\{ (\cos(2\pi - \varepsilon) - \cos x_0)^{(n+1)\alpha} (2\pi - \varepsilon - x_0) \left\| D_{x_0+;\cos}^{(n+1)\alpha} h \right\|_{\infty,[x_0,2\pi-\varepsilon]} + \right.$$

$$\left. (\cos x_0 - \cos(\pi + \varepsilon))^{(n+1)\alpha} (x_0 - \pi - \varepsilon) \left\| D_{x_0-;\cos}^{(n+1)\alpha} h \right\|_{\infty,[\pi+\varepsilon,x_0]} \right\}.$$

Proof. By Theorem 15. □

From now on $f(z)$, $z(t)$, $t \in (a,b)$, γ will be as in section 1. Introduction. Put $z(a) = u$, $z(b) = w$ and $z(c) = v$, where $u, w, v \in \mathbb{C}$, with $c \in [a,b]$.

We will use here $h(t) := f(z(t)) z'(t)$, $t \in [a,b]$.

In that case we will have

$$\left| \frac{1}{b-a} \int_a^b h(t)\, dt - h(c) \right| = \left| \frac{1}{b-a} \int_a^b f(z(t)) z'(t)\, dt - f(z(c)) z'(c) \right| \stackrel{(1)}{=}$$

$$\left| \frac{1}{b-a} \int_{\gamma_{u,w}} f(z)\, dz - f(v) z'(c) \right| \stackrel{(1)}{=} \left| \frac{1}{b-a} \int_\gamma f(z)\, dz - f(v) z'(c) \right|, \quad (44)$$

where $\gamma_{u,w} = \gamma$.

We have the following advanced generalized fractional complete $\mathbb{C}$-Ostrowski type inequalities:

Theorem 20. *Let* $g \in C^1([a,b])$ *and strictly increasing, such that* $g^{-1} \in C^1([g(a), g(b)])$, *and* $0 < \alpha < 1$, $n \in \mathbb{N}$, $f(z(\cdot)) z'(\cdot) \in C^1([a,b], \mathbb{C})$. *Let* $c \in [a,b]$ *be fixed. Assume that* $F_k^c := D_{c-;g}^{k\alpha} (f(z(\cdot)) z'(\cdot))$, *for* $k = 1, \ldots, n$, *fulfill* $F_k^c \in C^1([a,b], \mathbb{C})$ *and* $F_{n+1}^c \in C([a,c], \mathbb{C})$ *and* $\left(D_{c-;g}^{i\alpha} (f(z(\cdot)) z'(\cdot)) \right)(c) = 0$, $i = 1, \ldots, n$.

Similarly, we assume that $G_k^c := D_{c+;g}^{k\alpha} (f(z(\cdot)) z'(\cdot))$, *for* $k = 1, \ldots, n$, *fulfill* $G_k^c \in C^1([c,b], \mathbb{C})$ *and* $G_{n+1}^c \in C([c,b], \mathbb{C})$ *and* $\left(D_{c+;g}^{i\alpha} (f(z(\cdot)) z'(\cdot)) \right)(c) = 0$, $i = 1, \ldots, n$.
Then

$$\left| \frac{1}{b-a} \int_{\gamma_{u,w}} f(z)\, dz - f(v) z'(c) \right| \leq \frac{1}{(b-a)\Gamma((n+1)\alpha + 1)}. \quad (45)$$

$$\left\{ (g(b) - g(c))^{(n+1)\alpha} (b-c) \left\| D_{c+;g}^{(n+1)\alpha} (f(z(\cdot))z'(\cdot)) \right\|_{\infty,[c,b]} + \right.$$
$$\left. (g(c) - g(a))^{(n+1)\alpha} (c-a) \left\| D_{c-;g}^{(n+1)\alpha} (f(z(\cdot))z'(\cdot)) \right\|_{\infty,[a,c]} \right\}.$$

Proof. By Theorem 17. □

We continue with:

Theorem 21. Let $0 < \alpha < 1$, $n \in \mathbb{N}$, $f(z(\cdot))z'(\cdot) \in C^1([a,b],\mathbb{C})$, $c \in [a,b]$. Assume that $F_k^c := D_{c-;e^t}^{k\alpha}(f(z(\cdot))z'(\cdot))$, for $k = 1, \ldots, n$, fulfill $F_k^c \in C^1([a,c],\mathbb{C})$ and $F_{n+1}^c \in C([a,c],\mathbb{C})$ and $\left(D_{c-;e^t}^{i\alpha}(f(z(\cdot))z'(\cdot)) \right)(c) = 0$, $i = 1, \ldots, n$.

Similarly, we assume that $G_k^c := D_{c+;e^t}^{k\alpha}(f(z(\cdot))z'(\cdot))$, for $k = 1, \ldots, n$, fulfill $G_k^c \in C^1([c,b],\mathbb{C})$ and $G_{n+1}^c \in C([c,b],\mathbb{C})$ and $\left(D_{c+;e^t}^{i\alpha}(f(z(\cdot))z'(\cdot)) \right)(c) = 0$, $i = 1, \ldots, n$.

Then

$$\left| \frac{1}{b-a} \int_{\gamma_{u,w}} f(z)\,dz - f(v)z'(c) \right| \leq \frac{1}{(b-a)\Gamma((n+1)\alpha+1)}. \quad (46)$$

$$\left\{ \left(e^b - e^c\right)^{(n+1)\alpha} (b-c) \left\| D_{c+;e^t}^{(n+1)\alpha}(f(z(\cdot))z'(\cdot)) \right\|_{\infty,[c,b]} + \right.$$
$$\left. (e^c - e^a)^{(n+1)\alpha} (c-a) \left\| D_{c-;e^t}^{(n+1)\alpha}(f(z(\cdot))z'(\cdot)) \right\|_{\infty,[a,c]} \right\}.$$

Proof. By Theorem 18. □

Finally and additionally, we choose that $a = \pi + \varepsilon$, $b = 2\pi - \varepsilon$, where $\varepsilon > 0$ is small, and $c \in [\pi + \varepsilon, 2\pi - \varepsilon]$. So here it is $z(\pi + \varepsilon) = u$, $z(2\pi - \varepsilon) = w$ and $z(c) = v$, where $u, w, u \in \mathbb{C}$.

We present:

Theorem 22. Let $0 < \alpha < 1$, $n \in \mathbb{N}$, $f(z(\cdot))z'(\cdot) \in C^1([\pi+\varepsilon, 2\pi-\varepsilon],\mathbb{C})$, $\varepsilon > 0$ small, $c \in [\pi+\varepsilon, 2\pi-\varepsilon]$. Assume that $F_k^c := D_{c-;\cos}^{k\alpha}(f(z(\cdot))z'(\cdot))$, for $k = 1, \ldots, n$, fulfill $F_k^c \in C^1([\pi+\varepsilon,c],\mathbb{C})$ and $F_{n+1}^c \in C([\pi+\varepsilon,c],\mathbb{C})$ and $\left(D_{c-;\cos}^{i\alpha}(f(z(\cdot))z'(\cdot)) \right)(c) = 0$, $i = 1, \ldots, n$.

Similarly, we assume that $G_k^c := D_{c+;\cos}^{k\alpha}(f(z(\cdot))z'(\cdot))$, for $k = 1, \ldots, n$, fulfill $G_k^c \in C^1([c, 2\pi-\varepsilon],\mathbb{C})$ and $G_{n+1}^c \in C([c, 2\pi-\varepsilon],\mathbb{C})$ and $\left(D_{c+;\cos}^{i\alpha}(f(z(\cdot))z'(\cdot)) \right)(c) = 0$, $i = 1, \ldots, n$.

Then

$$\left| \frac{1}{\pi - 2\varepsilon} \int_{\gamma_{u,w}} f(z)\,dz - f(v)z'(c) \right| \leq \frac{1}{(\pi - 2\varepsilon)\Gamma((n+1)\alpha+1)}. \quad (47)$$

$$\left\{ (\cos(2\pi - \varepsilon) - \cos c)^{(n+1)\alpha} (2\pi - \varepsilon - c) \left\| D_{c+;\cos}^{(n+1)\alpha}(f(z(\cdot))z'(\cdot)) \right\|_{\infty,[c,2\pi-\varepsilon]} + \right.$$
$$\left. (\cos c - \cos(\pi + \varepsilon))^{(n+1)\alpha} (c - \pi - \varepsilon) \left\| D_{c-;\cos}^{(n+1)\alpha}(f(z(\cdot))z'(\cdot)) \right\|_{\infty,[\pi+\varepsilon,c]} \right\}.$$

Proof. By Theorem 19. □

REFERENCES

[1] Dragomir, S. S. 2019. "An extension of Opial's inequality to the complex integral". *RGMIA Res. Rep. Coll.* 22, Art. 9, 9 pp., rgmia.org/v22.php.

[2] Anastassiou, G. A. 2017. "Principles of General Fractional Analysis for Banach space valued functions". *Bulletin of Allahabad Math. Soc.* 32(1): 71–145.

[3] Mikusinski, J. 1978. *The Bochner integral.* New York: Academic Press.

[4] Shilov, G. E. 1996. *Elementary Functional Analysis.* New York: Dover Publications, Inc.

[5] Anastassiou, G. A. 2017. "A strong Fractional Calculus Theory for Banach space valued functions". *Nonlinear Functional Analysis and Applications* 22(3): 495–524.

[6] Anastassiou, G. A. 2017. "Strong Right Fractional Calculus for Banach space valued functions". *Revista Proyecciones* 36(1): 149–186.

[7] Anastassiou, G. A. 2017. "Strong mixed and generalized fractional calculus for Banach space valued functions". *Mat. Vesnik* 69(3): 176–191.

[8] Ostrowski, A. M. 1970. "On an integral inequality". *Aequat. Math.* 4: 358–373.

Chapter 4

IMPROVED QUALITATIVE ANALYSIS FOR NEWTON–LIKE METHODS WITH R–ORDER OF CONVERGENCE AT LEAST THREE IN BANACH SPACES

Ioannis K. Argyros[1,*] **_and Santhosh George_**[2,†]
[1]Department of Mathematical Sciences
Cameron University, Lawton, OK, US
[2]Department of Mathematical and Computational Sciences
National Institute of Technology, Karnataka, India

Abstract

The aim of this study is to extend the applicability of a certain family of Newton–like methods with R–order of convergence at least three. By using our new idea of restricted convergence, we find a more precise location where the iterates lie leading to smaller constants and functions than in earlier studies which in turn lead to a tighter semi–local convergence for these methods. This idea can be used on other iterative methods as well as in the local convergence analysis of these methods. Numerical examples further show the advantages of the new results over the ones in earlier studies.

Keywords: Newton–like methods, Banach space, semi–local convergence

AMS Subject Classification: 65G99, 65H10, 49M15, 47H17

1. Introduction

Let B_1, B_2 be Banach spaces and D be an open and convex subset of B_1.
The problem of finding a locally unique solution p of the equation

$$F(x) = 0 \tag{1}$$

---

[*]Corresponding Author's E-mail: iargyros@cameron.edu.
[†]E-mail: sgeorge@nitk.ac.in.

is very important, since many problems in Sciences, Mathematical Economics, Mathematical Biology and Engineering using Mathematical Modeling can be formulated like (1). It is well known that the solution p can be found in explicit form only in special cases. That is why, in practise we use some iterative method to generate a sequence approximating p.

Next, we introduce a class of iterative methods defined in [1] as

$$x_{n+1} = G(x_n) = x_n - H(L_F(x_n))F'(x_n)^{-1}F(x_n), \qquad (2)$$

where

$$H(L_F(x_n)) = I + \frac{1}{2}L_F(x_n) + \sum_{k \geq 2} A_k L_F(x_n)^k, \ \{A_k\}_{k \geq 2} \subset [0, \infty),$$

$\{A_k\}_{k \geq 2}$ is a non–increasing real sequence with $\sum_{k \geq 2} A_k t^k < \infty$ for $|t| < r$, $L_F(x_n)^k, k \in \mathbb{N}$ stands for $L_F(x)^k = L_F(x) \underset{k-times}{\circ \ldots \circ} L_F(x)$ is a linear operator in D, and $I = L_F(x)^0$. Most popular iterative methods of R–order at least three are special cases of method (2) [1–40].

Chebyshev

$$H(L_F(x_n)) = I + \frac{1}{2}L_F(x_n).$$

Chebyshev–Like method

$$H(L_F(x_n)) = I + \frac{1}{2}L_F(x_n) + \frac{1}{2}L_F(x_n)^2.$$

Euler

$$H(L_F(x_n)) = I + \frac{1}{2}L_F(x_n) + \sum_{k \geq 2}(-1)^k 2^{k+1}\frac{1}{2(k+1)}L_F(x_n)^k.$$

Halley

$$H(L_F(x_n)) = I + \frac{1}{2}L_F(x_n) + \sum_{k \geq 2}\frac{1}{2^k}L_F(x_n)^k.$$

Super–Halley

$$H(L_F(x_n)) = I + \frac{1}{2}L_F(x_n) + \sum_{k \geq 2}\frac{1}{2}L_F(x_n)^k.$$

Ostrowski

$$H(L_F(x_n)) = I + \frac{1}{2}L_F(x_n) + \sum_{k \geq 2}(-1)^k\frac{1}{2k}L_F(x_n)^k.$$

The Logarithmic Method

$$H(L_F(x_n)) = I + \frac{1}{2}L_F(x_n) + \sum_{k \geq 2}\frac{1}{(k+1)!}L_F(x_n)^k.$$

The Exponential Method

$$H(L_F(x_n)) = I + \frac{1}{2}L_F(x_n) + \sum_{k \geq 2}\frac{1}{k+1}L_F(x_n)^k.$$

Therefore, it is important to study method (2) that unifies all the above and other methods. It is well known that as the convergence order of iterative methods increases the convergence domain decreases. That is why we revisit methods (2) to provide under the same conditions as in [1] a finer convergence analysis using our ideas of center Lipschitz condition in combination with the restricted convergence region.

2. SEMI–LOCAL CONVERGENCE

The region of convergence for iterative methods is generally small. In practise this region depends on the: operator F, Fréchet derivatives of F, the initial point x_0 and some non–negative parameters depending on Lipschitz type conditions. As an example, we first consider the semi–local convergence for one-point iterative methods presented in [1]:

Case 1. Kantorovich Conditions

Let B_1, B_2 denote Banach spaces and let D be a non–empty, open and convex subset of B_1. Let F be also a twice continuously differentiable function Fréchet differentiable operator satisfying:

(A_1) There exists $x_0 \in D$ and $b > 0$ such that $\Gamma_0 = F'(x_0)^{-1} \in L(B_2, B_1)$ and $\|\Gamma_0\| \le b$

(A_2) $\|\Gamma_0 F(x_0)\| \le \eta$ for some $\eta \ge 0$

(A_3) There exists $M \ge 0$ such that $\|F''(x)\| \le M$ for all $x \in D$

(A_4) There exists $N \ge 0$ such that $\|F'''(x)\| \le N$ for all $x \in D$

Later [1], the hypothesis (A_4) was replaced by the weaker hypothesis (A_4').
There exists $K \ge 0$ such that $\|F''(x) - F''(y)\| \le K\|x - y\|$ for all $x, y \in D$.
It is very important to notice that M, N, K depend on D that is

$$M = M(D), N = N(D) \quad \text{and} \quad K = K(D). \tag{3}$$

Clearly, the existing results can be improved if these parameters are replaced by at least as small parameters as follows: It follows from (A_3) that:
$(\bar{A}_3)$ There exists $M_0 \ge 0$ such that $\|F'(x) - F'(x_0)\| \le M_0\|x - x_0\|$ for all $x \in D$.
Notice that

$$M_0 \le M \tag{4}$$

holds in general and $\frac{M}{M_0}$ can be arbitrarily large [9].

In the earlier studies the existence of $F'(x)^{-1}$ is established using (A_3) and the Banach lemma on invertible operators to arrive at the estimate

$$\|F'(x)^{-1}\| \le \frac{b}{1 - bM\|x - x_0\|} \tag{5}$$

provided that $x \in U(x_0, \frac{1}{bM})$. However, if one uses $(\bar{A}_3)$ instead of the less precise (A_3) one obtains

$$\|F'(x)^{-1}\| \le \frac{b}{1 - bM_0\|x - x_0\|} \tag{6}$$

provided that $x \in U(x_0, \frac{1}{bM_0})$. Then, the results in the literature can be improved if one uses (6) instead of (5) in the study of the semilocal convergence of these methods. The same technique can be used on other iterative methods as well as in the case of local convergence. It turns out that we can do even better if we use $(\bar{A}_3)$ to introduce at least as small parameters M and N as follows:

Define the set D_0 by $D_0 = D \cap U(x_0, \frac{1}{bM_0})$. Then, replace (A_3), (A_4) and (A'_4) by
$(\bar{A}_3)$ There exists $\bar{M} \geq 0$ such that $||F''(x)|| \leq \bar{M}$ for all $x \in D_0$
$(\bar{A}_4)$ There exists $\bar{N} \geq 0$ such that $||F'''(x)|| \leq \bar{N}$ for all $x \in D_0$
$(\bar{A}'_4)$ There exists $\bar{K} \geq 0$ such that $||F''(x) - F''(y)|| \leq \bar{K}||x-y||$ for all $x, y \in D_0$.
Then, we have that
$$\bar{M} \leq M \tag{7}$$
$$\bar{N} \leq N \tag{8}$$
and
$$\bar{K} \leq K \tag{9}$$
since $D_0 \subseteq D$. Notice that
$$\bar{M} = \bar{M}_0(D_0), \bar{N} = \bar{N}_0(D_0), \bar{K} = \bar{K}_0(D_0). \tag{10}$$

Then, by simply noticing that the iterates $\{x_n\}$ lie in D_0 which is a more precise location than D, one may rewrite all previous results using $\bar{M}, \bar{N}, \bar{K}$ instead of M, N, K, respectively using these modifications in the previous proofs. In the rest of this study we assume that F is a twice continuously Frechet–differentiable on D unless otherwise stated.

As in [1], it is convenient for the semi-local convergence analysis that follows to introduce some scalar functions: $v, h, \phi, f, g(t,s)$ by

$$v(t) = 2 \sum_{k \geq 2} A_k t^{k-2},$$

$$h(t) = 1 + \frac{1}{2}t(1 + tv(t)),$$

$$\phi(t) = 12 + 6t - 6h(t)(1 + 2t) + 3h^2(t)t(2t-1),$$

$$f(t) = \frac{1}{1 - th(t)},$$

and

$$g(t,s) = \frac{t^2}{2}[1 + (1+t)v(t) + \frac{t}{4}(1 + tv(t))^2 + \frac{s}{6}.$$

Next, we present the semi–local convergence of method (2).

Theorem 1. *Suppose: hypothesis* $(A_1), (A_2), (\bar{A}_3), (\bar{\bar{A}}_3)$ *and* (A'_4),

$$\bar{a}_0 = \bar{M}bn < r, \bar{a}_0\bar{h}(\bar{a}_0) \leq 1, \bar{b}_0 = \bar{K}\beta\eta^2 \leq \phi(\bar{a}_0) \tag{11}$$

and

$$U(x_0, \bar{R}) \subseteq D \tag{12}$$

hold, where
$$\bar{R} = \frac{h(\bar{a}_0)n}{1 - f(\bar{a}_0)g(\bar{a}_0, \bar{b}_0)} \quad (13)$$

Then, sequence $\{x_n\}$ generated for $x_0 \in D$ from method (2) is well–defined remains in $\bar{U}(x_0, \bar{R})$ for all $n = 0, 1, 2, \ldots$ and converges with R–order of convergence at least three to a point p which is the only solution of equation $F(x) = 0$ in $U(x_0, \frac{2}{M_0 b} - \bar{R}\eta) \cap D$. Moreover, the following a priori estimates hold

$$\|x_n - p\| \le h(\bar{a}_0 \bar{\gamma}^{\frac{3^n-1}{2}})\eta \frac{\bar{\gamma}^{\frac{3^n-1}{2}} \bar{\Delta}^n}{1 - \bar{\gamma}^{3^n} \bar{\Delta}} < (\bar{\gamma}^{\frac{1}{2}})^{3^n} \frac{\bar{R}}{(\bar{\gamma})^{\frac{1}{2}}}, \quad (14)$$

where $\bar{\gamma} = f(\bar{a}_0)g(\bar{a}_0, \bar{b}_0) \in (0, 1)$, and $\bar{\Delta} = \frac{1}{f(\bar{a}_0)}$.

Remark 2. Notice that we have in [1] $(A_1), (A_2), (A_3)$ and (A'_4),

$$a_0 = Mb_n < r, a_0 h(a_0) < 1, b_0 = K\beta\eta^2 < \phi(a_0) \quad (15)$$

and
$$U(x_0, R) \subseteq D \quad (16)$$

hold, where
$$R = \frac{h(a_0)n}{1 - f(a_0)g(a_0, b_0)}. \quad (17)$$

Moreover, the estimate corresponding to (14) is given by

$$\begin{aligned}\|x_n - p\| &\le h(a_0 \gamma^{\frac{3^n-1}{2}})\eta \frac{\gamma^{\frac{3^n-1}{2}} \Delta^n}{1 - \gamma^{3^n} \Delta} \\ &< (\gamma^{\frac{1}{2}})^{3^n} \frac{R}{\gamma^{\frac{1}{2}}}.\end{aligned} \quad (18)$$

Notice that in view of (4), (7)–(9)

$$\text{hypotheses (15)} \Longrightarrow \text{hypotheses (11)}, \quad (19)$$

$$\bar{R} \le R. \quad (20)$$

Our uniqueness ball $U(x_0, \frac{2}{M_0 b} - \bar{R})$ is at least as large as $U(x_0, \frac{2}{Mb} - R)$ given in [1] and the error bounds (14) are at least as precise as (18). These improvements are obtained under the same computational cost since in practise the computation of M (or K or N) requires the computation of $M_0, \bar{M}$ (or $\bar{K}$ or $\bar{N}$) as special cases. All the above justify the claims made at the introduction of this study (see also [9]).

Next, we present similar extensions of related works.

Case 2. Operators with w–conditioned second derivative. As in [1], consider a generalization of the Lipschitz and Hölder condition:

$(\bar{A}''_4) \|F'(x) - F''(y)\| \le \bar{w}(\|x - y\|)$ for all $x, y \in D_0$, where $\bar{w} : [0, \infty) \to [0, \infty)$ is continuous and nondecreasing with $\bar{w}(0) = 0$. We say that $\bar{w}$ is quasi–continuous if $\bar{w}(\theta z) \le \theta^q \bar{w}(z)$ for all $\theta[0, 1]$ and some $q \ge 0$.

($\bar{A}_5$) There exists a continuous and nondecreasing function $\bar{\phi}: [0,1] \to (0, \infty)$ such that $\bar{w}(\theta z) \leq \bar{\phi}(\theta)\bar{w}(z)$ for all $\theta \in [0,1]$ and $z \in (0, \infty)$.

Let $\bar{T} = \int_0^1 \bar{\phi}(\theta)(1-\theta)d\theta$ and $g(t,s) = \frac{t}{2}(tv(t) + h(t)^2 - 1) + \frac{s}{(q+2)(q+1)}$.

It us worth noticing that if $q = 1$ function g reduces to the corresponding one in Case 1.

Theorem 3. *Suppose: hypothesis* $(A_1), (A_2), (\bar{A}_3), (\bar{\bar{A}}_3), (\bar{A}_4'')$, $\bar{a}_0 = \bar{M}b\eta < r$, $\bar{a}_0 h(\bar{a}_0) < 1$, $\bar{b}_0 = b\bar{w}(\eta)\eta < \bar{K}_0$, *and* $U(x_0, \bar{R}\eta) \subseteq D$, *where*

$$\bar{K}_0 = \frac{1}{2}(q+2)(q+1)(2 + \bar{a}_0 - 4\bar{a}_0 h(\bar{a}_0) - \bar{a}_0 h(\bar{a}_0)^2 + 2\bar{a}_0^2 h(\bar{a}_0)^2 - \bar{a}_0^2 l(\bar{a}_0))$$

and

$$\bar{R} = \frac{h(\bar{a}_0)}{1 - f(\bar{a}_0)g(\bar{a}_0, \bar{b}_0)}.$$

Then, sequence $\{x_n\}$ *is generated for* $x_0 \in D$ *by method (2) is well defined, remains in* $\bar{U}(x_0, \bar{R})$ *for all* $n = 0, 1, 2, \ldots$, *and converges with* $R-$*order of convergence atleast* $q + 2$ *to a point* p *which is the only solution of equation* $F(x) = 0$ *in* $U(x_0, \frac{2}{M_0 b} - \bar{R}\eta) \cap D$. *Moreover, the following a priori estimates hold*

$$\|x_n - p\| \leq \bar{\gamma} \frac{(q+2)^n - 1}{q+1} \bar{\Delta}^n \frac{h(\bar{\gamma}^{\frac{(q+2)^n-1}{q+1}} \bar{a}_0)n}{1 - \bar{\gamma}(q+2)^n \bar{\Delta}} < \left(\bar{\gamma}^{\frac{1}{q+1}}\right)^{(q+2)^n} \frac{\bar{R}\eta}{\bar{\gamma}^{\frac{1}{q+1}}}$$

where $\bar{\gamma} = f(\bar{a}_0)^2 g(\bar{a}_0, \bar{b}_0) \in (0, 1)$ *and* $\bar{\Delta} = \frac{1}{f(\bar{a}_0)}$.

Remark 4. *The following stronger conditions were used in [1]:*

(A_4'') $\|F''(x) - F''(y)\| \leq w\|x - y\| \forall x, y \in D$ *where* w *is as* $\bar{w}$.

(A_5) *There exists a continuous and nondecreasing function* $\phi: [0,1] \to (0, \infty)$ *such that* $w(\theta z) \leq \phi(\theta)w(z)$ *for all* $\theta \in [0,1]$ *and* $z \in (0, \infty)$.

Notice that $\bar{w}(t) \leq w(t)$ *for all* $t \in [0, \frac{1}{bM}]$ *and* $\bar{\phi}(\theta) \leq \phi(\theta)$ *for all* $\theta \in [0,1]$.

Therefore, we obtain the advantages as already stated in Case 1.

Case 3. Convergence under weak conditions Suppose: (A_3') $\|F''(x_0)\| \leq \alpha$.

This hypothesis relaxes (A_3) which is useful in some cases [1].

Theorem 5. *Suppose: hypotheses* $(A_1), (A_2), (A_3), (A_3'), (\bar{A}_3), (\bar{A}_4'), U(x_0, \bar{s}*) \subseteq D$ *and* $\eta \leq \bar{\eta}_0 = \frac{1}{3\bar{k}^2}[\beta(\alpha^2 + \frac{2\bar{k}}{\beta})^{3/2} - \beta\alpha^3 - 3\bar{k}\alpha]$ *hold where* $\bar{s}*$ *is the smallest positive root of polynomial given by* $\bar{p}(t) = \frac{\bar{k}}{6}t^3 + \frac{\alpha}{2}t^2 - \frac{1}{\beta}t + \frac{\eta}{\beta}$ *and* s^{**} *denoting the larger positive root of* $\bar{p}$. *Then, sequence* $\{x_n\}$ *starting from* $x_0 \in D$, *generated by method (2) converges to a solution* p *of equation* $F(x) = 0$ *in* $\bar{U}(x_0, \bar{s}*)$. *Moreover, if* $\bar{s}* < s^{**}$, *the solution* p *is unique in* $\bar{U}(x_0, s^{**}) \cap D$. *Sequence* $\{x_n\}$ *has order of convergence at least three. If* $\bar{s}* = s^{**}$, *the solution* P *is unique in* $U(x_0, \bar{s}*)$.

Suppose $(\bar{A}_3'')$ $\|F''(x)\| \leq \bar{w}(\|x\|) \forall x \in D_0$ *where* $\bar{w}: [0, \infty) \to [0, \infty)$ *is a monotone continuous function with* $\bar{w}(0) \geq 0$. *The equation* $4t - 2h(\beta\eta\bar{\phi}(t)(c + \beta\eta\bar{\phi}(t))t + \eta) - \beta\eta\bar{\phi}(t)h(\beta\eta\bar{\phi}(t))^2(t - 2\eta) = 0$ *has atleast one positive solution where*

$$\bar{\phi}(t) = \begin{cases} \bar{w}(\|x_0\| + t), & \text{if } \bar{w} \text{ is nondecreasing,} \\ \bar{w}(\|x_0\| - t), & \text{if } \bar{w} \text{ is nonincreasing.} \end{cases}$$

Denote by $\bar{R}$ *the smallest solution of the preceding equation. If* $\bar{w}$ *is non-increasing then* $\bar{R} < \|x_0\|$. *Define function* $\tilde{g}(t) = h(t)(1 + \frac{t}{2}h(t)) - 1$.

Theorem 6. *Suppose: Hypothesis* $(A_1), (A_2), (\bar{\bar{A}}_3), (\bar{\bar{A}}_3'')$ $\bar{a}_0 h(\bar{a}_0) < 1$, $f(\bar{a}_0)^2 g(a_0) < 1$ *and* $\bar{U}(x_0, \bar{\bar{R}}) \subseteq D$ *hold. Then sequence* $\{x_n\}$ *starting from* $x_0 \in D$ *generated by method (2) is well–defined, remains in* $\bar{U}(x_0, \bar{\bar{R}})$ *converges to a solution* p *of equation* $F(x) = 0$ *in* $\bar{U}(x_0, \bar{\bar{R}})$. *Moreover, the solution* p *is unique in* $U(x_0, \bar{\bar{R}})$ *where* $\bar{\bar{R}}$ *is the largest positive solution of the equation.*

$$b \int_0^1 \int_0^1 \phi(s(R + \theta(\zeta - R))) ds (R + \theta(\zeta - R)) d\theta = 1.$$

Remark 7. *The advantages stated in Remark 2 carry out to:*

(i) *Theorem 5. Since* $\bar{K} \leq K$. *In particular, we have*

$$\begin{aligned} \eta_0 &\leq \bar{\eta}_0 \\ \bar{p}(t) &\leq p(t) \\ \bar{s}^* &\leq s^* \\ s^{**} &\leq \bar{s}^{**} \end{aligned}$$

where $p(t) = \frac{k}{6}t^3 + \frac{\alpha}{2}t^2 - \frac{1}{\beta}t + \frac{n}{\beta}$, s^*, s^{**} *are the positive roots of* p *and* $\eta_0 = \frac{1}{3k^2}[\beta(\alpha^2 + \frac{2k}{\beta})^{3/2} - \beta\alpha^3 - 3k\bar{\alpha}]$ *used in [1].*

(ii) *Theorem 6. We have that* $\bar{w}(t) \leq w(t)$ *for all* $t \in [0, t_0)$, *for some* $t_0 > 0$, $\bar{\phi}(t) \leq \phi(t)$ *where* w, ϕ *are the functions used in [1]:*

(A_3'') $\|F''(x)\| \leq w\|x\|$, *for all* $x \in D$.

3. REGIONS OF ACCESSIBILITY

By simply using $(\bar{A}_3), (\bar{\bar{A}}_3)$ instead of (A_3) in [1], we obtain:

Theorem 8. *Suppose: Hypothesis* $(A_1), (A_2), (\bar{A}_3), (\bar{\bar{A}}_3), U(x_0, \bar{R}_N) \subseteq D$ *and*

$$\bar{a} = \bar{M}bn < \frac{1}{2} \tag{21}$$

hold, where

$$\bar{R}_N = \frac{2(1 - \bar{a})}{2 - 3\bar{a}}\eta \tag{22}$$

Then, Newton's method converges to a solution z^* *of equation* $F(x) = 0$ *with R-order of convergence atleast two.*

Remark 9. *In [1] the convergence was established using* (A_3),

$$a = Mbn < \frac{1}{2} \tag{23}$$

and

$$R_N = \frac{2(1 - a)}{3 - 2a}\eta. \tag{24}$$

However, since $\bar{M} \leq M$, we have

$$a < \frac{1}{2} \implies \bar{a} < \frac{1}{2} \tag{25}$$

but not necessarily vice–verca unless $\bar{M} = M$. Moreover, we have that

$$R_N \leq \bar{R}_N. \tag{26}$$

In view of (25) and (26) if $\bar{M} < M$ then $\bar{a}, \bar{R}_N$ can replace a, R_N, respectively in the related results in [1]. Clearly, the new results if rewritten using the ones in [1] but with the above modifications will be atleast as good. In particular, as in [1] let us consider the hybrid method (HM) defined by $x_0 \in D$

$$\begin{aligned}
x_n &= x_{n-1} - F'(x_{n-1})^{-1} F(x_{n-1}) \text{ for all } n = 1, 2, \ldots, \bar{N}_0, \\
z_0 &= x_{N_0}, \\
y_{k-1} &= z_{k-1} - F'(z_{k-1})^{-1} F(z_{k-1}) \\
z_k &= y_{k-1} + \frac{1}{2} L(z_{k-1}) H(L(z_{k-1}))(y_{k-1} - z_{k-1}), k \in N
\end{aligned}$$

where x_0 is an initial point and L, H are defined as in [1]. Sequence (HM) can be written:

$$w_n = \begin{cases} x_n, & \text{if } n \leq \bar{N}_0 \\ z_n - \bar{N}_0, & \text{if } n > \bar{N}_0. \end{cases} \tag{27}$$

Then, we have:

Theorem 10. Suppose: hypothesis $(A_1), (A_2), (\bar{A}_3), (\bar{\bar{A}}_3), (21)$ and $\bar{U}(x_0, \bar{R}_N + \bar{R}) \subseteq D$ hold. Then, there exists $N_0 \in N$ such that sequence defined in (27) and starting at w_0 converges to a solution x^* of equation $F(x) = 0$ and $z^*, w_n \in \bar{U}(x_0, \bar{R}_N + \bar{R})$.

Theorem 11. Suppose: hypothesis of Theorem 10, 8 hold but not (11), for some $x_0 \in D$ satisfying $(A_1), (A_2), (\bar{A}_3), (\bar{\bar{A}}_3)$. Set $z_0 = x_{\bar{N}_0}$ where

$$\bar{N}_1 = 1 + \frac{\log r - \log \bar{a}}{\log f(\bar{a}) g(\bar{a})},$$

$$\bar{N}_2 = 1 + \left[\frac{-\log \bar{a} \phi(\bar{a})}{\log f(\bar{a}) g(\bar{a})} \right],$$

$$\bar{N}_3 = 1 + \left[\frac{\log \rho - \log \bar{a}}{\log f(\bar{a}) g(\bar{a})} \right],$$

$$\bar{N}_4 = 1 + \left[\frac{\log \rho - \log \bar{a}}{\log f(\bar{a}) g(\bar{a})} \right],$$

$$\bar{N}_0 = \max(\bar{N}_1, \bar{N}_2, \bar{N}_3, \bar{N}_4),$$

when they are positive or set them equal to zero. Now f, ϕ, g are as defined before Theorem 1, $\mu = \phi(\bar{\alpha}_{\bar{N}_3})$ ρ is the smallest positive solution of equation $\phi(t) = 0$ and $[x]$ is the integer part of real number x. Then z_0 satisfies (11).

Remark 12. *Clearly, Theorem 10 and Theorem 11 improve the corresponding ones in [1] and $\bar{N}_0 \leq N_0$.*

Remark 13. *The sufficient semi–local convergence hypothesis (21) can be weakened even further. Indeed $\bar{a}, \bar{M}, \bar{R}_N$ in Theorem 8 can be replaced respectively by [1]*

$$\tilde{a} = \tilde{M}\beta\eta < \frac{1}{2}$$

$$\tilde{R}_N = \left[1 + \frac{M_0\eta}{2(1-\delta)(1-M_0\eta)}\right]\eta$$

$$\tilde{M} = \frac{1}{8}(4M_0 + \sqrt{M_0\bar{M}} + \sqrt{M_0\bar{M} + 8M_0^2})$$

$$\delta = \frac{2\bar{M}}{\bar{M} + \sqrt{\bar{M}^2 + 8M_0\bar{M}}}$$

Notice that

$$a < \frac{1}{2} \implies \bar{a} < \frac{1}{2} \implies \tilde{a} < \frac{1}{2}$$

and $\frac{\tilde{a}}{a} \to 0$ as $\frac{M_0}{M} \to 0$.

That is the applicability of Newton's method which can be extended infinitely many times over the old Kantorovich hypothesis (23).

Then with $\tilde{a}, \tilde{R}$ replacing $\bar{a}, \bar{R}_N$ the preceding results can be improved even further as well as the ones that we do not include here but can be found in [1]. We leave this task to the motivated reader. These advantages can be extended in the local convergence case for these methods using the same idea.

REFERENCES

[1] Hernández, M. A. and Romero, N. 2016. *A qualitative analysis of a family of Newton–like iterative processes with R–order of convergence at least three*. Switzerland: Springer international publishing.

[2] Amat, S. and Busquier. S, (eds.). 2016. "Advances in iterative methods for non–linear equations". *SEMA SIMAI Springer Series* 10: 173–210.

[3] Amat, S. and Busquier, S. 2001. "Geometry and convergence of some third–order methods". *Southwest J. Pure Appl. Math* 2: 61–72.

[4] Argyros, I. K. and Magreñán, Á. A. 2018. *A contemporary study of iterative methods*. New York: Elsevier (Academic Press).

[5] Argyros, I. K. and Szidarovszky, F. 1993. *The theory and Applications of Iteration Methods*. FL: CRC Press, Boca Raton.

[6] Argyros, I. K., Chen D. and Qian, Q. 1994. "The Jarratt method in Banach space setting". *J.Comput. Appl. Math.* 51: 103–106.

[7] Argyros, I. K., George, S. and Thapa, N. 2018. *Mathematical modeling for the solution of equations and systems of equations with applications*, Volume–I. NY: Nova Publishes.

[8] Argyros, I. K., George, S. and Thapa, N. 2018. *Mathematical modeling for the solution of equations and systems of equations with applications*, Volume–II. New York: Nova Publishers.

[9] Argyros, I. K. and Hilout, S. 2012. "Weaker conditions for the convergence of Newton's method". *Journal of Complexity* 28: 364–387.

[10] Bruns, D. D. and Bailey, J. E. 1977. "Nonlinear feedback control for operating a nonisothermal CSTR near an unstable steady state". *Chem. Eng. Sci.* 32: 257–264.

[11] Conway, J. B. 1990. *A Course in Functional Analysis*. New York: Springer.

[12] Davis, H. T. 1962. *Introduction to Nonlinear Differential and Integral Equations*. New York: Dover.

[13] Ezquerro, J. A. and Hernández, M. A. 2003. "A uniparametric Halley–type iteration with free second derivative". *Int. J. Pure Appl. Math.* 6: 103–114.

[14] Ezquerro, J. A., Hernández, M. A. and Romero, N. 2018. "A modification of Cauchy's method for quadratic equations". *J. Math. Anal. Appl.* 339(2): 954–969.

[15] Ezquerro, J. A., Hernández, M. A., and Romero, N. 2010. "On some one–point hybrid iterative methods". *Nonlinear Anal. Ser. A Theory methods Appl.* 72: 587–601.

[16] Ezquerro, J. A., Hernández, M. A. and Romero, N. 2010. "Newton–like methods for operators with bounded second Frechet derivative". *Monografías del Seminario Matemático García Galdeano* 35: 137–144.

[17] Ezquerro, J. A., Hernández, M. A. and Romero, N. 2011. "Solving nonlinear integral equations of Fredholm type with high order iterative methods". *J. Comput. Appl. Math.* 236(6): 1449–1463.

[18] Gander, W. 1985. "On Halley's iteration method". *Am. Math. Mon.* 92: 131–134.

[19] Ganesh, M. and Joshi, M. C. 1991. "Numerical solvability of Hammerstein integral equations of mixed type". *IMA J. Numer. Anal.* 11: 21–31.

[20] Hairer, E. and Wanner, G. 1991. *Solving Ordinary Differential Equations II: Stiff and Differential–Algebraic Problems*. Berlin: Springer.

[21] Hernández, M. A. 2001. "The Newton method for operators with Holder continuous first derivative". *J. Optim. Theory Appl.* 109: 631–648.

[22] Hernández, M. A. 2001. "Chebyshev's approximation algorithms and applications". *Comput. Math. Appl.* 41: 433–445.

[23] Hernández, M. A. and Romero, N. 2005. "On a new multiparametric family of Newton–like methods". *Appl. Numer. Anal. Comput. Math.* 2: 78–88.

[24] Hernández, M. A. and Romero, N. 2005. "On a characterization of some Newton–like methods of R–order atleast three". J. Comput. Appl. Math. 183(1): 53-66.

[25] Hernández, M. A. and Romero, N. 2007. "General study of iterative processes of R–order atleast three under weak convergence conditions". *J. Optim. Theory. Appl.* 133: 163–177.

[26] Hernández, M. A. and Romero, N. 2007. "Application of iterative processes of R–order atleast three to operators with unbounded second derivative". *Appl. Math. Comput.* 185: 737–747.

[27] Hernández, M. A. and Romero, N. 2009. "Toward a unified theory for third R–order iterative methods for operators with unbounded second derivative". *Appl. Math. Comput.* 215(6): 2248–2261.

[28] Hernández, M. A. and Salanova, M. A. 1999. "Index of convexity and concavity: application to Halley method". *Appl. Math. Comput.* 103: 27–49.

[29] Jerome, J. W. and Varga, R. S. 1969. *Generalizations of Spline Functions and Applications to Nonlinear Boundary Value and Eigenvalue Problems. Theory and Applications of Spline Functions Academic*. New York.

[30] Kantorovich, L. V. and Akilov, G. P. 1982. *Functional Analysis*. Oxford: Pergamon Press.

[31] Keller, H. B. 1992. *Numerical Methods for Two–Point Boundary Value Problems*. New York: Dover Publications.

[32] Kneisi, K. 2001. " Julia sets for the super–Newton method, Cauchy's method and Halley's method". *Chaos* 11(2): 359–370.

[33] Magreñán, A. A. 2014. "Different anomalies in a Jarratt family of iterative root finding methods". *Appl. Math. Comput.* 233: 29–38.

[34] Magreñán, A. A. 2014. "A new tool to study real dynamics: The convergence plane". *Appl. Math. Comput.* 248: 29–38.

[35] Macnamee, J. M. 2007. *Numerical methods for Roots of Polynomials–Part I*. Amsterdam: Studies in Computational Mathematics, vol 14. Elsevier.

[36] Potra, F. A. and Ptak, V. 1984. *Nondiscrete Induction and Iterative Processes*. London: Pitman Advanced Publishing Program.

[37] Safiev, R. A. 1964. "On some iterative processes, Ž. Vyčcisl. Mat. Fiz". 4: 139–143. Translated into English by L.B.Rall as MRC Technical Summary Report, vol. 649, University of Wincosin–Madison (1966).

[38] Schröder, E. 1870. "Über unendlich viele Algotithmen zur Auflösung der Gleichugen". *Math. Ann.* 2: 317–365.

[39] Traub, J. F. 1964. *Iterative methods for the Solution of Equations.* Englewood Cliffs: Prentice Hall.

[40] Weerakoon, S. and Fernando, T. G. I. 2000. "A variant of Newton's method with accelerated thirs–order convergence". *Appl. Math. Lett.* 13: 87–93.

In: Understanding Banach Spaces
Editor: Daniel González Sánchez

ISBN: 978-1-53616-745-0
© 2020 Nova Science Publishers, Inc.

Chapter 5

DEVELOPMENTS ON THE CONVERGENCE REGION OF NEWTON–LIKE METHODS WITH GENERALIZED INVERSES IN BANACH SPACES

Ioannis K. Argyros[1,∗] *and Santhosh George*[2,†]
[1]Department of Mathematical Sciences
Cameron University, Lawton, OK, US
[2]Department of Mathematical and Computational Sciences
National Institute of Technology, Karnataka, India

Abstract

The convergence region of Newton–like methods involving Banach space valued mappings and generalized inverses is extended. To achieve this task, a region is found inside the domain of the mapping containing the iterates. Then, the semi–local as well as local convergence analysis is finer, since the new Lipschitz parameters are at least as small and in earlier work using the same information. We compare convergence criteria using numerical examples.

Keywords: Banach space, Newton–like methods, local, semi–local convergence, generalized inverses

AMS Subject Classification: 49M15, 47H17, 65G99, 37F50, 65N12

1. INTRODUCTION

Let $\mathbb{B}_1, \mathbb{B}_2$ be a Banach spaces, $\Omega \subseteq \mathbb{B}_1$ stand for a nonempty, convex and open set. Define $U(x, r) = \{y \in \mathbb{B}_1 : \|y - x\| < r\}, \bar{U}(x, r) = \{y \in \mathbb{B}_1 : \|y - x\| \leq r\}$ and $\mathcal{L}B(\mathbb{B}_1, \mathbb{B}_2) = \{R : \mathbb{B}_1 \to \mathbb{B}_2$ is bounded and linear$\}$. From now on by differentiable mapping we mean differentiable in the Fréchet sense. The computation of a solution x_* of

$$\triangle F(x) = 0, \qquad (1)$$

[∗]Corresponding Author's E-mail: iargyros@cameron.edu.
[†]E-mail: sgeorge@nitk.ac.in.

where $F : \Omega \to \mathbb{B}_2$ is differentiable, $\triangle \in \mathcal{L}B(\mathbb{B}_1, \mathbb{B}_2)$ constitutes a difficult task in general. However, we need to find x_* due to its relationship to application in various disciplines such as optimization, least squares, ill–posed problems, Mathematical Physics, Economics, Chemistry, Biology, Medicine and also in Engineering [1–13]. These problems are converted using Mathematical Modeling to the form of an equation like (1). Although we hope for an analytic or closed form solution, we resort to iterative methods since the former is difficult or impossible. In this work, we revisit the local an semilocal convergence study of Newton–like methods given as

$$x_{n+1} = x_n - A(x_n)^\# F(x_n) \qquad (2)$$

for all $n = 0, 1, 2, \ldots$ and $x_0 \in \Omega$ an initial point, $A(x_n)^\#$ stands for an outer inverse of $A(x_n)$, so that $A(x_n)^\# A(x_n) A(x_n)^\# = A(x_n)^\#$.

Our work is needed also for Newton methods for undetermined systems, least–squares problems and Gauss–Newton methods, as well as Newton–like methods for ill–posed problems in abstract spaces.

The background of method (2) and a Kantorovich–type semilocal convergence analysis of method (2) can be found in the elegant work in [9] (see also [13]). The convergence region of method (2) is small in general under the Kantorovich–like semilocal convergence criteria given in [9]. This limits the applicability of method (2), and its special cases.

The novelty of our work lies in the fact that we can extend the convergence region by locating a region inside $Dom(F)$ also containing the iterates x_n. Then, the Lipschitz parameters in that region are at least as tight as the ones in $Dom(F)$. This modification leads to: enlarged region of convergence, weaker semilocal convergence criteria, better estimates on $\|x_{n+1} - x_n\|$, $\|x_n - x_*\|$ and better uniqueness results. Another advantage of our technique is that these developments require Lipschitz parameters that are special cases of the old ones. That is these developments are obtained under the same computational cost as in [9]. This technique can be used to extend the applicability of other methods in an analogous way [1–13]. The local convergence analysis of method (2) not given in [9] is also studied.

The layout of the rest of the chapter is: The semi–local, local analysis appears in Section 2, Section 3 respectively. Section 4 contains numerical examples.

2. Semilocal Convergence

We present an extension of Theorem 3.1 in [9], which in turn extended a Theorem in [13] from inverses to using bounded outer inverses. To avoid repetitions, we refer the reader to [7, 9] for definitions and properties of generalized inverses.

Theorem 1. *Let* $F : \Omega \to \mathbb{B}_2$ *be differentiable, and let* $A(x) \in \mathcal{L}B(\mathbb{B}_1, \mathbb{B}_2)$ *be close to* $F'(x)$. *Suppose that there exist* $x_0 \in \Omega$, *a bounded outer inverse* $A^\#$ *of* $A(x_0)$, *and parameters* $\eta, L > 0, 0 < l < 1, M, L, \mu \geq 0$ *such that for each* $x \in \Omega$

$$\|A^\# F(x_0)\| \leq \eta, \qquad (3)$$
$$\|A^\# (A(x) - A)\| \leq L\|x - x_0\| + l. \qquad (4)$$

Let $\Omega_0 = \Omega \cap U(x_1, \frac{1-l}{L} - \eta)$, there exist parameters $K_0 = K_0(\Omega_0)$, $K = K(\Omega_0)$, $M = M(\Omega_0)$, $\mu = \mu(\Omega_0)$ such that for each $x, y \in \Omega_0$

$$\|A^{\#}(F'(x) - F'(x_0))\| \leq K_0\|x - x_0\|, \tag{5}$$
$$\|A^{\#}(F'(y) - F'(x))\| \leq K\|y - x\|, \tag{6}$$
$$\|A^{\#}(F'(x) - A(x))\| \leq M\|x - x_0\| + \mu, \tag{7}$$
$$b = \mu + l < 1, \tag{8}$$
$$L\eta + l < 1, \tag{9}$$
$$h = \sigma\eta \leq \frac{1}{2}(1 - b)^2, \tag{10}$$

and

$$\bar{U}(x_0, t_*) \subset D, \tag{11}$$

where

$$\sigma = \max\{K, M + L\} \tag{12}$$

and

$$t_* = \frac{1 - b - \sqrt{(1 - b)^2 - 2h}}{\sigma}. \tag{13}$$

Then, the following items hold.

1) Sequence $\{x_n\}$ generated by method (2) with $A(x_n)^{\#} = (I + A^{\#}(A(x_n) - A))^{-1}A^{\#}$ is well defined, remains in $U(x_1, t_* - \eta)$, and converges to a solution $x_* \in \bar{U}(x_1, t_* - \eta)$ of equation $A^{\#}F(x) = 0$.

2) The solution x_* is unique in

$$\tilde{U} \bigcap \{R(A^{\#}) + x_1\}$$

where $\tilde{U} = \begin{cases} \bar{U}(x_1, t_* - \eta) \cap \Omega & \text{if } h = \frac{1}{2}(1 - b)^2, \\ U(x_1, t_* - \eta) \cap \Omega & \text{if } h < \frac{1}{2}(1 - b)^2, \end{cases}$

$$x_1 = x_0 - A(x_0)^{\#}F(x_0),$$
$$R(A^{\#}) + x_1 = \{x + x_1 : x \in R(A^{\#})\},$$

and

$$t_{**} = \frac{1 - b + \sqrt{(1 - b)^2 - 2h}}{\sigma}. \tag{14}$$

Proof. Simply notice that the iterates x_n lie in Ω, which is such that $\Omega_0 \subseteq \Omega$. The rest is identical to the proof of Theorem 3.1 in [9] by simply exchanging Ω by Ω_0. □

Remark 2. (a) Denote by $K_1, M_1, \mu_1, L_1, l_1, h_1, b_1, \sigma_1, t_*^1, t_{**}^1$ the corresponding constants but depending on Ω (used in [9]) not Ω_0. Then, we have by $\Omega_0 \subseteq \Omega$ that

$$K_0 \leq K_1, \quad (15)$$
$$K \leq K_1, \quad (16)$$
$$M \leq M_1, \quad (17)$$
$$\mu \leq \mu_1, \quad (18)$$
$$l = l_1, \quad (19)$$
$$L = L_1, \quad (20)$$
$$b \leq b_1, \quad (21)$$
$$\sigma \leq \sigma_1, \quad (22)$$
$$t_* \leq t_*^1, \quad (23)$$
$$t_{**}^1 \leq t_{**} \quad (24)$$

and

$$h_1 = \sigma_1 \eta \leq \frac{1}{2}(1-b_1)^2 \Rightarrow h \leq \frac{1}{2}(1-b)^2 \quad (25)$$

but not necessarily vice–versa, unless if $b = b_1$. The new uniqueness results are more precise too.

Our parameters are also special cases of $K_1, \ldots, t_{**}^1$, respectively. Hence, the semi–local convergence criteria and uniqueness results are extended under the same computational effort.

(b) It turns out that the error estimates are improved too. Indeed, consider iterations $\{r_n\}, \{s_n\}, \{t_n\}$ defined by

$$r_0 = 0, r_1 = \eta, r_2 = r_1 + \frac{\frac{K_0}{2}(r_1 - r_0)^2 + \mu(r_1 - r_0)}{1 - Lr_1 - l},$$

for $n = 2, \ldots$

$$r_{n+1} = r_n + \frac{\frac{K}{2}(r_n - r_{n-1})^2 + (M(r_{n-1} - r_0) + \mu)(r_n - r_{n-1})}{1 - Lr_n - l},$$

$s_0 = 0, s_1 = \eta$, for $n = 1, 2, \ldots$

$$s_{n+1} = s_n + \frac{\frac{K}{2}(s_n - s_{n-1})^2 + (M(s_{n-1} - s_0) + \mu)(s_n - s_{n-1})}{1 - Ls_n - l},$$

$t_0 = 0$, for $n = 1, 2, \ldots$

$$t_{n+1} = t_n + \frac{s(t_n)}{g(t_n)},$$

where $s(t) = \frac{\sigma_1}{2}t^2 - (1-b_1)t + \eta$, and $g(t) = 1 - Lt - l$. Sequence $\{s_n\}$ and $\{t_n\}$ appeared in the proof of Theorem 3.1 [9]. In view of the proof of Theorem 3.1 [9] our Theorem 2.1, and their hypotheses, we have by a simply inductive argument that $\|x_{n+1} - x_n\| \leq r_{n+1} - r_n \leq s_{n+1} - s_n \leq t_{n+1} - t_n$, so our error bounds are tighter too.

(c) Let us specialize these results in the case of Newton's method, i.e., $A(x) = F'(x)$. Then, we set $M = \mu$, $M_1 = \mu_1$, $l = l_1 = 0$, $L_1 = L = K_0$, $b = b_1 = 0$, $h_1 = \sigma_1 \eta \leq \frac{1}{2} \Rightarrow h = \sigma \eta \leq \frac{1}{2}$ but not necessarily vice versa, unless if $\sigma = \sigma_1$, i.e., $K_1 = K$ (see also the numerical examples, where $K < K_1$). The rest of the results in [9] are improved along the same lines. The details are left to the motivated reader.

We wanted to present an extension using similar parameters, so we can compare the old Theorem 3.1 [9] to our Theorem 2.1. However, in our local convergence analysis, there is no result to compare it to in [9], so we use generalized ω–conditions.

3. LOCAL CONVERGENCE

We need some real functions and parameters. Let function $q_0 : [0, \infty) \to (-\infty, \infty)$ be increasing, continuous, and satisfying $q_0(0) = 0$. Assume

$$q_0(t) = 1$$

has a minimal positive zero, which we denote by ρ_0. Set $I = [0, \rho_0)$. Let $q : I \to (-\infty, \infty)$ be increasing, continuous, and satisfying $q(0) = 0$. Define functions p_0 and p on the interval I by

$$p_0(t) = \frac{\int_0^1 q((1-\tau)t)d\tau}{1 - q_0(t)}$$

and

$$p(t) = p_0(t) - 1.$$

By these definitions $p(0) = -1$ and $p(t) \to \infty$ for $t \to \rho_0^-$. The intermediate value theorem applied to function p defined on the interval I assures the existence of a minimal zero in $(0, \rho_0)$, which we denote by ρ_0. By the definition of ρ

$$0 \leq q_0(t) < 1 \tag{26}$$

and

$$0 \leq q(t) < 1 \quad \forall t \in [0, \rho). \tag{27}$$

Theorem 3. *Let $F : \Omega \to \mathbb{B}_2$ be a differentiable operator, and let $A(x) \in \mathcal{LB}(\mathbb{B}_1, \mathbb{B}_2)$ be close to $F'(x)$. Assume there exist $x_* \in \Omega$, a bounded outer inverse $A^\#$ of $A(x_*)$, and function $q_0 : [0, \infty) \to (-\infty, \infty)$ increasing, continuous and satisfying $q_0(0) = 0$ such that for each $x \in \Omega$*

$$\|A^\#(A(x) - A(x_*))\| \leq q_0(\|x - x_*\|). \tag{28}$$

Set $\Omega_1 = \Omega \cap U(x_*, \rho_0)$. There exist function $q : I \to (-\infty, \infty)$ increasing, continuous and satisfying $q(0) = 0$ such that for each $x, y \in \Omega_1$

$$\|A^\#(F'(x_* + \tau(x - x_*)) - A(x))\| \leq q((1-\tau)\|x - x_*\|) \tag{29}$$

for each $\tau \in [0, 1]$. Then, starting from $x_0 \in U(x_*, \rho) - \{x_*\}$, sequence $\{x_n\} \subset U(x_*, \rho)$, and converges to x_*, so that

$$\|x_{n+1} - x_*\| \leq \bar{P}_0(\|x_n - x_*\|)\|x_n - x_*\| \leq \|x_n - x_*\| < \rho. \tag{30}$$

where $\bar{P}_0 = \begin{cases} P_0, & n = 0 \\ P & n = 1, 2, \ldots \end{cases}$ Moreover, if there exists $\bar{\rho} \geq \rho$ such that

$$\int_0^1 q_0(\tau \bar{\rho}) d\tau < 1, \tag{31}$$

then, x_* is the only exact solution of equation of (1) in the set $\Omega_2 = \Omega \bigcap \bar{U}(x_*, \bar{\rho})$.

Proof. If $u \in U(x_*, \rho)$, then by the definition of ρ_0, ρ and (26)

$$\|A^\#(A(u) - A(x_*))\| \leq q_0(\|u - x_*\|) \leq q_0(\rho) < 1, \tag{32}$$

so $A(u)^\#$ exists, and

$$\|A(u)^\# A(x_*)\| \leq \frac{1}{1 - q_0(\|u - x_*\|)}. \tag{33}$$

In particular, for $u = x_0$, x_1 exists. We can write by method (2), (27), (29) and (33) in turn

$$\begin{aligned}
\|x_1 - x_*\| &= \|x_0 - A(x_0)^\# F(F(x_0))\| \\
&= \|A(x_0)^\#(A(x_0)(x_0 - x_*) - F(x_0) + F(x_*))\| \\
&= \|-A(x_0)^\# A(x_*) A(x_*)^\#(A(x_0)(x_0 - x_*) - F(x_0) + F(x_*))\| \\
&\leq \|A(x_0)^\# A(x_*)\| \|\int_0^1 A(x_*)^\#(F'(x_* - \tau(x_0 - x_*)) - A(x_0)) \\
&\quad d\tau (x_0 - x_*)\| \\
&\leq \frac{\int_0^1 q_0(\tau \|x_0 - x_*\|) d\tau \|x_0 - x_*\|}{1 - q_0(\|x_0 - x_*\|)} \\
&\leq \frac{\int_0^1 q(\tau \|x_0 - x_*\|) d\tau \|x_0 - x_*\|}{1 - q_0(\|x_0 - x_*\|)} \\
&\leq \|x_0 - x_*\| < \rho,
\end{aligned} \tag{34}$$

showing $x_1 \in U(x_*, \rho)$, and (30) for $n = 0$. The induction for (30) finishes, if x_m, x_{m+1} replace above x_0, x_1, respectively. It then follows from the estimation

$$\|x_{m+1} - x_*\| \leq a \|x_m - x_*\| < \rho, \tag{35}$$
$$a = q(\|x_0 - x_*\|) \in [0, 1]$$

that $x_{m+1} \in U(x_*, \rho)$, and $\lim_{m \to \infty} x_m = x_*$. The uniqueness part as standard is omitted [9]. $\square$

Remark 4. Let us specialize A, q_0, q, so that we can compare our results to earlier ones, in the case of Newton's method, since as noted earlier, we cannot find an analog of Theorem 26 in the literature. Choose $A(x) = F'(x), q_1(t) = c_1 t$, where c_1 is the Lipschitz constant on Ω. In this case the radius of convergence given by Rheinbnldt [11] and Traub [12] is

$$\rho_1 = \frac{2}{3c_1}. \tag{36}$$

In our case we have $q_0(t) = c_0 t$, $q(t) = ct$,

$$c_0 \leq c_1 \tag{37}$$
$$c \leq c_1, \tag{38}$$

and

$$\rho = \frac{2}{2c_0 + c}. \tag{39}$$

Then, by (36)–(39)

$$\rho_1 \leq \rho \tag{40}$$

Moreover, if $c_0 < c$, or $c < c_1$ then

$$\rho_1 < \rho. \tag{41}$$

In these case corresponding error bounds become

$$\|x_{n+1} - x_*\| \leq \frac{c_1 \|x_n - x_*\|^2}{\tau(1 - c_1\|x_n - x_*\|)} = e_n^1 \|x_n - x_*\| \tag{42}$$

and

$$\|x_{n+1} - x_*\| \leq \frac{\bar{c}\|x_n - x_*\|^2}{\tau(1 - c_0\|x_n - x_*\|)} = e_n \|x_n - x_*\| \tag{43}$$

where $\bar{c} == \begin{cases} c_0, & n = 0 \\ c & n = 1, 2, \ldots \end{cases}$ and $e_n < e_n^1$. Finally, for the uniqueness of the solution, we have $\Omega_2 = \Omega \cap \bar{U}(x_0, \frac{1}{c_0})$ whereas for the old one $\Omega_2^1 = \Omega \cap \bar{U}(x_0, \frac{1}{c_1})$ and $\Omega_2^1 \subset \Omega_2$. Hence the applicability has been extended in the local convergence of method (2) too.

4. NUMERICAL EXAMPLES

We set $A(x)^\# = F'(x)^{-1}$ in all examples for simplicity.

Example 5. *Let* $\mathbb{B}_1 = \mathbb{B}_2 = \mathbb{R}^3$, $\Omega = U(0, 1)$, $x^* = (0, 0, 0)^T$ *and define F on Ω by*

$$F(x) = F(x_1, x_2, x_3) = (e^{x_1} - 1, \frac{e-1}{2}x_2^2 + x_2, x_3)^T. \tag{44}$$

For the points $u = (u_1, u_2, u_3)^T$, the Fréchet derivative is given by

$$F'(u) = \begin{pmatrix} e^{u_1} & 0 & 0 \\ 0 & (e-1)u_2 + 1 & 0 \\ 0 & 0 & 1 \end{pmatrix}.$$

Using the norm of the maximum of the rows, Theorem 3.1 and since $F'(x^*) = diag(1, 1, 1)$, we can define parameters for method (2) by $c_0 = (e - 1) < c = e^{\frac{1}{e-1}} < c_1 = e$. Then, $\rho_1 = 0.2453 < \rho = 0.3827$.

Example 6. Let $\mathcal{B}_1 = \mathcal{B}_2 = C[0,1]$, and $\Omega = \overline{U}(0,1)$. Define function F on Ω by

$$F(\varphi)(x) = \varphi(x) - 5\int_0^1 x\theta\varphi(\theta)^3 d\theta. \tag{45}$$

We have that

$$F'(\varphi(\xi))(x) = \xi(x) - 15\int_0^1 x\theta\varphi(\theta)^2 \xi(\theta)d\theta, \text{ for each } \xi \in \Omega.$$

We get that $x^* = 0$, $c0 = 7.5 < c = c_1 = 15$. This $\rho_1 = 0.0444 < \rho = 0.0667$.

Example 7. Let $\mathbb{B}_1 = \mathbb{B}_2 = \mathbb{R}$, $\Omega = \bar{U}(x_0, 1-\varsigma)$, $x_0 = 1$ and $\varsigma \in [0, \frac{1}{2})$. Define function F on Ω by

$$F(x) = x^3 - \varsigma.$$

Then, we get by Theorem 2.1 and Remark 2.2 that for $\eta = \frac{1}{3}(1-\varsigma), \ell = \ell_1 = \mu = \mu_1 = M = M_1 = 0, L = K_0 = 3-\varsigma, K = \frac{2}{3(3-\varsigma)}(-2\varsigma^2 + 5\varsigma + 6) < K_1 = L_1 = 2(2-\varsigma)$. Then $h_1 > \frac{1}{2}$ for each $\varsigma \in [0, \frac{1}{2})$ so, there is no guarantee that $\{x_n\}$ converges to $x_* = \sqrt[3]{\varsigma}$ under the old assumptions. But in our case $h \leq \frac{1}{2}$ for each $\varsigma \in [0.3720452, 0.5)$. Hence, the applicability of the method is extended.

Example 8. Let $\mathbb{B}_1 = \mathbb{B}_2 = C[0,1], \Omega = \bar{U}(x^*, 1)$ and $\Omega^* = \{x \in C[0,1]; \|x\| \leq R\}$, such that $R > 0$ and F defined on Ω^* and given as

$$F(x)(s) = x(s) - f(s) - \xi \int_0^1 Q(s,t)x(t)^3\, dt, \quad x \in C[0,1], s \in [0,1],$$

where $f \in C[0,1]$ is a given function, ξ is a real constant and the kernel F is the Green function. In this case, for each $x \in \Omega^*$, $F'(x)$ is a linear operator defined on Ω^* by the following expression:

$$[G'(x)(v)](s) = v(s) - 3\xi \int_0^1 Q(s,t)x(t)^2 v(t)\, dt, \quad v \in C[0,1], s \in [0,1].$$

If we choose $x_0(s) = f(s) = 1$, it follows $\|I - G'(x_0)\| \leq 3|\xi|/8$. Thus, if $|\xi| < 8/3$, $G'(x_0)^{-1}$ is defined and

$$\|G'(x_0)^{-1}\| \leq \frac{8}{8-3|\xi|}, \quad \|G(x_0)\| \leq \frac{|\xi|}{8}, \quad \eta = \|G'(x_0)^{-1}G(x_0)\| \leq \frac{|\xi|}{8-3|\xi|}.$$

Moreover, we have $\ell = \ell_1 = \mu = \mu_1 = M = M_1 = 0, L = K_0 = 2.6, T = K_1 = 3.8, K = 1.38154$, so $h_1 = 0.76 > \frac{1}{2}$ but $h = 0.485085 < \frac{1}{2}$. Hence, the old criterion is not satisfied but the new guarantees convergence to x^*.

REFERENCES

[1] Amat, S., Argyros, I. K, Busquier, S., and Herńandez–Verón, M. A. 2018. "On two high–order families of frozen Newton–type methods". *Numerical Linear Algebra with Applications* 25(1).

[2] Argyros, I. K. 2007. *Computational theory of iterative methods*, volume 15. Elsevier.

[3] Argyros, I. K. and Hilout, S. 2012. "Weaker conditions for the convergence of Newton's method". *Journal of Complexity* 28(3): 364–387.

[4] Argyros, I. K. and Magreñán, Á. A. 2018. *A contemporary study of iterative methods*. New York: Elsevier (Academic Press).

[5] Argyros, I. K. and Magreñán, Á. A. 2017. *Iterative methods and their dynamics with applications*. New York: CRC Press.

[6] Ezquerro, J. A. and Herńandez–Verón, M. A. 2014. "How to improve the domain of starting points for Steffensen's method". *Studies in Applied Mathematics* 132(4): 354–380.

[7] Häussler, W. M. 1986. "A Kantorovich–type convergence analysis for the Gauss–Newton method". *Numer. Math.* 48: 119–125.

[8] Kantorovich, L. V. and Akilov, G. P. 1982. *Functional analysis*. Pergamon press.

[9] Nashed, M. Z. and Chen, X. 1993. "Convergence of Newton–like methods for singular operator equations using outer inverses". *Numer. Math.* 66: 235–257.

[10] Potra, F. A. and Pták, V. 1984. *Nondiscrete induction and iterative processes*, volume 103. Pitman Advanced Publishing Program.

[11] Rheinboldt, W. C. 1978. "An adaptive continuation process for solving systems of nonlinear equations". *Polish Academy of Science, Banach Ctr. Publ.* 3(1): 129–142.

[12] Traub, J. F. 1982. *Iterative methods for the solution of equations*. American Mathematical Soc.

[13] Yamamoto, T. 1987. "A convergence theorem for Newton–like methods in Banach spaces". *Numer. Math.* 51: 545–557.

In: Understanding Banach Spaces
Editor: Daniel González Sánchez
ISBN: 978-1-53616-745-0
© 2020 Nova Science Publishers, Inc.

Chapter 6

MODIFIED NEWTON–TYPE COMPOSITIONS FOR SOLVING EQUATIONS IN BANACH SPACES

Ioannis K. Argyros[1,*] *and Santhosh George*[2,†]
[1]Department of Mathematical Sciences
Cameron University, Lawton, OK, US
[2]Department of Mathematical and Computational Sciences
National Institute of Technology, Karnataka, India

Abstract

We compare the radii of convergence as well as the error bounds of two efficient sixth convergence order methods for solving Banach space valued operators. The convergence criteria invlove conditions on the first derivative. Earlier convergence criteria require the existence of derivatives up to order six. Therefore, our results extended the usage of these methods. Numerical examples complement the theoretical results.

Keywords: sixth convergence order method, Banach space, local convergence

AMS Subject Classification: 65F08, 37F50, 65N12

1. INTRODUCTION

Let $\mathcal{E}_1, \mathcal{E}_2$ be Banach spaces and $E \subset \mathcal{E}_1$ be an open, and convex set. $\mathcal{B}(\mathcal{E}_1, \mathcal{E}_2)$, denotes the space of bounded linear operators from $\mathcal{E}_1$ into $\mathcal{E}_2$. The set $T(x, a)$, is an open ball with center $x \in \mathcal{E}_1$ and of radius $a > 0$. One of the most important tasks in Computational Mathematics is to find a solution x_* of

$$\mathcal{F}(x) = 0, \qquad (1)$$

where $\mathcal{F} : E \longrightarrow \mathcal{E}_2$ is continuously differentiable in the Fréchet sense. This task is very imporatant to undertake, since many problems from different disciplines such

---

[*]Corresponding Author's E-mail: iargyros@cameron.edu.
[†]E-mail: sgeorge@nitk.ac.in.

as Applied Mathematics, Optimization, Mathematical Biology, Chemistry, Economics, Medicine, Physics, Engineering and related areas are formulated as in (1) by utilizing Mathematical modeling [1–27]. However, closed form solutions are hard to find, so iterative methods are proposed, where a sequence is generated approximating x_*. The most popular quadratically convergent method is Newton's given as

$$x_{n+1} = x_n - \mathcal{F}'(x_n)^{-1}\mathcal{F}(x) \text{ for all } n = 0, 1, 2, \ldots, \tag{2}$$

with $x_0 \in \mathcal{E}_1$. Many authors have proposed higher than two convergence order methods. Their convergence requires Taylor series and the existence of high order derivatives limiting the applicability of these methods. We use only conditions on the first derivative and for the convergence order, we use computational order of convergence (COC) or approximate computational order of convergence (ACOC) that do not require high convergence derivatives and in the case of ACOC not even knowledge of x_* (see Remark 2). That is how we extend the applicability of iterative methods. In particular we consider Newton–type methods defined as

$$\begin{aligned}
y_n &= x_n - \mathcal{F}'(x_n)^{-1}\mathcal{F}(x_n) \\
z_n &= x_n - \frac{1}{2}(\mathcal{F}(x_n)^{-1} + \mathcal{F}'(y_n)^{-1})\mathcal{F}(x_n) \\
x_{n+1} &= z_n - \frac{1}{2}(\mathcal{F}'(x_n)^{-1} + \mathcal{F}'(y_n)^{-1}\mathcal{F}'(x_n)\mathcal{F}'(y_n)^{-1})\mathcal{F}(z_n),
\end{aligned} \tag{3}$$

and

$$\begin{aligned}
y_n &= x_n - \frac{2}{3}\mathcal{F}'(x_n)^{-1}\mathcal{F}(x_n) \\
z_n &= x_n - (6\mathcal{F}'(y_n) + 2\mathcal{F}'(x_n))^{-1}(3\mathcal{F}'(y_n) + \mathcal{F}'(x_n))\mathcal{F}'(x_n)^{-1}\mathcal{F}(x_n) \\
x_{n+1} &= z_n - (\frac{3}{2}\mathcal{F}'(y_n)^{-1} - \frac{1}{2}\mathcal{F}'(x_n)^{-1})\mathcal{F}(z_n).
\end{aligned} \tag{4}$$

Methods (3), (4) were studied in [23, 27], respectively, when $\mathcal{E}_1 = \mathcal{E}_2 = \mathbb{R}^k$. Their convergence order was shown to be six in this case under hypotheses on the sixth derivative. But these hypotheses limits the applicability of these method.

As an academic example: Let $\mathcal{E}_1 = \mathcal{E}_2 = \mathbb{R}$, $E = [-\frac{1}{2}, \frac{3}{2}]$. Define $\mathcal{F}$ on E by

$$\mathcal{F}(x) = x^3 \log x^2 + x^5 - x^4$$

Then, we have $x_* = 1$, and

$$\mathcal{F}'(x) = 3x^2 \log x^2 + 5x^4 - 4x^3 + 2x^2,$$

$$\mathcal{F}''(x) = 6x \log x^2 + 20x^3 - 12x^2 + 10x,$$

$$\mathcal{F}'''(x) = 6 \log x^2 + 60x^2 - 24x + 22.$$

Obviously $\mathcal{F}'''(x)$ is not bounded on E. So, the convergence of methods (3), (4) are not guaranteed by the analysis in [23, 27].

The purpose of this chapter is twofold: First we provide a local convergence based on the first derivative. The analysis includes upper error estimates on $\|x_n - x_*\|$ as well as uniqueness results depending on generalized Lipschitz conditions not available in [23, 27]. Secondly, we compare the two methods with each other. Our technique can be used to study and improve the applicability of other iterative methods along the same lines [1–22, 24–27].

The rest of the chapter is lays out as follows: Section 2, Section 3 contains the local convergence of methods (3), (4), respectively. The theoretical results are tested using numerical examples in Section 4.

2. ANALYSIS OF METHOD (3)

Consider function $a_0 : S \longrightarrow S$ increasing, and continuous so that $a_0(0) = 0$ provided that $S = [0, \infty)$. Assume

$$a_0(t) = 1 \tag{5}$$

has a minimal positive zero ρ_0. Set $S_0 = [0, \rho_0)$. Let $a : S_0 \longrightarrow S$ be increasing, and continuous so that $a(0) = 0$. Consider functions b and $\bar{b}$ on S_0 as

$$b(t) = \frac{\int_0^1 a((1-\tau)t)d\tau}{1 - a_0(t)}$$

and

$$\bar{b}(t) = b(t) - 1.$$

These definitions give $\bar{b}(0) = -1$ and $\bar{b}(t) \longrightarrow \infty$ provided that $t \longrightarrow \rho_0^-$. Consider r_1 to be the minimal zero of $\bar{b}(t) = 0$ in $(0, \rho_0)$ which is assured by the mean value theorem. Assume

$$a_0(b(t)t) = 1 \tag{6}$$

has a minimal positive zero ρ_1. Let $S_1 = [0, \rho_2)$ with $\rho_2 = \min\{\rho_0, \rho_1\}$. Consider function $a_1 : S_1 \longrightarrow S$ increasing and continuous. Let functions c and $\bar{c}$ on S_1 as

$$c(t) = a_1(t) + \frac{1}{2} \frac{(a_0(t) + a_0(b(t))) \int_0^1 a_1(\tau t)d\tau}{(1 - a_0(t))(1 - a_0(b(t)))}$$

and

$$\bar{c}(t) = c(t) - 1.$$

These definitions give $\bar{c}(0) = -1$ and $\bar{c}(t) \longrightarrow \infty$ provided that $t \longrightarrow \rho_2^-$. Let r_2 to be the minimal zero of $\bar{c}(t) = 0$ in $(0, \rho_2)$. Assume

$$a_0(c(t)t) = 1 \tag{7}$$

has a minimal positive zero ρ_3. Let $S_2 = [0, \rho_4)$ with $\rho_4 = \min\{\rho_2, \rho_3\}$. Consider functions d and $\bar{d}$ on S_2 as

$$d(t) = \frac{\int_0^1 a((1-\tau)c(t)t)c(t)d\tau}{1 - a_0(c(t)t)}$$
$$+ \frac{(a_0(t) + a_0(c(t)))\int_0^1 a_1(\tau c(t)t)d\tau c(t)}{(1 - a_0(t))(1 - a_0(c(t)t))}$$
$$+ \frac{1}{2}\frac{1}{1 - a_0(b(t)t)}\left[\frac{a_0(t) + a_0(b(t)t)}{1 - a_0(b(t)t)}\right.$$
$$\left.\frac{a_0(c(t)t) + a_0(b(t)t)}{1 - a_0(c(t)t)}\right]\int_0^1 a_1(\tau c(t)t)d\tau$$

and

$$\bar{d}(t) = d(t) - 1.$$

These definitions give $\bar{d}(0) = -1$ and $\bar{d}(t) \longrightarrow \infty$ provided that $t \longrightarrow \rho_4^-$. Let r_3 stand for the minimal zero of $\bar{d}(t) = 0$ in $(0, \rho_4)$. Set a radius r as

$$r = \min\{r_j\}, \; j = 1, 2, 3. \tag{8}$$

We have

$$0 \leq a_0(t) < 1 \tag{9}$$
$$0 \leq a_0(b(t)t) < 1 \tag{10}$$
$$0 \leq a_0(c(t)t) < 1 \tag{11}$$
$$0 \leq b(t) < 1 \tag{12}$$
$$0 \leq c(t) < 1 \tag{13}$$

and

$$0 \leq d(t) < 1 \text{ for all } t \in [0, r). \tag{14}$$

Consider conditions (H):

(H1) $\mathcal{F}: E \longrightarrow \mathcal{E}_2$ is continuously differentiable. There exists $x_* \in E$ with $\mathcal{F}(x_*) = 0$, and $\mathcal{F}'(x_*)^{-1} \in \mathcal{B}(\mathcal{E}_2, \mathcal{E}_1)$.

(H2) $a_0 : S \longrightarrow S$ is increasing, continuous, $a_0(0) = 0$ so that for all $x \in E$

$$\|\mathcal{F}'(x_*)^{-1}(\mathcal{F}'(x_*) - \mathcal{F}'(x))\| \leq a_0(\|x_* - x\|).$$

Consider $Q_0 = E \cap T(x_*, \rho_0)$ with ρ_0 given in (5).

(H3) $a : [0, \rho_0) \longrightarrow I$, $a_1 : [0, \rho_0) \longrightarrow I$ are increasing, continuous, $a(0) = 0$ so that for all $x, y \in Q_0$

$$\|\mathcal{F}'(x_*)^{-1}(\mathcal{F}'(y) - \mathcal{F}'(x))\| \leq a(\|y - x\|)$$

and

$$\|\mathcal{F}'(x_*)^{-1}\mathcal{F}'(x)\| \leq a_1(\|x_* - x\|).$$

(H4) ρ_j, $j = 1, 2, 3, 4$ exist and $\bar{T}(x_*, r) \subseteq E$, with r defined in (8).

(H5) $\int_0^1 a_0(\tau r_*) d\tau < 1$ for some $r_* \geq r$. Consider $Q_1 = E \cap \bar{T}(x_*, r_*)$.

Theorem 1. *Assume conditions (H), and choose $x_0 \in T(x_*, r) - \{x_*\}$. Then, the following items hold*

$$\{x_n\} \subseteq T(x_*, r),$$

$$\lim_{n \to \infty} x_n = x_*,$$

$$\|y_n - x_*\| \leq b(\|x_n - x_*\|)\|x_n - x_*\| \leq \|x_n - x_*\| < r, \quad (15)$$

$$\|z_n - x_*\| \leq c(\|x_n - x_*\|)\|x_n - x_*\| \leq \|x_n - x_*\|, \quad (16)$$

$$\|x_{n+1} - x_*\| \leq d(\|x_n - x_*\|)\|x_n - x_*\| \leq \|x_n - x_*\|, \quad (17)$$

and the only solution of equation $\mathcal{F}(x) = 0$ in Q_1 is x_*.

Proof. Estimations (15)–(17) are shown by induction. Consider $x \in T(x_*, r)$. Then, by (H1), (H2), (8), and (9)

$$\|\mathcal{F}'(x_*)^{-1}(\mathcal{F}'(x_*) - \mathcal{F}'(x))\| \leq a_0(\|x_* - x\|) \leq a_0(r) < 1, \quad (18)$$

which in combination with the Lemma by Banach and invertible operators [19] $\mathcal{F}'(x)$ is invertible satisfying

$$\|\mathcal{F}'(x)^{-1}\mathcal{F}'(x_*)\| \leq \frac{1}{1 - a_0(\|x_* - x\|)}. \quad (19)$$

The first substep of method (3) also gives that y_0 is well defined. Then, by (8), (12), (H3), method (3), and (19) the following estimation is obtained

$$\begin{aligned}
\|y_0 - x_*\| &= \|\|x_0 - x_* - \mathcal{F}'(x_0)^{-1}\mathcal{F}(x_0)\| \\
&\leq \|\mathcal{F}'(x_0)^{-1}\mathcal{F}'(x_*)\| \\
&\quad \times \|\int_0^1 \mathcal{F}'(x_*)^{-1}(\mathcal{F}'(x_* + \tau(x_0 - x_*)) - \mathcal{F}'(x_0))d\tau(x_0 - x_*)\| \\
&\leq \frac{\int_0^1 a((1-\tau)\|x_0 - x_*\|)d\tau \|x_0 - x_*\|}{1 - a_)(\|x_0 - x_*\|)} \\
&= b(\|x_0 - x_*\|)\|x_0 - x_*\| \leq \|x_0 - x_*\| < r
\end{aligned} \quad (20)$$

showing $y_0 \in T(x_*, r)$ and estimation (15) for $n = 0$. The existence of z_0, x_1 follow by (18) for $x = y_0$ and (11). Using the second substep of method (3), (8)–(10), (13), (H3), (19) (for $x = x_0, y_0$) and the estimation

$$\begin{aligned}
\|\mathcal{F}'(x_*)^{-1}\mathcal{F}(x_0)\| &= \|\int_0^1 \mathcal{F}'(x_*)^{-1}\mathcal{F}'(x_* + \tau(x_0 - x_*))d\tau(x_0 - x_*)\| \\
&\leq \int_0^1 a_1(\tau\|x_0 - x_*\|)d\tau\|x_0 - x_*\|, \quad (21)
\end{aligned}$$

we obtain in turn

$$\begin{aligned}
\|z_0 - x_*\| &= \|(x_0 - x_* - \mathcal{F}'(x_0)^{-1}\mathcal{F}(x_0)) \\
&\quad + \frac{1}{2}\mathcal{F}'(x_0)^{-1}[(\mathcal{F}'(y_0) - \mathcal{F}'(x_*)) + (\mathcal{F}'(x_*) - \mathcal{F}(x_0))]\mathcal{F}'(y_0)^{-1}\mathcal{F}(x_0)\| \\
&\leq \left[b(\|x_0 - x_*\|) \right. \\
&\quad \left. + \frac{1}{2}\frac{(a_0(\|x_0 - x_*\|) + a_0(\|y_0 - x_*\|))\int_0^1 a_1(\tau\|x_0 - x_*\|)d\tau}{2(1 - a_0(\|x_0 - x_*\|))(1 - a_0(\|y_0 - x_*\|))}\right]\|x_0 - x_*\| \\
&\leq c(\|x_0 - x_*\|)\|x_0 - x_*\| \leq \|x_0 - x_*\| \quad (22)
\end{aligned}$$

implying $z_0 \in T(x_*, r)$, and (16) holds for $n = 0$. By the third substep of method (3), (8), (11), (14), (19) (for $x = z_0, x_0, y_0$), (H1)–(H3), (20), (21) (for $x = z_0$), and some algebtaic manipulations, we write in turn that

$$\begin{aligned}
\|x_1 - x_*\| &= \|(z_0 - x_* - F'(z_0)^{-1}F(z_0)) \\
&\quad + F'(x_0)^{-1}[(F'(z_0) - F'(x_*)) + (F'(x_*) - F'(z_0))]F'(z_0)^{-1}F(z_0) \\
&\quad - \frac{1}{2}F'(y_0)^{-1}\{[(F'(x_0) - F'(x_*)) + (F'(x_*) - F'(y_0))]F'(y_0)^{-1} \\
&\quad + [(F'(z_0) - F'(x_*)) + (F'(x_*) - F'(y_0))]F'(z_0)^{-1}\}F(z_0)\| \\
&\leq \frac{\int_0^1 a((1-\tau)c(t)t)d\tau c(t)}{1 - a_0(c(t)t)} \\
&\quad + \frac{(a_0(t) + a_0(c(t)t))\int_0^1 a_1(\tau c(t)t)d\tau c(t)}{(1 - a_0(t))(1 - a_0(c(t)t))} \\
&\quad + \frac{1}{2}\frac{1}{1 - a_0(b(t)t)}\left[\frac{a_0(t) + a_0(b(t)t)}{1 - a_0(b(t)t)} + \frac{a_0(c(t)t) + a_0(b(t)t)}{1 - a_0(c(t)t)}\right] \\
&\quad \times \int_0^1 a_1(\tau c(t)t)d\tau c(t) \\
&= d(t)t \leq t \quad (23)
\end{aligned}$$

for $\|x_0 - x_*\| \leq t$ leading to $x_1 \in T(x_*, r)$, and (17) for $n = 0$. Substitute x_0, y_0, z_0, x_1 by x_m, y_m, z_m, x_{m+1} in the above estimations to complete the induction for (15)–(17). Then, the estimation

$$\|x_{m+1} - x_*\| \leq \lambda\|x_m - x_*\| \leq r \quad (24)$$

for $\lambda = d(\|x_0 - x_*\|) \in [0,1)$ gives $x_{m+1} \in T(x_*, r)$ and $\lim_{m \to \infty} x_m = x_*$. Let $\mu = \int_0^1 \mathcal{F}'(y_* + \tau(x_* - y_*))d\tau$ for $y_* \in Q_1$, and $\mathcal{F}(y_*) = 0$. By (H5) and (H2), we get

$$\begin{aligned}
\|\mathcal{F}'(x_*)^{-1}(\mu - \mathcal{F}/(x_*))\| &\leq \int_0^1 a_0(\tau\|x_* - y_*\|)d\tau \\
&\leq \int_0^1 a_0(\tau - r_*)d\tau < 1,
\end{aligned}$$

so μ^{-1} exists, and from

$$0 = \mathcal{F}(x_*) - \mathcal{F}(y_*) = \mu(x_* - y_*), \ x_* = y_*.$$

□

Remark 2. (a) In view of (H2), we can write

$$\begin{aligned}\|\mathcal{F}'(x_*)^{-1}\mathcal{F}'(x)\| &= \|\mathcal{F}'(x_*)^{-1}[(\mathcal{F}'(x) - \mathcal{F}'(x_*)) + \mathcal{F}'(x_*)]\| \\ &\leq 1 + \|\mathcal{F}'(x_*)^{-1}(\mathcal{F}'(x) - \mathcal{F}'(x_*))\| \\ &\leq 1 + a_0(\|x - x_*\|),\end{aligned} \quad (25)$$

so the second condition in (H3) can be dropped, and we choose $a_1(t) = 1 + a_0(t)$.

(b) Suppose that $a_0(t) = L_0 t$, $a(t) = Lt$ and $a_1(t) = L_1 t$ for some nonnegative constants L_0, L, L_1. It follows from the definition of r and $r_A = \frac{2}{2L_0 + L}$ that $r < r_A$. That is, the radius of convergence r cannot be larger than the radius of convergence r_A of Newton's method obtained by us [2–11]. Suppose that L_1 is the Lipschitz constant on E. Then, we have $L_0 \leq L_1$ and $L \leq L_1$. Moreover, the radius of convergence for Newton's method given independently by Rheinboldt [25] and Traub [27] $r_{TR} = \frac{2}{3L_1}$ is such that $r_{TR} \leq r_A$.

(c) It is worth noticing that the studied methods are not changing when we use the conditions of the preceding Theorems instead of the stronger conditions used in earlier studies. Moreover, the preceding Theorems we can compute the computational order of convergence (COC) defined by

$$\xi = \ln\left(\frac{\|x_{n+1} - x_**\|}{\|x_n - x_*\|}\right) / \ln\left(\frac{\|x_n - x_*\|}{\|x_{n-1} - x_*\|}\right)$$

or the approximate computational order of convergence (ACOC) [9, 10]

$$\xi_1 = \ln\left(\frac{\|x_{n+1} - x_n\|}{\|x_n - x_{n-1}\|}\right) / \ln\left(\frac{\|x_n - x_{n-1}\|}{\|x_{n-1} - x_{n-2}\|}\right).$$

This way we obtain in practice the order of convergence.

(d) The results obtained here can be used for operators $\mathcal{F}$ satisfying autonomous differential equations [2, 4, 5] of the form

$$\mathcal{F}'(x) = G(\mathcal{F}(x))$$

where G is a continuous operator. Then, since $\mathcal{F}'(x^*) = G(\mathcal{F}(x^*)) = G(0)$, we can apply the results without actually knowing x_*. For example, let $\mathcal{F}(x) = e^x - 1$. Then, we can choose: $G(x) = x + 1$.

(e) The local results obtained here can be used for projection methods such as the Arnoldi's method, the generalized minimum residual method (GMRES), the generalized conjugate method(GCR) for combined Newton/finite projection methods and in connection to the mesh independence principle can be used to develop the cheapest and most efficient mesh refinement strategies [2–7, 19].

In a similar way, we improve the semi–local convergence analysis of method (3) given in [24]. The work is given in the next section.

3. ANALYSIS OF METHOD (4)

The analysis of method (4) is analogous to the one of method (3). Consider function $\varphi_0 : S \longrightarrow S$ increasing, and continuous so that $\varphi_0(0) = 0$. Assume

$$\varphi_0(t) = 1 \tag{26}$$

has a minimal positive zero λ_0. Set $P_0 = [0, \lambda_0)$. Consider functions $\varphi : P_0 \longrightarrow S$, $\varphi_1 : P_0 \longrightarrow S$ be increasing and continuous so that $\varphi(0) = 0$. Let functions ψ_1 and $\bar{\psi}_1$ on P_0 as

$$\psi_1(t) = \frac{\int_0^1 \varphi((1-\tau)t)d\tau + \frac{1}{3}\int_0^1 \varphi_1(\tau t)d\tau}{1 - \varphi_0(t)},$$

$$\bar{\psi}_1(t) = \psi_1(t) - 1.$$

Assume

$$\frac{\varphi_1(0)}{3} < 1. \tag{27}$$

By these definitions $\bar{\psi}_1(0) = \frac{\varphi_1(0)}{3} - 1 < 0$ and $\bar{\psi}_1(t) \longrightarrow \infty$ provided that $t \longrightarrow \lambda_0^-$. Let R_1 be the minimal zero of $\bar{\psi}_1(t) = 0$. Assume

$$p(t) = 1 \tag{28}$$

has a minimal positive zero λ_p, where $p(t) = \frac{1}{2}(3\varphi_0(\psi_1(t)t) + \varphi_0(t))$. Set $P_1 = [0, \lambda_1)$ with $\lambda_1 = \min\{\lambda_0, \lambda_p\}$. Consider functions ψ_2 and $\bar{\psi}_2$ on P_1 as

$$\psi_2(t) = \frac{\int_0^1 \varphi(\tau t)d\tau}{1 - \varphi_0(t)} + \frac{3(\varphi_0(t) + \varphi_0(\psi_1(t)t))\int_0^1 \varphi_1(\tau t)d\tau}{4(1-p(t))(1-\varphi_0(t))}$$

and

$$\bar{\psi}_2(t) = \psi_2(t) - 1.$$

By these definitions, we obtain $\bar{\psi}_2(0) = -1$, and as $t \longrightarrow \lambda_1^-$, $\bar{\psi}_2(t) \longrightarrow \infty$, so let R_2 be the minimal zero of $\bar{\psi}_2(t) = 0$. Assume

$$\varphi_0(\psi_1(t)t) = 1$$

$$\varphi_0(\psi_2(t)t) = 1$$

have minimal positive zeros λ_2, λ_3, respectively. Set $P_2 = [0, \lambda_4)$ with $\lambda_4 = \min\{\lambda_1, \lambda_2, \lambda_3\}$. Consider fucntions ψ_3 and $\bar{\psi}_3$ on P_2 as

$$\psi_3(t) = \frac{\int_0^1 \varphi(\tau \psi_2(t)t)d\tau \psi_2(t)}{1 - \varphi_0(\psi_2(t)t)}$$
$$+ \frac{(\varphi_0(\psi_1(t)t) + \varphi_0(\psi_2(t)t))\int_0^1 \varphi_1(\tau \psi_2(t)t)d\tau \psi_2(t)}{(1 - \varphi_0(\psi_2(t)t))(1 - \varphi_0(\psi_1(t)t))}$$
$$+ \frac{1}{2}\frac{(\varphi_0(\psi_1(t)t) + \varphi_0(t))\int_0^1 \varphi_1(\tau \psi_1(t)t)d\tau \psi_2(t)}{(1 - \varphi_0(t))(1 - \varphi_0(\psi_1(t)t))}$$

and
$$\bar{\psi}_3(t) = \psi_3(t) - 1.$$

It follows $\bar{\psi}_3(0) = -1$ and $\bar{\psi}_3(t) \longrightarrow \infty$ with $t \longrightarrow \lambda_4^-$. Let R_3 be the minimal zero of $\bar{\psi}_3(t) = 0$ in $(0, \lambda_4)$. Set

$$R = \min\{R_j\}, \ j = 1, 2, 3, \tag{29}$$

then

$$0 \leq \varphi_0(t) < 1 \tag{30}$$
$$0 \leq \varphi_0(\psi_1(t)t) < 1 \tag{31}$$
$$0 \leq \varphi_0(\psi_2(t)t) < 1 \tag{32}$$
$$0 \leq \psi_1(t) < 1 \tag{33}$$
$$0 \leq \psi_2(t) < 1 \tag{34}$$

and

$$0 \leq \psi_3(t) < 1 \tag{35}$$

for all $t \in [0, R)$. To obtain conditions needed for the local convergence of method (4) simply replace $a_0, a, a_1, r, r_*, b, c, d$ by $\varphi_0, \varphi, \varphi_1, R, R_*, \psi_1, \psi_2, \psi_3$, respectively and also assume (27) as condition (H6'). Denote these conditions by (H').

Theorem 3. *Assume conditions (H'), and choose $x_0 \in T(x_*, R) - \{x_*\}$. Then, the following items hold*

$$\{x_n\} \subseteq T(x_*, R),$$
$$\lim_{n \longrightarrow \infty} x_n = x_*,$$
$$\|y_n - x_*\| \leq \psi_1(\|x_n - x_*\|)\|x_n - x_*\| \leq \|x_n - x_*\| < R \tag{36}$$
$$\|z_n - x_*\| \leq \psi_2(\|x_n - x_*\|)\|x_n - x_*\| \leq \|x_n - x_*\| \tag{37}$$
$$\|x_{n+1} - x_*\| \leq \psi_3(\|x_n - x_*\|)\|x_n - x_*\| \leq \|x_n - x_*\| \tag{38}$$

and the only solution of equation $\mathcal{F}(x) = 0$ in Q_1 is x_*.

Proof. As in Theorem 1, we get

$$\|\mathcal{F}'(x)^{-1}\mathcal{F}'(x_*)\| \leq \frac{1}{1 - \varphi_0(\|x_* - x\|)} \tag{39}$$

for each $x \in T(x_*, R)$. Then, by the first substep of method (4),

$$\begin{aligned}
\|y_0 - x_*\| &= \|(x_0 - x_* - \mathcal{F}'(x_0)^{-1}\mathcal{F}(x_0)) + \frac{1}{3}\mathcal{F}'(x_0)^{-1}\mathcal{F}(x_0)\| \\
&\leq \frac{\int_0^1 \varphi((1-\tau)\|x_0 - x_*\|)d\tau\|x_0 - x_*\| + \frac{1}{3}\int_0^1 \varphi_1(\tau\|x_0 - x_*\|)d\tau\|x_0 - x_*\|}{1 - \varphi_0(\|x_0 - x_*\|)} \\
&= \psi_1(\|x_0 - x_*\|)\|x_0 - x_*\| \leq \|x_0 - x_*\| < R, \tag{40}
\end{aligned}$$

showing $y_0 \in T(x_*, R)$ and (36) for $n = 0$. We must show $(6\mathcal{F}'(y_0) + 2\mathcal{F}'(x_0))$ is invertible. By (H2'), we get

$$\begin{aligned}
&\|(2\mathcal{F}'(x_*))^{-1}[3\mathcal{F}'(y_0) + \mathcal{F}'(x_0) - 2\mathcal{F}'(x_*)]\| \\
&\leq \frac{1}{2}[3\|\mathcal{F}'(x_*)^{-1}(\mathcal{F}'(y_0) - \mathcal{F}'(x_*))\| \\
&\quad + \|\mathcal{F}'(x_*)^{-1}\mathcal{F}'(x_0) - \mathcal{F}'(x_*))\|] \\
&\leq \frac{1}{2}(3\varphi_0(\|y_0 - x_*\|) + \varphi_0(\|x_0 - x_*\|)) \\
&= p(\|x_0 - x_*\|) \leq p(R) < 1,
\end{aligned}$$

so

$$\|(6\mathcal{F}'(y_0) + 2\mathcal{F}'(x_0))^{-1}\mathcal{F}'(x_*)\| \leq \frac{3}{4(1 - p(\|x_0 - x_*\|))}. \tag{41}$$

That is z_0 exists. Then, we get

$$\begin{aligned}
\|z_0 - x_*\| &= \|(x_0 - x_* - \mathcal{F}'(x_0)^{-1}\mathcal{F}(x_0) \\
&\quad + [I - (6\mathcal{F}'(y_0) + 2\mathcal{F}'(x_0))^{-1}(3\mathcal{F}'(y_0) + \mathcal{F}'(x_0))]\mathcal{F}'(x_0)^{-1}\mathcal{F}(x_0)\| \\
&\leq \psi_1(\|x_0 - x_*\|)\|x_0 - x_*\| + \frac{3}{2}\|(3\mathcal{F}'(y_0) + \mathcal{F}'(x_0))^{-1}\mathcal{F}'(x_*)\| \\
&\quad \times [\|\mathcal{F}'(x_*)^{-1}((\mathcal{F}'(y_0) - \mathcal{F}'(x_*))\| + \|\mathcal{F}'(x_*)^{-1}(\mathcal{F}'(x_0) - \mathcal{F}'(x_*))\|] \\
&\quad \times \|\mathcal{F}'(x_0)^{-1}\mathcal{F}'(x_*)\|\|\mathcal{F}'(x_*)^{-1}\mathcal{F}(x_)\| \\
&\leq \left[\psi_1(\|x_0 - x_*\|) + \frac{3(\varphi_0(\|y_0 - x_*\|) + \varphi_0(\|x_0 - x_*\|))\int_0^1 \varphi_1(\tau\|x_0 - x_*\|)d\tau}{4(1 - p(\|x_0 - x_*\|))(1 - \varphi_0(\|x_0 - x_*\|))}\right] \\
&\quad \times \|x_0 - x_*\| \\
&= \psi_2(\|x_0 - x_*\|)\|x_0 - x_*\| \leq \|x_0 - x_*\|, \tag{42}
\end{aligned}$$

leading to $z_0 \in T(x_*, R)$ and (37) for $n = 0$. Notice that x_1 exists, since both $\mathcal{F}'(y_0)$ and $\mathcal{F}'(x_0)$ are invertible.

Next, we obtain

$$\begin{aligned}
\|x_1 - x_*\| &= \|(z_0 - x_* - \mathcal{F}'(z_0)^{-1}\mathcal{F}(z_0)) \\
&\quad + \mathcal{F}'(z_0)^{-1}[(\mathcal{F}'(y_0) - \mathcal{F}'(x_*)) + (\mathcal{F}'(x_*) - \mathcal{F}'(z_0))]\mathcal{F}'(y_0)^{-1}\mathcal{F}(z_0) \\
&\quad + \frac{1}{2}\mathcal{F}'(x_0)^{-1}(\mathcal{F}'(y_0) - \mathcal{F}'(x_0))\mathcal{F}'(y_0)^{-1}\mathcal{F}'(z_0)\| \\
&\leq \left[\psi_2(\|z_0 - x_*\|) + \frac{(\varphi_0(\|y_0 - x_*\|) + \varphi_0(\|z_0 - x_*\|))\int_0^1 \varphi_1(\tau\|z_0 - x_*\|)d\tau}{(1 - \varphi_0(\|z_0 - x_*\|))(1 - \varphi_0(\|y_0 - x_*\|))} \right. \\
&\quad \left. + \frac{1}{2}\frac{\varphi_0(\|y_0 - x_*\|) + \varphi_0(\|x_0 - x_*\|)\int_0^1 \varphi_1(\tau\|z_0 - x_0\|)d\tau}{(1 - \varphi_0(\|x_0 - x_*\|))(1 - \varphi_0(\|y_0 - x_*\|))}\right]\|z_0 - x_*\| \\
&\leq \psi_3(\|x_0 - x_*\|)\|x_0 - x_*\| \leq \|x_0 - x_*\| \tag{43}
\end{aligned}$$

leading to $x_1 \in T(x_*, R)$ and (38) for $n = 0$. The rest as identical to Theorem 1 is omitted. $\square$

Remark 4. *Similar comments as in Remark 2 can be made for method (4).*

4. NUMERICAL EXAMPLES

We present the following examples to test the convergence criteria.

Example 5. *Let $\mathcal{E}_1 = \mathcal{E}_1 = \mathbb{R}^3$, $E = U(0,1)$, $x^* = (0,0,0)^T$ and define $\mathcal{F}$ on E by*

$$\mathcal{F}(x) = \mathcal{F}(u_1, u_2, u_3) = \left(e^{u_1} - 1, \frac{e-1}{2}u_2^2 + u_2, u_3\right)^T. \tag{44}$$

For the points $u = (u_1, u_2, u_3)^T$, the Fréchet derivative is given by

$$\mathcal{F}'(u) = \begin{pmatrix} e^{u_1} & 0 & 0 \\ 0 & (e-1)u_2 + 1 & 0 \\ 0 & 0 & 1 \end{pmatrix}.$$

Using the norm of the maximum of the rows and since $\mathcal{F}'(x^*) = diag(1,1,1)$, we can define parameters $\varphi_0(t) = a_0(t) = (e-1)t$, $\varphi(t) = a(t) = e^{\frac{1}{e-1}}t$, $\varphi_1(t) = a_1(t) = e^{\frac{1}{e-1}}$. Then, we have

$$r_1 = 0.3827, \, r_2 = 2.5490, \, r_3 = 0.0475 = r$$

and

$$R_1 = 0.1150 = R, \, R_2 = 0.1756, \, R_3 = 0.1205.$$

Example 6. *Returning back to the motivational example at the introduction of this study, we have $\varphi_0(t) = a_0(t) = \varphi(t) = a(t) = 96.662907t$, $\varphi_1(t) = a_1(t) = 1.0631$. Then, we have*

$$r_1 = 0.0069, \, r_2 = 0.0162, \, r_3 = 0.0004 = r$$

and

$$R_1 = 0.1150 = R, \, R_2 = 0.1756, \, R_3 = 0.1205.$$

REFERENCES

[1] Amat, S., Argyros, I. K., Busquier, S. and Hernández–Verón, M. A. 2018. "On two high–order families of frozen Newton–type methods". *Numerical Linear Algebra with Applications* 25(1).

[2] Argyros, I. K. 2007. *Computational theory of iterative methods*, volume 15. Elsevier.

[3] Argyros, I. K. 2008. "On the semilocal convergence of a fast two–step Newton method". *Revista Colombiana de Matemáticas* 42(1): 15–24.

[4] Argyros, I. K., George, S. and Thapa, N. 2018. *Mathematical modeling for the solution of equations and systems of equations with applications*, volume–I. New York: Nova Publishers.

[5] Argyros, I. K., George, S. and Thapa, N. 2018. *Mathematical modeling for the solution of equations and systems of equations with applications*, volume–II. New York: Nova Publishers.

[6] Argyros, I. K., Cordero, A., Magreñán, Á. A. and Torregrosa, J. R. 2017. "Third–degree anomalies of Traubs method". *Journal of Computational and Applied Mathematics* 309: 511–521.

[7] Argyros, I. K. and Hilout, S. 2012. "Weaker conditions for the convergence of Newton's method". *Journal of Complexity* 28(3): 364–387.

[8] Argyros, I. K. and Hilout, S. 2013. *Computational methods in nonlinear analysis: efficient algorithms, fixed point theory and applications*. World Scientific.

[9] Argyros, I. K., Magreñán, Á. A., Orcos, L. and Sicilia, J. A. 2017. "Local convergence of a relaxed two–step Newton like method with applications". *Journal of Mathematical Chemistry* 55(7): 1427–1442.

[10] Argyros, I. K. and Magreñán, Á. A. 2018. *A contemporary study of iterative methods*. New York: Elsevier (Academic Press).

[11] Argyros, I. K. and Magreñán, Á. A. 2017. *Iterative methods and their dynamics with applications*. New York, USA: CRC Press.

[12] Cordero, A. and Torregrosa, J. R. 2015. "Low–complexity root–finding iteration functions with no derivatives of any order of convergence". *Journal of Computational and Applied Mathematics* 275: 502–515.

[13] Cordero, A., Torregrosa, J. R. and Vindel, P. 2012. "Study of the dynamics of third–order iterative methods on quadratic polynomials". *International Journal of Computer Mathematics* 89(13–14): 1826–1836.

[14] Ezquerro, J. A. and Hernández–Verón, M. A. 2014. "How to improve the domain of starting points for Steffensen's method". *Studies in Applied Mathematics* 132(4): 354–380.

[15] Ezquerro, J. A. and Hernández–Verón, M. A. 2017. "Majorizing sequences for nonlinear fredholm Hammerstein integral equations". *Studies in Applied Mathematics* 140(3).

[16] Ezquerro, J. A., Grau–Sánchez, M., Hernández–Verón, M. A. and Noguera, M. 2015. "A family of iterative methods that uses divided differences of first and second orders". *Numerical algorithms* 70(3): 571–589.

[17] Ezquerro, J. A., Hernández–Verón, M. A. and Velasco, A. I. 2015. "An analysis of the semilocal convergence for secant–like methods". *Applied Mathematics and Computation* 266: 883–892.

[18] Hernández–Verón, M. A., Martínez, E. and Teruel, C. 2017. "Semilocal convergence of a k–step iterative process and its application for solving a special kind of conservative problems". *Numerical Algorithms* 76(2): 309–331.

[19] Kantlorovich, L. V. and Akilov, G. P. 1982. *Functional analysis*. Pergamon Press.

[20] Magreñán, Á. A., Cordero, A., Gutiérrez, J. M. and Torregrosa, J. R. 2014. "Real qualitative behavior of a fourth–order family of iterative methods by using the convergence plane". *Mathematics and Computers in Simulation* 105: 49–61.

[21] Magreñán, Á. A. and Argyros, I. K. 2016. "Improved convergence analysis for Newton–like methods". *Numerical Algorithms* 71(4): 811–826.

[22] Magreñán, Á. A. and Argyros, I. K. 2014. "Two–step Newton methods". *Journal of Complexity* 30(4): 533–553.

[23] Ozban, A. Y. 2004. "Some variants of Newton's method". *Appl. Math. Lett.* 17, 677–682.

[24] Potra, F. A. and Pták, V. 1984. "Nondiscrete induction and iterative processes", volume 103. Pitman Advanced Publishing Program.

[25] Rheinboldt, W. C. 1978. "An adaptive continuation process for solving systems of nonlinear equations". *Polish Academy of Science, Banach Ctr. Publ.* 3(1): 129–142.

[26] Traub, J. F. 1982. *Iterative methods for the solution of equations.* American Mathematical Soc.

[27] Wang, X., Kou, J. and Gu, C. 2011. "Semilocal convergence of a sixth–order Jarratt method in Banach spaces". *Numer. Algor.* 57: 441–456.

In: Understanding Banach Spaces
Editor: Daniel González Sánchez
ISBN: 978-1-53616-745-0
© 2020 Nova Science Publishers, Inc.

Chapter 7

BALL CONVERGENCE FOR OPTIMAL DERIVATIVE FREE METHODS

Ioannis K. Argyros[1,*] *and Ramandeep Behl*[2,†]
[1]Department of Mathematical Sciences
Cameron University, Lawton, OK, US
[2]Department of Mathematics
King Abdulaziz University, Jeddah, Saudi Arabia

Abstract

We present a local convergence analysis of a two–step fourth order derivative free method in order to approximate a locally unique solution of nonlinear equation in a Banach space setting. In the earlier study [1], the order of convergence was shown using Taylor series expansions and hypotheses up to the fourth order derivative or even higher of the function involved. However, no derivative appears in the proposed scheme. That restrict the applicability of the this scheme because there are many practical situations where the calculations of fourth derivatives are expensive and/or it requires a great deal of time for them to be given or calculated or sometimes fourth derivative is unbounded or not exist. Therefore, we expand the applicability of the Cordero et al. (2013) scheme using only hypotheses on the first order derivative of function F. We also proposed the computable radii of convergence, error bounds based on the Lipschitz constants and the range of initial guess that tell us how close the initial guess should be required for granted convergence of methods. Numerical examples where earlier studies cannot apply to solve nonlinear equations but our results apply are also given in this study.

Keywords: two–step method with memory, local convergence, convergence order

AMS Subject Classification: 65D10, 65D99

---

[*]Corresponding Author's E-mail: iargyros@cameron.edu.
[†]E-mail: ramanbehl87@yahoo.in.

1. INTRODUCTION

In this study we are concerned with the problem of approximating a locally unique solution x^* of the equation of the form

$$F(x) = 0, \tag{1}$$

where F is a twice Fréchet differentiable function defined on a subset D of S ($S = \mathbb{R}$ or $S = \mathbb{C}$) with values in S.

We can say that either lack or intractability of their analytic solutions often forces researchers from the worldwide to resort to an iterative method. While, using these iterative methods researchers face the problems of slow convergence, non–convergence, divergence, inefficiency or failure (for detail please see Traub [2] and Petkovic et al. [3]). The convergence analysis of iterative methods is usually divided into two categories: semi–local and local convergence analysis. The semi–local convergence matter is, based on the information around an initial point, to give criteria ensuring the convergence of iteration procedures. A very important problem in the study of iterative procedures is the convergence domain. Therefore, it is very important to propose the radius of convergence of the iterative methods.

We study the local convergence analysis of an optimal two–step fourth order method defined for each $n = 0, 1, 2, \ldots$ by

$$\begin{aligned} y_n &= x_n - \frac{F(x_n)^2}{F(x_n + F(x_n)) - F(x_n)}, \\ x_{n+1} &= y_n - \frac{[x_n, x_n + F(x_n); F] F(y_n)}{[x_n, y_n; F][y_n, x_n + F(x_n); F]} \end{aligned} \tag{2}$$

where x_0 are initial point. Method (2) was studied in [1] using Taylor expansions and hypotheses reaching up to the fourth derivative of function F when $X = \mathbb{R}$. The hypotheses on the derivatives of F limit the applicability of method (2). As a motivational example, define function F on $S = \mathbb{R}$, $D = [-\frac{1}{\pi}, \frac{2}{\pi}]$ by

$$F(x) = \begin{cases} x^3 \log(\pi^2 x^2) + x^5 \sin\left(\frac{1}{x}\right), & x \neq 0, \\ 0, & x = 0. \end{cases}$$

Then, we have that

$$F'(x) = 2x^2 - x^3 \cos\left(\frac{1}{x}\right) + 3x^2 \log(\pi^2 x^2) + 5x^4 \sin\left(\frac{1}{x}\right),$$

$$F''(x) = -8x^2 \cos\left(\frac{1}{x}\right) + 2x(5 + 3\log(\pi^2 x^2)) + x(20x^2 - 1) \sin\left(\frac{1}{x}\right)$$

and

$$F'''(x) = \frac{1}{x}\left[(1 - 36x^2) \cos\left(\frac{1}{x}\right) + x\left(22 + 6\log(\pi^2 x^2) + (60x^2 - 9) \sin\left(\frac{1}{x}\right)\right)\right].$$

One can easily find that the function $F'''(x)$ is unbounded on $\mathbb{D}$ at the point $x = 0$. Hence, the results in [1], cannot apply to show the convergence of method (2) requiring hypotheses on the fourth derivative of function F or higher. Notice that, in–particular there is a plethora of iterative methods for approximating solutions of nonlinear equations [1–14]. These results show that initial guess should be close to the required root for the convergence of the corresponding methods. But, how close initial guess should be required for the convergence of the corresponding method? These local results give no information on the radius of the convergence ball for the corresponding method. The same technique can be used on other methods.

In the present work, we study the local convergence analysis of method (2) in the more general setting of a Banach space. Method (2) can be written as

$$y_n = x_n - A_n^{-1} F(x_n),$$
$$x_{n+1} = y_n - [y_n, x_n + F(x_n); F]^{-1}[x_n, x_n + F(x_n); F][x_n, y_n; F]^{-1} F(y_n), \quad (3)$$

where $A_n = \int_0^1 F'(x_n + \theta F(x_n)) d\theta$ and the divided difference $[x, y; F] = \int_0^1 F'(y_n + \theta(x - y)) d\theta (x - y)$ for each $x, y \in D$.

In this way, we expand the applicability of method (2) using only hypotheses up to the first order derivative of function F. We also proposed the computable radii of convergence and error bounds based on the Lipschitz constants. We further present the range of initial guesses x_0 that tell us how close the initial guess should be required for granted convergence of the method (2). This problem was not addressed in [1].

2. LOCAL CONVERGENCE

In this section, we shall define some scalar functions and parameters in order to present the local convergence of method (3) that follows.

Let $K > 0$, $L_0 > 0$, $L > 0$, $M \geq 1$, $M_0 > 0$ and $N \geq 1$ be given parameters. The convergence is based on some scalar functions and parameters. Let us also assume some functions p defined on the interval $[0, \infty)$ by

$$p(t) = L_0 \left(1 + \frac{M_0}{2}\right) t,$$

parameter r_p by

$$r_p = \frac{1}{\left(1 + \frac{M_0}{2}\right) L_0}$$

and functions g_1 and h_1 on the interval $[0, r_p)$ by

$$g_1(t) = \frac{L}{2(1 - L_0 t)} \left(1 + \frac{M_0 M}{1 - p(t)}\right) t,$$
$$h_1(t) = g_1(t) - 1.$$

We have that $h_1(0) = -1 < 0$, and $h_1(t) \to +\infty$ as $t \to r_p^-$. By the intermediate value theorem functions h_1 has zeros in the interval $(0, r_p)$. Further, let r_1 be the smallest such zero. Moreover, define functions q_0, h_{q_0}, q and h_q on the interval $[0, r_p)$ by

$$q_0(t) = K(1 + g_1(t))t,$$
$$h_{q_0}(t) = q_0(t) - 1,$$
$$q(t) = K(1 + M_0 + g_1(t))t$$
and
$$h_q(t) = q(t) - 1.$$

Then, we get that $h_{q_0}(0) = h_q(0) - 1 < 0$ and $h_{q_0}(t) \to \infty$, $h_q(t) \to +\infty$ as $t \to r_p^{-1}$. Denote by r_{q_0} and r_q the smallest such zero of function h_{q_0} and h_q, respectively on the interval $(0, r_p)$. Notice that $h_q(t) = h_{q_0}(t) + M_0 t$, then $h_q(r_{q_0}) = h_{q_0}(r_{q_0}) + M_0 r_{q_0} = M_0 r_{q_0} > 0$, so $r_q < r_{q_0}$. Furthermore, define functions g_2 and h_2 on the interval $[0, r_q)$ by

$$g_2(t) = \left(1 + \frac{MN}{(1 - q_0(t))(1 - q(t))}\right) g_1(t),$$

and

$$h_2(t) = g_2(t) - 1.$$

We have that $h_2(0) = -1 < 0$, and $h_2(t) \to +\infty$ as $t \to r_q^-$. Let r be the smallest such zero. Then, we have that for each $t \in [0, r)$

$$0 \le p(t) < 1, \tag{4}$$

$$0 \le g_1(t) < 1, \tag{5}$$

$$0 \le q_0(t) < 1, \tag{6}$$

$$0 \le q(t) < 1, \tag{7}$$

and

$$0 \le g_2(t) < 1. \tag{8}$$

Let $U(\gamma, \rho)$ and $\bar{U}(\gamma, \rho)$ stand, respectively for the open and closed balls in S with center $\gamma \in S$ and radius $\rho > 0$. Next, we present the local convergence analysis of method (3) using the preceding notations.

Theorem 1. *Let us consider $F : D \subset X \to X$ be a Fréchet–differentiable operator. Suppose that there exist a divided difference of order one $[\cdot, \cdot; F] : D^2 \to L(X)$, $x^* \in D$ and $L_0 > 0$ such that for each $x \in D$*

$$F(x^*) = 0, \quad F'(x^*) \in L(X), \tag{9}$$

$$\|F(x^*)^{-1}(F'(x) - F'(x^*))\| \le L_0 \|x - x^*\|. \tag{10}$$

Moreover, suppose that there exist $L > 0$, $K > 0$, $M_0 > 0$, $N \ge 1$ and $M \ge 1$ such that for each $x, y \in D \cap U\left(x^*, \frac{1}{L_0}\right)$

$$\|F'(x^*)^{-1}(F'(x) - F'(y))\| \le L \|x - y\|, \tag{11}$$

$$\|F'(x^*)^{-1}F'(x)\| \leq M, \tag{12}$$

$$\|F'(x)\| \leq M_0, \tag{13}$$

$$\|F'(x^*)^{-1}([x, y; F] - F'(x^*))\| \leq K(\|x - x^*\| + \|y - x^*\|), \tag{14}$$

$$\|F'(x^*)^{-1}[x, y; F]\| \leq N, \tag{15}$$

and

$$\bar{U}(x^*, (1 + M_0)r) \subseteq D, \tag{16}$$

where the radius of convergence r was defined previously. Then, the sequence $\{x_n\}$ generated for $x_0 \in U(x^*, r) - \{x^*\}$ by method (3) is well defined, remains in $U(x^*, r)$ for each $n = 0, 1, 2, \ldots$ and converges to x^*. Moreover, the following estimates hold

$$\|y_n - x^*\| \leq g_1(\|x_n - x^*\|)\|x_n - x^*\| < \|x_n - x^*\| < r \tag{17}$$

and

$$\|x_{n+1} - x^*\| \leq g_2(\|x_n - x^*\|)\|x_n - x^*\| < \|x_n - x^*\|, \tag{18}$$

where the "g" functions are defined previously. Furthermore, for $T \in \left[r, \frac{2}{L_0}\right)$, the point x^* is the only solution of equation $F(x) = 0$ in $\bar{U}(x^*, T) \cap D$.

Proof. We shall show estimates (17) and (18) hold with the help of mathematical induction. Notice that

$$\|x_0 + F(x_0) - x^*\| \leq \|x_0 - x^*\| + \left\|\int_0^1 F'(x^* + \theta(x_0 - x^*))(x_0 - x^*)d\theta\right\|$$
$$r + M_0\|x_0 - x^*\| \leq (1 + M_0)r,$$

so $x_0 + F(x_0) \in \bar{U}(x^*, (1 + M_0)r) \subset D$. Then, we must show that $A_0^{-1} \in L(X)$. Using (10), we get that

$$\|F'(x^*)^{-1}(A_0 - F'(x^*))\| \leq L_0 \int_0^1 \|x_0 - x^* + \theta F(x_0)\|d\theta$$
$$\leq \|x_0 - x^*\| + M_0 \int_0^1 \theta d\theta\|x_0 - x^*\| \tag{19}$$
$$= \left(1 + \frac{M_0}{2}\right)\|x_0 - x^*\| = p(\|x_0 - x^*\|)$$
$$< p(r) < 1.$$

It follows from (19) and the Banach Lemma on invertible functions [7, 14] that $A_0^{-1} \in L(X)$,

$$\|A_0^{-1}F'(x^*)\| \leq \frac{1}{1 - p(\|x_0 - x^*\|)} \tag{20}$$

and y_0 is well defined by the first sub step of method (3) for $n = 0$. Then, from method (3), we obtain the identity

$$y_0 - x^* = (x_0 - x^* - F'(x_0)^{-1}F(x_0)) + (F'(x_0)^{-1}F'(x^*))\left(F'(x^*)^{-1}(A_0 - F'(x_0))\right)$$
$$\times \left(A_0^{-1}F'(x^*)\right)\left(F'(x^*)^{-1}F(x_0)\right). \tag{21}$$

Using (5), (9), (10), (11), (12), (13), (20) and (21), we get turn that

$$\begin{aligned}
\|y_0 - x^*\| &\leq \|F'(x_0)^{-1}F'(x^*)\| \left\| \int_0^1 F'(x^*)^{-1} \left[F'(x^* + \theta(x_0 - x^*)) - F'(x_0) \right] (x_0 - x^*) d\theta \right\| \\
&\quad + \|F'(x_0)^{-1}F'(x^*)\| \|F'(x^*)^{-1}(A_0 - F'(x_0))\| \|A_0^{-1}F'(x^*)\| \|F'(x^*)^{-1}F(x_0)\| \\
&\leq \frac{L\|x_0 - x^*\|^2}{2(1 - L_0\|x_0 - x^*\|)} + \frac{LM_0 M \|x_0 - x^*\|^2}{2(1 - L_0\|x_0 - x^*\|)(1 - p(\|x_0 - x^*\|))} \\
&= g_1(\|x_0 - x^*\|)\|x_0 - x^*\| < \|x_0 - x^*\| < r,
\end{aligned} \qquad (22)$$

where we also used the estimates

$$\|F'(x_0)^{-1}F'(x^*)\| \leq \frac{1}{1 - L_0\|x_0 - x^*\|}, \qquad (23)$$

$$\|F'(x^*)^{-1}(A_0 - F'(x_0))\| \leq M\|x_0 - x^*\| \qquad (24)$$

and

$$\begin{aligned}
\|(F'(x^*)^{-1}(A_0 - F'(x_0)))\| &\leq LM_0 \|x_0 - x^*\| \int_0^1 \theta d\theta \\
&= \frac{LM_0}{2}\|x_0 - x^*\|.
\end{aligned} \qquad (25)$$

Estimate (22) shows (17) for $n = 0$ and $y_0 \in U(x^*, r)$. Then, using (6), (7), (13), (14) and (16), we obtain in turn that

$$\begin{aligned}
\|F'(x^*)([x_0, y_0; F] - F'(x^*))\| &\leq K(\|x_0 - x^*\| + \|y_0 - x^*\|) \\
&\leq K(1 + g_1(\|x_0 - x^*\|))\|x_0 - x^*\| \\
&= q_0(\|x_0 - x^*\|) < q_0(r) < 1
\end{aligned} \qquad (26)$$

and

$$\begin{aligned}
\|F'(x^*)([y_0, x_0 + F(x_0); F] - F'(x^*))\| &\leq K(\|y_0 - x^*\| + \|x_0 + F(x_0) - x^*\|) \\
&\leq K(\|y_0 - x^*\| + \|x_0 - x^*\| + \|F(x_0)\|) \\
&\leq K(g_1(\|x_0 - x^*\|)\|x_0 - x^*\| + \|x_0 - x^*\| \\
&\quad + M_0\|x_0 - x^*\|) \\
&= q(\|x_0 - x^*\|) < q(r) < 1.
\end{aligned} \qquad (27)$$

It follows from (26) and (27) that $[x_0, y_0; F]^{-1} \in L(X)$ and $[y_0, x_0 + F(x_0); F]^{-1} \in L(X)$, respectively

$$\|[x_0, y_0; F]^{-1} F'(x^*)\| \leq \frac{1}{1 - q_0(\|x_0 - x^*\|)} \qquad (28)$$

and

$$\|[y_0, x_0 + F(x_0); F]^{-1} F'(x^*)\| \leq \frac{1}{1 - q(\|x_0 - x^*\|)} \qquad (29)$$

and x_1 is well defined by the second sub step of method (3) for $n = 0$. Then, using the first sub step of method (3) for $n = 0$, the definition of r, (8), (15), (22), (24), (28) and (29), we get in turn that

$$\|x_1 - x^*\| \leq \|y_0 - x^*\| + \frac{MN\|y_0 - x^*\|}{(1 - q_0(\|x_0 - x^*\|))(1 - q(\|x_0 - x^*\|))}$$

$$\leq \left(1 + \frac{MN}{(1 - q_0(\|x_0 - x^*\|))(1 - q(\|x_0 - x^*\|))}\right) g_1(\|x_0 - x^*\|)\|x_0 - x^*\| \quad (30)$$

$$= g_2(\|x_0 - x^*\|)\|x_0 - x^*\| < \|x_0 - x^*\| < r,$$

which shows (18) for $n = 0$ and $x_1 \in U(x^*, r)$. By simply replacing x_0, y_0, x_1 by x_k, y_k, x_{k+1} in the preceding estimates we arrive at (14) and (15). Then notice that $\|x_{k+1} - x^*\| \leq \|x_k - x^*\| < r$, $c = g_2(r) \in [0, 1)$. Hence, we conclude that $\lim_{k \to \infty} x_k = x^*$ and $x_{k+1} \in U(x^*, r)$. Finally, to show the uniqueness part, let $y^* \in \bar{U}(x^*, T)$ be such that $F(y^*) = 0$. Set $Q = \int_0^1 F'(x^* + \theta(y^* - x^*))\, d\theta$. Then, using (15), we get that

$$\|F'(x^*)^{-1}(Q - F'(x^*))\| \leq L_0 \int_0^1 \theta \|x^* - y^*\|\, d\theta = \frac{L_0}{2} T < 1. \quad (31)$$

Hence, $Q^{-1} \in L(Y, X)$. Then, in view of the identity $F(y^*) - F(x^*) = Q(y^* - x^*)$, we conclude that $x^* = y^*$. □

Remark 2. (a) *In view of (12) and the estimate*

$$\|F'(x^*)^{-1} F'(x)\| = \|F'(x^*)^{-1}(F'(x) - F'(x^*)) + I\|$$
$$\leq 1 + \|F'(x^*)^{-1}(F'(x) - F'(x^*))\|$$
$$\leq 1 + L_0 \|x_0 - x^*\|$$

condition (14) can be dropped and M can be replaced by

$$M = M(t) = 1 + L_0 t$$

or $M = 2$, since $t \in [0, \frac{1}{L_0})$.

(b) *The results obtained here can be used for operators F satisfying the autonomous differential equation [7, 8] of the form*

$$F'(x) = P(F(x)),$$

where P is a known continuous operator. Since $F'(x^*) = P(F(x^*)) = P(0)$, we can apply the results without actually knowing the solution x^*. Let as an example $F(x) = e^x + 2$. Then, we can choose $P(x) = x - 2$.

(c) *The radius $r_A = \frac{2}{L_0 + L}$ was shown by us in [7, 8] to be the convergence radius for Newton's method under conditions (12) – (14). Radius r_A is at least as large as the convergence ball given by Rheinboldt [14] and Traub [2]*

$$r_R = \frac{2}{3L}.$$

Notice that for $L_0 < L$,
$$r_R < r_A.$$
Moreover,
$$\frac{r_R}{r_A} \to \frac{1}{3} \quad as \quad \frac{L_0}{L} \to 0.$$

Hence, r_A is at most three times larger than r_R. In the numerical examples we compare r to r_A and r_R.

(d) *It is worth noticing that method (2) is not changing if we use the conditions of Theorem 1 instead of the stronger conditions given in [1]. Moreover, for the error bounds in practice we can use the computational order of convergence (COC) [10]*

$$\xi = \frac{\ln \frac{\|x_{n+2}-x^*\|}{\|x_{n+1}-x^*\|}}{\ln \frac{\|x_{n+1}-x^*\|}{\|x_n-x^*\|}}, \quad \text{for each } n = 0, 1, 2, \ldots \tag{32}$$

or the approximate computational order of convergence (ACOC) [10]

$$\xi^* = \frac{\ln \frac{\|x_{n+2}-x_{n+1}\|}{\|x_{n+1}-x_n\|}}{\ln \frac{\|x_{n+1}-x_n\|}{\|x_n-x_{n-1}\|}}, \quad \text{for each } n = 1, 2, \ldots \tag{33}$$

This way we obtain in practice the order of convergence in a way that avoids the bounds involving estimates higher than the first Fréchet derivative.

3. NUMERICAL EXAMPLES AND APPLICATIONS

In order to demonstrate the convergence behavior of the scheme proposed in [1] and to check the validity of the theoretical results which we have proposed in section 2. For this purpose, we shall choose a variety of nonlinear equations and system of nonlinear equations which are mentioned in the following examples including motivational example.

First of all, we shall calculate the values of r_R, r_A, r_p, r_1, r_{q_0}, r_q and r to find the range of convergence domain, which are presented in the corresponding example up to 5 significant digits. However, we have calculate the values of these constants up to several number of significant digits. Now, we can choose an initial guess x_0 with in the range of convergence domain, which gives the guarantee for convergence of the iterative methods. Further, we will also want to verify the theoretical order of convergence of these methods for scalar equations on the basis of the results obtain from computational order of convergence and $\left|\frac{e_{n+1}}{e_n^p}\right|$. In the Tables 3, we mentioned the number of iteration indexes (n), approximated zeros (x_n), residual error of the corresponding function $(|F(x_n)|)$, errors $|e_n|$ (where $e_n = x_n - x^*$), $\left|\frac{e_{n+1}}{e_n^p}\right|$ and the asymptotic error constant $\eta = \lim_{n \to \infty} \left|\frac{e_{n+1}}{e_n^p}\right|$. In addition, we use the formulas proposed by Sánchez et al. in [10] to calculate the computational order of convergence by, which is given by

$$\xi = \frac{\ln \frac{|x_{n+2}-x^*|}{|x_{n+1}-x^*|}}{\ln \frac{|x_{n+1}-x^*|}{|x_n-x^*|}}, \quad \text{for each } n = 0, 1, 2, \ldots$$

or the approximate computational order of convergence (ACOC) [10]

$$\xi^* = \frac{ln\frac{|x_{n+2}-x_{n+1}|}{|x_{n+1}-x_n|}}{ln\frac{|x_{n+1}-x_n|}{|x_n-x_{n-1}|}}, \quad \text{for each } n = 1, 2, \ldots$$

We calculate the computational order of convergence, asymptotic error constant and other constants up to several number of significant digits (minimum 1000 significant digits) to minimize the round off error.

In the context of system of nonlinear equations, we also consider a nonlinear system in examples 3 – 4 to check the proposed theoretical results for nonlinear system. In this regards, we displayed the number of iteration indexes (n), residual error of the corresponding function ($\|F(x_n)\|$), error in the iterations $\|x_{n+1} - x_n\|$ in the Tables 1 and 2. Moreover, we use the above formulas namely, 32 and 33 to calculate the computational order of convergence for nonlinear system.

As we mentioned in the earlier paragraph that we calculate the values of all the constants and functional residuals up to several number of significant digits but due to the limited paper space, we display the values of x_n up to 15 significant digits. Further, the values of other constants namely, $\rho(COC)$ up to 5 significant digits and the values $\left|\frac{e_n}{e_{n-1}^p}\right|$ and η are up to 10 significant digits. Furthermore, the residual error in the function/system of nonlinear functions ($|F(x_n)|$ or $\|F(x_n)\|$) and the error ($|e_n|$ or $\|x_{n+1} - x_n\|$) are display up to 2 significant digits with exponent power which are mentioned in the following Tables corresponding to the test function. However, minimum 1000 significant digits are available with us for every value.

Furthermore, we consider the approximated zero of test functions when the exact zero is not available, which is corrected up to 1000 significant digits to calculate $\|x_n - x^*\|$. During the current numerical experiments with programming language Mathematica (Version 9), all computations have been done with multiple precision arithmetic, which minimize round-off errors.

Example 3. Let $\mathbb{X} = \mathbb{Y} = \mathbb{R}^3$, $\mathbb{D} = \bar{U}(0, 1)$. Define F on $\mathbb{D}$ for $v = (x, y, z)^T$ by

$$F(v) = \left(e^x - 1, \frac{e-1}{2}y^2 + y, z\right)^T. \tag{34}$$

Then the Fréchet-derivative is given by

$$F'(v) = \begin{bmatrix} e^x & 0 & 0 \\ 0 & (e-1)y + 1 & 0 \\ 0 & 0 & 1 \end{bmatrix}.$$

Notice that $x^* = (0, 0, 0)$, $F'(x^*) = F'(x^*)^{-1} = diag\{1, 1, 1\}$, $L_0 = e - 1$, $L = 1.789572397$, $K = \frac{L_0}{2}$, $M_0 = 2$ and $M = N = 2$. Hence, we further yield the following values

$$r_R = 0.24525, \quad r_A = 0.32495, \quad r_p = 0.29099, \quad r_1 = 0.091008, \tag{35}$$
$$r_{q_0} = 0.19280, \quad r_q = 0.17396, \quad r = 0.024056.$$

Further, error in the iterations, residual error in the involved function and COC (ρ) are given in the following Table 1 on the basis of initial guess $(0.022, 0.022, 0.023)$.

Table 1. Convergence behavior on example 3

Method	n	$\|F(x_n)\|$	$\|x_{n+1} - x_n\|$	ξ
3	0	$3.9e(-2)$	$3.9e(-2)$	
	1	$1.7e(-7)$	$1.7e(-7)$	
	2	$3.7e(-28)$	$3.7e(-28)$	3.8666
	3	$1.3e(-110)$	$1.3e(-110)$	3.9895

Example 4. Let $\mathbb{X} = \mathbb{Y} = \mathbb{R}^2$, $\mathbb{D} = \bar{U}(0, 1)$. Define F on $\mathbb{D}$ for $v = (x, y)^T$ by

$$F(v) = \left(x^2 - x - y^2 - 1, \, -\sin x + y\right)^T. \tag{36}$$

Then the Fréchet-derivative is given by

$$F'(v) = \begin{bmatrix} 2x - 1 & -2y \\ -\cos x & 1 \end{bmatrix}.$$

Notice that $x^* = (-0.8452567\ldots, -0.7481415\ldots)$, $L_0 = L = 9.864796414$, $K = \frac{L_0}{2}$, $M_0 = 8.37359$ and $M = N = 2$. Hence, we deduce the following values

$$\begin{aligned} r_R &= 0.067580, \quad r_A = 0.0675804, \quad r_p = 0.019544, \quad r_1 = 0.0069792, \\ r_{q_0} &= 0.016643, \quad r_q = 0.014129 \quad r = 0.0018644. \end{aligned} \tag{37}$$

Finally, residual error in the involved function, error in the iterations and COC (ρ) are given in the following Table 2 by considering initial guess $(0.843, 0.746)$.

Table 2. Convergence behavior on example 4

Methods	n	$\|F(x_n)\|$	$\|x_{n+1} - x_n\|$	ξ
3	0	$2.9e(-3)$	$3.1e(-3)$	
	1	$3.4e(-11)$	$3.9e(-11)$	
	2	$8.2e(-43)$	$9.1e(-43)$	4.0012
	3	$2.5e(-169)$	$2.8e(-169)$	3.9999

Example 5. Returning back to the motivation example at the introduction on this paper, we have $L = L_0 = \frac{2}{2\pi+1}(80+16\pi+(11+12\log 2)\pi^2)$, $K = \frac{L_0}{2}$, $M = N = 2$, $M_0 = \frac{2(2\pi+1)}{\pi^3}$ and our required zero is $x^* = \frac{1}{\pi}$. Then, we obtain the following values

$$\begin{aligned} r_R &= 0.0075648, \quad r_A = 0.0075648, \quad r_p = 0.0091888, \quad r_1 = 0.00463528, \\ r_{q_0} &= 0.0063529, \quad r_q = 0.0061232 \quad r = 0.0015011. \end{aligned} \tag{38}$$

Hence, iteration indexes n, residual error in the involved function, error e_n etc. are displayed in the following Table 2.

Table 3. Convergence behavior on example 5

Methods	n	x_n	$\|F(x_n)\|$	$\|e_n\|$	ξ	$\left\|\dfrac{e_n}{e_{n-1}^p}\right\|$	η
	0	0.316	$5.3e(-4)$	$2.3e(-3)$			
2	1	0.318309934916025	$1.1e(-8)$	$4.9e(-8)$		1711.659417	1594.830615
	2	0.318309886183791	$2.1(-27)$	$9.0e(-27)$	4.0066	1594.828247	
	3	0.318309886183791	$2.5e(-102)$	$1.0e(-101)$	4.0000	1594.830615	

REFERENCES

[1] Cordero, A., Hueso, J. L., Martínez, E. and Torregrosa, J. R. 2013. "A new technique to obtain derivative free optimal iterative methods for solving nonlinear equations". *J. Comput. Appl. Math.* 252: 95–102.

[2] Traub, J. F. 1964. *Iterative methods for the solution of equations*. Englewood Cliffs, N. J.: Prentice–Hall Series in Automatic Computation.

[3] 12 Petkovic, M. S., Neta, B., Petkovic, L. and Džunič, J. 2013. *Multipoint methods for solving nonlinear equations*. Elsevier.

[4] Amat, S., Busquier, S. and Plaza, S. 2005. "Dynamics of the King and Jarratt iterations". *Aequationes Math.* 69(3): 212–223.

[5] Amat, S., Busquier, S. and Plaza, S. 2010. "Chaotic dynamics of a third–order Newton–type method". *J. Math. Anal. Appl.* 366(1): 24–32.

[6] Amat, S., Hernández, M. A. and Romero, N. 2008. "A modified Chebyshev's iterative method with at least sixth order of convergence". *Appl. Math. Comput.* 206(1): 164–174.

[7] Argyros, I. K. 2008. *Convergence and Application of Newton–type Iterations*. Springer.

[8] Argyros, I. K. and Hilout, S. 2013. *Numerical methods in Nonlinear Analysis*. New Jersey: World Scientific Publ. Comp. New Jersey.

[9] Behl, R. and Motsa, S. S. 2015. "Geometric construction of eighth–order optimal families of Ostrowski's method". *T. Sci. W. J.* vol. 2015, Article ID 614612, 11 pages.

[10] Ezquerro, J. A. and Hernández, M. A. 2009. "New iterations of R–order four with reduced computational cost". *BIT Numer. Math.* 49: 325–342.

[11] Kanwar, V., Behl, R. and Sharma, K. K. 2011. "Simply constructed family of a Ostrowski's method with optimal order of convergence". *Comput. Math. Appli.* 62(11): 4021–4027.

[12] Magreñán, Á. A. 2014. "Different anomalies in a Jarratt family of iterative root-finding methods". *Appl. Math. Comput.* 233: 29–38.

[13] Magreñán, Á. A. 2014. "A new tool to study real dynamics: The convergence plane". *Appl. Math. Comput.* 248: 215–224.

[14] Rheinboldt, W. C. 1978. "An adaptive continuation process for solving systems of nonlinear equations". *Polish Academy of Science, Banach Ctr. Publ.* 3: 129–142.

In: Understanding Banach Spaces
Editor: Daniel González Sánchez

ISBN: 978-1-53616-745-0
© 2020 Nova Science Publishers, Inc.

Chapter 8

BALL CONVERGENCE FOR A DERIVATIVE FREE METHOD WITH MEMORY

Ioannis K. Argyros[1,*] *and Ramandeep Behl*[2,†]
[1]Department of Mathematical Sciences
Cameron University, Lawton, OK, US
[2]Department of Mathematics
King Abdulaziz University, Jeddah, Saudi Arabia

Abstract

In this chapter, we present a local convergence analysis of a derivative free methods of order $1 + \sqrt{2}$ in order to obtain a locally unique solution in a Banach space setting. Further, we use hypotheses only on the first order derivative of the involved function. On the other hand, in the earlier study Compas et al. [1] proved the order of convergence with help of Taylor series expansions and hypotheses up to the third order derivative or even higher of the function involved which put restriction on the applicability of the proposed scheme. However, the proposed scheme is derivative free. The principle aim of this study is not only to expand the applicability of the proposed scheme on those examples where earlier study fails to work but also expand the space of application and proposed the convergence domain. Moreover, a variety of concrete numerical examples are demonstrated to confirm the proposed theory in this chapter.

Keywords: two–step method with memory, local convergence, convergence order

AMS Subject Classification: 65D10, 65D99

1. INTRODUCTION

The construction of the fixed point iterative methods for obtaining the solution of scalar and system of nonlinear equations becomes one of the most interesting and challenging task of numerical analysis because these techniques have been applied in many diverse fields as

---

*Corresponding Author's E-mail: iargyros@cameron.edu.
†E-mail: ramanbehl87@yahoo.in.

engineering, economics, physics and applied science, etc. Due to the importance of this topic many researchers from the worldwide proposed several numerical methods, most of them are iterative in nature.

In this study, we are concerned with the problem of approximating a locally unique solution x^* of the equation of the form

$$F(x) = 0, \tag{1}$$

where F is a Fréchet differentiable operator defined on a subset $\mathbb{D}$ of a Banach space $\mathbb{X}$ with values in a Banach space $\mathbb{Y}$. Analytic methods to obtain the exact solution of such problems are almost non existent. Therefore, the researchers from the worldwide to resort to an iterative method. Generally, these iterative methods provide an approximated solution corrected up to a specified degree of accuracy which is further depend upon the methods and programming software namely, Fortran, Maple, Matlab, Mathematica, etc. While, using these iterative schemes, researchers have to face several problems and some of them are related to slow convergence, non–convergence, oscillation problem close to the initial guess, divergence, failure etc. (for the detail explanation of these problem please see [2–5]).

In the past and recent years, researchers proposed a plethora of one point and multi–point methods for obtaining the solutions of the scalar and system of nonlinear equations [1–38]. Generally, most of the researchers mentioned that for the granted convergence of there proposed methods initial approximation should be close to the required solution. But, how close the initial approximation should be required which gives the grantee for the convergence of the corresponding method? That's why these local results give no information on the radius of the convergence ball for the corresponding method. Therefore, a very important problem in the study of iterative procedures is the convergence domain and it is very important to propose the radius of convergence of the iterative methods.

Keeping these things in our mind, we study the local convergence analysis of two–step method with memory defined for each $n = 0, 1, 2, \ldots$ by

$$\begin{aligned} y_n &= x_n - [x_{n-1}, x_n; F]^{-1} F(x_n) \\ x_{n+1} &= y_n - [x_n, y_n; F]^{-1} F(y_n), \end{aligned} \tag{2}$$

where x_{-1}, x_0 are initial points and $[x, y; F] : \mathbb{D}^2 \to L(\mathbb{X}, \mathbb{Y})$ is a divided difference for operator F [2, 3]. Method (2) was introduced and studied in [1] in the special case when $\mathbb{X} = \mathbb{Y} = \mathbb{R}$. It was shown to be of order $1 + \sqrt{2}$ using Taylor series expansions and hypotheses up to the third derivative of function F [1]. These hypotheses on the derivatives of F limit the applicability of method (2). As a motivational example, define function F on $S = \mathbb{R}$, $D = [-\frac{1}{\pi}, \frac{2}{\pi}]$ by

$$F(x) = \begin{cases} x^3 \log(\pi^2 x^2) + x^5 \sin\left(\dfrac{1}{x}\right), & x \neq 0, \\ 0, & x = 0. \end{cases}$$

Then, we have that

$$F'(x) = 2x^2 - x^3 \cos\left(\frac{1}{x}\right) + 3x^2 \log(\pi^2 x^2) + 5x^4 \sin\left(\frac{1}{x}\right),$$

$$F''(x) = -8x^2 \cos\left(\frac{1}{x}\right) + 2x(5 + 3\log(\pi^2 x^2)) + x(20x^2 - 1)\sin\left(\frac{1}{x}\right)$$

and

$$F'''(x) = \frac{1}{x}\left[(1 - 36x^2)\cos\left(\frac{1}{x}\right) + x\left(22 + 6\log(\pi^2 x^2) + (60x^2 - 9)\sin\left(\frac{1}{x}\right)\right)\right].$$

One can easily find that the function $F'''(x)$ is unbounded on $\mathbb{D}$ at the point $x = 0$. Hence, the results in [1], cannot apply to show the convergence of method (2) requiring hypotheses on the third derivative of function F or higher. However, the proposed scheme (2) do not have a derivative of the involved function.

The principle aim of the present study not only expand the applicability of the proposed scheme on the above example where earlier study fails to work but also expand the space of application and proposed the convergence domain. In this regards, we use only hypotheses on the first derivative of operator F. We also proposed the computable radii of convergence and error bounds based on the Lipschitz constants. We further present the range of initial guesses x_0 that tell us how close the initial guess should be required for granted convergence of the method (2). These problems were not discussed in the earlier study proposed by Compas et al. in [1].

The chapter is organized as follows. The section 2 is devoted to the local convergence analysis of the method (2). We also provide a radius of convergence, computable error bounds and uniqueness result not given in the earlier studies using Taylor expansions. Several numerical examples are presented in the section 3. Finally, we mentioned the concluding remarks in the section 4.

2. LOCAL CONVERGENCE

In this section, we shall define some scalar functions and parameters in order to present the local convergence of method (2) that follows.

Let $L_0 > 0$, $L > 0$, $K_0 > 0$, $K > 0$ and $M \geq 1$ be given parameters. The local convergence analysis of method (2) is based on some scalars functions and parameters. Let

$$r_0 = \min\left\{\frac{1}{L_0}, \frac{1}{2K_0}\right\}. \tag{3}$$

Let us also assume some functions g_1 and h_1 defined on the interval $[0, r_0)$ by

$$g_1(t) = \frac{1}{2(1 - L_0 t)}\left(L + \frac{4KM}{1 - 2K_0 t}\right)t,$$

and

$$h_1(t) = g_1(t) - 1.$$

We have that $h_1(0) = -1 < 0$ and $h_1(t) \to +\infty$ as $t \to r_0^-$. Then, by the intermediate value theorem function h_1 have zeros in the interval $(0, r_0)$. Further, let r_1 be the smallest such zero on the interval $(0, r_0)$.

Moreover, define functions g_2 and h_2 on the interval $[0, r_0)$ by

$$g_2(t) = \left(1 + \frac{M}{1 - 2K_0 t}\right) g_1(t),$$

and

$$h_2(t) = g_2(t) - 1.$$

We get that $h_2(0) = -1 < 0$ and $h_2(t) \to +\infty$ $t \to r_0^-$. Denote by r_2 the smallest such zero. Define

$$r = \min\{r_1, r_2\}. \tag{4}$$

Then, we have that for each $t \in [0, r)$

$$0 \leq g_1(t) < 1 \tag{5}$$

and

$$0 \leq g_2(t) < 1. \tag{6}$$

Let $U(\gamma, \delta)$ and $\bar{U}(\gamma, \delta)$ stand, respectively for the open and closed balls in $\mathbb{X}$ with center $\gamma \in \mathbb{X}$ and of radius $\delta > 0$. Next, we present the local convergence analysis of method (2) using the preceding notations.

Theorem 1. *Let us consider $F : D \subset \mathbb{X} \to \mathbb{Y}$ be a Fréchet differentiable operator. Suppose, there exist a divided difference of order one $[\cdot, \cdot ; F] : D^2 \to L(\mathbb{X}, \mathbb{Y})$. Further, suppose that there exist $x^* \in D$, $L_0 > 0$ and $K_0 > 0$ such that for each $x \in D$*

$$F(x^*) = 0, \quad F'(x^*)^{-1} \in L(\mathbb{Y}, \mathbb{X}), \tag{7}$$

$$\|F(x^*)^{-1}\left(F'(x) - F'(x^*)\right)\| \leq L_0 \|x - x^*\| \tag{8}$$

and

$$\|F(x^*)^{-1}\left([x, y; F] - F'(x^*)\right)\| \leq K_0(\|x - x^*\| + \|y - x^*\|). \tag{9}$$

Moreover, suppose that there exist $L > 0$, $M \geq 1$ and $K > 0$ such that for each $x, y \in \bar{U}(x^*, r_0) \cap D$

$$\|F'(x^*)^{-1}\left(F'(x) - F'(y)\right)\| \leq L\|x - y\|, \tag{10}$$

$$|F'(x^*)^{-1}\left([x, y; F] - F'(x)\right)| \leq K\|x - y\|, \tag{11}$$

$$\|F'(x^*)^{-1} F'(x)\| \leq M, \tag{12}$$

and

$$\bar{U}(x^*, r) \subseteq D, \tag{13}$$

where the radii r_0 and r are given by (3) and (4), respectively. Then, the sequence $\{x_n\}$ generated by method (2) for $x_{-1}, x_0 \in U(x^*, r) - \{x^*\}$ with $x_{-1} \neq x_0$ is well defined, remains in $U(x^*, r)$ for each $n = 0, 1, 2, \ldots$ and converges to x^*. Moreover, the following estimates hold

$$\|y_n - x^*\| \leq g_1(r)\|x_n - x^*\| < \|x_n - x^*\| < r \tag{14}$$

and

$$\|x_{n+1} - x^*\| \leq g_2(r)\|x_n - x^*\| < \|x_n - x^*\|, \tag{15}$$

where the "g" functions are defined previously. Furthermore, for $T \in \left[r, \frac{2}{L_0}\right)$, the limit point x^* is the only solution of equation $F(x) = 0$ in $\bar{U}(x^*, r) \cap D$.

Proof. We shall show estimates (14) and (15) hold with the help of mathematical induction. Using hypotheses $x_0 \in U(x^*, r) - \{x^*\}$ and (8), we get that

$$\|F'(x^*)^{-1}(F'(x_0) - F'(x^*))\| \le L_0\|x - x^*\| < L_0 r < 1. \tag{16}$$

It follows from (16) and the Banach Lemma on invertible operators [2,4] that $F'(x_0)^{-1} \in L(\mathbb{Y}, \mathbb{X})$ and

$$\|F'(x_0)^{-1} F'(x^*)\| \le \frac{1}{1 - L_0\|x_0 - x^*\|}. \tag{17}$$

We also have by (3), (4) and (9) that

$$\|F'(x_0)^{-1}([x_{-1}, x_0; F] - F'(x^*))\| \le K_0(\|x_{-1} - x^*\| + \|x_0 - x^*\|) < 2K_0 r < 1. \tag{18}$$

Hence, we get by (18) that y_0 is well defined

$$\|[x_{-1}, x_0; F]^{-1} F'(x^*)\| \le \frac{1}{1 - K_0(\|x_{-1} - x^*\| + \|x_0 - x^*\|)} < \frac{1}{1 - 2K_0 r}. \tag{19}$$

Using the first sub step of method (2) for $n = 0$, (4), (5), (7) (10) – (13), (17) and (19), we get in turn from the identity

$$y_0 - x^* = (x_0 - x^* - F'(x_0)^{-1} F(x_0)) + (F'(x_0)^{-1} F'(x^*))(F'(x^*)^{-1}([x_{-1}, x_0; F] - F'(x_0)))$$
$$\times ([x_{-1}, x_0; F]^{-1} F'(x^*))(F'(x^*)^{-1} F'(x_0))$$

that

$$\|y_0 - x^*\| \le \|F'(x_0)^{-1} F'(x^*)\| \left\|\int_0^1 F'(x^*)^{-1}(F'(x^* + \theta(x_0 - x^*)) - F'(x_0))(x_0 - x^*) d\theta\right\|$$
$$+ \|F'(x_0)^{-1} F'(x^*)\| \|F'(x^*)^{-1}([x_{-1}, x_0; F] - F'(x_0))\|$$
$$\times \|[x_{-1}, x_0; F]^{-1} F'(x^*)\| \|F'(x^*)^{-1} F'(x_0)\| \tag{20}$$
$$\le \frac{L\|x_0 - x^*\|^2}{2(1 - L_0\|x_0 - x^*\|)} + \frac{K[\|x_{-1} - x^*\| + \|x_0 - x^*\|] M \|x_0 - x^*\|}{(1 - L_0\|x_0 - x^*\|)(1 - K_0(\|x_{-1} - x^*\| + \|x_0 - x^*\|))}$$
$$< g_1(r)\|x_0 - x^*\| < \|x_0 - x^*\| < r,$$

which shows (14) for $n = 0$ and $y_0 \in U(x^*, r)$, where we also used that

$$\|F'(x^*)^{-1} F'(x_0)\| = \left\|\int_0^1 F'(x^*)^{-1}(F'(x^* + \theta(x_0 - x^*)) - F'(x_0))(x_0 - x^*) d\theta\right\|$$
$$\le M\|x_0 - x^*\|, \tag{21}$$

since $\|x^* + \theta(x_0 - x^*) - x^*\| = \theta\|x_0 - x^*\| < r$. Next, we shall show that $[x_0, y_0; F]^{-1} \in L(Y, X)$. Using (4), (9), (14) (for $n = 0$) and (20), we get in turn that

$$\|F'(x^*)^{-1}([x_0, y_0; F] - F'(x^*))\| \le K_0(\|x_0 - x^*\| + \|y_0 - x^*\|)$$
$$\le K_0(1 + g_1(\|x_0 - x^*\|))\|x_0 - x^*\| \tag{22}$$
$$< 2K_0 r < 1.$$

In view of (22), we get that

$$\|[x_0, y_0; F]^{-1} F'(x^*)\| \le \frac{1}{1 - K_0(1 + g_1(\|x_0 - x^*\|))\|x_0 - x^*\|}$$
$$\le \frac{1}{1 - 2K_0 r} \tag{23}$$

and x_1 is well defined. Then, using the last sub–step of method (2) for $n = 0$, (4), (6), (10) – (12), (17), (20), and (23) that

$$\|x_1 - x^*\| \le \|y_0 - x^*\| + \frac{M\|y_0 - x^*\|}{1 - K_0(1 + g_1(\|x_0 - x^*\|))\|x_0 - x^*\|} \quad (24)$$
$$< g_2(r)\|x_0 - x^*\| < \|x_0 - x^*\| < r,$$

which shows (24) for $n = 0$ and $x_1 \in U(x^*, r)$. By simply replacing x_0, y_0, x_1 by x_k, y_k, x_{k+1} in the preceding estimates we arrive at (13) and (14). Then notice that $\|x_{k+1} - x^*\| < \|x_k - x^*\| < r$, $c = g_2(r) \in [0, 1)$. Hence, we conclude that $\lim_{k \to \infty} x_k = x^*$ and $x_{k+1} \in U(x^*, r)$. Finally, to show the uniqueness part, let $y^* \in \bar{U}(x^*, T)$ be such that $F(y^*) = 0$. Set $Q = \int_0^1 F'(x^* + \theta(y^* - x^*)) \, d\theta$. Then, using (14), we get that

$$\|F'(x^*)^{-1}(Q - F'(x^*))\| \le L_0 \int_0^1 \theta\|x^* - y^*\| d\theta = \frac{L_0}{2}T < 1. \quad (25)$$

Hence, $Q^{-1} \in L(Y, X)$. Then, in view of the identity $F(y^*) - F(x^*) = Q(y^* - x^*)$, we conclude that $x^* = y^*$. □

Remark 2. *(a) In view of (11) and the estimate*

$$\|F'(x^*)^{-1}F'(x)\| = \|F'(x^*)^{-1}(F'(x) - F'(x^*)) + I\|$$
$$\le 1 + \|F'(x^*)^{-1}(F'(x) - F'(x^*))\|$$
$$\le 1 + L_0\|x_0 - x^*\|$$

condition (13) can be dropped and M can be replaced by

$$M = M(t) = 1 + L_0 t$$

or $M = 2$, since $t \in [0, \frac{1}{L_0})$.

(b) The results obtained here can be used for operators F satisfying the autonomous differential equation [2, 3] of the form

$$F'(x) = P(F(x)),$$

where P is a known continuous operator. Since $F'(x^*) = P(F(x^*)) = P(0)$, we can apply the results without actually knowing the solution x^*. Let as an example $F(x) = e^x + 2$. Then, we can choose $P(x) = x - 2$.

(c) The radius $r_A = \frac{2}{L_0 + L}$ was shown by us in [2, 3] to be the convergence radius for Newton's method under conditions (11) – (13). Radius r_A is at least as large as the convergence ball given by Rheinboldt [38] and Traub [5]

$$r_R = \frac{2}{3L}.$$

Notice that for $L_0 < L$,
$$r_R < r_A.$$

Moreover,
$$\frac{r_R}{r_A} \to \frac{1}{3} \quad as \quad \frac{L_0}{L} \to 0.$$

Hence, r_A is at most three times larger than r_R. In the numerical examples we compare r to r_A and r_R.

(d) *It is worth noticing that method (2) is not changing if we use the conditions of Theorem 1 instead of the stronger conditions given in [1]. Moreover, for the error bounds in practice we can use the computational order of convergence (COC) [17]*

$$\rho = \frac{ln\frac{\|x_{n+2}-x^*\|}{\|x_{n+1}-x^*\|}}{ln\frac{\|x_{n+1}-x^*\|}{\|x_n-x^*\|}}, \quad \text{for each } n = 0, 1, 2, \ldots \quad (26)$$

or the approximate computational order of convergence (ACOC) [17]

$$\rho^* = \frac{ln\frac{\|x_{n+2}-x_{n+1}\|}{\|x_{n+1}-x_n\|}}{ln\frac{\|x_{n+1}-x_n\|}{\|x_n-x_{n-1}\|}}, \quad \text{for each } n = 1, 2, \ldots \quad (27)$$

This way we obtain in practice the order of convergence in a way that avoids the bounds involving estimates higher than the first Fréchet derivative. Notice that the solution x^* is not used to compute ρ^* in the formula (27).

3. NUMERICAL EXPERIMENTS

In this section, we consider a concrete variety of scalar and system of nonlinear equations which are given in the following examples including motivational example to illustrate the convergence behavior and the validation of the proposed study.

First of all, we shall calculate the values of r_R, r_A, r_0, r_1, r_2 and r for obtaining the range of convergence domain. The values of these constants are given in the corresponding example up to 5 significant digits. However, we have calculate the values of these constants up to several number of significant digits. With this convergence domain, we can choose any initial approximation x_0 from this domain which gives the guarantee of convergence of the proposed study.

In addition, we verify the theoretical order of convergence for scalar equation on the basis of the results obtain from $\left|\frac{e_{n+1}}{e_n^p}\right|$. In the Table 5, we displayed the number of iteration indexes (n), approximated zeros (x_n), residual error of the corresponding function $(|F(x_n)|)$, errors $|e_n|$ (where $e_n = x_n - x^*$), $\left|\frac{e_{n+1}}{e_n^p}\right|$ and the asymptotic error constant $\eta = \lim_{n \to \infty} \left|\frac{e_{n+1}}{e_n^p}\right|$. Moreover, we calculate the computational order of convergence [17] by using the following formulas

$$\xi = \frac{ln\frac{|x_{n+2}-x^*|}{|x_{n+1}-x^*|}}{ln\frac{|x_{n+1}-x^*|}{|x_n-x^*|}}, \quad \text{for each } n = 0, 1, 2, \ldots$$

or the approximate computational order of convergence (ACOC)

$$\xi^* = \frac{\ln\frac{|x_{n+2}-x_{n+1}|}{|x_{n+1}-x_n|}}{\ln\frac{|x_{n+1}-x_n|}{|x_n-x_{n-1}|}}, \quad \text{for each } n = 1, 2, \ldots$$

In order to minimize the round off error, we calculate the values of computational order of convergence, asymptotic error constant and other constants up to several number of significant digits (minimum 1000 significant digits).

In the context of system of nonlinear equations, we also consider a variety of nonlinear systems in examples 3 – 6 to further check the validity of theoretical results for nonlinear system. Therefore, we given the number of iteration indexes (n), residual error of the corresponding function ($\|F(x_n)\|$), error in the iterations $\|x_{n+1} - x_n\|$ in the Tables 1 – 4. Moreover, we also calculate the computational order of convergence for nonlinear system by using the above mentioned formulas namely, 26 and 27.

As we mentioned in the earlier paragraph that we calculate the values of all the constants and functional residuals in multi precision but due to the limited chapter space, we display the values of x_n up to 15 significant digits for nonlinear equations. Further, the values of other constants namely, $\xi(COC)$ up to 5 significant digits and the values $\left|\frac{e_n}{e_{n-1}^p}\right|$ and η are up to 10 significant digits. Moreover, the residual error in the function/system of nonlinear functions ($|F(x_n)|$ or $\|F(x_n)\|$) and the error ($|e_n|$ or $\|x_{n+1} - x_n\|$) are display up to 2 significant digits with exponent power in the following Tables corresponding to the test function. However, we calculate each value for minimum 1000 significant digits.

Moreover, we consider the approximated zero up to 1000 significant digits of test functions when the exact zero is not available to calculate $\|x_n - x^*\|$. During the current numerical experiments with programming language Mathematica (Version 9), all computations have been done with multiple precision arithmetic, which minimize round–off errors.

Example 3. *Let us consider the following boundary value problem [18]*

$$y'' + y^3 = 0, \; y(0) = 0, \; y(1) = 1. \tag{28}$$

Further, assume the partition of the interval $[0, 1]$, which is defined as follows

$$x_0 = 0 < x_1 < x_2 < \cdots < x_{n-1} < x_n = 1, \; x_{i+1} = x_i + h, \; h = \frac{1}{n}.$$

Let us define $y_0 = y(x_0) = 0$, $y_1 = y_1(x_1), \ldots, y_{n-1} = y(x_{n-1})$, $y_n = y(x_n) = 1$. We use the following way to discretization the second derivative, we get

$$y_k'' = \frac{y_{k-1} - 2y_k + y_{k+1}}{h^2}, \; k = 1, 2, 3, \ldots, n-1,$$

further yield a system of nonlinear equations of order $n - 1$

$$y_{k-1} - 2y_k + y_{k+1} + h^2 y_k^3 = 0, \; k = 1, 2, 3, \ldots, n-1. \tag{29}$$

Defines an operator $F : U(x^*, 1) \subseteq \mathbb{R}^7 \to \mathbb{R}^7$ for system (29). We solve this system for particular value $n = 7$ by considering the initial approximation $y(0) =$

$(1, 0, 1, 0, 1, 0)^T$. Notice that $x^* = (0.1506388\ldots, 0.3012079\ldots, 0.4512193\ldots, 0.5993558\ldots, 0.7430983\ldots, 0.8784667\ldots)$, $L_0 = L = 14.8611751406$, $K_0 = K = \frac{L}{2}$ and $M = 2$. Then, we obtain the following values

$$r_R = 0.044860, \quad r_A = 0.067289, \quad r_0 = 0.067289,$$
$$r_1 = 0.016264, \quad r_2 = 0.0068653 \quad r = 0.0068653. \tag{30}$$

Further, we display the residual error of the involved function, error in the iterations and COC (ξ) in the following Table 1 by assuming $x_0 = (0.155, 0.306, 0.456, 0.604, 0.748, 0.883)$ and $x_{-1} = (0.154, 0.305, 0.454, 0.605, 0.746, 0.880)$.

Table 1. Convergence behavior on example 3

Method	n	$\|F(x_n)\|$	$\|x_{n+1} - x_n\|$	ξ
	1	$1.2e(-9)^a$	$6.2e(-9)$	
2	2	$5.8e(-22)$	$3.0e(-21)$	
	3	$8.3e(-53)$	$4.2e(-52)$	2.5046

[a] $1.2e(-9)$ denotes 1.2×10^{-9}.

Example 4. Let $\mathbb{X} = \mathbb{Y} = \mathbb{R}^4$, $\mathbb{D} = \bar{U}(x^*, 1)$. Define F on $\mathbb{D}$ for $v = (x - 1, x_2, x_3, x_4)^T$ by

$$F(v) = (x_2 x_3 + x_4(x_1 + x_3), x_1 x_3 + x_4(x_1 + x_3), x_1 x_2 + x_4(x_1 + x_2), x_1 x_2 + x_1 x_3 + x_2 x_3 - 1)^T. \tag{31}$$

Then the Fréchet–derivative is given by

$$F'(v) = \begin{bmatrix} x_4 & x_3 & x_2 + x_4 & x_1 + x_3 \\ x_3 + x_4 & 0 & x_1 + x_4 & x_1 + x_3 \\ x_2 + x_4 & x_1 + x_4 & 0 & x_1 + x_2 \\ x_2 + x_3 & x_1 + x_3 & x_1 + x_2 & 0 \end{bmatrix}.$$

Notice that $x^* = (0.5773503\ldots, 0.5773503\ldots, 0.5773503\ldots, -0.2886751\ldots)$, $L_0 = L = 29.44485736$, $K_0 = K = 14.72242868$ and $M = 2$. Then, we obtain the following values

$$r_R = 0.0226412, \quad r_A = 0.033962, \quad r_0 = 0.033962,$$
$$r_1 = 0.093677, \quad r_2 = 0.0034650 \quad r = 0.0034650. \tag{32}$$

Further, residual error of the involved function, error in the iterations and COC (ξ) are given in the following Table 2, by considering the initial approximations $x_0 = (0.574, 0.574, 0.574, -0.291)$ and $x_{-1} = (0.580, 0.579, 0.580, -0.285)$.

Example 5. Let $\mathbb{X} = \mathbb{Y} = \mathbb{R}^3$, $\mathbb{D} = \bar{U}(x^*, 2.5)$. Define F on $\mathbb{D}$ for $v = (x, y, z)^T$ by

$$F(v) = \left(x^2 + y^2 + z^2 - 9, xyz - 1, x + y - z^2\right)^T. \tag{33}$$

Table 2. Convergence behavior on example 4

Method	n	$\|F(x_n)\|$	$\|x_{n+1} - x_n\|$	ξ
	1	$1.2e(-7)$	$7.0e(-8)$	
2	2	$2.3e(-17)$	$6.1e(-17)$	
	3	$1.8e(-40)$	$1.7e(-40)$	2.5991

Then the Fréchet–derivative is given by

$$F'(v) = \begin{bmatrix} 2x & 2y & 2z \\ yz & xz & xy \\ 1 & 1 & -2z \end{bmatrix}.$$

Notice that $x^* = (-2.090295\ldots, 2.140258\ldots, -0.2235251\ldots)$, $L_0 = L = 10.42490887$, $K_0 = K = 5.21245444$ *and* $M = 2$. *Then, we obtain the following values*

$$\begin{aligned} r_R &= 0.063949, \quad r_A = 0.095924, \quad r_0 = 0.095924, \\ r_1 &= 0.023184, \quad r_2 = 0.0097867 \quad r = 0.00978674. \end{aligned} \tag{34}$$

Hence, by considering $x_0 = (-2.08, 2.13, -0.210)$ *and* $x_{-1} = (-2.07, 2.12, -0.211)$, *we obtain residual error of the involved function, error in the iterations and COC* (ξ), *which are presented in the following Table 3.*

Table 3. Convergence behavior on example 5

Method	n	$\|F(x_n)\|$	$\|x_{n+1} - x_n\|$	ξ
	1	$6.9e(-6)$	$2.5e(-6)$	
2	2	$1.8e(-14)$	$3.7e(-15)$	
	3	$5.2e(-36)$	$8.6e(-37)$	2.4496

Example 6. *Let* $\mathbb{X} = \mathbb{Y} = \mathbb{R}^2$, $\mathbb{D} = \bar{U}(x^*, 1)$. *Define F on $\mathbb{D}$ for* $v = (x, y)^T$ *by*

$$F(v) = \left(x^2 - x - y^2 - 1, \ -\sin x + y\right)^T. \tag{35}$$

Then the Fréchet–derivative is given by

$$F'(v) = \begin{bmatrix} 2x - 1 & -2y \\ -\cos x & 1 \end{bmatrix}.$$

Notice that $x^* = (-0.8452567\ldots, -0.7481415\ldots)$, $L_0 = L = 7.902715832$, $K_0 = K = 3.951357916$ *and* $M = 2$. *Then, we obtain the following values*

$$\begin{aligned} r_R &= 0.0843592, \quad r_A = 0.12654, \quad r_0 = 0.12654, \\ r_1 &= 0.030584, \quad r_2 = 0.012910 \quad r = 0.012910. \end{aligned} \tag{36}$$

Moreover, the residual error in the involved function, error in the iterations and COC (ξ) are displayed in the following Table 4, by considering $x_0 = (-0.835, -0.738)$ and $x_{-1} = (-0.836, -0.736)$.

Table 4. Convergence behavior on example 6

Method	n	$\|F(x_n)\|$	$\|x_{n+1} - x_n\|$	ξ
2	1	$3.1e(-7)$	$1.4e(-7)$	
	2	$1.1e(-16)$	$1.6e(-16)$	
	3	$5.9e(-40)$	$5.1e(-40)$	2.6306

Example 7. *Returning back to the motivational example at the introduction on this chapter, we have $L = L_0 = \frac{2}{2\pi+1}(80 + 16\pi + (11 + 12\log 2)\pi^2)$, $K = K_0 = \frac{L}{2}$, $M = 2$ and our required zero is $x^* = \frac{1}{\pi}$.*

$$r_R = 0.0075648, \quad r_A = 0.011347, \quad r_0 = 0.011347,$$
$$r_1 = 0.0027426, \quad r_2 = 0.0011577 \quad r = 0.00115771. \tag{37}$$

Further, we presented the residual error of the involved function, error in the iterations and COC (ξ), etc. in the following Table 5 by assuming initial approximations $x_0 = 0.317$ and $x_{-1} = 0.319$.

Table 5. Convergence behavior of different cases on example 7

Method	n	x_n	$\|F(x_n)\|$	$\|e_n\|$	ξ	$\frac{e_n}{e_{n-1}^{(1+\sqrt{2})}}$	η
(2)	1	0.318309972751427	$2.0e(-8)$	$8.7e(-18)$			
	2	0.318309886183790	$1.7e(-16)$	$7.1e(-16)$		80.35679694	11.74502867
	3	0.318309886183791	$7.5e(-37)$	$3.2e(-36)$	2.5175	11.74502867	

REFERENCES

[1] Campos, B., Cordero, A., Torregrosa, J. R. and Vindel, P. "Memorizing the dynamical analysis of iterative methods". *Appl. Math. Comput.* to appear.

[2] Argyros, I. K. 2008. *Convergence and Application of Newton–type Iterations.* Springer.

[3] Argyros, I. K. and Hilout, S. 2013. *Numerical methods in Nonlinear Analysis.* New Jersey: World Scientific Publ. Comp.

[4] Petkovic, M. S., Neta, B., Petkovic, L. and Džunič, J. 2013. *Multipoint methods for solving nonlinear equations*. Elsevier.

[5] Traub, J. F. 1964. *Iterative methods for the solution of equations*. Englewood Cliffs, N. J.: Prentice–Hall Series in Automatic Computation.

[6] Amat, S., Busquier, S. and Plaza, S. 2005. "Dynamics of the King and Jarratt iterations". *Aequationes Math*. 69(3): 212–223.

[7] Amat, S., Busquier, S. and Plaza, S. 2010. "Chaotic dynamics of a third–order Newton–type method". *J. Math. Anal. Appl.* 366(1): 24–32.

[8] Amat, S., Hernández, M. A. and Romero, N. 2008. "A modified Chebyshev's iterative method with at least sixth order of convergence". *Appl. Math. Comput.* 206(1): 164–174.

[9] Behl, R. and Motsa, S. S. 2015. "Geometric construction of eighth–order optimal families of Ostrowski's method". *T. Sci. W. J.*, vol. 2015, Article ID 614612, 11 pages.

[10] Chicharro, F., Cordero, A. and Torregrosa, J. R. 2013. "Drawing dynamical and parameters planes of iterative families and methods". *The Scientific World Journal Volume*. Article ID 780153.

[11] Chun, C. 1990. "Some improvements of Jarratt's method with sixth–order convergence". *Appl. Math. Comput.* 190(2): 1432–1437.

[12] Chun, C. 2007. "Some improvements of Jarratts method with sixth–order convergence". *Appl. Math. Comput.* 190: 1432–1437.

[13] Cordero, A., García–Maimó, J., Torregrosa, J. R., Vassileva, M. P. and Vindel, P. 2013. "Chaos in King's iterative family". *Appl. Math. Lett.* 26: 842–848.

[14] Cordero, A., Torregrosa, J. R. and Vindel, P. 2013. "Dynamics of a family of Chebyshev–Halley type methods". *Appl. Math. Comput.* 219: 8568–8583.

[15] Cordero, A. and Torregrosa, J. R. 2007. "Variants of Newton's method using fifth–order quadrature formulas". *Appl. Math. Comput.* 190: 686–698.

[16] Ezquerro, J. A. and Hernández, M. A. 2000. "Recurrence relations for Chebyshev–type methods". *Appl. Math. Optim.* 41(2): 227–236.

[17] Ezquerro, J. A. and Hernández, M. A. 2009. "New iterations of R–order four with reduced computational cost". *BIT Numer. Math.* 49: 325–342.

[18] Grau–Sánchez, M., Peris, J. M. and Gutiérrez, J. M. 2007. "Accelerated iterative methods for finding solutions of a system of nonlinear equations". *Appl. Math. Comput.* 190(2): 1815–1823.

[19] Gutiérrez, J. M. and Hernández, M. A. 1998. "Recurrence relations for the super–Halley method". *Computers Math. Applic.* 36(7):1–8.

[20] Hernández, M. A. 2001. "Chebyshev's approximation algorithms and applications". *Computers Math. Applic.* 41(3–4): 433–455.

[21] Hernández, M. A. and Salanova, M. A. 1999. "Sufficient conditions for semilocal convergence of a fourth order multipoint iterative method for solving equations in Banach spaces". *Southwest J. Pure Appl. Math.* 1: 29–40.

[22] Herceg, D. and Herceg, Dj. 2013. "Means based modifications of Newtons method for solving nonlinear equations". *Appl. Math. Comput.* 219(11): 6126–6133.

[23] Herceg, D. and Herceg, Dj. 2013. "Third–order modifications of Newtons method based on Stolarsky and Gini means". *J. Comput. Appl. Math.* 245: 53–61.

[24] Herceg, D. and Herceg, Dj. 2014. "Sixth–order modifications of Newtons method based on Stolarsky and Gini means". *J. Comput. Appl. Math.* 267: 244–253.

[25] Homeier, H. H. H. 2005. "On Newton–type methods with cubic convergence". *J. Comput. Appl. Math.* 176: 425–432.

[26] Jarratt, P. 1966. "Some fourth order multipoint methods for solving equations". *Math. Comput.* 20(95): 434–437.

[27] Kou, J. and Wang, X. 2012. "Semilocal convergence of a modified multi–point Jarratt method in Banach spaces under general continuity conditions". *Numer. Algorithms* 60: 369–390.

[28] Kou, J. 2007. "On Chebyshev–Halley methods with sixth–order convergence for solving non–linear equations". *Appl. Math. Comput.* 190: 126–131.

[29] Kou, J., Li, Y. and Wang, X. 2007. "Third–order modification of Newton's method". "*J. Comput. Appl. Math*". 205: 1–5.

[30] Li, D., Liu, P. and Kou, J. 2014. "An improvement of the Chebyshev–Halley methods free from second derivative". *Appl. Math. Comput.* 235: 221–225.

[31] Lukić, T. and Ralević, N. M. 2008. "Geometric mean Newtons method for simple and multiple roots". *Appl. Math. Lett.* 21: 30–36.

[32] Magreñán, Á. A. 2014. "Different anomalies in a Jarratt family of iterative root–finding methods". *Appl. Math. Comput.* 233: 29–38.

[33] Magreñán, Á. A. 2014. "A new tool to study real dynamics: The convergence plane". *Appl. Math. Comput.* 248: 215–224.

[34] Neta, B. 1979. "A sixth order family of methods for nonlinear equations". *Int. J. Comput. Math.* 7: 157–161.

[35] Ozban, A. Y. 2004. "Some new variants of Newtons method". *Appl. Math. Lett.* 17: 677–682.

[36] Parhi, S. K. and Gupta, D. K. 2007. "Recurrence relations for a Newton–like method in Banach spaces". *J. Comput. Appl. Math.* 206(2): 873–887.

[37] Parhi, S. K. and Gupta, D. K. 2008. "A sixth order method for nonlinear equations". *Appl. Math. Comput.* 203: 50–55.

[38] Rheinboldt, W. C. 1978. "An adaptive continuation process for solving systems of nonlinear equations". *Polish Academy of Science, Banach Ctr. Publ.* 3: 129–142.

In: Understanding Banach Spaces
Editor: Daniel González Sánchez

ISBN: 978-1-53616-745-0
© 2020 Nova Science Publishers, Inc.

Chapter 9

WEAKER CONVERGENCE CONDITIONS OF AN ITERATIVE METHOD FOR NONLINEAR ILL–POSED EQUATIONS

Ioannis K. Argyros[1,*] *and Santhosh George*[2,†]
[1]Department of Mathematical Sciences
Cameron University, Lawton, OK, US
[2]Department of Mathematical and Computational Sciences
National Institute of Technology, Karnataka, India

Abstract

In this chapter we expand the applicability of an iterative method which converges to the unique solution x_α^δ of the method of Lavrentiev regularization, i.e., $F(x) + \alpha(x - x_0) = y^\delta$, approximating the solution $\hat{x}$ of the ill–posed problem $F(x) = y$ where $F : D(F) \subseteq X \longrightarrow X$ is a nonlinear monotone operator defined on a real Hilbert space X. We use a center–Lipschitz instead of a Lipschitz condition used in [1–3]. The convergence analysis and the stopping rule are based on the majorizing sequence. The choice of the regularization parameter is the crucial issue. We show that the adaptive scheme considered by Perverzev and Schock [4] for choosing the regularization parameter can be effectively used here for obtaining order optimal error estimate. Numerical examples are presented to show that older convergence conditions [1–3] are not satisfied but the new ones are satisfied.

Keywords: ill–posed problems, local–semilocal convergence, Fréchet derivative

AMS Subject Classification: 65G9, 65H10

1. Introduction

This chapter extends the analysis of the Lavrentiev regularization for nonlinear ill–posed problems

$$F(x) = y, \qquad (1)$$

---

[*]Corresponding Author's E-mail: iargyros@cameron.edu.
[†]E-mail: sgeorge@nitk.ac.in.

where $F : D(F) \subseteq X \to X$ is a nonlinear monotone operator considered in [5]. Recall that F is a monotone operator if it satisfies the relation

$$\langle F(x_1) - F(x_2), x_1 - x_2 \rangle \geq 0, \qquad \forall x_1, x_2 \in D(F). \tag{2}$$

Here X be a real Hilbert space with inner product $\langle .,. \rangle$ and norm $\|.\|$. Let $U(x, R)$ and $\overline{U(x, R)}$, stand respectively, for the open and closed ball in X with center x and radius $R > 0$. Let also $L(X)$ be the space of all bounded linear operators from X into itself.

We assume that (1) has a solution, namely $\hat{x}$ and that F possesses a locally uniformly bounded Fréchet derivative $F'(.)$ in a ball around $\hat{x} \in X$.

In application, usually only noisy data y^δ are available, such that

$$\|y - y^\delta\| \leq \delta. \tag{3}$$

Then the problem of recovery of $\hat{x}$ from noisy equation $F(x) = y^\delta$ is ill–posed.

In [5], we considered an iterative regularization method;

$$x_{n+1,\alpha}^\delta = x_{n,\alpha}^\delta - (F'(x_0) + \alpha I)^{-1}(F(x_{n,\alpha}^\delta) - y^\delta + \alpha(x_{n,\alpha}^\delta - x_0)), \tag{4}$$

and proved that $(x_{n,\alpha}^\delta)$ converges linearly to the unique solution x_α^δ of

$$F(x) + \alpha(x - x_0) = y^\delta. \tag{5}$$

It is known (cf. [6], Theorem 1.1) that the equation (5) has a unique solution x_α^δ for $\alpha > 0$, provided F is Fréchet differentiable and monotone in the ball $B_r(\hat{x}) \subset D(F)$ with radius $r = \|\hat{x} - x_0\| + \delta/\alpha$. However the regularized equation (5) remains nonlinear and one may have difficulties in solving them numerically.

Many authors (see [3, 7–10]) considered iterative regularization methods for obtaining stable approximate solutions for (5). Recall ([2]) that, an iterative method with iterations defined by

$$x_{k+1}^\delta = \Phi(x_0^\delta, x_1^\delta, \ldots, x_k^\delta; y^\delta),$$

where $x_0^\delta := x_0 \in D(F)$ is a known initial approximation of $\hat{x}$, for a known function Φ together with a stopping rule which determines a stopping index $k_\delta \in \mathbb{N}$ is called an iterative regularization method if $\|x_{k_\delta}^\delta - \hat{x}\| \to 0$ as $\delta \to 0$.

In [3], Bakushinskii and Smirnova considered the iteratively regularized Lavrentiev method

$$x_{k+1}^\delta = x_k^\delta - (A_k^\delta + \alpha_k I)^{-1}(F(x_k^\delta) - y^\delta + \alpha_k(x_k^\delta - x_0)), \quad k = 0, 1, 2, \ldots, \tag{6}$$

where $A_k^\delta := F'(x_k^\delta)$ and $\{\alpha_k\}$ is a sequence of positive real numbers such that $\lim_{k \to \infty} \alpha_k = 0$. In fact, the stopping index k_δ in [3] was chosen according to the discrepancy principle

$$\|F(x_{k_\delta}^\delta) - y^\delta\| \leq \tau\delta < \|F(x_k^\delta) - y^\delta\|, \qquad 0 \leq k < k_\delta$$

for some $\tau > 1$ and showed that $x_{k_\delta}^\delta \to \hat{x}$ as $\delta \to 0$ under the following assumptions:

- There exists $L > 0$ such that $\|F'(x) - F'(y)\| \leq L\|x - y\|$ for all $x, y \in D(F)$.

- There exists $p > 0$ such that

$$\frac{\alpha_k - \alpha_{k+1}}{\alpha_k \alpha_{k+1}} \leq p, \quad \forall k \in \mathbb{N}, \tag{7}$$

- $\sqrt{(2+L\sigma)}\|x_0 - \hat{x}\|td \leq \sigma - 2\|x_0 - \hat{x}\|t \leq d\alpha_0$, where $\sigma := (\sqrt{\tau} - 1)^2, t := p\alpha_0 + 1$ and $d = 2(t\|x_0 - \hat{x}\| + p\sigma)$.

However, no error estimate for $\|x_{k_\delta}^\delta - \hat{x}\|$ was given in [3]. Later in [2], Mahale and Nair considered method (6) and obtained an error estimate for $\|x_{k_\delta}^\delta - \hat{x}\|$ under a weaker condition than (7). Precisely they choose the stopping index k_δ as the first nonnegative integer such that x_k^δ in (7) is defined for each $k \in \{0, 1, 2, \ldots, k_\delta\}$ and

$$\|\alpha_{k_\delta}(A_{k_\delta}^\delta + \alpha_{k_\delta}I)^{-1}(F(x_{k_\delta}) - y^\delta)\| \leq c_0 \text{ with } c_0 > 4.$$

In [2], Mahale and Nair showed that $x_{k_\delta}^\delta \to \hat{x}$ as $\delta \to 0$ and obtained an optimal order error estimate for $\|x_{k_\delta}^\delta - \hat{x}\|$ under the following assumptions:

Assumption 1. *There exists $r > 0$ such that $U(\hat{x}, r) \subseteq D(F)$ and F is Fréchet differentiable at all $x \in U(\hat{x}, r)$.*

Assumption 2. *(cf. [11], Assumption 3) There exists a constant $k_0 > 0$ such that for every $x, u \in U(\hat{x}, r))$ and $v \in X$ there exists an element $\Phi(x, u, v) \in X$ such that $[F'(x) - F'(u)]v = F'(u)\Phi(x, u, v), \|\Phi(x, u, v)\| \leq k_0 \|v\|\|x - u\|.$*

Assumption 3. *There exists a continuous, strictly monotonically increasing function $\varphi : (0, a] \to (0, \infty)$ with $a \geq \|F'(\hat{x})\|$ satisfying*

1. $\lim_{\lambda \to 0} \varphi(\lambda) = 0$

2. *for $\alpha \leq 1, \varphi(\alpha) \geq \alpha$*

3. $\sup_{\lambda \geq 0} \frac{\alpha\varphi(\lambda)}{\lambda + \alpha} \leq c_\varphi \varphi(\alpha), \quad \forall \lambda \in (0, a]$

4. *there exists $w \in X$ such that*

$$x_0 - \hat{x} = \varphi(F'(\hat{x}))w. \tag{8}$$

Assumption 4. *There exists a sequence $\{\alpha_k\}$ of positive real numbers such that $\lim_{k \to \infty} \alpha_k = 0$ and there exists $\mu > 1$ such that*

$$1 \leq \frac{\alpha_k}{\alpha_{k+1}} \leq \mu, \quad \forall k \in \mathbb{N}. \tag{9}$$

Note that (9) is weaker than (7).

In [1] motivated by iteratively regularized Lavrentiev method (see, [3], [2]), we showed the quadratic convergence of the method defined by

$$x_{n+1,\alpha}^\delta = x_{n,\alpha}^\delta - (F'(x_{n,\alpha}^\delta) + \alpha I)^{-1}(F(x_{n,\alpha}^\delta) - y^\delta + \alpha(x_{n,\alpha}^\delta - x_0)), \tag{10}$$

where $x_{0,\alpha}^\delta := x_0$ is a starting point of the iteration. Let $R_\alpha(x) = F'(x) + \alpha I$ and

$$G(x) = x - R_\alpha(x)^{-1}(F(x_{n,\alpha}^\delta) - y^\delta + \alpha(x_{n,\alpha}^\delta - x_0)). \tag{11}$$

Note that with the above notation $x_{n+1,\alpha}^\delta = G(x_{n,\alpha}^\delta)$. The assumptions used instead of Assumption 1 and Assumption 2 in [1] are, respectively.

Assumption 5. *There exists $r > 0$ such that $U(x_0, r) \cup U(\hat{x}, r)) \subseteq D(F)$ and F is Fréchet differentiable at all $x \in U(x_0, r) \cup U(\hat{x}, r))$.*

Assumption 6. *There exists a constant $k_0 > 0$ such that for every $x, u \in U(x_0, r) \cup U(\hat{x}, r))$ and $v \in X$ there exists an element $\Phi(x, u, v) \in X$ satisfying $[F'(x) - F'(u)]v = F'(u)\Phi(x, u, v), \|\Phi(x, u, v)\| \le k_0 \|v\| \|x - u\|$.*

The second condition in Assumption 6 is essentially a Lipschitz–type condition. However, it is in general very difficult to verify or may not even be satisfied [12]–[16]. In order for us to expand the applicability of the method, we consider the following weaker assumptions.

Assumption 7. *There exists $r > 0$ such that $U(x_0, r) \subseteq D(F)$ and F is Fréchet differentiable at all $x \in U(x_0, r)$.*

Assumption 8. *Let $x_0 \in X$ be fixed. There exists a constant $L > 0$ such that for every $x, u \in U(x_0, r) \subseteq D(F)$ and $v \in X$ there exists an element $\Phi(x, u, v) \in X$ satisfying $[F'(x) - F'(u)]v = F'(u)\Phi(x, u, v), \|\Phi(x, u, v)\| \le L\|v\|(\|x - x_0\| + \|u - x_0\|)$.*

Note that in view of the estimate

$$\|x - u\| \le \|x - x_0\| + \|x_0 - u\|,$$

Assumption 6 implies Assumption 8 with $k_0 = L$ but not necessarily vice versa(see numerical examples at the end of the chapter). Throughout this chapter we assume that the operator F satisfies Assumptions 7 and 8.

Remark 9. *If Assumptions 7 and 8 are fulfilled only for all $x, u \in U(x_0, r) \cap Q \ne \emptyset$, where Q is a convex closed a priori set for which $\hat{x} \in Q$, then we can modify the method (10) in the following way:*

$$x_{n+1,\alpha}^\delta = P_Q(G(x_{n,\alpha}^\delta))$$

to obtain the same estimates in this chapter; here P_Q is the metric projection onto the set Q.

The plan of this chapter is as follows. In Section 2, we prove the convergence of the method and in Section 3, we give error bounds under source conditions. Section 4 deals with the starting point and algorithm. Finally, we give some numerical examples in concluding Section 5.

2. CONVERGENCE ANALYSIS

We use a majorizing sequence for proving our results. Recall (see [12], Definition 1.3.11) that a nonnegative sequence $\{t_n\}$ is said to be a majorizing sequence of a sequence $\{x_n\}$ in X if

$$\|x_{n+1} - x_n\| \leq t_{n+1} - t_n, \quad \forall n \geq 0.$$

In the convergence analysis we will use the following Lemma on majorization, which is a reformulation of Lemma 1.3.12 in [12].

Lema 10. *(cf. [17], Lemma 2.1) Let $\{t_n\}$ be a majorizing sequence for $\{x_n\}$ in X. If $\lim_{n\to\infty} t_n = t^*$, then $x^* = \lim_{n\to\infty} x_n$ exists and*

$$\|x^* - x_n\| \leq t^* - t_n, \quad \forall n \geq 0. \tag{12}$$

We need an auxillary result on majorizing sequences for method (10).

Lema 11. *Suppose that there exist non–negative numbers L, η such that*

$$16L\eta \leq 1. \tag{13}$$

Let

$$q = \frac{1 - \sqrt{1 - 16L\eta}}{2}. \tag{14}$$

Then, scalar sequence $\{t_n\}$ given by

$$t_0 = 0,\ t_1 = \eta,\ t_{n+1} = t_n + \frac{L}{2}(5t_n + 3t_{n-1})(t_n - t_{n-1}) \text{ for each } n = 1, 2, \ldots \tag{15}$$

is well defined, nondecreasing, bounded from above by t^{**} given by

$$t^{**} = \frac{\eta}{1 - q} \tag{16}$$

and converges to its unique least upper bound t^* which satisfies

$$\eta \leq t^* \leq t^{**}. \tag{17}$$

Moreover the following estimates hold for each $n = 1, 2, \ldots$.

$$t_{n+1} - t_n \leq q(t_n - t_{n+1}) \tag{18}$$

and

$$t^* - t_n \leq \frac{q^n}{1 - q}\eta. \tag{19}$$

Proof. Note that $q \in (0, 1)$. We shall show using mathematical induction that

$$\frac{L}{2}(5t_m + 3t_{m-1}) \leq q. \tag{20}$$

Estimate is true for $m = 1$ by the definition of sequence $\{t_n\}$ and (13). Then, we have by (15) that $t_2 - t_1 \leq q(t_1 - t_0)$ and $t_2 \leq \eta + q\eta = (1+q)\eta = \frac{1-q^2}{1-q}\eta < \frac{\eta}{1-q} = t^{**}$. Let us assume (20) holds for all integers smaller or equal to m. Hence, we get that

$$t_{m+1} - t_m \leq q(t_m - t_{m-1}). \tag{21}$$

and

$$t_{m+1} \leq \frac{1-q^{m+1}}{1-q}\eta. \tag{22}$$

We shall show that (20) holds for m replaced by $m+1$. Using (21) and (22), estimates (20) shall be true if

$$\frac{L}{2}[5(\frac{1-q^m}{1-q}) + 3(\frac{1-q^{m-1}}{1-q})]\eta \leq q. \tag{23}$$

Estimate (23) motivates us to define recurrent functions f_m on $[0, 1)$ by

$$f_m(t) = L[5(1 + t + \cdots + t^{m-1}) + 3(1 + t + \cdots + t^{m-2})]\eta - 2t. \tag{24}$$

We need n relationship between two consecutive functions f_m. Using (24) we get that

$$f_{m+1}(t) = f_m(t) + (5t + 3)L\eta t^{m-1} \geq f_m(t). \tag{25}$$

Define function f_∞ on $[0, 1)$ by

$$f_\infty(t) = \lim_{m \to \infty} f_m(t). \tag{26}$$

Then, using (24) we get that

$$f_\infty(t) = \frac{8L}{1-t}\eta - 2t. \tag{27}$$

Evidently, (23) is true if

$$f_\infty(t) \leq 0, \tag{28}$$

since

$$f_m(q) \leq f_{m+1}(q) \leq \cdots \leq f_\infty(q). \tag{29}$$

But (28) is true by (23) and (24). The induction for (20) is complete. Therefore, sequence $\{t_n\}$ is nondecreasing, bounded from above by t^{**} and such it converges to some t^* which satisfies (17). The proof of the Lemma is complete. □

Lema 12. ([1], Lemma 2.3) For $u, v \in B_{r_0}(x_0)$

$$F(u) - F(v) - F'(u)(u-v) = F'(u)\int_0^1 \Phi(v + t(u-v), u, u-v)dt.$$

Here after we assume that for $r > 0$, $L > 0$ and q given in (14):

$$\frac{\delta}{\alpha} < \eta \leq \min\left\{\frac{1}{16L}, r(1-q)\right\} \tag{30}$$

and $\|\hat{x} - x_0\| \leq \rho$, where

$$\rho \leq \frac{1}{L}(\sqrt{1 + 2L(\eta - \delta/\alpha)} - 1) \tag{31}$$

Remark 13. *Note that (30) and (31) imply*

$$\frac{L}{2}\rho^2 + \rho + \frac{\delta}{\alpha} \leq \eta \leq \min\{\frac{1}{16L}, r(1-q)\}. \tag{32}$$

Theorem 14. *Suppose Assumption 8 holds. Let the assumptions in Lemma 11 are satisfied with η as in (32). Then the sequence $\{x_{n,\alpha}^\delta\}$ defined in (11) is well defined and $x_{n,\alpha}^\delta \in U(x_0, t^*)$ for all $n \geq 0$. Further $\{x_{n,\alpha}^\delta\}$ is a Cauchy sequence in $U(x_0, t^*)$ and hence converges to $x_\alpha^\delta \in \overline{U(x_0, t^*)} \subset U(x_0, t^{**})$ and $F(x_\alpha^\delta) = y^\delta + \alpha(x_0 - x_\alpha^\delta)$.*

Moreover, the following estimates hold for all $n \geq 0$:

$$\|x_{n+1,\alpha}^\delta - x_{n,\alpha}^\delta\| \leq t_{n+1} - t_n, \tag{33}$$

$$\|x_{n,\alpha}^\delta - x_\alpha^\delta\| \leq t^* - t_n \leq \frac{q^n \eta}{1-q} \tag{34}$$

and

$$\|x_{n+1,\alpha}^\delta - x_{n,\alpha}^\delta\| \leq \frac{L}{2}[5\|x_{n,\alpha}^\delta - x_{0,\alpha}^\delta\| + 3\|x_{n-1,\alpha}^\delta - x_{0,\alpha}^\delta\|]\|x_{n,\alpha}^\delta - x_{n-1,\alpha}^\delta\|. \tag{35}$$

Proof. First we shall prove that

$$\|x_{n+1,\alpha}^\delta - x_{n,\alpha}^\delta\| \leq \frac{L}{2}[5\|x_{n,\alpha}^\delta - x_{0,\alpha}^\delta\| + 3\|x_{n-1,\alpha}^\delta - x_{0,\alpha}^\delta\|]\|x_{n,\alpha}^\delta - x_{n-1,\alpha}^\delta\|. \tag{36}$$

With G as in (11), we have for $u, v \in B_{t^*}(x_0)$,

$$\begin{aligned}
G(u) - G(v) &= u - v - R_\alpha(u)^{-1}[F(u) - y^\delta + \alpha(u - x_0)] + R_\alpha(v)^{-1} \\
&\quad \times [F(v) - y^\delta + \alpha(v - x_0)] \\
&= u - v - [R_\alpha(u)^{-1} - R_\alpha(v)^{-1}](F(v) - y^\delta + \alpha(v - x_0)) \\
&\quad - R_\alpha(u)^{-1}(F(u) - F(v) + \alpha(u - v)) \\
&= R_\alpha(u)^{-1}[F'(u)(u - v) - (F(u) - F(v))] \\
&\quad - R_\alpha(u)^{-1}[F'(v) - F'(u)]R_\alpha(v)^{-1}(F(v) - y^\delta + \alpha(v - x_0)) \\
&= R_\alpha(u)^{-1}[F'(u)(u - v) - (F(u) - F(v))] \\
&\quad - R_\alpha(u)^{-1}[F'(v) - F'(u)](v - G(v)) \\
&= R_\alpha(u)^{-1}[F'(u)(u - v) + \int_0^1 (F'(u + t(v - u))(v - u))dt] \\
&\quad - R_\alpha(u)^{-1}[F'(v) - F'(u)](v - G(v)) \\
&= \int_0^1 R_\alpha(u)^{-1}[(F'(u + t(v - u)) - F'(u))(v - u)dt] \\
&\quad - R_\alpha(u)^{-1}[F'(v) - F'(u)](v - G(v)).
\end{aligned}$$

The last, but one step follows from the Lemma 10. So by Assumption 8 and the estimate

$\|R_\alpha(u)^{-1}F'(u)\| \leq 1$, we have

$$\begin{aligned}
\|G(u) - G(v)\| &\leq L\int_0^1 [\|u + t(v-u) - x_0\| + \|u - x_0\|]dt\|v - u\| \\
&\quad + L[\|v - x_0\| + \|u - x_0\|]\|v - u\| \\
&\leq \frac{L}{2}[3\|u - x_0\| + \|v - x_0\|] + L[\|v - x_0\| + \|u - x_0\|] \\
&\leq \frac{L}{2}[5\|u - x_0\| + 3\|v - x_0\|]\|v - G(u)\|.
\end{aligned} \quad (37)$$

Now by taking $u = x_{n,\alpha}^\delta$ and $v = x_{n-1,\alpha}^\delta$ in (37), we obtain (36).

Next we shall prove that the sequence (t_n) defined in Lemma 10 is a majorizing sequence of the sequence $(x_{n,\alpha}^\delta)$. Note that $F(\hat{x}) = y$, so by Lemma 11,

$$\begin{aligned}
\|x_{1,\alpha}^\delta - x_0\| &= \|R_\alpha(x_0)^{-1}(F(x_0) - y^\delta)\| \\
&= \|R_\alpha(x_0)^{-1}(F(x_0) - y + y - y^\delta)\| \\
&= \|R_\alpha(x_0)^{-1}(F(x_0) - F(\hat{x}) - F'(x_0)(x_0 - \hat{x}) \\
&\quad + F'(x_0)(x_0 - \hat{x}) + y - y^\delta)\| \\
&\leq \|R_\alpha(x_0)^{-1}(F(x_0) - F(\hat{x}) - F'(x_0)(x_0 - \hat{x}))\| \\
&\quad + \|R_\alpha(x_0)^{-1}F'(x_0)(x_0 - \hat{x})\| + \|R_\alpha(x_0)^{-1}(y - y^\delta)\| \\
&\leq \|R_\alpha(x_0)^{-1}F'(x_0)\int_0^1 \Phi(\hat{x} + t(x_0 - \hat{x}), x_0, (x_0 - \hat{x}))dt\| \\
&\quad + \|R_\alpha(x_0)^{-1}F'(x_0))(x_0 - \hat{x})\| + \frac{\delta}{\alpha} \\
&\leq \frac{L}{2}\|x_0 - \hat{x}\|^2 + \|x_0 - \hat{x}\| + \frac{\delta}{\alpha} \\
&\leq \frac{L}{2}\rho^2 + \rho + \frac{\delta}{\alpha} \\
&\leq \eta = t_1 - t_0.
\end{aligned}$$

Assume that $\|x_{i+1,\alpha}^\delta - x_{i,\alpha}^\delta\| \leq t_{i+1} - t_i$ for all $i \leq k$ for some k. Then

$$\begin{aligned}
\|x_{k+1,\alpha}^\delta - x_0\| &\leq \|x_{k+1,\alpha}^\delta - x_{k,\alpha}^\delta\| + \|x_{k,\alpha}^\delta - x_{k-1,\alpha}^\delta\| + \cdots + \|x_{1,\alpha}^\delta - x_0\| \\
&\leq t_{k+1} - t_k + t_k - t_{k-1} + \cdots + t_1 - t_0 \\
&= t_{k+1} \leq t^*.
\end{aligned}$$

So, $x_{i+1,\alpha}^\delta \in B_{t^*}(x_0)$ for all $i \leq k$, and hence, by (36),

$$\begin{aligned}
\|x_{k+2,\alpha}^\delta - x_{k+1,\alpha}^\delta\| &\leq \frac{L}{2}[5\|x_{n,\alpha}^\delta - x_{0,\alpha}^\delta\| + 3\|x_{n-1,\alpha}^\delta - x_{0,\alpha}^\delta\|]\|x_{n,\alpha}^\delta - x_{n-1,\alpha}^\delta\| \\
&\leq \frac{L}{2}(5t_{k+1} + 3t_{k-1}) = t_{k+2} - t_{k+1}.
\end{aligned}$$

Thus by induction $\|x_{n+1,\alpha}^\delta - x_{n,\alpha}^\delta\| \leq t_{n+1} - t_n$ for all $n \geq 0$ and hence $\{t_n\}, n \geq 0$ is a majorizing sequence of the sequence $\{x_{n,\alpha}^\delta\}$. In particular $\|x_{n,\alpha}^\delta - x_0\| \leq t_n \leq t^*$, i.e.,

$x_{n,\alpha}^\delta \in U(x_0, t^*)$, for all $n \geq 0$. So, $\{x_{n,\alpha}^\delta\}, n \geq 0$ is a Cauchy sequence and converges to some $x_\alpha^\delta \in \overline{U(x_0, t^*)} \subset U(x_0, t^{**})$ and by Lemma 12

$$\|x_\alpha^\delta - x_{n,\alpha}^\delta\| \leq t^* - t_n \leq \frac{q^n \eta}{1-q}.$$

To prove (35), we observe that $G(x_\alpha^\delta) = x_\alpha^\delta$, so (35) follows from (37), by taking $u = x_{n,\alpha}^\delta$ and $v = x_\alpha^\delta$ in (37). Now by letting $n \to \infty$ in (10) we obtain $F(x_\alpha^\delta) = y^\delta + \alpha(x_0 - x_\alpha^\delta)$.

This completes the proof of the Theorem. $\square$

Remark 15. *The convergence order of the method is two [1], under Assumption 6. In Theorem 14 the error bounds are too pessimistic. That is why in practice we shall use the computational order of convergence (COC) (see eg. [15]) defined by*

$$\varrho \approx \ln\left(\frac{\|x_{n+1} - x_\alpha^\delta\|}{\|x_n - x_\alpha^\delta\|}\right) / \ln\left(\frac{\|x_n - x_\alpha^\delta\|}{\|x_{n-1} - x_\alpha^\delta\|}\right).$$

The (COC) ϱ will then be close to 2 which is the order of convergence of the method.

3. ERROR BOUNDS UNDER SOURCE CONDITIONS

The objective of this section is to obtain an error estimate for $\|x_{n,\alpha}^\delta - \hat{x}\|$ under

Assumption 16. *There exists a continuous, strictly monotonically increasing function $\varphi : (0, a] \to (0, \infty)$ with $a \geq \|F'(x_0)\|$ satisfying;*

(i) $\lim_{\lambda \to 0} \varphi(\lambda) = 0$,

(ii) $\sup_{\lambda \geq 0} \frac{\alpha \varphi(\lambda)}{\lambda + \alpha} \leq \varphi(\alpha) \quad \forall \lambda \in (0, a]$ and

(iii) there exists $v \in X$ with $\|v\| \leq 1$ (cf. [18]) such that

$$x_0 - \hat{x} = \varphi(F'(x_0))v.$$

Proposition 17. *Let $F : D(F) \subseteq X \to X$ be a monotone operator in X. Let x_α^δ be the solution of (5) and $x_\alpha := x_\alpha^0$. Then*

$$\|x_\alpha^\delta - x_\alpha\| \leq \frac{\delta}{\alpha}.$$

Proof. The result follows from the monotonicity of F and the relation;

$$F(x_\alpha^\delta) - F(x_\alpha) + \alpha(x_\alpha^\delta - x_\alpha) = y^\delta - y.$$

$\square$

Theorem 18. *(cf. [11], Proposition 4.1 or [6], Theorem 3.3) Suppose that Assumption 7, 8 and hypotheses of Proposition 17 hold. Let $\hat{x} \in D(F)$ be a solution of (1). Then, the following assertion holds*

$$\|x_\alpha - \hat{x}\| \leq (Lr + 1)\varphi(\alpha).$$

Theorem 19. *Suppose hypotheses of Theorem 14 and Theorem 18 hold. Then, the following assertion holds*

$$\|x_{n,\alpha}^\delta - \hat{x}\| \leq \frac{q^n \eta}{1-q} + c_1 \left(\varphi(\alpha) + \frac{\delta}{\alpha} \right)$$

where $c_1 = \max\{1, (L_0 r + 1)\}$.

Let

$$\bar{c} := \max \left\{ \frac{\eta}{1-q} + 1, (L_0 r + 2) \right\}, \tag{38}$$

and let

$$n_\delta := \min\{n : q^n \leq \frac{\delta}{\alpha}\}. \tag{39}$$

Theorem 20. *Let $\bar{c}$ and n_δ be as in (38) and (39) respectively. Suppose that hypothese of Theorem 19 hold. Then, the following assertions hold*

$$\|x_{n_\delta,\alpha}^\delta - \hat{x}\| \leq \bar{c} \left(\varphi(\alpha) + \frac{\delta}{\alpha} \right). \tag{40}$$

Note that the error estimate $\varphi(\alpha) + \frac{\delta}{\alpha}$ in (32) is of optimal order if $\alpha := \alpha_\delta$ satisfies, $\varphi(\alpha_\delta)\alpha_\delta = \delta$.

Now using the function $\psi(\lambda) := \lambda \varphi^{-1}(\lambda), 0 < \lambda \leq a$ we have $\delta = \alpha_\delta \varphi(\alpha_\delta) = \psi(\varphi(\alpha_\delta))$, so that $\alpha_\delta = \varphi^{-1}(\psi^{-1}(\delta))$. In view of the above observations and (32) we have the following.

Theorem 21. *Let $\psi(\lambda) := \lambda \varphi^{-1}(\lambda)$ for $0 < \lambda \leq a$, and the assumptions in Theorem 20 hold. For $\delta > 0$, let $\alpha := \alpha_\delta = \varphi^{-1}(\psi^{-1}(\delta))$ and let n_δ be as in (39). Then*

$$\|x_{n_\delta,\alpha}^\delta - \hat{x}\| = O(\psi^{-1}(\delta)).$$

In this section, we present a parameter choice rule based on the balancing principle studied in [19], [4], [20]. In this method, the regularization parameter α is selected from some finite set

$$D_M(\alpha) := \{\alpha_i = \mu^i \alpha_0, i = 0, 1, \ldots, M\}$$

where $\mu > 1, \alpha_0 > 0$ and let

$$n_i := \min \left\{ n : e^{-\gamma_0 n} \leq \frac{\delta}{\alpha_i} \right\}.$$

Then for $i = 0, 1, \ldots, M$, we have

$$\|x_{n_i,\alpha_i}^\delta - x_{\alpha_i}^\delta\| \leq c \frac{\delta}{\alpha_i}, \quad \forall i = 0, 1, \ldots, M.$$

Let $x_i := x_{n_i,\alpha_i}^\delta$. The parameter choice strategy that we are going to consider in this chapter, we select $\alpha = \alpha_i$ from $D_M(\alpha)$ and operate only with corresponding x_i, $i = 0, 1, \cdots, M$. Proof of the following theorem is analogous to the proof of Theorem 3.1 in [11].

Theorem 22. *(cf. [11], Theorem 3.1)* Assume that there exists $i \in \{0, 1, 2, \ldots, M\}$ such that $\varphi(\alpha_i) \leq \frac{\delta}{\alpha_i}$. Suppose the hypotheses of Theorem 19 and Theorem 20 hold and let

$$l := \max\left\{i : \varphi(\alpha_i) \leq \frac{\delta}{\alpha_i}\right\} < M,$$

$$k := \max\left\{i : \|x_i - x_j\| \leq 4\bar{c}\frac{\delta}{\alpha_j}, \quad j = 0, 1, 2, \ldots, i\right\}.$$

Then $l \leq k$ and
$$\|\hat{x} - x_k\| \leq c\psi^{-1}(\delta)$$

where $c = 6\bar{c}\mu$.

4. Implementation of Adaptive Choice Rule

The main goal of this section is to provide a starting point for the iteration approximating the unique solution x_α^δ of (5) and then to provide an algorithm for the determination of a parameter fulfilling the balancing principle. The choice of the starting point involves the following steps:

- For $q = \frac{1-\sqrt{1-16L\eta}}{2}$ choose $0 < \alpha_0 < 1$ and $\mu > 1$.

- Choose η such that η satisfies (13).

- Choose ρ such that ρ satisfies (31).

- Choose $x_0 \in D(F)$ such that $\|x_0 - \hat{x}\| \leq \rho$.

- Choose the parameter $\alpha_M = \mu^M \alpha_0$ big enough with $\mu > 1$, not too large.

- Choose n_i such that $n_i = \min\{n : q^n \leq \frac{\delta}{\alpha_i}\}$.

Finally the adaptive algorithm associated with the choice of the parameter specified in Theorem 22 involves the following steps:

4.1. Algorithm

- Set $i \leftarrow 0$.

- Solve $x_i := x_{n_i,\alpha_i}^\delta$ by using the iteration (10).

- If $\|x_i - x_j\| > 4c\frac{\sqrt{\delta}}{\mu^j}$, $j \leq i$, then take $k = i - 1$.

- Set $i = i + 1$ and return to Step 2.

5. NUMERICAL EXAMPLES

In this section we give an example for illustrating the algorithm considered in the above section. We apply the algorithm by choosing a sequence of finite dimensional subspace (V_n) of X with dim $V_n = n + 1$. Precisely we choose V_n as the space of linear splines in a uniform grid of $n + 1$ points in $[0, 1]$.

Example 1. *(see [11], section 4.3) Let $F : D(F) \subseteq L^2(0,1) \longrightarrow L^2(0,1)$ defined by*

$$F(u) := \int_0^1 k(t,s)u^3(s)ds,$$

where

$$k(t,s) = \begin{cases} (1-t)s, & 0 \leq s \leq t \leq 1, \\ (1-s)t, & 0 \leq t \leq s \leq 1. \end{cases}$$

Then for all $x(t), y(t) : x(t) > y(t)$:

$$\langle F(x) - F(y), x - y \rangle = \int_0^1 \left[\int_0^1 k(t,s)(x^3 - y^3)(s)ds \right]$$

$$\times (x-y)(t)dt \geq 0.$$

Thus the operator F is monotone. The Fréchet derivative of F is given by

$$F'(u)w = 3\int_0^1 k(t,s)u^2(s)w(s)ds. \tag{41}$$

Note that for $u, v > 0$,

$$[F'(v) - F'(u)]w = 3\int_0^1 k(t,s)(v^2(s) - u^2(s))ds$$
$$:= F'(u)\Phi(v,u,w)$$

where $\Phi(v, u, w) = \frac{[v^2(s) - u^2(s)]w(s)}{u^2(s)}$.
Observe that

$$\Phi(v,u,w) = \frac{[v^2(s) - u^2(s)]w(s)}{u^2(s)} = \frac{[u(s)+v(s)][v(s)-u(s)]w(s)}{u^2(s)}.$$

So Assumption 2 satisfies with $k_0 \geq \left\| \frac{[u(s)+v(s)]}{u^2(s)} \right\|$. In our computation, we take $f(t) = \frac{6\sin(\pi t) + \sin^3(\pi t)}{9\pi^2}$ and $f^\delta = f + \delta$. Then the exact solution

$$\hat{x}(t) = \sin(\pi t).$$

We use

$$x_0(t) = \sin(\pi t) + \frac{3[t\pi^2 - t^2\pi^2 + \sin^2(\pi t)]}{4\pi^2}$$

as our initial guess, so that the function $x_0 - \hat{x}$ satisfies the source condition

$$x_0 - \hat{x} = \varphi(F'(x_0))\frac{\hat{x}}{4x_0}$$

where $\varphi(\lambda) = \lambda$.

For the operator $F'(.)$ defined in (41), $\varepsilon_h = O(n^{-2})$ (cf. [21]). Thus we expect to obtain the rate of convergence $O((\delta + \varepsilon_h)^{\frac{1}{2}})$.

We choose $\alpha_0 = (1.1)(\delta + \varepsilon_h)$, $\mu = 1.1$, $\rho = 0.11$, and $q = 0.46$. The results of the computation are presented in Table 1. The plots of the exact solution and the approximate solution obtained are given in Figures 1 and 2.

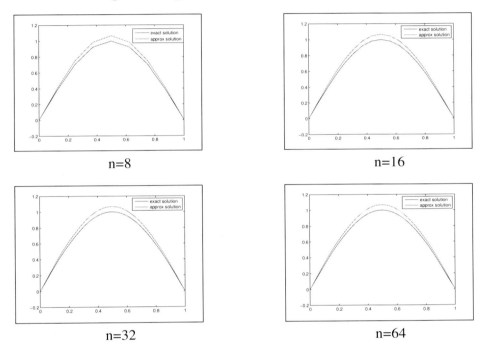

Figure 1. Curves of the exact and approximate solutions.

Example 2. Let $X = Y = \mathbb{R}$, $D = [0, \infty)$, $x_0 = 1$ and define function F on D by

$$F(x) = \frac{x^{1+\frac{1}{i}}}{1+\frac{1}{i}} + c_1 x + c_2, \qquad (42)$$

where c_1, c_2 are real parameters and $i > 2$ an integer. Then $F'(x) = x^{1/i} + c_1$ is not Lipschitz on D. However central Lipschitz condition $(C2)'$ holds for $L_0 = 1$.

Indeed, we have

$$\|F'(x) - F'(x_0)\| = |x^{1/i} - x_0^{1/i}|$$
$$= \frac{|x - x_0|}{x_0^{\frac{i-1}{i}} + \cdots + x^{\frac{i-1}{i}}}$$

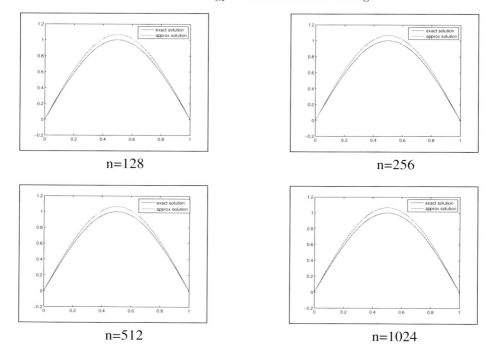

Figure 2. Curves of the exact and approximate solutions.

Table 1. Iterations and corresponding error estimates

n	k	n_k	$\delta + \varepsilon_h$	α	$\|x_k - \hat{x}\|$	$\frac{\|x_k - \hat{x}\|}{(\delta + \varepsilon_h)^{1/2}}$
8	2	2	0.0200	0.0267	0.0359	0.2538
16	2	2	0.0200	0.0266	0.0432	0.3055
32	2	2	0.0200	0.0266	0.0450	0.3183
64	2	2	0.0200	0.0266	0.0455	0.3215
128	17	3	0.0200	0.1112	0.0456	0.3223
256	17	3	0.0200	0.1112	0.0456	0.3225
512	30	3	0.0200	0.3839	0.0456	0.3226
1024	30	3	0.0200	0.3839	0.0456	0.3226

so
$$\|F'(x) - F'(x_0)\| \leq L_0|x - x_0|.$$

Example 3. ([22, Example 7.5]) We consider the integral equations

$$u(s) = f(s) + \lambda \int_a^b G(s,t) u(t)^{1+1/n} dt, \quad n \in \mathbb{N}. \tag{43}$$

Here, f is a given continuous function satisfying $f(s) > 0, s \in [a,b], \lambda$ is a real number, and the kernel G is continuous and positive in $[a, b] \times [a, b]$.

For example, when $G(s, t)$ is the Green kernel, the corresponding integral equation is equivalent to the boundary value problem

$$u'' = \lambda u^{1+1/n}$$
$$u(a) = f(a), u(b) = f(b).$$

These type of problems have been considered in [12]–[16].

Equation of the form (43) generalize equations of the form

$$u(s) = \int_a^b G(s,t) u(t)^n dt \qquad (44)$$

studied in [12]–[16]. Instead of (43) we can try to solve the equation $F(u) = 0$ where

$$F : \Omega \subseteq C[a,b] \to C[a,b], \Omega = \{u \in C[a,b] : u(s) \geq 0, s \in [a,b]\},$$

and

$$F(u)(s) = u(s) - f(s) - \lambda \int_a^b G(s,t) u(t)^{1+1/n} dt.$$

The norm we consider is the max–norm.

The derivative F' is given by

$$F'(u)v(s) = v(s) - \lambda(1 + \frac{1}{n}) \int_a^b G(s,t) u(t)^{1/n} v(t) dt, \quad v \in \Omega.$$

First of all, we notice that F' does not satisfy a Lipschitz–type condition in Ω. Let us consider, for instance, $[a,b] = [0,1], G(s,t) = 1$ and $y(t) = 0$. Then $F'(y)v(s) = v(s)$ and

$$\|F'(x) - F'(y)\| = |\lambda|(1 + \frac{1}{n}) \int_a^b x(t)^{1/n} dt.$$

If F' were a Lipschitz function, then

$$\|F'(x) - F'(y)\| \leq L_1 \|x - y\|,$$

or, equivalently, the inequality

$$\int_0^1 x(t)^{1/n} dt \leq L_2 \max_{x \in [0,1]} x(s), \qquad (45)$$

would hold for all $x \in \Omega$ and for a constant L_2. But this is not true. Consider, for example, the functions

$$x_j(t) = \frac{t}{j}, \quad j \geq 1, \ t \in [0,1].$$

If these are substituted into (45)

$$\frac{1}{j^{1/n}(1 + 1/n)} \leq \frac{L_2}{j} \Leftrightarrow j^{1-1/n} \leq L_2(1 + 1/n), \ \forall j \geq 1.$$

This inequality is not true when $j \to \infty$.

Therefore, Assumption 6 is not satisfied in this case. However, Assumption 8 holds. To show this, let $x_0(t) = f(t)$ and $\gamma = \min_{s \in [a,b]} f(s), \alpha > 0$ Then for $v \in \Omega$,

$$\begin{aligned} \|[F'(x) - F'(x_0)]v\| &= |\lambda|(1 + \frac{1}{n}) \max_{s \in [a,b]} |\int_a^b G(s,t)(x(t)^{1/n} - f(t)^{1/n})v(t)dt| \\ &\leq |\lambda|(1 + \frac{1}{n}) \max_{s \in [a,b]} G_n(s,t) \end{aligned}$$

where

$$G_n(s,t) = \frac{G(s,t)|x(t) - f(t)|}{x(t)^{(n-1)/n} + x(t)^{(n-2)/n}f(t)^{1/n} + \cdots + f(t)^{(n-1)/n}}.$$

Hence,

$$\begin{aligned} \|[F'(x) - F'(x_0)]v\| &= \frac{|\lambda|(1+1/n)}{\gamma^{(n-1)/n}} \max_{s \in [a,b]} \int_a^b G(s,t)v(t)dt \|x - x_0\| \\ &\leq L_0 \|v\| \|x - x_0\|, \end{aligned}$$

where $L_0 = \frac{|\lambda|(1+1/n)}{\gamma^{(n-1)/n}} N$ and $N = \max_{s \in [a,b]} \int_a^b G(s,t)dt$. *Then by the following inequality;*

$$\|[F'(x) - F'(u)]v\| \leq \|[F'(x) - F'(x_0)]v\| + \|[F'(x_0) - F'(u)]v\| \leq L_0\|v\|(\|x - x_0\| + \|u - x_0\|)$$

Assumption 8 holds for sufficiently small λ.

REFERENCES

[1] George, S. and Elmahdy, A. I. 2012. "A quadratic convergence yielding iterative method for nonlinear ill–posed operator equations". *Comput. Methods Appl. Math.* 12(1): 32–45.

[2] Mahale, P. and Nair, M. T. 2009. "Iterated Lavrentiev regularization for nonlinear ill–posed problems". *ANZIAM Journal.* 51: 191–217.

[3] Bakushinsky, A. and Smirnova, A. 2005. "On application of generalized discrepancy principle to iterative methods for nonlinear ill–posed problems". *Numer. Funct. Anal. Optim.* 26: 35–48.

[4] Perverzev, S. V. and Schock, E. 2005. "On the adaptive selection of the parameter in regularization of ill–posed problems". *SIAM J. Numer. Anal.* 43: 2060–2076.

[5] Argyros, I. K. and George, S. "Expanding the applicability of an iterative regularization method for ill–posed problems" (communicated).

[6] Tautanhahn, U. 2002. "On the method of Lavrentiev regularization for nonlinear ill–posed problems". *Inverse Problems.* 18: 191–207.

[7] Bakushinskii, A. B. 1992. "The problem of convergence of the iteratively regularized Gauss–Newton method". *Comput. Math. Math. Phys.* 32: 1353–1359.

[8] Bakushinskii, A. B. 1995. "Iterative methods without saturation for solving degenerate nonlinear operator equations". *Dokl. Akad. Nauk.* 344: 7–8.

[9] Bakushinsky, A. B. and Kokurin, M. Yu. 2004. *Iterative methods for approximate solution of inverse problems*. New York: Mathematics and Its Applications, Volume 577, Springer, Dordrecht.

[10] Kaltenbacher, B., Neubauer, A. and Scherzer, O. 2008. *Iterative regularization methods for nonlinear ill–posed problems*. Berlin: Radon Series on Computational and Applied Mathematics, volume 6. Walter de Gruyter GmbH & Co. KG.

[11] Semenova, E. V. 2010. "Lavrentiev regularization and balancing principle for solving ill–posed problems with monotone operators". *Comput. Methods Appl. Math.* 4: 444–454.

[12] Argyros, I. K. 2008. *Convergenve and Applications of Newton–type Iterations*. New York: Springer.

[13] Argyros, I. K. 2007. "Approximating solutions of equations using Newton's method with a modified Newton's method iterate as a starting point". *Rev. Anal. Numer. Theor. Approx.* 36: 123–138.

[14] Argyros, I. K. 2011. "A Semilocal convergence for directional Newton methods". *Math. Comput. (AMS).* 80: 327–343.

[15] Argyros, I. K. and Hilout, S. 2012. "Weaker conditions for the convergence of Newton's method". *J. Complexity.* 28: 364–387.

[16] Argyros, I. K., Cho, Y. J. and Hilout, S. 2012. *Numerical methods for equations and its applications*. New York: CRC Press, Taylor and Francis.

[17] George, S. and Elmahdy, A. I. 2010. "An analysis of Lavrentiev regularization for nonlinear ill–posed problems using an iterative regularization method". *Int. J. Comput. Appl. Math.* 5(3): 369–381.

[18] Nair, M. T. and Ravishankar, P. 2008. "Regularized versions of continuous Newton's method and continuous modified Newton's method under general source conditions". *Numer. Funct. Anal. Optim.* 29(9–10): 1140–1165.

[19] Mathe, P. and Perverzev, S. V. 2003. "Geometry of linear ill–posed problems in variable Hilbert scales". *Inverse Problems.* 19(3): 789–803.

[20] Bauer, F. and Hohage, T. 2005. "A Lepskij–type stopping rule for regularized Newton methods". *Inverse Problems.* 21(6): 1975–1991.

[21] Groetsch, C. W., King, J. T. and Murio, D. 1982. "Asymptotic analysis of a finite element method for Fredholm equations of the first kind", in *Treatment of Integral Equations by Numerical Methods*. London: C. T. H. Baker and G. F. Miller, Academic Press.

[22] Argyros, I. K., Cho, Y. J. and George, S. 2013. "Expanding the applicability of Lavrentiev regularization method for ill–posed problems". *Boundary Value Problems*. 114.

In: Understanding Banach Spaces
Editor: Daniel González Sánchez

ISBN: 978-1-53616-745-0
© 2020 Nova Science Publishers, Inc.

Chapter 10

BALL CONVERGENCE THEOREM FOR A FIFTH–ORDER METHOD IN BANACH SPACES

Ioannis K. Argyros[1,*] *and Santhosh George*[2,†]
[1]Department of Mathematical Sciences
Cameron University, Lawton, OK, US
[2]Department of Mathematical and Computational Sciences
National Institute of Technology, Karnataka, India

Abstract

We present a local convergence analysis for a fifth–order method in order to approximate a solution of a nonlinear equation in a Banach space. Our sufficient convergence conditions involve only hypotheses on the first Fréchet–derivative of the operator involved. Earlier studies use hypotheses up to the fourth Fréchet–derivative [1]. Hence, the applicability of these methods is expanded under weaker hypotheses and less computational cost for the constants involved in the convergence analysis. Numerical examples are also provided in this study.

Keywords: high convergence order method, Banach space, local convergence, Fréchet–derivative

AMS Subject Classification: 65D10, 65D99

1. Introduction

Many problems in computational sciences and other disciplines can be brought in the form of the nonlinear equation

$$F(x) = 0, \qquad (1)$$

where F is a Fréchet–differentiable operator defined on a subset D of a Banach space X with values in a Banach space Y. In this study we are concerned with the problem of approximating a solution x^* of equation (1). In Numerical Functional Analysis, for finding

---

[*]Corresponding Author's E-mail: iargyros@cameron.edu.
[†]E-mail: sgeorge@nitk.ac.in.

solution of (1) is essentially connected to Newton–like methods [1–23]. The Newton–like methods are usually studied based on: semi–local and local convergence. The semi–local convergence matter is, based on the information around an initial point, to give conditions ensuring the convergence of the iterative procedure; while the local one is, based on the information around a solution, to find estimates of the radii of convergence balls. There exist many studies which deal with the local and semilocal convergence analysis of Newton–like methods such as [1–23].

We present a local convergence analysis for a fifth–order method defined for each $n = 0, 1, 2, \ldots$ by

$$\begin{aligned} y_n &= x_n - F'(x_n)^{-1} F(x_n), \\ z_n &= y_n - 5 F'(x_n)^{-1} F(y_n), \\ x_{n+1} &= z_n - \frac{1}{5} F'(x_n)^{-1} (-16 F(y_n) + F(z_n)), \end{aligned} \quad (2)$$

where x_0 is an initial point. The convergence the method (2) was studied in [1]. The convergence in [1] was studied under the assumptions that $F^{(i)}, i = 1, 2, 3, 4$ are bounded.

Similar assumptions have been used by several authors [2]–[23], on other high convergence order methods. These assumptions however are very restrictive. As a motivational example, let us define function f on $D = [-\frac{1}{2}, \frac{5}{2}]$ by

$$f(x) = \begin{cases} x^3 \ln x^2 + x^5 - x^4, & x \neq 0 \\ 0, & x = 0 \end{cases}$$

Choose $x^* = 1$. We have that

$$\begin{aligned} f'(x) &= 3x^2 \ln x^2 + 5x^4 - 4x^3 + 2x^2, \ f'(1) = 3, \\ f''(x) &= 6x \ln x^2 + 20x^3 - 12x^2 + 10x \\ f'''(x) &= 6 \ln x^2 + 60x^2 - 24x + 22. \end{aligned}$$

Then, obviously, function f''' is unbounded on D. In the present paper we only use hypotheses on the first Fréchet derivative (see conditions (16)–(21)). This way we expand the applicability of method (2).

The rest of the chapter is organized as follows. The local convergence of method (2) is given in Section 2, whereas the numerical examples are given in Section 3. Some comments are given in the concluding section 4.

2. Local Convergence Analysis

In this section we present the local convergence analysis of method (2). Denote by $U(v, \rho), \bar{U}(v, \rho)$ the open and closed balls, respectively, in X of center v and radius $\rho > 0$.

Let $L_0 > 0, L > 0$ and $M > 0$ be given parameters. It is convenient for the local convergence analysis of method (2) that follows to define functions on the interval $[0, \frac{1}{L_0})$

Ball Convergence Theorem for a Fifth–Order Method in Banach Spaces

by

$$g(t) = \frac{Lt}{2(1 - L_0 t)},$$

$$g_1(t) = [1 + \frac{5M}{1 - L_0 t}]g(t),$$

$$g_2(t) = g_1(t) + \frac{M(16g(t) + g_1(t))}{5(1 - L_0 t)}$$

$$= \left[1 + \frac{5M}{1 - L_0 t} + \frac{M(16 + t)}{5(1 - L_0 t)}\right] g(t),$$

$$h_1(t) = (1 - L_0 t + 5M)Lt - 2(1 - L_0 t)^2,$$

$$h_2(t) = (5(1 - L_0 t + 5M) + M(16 + t))Lt - 10(1 - L_0 t)^2$$

and parameter

$$r = \frac{2}{2L_0 + L}, \tag{3}$$

We have by the choice of r that

$$0 \leq g(t) < 1 \text{ for each } t \in [0, r). \tag{4}$$

Using the definition of function h_1 we get that $h_1(0) = -2 < 0$ and $h_1(\frac{1}{L_0}) = \frac{5ML}{L_0} > 0$. If follows from the intermediate value theorem that function h_1 has zeros in the interval $(0, \frac{1}{L_0})$. Denote by r_1 the smallest such zero. We also have that

$$h_1(t) = 2(1 - L_0 t)^2[(1 + \frac{5M}{1 - L_0 t})g(t) - 1]$$

so

$$h_1(r) = 10M(1 - L_0 r) > 0 \tag{5}$$

since $g(r) - 1 = 0$ and $r < \frac{1}{L_0}$. Then, we have by (3), (4) and (5) that

$$0 < r_1 < r, \tag{6}$$

and

$$0 \leq g_1(t) < 1 \text{ for each } t \in [0, r_1), \tag{7}$$

Similarly, we have that $h_2(0) = -10 < 0$ and $h_2(\frac{1}{L_0}) = (41 + \frac{1}{L_0})\frac{ML}{L_0} > 0$. Then, function h_2 has zeros in the interval $(0, \frac{1}{L_0})$. Denote by r_2 the smallest such zero. Then, again we have that

$$h_2(r_1) > 0 \tag{8}$$

since $g_1(r_1) - 1 = 0$, $r_1 < \frac{1}{L_0}$ and $L_0 r_1 < 1$. Then, we have that

$$r_2 < r_1 < r, \tag{9}$$

$$0 \leq g(t) < 1, \tag{10}$$

$$0 \leq g_1(t) < 1, \tag{11}$$

and

$$0 \leq g_2(t) < 1 \tag{12}$$

for each $t \in [0, r_2)$.

Next, we present the local convergence analysis of method (2) using the above notation.

Theorem 1. *Let $F : D \subseteq X \to Y$ be a Fréchet–differentiable operator. Suppose that there exist $x^* \in D$, parameters $L_0 > 0, L > 0$ and $M > 0$ such that for each $x, y \in D$*

$$F(x^*) = 0, \quad F'(x^*)^{-1} \in L(Y, X), \tag{13}$$

$$\|F'(x^*)^{-1}(F'(x) - F'(x^*))\| \leq L_0 \|x - x^*\|, \tag{14}$$

$$\|F'(x^*)^{-1}(F'(x) - F'(y))\| \leq L \|x - y\|, \tag{15}$$

$$\|F'(x^*)^{-1} F'(x)\| \leq M \tag{16}$$

and

$$\bar{U}(x^*, r_2) \subseteq D, \tag{17}$$

where r_2 is defined above Theorem 1. Then, sequence $\{x_n\}$ generated by method (2) for $x_0 \in U(x^*, r_2)$ is well defined, remains in $U(x^*, r_2)$ for each $n = 0, 1, 2, \ldots$ and converges to x^*. Moreover, the following estimates hold for each $n = 0, 1, 2, \ldots$,

$$\|y_n - x^*\| \leq g(\|x_n - x^*\|)\|x_n - x^*\| < \|x_n - x^*\| < r_2, \tag{18}$$

$$\|z_n - x^*\| \leq g_1(\|x_n - x^*\|)\|x_n - x^*\| < \|x_n - x^*\|, \tag{19}$$

and

$$\|x_{n+1} - x^*\| \leq g_2(\|x_n - x^*\|)\|x_n - x^*\| < \|x_n - x^*\| \tag{20}$$

where the "g" functions are defined above Theorem 1. Furthermore, suppose that there exists $R \in [r_2, \frac{2}{L_0})$ such that $\bar{U}(x^*, R) \subset D$, then the limit point x^* is the only solution of equation $F(x) = 0$ in $\bar{U}(x^*, R)$.

Proof. Using (15), the definition of r_2 and the hypothesis $x_0 \in U(x^*, r_2)$, we have that

$$\|F'(x^*)^{-1}(F'(x_0) - F'(x^*))\| \leq L_0 \|x_0 - x^*\| < L_0 r_2 < 1. \tag{21}$$

It follows from (21) and the Banach Lemma on invertible operators [4,5] that $F'(x_0)^{-1} \in L(Y, X)$ and

$$\|F'(x_0)^{-1} F'(x^*)\| \leq \frac{1}{1 - L_0 \|x_0 - x^*\|} < \frac{1}{1 - L_0 r_2}. \tag{22}$$

Hence, y_0 is well defined by method (2). Using the first substep in method (2) for $n = 0$, we get that

$$\begin{aligned} y_0 - x^* &= x_0 - x^* - F'(x_0)^{-1} F(x_0) \\ &= -F'(x_0)^{-1} F'(x^*) \int_0^1 F'(x^*)^{-1} \\ &\quad \times [F'(x^* + t(x_0 - x^*)) - F'(x_0)](x_0 - x^*) dt. \end{aligned} \tag{23}$$

It follows from (15), (21) and (23) that

$$\begin{aligned}
\|x_0 - x^* - F'(x_0)^{-1}F(x_0)\| &\leq \|F'(x_0)^{-1}F'(x^*)\| \\
&\quad \left\| \int_0^1 [F'(x^* + t(x_0 - x^*)) - F'(x_0)](x_0 - x^*)dt \right\| \\
&\leq \frac{L\|x_0 - x^*\|^2}{2(1 - L_0\|x_0 - x^*\|)} \\
&\leq \frac{Lr_2}{2(1 - L_0 r_2)}\|x_0 - x^*\| < \|x_0 - x^*\| < r_2, \quad (24)
\end{aligned}$$

which shows (18) for $n = 0$. Using (16) we have that

$$F(x_0) = F(x_0) - F(x^*) = \int_0^1 F'(x^* + \theta(x_0 - x^*))(x_0 - x^*)d\theta$$

so,

$$\|F'(x^*)^{-1}F(x_0)\| \leq M\|x_0 - x^*\|, \quad (25)$$

since $\|x^* - (x^* + \theta(x_0 - x^*))\| = |\theta|\|x_0 - x^*\| < r_2$, i.e., $x^* + \theta(x_0 - x^*) \in U(x^*, r_2)$ for each $\theta \in [0, 1]$. Using (14) and the definition of r_2, we get in turn that

$$\begin{aligned}
\|(2F'(x^*)^{-1})(F'(x_0) + F'(y_0) - 2F'(x^*))\| &\leq \frac{1}{2}(\|F'(x^*)^{-1}(F'(x_0) - F'(x^*))\| \\
&\quad + \|F'(x^*)^{-1}(F'(y_0) - F'(x^*))\|) \\
&\leq \frac{L_0}{2}(\|x_0 - x^*\| + \|y_0 - x^*\|) \\
&< \frac{L_0}{2}(\|x_0 - x^*\| + \|x_0 - x^*\|) \\
&= L_0 r_2 < 1. \quad (26)
\end{aligned}$$

It follows from (26) and Banach lemma on invertible operators that $(F'(x_0) + F'(y_0))^{-1} \in L(Y, X)$ and

$$\begin{aligned}
\|(2(F'(x_0) + F'(y_0)))^{-1}F'(x^*)\| &= \|(F'(x_0) + F'(y_0))^{-1}(2F'(x^*))\| \\
&\leq \frac{1}{1 - \frac{L_0}{2}(\|x_0 - x^*\| + \|y_0 - x^*\|)} \\
&\leq \frac{1}{1 - \frac{L_0}{2}(1 + g(\|x_0 - x^*\|))\|x_0 - x^*\|} \\
&\leq \frac{1}{1 - \frac{L_0}{2}(1 + g(r_2))r_2}. \quad (27)
\end{aligned}$$

It also follows that z_0 is well defined by the second step of method (2) for $n = 0$. Then, we have from the second step of method (2), (9)–(11), (22), (24) and (25) that

$$\begin{aligned}
\|z_0 - x^*\| &\leq \|y_0 - x^*\| + 5\|F'(x_0)^{-1}F'(x^*)\|(\|F'(x^*)^{-1}F(y_0)\|) \\
&\leq \|y_0 - x^*\| + \frac{5M\|y_0 - x^*\|}{1 - L_0\|x_0 - x^*\|} \\
&= [1 + \frac{5M}{1 - L_0\|x_0 - x^*\|}]\|y_0 - x^*\| \\
&\leq (1 + \frac{5M}{1 - L_0\|x_0 - x^*\|})g(\|x_0 - x^*\|)\|x_0 - x^*\| \\
&= g_1(\|x_0 - x^*\|)\|x_0 - x^*\| \\
&< \|x_0 - x^*\| < r_2,
\end{aligned} \quad (28)$$

which shows (19) for $n = 0$. We also have by the third step of method (2) for $n = 0$ and (28) that x_1 is well defined. Then, using method (2) for $n = 0$, (10), (13), (23), (26) (for x_0 replaced by z_0), (27) and (28) we obtain in turn that

$$\begin{aligned}
\|x_1 - x^*\| &\leq \|z_0 - x^*\| \\
&\quad + \frac{1}{5}\|F'(x_0)^{-1}F'(x^*)\|(16\|F'(x^*)^{-1}F'(y_0)\|) \\
&\quad + \|F'(x^*)^{-1}F'(z_0)\| \\
&\leq \|z_0 - x^*\| + \frac{M(16\|y_0 - x^*\| + \|z_0 - x^*\|)}{5(1 - L_0\|x_0 - x^*\|)} \\
&\leq g(\|x_0 - x^*\|)\|x_0 - x^*\| + \frac{M(16g(\|x_0 - x^*\|) + g_1(\|x_0 - x^*\|))\|x_0 - x^*\|}{5(1 - L_0\|x_0 - x^*\|)} \\
&= g_2(\|x_0 - x^*\|)\|x_0 - x^*\| < \|x_0 - x^*\| < r_2,
\end{aligned} \quad (29)$$

which shows (21) for $n = 0$. By simply replacing x_0, y_0, z_0, x_1 by x_k, y_k, z_k, x_{k+1} in the preceding estimates we arrive at estimate (18)–(20). Using the estimate $\|x_{k+1} - x^*\| < \|x_k - x^*\| < r_2$ we deduce that $x_{k+1} \in U(x^*, r_2)$ and $\lim_{k \to \infty} x_k = x^*$.

Finally, to show the uniqueness part, let $T = \int_0^1 F'(y^* + t(x^* - y^*))dt$ for some $y^* \in \bar{U}(x^*, R)$ with $F(y^*) = 0$. Using (14) and the estimate

$$\begin{aligned}
\|F'(x^*)^{-1}(T - F'(x^*))\| &\leq \int_0^1 L_0\|y^* + t(x^* - y^*) - x^*\|dt \\
&\leq \int_0^1 (1 - t)\|x^* - y^*\|dt \leq \frac{L_0}{2}R < 1,
\end{aligned}$$

it follows that T^{-1} exists. Then, from the identity $0 = F(x^*) - F(y^*) = T(x^* - y^*)$, we deduce that $x^* = y^*$. $\square$

Remark 2. 1. *In view of (14) and the estimate*

$$\begin{aligned}
\|F'(x^*)^{-1}F'(x)\| &= \|F'(x^*)^{-1}(F'(x) - F'(x^*)) + I\| \\
&\leq 1 + \|F'(x^*)^{-1}(F'(x) - F'(x^*))\| \leq 1 + L_0\|x - x^*\|
\end{aligned}$$

condition (16) can be dropped and M can be replaced by

$$M(t) = 1 + L_0 t.$$

2. The results obtained here can be used for operators F satisfying autonomous differential equations [4] of the form

$$F'(x) = P(F(x))$$

where P is a continuous operator. Then, since $F'(x^*) = P(F(x^*)) = P(0)$, we can apply the results without actually knowing x^*. For example, let $F(x) = e^x - 1$. Then, we can choose: $P(x) = x + 1$.

3. The local results obtained here can be used for projection methods such as the Arnoldi's method, the generalized minimum residual method (GMRES), the generalized conjugate method(GCR) for combined Newton/finite projection methods and in connection to the mesh independence principle can be used to develop the cheapest and most efficient mesh refinement strategies [4, 5].

4. The radius r given by (3) was shown by us to be the convergence radius of Newton's method [4, 5]

$$x_{n+1} = x_n - F'(x_n)^{-1} F(x_n) \text{ for each } n = 0, 1, 2, \ldots \tag{30}$$

under the conditions (14) and (15). It follows from (3) and (9) that the convergence radius r_2 of the method (2) cannot be larger than the convergence radius r of the second order Newton's method (30). As already noted in [4, 5] r is at least as large as the convergence ball given by Rheinboldt [22]

$$r_R = \frac{2}{3L}. \tag{31}$$

In particular, for $L_0 < L$ we have that

$$r_R < r$$

and

$$\frac{r_R}{r} \to \frac{1}{3} \text{ as } \frac{L_0}{L} \to 0.$$

That is our convergence ball r is at most three times larger than Rheinboldt's. The same value for r_R was given by Traub [23].

5. It is worth noticing that method (2) is not changing when we use the conditions of Theorem 1 instead of the stronger conditions used in [1]. Moreover, we can compute the computational order of convergence (COC) defined by

$$\xi = \ln\left(\frac{\|x_{n+1} - x^*\|}{\|x_n - x^*\|}\right) / \ln\left(\frac{\|x_n - x^*\|}{\|x_{n-1} - x^*\|}\right)$$

or the approximate computational order of convergence

$$\xi_1 = \ln\left(\frac{\|x_{n+1} - x_n\|}{\|x_n - x_{n-1}\|}\right) / \ln\left(\frac{\|x_n - x_{n-1}\|}{\|x_{n-1} - x_{n-2}\|}\right).$$

This way we obtain in practice the order of convergence in a way that avoids the bounds involving estimates using estimates higher than the first Fréchet derivative of operator F.

3. NUMERICAL EXAMPLES

We present two numerical examples in this section.

Example 3. *Let $X = Y = \mathbb{R}^3, D = \bar{U}(0,1)$. Define F on D for $v = (x,y,z)$ by*

$$F(v) = (e^x - 1, \frac{e^x - 1}{2}y^2 + y, z)). \tag{32}$$

Then, the Fréchet derivative is given by

$$F'(v) = \begin{bmatrix} e^x & 0 & 0 \\ 0 & (e-1)y + 1 & 0 \\ 0 & 0 & 1 \end{bmatrix}.$$

Notice that $x^* = (0,0,0)$, $F'(x^*) = F'(x^*)^{-1} = diag\{1,1,1\}$, $L_0 = e - 1 < L = e$, $M = e$. Then, we have

$$r_2 = 0.0286 < r_1 = 0.0622 < r = 0.3249.$$

Example 4. *Returning back to the motivational example at the introduction of this study, we have $L_0 = L = N = 146.6629073$, $M = 101.5578008$. Then we have*

$$r_2 = 1.6277e - 005 < r_1 = 2.7015e - 005 < r = 0045.$$

REFERENCES

[1] Arroyo, V., Cordero, A. and Torregrosa, J. R. 2011. "Approximation of artificial intelligence satellites preliminary orbits: The efficiency challenge". *Math. Comput. Modelling.* 54: 1802–1807.

[2] Amat, S., Busquier, S. and Gutiérrez, J. M. 2003. "Geometric construction of iterative functions to solve nonlinear equations". *J. Comput. Appl. Math.* 157: 197–205.

[3] Amat, S., Hernández, M. A. and Romero, N. 2008. "A modified Chebyshev's iterative method with at least sixth order of convergence". *Appl. Math. Comput.* 206:(1): 164–174.

[4] Argyros, I. K. 2008. *Convergence and Application of Newton–type Iterations*. Springer.

[5] Argyros, I. K. and Hilout, S. 2013. *Computational methods in nonlinear Analysis*. New Jersey, USA: World Scientific Publ. Co.

[6] Candela, V. and Marquina, A. 1990. "Recurrence relations for rational cubic methods I: The Halley method". *Computing* 44: 169–184.

[7] Cordero, A. and Torregrosa, J. R. 2007. "Variants of Newton's method using fifth order quadrature formulas". *Appl. Math. Comput.* 190: 686–698.

[8] Chun, C. 1990. "Some improvements of Jarratt's method with sixth–order convergence". *Appl. Math. Comput.* 190(2): 1432–1437.

[9] Ezquerro, J. A. and Hernández, M. A. 2003. "A uniparametric Halley–type iteration with free second derivative". *Int. J. Pure and Appl. Math.* 6(1): 99–110.

[10] Ezquerro, J. A. and Hernández, M. A. 2009. "New iterations of R–order four with reduced computational cost". *BIT Numer. Math.* 49: 325–342.

[11] Ezquerro, J. A. and Hernández, M. A. 2005. "On the R–order of the Halley method". *J. Math. Anal. Appl.* 303: 591–601.

[12] Gutiérrez, J. M. and Hernández, M. A. 1998. "Recurrence relations for the super–Halley method". *Computers Math. Applic.* 36(7): 1–8.

[13] Ganesh, M. and Joshi, M. C. 1991. "Numerical solvability of Hammerstein integral equations of mixed type". *IMA J. Numer. Anal.* 11: 21–31.

[14] Hernández, M. A. 2001. "Chebyshev's approximation algorithms and applications". *Computers Math. Applic.* 41(3–4): 433–455.

[15] Hernández, M. A. and Salanova, M. A. 1999. "Sufficient conditions for semilocal convergence of a fourth order multipoint iterative method for solving equations in Banach spaces". *Southwest J. Pure Appl. Math.* 1: 29–40.

[16] Jarratt, P. 1966. "Some fourth order multipoint methods for solving equations". *Math. Comput.* 20(95): 434–437.

[17] Kou, J. and Li, Y. 2007. "An improvement of the Jarratt method". *Appl. Math. Comput.* 189: 1816–1821.

[18] Parhi, S. K. and Gupta, D. K. 2010. "Semilocal convergence of a Stirling–like method in Banach spaces". *Int. J. Comput. Methods.* 7(2): 215–228.

[19] Parhi, S. K. and Gupta, D. K. 2007. "Recurrence relations for a Newton–like method in Banach spaces". *J. Comput. Appl. Math.* 206(2): 873–887.

[20] Rall, L. B. 1979. *Computational solution of nonlinear operator equations*. New York: Robert E. Krieger.

[21] Ren, H., Wu, Q. and Bi, W. 2009. "New variants of Jarratt method with sixth–order convergence". *Numer. Algorithms* 52(4): 585–603.

[22] Rheinboldt, W. C. 1978. An adaptive continuation process for solving systems of non-linear equations, In: *Mathematical models and numerical methods* (A.N.Tikhonov et al. eds.) pub.3, (19), 129–142. Warsaw Poland: Banach Center.

[23] Traub, J. F. 1964. *Iterative methods for the solution of equations.* New Jersey USA: Prentice Hall Englewood Cliffs.

In: Understanding Banach Spaces
Editor: Daniel González Sánchez

ISBN: 978-1-53616-745-0
© 2020 Nova Science Publishers, Inc.

Chapter 11

EXTENDED CONVERGENCE OF KING–WERNER–LIKE METHODS WITHOUT DERIVATIVES

Ioannis K. Argyros[1,*] *and Santhosh George*[2,†]
[1]Department of Mathematical Sciences
Cameron University, Lawton, OK, US
[2]Department of Mathematical and Computational Sciences
National Institute of Technology, Karnataka, India

Abstract

We provide a semilocal as well as a local convergence analysis of some efficient King–Werner–like methods of order $1+\sqrt{2}$ free of derivatives for Banach space valued operators. We use our new idea of the restricted convergence region to find a smaller subset than before containing the iterates. Consequently the resulting Lipschitz parameters are smaller than in earlier works. Hence, to a finer convergence analysis is obtained. The extensions involve no new constants, since the new ones specialize to the ones in previous works. Examples are used to test the convergence criteria.

Keywords: semilocal and local convergence analysis, Werner's method, Fréchet–derivative, King's method, Banach space

AMS Subject Classification: 65H10, 65G99, 49M15, 65J15

1. INTRODUCTION

Denote by $\mathbb{E}_1, \mathbb{E}_2$ Banach spaces and let $\Omega \subset \mathbb{E}_1$ be convex. Set $U(w, \mu) = \{x \in \mathbb{E}_1 : \|x - w\| < \mu, \mu > 0\}$ and denote by $\bar{U}(w, \mu)$ the closure of $U(w, \mu)$. Consider $\mathcal{H} : \Omega \subset \mathbb{E}_1 \longrightarrow \mathbb{E}_2$ to be differentiable in the Fréchet sense. Mathematical modeling can be used to solve problems in diverse disciplines. This approach requires solving

$$\mathcal{H}(x) = 0, \qquad (1)$$

---

[*]Corresponding Author's E-mail: iargyros@cameron.edu.
[†]E-mail: sgeorge@nitk.ac.in.

but this is a difficult task in general. To find a solution x_* in closed form is hard, and can be achieved in some cases. Hence, authors resort to iterative procedures generating sequences converging to x_*. Newton's method is a widely used quadratically convergent method that is widely used. Third convergent order methods have been used such as Chebyshev or Halley but these involve the expensive $\mathcal{H}''$ at each step, which are useful when $\mathcal{H}''$ is constant or diagonal with blocks, and not expensive or in some problems that are stiff [1]. That explains why authors propose alternative procedures free of first or second derivative.

Werner in [2, 3] considered King's method [4] as:

For $x_0, y_0 \in \Omega$, let

$$\begin{aligned} x_{n+1} &= x_n - \mathcal{H}'(\tfrac{x_n+y_n}{2})^{-1}\mathcal{H}(x_n), \\ y_{n+1} &= x_{n+1} - \mathcal{H}'(\tfrac{x_n+y_n}{2})^{-1}\mathcal{H}(x_{n+1}) \end{aligned} \qquad (2)$$

for each $n = 0, 1, 2, \ldots$ and $\mathbb{E}_1 = \mathbb{R}^i, \mathbb{E}_2 = \mathbb{R}$ using

(H_0) $x_* \in \Omega$ with $\mathcal{H}(x_*) = 0$;
(H_1) $\mathcal{H} \in C^{2,a}(\Omega), a \in (0, 1]$;
(H_2) $\mathcal{H}'(x)^{-1} \in L(\mathbb{E}_2, \mathbb{E}_1)$ and $\|\mathcal{H}'(x)^{-1}\| \leq \Gamma$;
(H_3)

$$\|\mathcal{H}'(y) - \mathcal{H}'(x)\| \leq L_1 \|y - x\|$$

for each $x, y \in \Omega$;

(H_4)

$$\|\mathcal{H}''(y) - \mathcal{H}''(x)\| \leq L_{2,a} \|y - x\|^a$$

for each $x, y \in \Omega$;

(H_5) $U(x_*, b) \subseteq \Omega$ with

$$b = \min\{b_0, \rho_0, \rho_1\},$$

$$b_0 = \frac{2}{L_1 \Gamma}$$

and ρ_0, ρ_1 satisfy [5, p. 337]

$$\begin{aligned} Bv^{1+a} + Aw &= 1 \\ 2Av^2 + Avw &= w, \\ A = \tfrac{1}{2}\Gamma L_1, \quad B &= \tfrac{\Gamma L_{2,a}}{4(a+1)(a+2)}. \end{aligned}$$

The convergence order is $1 + \sqrt{2}$. But some assumptions may not hold. Consider function $\mathcal{H} : [-1, 1] \to (-\infty, \infty)$ as

$$\mathcal{H}(x) = x^2 \ln x^2 + c_1 x^2 + c_2 x + c_3, \quad f(0) = c_3,$$

where c_1, c_2, c_3 are constants. Notice that $\lim_{x \to 0} x^2 \ln x^2 = 0$, $\lim_{x \to 0} x \ln x^2 = 0$, $\mathcal{H}'(x) = 2x \ln x^2 + 2(c_1 + 1)x + c_2$ and $\mathcal{H}''(x) = 2(\ln x^2 + 3 + c_1)$, violating (H_4).

Later, for $\mathbb{E}_1 = \mathbb{E}_2 = \mathbb{R}$ McDougall et al. in [6] proposed:

$$\begin{aligned} y_0 &= x_0 \\ x_1 &= x_0 - \mathcal{H}'(\tfrac{x_0+y_0}{2})^{-1}\mathcal{H}(x_0), \\ y_n &= x_n - \mathcal{H}'(\tfrac{x_{n-1}+y_{n-1}}{2})^{-1}\mathcal{H}(x_n), \\ x_{n+1} &= x_n - \mathcal{H}'(\tfrac{x_n+y_n}{2})^{-1}\mathcal{H}(x_n) \end{aligned} \tag{3}$$

for $x_0 \in \Omega$. Method (3) is of order $1 + \sqrt{2}$, and allows the best possible value of the derivative to be utilized in the last substep.

In this chapter semilocal convergence analysis of method

$$\begin{aligned} x_{n+1} &= x_n - A_n^{-1}\mathcal{H}(x_n) \\ y_{n+1} &= x_{n+1} - A_n^{-1}\mathcal{H}(x_{n+1}), \end{aligned} \tag{4}$$

is studied, where $A_n = [x_n, y_n; \mathcal{H}]$ and $[x, y; \mathcal{H}]$ is divided difference of order one [6] with

$$[x, y; \mathcal{H}](y - x) = \mathcal{H}(y) - \mathcal{H}(x) \quad for\ each\ x, y \in \Omega\ with\ y \neq x, \tag{5}$$

and $\mathcal{H}'(x) = [x, x; \mathcal{H}]$ for each $x \in \Omega$ if $\mathcal{H}$ is differentiable on Ω.

Method (4) is a also of order $1 + \sqrt{2}$ but has the advantage of avoiding the derivatives.

The analysis is presented in Section 2 and Section 3, respectively. The results are tested using numerical examples and favor our results.

2. SEMILOCAL CONVERGENCE

The convergence uses:

Lemma 1. *[7] Let $L_0 > 0$, $L > 0$, $s_0 \geq 0$, $t_1 \geq 0$. Let α stand for the unique zero in $(0, 1)$ of cubic polynomial p given as*

$$p(t) = L_0 t^3 + L_0 t^2 + 2Lt - 2L. \tag{6}$$

Assume that

$$0 < \frac{L(t_1 + s_0)}{1 - L_0(t_1 + s_1 + s_0)} \leq \alpha \leq 1 - \frac{2L_0 t_1}{1 - L_0 s_0}, \tag{7}$$

where

$$s_1 = t_1 + L(t_1 + s_0)t_1. \tag{8}$$

Then, sequence $\{t_n\}$

$$\begin{aligned} t_0 &= 0,\ s_{n+1} = t_{n+1} + \frac{L(t_{n+1}-t_n+s_n-t_n)(t_{n+1}-t_n)}{1-L_0(t_n-t_0+s_n+s_0)}, &\text{for each } n = 1, 2, \ldots, \\ t_{n+2} &= t_{n+1} + \frac{L(t_{n+1}-t_n+s_n-t_n)(t_{n+1}-t_n)}{1-L_0(t_{n+1}-t_0+s_{n+1}+s_0)}, &\text{for each } n = 0, 1, 2, \ldots \end{aligned} \tag{9}$$

increasing converges to some $t^\star$ so that

$$t_1 \leq t^\star \leq t^{\star\star}, \tag{10}$$

$$t^{\star\star} = \frac{t_1}{1 - \alpha}. \tag{11}$$

so that
$$s_n - t_n \leq \alpha(t_n - t_{n-1}) \leq \alpha^n(t_1 - t_0), \tag{12}$$
$$t_{n+1} - t_n \leq \alpha(t_n - t_{n-1}) \leq \alpha^n(t_1 - t_0) \tag{13}$$
and
$$t_n \leq s_n \tag{14}$$
for each $n = 1, 2, \ldots$

Theorem 2. *[7] Let $\mathcal{H} : \Omega \subset \mathbb{E}_1 \to \mathbb{E}_2$ and $[.,.,;\mathcal{H}]$ a divided difference for operator $\mathcal{H}$ on $\Omega \times \Omega$. Assume hypotheses of Lemma 2.1 hold, and there exist $x_0, y_0 \in \Omega$, $L_0 > 0$, $L > 0$, $s_0 \geq 0$, $t_1 \geq 0$, so for $x, y \in \Omega$*
$$A_0^{-1} \in L(\mathbb{E}_2, \mathbb{E}_1) \tag{15}$$
$$\|A_0^{-1}\mathcal{H}(x_0)\| \leq t_1, \tag{16}$$
$$\|x_0 - y_0\| \leq s_0, \tag{17}$$
Set $\Omega_0 = \Omega \cap U(x_0, \frac{1}{2L_0})$. For each $x, y, z, v \in \Omega_0$
$$\|A_0^{-1}([x, y; \mathcal{H}] - A_0)\| \leq L_0(\|x - x_0\| + \|y - y_0\|), \tag{18}$$
$$\|A_0^{-1}([x, y; \mathcal{H}] - [z, v; \mathcal{H}])\| \leq L(\|x - z\| + \|y - v\|), \tag{19}$$
and
$$\overline{U}(x_0, t^\star) \subseteq \Omega. \tag{20}$$
Then, $\{x_n\} \subset \overline{U}(x_0, t^\star)$ and converges to a unique solution $x_* \in \overline{U}(x_0, t^\star)$, $\lim_{n \to \infty} x_n = x_* \in \overline{U}(x_0, t^\star)$, $\mathcal{H}(x_*) = 0$, so that
$$\|x_n - x_*\| \leq t^\star - t_n. \tag{21}$$

Furthermore, assume
$$L_0(t^\star + R - s_0) < 1, \tag{22}$$
for $R > t^*$ then, x_* is unique in $\Omega_1 = \Omega \cap \bar{U}(x_0, R)$.

Proof. The proof is similar to the ones in [7] but Ω_0 replaces Ω. □

Remark 3. *(a) The limit point $t^\star$ can be switched by $t^{\star\star}$ provided in closed form by (11) in Theorem 2.*
(b) By (19) and (18)
$$L_0 \leq L$$
and $\frac{L}{L_0}$ can be arbitrarily large [9]. Next, define the tighter (than $\{t_n\}$) majorizing sequence $\{\bar{t}_n\}$ (for $\{x_n\}$) as
$$\begin{aligned}&\bar{t}_0 = 0, \bar{t}_1 = t_1, \bar{s}_0 = s_0, \bar{s}_1 = \bar{t}_1 + L_0(\bar{t}_1 + \bar{s}_0)\bar{t}_1, \\ &\bar{s}_{n+1} = \bar{t}_{n+1} + \frac{L(\bar{t}_{n+1} - \bar{t}_n + \bar{s}_n - \bar{t}_n)(\bar{t}_{n+1} - \bar{t}_n)}{1 - L_0(\bar{t}_n - \bar{t}_0 + \bar{s}_n + \bar{s}_0)} \quad \text{for each } n = 1, 2, \ldots\end{aligned} \tag{23}$$

and
$$\bar{t}_{n+2} = \bar{t}_{n+1} + \frac{L(\bar{t}_{n+1}-\bar{t}_n+\bar{s}_n-\bar{t}_n)(\bar{t}_{n+1}-\bar{t}_n)}{1-L_0(\bar{t}_{n+1}-\bar{t}_0+\bar{s}_{n+1}+\bar{s}_0)} \quad for\ each\ n = 0, 1, \ldots$$

Then, by induction
$$\bar{t}_n \leq t_n, \tag{24}$$
$$\bar{s}_n \leq s_n, \tag{25}$$
$$\bar{t}_{n+1} - \bar{t}_n \leq t_{n+1} - t_n, \tag{26}$$
$$\bar{s}_n - \bar{t}_n \leq s_n - t_n \tag{27}$$

and
$$\bar{t}^\star = \lim_{n \to \infty} \bar{t}_n \leq t^\star.$$

If $L_0 < L$, (24)–(27) are strict for $n \geq 2$, $n \geq 1$, $n \geq 1$, $n \geq 1$, respectively. Clearly, sequence $\{\bar{t}_n\}$ increasing converges to $\bar{t}^\star$ under the hypotheses of Lemma 1 and can replace $\{t_n\}$ as a majorizing sequence for $\{x_n\}$ in Theorem 2.

(c) In [7] the stronger than (18) condition for each $x, y, v, z \in \Omega$
$$\|A_0^{-1}([x,y;\mathcal{H}] - [z,v;\mathcal{H}])\| \leq N(\|x-z\| + \|y-v\|)$$

was used. But
$$L \leq N$$
and
$$L_0 \leq N,$$
since $\Omega_0 \subseteq \Omega$. Therefore, we have extended the applicability of method, since the old sufficient convergence criterion
$$0 < \frac{N(t_1+s_0)}{1-L_0(t_1+s_1=s_0)} \leq \bar{\alpha} \leq 1 - \frac{2L_0 t_1}{1-L_0 s_0} \tag{28}$$

implies (7) but not necessarily vice versa, where $\bar{\alpha}$ is the only root in $(0,1)$ of
$$\bar{p}(t) = L_0 t^3 + L_0 t^2 + 2Nt - 2N. \tag{29}$$

Clearly the new majorizing sequences are tighter than the previous ones too.

3. LOCAL CONVERGENCE

Theorem 4. Let $\mathcal{H}, [.,.;\mathcal{H}]$ be as in Theorem 2. Assume there exist $x_* \in \Omega$, $l_0 > 0$ and $l > 0$ such that for each $x, y \in \Omega$
$$\mathcal{H}(x_*) = 0, \ \mathcal{H}'(x_*)^{-1} \in L(\mathbb{E}_2, \mathbb{E}_1), \tag{30}$$

Set $\Omega_0 = \Omega \cap U(x_*, \frac{1}{2l_0})$, for each $x, y, z, v \in \Omega_0$
$$\|\mathcal{H}'(x_*)^{-1}([x,y;\mathcal{H}] - \mathcal{H}'(x_*))\| \leq l_0(\|x-x_*\| + \|y-x_*\|) \tag{31}$$

$$\|\mathcal{H}'(x_*)^{-1}([x, y; \mathcal{H}] - [z, u; \mathcal{H}])\| \leq l(\|x - z\| + \|y - u\|), \tag{32}$$

and

$$\overline{U}(x_*, \rho) \subseteq \Omega, \tag{33}$$

where

$$\rho = \frac{1}{(1 + \sqrt{2})l + 2l_0}. \tag{34}$$

Then, sequence $\{x_n\} \subset \overline{U}(x_*, \rho)$, $\lim_{n \to \infty} x_n = x_*$, the order is $1 + \sqrt{2}$ at least, if $x_0, y_0 \in U(x_*, \rho)$, and the following items hold

$$\|x_{n+2} - x_*\| \leq \frac{\sqrt{2} - 1}{\rho^2} \|x_{n+1} - x_*\|^2 \|x_n - x_*\| \tag{35}$$

and

$$\|x_n - x_*\| \leq \left(\frac{\sqrt{\sqrt{2} - 1}}{\rho}\right)^{F_n - 1} \|x_1 - x_*\|^{F_n} \tag{36}$$

where F_n is Fibonacci sequence given as $F_1 = F_2 = 1$ and $F_{n+2} = 2F_{n+1} + F_n$.

The proof is similar to the one in [7] but Ω_0 replaces Ω.

Proof. Using the hypotheses that $x_0, y_0 \in U(x_*, \rho)$, (31) and (34) we have

$$\begin{aligned} \|\mathcal{H}'(x_*)^{-1}(\mathcal{H}'(x_*) - A_0)\| &= \|\mathcal{H}'(x_*)^{-1}(\mathcal{H}'(x_*) - [x_0, y_0; \mathcal{H}])\| \\ &\leq l_0(\|x_* - x_0\| + \|x_* - y_0\|) \\ &< 2l_0\rho = \frac{2l_0}{(1+\sqrt{2})l + 2l_0} < 1. \end{aligned} \tag{37}$$

By (37), and the Banach lemma on invertible operators [8] $A_0^{-1} \in L(\mathbb{E}_2, \mathbb{E}_1)$ and

$$\|A_0^{-1} \mathcal{H}'(x_*)\| \leq \frac{1}{1 - l_0(\|x_* - x_0\| + \|x_* - y_0\|)} < \frac{1}{1 - 2l_0\rho} = \frac{(1 + \sqrt{2})l + 2l_0}{(1 + \sqrt{2})l}. \tag{38}$$

By the first substep of (4) and (30),

$$\begin{aligned} x_1 - x_* &= x_0 - x_* - A_0^{-1}\mathcal{H}(x_0) = A_0^{-1}[A_0(x_0 - x_*) - (\mathcal{H}(x_0) - \mathcal{H}(x_*))] \\ &= -A_0^{-1}\mathcal{H}'(x_*)\mathcal{H}'(x_*)^{-1}([x_0, x_*; \mathcal{H}] - [x_0, y_0; \mathcal{H}])(x_0 - x_*). \end{aligned} \tag{39}$$

Next, by (32), (34), (38) and (39)

$$\begin{aligned} \|x_1 - x_*\| &\leq \frac{l\|y_0 - x_*\|}{1 - l_0(\|y_0 - x_*\| + \|x_0 - x_*\|)} \|x_0 - x_*\| \\ &\leq \frac{l\rho}{1 - 2l_0\rho} \|x_0 - x_*\| = (\sqrt{2} - 1)\|x_0 - x_*\| < \rho, \end{aligned} \tag{40}$$

so, $x_1 \in U(x_*, \rho)$.

Then, by (4), (30)

$$\begin{aligned} y_1 - x_* &= x_1 - x_* - A_0^{-1}\mathcal{H}(x_1) \\ &= A_0^{-1}\mathcal{H}'(x_*)\mathcal{H}'(x_*)^{-1}([x_0, y_0; \mathcal{H}] - [x_1, x_*; \mathcal{H}])(x_1 - x_*). \end{aligned} \tag{41}$$

Using (32), (34), (38), (40) and (41) that

$$
\begin{aligned}
\|y_1 - x_*\| &\leq \|A_0^{-1}\mathcal{H}'(x_*)\|\|\mathcal{H}'(x_*)^{-1}([x_0,y_0;\mathcal{H}] - [x_1,x_*;\mathcal{H}])\|\|x_1 - x_*\| \\
&\leq \frac{l(\|x_1-x_0\|+\|y_0-x_*\|)}{1-l_0(\|x_*-x_0\|+\|x_*-y_0\|)}\|x_1 - x_*\| \\
&\leq \frac{l(\|x_1-x_*\|+\|x_0-x_*\|+\|y_0-x_*\|)}{1-l_0(\|x_*-x_0\|+\|x_*-y_0\|)}\|x_1 - x_*\| \\
&\leq \frac{l([\frac{l\rho}{1-2l_0\rho}+1]\|x_0-x_*\|+\|y_0-x_*\|)}{1-2l_0\rho}\|x_1 - x_*\| \\
&\leq \frac{l\rho(\frac{l\rho}{1-2l_0\rho}+2)}{1-2l_0\rho}\|x_1 - x_*\| \\
&= \|x_1 - x_*\| < \rho,
\end{aligned}
\tag{42}
$$

which shows $y_1 \in U(x_*,\rho)$. Switch x_0, y_0, x_1, y_1 and A_0 for $x_k, y_k, x_{k+1}, y_{k+1}$ and A_k in the estimations above to obtain as in (40) and (42)

$$
\begin{aligned}
\|x_{k+1} - x_*\| &\leq \frac{l\|y_k-x_*\|}{1-l_0(\|y_k-x_*\|+\|x_k-x_*\|)}\|x_k - x_*\| \\
&\leq \frac{l\rho}{1-2l_0\rho}\|x_k - x_*\| = (\sqrt{2}-1)\|x_k - x_*\| < \rho
\end{aligned}
\tag{43}
$$

and

$$
\begin{aligned}
\|y_{k+1} - x_*\| &\leq \frac{l(\|x_{k+1}-x_*\|+\|x_k-x_*\|+\|y_k-x_*\|)}{1-l_0(\|x_k-x_*\|+\|y_k-x_*\|)}\|x_{k+1} - x_*\| \\
&\leq \frac{l([\frac{l\rho}{1-2l_0\rho}+1]\|x_k-x_*\|+\|y_k-x_*\|)}{1-2l_0\rho}\|x_{k+1} - x_*\| \\
&\leq \frac{l\rho(\frac{l\rho}{1-2l_0\rho}+2)}{1-2l_0\rho}\|x_{k+1} - x_*\| \\
&= \|x_{k+1} - x_*\| < \rho
\end{aligned}
\tag{44}
$$

and by (43) and (44)

$$
\begin{aligned}
\|y_{k+2} - x_*\| &\leq \frac{l(\|x_{k+2}-x_*\|+\|x_{k+1}-x_*\|+\|y_{k+1}-x_*\|)}{1-l_0(\|x_{k+1}-x_*\|+\|y_{k+1}-x_*\|)}\|x_{k+2} - x_*\| \\
&\leq \frac{(\frac{l\rho}{1-2l_0\rho}+2)l\|x_{k+1}-x_*\|}{1-2l_0\rho}\|x_{k+2} - x_*\| = \frac{1}{\rho}\|x_{k+1} - x_*\|\|x_{k+2} - x_*\|
\end{aligned}
\tag{45}
$$

so

$$
\begin{aligned}
\|x_{k+3} - x_*\| &\leq \frac{l\|y_{k+2}-x_*\|}{1-l_0(\|y_{k+2}-x_*\|+\|x_{k+2}-x_*\|)}\|x_{k+2} - x_*\| \\
&\leq \frac{l}{1-2l_0\rho}\frac{1}{\rho}\|x_{k+2}-x_*\|^2\|x_{k+1}-x_*\| \\
&= \frac{\sqrt{2}-1}{\rho^2}\|x_{k+2}-x_*\|^2\|x_{k+1}-x_*\|.
\end{aligned}
\tag{46}
$$

holds for each $k = 0, 1, \ldots$, leading to (35).

Consider $\{u_k\}$ for $k = 0, 1, \ldots$ as

$$u_k = \frac{\sqrt{\sqrt{2}-1}}{\rho}\|x_k - x_*\| \tag{47}$$

Then, we have from (46) and (47) that

$$u_{k+3} \leq u_{k+2}^2 u_{k+1}. \tag{48}$$

holds for each $k = 0, 1, \ldots$.

By (48) we have

$$u_k \leq u_1^{F_k} \tag{49}$$

so
$$\|x_k - x_*\| \leq \left(\frac{\sqrt{\sqrt{2}-1}}{\rho}\right)^{F_k-1} \|x_1 - x_*\|^{F_k} \tag{50}$$

and (36) holds in [8] they got

$$F_k \geq (\sqrt{2}-1)^2(1+\sqrt{2})^k, \quad k \geq 1, \tag{51}$$

so
$$\|x_k - x_*\| \leq \left(\frac{\sqrt{\sqrt{2}-1}}{\rho}\right)^{(\sqrt{2}-1)^2(1+\sqrt{2})^k-1} \|x_1 - x_*\|^{(\sqrt{2}-1)^2(1+\sqrt{2})^k} \tag{52}$$

so $\lim_{k \to \infty} x_k = x_*$ with order of $1 + \sqrt{2}$ at least. $\square$

Remark 5. *(a) If $\mathbb{E}_1 = \mathbb{E}_2 = \mathbb{R}$, in [8] they provide a radius of convergence for (4) as*

$$\rho_\star = \frac{s^\star}{\mu}, \tag{53}$$

with $s^\star \approx 0.55279$ $\mu > 0$ an upper bound for $\|\mathcal{H}(x_*)^{-1}\mathcal{H}''(x)\|$ in view (32)

$$\|\mathcal{H}'(x_*)^{-1}(\mathcal{H}'(y) - \mathcal{H}'(x))\| \leq 2l\|y - x\| \quad \text{for any } x, y \in \Omega. \tag{54}$$

Let $l = \frac{\mu}{2}$ and $l_0 = l$ and use (34)

$$\rho = \frac{2}{(3+\sqrt{2})\mu} = \frac{2(3-\sqrt{2})}{5\mu} \approx \frac{0.63432}{\mu} > \frac{s^\star}{\mu} = \rho_\star. \tag{55}$$

(b) By (32), and (31), we have

$$l_0 \leq l \tag{56}$$

and $\frac{l_0}{l}$ can be arbitrarily small [1, 7, 9–11].

(c) In [7], we use instead of (31) the stronger condition for each $x, y, z, v \in \Omega$

$$\|\mathcal{H}'(x_*)^{-1}([x, y; \mathcal{H}] - [z, v; \mathcal{H}])\| \leq \ell_1(\|x - z\| + \|y - v\|). \tag{57}$$

Then, we have by (31) and (57) that

$$\ell_0 \leq \ell_1$$

and

$$\ell \leq \ell_1$$

since $\Omega_0 \subseteq \Omega$. Hence, new radius is at least as large as in [7]. Error bounds are also tighter.

4. NUMERICAL EXAMPLES

Example 6. For $\mathbb{E}_1 = \mathbb{E}_2 = \mathbb{R}$, $\Omega = (-1, 1)$ and $x_* = 0$ Consider $\mathcal{H}$ on Ω as

$$\mathcal{H}(x) = e^x - 1. \tag{58}$$

We obtain in turn

$$\begin{aligned}|\mathcal{H}'(x_*)^{-1}([x,y;\mathcal{H}] - [z,u;\mathcal{H}])| &= |\int_0^1 (\mathcal{H}'(\tau x + (1-\tau)y) - \mathcal{H}'(\tau z + (1-\tau)u))d\tau| \\ &= |\int_0^1 \int_0^1 (\mathcal{H}''(\lambda(\tau x + (1-\tau)y) + (1-\lambda)(\tau z + (1-\tau)u))(\tau x + (1-\tau)y \\ &\quad -(\tau z + (1-\tau)u))d\lambda d\tau| \\ &= |\int_0^1 \int_0^1 (e^{\lambda(\tau x+(1-\tau)y)+(1-\lambda)(\tau z+(1-\tau)u)}(\tau x + (1-\tau)y - (\tau z + (1-\tau)u))d\lambda d\tau| \\ &\leq \int_0^1 e|\tau(x-z) + (1-\tau)(y-u)|d\tau \\ &\leq \tfrac{e}{2}(|x-z| + |y-u|).\end{aligned} \tag{59}$$

Then, $\ell = \tfrac{1}{2}e^{\frac{1}{e-1}}$, $\ell_0 = \tfrac{e-1}{2}$, $\ell_1 = \tfrac{e}{2}$.
$\rho = 0.2578$, $\rho 1 = 0.2000$.

Example 7. For $\mathbb{E}_1 = \mathbb{E}_2 = C[0,1]$, and $D = \overline{U}(0,1)$, $x_*(s) = 0$ Consider function $\mathcal{H}$ on Ω, as

$$\mathcal{H}(x)(s) = x(s) - 5\int_0^1 stx^3(t)dt, \tag{60}$$

and the divided difference of $\mathcal{H}$ as

$$[x, y; \mathcal{H}] = \int_0^1 \mathcal{H}'(tx + (1-t)y)dt, \tag{61}$$

we get

$$[\mathcal{H}'(x)y](s) = y(s) - 15\int_0^1 stx^2(t)y(t)dt, \quad \text{for each } y \in \Omega. \tag{62}$$

We get $l_0 = 3.75$ and $l = \ell_1 = 7.5$ [7], so

$$\rho = \frac{1}{(1+\sqrt{2})l + 2l_0} \approx 0.039052429, \tag{63}$$

and

$$\bar{\rho} = \frac{1}{(1+\sqrt{2})l_1 + 2l} \approx 0.030205456, \tag{64}$$

so $\rho > \bar{\rho}$.

Example 8. For $\mathbb{E}_1 = \mathbb{E}_2 = C[0,1]$ and $\Omega = U(0,r)$ $r > 1$, consider $\mathcal{H}$ on Ω by

$$\mathcal{H}(x)(s) = x(s) - y(s) - \mu \int_0^1 G(s,t)x^3(t)dt, \quad x \in C[0,1], \ s \in [0,1].$$

with $y \in C[0,1]$ is given, μ is a real parameter, and the Kernel G is the Green's function

$$G(s,t) = \begin{cases} (1-s)t & \text{if } t \leq s \\ s(1-t) & \text{if } s \leq t, \end{cases}$$

so

$$(F'(x)(w))(s) = w(s) - 3\mu \int_0^1 G(s,t)x^2(t)w(t)dt, \quad w \in C[0,1], \ s \in [0,1].$$

For $x_0(s) = y_0(s) = y(s) = 1$ and $|\mu| < \frac{8}{3}$, we obtain

$$\|I - A_0\| \leq \tfrac{3}{8}\mu, \quad A_0^{-1} \in L(Y, X),$$
$$\|A_0^{-1}\| \leq \tfrac{8}{8-3|\mu|}, \quad s_0 = 0, \quad t_1 = \tfrac{|\mu|}{8-3|\mu|}, \quad L_0 = \tfrac{3(1+r)|\mu|}{2(8-3|\mu|)},$$

and

$$L = \frac{3r|\mu|}{8 - 3|\mu|}.$$

Set $r = 3$ and $\mu = \frac{1}{2}$, to obtain

$$t_1 = 0.076923077, \quad L_0 \approx 0.461538462, \quad L \approx 0.631279217$$

and

$$\frac{L(t_1 + s_0)}{1 - L_0(t_1 + s_1 + s_0)} \approx 0.057441746, \quad \alpha \approx 0.711345739, \quad 1 - \frac{2L_0 t_1}{1 - L_0 s_0} \approx 0.928994083.$$

Hence, Theorem 2 applies.

REFERENCES

[1] Argyros, I. K. 2007. *Computational theory of iterative methods, Series: Studies in Computational Mathematics 15, Editors, C.K. Chui and L. Wuytack*. New York, USA: Elsevier Publ. Co.

[2] Werner, W. 1979. "Uber ein Verfahren der Ordnung $1 + \sqrt{2}$ zur Nullstellenbestimmung". *Numer. Math.* 32: 333–342.

[3] Werner, W. 1982. "Some supplementary results on the $1 + \sqrt{2}$ order method for the solution of nonlinear equations. *Numer. Math.* 38: 383–392.

[4] King, R. F. 1972. "Tangent methods for nonlinear equations". *Numer. Math.* 18: 298–304.

[5] Traub, J. F. 1984. *Iterative Methods for the Solution of Equations*. Englewood Cliffs: Prentice Hull.

[6] McDougall, T. J. and Wotherspoon, S. J. 2014. "A simple modification of Newton's method to achieve convergence of order $1 + \sqrt{2}$". *Appl. Math. Lett.* 29: 20–25.

[7] Argyros, I. K. and Ren, H. 2015. "On the convergence of King–Werner–type methods of order $1 + \sqrt{2}$ free of derivatives". *Appl. Math. Comput.* 256: 148–159.

[8] Kantorovich, L.V. and Akilov, G. P. 1982. *Functional Analysis*. Oxford: Pergamon Press.

[9] Argyros, I. K. and Hilout, S. 2012. "Weaker conditions for th convergence of Newton's method" *J. Complexity* 28: 364–387.

[10] Argyros, I. K. and Magreñán, A. A. 2017. *Iterative methods and their dynamics with applications*. New York, USA: CRC Press, New York.

[11] Argyros, I. K. and Magreñán, A. A. 2018. *A contemporary study of iterative methods*. New York: Elsevier (Academic Press).

In: Understanding Banach Spaces
Editor: Daniel González Sánchez

ISBN: 978-1-53616-745-0
© 2020 Nova Science Publishers, Inc.

Chapter 12

ON AN EIGHTH ORDER STEFFENSEN–TYPE SOLVER FREE OF DERIVATIVES

Ioannis K. Argyros[1,*] *and Santhosh George*[2,†]
[1]Department of Mathematical Sciences
Cameron University, Lawton, OK, US
[2]Department of Mathematical and Computational Sciences
National Institute of Technology, Karnataka, India

Abstract

We expand the applicability of an eighth convergence order Steffensen–type solver for equations involving Banach space valued operators using only the first order derivative in contrast to earlier works using derivatives of order five which do not appear in the method, and in the special case of the i–dimensional Euclidean space.

Keywords: Banach space, Steffensen–type method, convergence order, local convergence

AMS Subject Classification: 65F08, 37F50, 65N12

1. INTRODUCTION

We solve equations
$$F(x) = 0, \tag{1}$$
where $F : \Omega \subset B \longrightarrow B$ is a continuous operator, B is a Banach space, and Ω nonempty open set using an iterative method. The method is of order eight [1] using only divided differences, derivatives up to the order five and Taylor expansions in the special case when $\mathcal{B} = \mathbb{R}^i$ and $Q(x) = (f_1^m(x), f_2^m(x), \ldots, f_i^m(x)), \quad m \geq 2$. We study this method, but

---
[*]Corresponding Author's E-mail: iargyros@cameron.edu.
[†]E-mail: sgeorge@nitk.ac.in.

in the more general setting of Banach space valued operators:

$$y_n = x_n - [x_n + Q(x_n), x_n, F]^{-1} F(x_n)$$
$$z_n = y_n - \left[\frac{13}{4}I - A_n\left(\frac{7}{2}I - \frac{5}{4}A_n\right)\right][x_n + Q(x_n), x_n, F]^{-1} F(y_n) \quad (2)$$
$$x_{n+1} = z_n - \left[\frac{7}{2}I - A_n\left(4I - \frac{3}{2}A_n\right)\right][x_n + Q(x_n), x_n, F]^{-1} F(z_n),$$

where $x_0 \in \Omega$ is an initial point, $Q : \Omega \longrightarrow B$ is a continuous operator,

$$A_n = [x_n + Q(x_n), x_n, F]^{-1}[y + Q(y_n), y_n; F],$$

$[\cdot, \cdot; F] : \Omega \times \Omega \to \mathscr{L}(\mathcal{B}, \mathcal{B})$, is a divided difference of order one with

$$[x, y; F](x - y) = F(x) - F(y) \quad \text{for each } x, y \in D \quad \text{with } x \neq y$$

and

$$[x, x, F] = F'(x) \quad \text{for each } x \in D,$$

if F is differentiable at x. (similarly for Q), where Q is as F.

As an academic example: Let $\mathcal{B}_1 = \mathcal{B}_2 = \mathbb{R}$, $\Omega = [-\frac{1}{2}, \frac{3}{2}]$. Define $\mathcal{G}$ on Ω by

$$\mathcal{G}(x) = x^3 \log x^2 + x^5 - x^4$$

Then, we have $x_* = 1$, and

$$\mathcal{G}'(x) = 3x^2 \log x^2 + 5x^4 - 4x^3 + 2x^2,$$

$$\mathcal{G}''(x) = 6x \log x^2 + 20x^3 - 12x^2 + 10x,$$

$$\mathcal{G}'''(x) = 6 \log x^2 + 60x^2 = 24x + 22.$$

Obviously $\mathcal{G}'''(x)$ is not bounded on Ω. So, the convergence of solver (2) not guaranteed by the analysis in [2].

Other problems with the usage of solver (2) are: no information on how to choose x_0; bounds on $\|x_n - x_*\|$ and information on the location of x_*. All these are addressed in this chapter by only using conditions on the first derivative, and in the more general setting of Banach space valued operators. That is how, we expand the applicability of solver (2). To avoid the usage of Taylor series and high convergence order derivatives, we rely on the computational order of convergence (COC) or the approximate computational order of convergence (ACOC) [3].

The layout of the rest of the chapter includes: the local convergence in Section 2, and the example in Section 3.

2. LOCAL CONVERGENCE

Set $S = [0, \infty)$. Certain real functions and parameters appearing in the local convergence analysis of method (1) are introduced. Let $W_0 : S \times S \to S$, $W_1 : S \to S$ be continuous and increasing functions with $W_0(0, 0) = 0$. Suppose that equation

$$W_0(W_1(t)\, t, t) = 1 \tag{3}$$

has at least one positive solution. Denote by ρ_0 the smallest such solution. Set $S_0 = [0, \rho_0)$. Let also $W : S_0 \times S_0 \to S$, $V : S_0 \to S$ and $W_2 : S_0 \to S$ be continuous and increasing functions with $W(0, 0) = 0$. Define functions g_1 and $\overline{g_1}$ in the interval S_0 by $g_1(t) = \frac{W(W_1(t)\, t, t)}{a(t)}$, and $\overline{g_1}(t) = g_1(t) - 1$, where $a(t) = 1 - W_0(W_1(t)\, t, t)$. We have $\overline{g_1}(0) = -1$ and $\overline{g_1}(t) \to \infty$ for $t \to \rho_0^-$. It follows from the intermediate value theorem that equation $\overline{g_1}(t) = 0$ has at least one solution in $(0, \rho_0)$. Denote by r_1 the smallest such solution.

Suppose that equation

$$W_0(W_1(g_1(t)\, t)\, g_1(t)\, t, g_1(t)\, t) = 1 \tag{4}$$

has at least one positive solution. Denote by ρ_1 the smallest such solution. Set $\rho_2 = \min\{\rho_0, \rho_1\}$ and $S_1 = [0, \rho_2)$. Define functions g_2 and $\overline{g_2}$ on the interval S_1 by

$$g_2(t) = \{g_1(g_1(t)\, t) + \frac{d(t)}{a(t)\, b(t)} V(g_1(t)\, t) + \frac{1}{4}[4\, h(t) + 5\, h^2(t)] \frac{V(g_1(t)\, t)}{a(t)}\} g_1(t)$$

and

$$\overline{g_2}(t) = g_2(t) - 1,$$

where

$$b(t) = 1 - W_0(W_1(g_1(t)\, t)\, g_1(t)\, t, g_1(t)\, t),$$
$$d(t) = W_0(W_1(t)\, t, t) + W_0(W_1(g_1(t)\, t), g_1(t)\, t)$$

and

$$h(t) = \frac{d(t)}{a(t)}.$$

Then, we also get $\overline{g_2}(0) = -1$ and $\overline{g_2}(t) \to \infty$ as $t \to \rho_2^-$. Denote by r_2 the smallest solution of equation $\overline{g_2}(t) = 0$ in $(0, \rho_2)$.

Suppose that equation

$$W_0(W_1(g_2(t)\, t)\, g_2(t)\, t, g_2(t)\, t) = 1 \tag{5}$$

has at least one positive solution. Denote by ρ_3 the smallest such solution. Set $\rho = \min\{\rho_2, \rho_3\}$ and $S_2 = [0, \rho)$. Define functions g_3 and $\overline{g_3}$ on the interval S_2 by

$$g_3(t) = \{g_1(g_2(t)\, t) + \frac{e(t)\, V(g_2(t)\, t)}{a(t)\, c(t)} + \frac{1}{2}[2\, h(t) + 3\, h^2(t)] \frac{V(g_2(t)\, t)}{a(t)}\} g_2(t)$$

and
$$\bar{g}_3(t) = g_3(t) - 1,$$
where
$$c(t) = 1 - W_0(W_2(g_2(t)t)g_2(t)t, g_2(t)t),$$
and
$$e(t) = W_0(W_2(t)t,t) + W_0(W_2(g_2(t)t), g_2(t)t).$$

We obtain again $\bar{g}_3(0) = -1$ and $\bar{g}_3(t) \to \infty$ as $t \to \rho^-$. Denote by r_3 the smallest solution of equation $\bar{g}_3(t) = 0$ in $(0, \rho)$. Define a radius of convergence r by

$$r = \min\{r_j\}, \quad j = 1, 2, 3. \tag{6}$$

Then, we have that for all $t \in [0, r)$

$$a(t) > 0, \tag{7}$$
$$b(t) > 0, \tag{8}$$
$$c(t) > 0, \tag{9}$$
$$d(t) \geq 0, \tag{10}$$
$$e(t) \geq 0, \tag{11}$$
$$h(t) \geq 0. \tag{12}$$

and
$$o \leq g_j(t) < 1. \tag{13}$$

In order to study the local convergence of method (2), we need to rewrite the three substeps.

Lemma 1. *Suppose that method (2) is well–defined for each $n = 0, 1, 2, \ldots$, and $[x + Q(x), x; F]^{-1} \in \mathscr{L}(\mathcal{B}_2, \mathcal{B}_1)$. Then, the following assertions hold*

$$y_n - x_* = [x_n + Q(x_n), x_n; F]^{-1} ([x_n + Q(x_n), x_n; F] - [x_n, x_*; F]) (x_n - x_*) \tag{14}$$

$$z_n - x_* = y_n - x_* + [y_n + Q(y_n), y_n; F]^{-1} F(y_n) + ([y_n + Q(y_n), y_n; F]^{-1}$$
$$- [x_n + Q(x_n), x_n; F]^{-1}) F(y_n) - \frac{1}{4}[4(I - A_n) + 5(I - A_n)^2] \tag{15}$$
$$\times [x_n + Q(x_n), x_n; F]^{-1}) F(y_n)$$

and

$$x_{n+1} - x_* = z_n - x_* - [z_n + Q(z_n), z_n; F]^{-1} F(z_n) + ([z_n + Q(z_n), z_n; F]^{-1}$$
$$- [x_n + Q(x_n), x_n; F]^{-1}) F(z_n) - \frac{1}{2}[2(I - A_n) + 3(I - A_n)^2] \tag{16}$$
$$\times [x_n + Q(x_n), x_n; F]^{-1} F(z_n)$$

Proof. We have in turn by the first substep of method (2) and the definition of the divided difference

$$y_n - x_* = x_n - x_* - [x_n + Q(x_n), x_n; F]^{-1} F(x_n)$$
$$= [x_n + Q(x_n), x_n; F]^{-1} ([x_n + Q(x_n), x_n; F] - [x_n, x_*; F]) (x_n - x_*)$$

which shows (14). Then similarly from the second substep of method (2)

$$z_n - x_* = y_n - x_* - [y_n + Q(y_n), y_n; F]^{-1} F(y_n) + ([y_n + Q(y_n), y_n; F]^{-1}$$
$$- [x_n + Q(x_n), x_n; F]^{-1}) F(y_n) - \frac{1}{4}(9\,I - 14\,A_n + 5\,A_n^2)$$
$$[x_n + Q(x_n), x_n; F]^{-1} F(y_n)$$
$$= y_n - x_* - [y_n + Q(y_n), y_n; F]^{-1} F(y_n) + ([y_n + Q(y_n), y_n; F]^{-1}$$
$$- [x_n + Q(x_n), x_n; F]^{-1}) F(y_n) - \frac{1}{4}[4(I - A_n) + 5(I - A_n)^2]$$
$$\times [x_n + Q(x_n), x_n; F]^{-1} F(y_n)$$

which shows (15).

Finally, from the third substep of method (2), we obtain in turn that

$$x_{n+1} - x_* = z_n - x_* - [z_n + Q(z_n), z_n; F]^{-1} F(z_n) + ([z_n + Q(z_n), z_n; F]^{-1}$$
$$- [x_n + Q(x_n), x_n; F]^{-1}) F(z_n) - [\frac{5}{2}I - A_n (4\,I - \frac{3}{2}A_n)]$$
$$\times [x_n + Q(x_n), x_n; F]^{-1}) F(z_n)$$
$$= z_n - x_* - [z_n + Q(z_n), z_n; F]^{-1} F(z_n) + ([z_n + Q(z_n), z_n; F]^{-1}$$
$$- [x_n + Q(x_n), x_n; F]^{-1}) F(z_n) - \frac{1}{2}[2(I - A_n) + 3(I - A_n)^2]$$
$$\times [x_n + Q(x_n), x_n; F]^{-1}) F(z_n),$$

which completes the proof. □

By $U(u, \lambda), \overline{U}(u, \lambda)$ we denote the open and closed balls in $\mathcal{B}_1$, respectively with center $u \in \mathcal{B}_1$ and of radius $\lambda > 0$.

The local convergence analysis is based on the conditions (A)

(a_1) $F : \Omega \to \mathcal{B}$ is continuously differentiable in the sense of Fréchet, $[\cdot, \cdot; F] : \Omega \times \Omega \to \mathscr{L}(\mathcal{B}, \mathcal{B})$, $[\cdot, \cdot; F] : \Omega \times \Omega \to \mathscr{L}(\mathcal{B}, \mathcal{B})$, are divided differences of order one and there exists $x_* \in \Omega$ such that $F(x_*) = Q(x_*) = 0$, and $F'(x_*)^{-1} \in \mathscr{L}(\mathcal{B}, \mathcal{B})$.

(a_2) There exist continuous and increasing functions $W_0 : S \times S \to S$ and $W_1 : S \to S$ with $W_0(0, 0) = 0$, so that for each $x \in \Omega$

$$\|F'(x_*)^{-1}([x + Q(x), x; F] - F'(x_*))\| \le W_0 (\|x + F(x) - x_*\|, \|x - x_*\|).$$

and

$$\|I + [x, x_*; F]\| \le W_1 (\|x - x_*\|).$$

Set $\Omega_0 = \Omega \cap U(x_*, \rho_0)$, where ρ_0 is given in (3).

(a_3) There exist continuous and increasing functions $W_0 : S_0 \times S_0 \to S$ and $W_2 : S_0 \to S$ and $V : S_0 \to S$ sush that for each $x \in \Omega_0$

$$\|F'(x_*)^{-1}([x+Q(x), x; F] - [x, x_*; F])\| \leq W(\|FG(x)\|, \|x - x_*\|).$$

$$\|Q(x)\| \leq W_2(\|x - x_*\|) \|x - x_*\|$$

and

$$\|F'(x_*)^{-1}[x, x_*; F]\| \leq V(\|x - x_*\|).$$

(a_4) $\overline{U}(x_*, \bar{r}) \subset \Omega$, $\bar{r} = W_1(r)r$, ρ_0, ρ_1 and ρ_3 given by (3) – (5), respectively exist and r is defined in (6).

(a_5) There exists $r_* \geq r$ such that $V_0(r_*) < 1$, where function $V_0 : S_0 \to S$ is continuous and increasing with $V_0(0) = 0$.

Set $\Omega_1 = \Omega \cap U(x_*, r_*)$.

By Lemma 1, we can use the notations

$$a_n = 1 - W_0\left(W_1(\|x_n - x_*\|)\|x_n - x_*\|, \|x_n - x_*\|\right)$$
$$b_n = 1 - W_0\left(W_1(g_1(\|x_n - x_*\|)\|x_n - x_*\|), g_1(\|x_n - x_*\|)\|x_n - x_*\|\right)$$
$$c_n = 1 - W_0\left(W_1(g_2(\|x_n - x_*\|)\|x_n - x_*\|)g_2(\|x_n - x_*\|)\|x_n - x_*\|,\right.$$
$$\left. g_2(\|x_n - x_*\|)\|x_n - x_*\|\right)$$
$$d_n = W_0\left(W_1(\|x_n - x_*\|)\|x_n - x_*\|, \|x_n - x_*\|\right)$$
$$\quad + W_0\left(W_1(g_1(\|x_n - x_*\|)\|x_n - x_*\|), g_1(\|x_n - x_*\|)\|x_n - x_*\|\right)$$
$$e_n = W_0\left(W_1(\|x_n - x_*\|)\|x_n - x_*\|, \|x_n - x_*\|\right)$$
$$\quad + W_0\left(W_1(g_2(\|x_n - x_*\|)\|x_n - x_*\|), g_2(\|x_n - x_*\|)\|x_n - x_*\|\right)$$

and

$$h_n = \frac{d_n}{a_n}.$$

Next, we present the local convergence analysis of method (2) using the conditions (A) and the preceding notation.

Theorem 2. *Suppose that the conditions (A) hold. Then, the sequence $\{x_n\}$ generated for $x_0 \in U(x_*, r) - \{x_*\}$ is well defined, remains in $U(x_*, r)$ for each $n = 0, 1, 2, \ldots$ and converges to x_*, so that*

$$\|y_n - x_*\| \leq g_1(\|x_n - x_*\|)\|x_n - x_*\| \leq \|x_n - x_*\| < r \tag{17}$$
$$\|z_n - x_*\| \leq g_2(\|x_n - x_*\|)\|x_n - x_*\| \leq \|x_n - x_*\|, \tag{18}$$

and

$$\|x_{n+1} - x_*\| \leq g_3(\|x_n - x_*\|)\|x_n - x_*\| \leq \|x_n - x_*\|, \tag{19}$$

where functions g_j are given previously and r is defined in (6).

Proof. We shall show estimates (17)–(19) using mathematical induction. Let $x \in U(x_*, r) - \{x_*\}$. By (a_1), (a_2), (a_4) and (6), we obtain in turn

$$\begin{aligned}
\|F'(x_*)^{-1}\left([x + Q(x), x; F] - F'(x_*)\right)\| &\leq W_0\left(\|x + G(x) - x_*\|, \|x - x_*\|\right) \\
&\leq W_0\left(\|(I + [x, x_*])(x - x_*)\|, \|x - x_*\|\right) \\
&\leq W_0\left(W_1(\|x - x_*\|)\|x - x_*\|, \|x - x_*\|\right) \\
&< W_0\left(W_1(r)r, r\right) r < 1,
\end{aligned} \quad (20)$$

where we also used that

$$\|x + F(x) - x_*\| = \|(I + [x, x_*; F])(x - x_*)\|$$
$$\leq W_1(\|x - x_*\|)\|x - x_*\| \leq W_1(r)r = \bar{r},$$

so $x + F(x) - x_* \in U(x_*, \bar{r})$. It follows from (20) and the Banach perturbation Lemma [1,4,5], that $[x + Q(x), x; F]^{-1} \in \mathscr{L}(\mathcal{B}, \mathcal{B})$, and

$$\|[x + Q(x), x; F]^{-1} F'(x_*)\| \leq \frac{1}{a(\|x - x_*\|)}, \quad (21)$$

and y_0, z_0, x_1 are well defined by method (2) for $n = 0$.

Then, by (6), (13) (for $j = 1$), (14), (21) (for $x = x_0$), and (a_3), we get in turn that

$$\begin{aligned}
\|y_0 - x_*\| &\leq \|[x_0 + Q(x_0), x_0; F]^{-1} F'(x_*)\| \|F'(x_*)^{-1}([x_0 + Q(x_0), x_0; F] \\
&\quad - [x_0, x_*; F])\| \|x_0 - x_*\| \\
&\leq \frac{W(W_1(\|x_0 - x_*\|)\|x_0 - x_*\|, \|x_0 - x_*\|) \|x_0 - x_*\|}{a(\|x_0 - x_*\|)} \\
&= g_1(\|x_0 - x_*\|) \|x_0 - x_*\| \leq \|x_0 - x_*\| < r,
\end{aligned} \quad (22)$$

so (17) holds for $n = 0$ and $y_0 \in U(x_*, r)$. We need the estimate obtained using (a_2) and (20)

$$\begin{aligned}
\|I - A_0\| &= \|([x_0 + Q(x_0), x_0; F]^{-1} F'(x_*))(F'(x_*)^{-1}[([x_0 + Q(x_0), x_0; F] - F'(x_*)) \\
&\quad + (F'(x_*) - [y_0 + Q(y_0), y_0; F])])\| \\
&\leq \frac{d(\|x_0 - x_*\|)}{a(\|x_0 - x_*\|)}
\end{aligned} \quad (23)$$

and the estimate using (a_1) and (a_3) that

$$F(x) - F(x_*) = [x, x_*; F](x - x_*),$$

so

$$\|F'(x_*)[x, x_*; F]\| \leq V(\|x - x_*\|). \quad (24)$$

Then, by (6), (13) (for $j = 2$), (15) and (21) (for $x = y_0$) – (24), we have in turn that

$$\begin{aligned}
\|z_0 - x_*\| &\leq \{g_1(g_1(\|x_0 - x_*\|)\|x_0 - x_*\|) \\
&\quad + \frac{d(\|x_0 - x_*\|) V(g_1(\|x_0 - x_*\|)\|x_0 - x_*\|)}{a(\|x_0 - x_*\|) b(\|x_0 - x_*\|)} \\
&\quad + \frac{1}{4}[4\frac{d(\|x_0 - x_*\|)}{a(\|x_0 - x_*\|)} + 5\frac{d(\|x_0 - x_*\|)}{a(\|x_0 - x_*\|)}]\frac{V(g_1(\|x_0 - x_*\|))\|x_0 - x_*\|}{a(\|x_0 - x_*\|)}\} \\
&\quad \times g_1(\|x_0 - x_*\|) \\
&= g_2(\|x_0 - x_*\|)\|x_0 - x_*\|,
\end{aligned} \qquad (25)$$

so (18) holds for $n = 0$ and $z_0 \in U(x_*, r)$. Next, from (6), (13) (for $j = 3$), (21) (for $x = z_0$), (22)–(25), we obtain in turn that

$$\begin{aligned}
\|x_1 - x_*\| &\leq \{g_1(g_2(\|x_0 - x_*\|)\|x_0 - x_*\|) + \frac{e(\|x_0 - x_*\|) V(g_2(\|x_0 - x_*\|)\|x_0 - x_*\|)}{a(\|x_0 - x_*\|) c(\|x_0 - x_*\|)} \\
&\quad + \frac{1}{2}[2\frac{d(\|x_0 - x_*\|)}{a(\|x_0 - x_*\|)} + 3(\frac{d(|x_0 - x_*\|)}{a(|x_0 - x_*\|)})^2]\frac{V(g_2(\|x_0 - x_*\|)\|x_0 - x_*\|)}{a(t)}\} \\
&\quad \times g_2(\|x_0 - x_*\|) \\
&= g_3(\|x_0 - x_*\|)\|x_0 - x_*\|,
\end{aligned} \qquad (26)$$

so (19) holds for $n = 0$ and $x_1 \in U(x_*, r)$. The induction for estimates (17)–(19) is terminated, if we replace x_0, y_0, z_0, x_1 by x_m, y_m, z_m, x_{m+1} in the preceding computations. Then, from the estimate

$$\|x_{m+1} - x_*\| \leq q\|x_m - x_*\| < r, \quad q = g_3(\|x_0 - x_*\|) \in [0, 1), \qquad (27)$$

we get that $\lim_{m \to \infty} x_m = x_*$ and $x_{n+1} \in U(x_*, r)$.

Finally to show the uniqueness part of the proof, let $G = [y_*, x_*; F]$ for some $y_* \in \Omega_1$ with $F(y_*) = 0$. Using (a_5), we get in turn that

$$\|F'(x_*)^{-1}(G - F'(x_*))\| \leq V_0(\|y_* - x_*\|) \leq V(r_*) < 1,$$

So $G^{-1} \in \mathcal{L}(\mathcal{B}, \mathcal{B})$. Then from the identity

$$0 = F(y_*) - F(x_*) = G(y_* - x_*),$$

obtain that $x_* = y_*$. $\square$

Remark 1. *It is worth noticing that solvers (2) is not changing when we use the conditions of the preceding Theorem instead of the stronger conditions used in [2]. Moreover, we can compute the computational order of convergence (COC) defined as*

$$\xi = \ln\left(\frac{\|x_{n+1} - x_*\|}{\|x_n - x_*\|}\right) / \ln\left(\frac{\|x_n - x_*\|}{\|x_{n-1} - x_*\|}\right)$$

or the approximate computational order of convergence (ACOC) [3]

$$\xi_1 = \ln\left(\frac{\|x_{n+1} - x_n\|}{\|x_n - x_{n-1}\|}\right) / \ln\left(\frac{\|x_n - x_{n-1}\|}{\|x_{n-1} - x_{n-2}\|}\right).$$

This way we obtain in practice the order of convergence, but not higher order derivatives are used.

REFERENCES

[1] Argyros, I. K., Ezquerro, J. A., Gutiérrez, J. M., Hernández, M. A. and Hilout, S. 2011. "On the semilocal convergence of efficient Chebyshev–Secant–type solvers". *J. Comput. Appl. Math.* 235: 3195–3206.

[2] Amiri, A., Cordero, A., Darvishi, M. T. and Torregrosa, J. R. 2018. "Presenting the order of convergence: Low complexity Jacobian–free iterative schemes for solving nonlinear systems". *J. Comput. Appl. Math.* 337: 87–97.

[3] Cordero, A., Hueso, J. L., Martínez, E. and Torregrosa, J. R. 2010. "A modified Newton–Jarratt's composition". *Numer. Algor.* 55: 87–99.

[4] Osrowski, A. M. 1960. *Solution of equations and systems of equations.* New York: Academic Press, New York.

[5] Potra, F. A. and Pták, V. 1984. *Nondiscrete induction and iterative processes. Volume 103.* Pitman Advanced Publishing Program.

In: Understanding Banach Spaces
Editor: Daniel González Sánchez
ISBN: 978-1-53616-745-0
© 2020 Nova Science Publishers, Inc.

Chapter 13

LOCAL CONVERGENCE OF OSADA'S METHOD FOR FINDING ZEROS WITH MULTIPLICITY

Ioannis K. Argyros[1,*] *and Santhosh George*[2,†]
[1]Department of Mathematical Sciences
Cameron University, Lawton, OK, US
[2]Department of Mathematical and Computational Sciences
National Institute of Technology, Karnataka, India

Abstract

We provide an extended local convergence of Osada's method for approximating a zero of a nonlinear equation with multiplicity m, where m is a natural number. The new technique provides a tighter convergence analysis under the same computational cost as in earlier works. This technique can be used on other iterative methods too. Numerical examples further validate the theoretical results.

Keywords: inexact method, radius of convergence, divided difference, derivative, zero with multiplicity

AMS Subject Classification: 49M15, 65D10, 65G99, 47H09

1. INTRODUCTION

A plethora of problems can be written as

$$F(x) = 0,$$

using mathematical modeling, where function $F : \Omega \subseteq \mathcal{E}_1 \longrightarrow \mathcal{E}_2$ is sufficiently many times differentiable, and $\Omega, \mathcal{E}_1, \mathcal{E}_2$ are convex subsets in $\mathbb{R}$. In the present study, we pay attention to the case of a solution x^* with multiplicity $m > 1$, namely, $F(x^*) = 0, F^{(i)}(x^*) = 0$ for $i = 1, 2, \ldots, m-1$, and $F^{(m)}(x^*) \neq 0$. The determination of solutions of multiplicity m is of great interest. As an example, in the study of electron trajectories, when the electron

---

[*]Corresponding Author's E-mail: iargyros@cameron.edu.
[†]E-mail: sgeorge@nitk.ac.in.

reaches a plate of zero speed, the function distance from the electron to the plate has a solution of multiplicity two. Moreover, the multiplicity of solutions appears in connection to Van Der Waals equation of state and other phenomena. The convergence order of iterative methods decreases, if the equation has solutions of multiplicity m. Modifications in the iterative function are needed to improve the order of convergence.

We present the Osada's method (OM) defined for each $n = 0, 1, 2, \ldots$ as

$$x_{n+1} = x_n - \frac{1}{2}m(m+1)\frac{F(x_n)}{F'(x_n)} + \frac{1}{2}(m-1)^2, \tag{1}$$

where x_0 is an initial point. Our technique can be used in an analogous way on other methods:

Newton's Method:

$$x_{n+1} = x_n - \frac{F(x_n)}{F'(x_n)}.$$

Modified Newton's Method:

$$x_{n+1} = x_n - m\frac{F(x_n)}{F'(x_n)}.$$

Traub method and choose:

$$\xi_n = m\frac{F(x_n)}{F'(x_n)} - \frac{m(3-m)}{2}\frac{F(x_n)}{F'(x_n)} - \frac{m^2}{2}\frac{F(x_n)^2 F''(x_n)}{F'(x_n)^3}.$$

We shall determine a radius of convergence which will assist us obtain x_0 so that $\lim_{n \to \infty} x_n = x_*$. In particular, Alarcon et al. [1] used in their local convergence analysis instead of the usual conditions [1–6]:

$$|(F^{(m)}(x^*))^{-1}(F^{(m+1)}(y) - F^{(m+1)}(x))| \leq K_0|y-x|^p \tag{2}$$

$$|F^{(m)}(x^*)^{-1}F^{(m+1)}(x)| \leq K_m. \tag{3}$$

$$|(F^{(m)}(\gamma))^{-1}F^{(m+1)}(x)| \leq \bar{b},\ x \in \Omega,\ \bar{b} > 0. \tag{4}$$

For the method that uses second derivative we need to add another assumption as follows:

$$|(F^{(m)}(\gamma))^{-1}F^{(m+2)}(x)| \leq \bar{c},\ x \in \Omega,\ \bar{c} > 0. \tag{5}$$

This way the proofs were simplified. More advantages can be found in [1].

In our chapter we extend further the applicability of OM by first considering the center–Lipschitz conditions for each $x \in \Omega$

$$|F'(x^*)^{-1}(F'(x) - F'(x^*))| \leq a_1|x-x^*|, a_1 > 0 \tag{6}$$

and

$$|F'(x^*)^{-1}(F''(x) - F''(x^*))| \leq a_2|x-x^*|, a_2 > 0. \tag{7}$$

Set $a = \max\{a_1, a_2\}$ and $\Omega_0 = \Omega \cap S(x^*, \frac{1}{a})$, where $S(x, \rho)$ is the sphere in $\mathcal{E}_1$ with center $x \in \mathcal{E}_1$ and of radius $\rho > 0$. By $\bar{S}(x, \rho)$, we denote the closure of $S(x, \rho)$. It follows from (6) and (7) that $F'(x) \neq 0, F''(x) \neq 0$ for each $x \in \Omega_0$, so OM is well defined as long as $x_n \in \Omega_0$ for each $n = 0, 1, 2, \ldots$ Then, we consider the restricted convergence conditions, corresponding to (4) and (5), given, respectively by

$$|F'(x^*)^{-1}F^{(m+1)}(x)| \leq b, x \in \Omega_0 \tag{8}$$

and

$$|(F^{(m)}(\gamma))^{-1}F^{(m+2)}(x)| \leq c, \ x \in \Omega. \tag{9}$$

Notice that $b = b(\Omega_0)$ and $c = c(\Omega_0)$. It follows from (4), (5), (8) and (9) that

$$b \leq \bar{b} \tag{10}$$

and

$$c \leq \bar{c}, \tag{11}$$

since $\Omega_0 \subseteq \Omega$. Therefore, b, c can replace $\bar{b}, \bar{c}$ in the proofs in [1]. This way, we obtain the improvements if strict inequality holds in (10) and (11) (I):

(i1) Larger radius of convergence. That is we have more initial points.

(i2) More precise estimation on $|x_n - x^*|$, so fewer iterates are needed to arrive at a desired error tolerance.

(i3) The uniqueness sphere is smaller.

The improvements are obtained using constants a and b which are special cases of $\bar{a}$ and $\bar{b}$, so no additional cost is involved in obtaining them.

2. LOCAL CONVERGENCE

We present the results without their proofs, since they are obtained by simply replacing $\bar{a}, \bar{b}$ by a, b respectively in the proofs in [1].

Lemma 1. *Assume function $F : \Omega \longrightarrow \mathbb{R}$ is sufficiently many times differentiable, and p be a zero of multiplicity $m (m > 1)$ for F. Then,*

$$F(x) = (x - x^*)^m h(x), \ h(x^*) \neq 0 \tag{12}$$

$$h(x^*) = \frac{F^{(m)}(x^*)}{m!}, \tag{13}$$

and

$$h^{(i)}(x) = \frac{1}{(m-1)!} \int_0^1 F^{(m+1)}(x^* + \tau(x - x^*))\tau^i(1-\tau)^{m-1} d\tau, \tag{14}$$

where

$$h(x) = \frac{F^{(m)}(x^*)}{m!} + \frac{1}{(m-1)!} \int_0^1 [F^{(m)}(x^* + \tau(x-x^*)) - F^{(m)}(x^*)](1-\tau)^{m-1} d\tau. \tag{15}$$

Lemma 2. *Suppose that conditions (8) and (9) hold. Set* $I_0 = (x^* - r_0^0, p + r_0^0)$, *where* $r_0^0 = \frac{m+1}{k}$ *and* $e_0 = x_0 - x^*$ *for* $x_0 \in I_0$. *Then, the following items hold*

$$|h(x^*)h(x_0)| \leq \frac{m+1+b|e_0|}{m+1} \tag{16}$$

$$|h(x^*)^{-1}h'(x_0)| \leq \frac{b}{m+1} \tag{17}$$

$$|h(x_0)^{-1}h(x^*)| \leq \frac{m+1}{m+1-b|e_0|} \tag{18}$$

$$|h(x_0)^{-1}h'(x_0)| \leq \frac{b}{m+1-b|e_0|} \tag{19}$$

$$|h(x_0)^{-1}h''(x_0)| \leq \frac{2c}{(m+2)(m+1-b|e_0|)}. \tag{20}$$

$$f_1(t) = \frac{2bc}{m^2(m^2-1)(m+2)(m+1-bt)}t^3 + \frac{2[(m+2)b^2 + (m+1-bt)c]}{m(m^2-1)(m+2)(m+1-bt)}t^2$$
$$+ \frac{(m^2+2m-1)b}{m(m^2-1)}t. \tag{21}$$

$$f_2(t) = \frac{\psi(t)}{1 - f_1(t)}, \tag{22}$$

where

$$\psi(t) = \frac{4bct^3 + [(m+1)^2(m+2)b^2 + 2m(m-1)(m+1-bt)]t^2}{2m^2(m^2-1)(m+2)(m+1-bt)}. \tag{23}$$

Based on the two Lemmas, we have:

Theorem 3. *Suppose: Conditions of Lemma 2 hold and* r_j, $j = 1, 2$ *are the unique positive solutions of equations* $h_j(t) = f_j(t) - 1$, *where the functions* f_j *are defined previously. Set* $r = r_2$. *Then, starting from* $x_0 \in T_r = (x^* - r, x^* + r)$, *sequence* $\{x_n\}$ *generated by OM is well defined* $x_n \in I_r$ *for each* $n = 0, 1, 2, \ldots$, $\lim_{n \to \infty} x_n = x^*$, *and*

$$|e_{n+1}| \leq \frac{|e_n|^3}{r^3}, \tag{24}$$

where $e_n = x_n - x^*$ *so the order of convergence of OM is three.*

Remark 4. *If* $b = \bar{b}$ *and* $c = \bar{c}$, *then our results become the ones in [1]. But, i.e., if strict inequality holds in (10) or (11), then the new results improve the ones in [1]. Let us denote by* $\bar{\psi}, \bar{h}, \bar{f}_j, \bar{r}_j$, *the old functions and parameters defined as the non bar functions and parameters by replacing* a, b *by* $\bar{a}, \bar{b}$, *respectively. Indeed, the bounds our Lemma 2 are tighter than the ones in Lemma 2 in [1];*

$$\psi(t) \leq \bar{\psi}(t) \tag{25}$$
$$f_j(t) \leq \bar{f}_j, \tag{26}$$
$$h_j(t) \leq \bar{h}_j(t), \tag{27}$$

so

$$\bar{r} \leq r \tag{28}$$

(see also the numerical example).

3. NUMERICAL EXAMPLE

We show that (10) and (11) hold as strict inequalities, so we obtain the improvements (I).

Example 5. *Let $\mathcal{E}_1 = \mathcal{E}_2 = \mathbb{R}, \Omega = [-1,1]$. Consider function F on Ω as*

$$F(x) = e^x - x - 1.$$

Then, we get $x^* = 0, m = 2, a_1 = a_2 = a = e - 1, c = b = e^{\frac{1}{a}}$, and $\bar{b} = \bar{c} = e$. Notice that $c < \bar{c}$. Then,

$$r = 0.4708186722060214712826109462184 8,$$

$$\bar{r} = 0.2550027089102493982331054667156 4.$$

and

$$\bar{r} < r.$$

REFERENCES

[1] Alarcón, D., Hueso, J. L. and Martínez, E. *An alternative analysis for the local convergence of iterative methods for multiple roots including when the multiplicity is unknown.* Submitted.

[2] Amat, S., Hernández, M. A. and Romero, N. 2012. "Semilocal convergence of a sixth order iterative method for quadratic equations". *Applied Numerical Mathematics* 62: 833-841.

[3] Osada, N. 1994. "An optimal mutiple root-finding method of order three". *J. Comput. Appl. Math.* 52: 131–133.

[4] Petkovic, M. S, Neta, B., Petkovic, L. and Džunič, J. 2013. *Multipoint methods for solving nonlinear equations.* Elsevier.

[5] Zhou, X. and Song, Y. 2011. "Convergence radius of Osada's method for multiple roots under Hölder and Center–Hölder continous conditions". *ICNAAM 2011(Greece), AIP Conf. Proc.* 1389: 1836–1839.

[6] Zhou, X., Chen, X. and Song, Y. 2014. "On the convergence radius of the modified Newton method for multiple roots under the center– Hölder condition". *Numer. Algor.* 65(2): 221–232.

In: Understanding Banach Spaces
Editor: Daniel González Sánchez

ISBN: 978-1-53616-745-0
© 2020 Nova Science Publishers, Inc.

Chapter 14

EXPANDING THE APPLICABILITY OF AN EIGHTH–ORDER METHOD IN BANACH SPACE UNDER WEAK CONDITIONS

Ioannis K. Argyros[1,], Ramandeep Behl[2,†],*
Daniel González Sánchez[3,‡] and S. S. Motsa[2,§]
[1]Department of Mathematical Sciences
Cameron University, Lawton, OK, US
[2]School of Mathematics, Statistics and Computer Sciences
University of KwaZulu–Natal, Pietermaritzburg, South Africa
[3]Escuela de Ciencias Físicas y Matemáticas
Universidad de Las Américas, Quito, Ecuador

Abstract

We present a local convergence analysis for an eighth–order convergent method in order to approximate a locally unique solution of nonlinear equation in a Banach space setting. In contrast to the earlier studies using hypotheses up to the seventh Fréchet–derivative, we only use hypotheses on the first–order Fréchet–derivative and Lipschitz constants. This way, we not only expand the applicability of these methods but also proposed the computable radius of convergence of these methods. Finally, a variety of concrete numerical examples demonstrate that our results even apply to solve those nonlinear equations where earlier studies cannot apply.

Keywords: iterative method, local convergence, Banach space, Lipschitz constant, order of convergence

AMS Subject Classification: 65G99, 65H10, 47J25, 47J05

---

[*]E-mail: iargyros@cameron.edu.
[†]E-mail: ramanbehl87@yahoo.in.
[‡]Corresponding Author's E-mail: daniel.gonzalez.sanchez@udla.edu.ec.
[§]E-mail: motsas1@ukzn.ac.za.

1. INTRODUCTION

One of the most basic and important problems in Numerical analysis concerns with approximating a locally unique solution x^* of the equation of the form

$$F(x) = 0, \qquad (1)$$

where $F : D \subset X \to Y$ is a Fréchet–differentiable operator, X, Y are Banach spaces and D is a convex subset of X. Let us also denote $L(X, Y)$ as the space of bounded linear operators from X to Y.

The problem of finding approximate unique solution x^* is very important, since the problems can be reduced to equation (1) using mathematical modeling [1–8]. However, it is not always possible to find the solution x^* in a closed form. Therefore, most of the methods are iterative to solve such type of problems. The convergence analysis of iterative methods is usually divided into two categories: semi–local and local convergence analysis. The semi–local convergence matter is, based on the information around an initial point, to give criteria ensuring the convergence of iteration procedures. A very important problem in the study of iterative procedures is the convergence domain. Therefore, it is very important to propose the radius of convergence of the iterative methods.

We study the local convergence of the three step eighth–order convergent method defined for each $n = 0, 1, 2, \ldots$ by

$$\begin{aligned}
y_n &= x_n - \frac{2}{3} F'(x_n)^{-1} F(x_n), \\
z_n &= x_n - \left(I + \frac{21}{8} F'(x_n)^{-1} F(y_n) - \frac{9}{2} \left(F'(x_n)^{-1} F(y_n) \right)^2 + \frac{15}{8} \left(F'(x_n)^{-1} F(y_n) \right)^3 \right) \\
&\quad \times F'(x_n)^{-1} F(x_n), \\
w_n &= z_n - \left(3I - \frac{5}{2} F'(x_n)^{-1} F(y_n) + \frac{1}{2} \left(F'(x_n)^{-1} F(y_n) \right)^2 \right) F'(x_n)^{-1} F(z_n), \\
x_{n+1} &= w_n - \left(3I - \frac{5}{2} F'(x_n)^{-1} F(y_n) + \frac{1}{2} \left(F'(x_n)^{-1} F(y_n) \right)^2 \right) F'(x_n)^{-1} F(w_n),
\end{aligned} \qquad (2)$$

where $x_0 \in D$ is an initial point and F' denotes the first–order Fréchet–derivative of operator F. The eighth–order of convergence was shown in [1] using the Taylor series expansion, certain hypotheses on the function F and hypotheses reaching up to the seventh–order derivative of the involved function F although only first order derivative appears in the proposed scheme (2). The semilocal convergence hypotheses were given in [5] under Hölder condition. The hypotheses on the derivatives of F restrict the applicability of method (2). As a motivational example, define function F on $\mathbb{X} = \mathbb{Y} = \mathbb{R}$, $D = [-\frac{5}{2}, \frac{1}{2}]$ by

$$F(x) = \begin{cases} x^3 \ln x^2 + x^5 - x^4, & x \neq 0, \\ 0, & x = 0. \end{cases}$$

Then, we have that

$$F'(x) = 3x^2 \ln x^2 + 5x^4 - 4x^3 + 2x^2,$$
$$F''(x) = 6x \ln x^2 + 20x^3 - 12x^2 + 10x$$

and
$$F'''(x) = 6\ln x^2 + 60x^2 - 24x + 22.$$

Then, obviously the third–order derivative of the involved function $F'''(x)$ is not bounded on D. Notice that, in particular there is a plethora of iterative methods for approximating solutions of nonlinear equations [1–22]. These results show that initial guess should be close to the required root for the convergence of the corresponding methods. But, how close initial guess should be required for the convergence of the corresponding method? These local results give no information on the radius of the ball convergence for the corresponding method. We address this question for method (2) in the next section 2.

In the present study, we expand the applicability of method (2) by using only hypotheses on the first–order derivative of function F and generalized Lipschitz conditions. Moreover, we we will avoid to use Taylor series expansions and use Lipschitz parameters. In this way, there is no need to use the higher–order derivatives to show the convergence of the scheme (2).

The rest of the chapter is organized as follows: in Section 2 contains the local convergence analysis of method (2). The numerical examples appear in the concluding Section 3.

2. LOCAL CONVERGENCE

We present the local convergence analysis of method (2) in this section by using some scalar functions and parameters. Let v, w_0, w be nondecreasing continuous functions defined on the interval $[0, \infty)$ with values in $[0, \infty)$ and $w_0(0) = w(0) = 0$. Define

$$r_0 = \sup\{t \geq 0 : w_0(t) < 1\}. \tag{3}$$

Moreover, define functions $g_i, h_i, i = 1, 2, 3, 4$ on the interval $[0, r_0)$ by

$$g_1(t) = \frac{\int_0^1 w((1-\theta)t)d\theta + \frac{1}{3}\int_0^1 v(\theta t)d\theta}{1 - w_0(t)},$$

$$g_2(t) = \frac{\int_0^1 w((1-\theta)t)d\theta}{1 - w_0(t)} + \frac{21}{8}\frac{\int_0^1 v(\theta t)d\theta \times \int_0^1 v(\theta g_1(t)t)d\theta g_1(t)t}{(1 - w_0(t))^2}$$

$$+ \frac{9}{2}\frac{\int_0^1 v(\theta t)d\theta \times \left(\int_0^1 v(\theta g_1(t)t)d\theta g_1(t)t\right)^2}{(1 - w_0(t))^3}$$

$$+ \frac{15}{8}\frac{\int_0^1 v(\theta t)d\theta \times \left(\int_0^1 v(\theta g_1(t)t)d\theta g_1(t)t\right)^3}{(1 - w_0(t))^4},$$

$$g_3(t) = g_2(t) + \frac{\int_0^1 v(\theta g_2(t)t)d\theta g_2(t)}{1-w_0(t)} \times \left(3 + \frac{5}{2}\frac{\int_0^1 v(\theta g_1(t)t)d\theta g_1(t)t}{1-w_0(t)}\right.$$
$$\left. + \frac{1}{2}\left(\frac{\int_0^1 v(\theta g_1(t)t)d\theta g_1(t)t}{1-w_0(t)}\right)^2\right),$$

$$g_4(t) = g_3(t) + \frac{\int_0^1 v(\theta g_3(t)t)d\theta g_3(t)}{1-w_0(t)} \times \left(3 + \frac{5}{2}\frac{\int_0^1 v(\theta g_1(t)t)d\theta g_1(t)t}{1-w_0(t)}\right.$$
$$\left. + \frac{1}{2}\left(\frac{\int_0^1 v(\theta g_1(t)t)d\theta g_1(t)t}{1-w_0(t)}\right)^2\right),$$

and
$$h_i = g_i(t) - 1.$$

Suppose that
$$v(0) < 3. \tag{4}$$

We have that $h_1(0) = \frac{v(0)}{3} - 1 < 0$, $h_j(0) = -1 < 0$, $j = 2, 3, 4$ and $h_i(t) \to +\infty$ as $t \to r_0^-$. Then, by the intermediate value theorem, we know that the functions h_i have zeros in the interval $(0, r_0)$. Denote by r_i, respectively the the smallest such zero. Define the radius of convergence r by
$$r = \min\{r_i\}. \tag{5}$$

Then, we have that
$$0 < r < r_0 \tag{6}$$
and for each $t \in [0, r)$
$$0 \le w_0(t) < 1 \tag{7}$$
and
$$0 \le g_i(t) < 1. \tag{8}$$

Let $U(\gamma, \rho)$, $\bar{U}(\gamma, \rho)$, stand respectively for the open and closed balls in X with center $\gamma \in X$ and of radius $\rho > 0$. Next, we present the local convergence analysis of method (2) using the preceding notations.

Theorem 1. *Let $F : D \subseteq X \to Y$ be a differentiable operator. Let v, w_0, $w : [0, \infty) \to [0, \infty)$ be nondecreasing continuous functions with $w_0(0) = w(0) = 0$ so that (4) holds and r_0 be given by (3). Suppose that there exists $x^* \in D$ such that for each $x \in D$*
$$F(x^*) = 0, \quad F'(x^*)^{-1} \in L(Y, X), \tag{9}$$
$$\|F'(x^*)^{-1}(F'(x) - F'(x^*))\| \le w_0(\|x - x^*\|). \tag{10}$$

Moreover, suppose that for each $x, y \in D_0 = D \cap U(x^*, r_0)$
$$\|F'(x^*)^{-1}(F'(x) - F'(y))\| \le w(\|x - y\|), \tag{11}$$
$$\|F'(x^*)^{-1}F'(x)\| \le v(\|x - x^*\|), \tag{12}$$

and
$$\bar{U}(x^*, r) \subseteq D, \tag{13}$$

Expanding the Applicability of an Eighth–Order Method ... 157

where the radius of convergence r is defined by (5). Then, sequence $\{x_n\}$ generated for $x_0 \in U(x^*, r) - \{x^*\}$ by method (2) is well defined, remains in $U(x^*, r)$ for each $n = 0, 1, 2, \ldots$ and converges to x^*. Moreover, the following estimates hold

$$\|y_n - x^*\| \leq g_1(\|x_n - x^*\|)\|x_n - x^*\| \leq \|x_n - x^*\| < r, \tag{14}$$

$$\|z_n - x^*\| \leq g_2(\|x_n - x^*\|)\|x_n - x^*\| \leq \|x_n - x^*\|, \tag{15}$$

$$\|z_n - x^*\| \leq g_3(\|x_n - x^*\|)\|x_n - x^*\| \leq \|x_n - x^*\| \tag{16}$$

and

$$\|x_{n+1} - x^*\| \leq g_4(\|x_n - x^*\|)\|x_n - x^*\| \leq \|x_n - x^*\|, \tag{17}$$

where the functions g_i are defined above the theorem. Furthermore, if

$$\int_0^1 w_0(\theta R) d\theta < 1 \text{ for } R \geq r, \tag{18}$$

then, the point x^* is the only solution of equation $F(x) = 0$ in $D_1 := D \cap \bar{U}(x^*, R)$.

Proof. We shall show sequences $\{x_n\}$ is well defined and converges to x^* so that estimates (14)–(17) hold with the help mathematical induction. By hypotheses $x_0 \in U(x^*, r) - \{x^*\}$, (3), (5) and (10), we have that

$$\|F'(x^*)^{-1}(F'(x_0) - F'(x^*))\| \leq w_0(\|x_0 - x^*\|) < w_0(r) < 1. \tag{19}$$

It follows from (19) and the Banach Lemma on invertible operators [2, 3] that $F'(x_0)^{-1} \in L(Y, X)$, y_0, z_0, w_0, x_1 are well defined and

$$\|F'(x_0)^{-1} F'(x^*)\| \leq \frac{1}{1 - w_0(\|x_0 - x^*\|)}. \tag{20}$$

We can write by (9)

$$F(x_0) = F(x_0) - F(x^*) = \int_0^1 F'(x^* + \theta(x_0 - x^*))(x_0 - x^*) d\theta. \tag{21}$$

Notice that $\|x^* + \theta(x_0 - x^*) - x^*\| = \theta\|x_0 - x^*\| < r$, so $x^* + \theta(x_0 - x^*) \in U(x^*, r)$ for each $\theta \in [0, 1]$.

By (12) and (21), we obtain that

$$\|F'(x^*)^{-1} F(x_0)\| \leq \int_0^1 v(\theta\|x_0 - x^*\|)\|x_0 - x^*\| d\theta. \tag{22}$$

In view (5), (6), (7) (for $i = 1$), (8), (11), (20) and (22), we obtain that

$$\|y_0 - x^*\| = \left\|(x_0 - x^* - F'(x_0)^{-1} F(x_0)) + \frac{1}{3} F'(x_0)^{-1} F(x_0)\right\|$$

$$\leq \|F'(x_0)^{-1} F(x^*)\| \left\|\int_0^1 F'(x^*)^{-1} (F'(x^* + \theta(x_0 - x^*)) - F'(x_0)) (x_0 - x^*)\right\|$$

$$+ \frac{1}{3} \|F'(x_0)^{-1} F(x^*)\| \|F'(x^*)^{-1} F(x_0)\|$$

$$\leq \frac{\int_0^1 w((1-\theta)\|x_0 - x^*\|) d\theta \|x_0 - x^*\| + \frac{1}{3} \int_0^1 v(\theta\|x_0 - x^*\|) d\theta \|x_0 - x^*\|}{1 - w_0(\|x_0 - x^*\|)}$$

$$= g_1(\|x_0 - x^*\|)\|x_0 - x^*\| \leq \|x_0 - x^*\| < r,$$

$$\tag{23}$$

which shows (14) for $n = 0$ and $y_0 \in U(x^*, r)$.

Then, as in (22), we have

$$\|F'(x^*)^{-1}F(y_0)\| \leq \int_0^1 v(\theta\|y_0 - x^*\|)\|y_0 - x^*\|d\theta g_1(\|x_0 - x^*\|)\|x_0 - x^*\|. \quad (24)$$

Using the second sub step of method (2) for $n = 0$, (5), (6), (7) (for $i = 2$), (20), (23) and (24), we get in turn that

$$\begin{aligned}
\|z_0 - x^*\| &= \|x_0 - x^* - F'(x_0)^{-1}F(x_0)\| + \frac{21}{8}\|F'(x_0)^{-1}F(x^*)\|^2 \|F'(x^*)^{-1}F(y_0)\| \\
&\quad \times \|F'(x^*)^{-1}F(x_0)\| \\
&\quad + \frac{9}{2}\|F'(x_0)^{-1}F(x^*)\|^3 \|F'(x^*)^{-1}F(y_0)\|^2 \|F'(x^*)^{-1}F(x_0)\| \\
&\quad + \frac{15}{8}\|F'(x_0)^{-1}F(x^*)\|^4 \|F'(x^*)^{-1}F(y_0)\|^3 \|F'(x^*)^{-1}F(x_0)\| \\
&\leq \frac{\int_0^1 w((1-\theta)\|x_0 - x^*\|)d\theta\|x_0 - x^*\|}{1 - w_0(\|x_0 - x^*\|)} \\
&\quad + \frac{21}{8}\frac{\int_0^1 v(\theta\|x_0 - x^*\|)d\theta\|x_0 - x^*\| \int_0^1 v(\theta\|y_0 - x^*\|)d\theta\|y_0 - x^*\|}{(1 - w_0(\|x_0 - x^*\|))^2}, \\
&\quad + \frac{9}{2}\frac{\int_0^1 v(\theta\|x_0 - x^*\|)d\theta\|x_0 - x^*\| \left(\int_0^1 v(\theta\|y_0 - x^*\|)d\theta\|y_0 - x^*\|\right)^2}{(1 - w_0(\|x_0 - x^*\|))^3} \\
&\quad + \frac{15}{8}\frac{\int_0^1 v(\theta\|x_0 - x^*\|)d\theta\|x_0 - x^*\| \left(\int_0^1 v(\theta\|y_0 - x^*\|)d\theta\|y_0 - x^*\|\right)^3}{(1 - w_0(\|x_0 - x^*\|))^4}, \\
&= g_2(\|x_0 - x^*\|)\|x_0 - x^*\| \leq \|x_0 - x^*\| < r,
\end{aligned} \quad (25)$$

which shows (15), holds for $n = 0$ and $z_0 \in U(x^*, r)$. We also have as in (22) that

$$\|F'(x^*)^{-1}F(z_0)\| \leq \int_0^1 v(\theta\|z_0 - x^*\|)\|z_0 - x^*\|d\theta. \quad (26)$$

Similarly, by (2), (5), (7) (for $i = 3$), (20), (23)–(26), we have in turn that

$$\begin{aligned}
\|w_0 - x^*\| &= \|z_0 - x^*\| + \left(3\|I\| + \frac{5}{2}\|F'(x_0)^{-1}F(y_0)\| \right. \\
&\quad \left. + \frac{1}{2}\|F'(x_0)^{-1}F(x^*)\|^2 \|F'(x^*)^{-1}F(y_0)\|^2\right) \\
&\quad \times \|F'(x_0)^{-1}F'(x^*)\| \|F'(x^*)^{-1}F(z_0)\| \\
&\leq g_2(\|x_0 - x^*\|)\|x_0 - x^*\| + \frac{\int_0^1 v(\theta\|z_0 - x^*\|)d\theta\|z_0 - x^*\|}{1 - w_0(\|x_0 - x^*\|)} \\
&\quad \times \left(3 + \frac{5}{2}\frac{\int_0^1 v(\theta\|y_0 - x^*\|)d\theta\|y_0 - x^*\|}{1 - w_0(\|x_0 - x^*\|)}\right. \\
&\quad \left. + \frac{1}{2}\left(\frac{\int_0^1 v(\theta\|y_0 - x^*\|)d\theta\|y_0 - x^*\|}{1 - w_0(\|x_0 - x^*\|)}\right)^2\right) \\
&= g_3(\|x_0 - x^*\|)\|x_0 - x^*\| \leq \|x_0 - x^*\| < r,
\end{aligned} \quad (27)$$

which shows (16) for $n = 0$ and $w_0 \in U(x^*, r)$.

We also have that

$$\|F'(x^*)^{-1}F(w_0)\| \leq \int_0^1 v(\theta\|w_0 - x^*\|)\|w_0 - x^*\|d\theta. \tag{28}$$

Next, by (5), (7) (for $i = 4$), (20), (22), (23), and (25)–(28), we get in turn that

$$\|x_1 - x^*\| \leq \|w_0 - x^*\| + \|F'(x_0)^{-1}F'(x^*)\| \|F'(x^*)^{-1}F(w_0)\|$$
$$\times \left[3\|I\| + \frac{5}{2}\frac{\int_0^1 v(\theta\|y_0 - x^*\|)d\theta\|y_0 - x^*\|}{1 - w_0(\|x_0 - x^*\|)}\right.$$
$$\left. + \frac{1}{2}\left(\frac{\int_0^1 v(\theta\|y_0 - x^*\|)d\theta\|y_0 - x^*\|}{1 - w_0(\|x_0 - x^*\|)}\right)^2\right] \tag{29}$$
$$= g_4(\|x_0 - x^*\|)\|x_0 - x^*\| \leq \|x_0 - x^*\| < r,$$

which shows (17) for $n = 0$ and $x_1 \in U(x^*, r)$. By simply replacing x_0, y_0, z_0, w_0, x_1 by x_k, y_k, z_k, w_k, x_{k+1} in the preceding estimates we arrive at (14)–(17). In view of the estimates

$$\|x_{k+1} - x^*\| \leq c\|x_k - x^*\| < r, \ c = g_4(\|x_0 - x^*\|) \in [0, 1), \tag{30}$$

we deduce that $\lim_{k \to \infty} x_k = x^*$ and $x_{k+1} \in U(x^*, r)$. Finally, to show the uniqueness part, let $y^* \in D_1$ with $F(y^*) = 0$. Define $Q = \int_0^1 F'(x^* + \theta(x^* - y^*))d\theta$. Using (7) and (14), we get that

$$\|F'(x^*)^{-1}(Q - F'(x^*))\| \leq \|\int_0^1 w_0(\theta\|y^* - x^*\|)d\theta$$
$$\leq \int_0^1 w_0(\theta R)d\theta < 1. \tag{31}$$

It follows from (31) that Q is invertible. Then, in view of the identity

$$0 = F(x^*) - F(y^*) = Q(x^* - y^*), \tag{32}$$

we conclude that $x^* = y^*$. $\square$

Remark 2. (a) *It follows from (7) that condition (9) can be dropped and be replaced by*

$$v(t) = 1 + w_0(t) \ or \ v(t) = 1 + w_0(r_0), \tag{33}$$

since,

$$\|F'(x^*)^{-1}\left[(F'(x) - F'(x^*)) + F'(x^*)\right]\| = 1 + \|F'(x^*)^{-1}(F'(x) - F'(x^*))\|$$
$$\leq 1 + w_0(\|x - x^*\|)$$
$$= 1 + w_0(t) \ for \ \|x - x^*\| \leq r_0. \tag{34}$$

(b) *If the function w_0 is strictly increasing, then we can choose*

$$r_0 = w_0^{-1}(1) \tag{35}$$

instead of (3).

(c) If w_0, w, v are constants functions, then

$$r_1 = \frac{2}{2w_0 + w} \tag{36}$$

and

$$r \leq r_1. \tag{37}$$

Therefore, the radius of convergence r can be larger than the radius of convergence r_1 for Newton's method

$$x_{n+1} = x_n - F'(x_n)^{-1} F(x_n). \tag{38}$$

Notice also that the earlier radius of convergence given independently by Rheindoldt [22] and Traub [8] is

$$r_{TR} = \frac{2}{3w_1} \tag{39}$$

and by Argyros [2, 3]

$$r_A = \frac{2}{2w_0 + w_1}, \tag{40}$$

where w_1 is the Lipschitz constant for (8) on D. But, we have

$$w \leq w_1, \ w_0 \leq w_1, \tag{41}$$

so

$$r_{TR} \leq r_A \leq r_1 \tag{42}$$

and

$$\frac{r_{TR}}{r_A} \to \frac{1}{3} \quad \text{as} \quad \frac{w_0}{w} \to 0. \tag{43}$$

The radius of convergence q used in [1] is smaller than the radius r_{DS} given by Dennis and Schabel [3]

$$q < r_{SD} = \frac{1}{2w_1} < r_{TR}. \tag{44}$$

However, q can not be computed using the Lipschitz constants.

(d) The order of convergence of method (2) was shown in [1] using hypotheses up to the fifth–order derivative of F. We have only used hypotheses on the first–order derivative of F. The order of convergence can be determined by using the following formula for the computational order of convergence COC given by

$$\xi = \frac{ln\frac{\|x_{n+2}-x^*\|}{\|x_{n+1}-x^*\|}}{ln\frac{\|x_{n+1}-x^*\|}{\|x_n-x^*\|}}, \quad \text{for each } n = 0, 1, 2, \ldots \tag{45}$$

or the approximate computational order of convergence (ACOC) [19], given by

$$\xi^* = \frac{ln\frac{\|x_{n+2}-x_{n+1}\|}{\|x_{n+1}-x_n\|}}{ln\frac{\|x_{n+1}-x_n\|}{\|x_n-x_{n-1}\|}}, \quad \text{for each } n = 1, 2, \ldots \tag{46}$$

that do not require derivatives. Notice also that ξ^* does not even require the knowledge of exact root x^*.

(e) *The results obtained here can be used for operators F satisfying the autonomous differential equation [2, 3] of the form*

$$F'(x) = P(F(x)) \tag{47}$$

where P is a known continuous operator. Since $F'(x^*) = P(F(x^*)) = P(0)$, we can apply the results without actually knowing the solution x^*. Let as an example $F(x) = e^x - 1$. Then, we can choose $P(x) = x + 1$.

3. NUMERICAL EXAMPLES AND APPLICATIONS

In this section, we shall demonstrate the theoretical results which we have proposed in the section 2. Therefore, we consider four numerical examples in this section, which are defined as follows:

Example 3. *Let $X = Y = C[0, 1]$ and consider the nonlinear integral equation of the mixed Hammerstein–type [5, 16], defined by*

$$x(s) = 1 + \int_0^1 G(s, t)\left(x(t)^{\frac{3}{2}} + \frac{x(t)^2}{2}\right) dt \tag{48}$$

where the kernel G is the Green's function defined on the interval $[0, 1] \times [0, 1]$ by

$$F(s, t) = \begin{cases} (1-s)t, & t \leq s, \\ s(1-t), & s \leq t. \end{cases} \tag{49}$$

The solution $x^*(s) = 0$ is the same as the solution of equation (1), where $F :\subseteq C[0, 1] \to C[0, 1]$ defined by

$$F(x)(s) = x(s) - \int_0^t G(s, t)\left(x(t)^{\frac{3}{2}} + \frac{x(t)^2}{2}\right) dt. \tag{50}$$

Notice that

$$\left\|\int_0^t G(s, t) dt\right\| \leq \frac{1}{8}. \tag{51}$$

Then, we have that

$$F'(x)y(s) = y(s) - \int_0^t G(s, t)\left(\frac{3}{2}x(t)^{\frac{1}{2}} + x(t)\right) dt,$$

so since $F'(x^*(s)) = I$,

$$\|F'(x^*)^{-1}(F'(x) - F'(y))\| \leq \frac{1}{8}\left(\frac{3}{2}\|x - y\|^{\frac{1}{2}} + \|x - y\|\right). \tag{52}$$

Therefore, we can choose

$$w_0(t) = w(t) = \frac{1}{8}\left(\frac{3}{2}t^{\frac{1}{2}} + t\right)$$

and by Remark 2 (a)
$$v(t) = 1 + w_0(t).$$

The results in [1,5] can not be used to solve this problem, since F' is not Lipschitz. However, our results can apply.

Example 4. *Suppose that the motion of an object in three dimensions is governed by system of differential equations*
$$\begin{aligned} f_1'(x) - f_1(x) - 1 &= 0 \\ f_2'(y) - (e-1)y - 1 &= 0 \\ f_3'(z) - 1 &= 0 \end{aligned} \tag{53}$$

with x, y, $z \in \Omega$ for $f_1(0) = f_2(0) = f_3(0) = 0$. then, the solution of the system is given for $w = (x, y, z)^T$ by function $F := (f_1, f_2, f_3) : \Omega \to \mathbb{R}^3$ defined by

$$F(v) = \left(e^x - 1, \frac{e-1}{2}y^2 + y, z\right)^T. \tag{54}$$

Then the Fréchet–derivative is given by

$$F'(v) = \begin{bmatrix} e^x & 0 & 0 \\ 0 & (e-1)y + 1 & 0 \\ 0 & 0 & 1 \end{bmatrix}.$$

Then, we have that $w_0(t) = L_0 t$, $w(t) = Lt$, $w_1(t) = L_1 t$, $w_0 = L_0$, $w_1 = L_1$ and $v(t) = M$, where $L_0 = e - 1 < L = e^{\frac{1}{L_0}} = 1.789572397$, $L_1 = e$ and $M = e^{\frac{1}{L_0}} = 1.7896$. The parameters using method (2) are:

$r_0 = 0.13333$, $r_1 = 0.066667$, $r_2 = 0.010381$, $r_3 = 0.0055809$, $r_4 = 0.0039782$,
$r_{TR} = 0.24525$, $r_A = 0.32495$, $r_{DS} = 0.18394$,

so
$$r = 0.0039782.$$

Example 5. *Let $A_1 = A_2 = C[0, 1]$, be the space of continuous functions defined on the interval $[0, 1]$ and be equipped with max norm. Let $\Omega = \bar{U}(0, 1)$ and $B(x) = F''(x)$ for each $x \in \Omega$. Define F on Ω*

$$F(\varphi)(x) = \phi(x) - 5\int_0^1 x\theta\varphi(\theta)^3 d\theta. \tag{55}$$

We have that

$$F'(\varphi(\xi))(x) = \xi(x) - 15\int_0^1 x\theta\varphi(\theta)^2 \xi(\theta) d\theta, \text{ for each } \xi \in \Omega. \tag{56}$$

Then, we have that $x^* = 0$, $L_0 = 7.5$, $L_1 = L = 15$ and $M = 2$. The parameters using method (2) for $w_0(t) = L_0 t$, $v(t) = 2 = M$, $w(t) = Lt$, $w_1 = L$ and $w_0 = L_0$ are:

$r_0 = 0.13333$, $r_1 = 0.022222$, $r_2 = 0.030157$, $r_3 = 0.0077783$, $r_4 = 0.0013404$,
$r_{TR} = 0.044444$, $r_A = 0.066667$, $r_{DS} = 0.033333$,

so
$$r = 0.0013404.$$

Example 6. *Returning back to the motivation example at the introduction on this paper, we have* $L = L_0 = 96.662907$ *and* $M = 2$. *The parameters using method (2) for* $w_0(t) = L_0 t$, $v(t) = 2 = M$, $w(t) = Lt$, $w_1(t) = L$ *and* $w_0 = L_0$ *are:*

$r_0 = 0.010345$, $r_1 = 0.0022989$, $r_2 = 0.0053004$, $r_3 = 0.0017053$, $r_4 = 0.00033322$,

$r_{TR} = 0.0068968$, $r_A = 0.0068968$, $r_{DS} = 0.0051726$,

so
$$r = 0.00033322.$$

REFERENCES

[1] Abbasbandy, S., Bakhtiari, P., Cordero, A., Torregrosa, J. R. and Lofti, T. "New efficient methods for solving nonlinear systems of equations with arbitrary even order". *Appl. Math. Comput.* (to appear).

[2] Argyros, I. K. 2008. *Convergence and Application of Newton–type Iterations.* Springer.

[3] Argyros, I. K. and Hilout, S. 2013. *Computational Methods in Nonlinear Analysis.* New Jersey: World Scientific Publ. Comp.

[4] Cordero, A., Torregrosa, J. R. and Vassileva, M. P. "Increasing the order of convergence of iterative schemes for solving nonlinear system". *J. Comput. Appl. Math.* 252: 86–94.

[5] Hernández, M. A. and Martínez, E. 2015. "On the semilocal convergence of a three steps Newton–type process under mild convergence conditions". *Numer. Algor.* 70: 377–392.

[6] Potra, F. A. and Pták, V. 1984. *Nondiscrete introduction and iterative process. Research Notes in Mathematics, 103.* Boston: Pitman.

[7] Sharma, J. R., Ghua, R. K. and Sharma, R. 2013. "An efficient fourth–order weighted–Newton method for system of nonlinear equations". *Numer. Algor.* 62: 307–323.

[8] Traub, J. F. 1964. *Iterative methods for the solution of equations.* Englewood Cliffs, N. J.: Prentice–Hall Series in Automatic Computation.

[9] Amat, S., Busquier, S., Plaza, S. and Gutiérrez, J. M. 2003. "Geometric constructions of iterative functions to solve nonlinear equations". *J. Comput. Appl. Math.* 157: 197–205.

[10] Amat, S., Busquier, S. and Plaza, S. 2005. "Dynamics of the King and Jarratt iterations". *Aequationes Math.* 69(3): 212–223.

[11] Amat, S., Hernández, M. A. and Romero, N. 2008. "A modified Chebyshev's iterative method with at least sixth order of convergence". *Appl. Math. Comput.* 206(1): 164–174.

[12] Argyros, I. K. and Magreñán, Á. A. 2015. "Ball convergence theorems and the convergence planes of an iterative methods for nonlinear equations". *SeMA* 71(1): 39–55.

[13] Argyros, I. K. and George, S. "Local convergence of some higher–order Newton–like method with frozen derivative". *SeMA* doi:10.1007/s40324-015-00398-8.

[14] Cordero, A. and Torregrosa, J. R. 2007. "Variants of Newton's method using fifth–order quadrature formulas". *Appl. Math. Comput.* 190: 686–698.

[15] Cordero, A. and Torregrosa, J. R. 2006. "Variants of Newton's method for functions of several variables". *Appl. Math. Comput.* 183: 199–208.

[16] Ezquerro, J. A. and Hernández, M. A. 2009. "New iterations of R–order four with reduced computational cost". *BIT Numer. Math.* 49: 325– 342.

[17] Ezquerro, J. A. and Hernández, M. A. 2003. "A uniparametric Halley type iteration with free second derivative". *Int. J. Pure and Appl. Math.* 6(1): 99–110.

[18] Gutiérrez, J. M. and Hernández, M. A. 1998. "Recurrence realtions for the super–Halley method". *Comput. Math. Appl.* 36: 1–8.

[19] Kou, J. 2007. "A third–order modification of Newton method for systems of nonlinear equations". *Appl. Math. Comput.* 191: 117–121.

[20] Montazeri, H., Soleymani, F., Shateyi, S. and Motsa, S. S. 1975. "On a new method for computing the numerical solution of systems of nonlinear equations". *J. Appl. Math.* Article ID 751975.

[21] Petkovic, M .S., Neta, B., Petkovic, L. and Džunič, J. 2013. *Multipoint methods for solving nonlinear equations*. Elsevier.

[22] Rheinboldt, W. C. 1978. *An adaptive continuation process for solving systems of nonlinear equations*. Polish Academy of Science, Banach Ctr. Publ. 3: 129–142.

In: Understanding Banach Spaces
Editor: Daniel González Sánchez

ISBN: 978-1-53616-745-0
© 2020 Nova Science Publishers, Inc.

Chapter 15

LOCAL CONVERGENCE FOR THREE–STEP EIGHTH– ORDER METHOD UNDER WEAK CONDITIONS

Ioannis K. Argyros[1,∗], Ramandeep Behl[2,†] and Daniel González Sánchez[3,‡]
[1]Department of Mathematical Sciences
Cameron University, Lawton, OK, US
[2]School of Mathematics, Statistics and Computer Sciences
University of KwaZulu–Natal, Pietermaritzburg, South Africa
[3]Escuela de Ciencias Físicas y Matemáticas
Universidad de Las Américas, Quito, Ecuador

Abstract

We present a local convergence analysis for an eighth–order convergent method in order to find a solution of nonlinear equation in a Banach space setting. In contrast to the earlier studies using hypotheses up to the seventh Fréchet–derivative, we only use hypotheses on the first–order Fréchet–derivative and Lipschitz constants. This way, we not only expand the applicability of these methods but also proposed the computable radius of convergence of these methods. Finally, a variety of concrete numerical examples demonstrate that our results even apply to solve those nonlinear equations where earlier studies cannot apply.

Keywords: iterative method, local convergence, Banach space, Lipschitz constant, order of convergence

AMS Subject Classification: 65G99, 65H10, 47J25, 47J05, 65D10, 65D99

[∗]E-mail: iargyros@cameron.edu.
[†]E-mail: ramanbehl87@yahoo.in.
[‡]Corresponding Author's E-mail: daniel.gonzalez.sanchez@udla.edu.ec.

1. INTRODUCTION

One of the most basic and important problems in Numerical analysis concerns with finding a locally unique solution x^* of the equation of the form

$$F(x) = 0, \tag{1}$$

where $F : D \subset X \to Y$ is a Fréchet–differentiable operator, X, Y are Banach spaces and D is a convex subset of X. Let us also denote $L(X, Y)$ as the space of bounded linear operators from X to Y.

The problem of finding x^* is very important, since many problems can be reduced to equation (1) using mathematical modeling [1–8]. However, it is not always possible to find the solution x^* in a closed form. Therefore, most of the methods are iterative to solve such type of problems. The convergence analysis of iterative methods is usually divided into two categories: semi–local and local convergence analysis. The semi–local convergence matter is, based on the information around an initial point, to give criteria ensuring the convergence of iteration procedures. A very important problem in the study of iterative procedures is the convergence domain. Therefore, it is very important to propose the radius of convergence of the iterative methods.

We study the local convergence of the three–step eighth–order convergent method defined for each $n = 0, 1, 2, \ldots$ by

$$\begin{aligned}
y_n &= x_n - F'(x_n)^{-1} F(x_n), \\
z_n &= \phi_1(x_n, y_n) \\
z_n^{(1)} &= z_n - \phi(x_n, y_n) F(z_n), \\
&\vdots \\
z_n^{(m-1)} &= z_n^{(m-2)} - \phi(x_n, y_n) F(z_n^{m-2}), \\
x_{n+1} &= z_n^{(m-1)} - \phi(x_n, y_n) F(z_n^{m-1}),
\end{aligned} \tag{2}$$

where $x_0 \in D$ is an initial point, $z_n = \phi_1(x_n, y_n)$ is a λ–order method (for $\lambda \geq 1$) and

$$\phi(s, t) = \frac{1}{3} \left\{ 4[3F'(t) - F'(s)]^{-1} + F'(s)^{-1} \right\}.$$

F' denotes the first–order Fréchet–derivative of operator F. The eighth–order of convergence was shown in [8] using the Taylor series expansion, certain hypotheses on the function F and hypotheses reaching up to the seventh–order derivative of the involved function F although only first–order derivative appears in the proposed scheme (2). The semilocal convergence hypotheses were given in [4] under Hölder condition. The hypotheses on the derivatives of F restrict the applicability of method (2). As a motivational example, define function F on $\mathbb{X} = \mathbb{Y} = \mathbb{R}$, $D = [-\frac{5}{2}, \frac{1}{2}]$ by

$$F(x) = \begin{cases} x^3 \ln x^2 + x^5 - x^4, & x \neq 0, \\ 0, & x = 0. \end{cases}$$

Then, we have that

$$F'(x) = 3x^2 \ln x^2 + 5x^4 - 4x^3 + 2x^2,$$

$$F''(x) = 6x \ln x^2 + 20x^3 - 12x^2 + 10x$$

and

$$F'''(x) = 6 \ln x^2 + 60x^2 - 24x + 22.$$

Then, obviously the third–order derivative of the involved function $F'''(x)$ is not bounded on D. Notice that, in–particular there is a plethora of iterative methods for approximating solutions of nonlinear equations [1–22]. These results show that initial guess should be close to the required root for the convergence of the corresponding methods. But, how close initial guess should be required for the convergence of the corresponding method? These local results give no information on the radius of the ball convergence for the corresponding method. We address this question for method (2) in the next section.

In the present study, we expand the applicability of method (2) by using only hypotheses on the first–order derivative of function F and generalized Lipschitz conditions. Moreover, we will avoid using Taylor series expansions and use Lipschitz parameters. In this way, there is no need to use the higher–order derivatives to show the convergence of the scheme (2).

The rest of the chapter is organized as follows: in section 2 contains the local convergence analysis of method (2). The numerical examples appear in the concluding Section 3.

2. LOCAL CONVERGENCE

In this section, we present the local convergence analysis of method (2), by using some scalar functions and parameters. Let $v, w_0, w : [0, +\infty) \to [0, +\infty)$ be nondecreasing continuous functions. In addition, let us consider $\phi : [0, +\infty)^2 \to [0, +\infty)$ be a nondecreasing continuous and $\phi_1 : D \times D \to X$ be continuous function such that $w_0(0) = w(0) = 0$. Suppose that equation

$$w_0(t) = 1 \qquad (3)$$

has at least one positive solution. Denote by r_0 the smallest such solution. Define functions g_1, g_2, h_1 and h_2 on the interval $[0, r_0)$ by

$$g_1(t) = \frac{\int_0^1 w((1-\theta)t)d\theta}{1 - w_0(t)},$$

$$g_2(t) = \phi(t, g_1(t)t)t^{\lambda-1},$$

$$h_1 = g_1(t) - 1$$

and

$$h_2(t) = g_2(t) - 1.$$

We have that $h_1(0) = h_2(0) = -1$ and $h_1(t) \to +\infty$, $h_2(t) \to +\infty$ as $t \to r_0^-$. Then, by the intermediate value theorem, we know that the functions h_1 and h_2 have zeros in the interval $(0, r_0)$. Denote by r_1 and r_2, respectively the the smallest such zero of the function h_1 and h_2. Define functions p and h_p on the interval $[0, r_0)$ by

$$p(t) = \frac{1}{2}\left[3w_0(g_1(t)t) + w_0(t)\right],$$

$$h_p(t) = p(t) - 1.$$

We have again that $h_p(0) = -1 < 0$ and $h_p(t) \to +\infty$ as $t \to r_0^-$. Let us denote r_p to be the smallest zero of the function h_p in the interval $(0, r_0)$. Let $\bar{r} = \min\{r_0, r_p\}$. Define functions $g^{(i)}$, $h^{(i)}$, $i = 1, 2, \ldots, m$ by

$$g^{(i)}(t) = \left(1 + q(t)\int_0^1 v(\theta g^{(i-1)}(t)t)d\theta\right)^{i-1} g_2(t)$$

and

$$h^{(i)} = g^{(i)}(t) - 1,$$

where

$$q(t) = \frac{1}{3}\left(\frac{2}{1 - p(t)} + \frac{1}{1 - w_0(t)}\right).$$

We also get that $h^{(i)}(0) = -1$ and $h^{(i)}(t) \to +\infty$ as $t \to \bar{r}^-$. Denote by $r^{(i)}$ be the smallest zeros of corresponding function $h^{(i)}$ on the interval $(0, \bar{r})$.

Define the radius of convergence r by

$$r = \min\{r_1, r_2, r^{(i)}\}. \tag{4}$$

Then, we have that

$$0 < r < r_0 \tag{5}$$

and for each $t \in [0, r)$

$$0 \leq g_1(t) < 1, \tag{6}$$

$$0 \leq g_2(t) < 1, \tag{7}$$

$$0 \leq p(t) < 1, \tag{8}$$

$$0 \leq q(t) \tag{9}$$

and

$$0 \leq g^{(i)}(t) < 1. \tag{10}$$

Let $U(\xi, \rho)$, $\bar{U}(\xi, \rho)$, stand respectively for the open and closed balls in X with center $\xi \in X$ and of radius $\rho > 0$. Next, we present the local convergence analysis of method (2) using the preceding notations.

Theorem 1. *Let $F : D \subseteq X \to Y$ be a Fréchet–differentiable operator. Let v, w_0, w : $[0, \infty) \to [0, \infty)$ be nondecreasing continuous functions, $\phi : [0, +\infty)^2 \to [0, +\infty)$ be a nondecreasing continuous and $\phi_1 : D \times D \to X$ be continuous function such that $w_0(0) = w(0) = 0$ and parameter r_0 be defined by (3). Suppose that there exists $x^* \in D$ such that for each $x \in D$*

$$F(x^*) = 0, \quad F'(x^*)^{-1} \in L(Y, X) \tag{11}$$

and
$$\|F'(x^*)^{-1}(F'(x) - F'(x^*))\| \leq w_0(\|x - x^*\|). \tag{12}$$

Moreover, suppose that for each $x, y \in D_0 = D \cap U(x^*, r_0)$

$$\|F'(x^*)^{-1}(F'(x) - F'(y))\| \leq w(\|x - y\|), \tag{13}$$

$$\|F'(x^*)^{-1}F'(x)\| \leq v(\|x - x^*\|), \tag{14}$$

$$\|\phi_1(x, y) - x^*\| \leq \phi(\|x - x^*\|, \|y - x^*\|)\|x - x^*\|^\lambda \tag{15}$$

and
$$\bar{U}(x^*, r) \subseteq D, \tag{16}$$

where the radius of convergence r is defined by (5). Then, sequence $\{x_n\}$ generated for $x_0 \in U(x^*, r) - \{x^*\}$ by method (2) is well defined, remains in $U(x^*, r)$ for each $n = 0, 1, 2, \ldots$ and converges to x^*. Moreover, the following estimates hold

$$\|y_n - x^*\| \leq g_1(\|x_n - x^*\|)\|x_n - x^*\| \leq \|x_n - x^*\| < r, \tag{17}$$

$$\|z_n - x^*\| \leq g_2(\|x_n - x^*\|)\|x_n - x^*\| \leq \|x_n - x^*\|, \tag{18}$$

$$\|z_n^{(i)} - x^*\| \leq g^{(i)}(\|x_n - x^*\|)\|x_n - x^*\| \leq \|x_n - x^*\|, \ i = 1, 2, \ldots, m-1 \tag{19}$$

and
$$\|x_{n+1} - x^*\| \leq g^{(m)}(\|x_n - x^*\|)\|x_n - x^*\| \leq \|x_n - x^*\|, \tag{20}$$

where the functions g are defined previously. Furthermore, if

$$\int_0^1 w_0(\theta R)d\theta < 1 \ \text{for} \ R \geq r, \tag{21}$$

then, the point x^* is the only solution of equation $F(x) = 0$ in $D_1 := D \cap \bar{U}(x^*, R)$.

Proof. Using mathematical induction, we shall show that the sequence $\{x_n\}$ is well defined and converges to x^* so that estimates (17) – (20) are satisfied. Using hypothesis $x_0 \in U(x^*, r) - \{x^*\}$, (3), (4) and (12), we have that

$$\|F'(x^*)^{-1}(F'(x_0) - F'(x^*))\| \leq w_0(\|x_0 - x^*\|) < w_0(r) < 1. \tag{22}$$

It follows from (22) and the Banach Lemma on invertible operators [1,2] that $F'(x_0)^{-1} \in L(Y, X)$, y_0, z_0 are well defined and

$$\|F'(x_0)^{-1}F'(x^*)\| \leq \frac{1}{1 - w_0(\|x_0 - x^*\|)}. \tag{23}$$

In view of (2), (4), (6), (13) and (23), we get in turn that

$$\|y_0 - x^*\| = \|x_0 - x^* - F'(x_0)^{-1}F(x_0)\|$$
$$\leq \|F'(x_0)^{-1}F'(x^*)\| \left\| \int_0^1 F'(x^*)^{-1}\left(F'(x^* + \theta(x_0 - x^*)) - F'(x_0)\right)(x_0 - x^*)d\theta \right\|$$
$$\leq \frac{\int_0^1 w((1-\theta)\|x_0 - x^*\|)d\theta \|x_0 - x^*\|}{1 - w_0(\|x_0 - x^*\|)}$$
$$= g_1(\|x_0 - x^*\|)\|x_0 - x^*\| \leq \|x_0 - x^*\| < r, \tag{24}$$

which implies (17) for $n = 0$ and $y_0 \in U(x^*, r)$. Then, by (2), (4), (7), (15) and (24), we get that

$$\begin{aligned}
\|z_0 - x^*\| &= \|\phi_1(x_0, y_0) - x^*\| \\
&\leq \psi(\|x_0 - x^*\|, \|y_0 - x^*\|)\|x_0 - x^*\|^\lambda \\
&\leq \psi(\|x_0 - x^*\|, g_1(\|x_0 - x^*\|)\|x_0 - x^*\|)\|x_0 - x^*\|^\lambda \\
&= g_2(\|x_0 - x^*\|)\|x_0 - x^*\| \leq \|x_0 - x^*\| < r,
\end{aligned} \quad (25)$$

which implies (18) for $n = 0$ and and $z_0 \in U(x^*, r)$. We can write by (11)

$$F(z_0) = F(z_0) - F(x^*) = \int_0^1 F'(x^* + \theta(z_0 - x^*)) d\theta (z_0 - x^*). \quad (26)$$

Notice that $\|x^* + \theta(z_0 - x^*) - x^*\| = \theta\|z_0 - x^*\| < r$, so $x^* + \theta(z_0 - x^*) \in U(x^*, r)$ for each $\theta \in [0, 1]$. Then, from (14), (25) and (26), we get that

$$\begin{aligned}
\|F'(x^*)^{-1} F(z_0)\| &\leq \int_0^1 v(\theta\|z_0 - x^*\|)\|z_0 - x^*\| d\theta \\
&\leq \int_0^1 v(\theta g_1(\|x_0 - x^*\|)\|x_0 - x^*\|) d\theta g_1(\|x_0 - x^*\|)\|x_0 - x^*\|.
\end{aligned} \quad (27)$$

We must show that $\phi(x_0, y_0) \neq 0$. Using (4), (8), (12) and (24), we obtain in turn that

$$\begin{aligned}
\left\| (2F'(x^*))^{-1} \left[3(F'(y_0) - F'(x^*)) + (F'(x^*) - F'(x_0)) \right] \right\| \\
\leq \frac{1}{2}(3w_0(\|y_0 - x^*\|) + w_0(\|x_0 - x^*\|)) \\
\leq \frac{1}{2}(3w_0(g_1(\|x_0 - x^*\|)\|x_0 - x^*\|) + w_0(\|x_0 - x^*\|)) \\
= p(\|x_0 - x^*\|) < p(r) < 1,
\end{aligned} \quad (28)$$

so $z_0, z_0^{(1)}, \ldots, z_0^{(m-1)}, x_1$ exist

$$\left\| \left[3F'(y_0) - F'(x^*) \right]^{-1} F'(x^*) \right\| \leq \frac{1}{2(1 - p(\|x_0 - x^*\|))} \quad (29)$$

and

$$\|\phi(x, y_0) F'(x^*)\| \leq \frac{1}{3} \left(\frac{2}{1 - p(\|x_0 - x^*\|)} + \frac{1}{1 - w_0(\|x_0 - x^*\|)} \right). \quad (30)$$

Using (2), (4), (7), (8), (10) (for $i = 2$), (27) and (30), we get in turn that

$$\begin{aligned}
\|z_0^{(1)} - x^*\| &= \|z_0 - x^*\| + \|\phi(x_0, y_0) F'(x^*)\| \|F'(x^*)^{-1} F(z_0)\| \\
&\leq \left(1 + \|\phi(x_0, y_0) F'(x^*)\| \int_0^1 v(\theta\|z_0 - x^*\|) d\theta \right) \|z_0 - x^*\| \\
&\leq g^{(1)}(\|x_0 - x^*\|)\|x_0 - x^*\| \leq \|x_0 - x^*\| < r,
\end{aligned} \quad (31)$$

so (19) holds for $n = 0$, $i = 1$ and $z_0^{-1} \in U(x^*, r)$. In an analogous way, we obtain for $i = 2, 3, \ldots, m - 1$ that

$$\|z_0^{(i-1)} - x^*\| = \|z_0^{(i-2)} - x^*\| + q(\|x_0 - x^*\|) \int_0^1 v(\theta\|z_0^{(i-2)} - x^*\|)d\theta \|z_0 - x^*\|$$
$$\leq g^{(i-1)}(\|x_0 - x^*\|)\|x_0 - x^*\| \leq \|x_0 - x^*\| < r, \quad (32)$$

which implies (19) holds for $n = 0$, $i = 1, 2, \ldots, m - 1$ and $z_0^m \in U(x^*, r)$. Then, for the last sub step of method (2), (4), (10) (for $i = m$) and the proceeding estimates, we obtain that

$$\|x_1 - x^*\| \leq \|z_0^{(m-1)} - x^*\| + \|\phi(x_0, y_0)F'(x^*)\|\|F'(x^*)^{-1}F(z_0^{(m-1)})\|$$
$$\leq \left(1 + q(\|x_0 - x^*\|) \int_0^1 v(\theta\|z_0^{(m-1)} - x^*\|)d\theta\right) \|z_0^{(m-1)} - x^*\| \quad (33)$$
$$= g^{(m)}(\|x_0 - x^*\|)\|x_0 - x^*\| \leq \|x_0 - x^*\| < r,$$

which shows (20) for $n = 0$ and $x_1 \in U(x^*, r)$. By simply replacing $x_0, y_0, z_0, z_0^{(i)}$, $i = 1, 2, \ldots, m$, x_1 by $x_k, y_k, z_k, z_k^{(i)}, x_{k+1}$ in the preceding estimates we arrive at (17)–(20). Using the estimates

$$\|x_{k+1} - x^*\| \leq c\|x_k - x^*\| < r, \ c = g^{(m)}(\|x_0 - x^*\|) \in [0, 1), \quad (34)$$

we deduce that $\lim_{k \to \infty} x_k = x^*$ and $x_{k+1} \in U(x^*, r)$. Finally, to show the uniqueness part, let $y^* \in D_1$ with $F(y^*) = 0$. Define $Q = \int_0^1 F'(x^* + \theta(x^* - y^*))d\theta$. Using (7) and (14), we get that

$$\|F'(x^*)^{-1}(Q - F'(x^*))\| \leq \|\int_0^1 w_0(\theta\|y^* - x^*\|)d\theta$$
$$\leq \int_0^1 w_0(\theta R)d\theta < 1. \quad (35)$$

It follows from (39) that Q is invertible. Then, in view of the identity

$$0 = F(x^*) - F(y^*) = Q(x^* - y^*), \quad (36)$$

we conclude that $x^* = y^*$. □

Remark 2. (a) *It follows from (12) that condition (14) can be dropped and be replaced by*

$$v(t) = 1 + w_0(t) \text{ or } v(t) = 1 + w_0(r_0), \quad (37)$$

since,

$$\|F'(x^*)^{-1}\left[(F'(x) - F'(x^*)) + F'(x^*)\right]\| = 1 + \|F'(x^*)^{-1}(F'(x) - F'(x^*))\|$$
$$\leq 1 + w_0(\|x - x^*\|)$$
$$= 1 + w_0(t) \ \text{ for } \|x - x^*\| \leq r_0. \quad (38)$$

(b) If the function w_0 is strictly increasing, then we can choose
$$r_0 = w_0^{-1}(1) \tag{39}$$
instead of (3).

(c) If w_0, w, v are constant functions, then
$$r_1 = \frac{2}{2w_0 + w} \tag{40}$$
and
$$r \leq r_1. \tag{41}$$

Therefore, the radius of convergence r can not be larger than the radius of convergence r_1 for Newton's method
$$x_{n+1} = x_n - F'(x_n)^{-1} F(x_n). \tag{42}$$

Notice also that the earlier radius of convergence given independently by Rheinboldt [22] and Traub [7] is
$$r_{TR} = \frac{2}{3w_1} \tag{43}$$
and by Argyros [1, 2]
$$r_A = \frac{2}{2w_0 + w_1}, \tag{44}$$
where w_1 is the Lipschitz constant for (8) on D. But, we have
$$w \leq w_1, \ w_0 \leq w_1, \tag{45}$$
so
$$r_{TR} \leq r_A \leq r_1 \tag{46}$$
and
$$\frac{r_{TR}}{r_A} \to \frac{1}{3} \quad \text{as} \quad \frac{w_0}{w} \to 0. \tag{47}$$

The radius of convergence q used in [8] is smaller than the radius r_{DS} given by Dennis and Schnabel
$$q < r_{SD} = \frac{1}{2w_1} < r_{TR}. \tag{48}$$
However, q can not be computed using the Lipschitz constants.

(d) The order of convergence of method (2) was shown in [8] using hypotheses up to the fifth–order derivative of F. We have only used hypotheses on the first–order derivative of F. The order of convergence can be determined by using the following formula for the computational order of convergence COC given by
$$\xi = \frac{ln \frac{\|x_{n+2} - x^*\|}{\|x_{n+1} - x^*\|}}{ln \frac{\|x_{n+1} - x^*\|}{\|x_n - x^*\|}}, \quad \text{for each } n = 0, 1, 2, \ldots \tag{49}$$

or the approximate computational order of convergence (ACOC) [19], given by

$$\xi^* = \frac{\ln \frac{\|x_{n+2}-x_{n+1}\|}{\|x_{n+1}-x_n\|}}{\ln \frac{\|x_{n+1}-x_n\|}{\|x_n-x_{n-1}\|}}, \quad \text{for each } n = 1, 2, \ldots \qquad (50)$$

that do not require derivatives. Notice also that ξ^* does not even require the knowledge of the solution x^*.

(e) The results obtained here can be used for operators F satisfying the autonomous differential equation [1, 2] of the form

$$F'(x) = P(F(x)) \qquad (51)$$

where P is a known continuous operator. Since $F'(x^*) = P(F(x^*)) = P(0)$, we can apply the results without actually knowing the solution x^*. Let as an example $F(x) = e^x - 1$. Then, we can choose $P(x) = x + 1$.

3. NUMERICAL EXAMPLES AND APPLICATIONS

We shall demonstrate the theoretical results which we have proposed in the section 2. In particular, we consider four numerical examples in this section, which are defined as follows:

Example 3. Let $X = Y = C[0, 1]$ and consider the nonlinear integral equation of the mixed Hammerstein–type [4, 16], defined by

$$x(s) = 1 + \int_0^1 G(s, t) \left(x(t)^{\frac{3}{2}} + \frac{x(t)^2}{2} \right) dt \qquad (52)$$

where the kernel G is the Green's function defined on the interval $[0, 1] \times [0, 1]$ by

$$F(s, t) = \begin{cases} (1-s)t, & t \leq s, \\ s(1-t), & s \leq t. \end{cases} \qquad (53)$$

The solution $x^*(s) = 0$ is the same as the solution of equation (1), where $F :\subseteq C[0, 1] \to C[0, 1]$ defined by

$$F(x)(s) = x(s) - \int_0^t G(s, t) \left(x(t)^{\frac{3}{2}} + \frac{x(t)^2}{2} \right) dt. \qquad (54)$$

Notice that

$$\left\| \int_0^t G(s, t) dt \right\| \leq \frac{1}{8}. \qquad (55)$$

Then, we have that

$$F'(x)y(s) = y(s) - \int_0^t G(s, t) \left(\frac{3}{2} x(t)^{\frac{1}{2}} + x(t) \right) dt,$$

so since $F'(x^*(s)) = I$,

$$\|F'(x^*)^{-1}(F'(x) - F'(y))\| \leq \frac{1}{8}\left(\frac{3}{2}\|x-y\|^{\frac{1}{2}} + \|x-y\|\right). \quad (56)$$

Therefore, we can choose

$$w_0(t) = w(t) = \frac{1}{8}\left(\frac{3}{2}t^{\frac{1}{2}} + t\right)$$

and by Remark 2 (a)

$$v(t) = 1 + w_0(t).$$

The results in [4,8] can not be used to solve this problem, since F' is not Lipschitz. However, our results can apply.

Example 4. Suppose that the motion of an object in three dimensions is governed by system of differential equations

$$\begin{aligned} f_1'(x) - f_1(x) - 1 &= 0 \\ f_2'(y) - (e-1)y - 1 &= 0 \\ f_3'(z) - 1 &= 0 \end{aligned} \quad (57)$$

with x, y, $z \in \Omega$ for $f_1(0) = f_2(0) = f_3(0) = 0$. then, the solution of the system is given for $w = (x, y, z)^T$ by function $F := (f_1, f_2, f_3) : \Omega \to \mathbb{R}^3$ defined by

$$F(v) = \left(e^x - 1, \frac{e-1}{2}y^2 + y, z\right)^T. \quad (58)$$

Then the Fréchet–derivative is given by

$$F'(v) = \begin{bmatrix} e^x & 0 & 0 \\ 0 & (e-1)y + 1 & 0 \\ 0 & 0 & 1 \end{bmatrix}.$$

Then, we have that $w_0(t) = L_0 t$, $w(t) = Lt$, $w_1(t) = L_1 t$, $w_0 = L_0$, $w_1 = L_1$ and $v(t) = M$, where $L_0 = e - 1 < L = e^{\frac{1}{L_0}} = 1.789572397$, $L_1 = e$ and $M = e^{\frac{1}{L_0}} = 1.7896$. Finally, we obtain

$$r = 0.0039782.$$

Example 5. Let $A_1 = A_2 = C[0, 1]$, be the space of continuous functions defined on the interval $[0, 1]$ and be equipped with max norm. Let $\Omega = \bar{U}(0, 1)$ and $B(x) = F''(x)$ for each $x \in \Omega$. Define F on Ω

$$F(\varphi)(x) = \phi(x) - 5\int_0^1 x\theta\varphi(\theta)^3 d\theta. \quad (59)$$

We have that

$$F'(\varphi(\xi))(x) = \xi(x) - 15\int_0^1 x\theta\varphi(\theta)^2 \xi(\theta) d\theta, \text{ for each } \xi \in \Omega. \quad (60)$$

Then, we have that $x^* = 0$, $L_0 = 7.5$, $L_1 = L = 15$, $M = 2$ and obtain $r = 0.0013404$ using method (2) for $w_0(t) = L_0 t$, $v(t) = 2 = M$, $w(t) = Lt$, $w_1 = L$ and $w_0 = L_0$.

Example 6. *Returning back to the motivation example at the introduction on this paper, we have* $L = L_0 = 96.662907$, $M = 2$ *and obtain* $r = 0.001$ *using method* (2) *for* $w_0(t) = L_0 t$, $v(t) = 2 = M$, $w(t) = Lt$, $w_1(t) = L$ *and* $w_0 = L_0$.

REFERENCES

[1] Argyros, I. K. 2008. *Convergence and Application of Newton–type Iterations.* Springer.

[2] Argyros, I. K. and Hilout, S. 2013. *Computational Methods in Nonlinear Analysis.* New Jersey: World Scientific Publ. Comp.

[3] Cordero, A., Torregrosa, J. R. and Vassileva, M. P. 2012. "Increasing the order of convergence of iterative schemes for solving nonlinear system". *J. Comput. Appl. Math.* 252: 86–94.

[4] Hernández, M. A. and Martínez, E. 2015. "On the semilocal convergence of a three steps Newton–type process under mild convergence conditions". *Numer. Algor.* 70: 377–392.

[5] Potra, F. A. and Pták, V. 1984. *Nondiscrete introduction and iterative process.* Boston: Research Notes in Mathematics, 103, Pitman.

[6] Sharma, J. R., Ghua, R. K. and Sharma, R. 2013. "An efficient fourth–order weighted–Newton method for system of nonlinear equations". *Numer. Algor.* 62: 307–323.

[7] Traub, J. F. 1964. *Iterative methods for the solution of equations, Prentice–Hall Series in Automatic Computation.* New Jersey: Englewood Cliffs.

[8] Xiao, X. and Yin, H. "Achieving higher order of convergence for solving systems of nonlinear equations". *Appl. Math. Comput.* (to appear).

[9] Amat, S., Busquier, S., Plaza, S. and Gutiérrez, J. M. 2003. "Geometric constructions of iterative functions to solve nonlinear equations". *J. Comput. Appl. Math.* 157: 197–205.

[10] Amat, S., Busquier, S. and Plaza, S. 2005. "Dynamics of the King and Jarratt iterations". *Aequationes Math.* 69(3): 212–223.

[11] Amat, S., Hernández, M. A. and Romero, N. 2018. "A modified Chebyshev's iterative method with at least sixth order of convergence". *Appl. Math. Comput.* 206(1): 164–174.

[12] Argyros, I. K. and Magreñán, Á. A. 2015. "Ball convergence theorems and the convergence planes of an iterative methods for nonlinear equations". *SeMA* 71(1): 39–55.

[13] Argyros, I. K. and George, S. "Local convergence of some higher–order Newton–like method with frozen derivative". *SeMa* doi:10.1007/s40324-015-00398-8.

[14] Cordero, A. and Torregrosa, J. R. 2007. "Variants of Newton's method using fifth–order quadrature formulas". *Appl. Math. Comput.* 190: 686–698.

[15] Cordero, A. and Torregrosa, J. R. 2006. "Variants of Newton's method for functions of several variables". *Appl. Math. Comput.* 183: 199–208.

[16] Ezquerro, J. A. and Hernández, M. A. 2009. "New iterations of R–order four with reduced computational cost". *BIT Numer. Math.* 49: 325–342.

[17] Ezquerro, J. A. and Hernández, M. A. 2003. "A uniparametric Halley type iteration with free second derivative". *Int. J. Pure and Appl. Math.* 6(1): 99–110.

[18] Gutiérrez, J. M. and Hernández, M. A. 1998. "Recurrence realtions for the super–Halley method". *Comput. Math. Appl.* 36: 1–8.

[19] Kou, J. 2007. "A third–order modification of Newton method for systems of nonlinear equations". *Appl. Math. Comput.* 191: 117–121.

[20] Montazeri, H., Soleymani, F., Shateyi, S. and Motsa, S. S. 2012. "On a new method for computing the numerical solution of systems of nonlinear equations". *J. Appl. Math.* Article ID 751975.

[21] Petkovic, M. S., Neta, B., Petkovic, L. and Džunič, J. 2013. *Multipoint methods for solving nonlinear equations.* Elsevier.

[22] Rheinboldt, W. C. 1978. *An adaptive continuation process for solving systems of nonlinear equations.* Polish Academy of Science, Banach Ctr. Publ. 3. 129–142.

In: Understanding Banach Spaces
Editor: Daniel González Sánchez

ISBN: 978-1-53616-745-0
© 2020 Nova Science Publishers, Inc.

Chapter 16

BALL CONVERGENCE FOR A THREE–STEP ONE PARAMETER EFFICIENT METHOD IN BANACH SPACE UNDER GENERALIZED CONDITIONS

Ioannis K. Argyros[1,*], *Ramandeep Behl*[2,†] *and Daniel González Sánchez*[3,‡]
[1]Department of Mathematical Sciences
Cameron University, Lawton, OK, US
[2]School of Mathematics, Statistics and Computer Sciences
University of KwaZulu–Natal, Pietermaritzburg, South Africa
[3]Escuela de Ciencias Físicas y Matemáticas
Universidad de Las Américas, Quito, Ecuador

Abstract

A local convergence analysis for an eighth–order convergent method is given in order to approximate a locally unique solution of a nonlinear equation for Banach space valued operators. In contrast to the earlier studies using hypotheses up to the seventh Fréchet–derivative, we only use hypotheses on the first–order Fréchet–derivative and Lipschitz constants. Hence, we not only expand the applicability of these methods but also proposed the computable radius of convergence of these methods. Finally, numerical examples demonstrate that our results apply to solve nonlinear equations but results in earlier studies cannot apply.

Keywords: iterative method, local convergence, Banach space, Lipschitz constant, order of convergence

AMS Subject Classification: 65G99, 65H10, 47J25, 47J05, 65D10, 65D99

---

[*]E-mail: iargyros@cameron.edu.
[†]E-mail: ramanbehl87@yahoo.in.
[‡]Corresponding Author's E-mail: daniel.gonzalez.sanchez@udla.edu.ec.

1. INTRODUCTION

One of the most basic and important problems in Numerical analysis concerns with approximating a locally unique solution x^* of the equation of the form

$$F(x) = 0, \tag{1}$$

where $F : \mathbb{D} \subset \mathbb{X} \to \mathbb{Y}$ is a Fréchet–differentiable operator, $\mathbb{X}$, $\mathbb{Y}$ are Banach spaces and $\mathbb{D}$ is a convex subset of $\mathbb{X}$. Let us also denote $L(\mathbb{X}, \mathbb{Y})$ as the space of bounded linear operators from $\mathbb{X}$ to $\mathbb{Y}$.

The location of unique solution x^* is very important, since a lot of problems can be reduced to equation (1) using mathematical modeling [1–8]. However, it is not always possible to find the solution x^* in a closed form. Therefore, most of the methods are iterative to solve such type of problems. The convergence analysis of iterative methods is usually divided into two categories: semi–local and local convergence analysis. The semi–local convergence matter is, based on the information around an initial point, to give criteria ensuring the convergence of iteration procedures. A very important problem in the study of iterative procedures is the convergence domain. Therefore, it is very important to propose the radius of convergence of the iterative methods.

We study the local convergence of the three–step eighth–order convergent method defined for each $n = 0, 1, 2, \ldots$ by

$$\begin{aligned} y_n &= x_n - \alpha F'(x_n)^{-1} F(x_n), \\ z_n &= y_n - F'(x_n)^{-1} F(y_n), \\ x_{n+1} &= z_n - \left[\frac{1}{\alpha} F'(y_n)^{-1} - \left(1 - \frac{1}{\alpha}\right) F'(x_n)^{-1} \right] F(z_n), \end{aligned} \tag{2}$$

where $x_0 \in \mathbb{D}$ is an initial point, for $\alpha \neq \pm 1$ the method reaches at fourth–order. The order of convergence was shown in [6] using Taylor series expansions and hypotheses reaching up to the fourth–order derivative of the involved function. The hypotheses on the derivatives of F restrict the applicability of method (2). As a motivational example, define function F on $\mathbb{X} = \mathbb{Y} = \mathbb{R}$, $\mathbb{D} = [-\frac{5}{2}, \frac{1}{2}]$ by

$$F(x) = \begin{cases} x^3 \ln x^2 + x^5 - x^4, & x \neq 0, \\ 0, & x = 0. \end{cases}$$

Then, we have that

$$F'(x) = 3x^2 \ln x^2 + 5x^4 - 4x^3 + 2x^2,$$
$$F''(x) = 6x \ln x^2 + 20x^3 - 12x^2 + 10x$$

and

$$F'''(x) = 6 \ln x^2 + 60x^2 - 24x + 22.$$

Then, obviously the third–order derivative of the involved function $F'''(x)$ is not bounded on $\mathbb{D}$. Notice that, in–particular there is a plethora of iterative methods for approximating solutions of nonlinear equations [1–23]. These results show that initial guess should be close to the required root for the convergence of the corresponding methods. But, how close

Ball Convergence for a Three–Step One Parameter Efficient Method ...

initial guess should be required for the convergence of the corresponding method? These local results give no information on the radius of the ball convergence for the corresponding method. We address this question for method (2) in the next section 2.

In the present study, we expand the applicability of method (2) by using only hypotheses on the first–order derivative of function F and generalized Lipschitz conditions. Moreover, we we will avoid to use Taylor series expansions and use Lipschitz parameters. In this way, there is no need to use the higher–order derivatives to show the convergence of the scheme (2).

The rest of the chapter is organized as follows: in section 2 contains the local convergence analysis of method (2). The numerical examples appear in the concluding Section 3.

2. LOCAL CONVERGENCE

The local convergence analysis that follows is based on some scalar functions and parameters. Let v, w_0, w, be increasing continuous functions defined on the interval $[0, +\infty)$ with values in $[0, +\infty)$ with $w_0(0) = w(0) = 0$ and $\alpha \in S - \{0\}$, $S = \mathbb{R}$ or $\mathbb{C}$. Define parameter r_0 to be the smallest positive solution of the following equation

$$w_0(t) = 1. \qquad (3)$$

Moreover, define functions g_1 and h_1 on the interval $[0, r_0)$ by

$$g_1(t) = \frac{\int_0^1 w((1-\theta)t)d\theta + |1-\alpha| \int_0^1 v(\theta t)d\theta}{1 - w_0(t)},$$

and

$$h_1 = g_1(t) - 1.$$

Suppose that

$$|1-\alpha|v(0) < 1. \qquad (4)$$

We have by (4) that $h_1(0) = |1-\alpha|v(0) - 1 < 0$ and $h_1(t) \to +\infty$ as $t \to r_0^-$. Then, by the intermediate value theorem, we know that the functions h_1 has zeros in the interval $(0, r_0)$. Denote by r_1 the smallest such zero of the function h_1 and h_q. Furthermore, define functions g_2, h_2, g_3 and h_3

$$g_2(t) = \frac{\int_0^1 w((1-\theta)g_1(t)t)d\theta}{1 - w_0(g_1(t)t)} + \frac{w((1+g_1(t))t) \int_0^1 v(\theta g_1(t)t)d\theta}{(1-w_0(t))(1-w_0(g_1(t)t))} g_1(t), \ t \in [0, \bar{r}_0)$$

$$h_2(t) = g_2(t) - 1,$$

$$g_3(t) = \frac{\int_0^1 w((1-\theta)g_2(t)t)d\theta}{1 - w_0(g_2(t)t)} + \frac{w((1+g_1(t))t) \int_0^1 v(\theta g_2(t)t)d\theta}{|\alpha|(1-w_0(t))(1-w_0(g_1(t)t))},$$

$$+ \frac{w((1+g_1(t))t) \int_0^1 v(\theta g_2(t)t)d\theta}{|\alpha|(1-w_0(t))(1-w_0(g_2(t)t))} g_2(t), \ t \in [0, \bar{\bar{r}}_0)$$

and

$$h_3(t) = g_3(t) - 1,$$

where $\bar{r}_0$ and $\bar{\bar{r}}_0$ are the smallest zeros of the functions $w_0(g_1(t)t) - 1$ and $w_0(g_2(t)t) - 1$, respectively in the interval $[0, r_0)$. We have again that $h_2(0) = -h_3(0) = -1 < 0$ and $h_2(t) \to +\infty$, $h_3(t) \to +\infty$ as $t \to r_0^-$. Let us denote r_2 and r_3 be the smallest zero of the functions h_2 and h_3, respectively on the interval $(0, r_0)$. Finally, define the radius of convergence r by

$$r = \min\{r_i\}, \quad i = 1, 2, 3. \tag{5}$$

Then, we have

$$0 < r < r_0, \tag{6}$$

and for each $t \in [0, r)$

$$0 \leq g_i(t) < 1, \tag{7}$$

$$0 \leq w_0(t) < 1, \tag{8}$$

$$0 \leq w_0(g_1(t)t) < 1, \tag{9}$$

and

$$0 \leq w_0(g_2(t)t) < 1. \tag{10}$$

Let $U(z, \rho)$, $\bar{U}(z, \rho)$, stand respectively for the open and closed balls in X with center $z \in \mathbb{X}$ and of radius $\rho > 0$. Next, we present the local convergence analysis of method (2) using the preceding notations.

Theorem 1. *Let $F : \mathbb{D} \subseteq \mathbb{X} \to \mathbb{Y}$ be a continuously Fréchet–differentiable operator. Let $v, w_0, w : [0, \infty) \to [0, \infty)$ be increasing continuous functions with $w_0(0) = w(0) = 0$ and let $r_0 \in [0, \infty)$, $\alpha \in S - \{0\}$ be such that (3) and (4) are satisfied. Suppose that there exists $x^* \in \mathbb{D}$ such that for each $x \in \mathbb{D}$,*

$$F(x^*) = 0, \quad F'(x^*)^{-1} \in L(\mathbb{Y}, \mathbb{X}) \tag{11}$$

and

$$\|F'(x^*)^{-1}(F'(x) - F'(x^*))\| \leq w_0(\|x - x^*\|). \tag{12}$$

Moreover, suppose that for each $x, y \in D_0 := D \cap U(x^*, r_0)$

$$\|F'(x^*)^{-1}(F'(x) - F'(y))\| \leq w(\|x - y\|), \tag{13}$$

$$\|F'(x^*)^{-1}F'(x)\| \leq v(\|x - x^*\|), \tag{14}$$

and

$$\bar{U}(x^*, r) \subseteq \mathbb{D}, \tag{15}$$

where the radius of convergence r is defined by (5). Then, sequence $\{x_n\}$ generated for $x_0 \in U(x^*, r) - \{x^*\}$ by method (2) is well defined, remains in $U(x^*, r)$ for each $n = 0, 1, 2, \ldots$ and converges to x^*. Moreover, the following estimates hold

$$\|y_n - x^*\| \leq g_1(\|x_n - x^*\|)\|x_n - x^*\| \leq \|x_n - x^*\| < r, \tag{16}$$

$$\|z_n - x^*\| \leq g_2(\|x_n - x^*\|)\|x_n - x^*\| \leq \|x_n - x^*\| \tag{17}$$

and

$$\|x_{n+1} - x^*\| \leq g_3(\|x_n - x^*\|)\|x_n - x^*\| \leq \|x_n - x^*\|, \tag{18}$$

where the functions g_i, $i = 1, 2, 3$ are defined above the Theorem. Furthermore, if

$$\int_0^1 w_0(\theta R) d\theta < 1, \text{ for } R \geq r, \tag{19}$$

then the point x^* is the only solution of equation $F(x) = 0$ in $D_1 := \mathbb{D} \cap \bar{U}(x^*, R)$.

Proof. We shall show using mathematical induction that the sequences $\{x_n\}$ is well defined in $U(x^*, r)$ and converges to x^*. By the hypothesis $x_0 \in U(x^*, r) - \{x^*\}$, (3), (5) and (12), we have that

$$\|F'(x^*)^{-1}(F'(x_0) - F'(x^*))\| \leq w_0(\|x_0 - x^*\|) < w_0(r) < 1. \tag{20}$$

In view of (20) and the Banach Lemma on invertible operators [1,2] $F'(x_0)^{-1} \in L(\mathbb{Y}, \mathbb{X})$, y_0 and z_0 are well defined by the first two sub steps of method (2) and

$$\|F'(x_0)^{-1} F'(x^*)\| \leq \frac{1}{1 - w_0(\|x_0 - x^*\|)}. \tag{21}$$

Using the first sub step of method (2), (5), (7) (for $i = 1$), (11), (13), (14) and (21), we obtain in turn that

$$\begin{aligned}
\|y_0 - x^*\| &= \|(x_0 - x^* - F'(x_0)^{-1} F(x_0)) + (1 - \alpha) F'(x_0)^{-1} F(x_0)\| \\
&\leq \|F'(x_0)^{-1} F(x^*)\| \left\| \int_0^1 F'(x^*)^{-1}(F'(x^* + \theta(x_0 - x^*)) - F'(x_0))(x_0 - x^*) d\theta \right\| \\
&\quad + |1 - \alpha| \|F'(x_0)^{-1} F'(x^*)\| \|F'(x_0)^{-1} F(x_0)\| \\
&\leq \frac{\int_0^1 w((1-\theta)\|x_0 - x^*\|) d\theta \|x_0 - x^*\| + |1 - \alpha| \int_0^1 v(\theta \|x_0 - x^*\|) d\theta \|x_0 - x^*\|}{1 - w_0(\|x_0 - x^*\|)} \\
&= g_1(\|x_0 - x^*\|) \|x_0 - x^*\| \leq \|x_0 - x^*\| < r,
\end{aligned} \tag{22}$$

which implies (16) for $n = 0$ and $y_0 \in U(x^*, r)$. We can write by the second sub step of method (2)

$$z_0 - x^* = y_0 - x^* - F'(y_0)^{-1} F(y_0) + F'(y_0)^{-1} (F'(y_0) - F'(x_0)) F'(x_0)^{-1} F(y_0). \tag{23}$$

Notice that since $y_0 \in U(x^*, r)$ with $y_0 = x_0$ in (20), $F'(y_0)^{-1} \in L(\mathbb{Y}, \mathbb{X})$ and x_1 are well defined and

$$\begin{aligned}
\|F'(y_0)^{-1} F'(x^*)\| &\leq \frac{1}{1 - w_0(\|y_0 - x^*\|)} \\
&\leq \frac{1}{1 - w_0(g_1(\|x_0 - x^*\|) \|x_0 - x^*\|)} \leq \frac{1}{1 - w_0(\|x_0 - x^*\|)}.
\end{aligned} \tag{24}$$

By (5), (6) (for $i = 2$), (9), (11), (13), (14), (21)–(24), we get in turn that

$$\begin{aligned}
\|z_0 - x^*\| &= \|y_0 - x^* - F'(y_0)^{-1}F(y_0))\| \\
&\quad + \|F'(y_0)^{-1}F'(x^*)\|\,|F'(x^*)^{-1}(F'(x_0) - F'(y_0))\| \\
&\quad \times \|F'(x_0)^{-1}F'(x^*)\|\|F'(x^*)^{-1}F'(y_0)\| \\
&\leq \frac{\int_0^1 w((1-\theta)\|y_0 - x^*\|)d\theta\|y_0 - x^*\|}{1 - w_0(\|y_0 - x^*\|)} \\
&\quad + \frac{w(\|y_0 - x^*\|)\int_0^1 v(\theta\|y_0 - x^*\|)d\theta\|y_0 - x^*\|}{(1 - w_0(\|x_0 - x^*\|))(1 - w_0(\|y_0 - x^*\|))} \\
&\leq g_2(\|x_0 - x^*\|)\|x_0 - x^*\| \leq \|x_0 - x^*\| < r,
\end{aligned} \quad (25)$$

which implies (17) for $n = 0$ and $z_0 \in U(x^*, r)$.

Then, by the third sub step of method (2), (5), (7) (for $i = 3$), (9), (10), (21)–(25) (for $y_0 = z_0$), we have in turn that

$$\begin{aligned}
\|x_1 - x^*\| &\leq \|z_0 - x^* - F'(z_0)^{-1}F(z_0)\| + \frac{1}{|\alpha|}\|F'(y_0)^{-1}(F'(x_0) - F'(y_0)) \\
&\quad \times F'(x_0)^{-1}F'(x^*)\|\|F'(x^*)^{-1}F(z_0)\| \\
&\quad + \|F'(z_0)^{-1}(F'(x_0) - F'(z_0))F'(x_0)^{-1}F'(x^*)\|\|F'(x^*)^{-1}F(z_0)\| \\
&\leq \frac{\int_0^1 w((1-\theta)\|z_0 - x^*\|)\|z_0 - x^*\|d\theta\|x_0 - x^*\|}{1 - w_0(\|x_0 - x^*\|)} \\
&\quad + \frac{1}{|\alpha|}\frac{w(\|y_0 - x^*\|)\int_0^1 v(\theta\|z_0 - x^*\|)d\theta\|z_0 - x^*\|}{(1 - w_0(\|y_0 - x^*\|))(1 - w_0(\|x_0 - x^*\|))} \\
&\quad + \frac{w(\|z_0 - x^*\|)\int_0^1 v(\theta\|z_0 - x^*\|)d\theta\|z_0 - x^*\|}{(1 - w_0(\|z_0 - x^*\|))(1 - w_0(\|x_0 - x^*\|))} \\
&\leq g_3(\|x_0 - x^*\|)\|x_0 - x^*\| \leq \|x_0 - x^*\| < r,
\end{aligned} \quad (26)$$

which shows (18) for $n = 0$ and $z_0 \in U(x^*, r)$. By simply replacing x_0, y_0, x_1 by x_k, y_k, x_{k+1} in the preceding estimates we arrive at (16) – (18). Then, in view of the estimates

$$\|x_{k+1} - x^*\| \leq c\|x_k - x^*\| < r, \quad c = g_3(\|x_0 - x^*\|) \in [0, 1), \quad (27)$$

we deduce that $\lim_{k \to \infty} x_k = x^*$ and $x_{k+1} \in U(x^*, r)$. Finally, to show the uniqueness part, let $y^* \in D_1$ with $F(y^*) = 0$. Define $Q = \int_0^1 F'(x^* + \theta(x^* - y^*))d\theta$. Using (12) and (19), we get that

$$\begin{aligned}
\|F'(x^*)^{-1}(Q - F'(x^*))\| &\leq \|\int_0^1 w_0(\theta\|y^* - x^*\|)d\theta \\
&\leq \int_0^1 w_0(\theta R)d\theta < 1.
\end{aligned} \quad (28)$$

It follows from (28) that Q is invertible. Then, in view of the identity

$$0 = F(x^*) - F(y^*) = Q(x^* - y^*), \quad (29)$$

we conclude that $x^* = y^*$. $\square$

Remark 2. (a) It follows from (12) that condition (14) can be dropped and be replaced by
$$v(t) = 1 + w_0(t) \text{ or } v(t) = 1 + w_0(r_0), \tag{30}$$
since,
$$\|F'(x^*)^{-1}\left[(F'(x) - F'(x^*)) + F'(x^*)\right]\| = 1 + \|F'(x^*)^{-1}(F'(x) - F'(x^*))\|$$
$$\leq 1 + w_0(\|x - x^*\|)$$
$$= 1 + w_0(t) \quad \text{for } \|x - x^*\| \leq r_0. \tag{31}$$

(b) If the function w_0 is strictly increasing, then we can choose
$$r_0 = w_0^{-1}(1) \tag{32}$$
instead of (3).

(c) If w_0, w, v are constants functions (the proof of Theorem 1 goes through for this case $+\infty$), then
$$r_1 = \frac{2}{2w_0 + w} \tag{33}$$
and
$$r \leq r_1. \tag{34}$$

Therefore, the radius of convergence r can be larger than the radius of convergence r_1 for Newton's method
$$x_{n+1} = x_n - F'(x_n)^{-1} F(x_n). \tag{35}$$

Notice also that the earlier radius of convergence given independently by Rheindoldt [22] and Traub [8] is
$$r_{TR} = \frac{2}{3w_1} \tag{36}$$
and by Argyros [1, 2]
$$r_A = \frac{2}{2w_0 + w_1}, \tag{37}$$
where w_1 is the Lipschitz constant for (8) on D. But, we have
$$w \leq w_1, \ w_0 \leq w_1, \tag{38}$$
so
$$r_{TR} \leq r_A \leq r_1 \tag{39}$$
and
$$\frac{r_{TR}}{r_A} \to \frac{1}{3} \quad \text{as} \quad \frac{w_0}{w} \to 0. \tag{40}$$

The radius of convergence q used in [23] is smaller than the radius r_{DS} given by Dennis and Schabel [2]
$$q < r_{SD} = \frac{1}{2w_1} < r_{TR}. \tag{41}$$
However, q can not be computed using the Lipschitz constants.

(d) *The order of convergence of method (2) was shown in [23] using hypotheses up to the fifth–order derivative of F. We have only used hypotheses on the first–order derivative of F. The order of convergence can be determined by using the following formula for the computational order of convergence COC given by*

$$\xi = \frac{\ln \frac{\|x_{n+2} - x^*\|}{\|x_{n+1} - x^*\|}}{\ln \frac{\|x_{n+1} - x^*\|}{\|x_n - x^*\|}}, \quad \text{for each } n = 0, 1, 2, \ldots \tag{42}$$

or the approximate computational order of convergence (ACOC) [19], given by

$$\xi^* = \frac{\ln \frac{\|x_{n+2} - x_{n+1}\|}{\|x_{n+1} - x_n\|}}{\ln \frac{\|x_{n+1} - x_n\|}{\|x_n - x_{n-1}\|}}, \quad \text{for each } n = 1, 2, \ldots \tag{43}$$

that do not require derivatives. Notice also that ξ^* does not even require the knowledge of exact root x^*.

(e) *The results obtained here can be used for operators F satisfying the autonomous differential equation [1, 2] of the form*

$$F'(x) = P(F(x)) \tag{44}$$

where P is a known continuous operator. Since $F'(x^*) = P(F(x^*)) = P(0)$, we can apply the results without actually knowing the solution x^*. Let as an example $F(x) = e^x - 1$. Then, we can choose $P(x) = x + 1$.

(f) *In view of the estimates*

$$\begin{aligned}
\|F'(x^*)^{-1}\big(F'(x_0) - F'(y_0)\big)\| &= \|F'(x^*)^{-1}(F'(x_0) - F'(x^*))\| \\
&\quad + \|F'(x^*)^{-1}\big(F'(y_0) - F'(x_0)\big)\| \\
&\leq w_0(\|x_0 - x^*\|) + w_0(\|y_0 - x^*\|) \\
&\leq w_0(\|x_0 - x^*\|) + w_0(g_1(\|x_0 - x^*\|)\|x_0 - x^*\|) \\
&\leq w_0(r) + w_0(g_1(r)r)
\end{aligned} \tag{45}$$

and similarly

$$\begin{aligned}
\|F'(x^*)^{-1}\big(F'(x_0) - F'(z_0)\big)\| &\leq w_0(\|x_0 - x^*\|) + w_0(\|z_0 - x^*\|) \\
&\leq w_0(r) + w_0(g_2(r)r),
\end{aligned} \tag{46}$$

we can replace the terms $w((1+g_1(t))t)$, $w((1+g_2(t))t)$ in the definition of functions g_2 and g_3 by $w_0(t) + w_0(g_1(t)t)$, $w_0(t) + w_0(g_2(t)t)$, respectively. If

$$w_0(t) \leq w(t), \ t \in [0, \ r_0)$$

and say w_0, w are constants, then the new functions g_2 and g_3 are tighter than the old one leading to larger r and tighter error bounds on the distances $\|x_n - x^*\|$ (if $w_0 < w$).

3. Numerical Examples and Applications

In this section, we shall demonstrate the theoretical results which we have proposed in the section 2. Therefore, we consider four numerical examples in this section, which are defined as follows:

Example 3. Let $\mathbb{X} = \mathbb{Y} = C[0, 1]$ and consider the nonlinear integral equation of the mixed Hammerstein–type [4, 16], defined on $\mathbb{D} = \bar{U}(0, 1)$ by

$$x(s) = 1 + \int_0^1 G(s, t)\left(x(t)^{\frac{3}{2}} + \frac{x(t)^2}{2}\right) dt \qquad (47)$$

where the kernel G is the Green's function defined on the interval $[0, 1] \times [0, 1]$ by

$$G(s, t) = \begin{cases} (1-s)t, & t \leq s, \\ s(1-t), & s \leq t. \end{cases} \qquad (48)$$

The solution $x^*(s) = 0$ is the same as the solution of equation (1), where $F : \mathbb{D} \subseteq C[0, 1] \to C[0, 1]$ defined by

$$F(x)(s) = x(s) - \int_0^t G(s, t)\left(x(t)^{\frac{3}{2}} + \frac{x(t)^2}{2}\right) dt. \qquad (49)$$

Notice that

$$\left\|\int_0^t G(s, t) dt\right\| \leq \frac{1}{8}. \qquad (50)$$

Then, we have that

$$F'(x)y(s) = y(s) - \int_0^t G(s, t)\left(\frac{3}{2}x(t)^{\frac{1}{2}} + x(t)\right) dt,$$

so since $F'(x^*(s)) = I$,

$$\|F'(x^*)^{-1}(F'(x) - F'(y))\| \leq \frac{1}{8}\left(\frac{3}{2}\|x - y\|^{\frac{1}{2}} + \|x - y\|\right). \qquad (51)$$

Therefore, we can choose

$$w_0(t) = w(t) = \frac{1}{8}\left(\frac{3}{2}t^{\frac{1}{2}} + t\right)$$

and by Remark 2 (a)

$$v(t) = 1 + w_0(t).$$

The results in [6] can not be used to solve this problem, since F' is not Lipschitz. However, our results can apply.

Example 4. *Suppose that the motion of an object in three dimensions is governed by system of differential equations*

$$f'_1(x) - f_1(x) - 1 = 0$$
$$f'_2(y) - (e-1)y - 1 = 0 \qquad (52)$$
$$f'_3(z) - 1 = 0$$

with x, y, $z \in \mathbb{D} = \bar{U}(0,1)$ for $f_1(0) = f_2(0) = f_3(0) = 0$. then, the solution of the system is given for $w = (x, y, z)^T$ by function $F := (f_1, f_2, f_3) : \mathbb{D} \to \mathbb{R}^3$ defined by

$$F(v) = \left(e^x - 1, \frac{e-1}{2}y^2 + y, z\right)^T. \qquad (53)$$

Then the Fréchet–derivative is given by

$$F'(v) = \begin{bmatrix} e^x & 0 & 0 \\ 0 & (e-1)y+1 & 0 \\ 0 & 0 & 1 \end{bmatrix}.$$

Then, we have that $w_0(t) = L_0 t$, $w(t) = Lt$, $w_1(t) = L_1 t$, $w_0 = L_0$, $w_1 = L_1$ and $v(t) = M$, where $L_0 = e - 1 < L = e^{\frac{1}{L_0}} = 1.789572397$, $L_1 = e$ and $M = e^{\frac{1}{L_0}} = 1.7896$. Then, we get

$$r = 0.0039782.$$

Example 5. *Let* $\mathbb{X} = \mathbb{Y} = C[0, 1]$, *be the space of continuous functions defined on the interval* $[0, 1]$ *and be equipped with max norm. Let* $\mathbb{D} = \bar{U}(0, 1)$ *and* $B(x) = F''(x)$ *for each* $x \in \mathbb{D}$. *Define* F *on* $\mathbb{D}$

$$F(\varphi)(x) = \phi(x) - 5\int_0^1 x\theta\varphi(\theta)^3 d\theta. \qquad (54)$$

We have that

$$F'(\varphi(\xi))(x) = \xi(x) - 15\int_0^1 x\theta\varphi(\theta)^2\xi(\theta)d\theta, \text{ for each } \xi \in \mathbb{D}. \qquad (55)$$

Then, we have that $x^* = 0$, $L_0 = 7.5$, $L_1 = L = 15$ and $M = 2$. Using method (2) for $w_0(t) = L_0 t$, $v(t) = 2 = M$, $w(t) = Lt$, $w_1 = L$ and $w_0 = L_0$, we get

$$r = 0.0013404.$$

Example 6. *Returning back to the motivation example at the introduction on this chapter, we have* $L = L_0 = 96.662907$ *and* $M = 2$. *Using method* (2) *for* $w_0(t) = L_0 t$, $v(t) = 2 = M$, $w(t) = Lt$, $w_1(t) = L$ *and* $w_0 = L_0$, *we can choose*

$$r = 0.0009.$$

REFERENCES

[1] Argyros, I. K. 2008. *Convergence and Application of Newton–type Iterations*. Springer.

[2] Argyros, I. K. and Hilout, S. 2013. *Computational Methods in Nonlinear Analysis*. New Jersey: World Scientific Publ. Comp.

[3] Cordero, A., Torregrosa, J. R. and Vassileva, M. P. 2012. "Increasing the order of convergence of iterative schemes for solving nonlinear system". *J. Comput. Appl. Math.* 252: 86–94.

[4] Hernández, M. A. and Martinez, E. 2015. "On the semilocal convergence of a three–steps Newton–type process under mild convergence conditions". *Numer. Algor.* 70: 377–392.

[5] Potra, F. A. and Pták, V. 1984. *Nondiscrete introduction and iterative process, Research Notes in Mathematics*. 103 Boston MA: Pitman.

[6] Singh, S., Gupta, D. K., Martínez, E. and Hueso, J. L. "Local convergence of a parameter based iteration with Hölder continuous derivative in Banach spaces". *Appl. Math. Comput.* (to appear).

[7] Sharma, J. R., Ghua, R. K. and Sharma, R. 2013. "An efficient fourth–order weighted–Newton method for system of nonlinear equations". *Numer. Algor.* 62: 307–323.

[8] Traub, J. F. 1964. "Iterative methods for the solution of equations, Prentice–Hall Series in Automatic Computation". New Jersey: Englewood Cliffs.

[9] Amat, S., Busquier, S., Plaza, S. and Gutiérrez, J. M. 2003. "Geometric constructions of iterative functions to solve nonlinear equations". *J. Comput. Appl. Math.* 157: 197–205.

[10] Amat, S., Busquier, S. and Plaza, S. 2005. "Dynamics of the King and Jarratt iterations". *Aequationes Math.* 69(3): 212–223.

[11] Amat, S., Hernández, M. A. and Romero, N. 2008. "A modified Chebyshev's iterative method with at least sixth order of convergence". *Appl. Math. Comput.* 206(1): 164–174.

[12] Argyros, I. K. and Magreñán, Á. A. 2015. "Ball convergence theorems and the convergence planes of an iterative methods for nonlinear equations". *SeMA* 71(1): 39–55.

[13] Argyros, I. K. and George, S. "Local convergence of some higher–order Newton–like method with frozen derivative". *SeMA* doi:10.1007/s40324-015-00398-8.

[14] Cordero, A. and Torregrosa, J. R. 2007. "Variants of Newton's method using fifth–order quadrature formulas". *Appl. Math. Comput.* 190: 686–698.

[15] Cordero, A. and Torregrosa, J. R. 2006. "Variants of Newton's method for functions of several variables". *Appl. Math. Comput.* 183: 199–208.

[16] Ezquerro, J. A. and Hernández, M. A. 2009. "New iterations of R–order four with reduced computational cost". *BIT Numer. Math.* 49: 325–342.

[17] Ezquerro, J. A. and Hernández, M. A. 2003. "A uniparametric halley type iteration with free second derivative". *Int. J. Pure and Appl. Math.* 6(1): 99–110.

[18] Gutiérrez, J. M. and Hernández, M. A. 1998. "Recurrence realtions for the super–Halley method". *Comput. Math. Appl.* 36: 1–8.

[19] Kou, J. 2007. "A third–order modification of Newton method for systems of nonlinear equations". *Appl. Math. Comput.* 191: 117–121.

[20] Montazeri, H., Soleymani, F., Shateyi, S. and Motsa, S. S. 2012. "On a new method for computing the numerical solution of systems of nonlinear equations". *J. Appl. Math.* Article ID 751975.

[21] Petkovic, M. S., Neta, B., Petkovic, L. and Džunič, J. 2013. *Multipoint methods for solving nonlinear equations.* Elsevier.

[22] Rheinboldt, W. C. 1978. "An adaptive continuation process for solving systems of nonlinear equations". *Polish Academy of Science, Banach Ctr. Publ.* 3: 129–142.

[23] Xiao, X. Y. and Yin, H. W. 2015. "Increasing the order convergence for iterative methods to solve nonlinear systems". *Calcolo.* DOI 10.100071510092-015-0149-9

In: Understanding Banach Spaces
Editor: Daniel González Sánchez
ISBN: 978-1-53616-745-0
© 2020 Nova Science Publishers, Inc.

Chapter 17

ON THE CONVERGENCE OF NEWTON–MOSER METHOD FROM DATA AT ONE POINT

José M. Gutiérrez[*] *and Miguel A. Hernández–Verón*[†]
Department of Mathematics and Computer Sciences
University of La Rioja, Logroño, Spain

Abstract

In this chapter we analyze both the local and semilocal convergence of Newton–Moser method, a variant of Newton's method for solving nonlinear equations. We develop this analysis by following the γ–theory and α–theory respectively. These two techniques, introduced by Smale in 1986, establish the convergence of an iterative method just by giving conditions at one point: the solution in the γ–theory and the starting point of the method in the α–theory. Some applications to the solution of integral equations are given.

Keywords: Newton–Moser method, local convergence theory, nonlinear equations, Banach spaces

AMS Subject Classification: 65J15

1. Introduction

In this chapter we analyze an iterative method, first explored by J. Moser [1] for investigating the stability of the N–body problem in Celestial Mechanics. Next the method has been generalized for numerically solving nonlinear equations in a general sense. In order to consider the more general case, let us consider a nonlinear equation

$$F(x) = 0, \qquad (1)$$

where F is an operator defined between two Banach spaces X and Y. Let us assume that x^* is a simple root of (1).

---

[*]Corresponding Author's E-mail: jmguti@unirioja.es.
[†]E-mail: mahernan@unirioja.es.

Initially, Moser proposed the following method

$$\begin{cases} x_{n+1} = x_n - A_n F(x_n), & n \geq 0, \\ A_{n+1} = A_n - A_n(F'(x_n)A_n - I), & n \geq 0, \end{cases} \quad (2)$$

for a given $x_0 \in X$, a given $A_0 \in \mathcal{L}(Y, X)$, the set of linear operators from Y to X, and where I is the identity operator in X. Note that the first equation is similar to Newton's method, but replacing the operator $F'(x_n)^{-1}$ by a linear operator A_n. The second equation is one step of Newton's method applied to equation $g_n(A) = 0$ where $g_n : \mathcal{L}(Y, X) \to \mathcal{L}(X, Y)$ is defined by $g_n(A) = A^{-1} - F'(x_n)$. So $\{A_n\}$ gives us an approximation of $F'(x_n)^{-1}$.

Note that with this procedure it is not necessary to calculate inverse operators, as it happens in Newton's method, $x_{n+1} = x_n - F'(x_n)^{-1}F(x_n)$, $n \geq 0$. Equivalently, it is not necessary to solve a linear equation at each iteration. Although this is one of the main advantages of this procedure, method (2) is unsatisfactory from a numerical point of view because it has not quadratic convergence but it uses the same amount of information per step as Newton's method. To overcome this disadvantage, some authors ([2], [3]) have proposed the following variant of method (2):

$$\begin{cases} x_{n+1} = x_n - B_n F(x_n), & n \geq 0, \\ B_{n+1} = 2B_n - B_n F'(x_{n+1})B_n, & n \geq 0, \end{cases} \quad (3)$$

where x_0 is a given point in X and B_0 is a given linear operator from Y to X. Observe that in (3) $F'(x_{n+1})$ appears instead of $F'(x_n)$.

The method (3) exhibits several attractive features. First, as in the case of method (2), it also avoids the calculus of inverse operators. Second, it has quadratic convergence, the same as Newton's method. Third, in addition to solve the nonlinear equation (1), the method produces successive approximations $\{B_n\}$ to the value of $F'(x^*)^{-1}$, being x^* a solution of (1).

The study of the convergence of method (3) has been made for different authors and with different techniques. For instance, Zehnder [4] or Petzeltova [5] have studied the convergence of the method under Kantorovich–type conditions [6], that is, under the classical assumption that the derivative F' is Lipschitz continuous around the solution x^*. In [7] or [8] the semilocal convergence of Newton–Moser method is established from a system of recurrence relations and in [9] a new semilocal convergence theorem is given by following the patterns of the α–theory introduced by Smale ([10], [11]) and later generalized by Wang and Zhao [12].

As far as we know, the case of the local convergence of method (3) has been less explored. For instance, Hald [2] gives a local convergence result in the scalar case and provides a generalization to systems of nonlinear equations, based on Kantorovich's theorem. In this paper we complete this study with a new local convergence result for method (3) that uses the ideas of the γ–theory also introduced by Smale ([10], [11]).

The chapter is organized as follows. In Section 2 we give a new local convergence result for Newton–Moser method by using the aforementioned γ–theory. Next, in Section 3 we remind some semilocal convergence results obtained previously by different authors using the α–theory of Smale. We finish in Section 4 with an application to nonlinear integral equations.

2. γ–THEORY FOR NEWTON–MOSER METHOD

The context for developing the known as γ–theory is an equation (1) given by on operator F defined between two Banach spaces X and Y that is analytic in a neighborhood of the solution x^*, let us say in $B(x^*, R)$ with $R > 0$. In other words, $F(x)$ can be written as a Taylor's series around x^*:

$$F(x) = \sum_{k=1}^{\infty} \frac{1}{k!} F^{(k)}(x^*)(x - x^*)^k,$$

for $x \in B(x^*, R)$.

We also assume that $F'(x^*)$ is invertible and we denote $\Gamma = F'(x^*)^{-1}$. We introduce the following constant γ that will play an important role in the convergence properties of the considered iterative method for solving equation (1):

$$\gamma = \sup_{k \geq 2} \left\| \frac{1}{k!} \Gamma F^{(k)}(x^*) \right\|^{1/(k-1)}. \tag{4}$$

For instance, in the case of Newton's method it is proved (see the comprehensive book of Dedieu [13], for instance) that when

$$\|x_0 - x^*\| \gamma \leq \frac{3 - \sqrt{7}}{2} = 0.17712\ldots$$

the Newton's sequence starting at x_0 converges to x^* with quadratic convergence. There are many improvements of the initially posed Smale's theory for Newton's method, as well as applications to other one–point third–order iterative methods (as the reader can see in [14]).

The γ–theory for a given iterative method describes the radius of a ball included in the basin of attraction of the solution x^*. For the case of Newton–Moser method (3) this radius could be influenced by another important parameter, that measures the distance between B_0, the initial seed of the sequence of operators B_n, and the inverse of $F'(x_0)$.

We introduce the sequences

$$e_n = \|x_n - x^*\|, \quad \beta_n = \|I - B_n F'(x_n)\|, \quad \gamma_n = \|I - B_n F'(x^*)\|,$$

$$s_n = \beta_n + 9\gamma e_n$$

to estate the following lemma.

Lemma 1. *If $x_0 \in B(x^*, R)$ and $s_0 < 1/2$, then $s_{n+1} \leq 2s_n^2$ for $n \geq 0$.*

Proof. Let us assume that $x_n \in B(x^*, R)$ and $s_n < 1/2$. We will then show that $x_{n+1} \in B(x^*, R)$ and $s_{n+1} \leq 2s_n^2$ and, therefore $s_{n+1} < 1/2$. As an application of Taylor's series we have

$$F(x_n) = F'(x^*)(x_n - x^*) + \sum_{k=2}^{\infty} \frac{1}{k!} F^{(k)}(x^*)(x - x^*)^k$$

and then,

$$\begin{aligned} x_{n+1} - x^* &= x_n - x^* - B_n F(x_n) = x_n - x^* - B_n F'(x^*)\Gamma F(x_n) \\ &= (I - B_n F'(x^*))(x_n - x^*) - B_n F'(x^*) \sum_{k=2}^{\infty} \frac{1}{k!} \Gamma F^{(k)}(x^*)(x_n - x^*)^k. \end{aligned}$$

Taking norms in this equality, keeping in mind $\|B_n F'(x^*)\| \leq 1 + \gamma_n$ and $\gamma e_n < 1$ we obtain

$$e_{n+1} \leq \gamma_n e_n + (1 + \gamma_n) \sum_{k=2}^{\infty} \gamma^{k-1} e_n^k \tag{5}$$

$$\leq \gamma_n e_n + \frac{1 + \gamma_n}{1 - \gamma e_n} \gamma e_n^2. \tag{6}$$

In addition, we also have

$$\begin{aligned} \gamma_n &\leq \|I - B_n F'(x_n)\| + \|B_n F'(x^*)\| \|\Gamma[F'(x_n) - F'(x^*)]\| \\ &\leq \beta_n + (1 + \gamma_n) \sum_{k=2}^{\infty} \frac{1}{(k-1)!} \|\Gamma F^{(k)}(x^*)\| e_n^{k-1} \\ &\leq \beta_n + (1 + \gamma_n) \sum_{k=2}^{\infty} k \gamma^{k-1} e_n^{k-1} \\ &= \beta_n + (1 + \gamma_n) \frac{2 - \gamma e_n}{(1 - \gamma e_n)^2} \gamma e_n. \end{aligned}$$

As $\gamma_n \leq \beta_n \leq s_n < 1/2$ and $\gamma e_n \leq s_n < 1/2$ the previous inequality can be written

$$\gamma_n \leq \beta_n + 9\gamma e_n = s_n. \tag{7}$$

Even more, as $\gamma_n \leq s_n < 1/2$, from (6) we deduce

$$e_{n+1} \leq \left(\gamma_n + \frac{1 + \gamma_n}{1 - \gamma e_n} \gamma e_n\right) e_n < (\beta_n + 9e_n) e_n = s_n e_n. \tag{8}$$

Finally, bearing in mind (7) and (8) and the inequalities $\gamma_n \leq \beta_n \leq s_n < 1/2$, we have

$$\begin{aligned} \beta_{n+1} &= \|I - B_{n+1} F'(x_{n+1})\| = \|I - B_n F'(x_{n+1})\|^2 \\ &= \|I - B_n F'(x^*) + B_n[F'(x^*) - F'(x_{n+1})]\|^2 \\ &\leq \left(\gamma_n + (1 + \gamma_n) \frac{2 - \gamma e_{n+1}}{(1 - \gamma e_{n+1})^2} \gamma e_{n+1}\right)^2 \\ &\leq \left(\beta_n + \frac{3}{2} \frac{2 - s_n \gamma e_n}{(1 - s_n \gamma e_n)^2} s_n \gamma e_n\right)^2 \\ &\leq \left(\beta_n + \frac{3}{2} \frac{2 - \gamma e_n}{(1 - \gamma e_n)^2} \gamma e_n\right)^2 \\ &\leq (\beta_n + 9\gamma e_n)^2 = s_n^2. \end{aligned}$$

So, by this last inequality, (8) and $3\lambda e_n/2 \le s_n$ we obtain the desired result

$$s_{n+1} = \beta_{n+1} + 9\gamma e_{n+1} \le s_n^2 + 9 s_n e_n = 2 s_n^2.$$

□

With this technical lemma, we can establish some convergence results of the sequence $\{x_n\}$ for x_0 close enough to the solution. In addition, we can also prove the convergence of the operators $\{B_n\}$ to the inverse $\Gamma = F'(x^*)^{-1}$.

Theorem 2. *If $x_0 \in B(x^*, R)$ and $s_0 < 1/2$, then*

$$e_n \le \frac{(2s_0)^{2^n}}{3\lambda 2^n}, \quad \beta_n \le \frac{(2s_0)^{2^n}}{4}, \quad \gamma_n \le \frac{(2s_0)^{2^n}}{2}.$$

Proof. By (8) we have $e_n \le s_{n-1} s_{n-2} \cdots s_1 s_0 e_0$. After the change of notation, $q_n = 2 s_n$, Lemma 1 can be written as $q_{n+1} \le q_n^2$. Consequently, $q_n \le (q_0)^{2^n}$. Then

$$s_{n-1} s_{n-2} \cdots s_1 s_0 = \frac{1}{2^n} q_0^{1+2+2^2+\cdots+2^{n-1}}$$

and

$$e_n \le \frac{1}{2^n} q_0^{2^n - 1} = \frac{1}{2^n q_0} q_0^{2^n} = \frac{1}{2^{n+1}} q_0^{2^n} \frac{e_0}{s_0}.$$

As

$$\frac{e_0}{s_0} = \frac{e_0}{\beta_0 + 3\lambda e_0/2} \le \frac{2}{3\lambda},$$

we deduce the first inequality.

The second inequality follows directly from Lemma 1. In fact, as $\beta_{n+1} \le s_n^2$,

$$\beta_n \le \frac{1}{4} q_{n-1}^2 \le \frac{1}{4} \left(q_0^{2^{n-1}} \right)^2 = \frac{(2 s_0)^{2^n}}{4}$$

Finally, the third inequality follows immediately from (7) and the recurrence for $\{q_n\}$:

$$\gamma_n \le s_n = \frac{q_n}{2} \le \frac{q_0^{2^n}}{2} = \frac{(2 s_0)^{2^n}}{2}.$$

□

3. α-THEORY FOR NEWTON–MOSER METHOD

The α-theory is a semilocal convergence theory introduced by Smale ([10], [11]), is an alternative to Kantorovich theory [6]. The main difference with the γ-theory presented in the previous section is that conditions on the starting point x_0 of a given iterative method are assumed instead on the solution x^*. In a few words, if x_0 is an initial value such that the sequence $\{x_n\}$ satisfies

$$\|x_n - x^*\| \le \left(\frac{1}{2}\right)^{2^n - 1} \|x_0 - x^*\|,$$

where x^* is a solution of an equation (1), then x_0 is said to be an approximate zero of F. The following conditions, introduced by Smale ([10], [11]) guarantee that x_0 is an approximated zero of Newton's method:

$$\begin{aligned} &i) \quad \|F'(x_0)^{-1}F(x_0)\| \leq \beta, \\ &ii) \quad \sup_{k \geq 2}\left(\frac{1}{k!}\|F'(x_0)^{-1}F^{(k)}(x_0)\|\right)^{1/(k-1)} \leq \gamma, \\ &iii) \quad \alpha = \beta\gamma \leq 3 - 2\sqrt{2}. \end{aligned} \quad (9)$$

This type of conditions have been considered for the study of the semilocal convergence of other iterative methods (see [14] for instance). In the case of Newton–Moser method (3), we can find in [8] a system of recurrence relations that establish its semilocal convergence under estimations at one point. In [9] a semilocal convergence theorem for Newton–Moser method (3) has been proved under conditions more general than the given in (9). These conditions were first assumed by Wang and Zhao [12] for Newton's method. Actually, in [9] the following conditions are assumed:

$$\begin{aligned} &i) \quad \|B_0 F(x_0)\| \leq \gamma_0, \\ &ii) \quad \|I - B_0 F'(x_0)\| \leq \beta < 1, \\ &iii) \quad \|B_0 F^{(j)}(x_0)\| \leq \gamma_j, \text{ for } j \geq 2, \\ &iv) \quad \text{there exists } R > 0 \text{ such that the series} \\ &\qquad \sum_{j \geq 2} \gamma_j t^j/j! \text{ is convergent for } t \in [0, R), \\ &v) \quad f(\hat{t}) < 0, \end{aligned} \quad (10)$$

where $\hat{t}$ is the absolute minimum of the function

$$f(t) = \gamma_0 + (\beta - 1)t + \sum_{j \geq 2} \frac{1}{j!}\gamma_j t^j, \quad t \geq 0. \quad (11)$$

In addition, the following scalar sequence is constructed:

$$\begin{cases} t_0 = 0, \quad b_0 = -1, \\ t_{n+1} = t_n - b_n f(t_n), \\ b_{n+1} = 2b_n - b_n f'(t_{n+1})b_n. \end{cases} \quad (12)$$

Condition $v)$ in (10) says that function $f(t)$ defined in (10) has at least one positive root. Denote t^* as the smallest positive solution of $f(t) = 0$. With the rest of conditions in (10), (11), (12), it is easy to prove that $\{t_n\}$ is an increasing monotone sequence to t^* and

$$\|x_{n+1} - x_n\| \leq t_{n+1} - t_n, \quad n \geq 0. \quad (13)$$

Consequently, as $\{t_n\}$ is a convergent sequence and $\{x_n\}$ is a sequence defined in a Banach space, $\{x_n\}$ converges to a limit x^*, that can be shown it is a solution of the nonlinear equation (1).

The details and proofs of the aforementioned comments can be seen in [8]. They are summarized in the following results.

Theorem 3. *Let us consider the scalar sequences $\{t_n\}$ and $\{b_n\}$ defined in (12). Then the following relations hold:*

1. $b_n < 0$.

2. $b_n f'(t_n) < 1$.

3. $t_n < t_{n+1} < t^*$, *where t^* is the smallest positive root of* (11).

Theorem 4. *Under conditions* (10), *the scalar sequence $\{t_n\}$ defined in* (12) *is a majorizing function for $\{x_n\}$ defined in* (3), *that is,*

$$\|x_{n+1} - x_n\| \leq t_{n+1} - t_n, \; n \geq 0. \tag{14}$$

Consequently, $\{x_n\}$ converges to a limit x^*. In addition, if $\|B_0\| \leq 1$, x^* is a solution of (1), *that is $F(x^*) = 0$.*

4. APPLICATION TO FREDHOLM INTEGRAL EQUATIONS

The semilocal convergence theorem can be seen as an existence of solution result for a nonlinear equation. In this section we consider the following integral equation:

$$x(t) = z(t) + \lambda \int_a^b k(t,s) H(x(s))\, ds, \quad t \in [a,b],$$

where z is a given continuous function, H is an analytic function, k is a kernel continuous in its two variables and λ is a real parameter. This equation can be written as a equation $F(x) = 0$, where $F : X \to X$ is an operator defined on $X = C[a,b]$, the space of continuous functions in the interval $[a,b]$. The expression of such operator is the following:

$$F(x)(t) = x(t) - z(t) - \lambda \int_a^b k(t,s) H(\phi(s))\, ds, \quad t \in [a,b]. \tag{15}$$

In the space of continuous functions in $[a,b]$ we consider the max–norm:

$$\|g\| = \max_{t \in [a,b]} |g(t)|, \; g \in C[a,b].$$

With this norm, we obtain the following bound for the kernel k:

$$\|k\| \leq \max_{t \in [a,b]} \int_a^b |k(t,s)|\, ds.$$

In [15] Newton's method has been considered for studying the solution of (15). The two main problems of using Newton's method for solving a nonlinear equation is the choice of the initial approximation x_0 and the calculus of the inverses $F'(x_k)^{-1}$ (or the corresponding solution of a linear equation) at each step. In [15] the initial approximation is chosen as $x_0(t) = z(t)$ and then it is established a set of values for the parameter λ in order equation (15) has a solution. An estimate for the norm of $F'(x_0)^{-1}$ is also given.

Now, in this section we use Newton–Moser method (3) for studying the solution of (15). We consider the same choice for the initial approximation, that is $x_0(t) = z(t)$, but the calculus of $F'(x_0)^{-1}$ it is not required now.

To construct the majorizing function (11) we need to calculate the parameters γ_0, β and γ_j, $j \geq 2$, given in (10), by taking as starting point the function $x_0 = z$. The derivatives of order j of (15) are j–linear operators from the space X^j on X given by:

$$F'(x)[y_1](t) = y_1(t) - \lambda \int_a^b k(t,s) H'(x(s)) y_1(s)\, ds,$$

$$F^{(j)}(x)[y_1, \ldots, y_j](t) = -\lambda \int_a^b k(t,s) H^{(j)}(x(s)) y_1(t) \cdots y_j(t)\, ds, \quad j \geq 2.$$

Now we consider a particular integral equation of type (15). We take $x_0(t) = z(t)$ and $B_0 = I$, the identity operator, as starting values for Newton–Moser method (3) and we study the existence of solutions for the corresponding majorizing equation $f(t) = 0$, with f defined in (11). Notice that different convergence results could be obtained under different choices for $x_0(t)$ and B_0.

Let us consider the integral equation

$$x(t) = 1 + \lambda \int_0^1 \cos(\pi s t) x(s)^m\, ds, \quad m \in \mathbb{N},\ t \in [0,1],$$

or equivalently, the nonlinear equation

$$F(x)(t) = x(t) - 1 - \lambda \int_0^1 \cos(\pi s t) x(s)^m\, ds. \qquad (16)$$

We take $x_0(t) = 1$ for all $t \in [0,1]$ and $B_0 = I$. Then, $\gamma_0 = |\lambda|$, $\beta = m|\lambda|$ and

$$\gamma_j = \begin{cases} |\lambda| m(m-1) \cdots (m-j+1) & \text{if } 2 \leq j \leq m \\ 0 & \text{if } j > m. \end{cases}$$

Consequently the majorizing function (11) is given by

$$f(t) = |\lambda| + (m|\lambda| - 1)t + |\lambda| \sum_{j=2}^m \binom{m}{j} t^j$$

$$= |\lambda|(1+t)^m - t.$$

If $m|\lambda| < 1$ this function has an absolute minimum

$$\hat{t} = \sqrt[m-1]{\frac{1}{m|\lambda|}} - 1$$

and, in addition, $f(\hat{t}) < 0$.

Then, according with the results of the previous section, we have established a result on the existence of solution for equations (16). In fact, if

$$|\lambda| < \frac{1}{m}$$

the integral equation (16) has a solution. In addition, this solution can be approximated by using Newton–Moser method (3) starting with $x_0(t) = 1$ and $B_0 = I$.

For instance, if we consider $m = 5$ and $\lambda = 1/20$ then, the iterations of Newton–Moser method are given by

$$\begin{aligned} x_1(t) &= 1 + 0.015915493 t^{-1} \sin(\pi t), \\ x_2(t) &= 1 + 0.017615759 t^{-1} \sin(\pi t), \\ x_3(t) &= 1 + 0.017633935 t^{-1} \sin(\pi t), \\ x_4(t) &= 1 + 0.017633938 t^{-1} \sin(\pi t). \end{aligned}$$

REFERENCES

[1] Moser, J. 1973. *Stable and random motions in dynamical systems with special emphasis on celestial mechanics. Herman Weil Lectures, Annals of Mathematics Studies.* 77. New Jersey: Princeton Univ. Press.

[2] Hald, O. H. 1975. "On a Newton–Moser type method". *Numer. Math.* 23: 411–425.

[3] Ulm, S. 1967. "On iterative methods with successive approximation of the inverse operator (Russian)". *Izv. Akad Nauk Est. SSR.* 16: 403–411.

[4] Zehnder, E. J. 1974. "A remark about Newton's method". *Comm. Pure Appl. Math.* 27: 361–366.

[5] Petzeltova, H. 1980. "Remark on Newton–Moser type method". *Commentationes Mathematicae Universitatis Carolinae.* 21: 719–725.

[6] Kantorovich, L. V. and G. P. Akilov. 1982. *Functional analysis.* Oxford: Pergamon Press.

[7] Ezquerro, J. A. and Hernández, M. A. 2008. "The Ulm method under mild differentiability conditions". *Numer. Math.* 109: 193–207.

[8] Gutiérrez, J. M., Hernández, M. A. and Romero, N. 2008. "A note on a modification of Moser's method". *Journal of Complexity.* 24: 185–197.

[9] Gutiérrez, J. M., Hernández, M. A. and Romero, N. 2010. "α–theory for Newton–Moser method". *Monografías Matemáticas García de Galdeano.* 35: 155–162.

[10] Smale, S. 1986. *Newton's method estimates from data at one point, in: R. Ewing, K. Gross and C. Martin (eds.): The Merging of Disciplines: New Directions in Pure, Applied and Computational Mathematics.* 185–196. New York: Springer–Verlag.

[11] Smale, S. 1987. *The fundamental theory for solving equations, in proceeding of the International Congress of Mathematicians.* Providence, RI: AMS, 185.

[12] Wang, D. R. and Zhao, F. G. 1995. "The theory of Smale's point estimation and its applications". *J. Comput. Appl. Math.* 60: 253–269.

[13] Dedieu, J. P. 2006. *Points fixes, zéros et la méthode de Newton*. Berlin Heidelberg: Springer–Verlag.

[14] Petkovic, M. 2008. *Point Estimation of Root Finding Methods*. Lecture Notes in Mathematics, Springer.

[15] Gutiérrez, J. M., Hernández, M. A. and Salanova, M. A. 2004. "α–theory for nonlinear Fredholm integral equations". *Grazer Mathematische Berichte*. 346: 187–196.

In: Understanding Banach Spaces
Editor: Daniel González Sánchez

ISBN: 978-1-53616-745-0
© 2020 Nova Science Publishers, Inc.

Chapter 18

APPROXIMATING INVERSE OPERATORS BY A FOURTH–ORDER ITERATIVE METHOD

J. A. Ezquerro[*], *Miguel A. Hernández–Verón*[†] *and J. L. Varona*[‡]
Department of Mathematics and Computer Sciences
University of La Rioja, Logroño, Spain

Abstract

From a known uniparametric family of one–point third–order iterative methods, we obtain a fourth–order iterative method to approximate inverse operators without using inverse operators, analyse the convergence of the method and illustrate the analysis with two numerical examples.

Keywords: inverse operator, iterative method, convergence, integral equation

AMS Subject Classification: 65J17, 65J22

1. INTRODUCTION

When we are interested in solving a nonlinear equation $F(x) = 0$, where $F : \Omega \subseteq X \to Y$ is an operator defined on a nonempty open convex domain Ω of a Banach space X with values in a Banach space Y, we usually turn to one–point iterative methods, which are of form

$$x_{n+1} = \Phi(x_n), \quad n \geq 0, \quad \text{for given } x_0, \qquad (1)$$

and, among these, we generally use the well–known Newton's method,

$$x_{n+1} = x_n - [F'(x_n)]^{-1} F(x_n), \quad n \geq 0, \quad \text{for given } x_0, \qquad (2)$$

since it is an efficient method as a consequence of its quadratic order of convergence and its low operational cost.

---

[*]Corresponding Author's E-mail: jezquer@unirioja.es.
[†]E-mail: mahernan@unirioja.es.
[‡]E-mail: jvarona@unirioja.es.

When choosing an iterative method, we must pay special attention to two aspects: the speed of convergence, which is measured through the order of convergence, and the operational cost of the method. For one–point iterative methods of form (1), it is well–known that the order of convergence depends explicitly on the derivatives of the operator F involved, so that if iteration (1) has order of convergence p, (1) depends on the first $p-1$ derivatives of F (p. 98 of [1], Theorem 5.3). As a consequence of this fact, when we want to solve nonlinear systems or infinite dimensional problems, the operational cost of using derivatives of high orders is high, so that it is common to resort to Newton's method because of its efficiency. However, for certain operators F, we may be interested in using iterative methods with higher order of convergence, as, for example, in solving quadratic equations ([2,3]).

If we now pay attention to one–point third–order iterative methods and, in particular, to Chebyshev's method,

$$x_{n+1} = x_n - \left(I + \frac{1}{2}L_F(x_n)\right)[F'(x_n)]^{-1}F(x_n), \quad n \geq 0, \quad \text{for given } x_0,$$

where I is the identity operator on X and $L_F(x) = [F'(x)]^{-1}F''(x)[F'(x)]^{-1}F(x)$, $x \in X$, provided that $[F'(x_n)]^{-1}$ exists at each step, we see that the only inverse operator that we have to calculate in each step is $[F'(x_n)]^{-1}$, which is the same inverse that calculate in each step of Newton's method. However, the algorithm of other third–order one–point iterative methods, as for example the Halley and super-Halley methods, involves the calculation of other different inverse operators. So, from this point of view, we can see Chebyshev's method as the third–order one–point iterative method with less operational cost.

Following the last two ideas, it is introduced in [4] and used in [5] the following uniparametric family of one-point iterative methods

$$x_{n+1} = x_n - \left(I + \frac{1}{2}L_F(x_n) + \alpha L_F(x_n)^2\right)[F'(x_n)]^{-1}F(x_n), \quad n \geq 0, \quad \text{for given } x_0, \tag{3}$$

with $\alpha \in [0, 1/2]$, that only needs the calculation of the inverse $[F'(x_n)]^{-1}$ in each step and has an important peculiarity: for the value of the parameter $\alpha = 1/2$, the order of convergence is four when it is applied to solve quadratic equations. Observe that (3) is reduced to Chebyshev's method for $\alpha = 0$. In [4], we can see that iterations of (3) have a faster convergence than Chebyshev's method.

The origin of iterations given in (3) is a Gander's result [6] on third–order one–point iterative methods in the real case. In [6], Gander proves that, for a real function f of real variable, the iteration of form

$$x_{n+1} = x_n - \psi\left(L_f(x_n)\right)\frac{f(x_n)}{f'(x_n)}, \quad n \geq 0, \quad \text{for given } x_0,$$

with $L_f(x) = \frac{f(x)f''(x)}{f'(x)^2}$, is of order three if ψ is a function such that $\psi(0) = 1$, $\psi'(0) = 1/2$ and $|\psi''(0)| < +\infty$. For (3), we consider $\psi(t) = \left(1 + \frac{1}{2}L_f(t) + \alpha L_f(t)^2\right)$ with a finite Taylor's series and its generalization to Banach spaces. It is obvious that there are other functions ψ that led to third–order one–point iterative methods. But, for instance, the functions that originate the Halley or super–Halley methods does not have a non finite Taylor's series, so that the operational cost is higher.

On the other hand, as we can see in [7], approximation of inverse operators often appears in mathematics, mechanics, physics, electronics, meteorology, geophysics, and other branches of the natural sciences. Also, they play an important role in solving nonlinear evolution equations of mathematical physics and its interest is increasing permanently. In this work, we ask if in the same way that we obtain a method of order four of family (3) to solve quadratic equations, we can obtain a method of order four to approximate inverse operators with the important characteristic of not using inverse operators. The answer is yes and we see that (3) with $\alpha = 1/4$ has order of convergence four when it is applied to the approximation of inverse operators without using inverse operators.

2. PRELIMINAIRES

We consider operator equations $Tz = \varphi$ for operators between Banach spaces X and Y. We are interested in approximating solutions of $Tz = \varphi$, so that there is a solution $x = T^{-1}(\varphi)$ if φ is in the domain of T^{-1}, by means of iterative methods that do not use inverse operators in their algorithms. To approximate the inverse operator T^{-1} we use the well-known Newton and Chebyshev methods and iteration (3).

For the last, we consider the set $GL(X,Y) = \{T \in \mathcal{L}(X,Y) : T^{-1} \text{ exists}\}$, where $\mathcal{L}(X,Y)$ is the set of bounded linear operators from the Banach space X into the Banach space Y, and approximate an operator T^{-1} with $T \in GL(X,Y)$. Then, we choose $\mathcal{F}(Z) = Z^{-1} - T$, where $\mathcal{F} : GL(Y,X) \to \mathcal{L}(X,Y)$, so that T^{-1} is the solution of $\mathcal{F}(Z) = 0$.

We first consider Newton's and Chebyshev's methods and see that they approximate inverse operators without using any inverse operator in their algorithm. From this peculiarity of both methods, our interest focuses then on approximating T^{-1} by iteration (3) and without using inverse operators. After that, we will prove that iteration (3) with $\alpha = 1/4$ has order of convergence four.

2.1. Newton's Method

If we apply Newton's method,

$$Z_{n+1} = Z_n - [\mathcal{F}(Z_n)]^{-1}\mathcal{F}(Z_n), \quad n \geq 0, \quad \text{for given } Z_0 \in GL(Y,X),$$

to the problem of approximating T^{-1}, it is clear that we can avoid the use of inverse operators for approximating Z_{n+1}, just write the last algorithm in the form

$$\mathcal{F}(Z_n) + \mathcal{F}'(Z_n)(Z_{n+1} - Z_n) = 0, \quad n \geq 0,$$

and calculate $\mathcal{F}'(Z_n)$.

So, given $Z \in GL(Y,X)$, as Z^{-1} exists,

$$\|I - Z^{-1}(Z + \beta P)\| \leq \|Z^{-1}\|\|\beta P\|,$$

so that we obtain $Z + \beta P \in GL(Y,X)$ and $\|\beta P\| < \frac{1}{\|Z^{-1}\|}$ with $P \in GL(Y,X)$ if

$$0 < \beta < \frac{1}{\|P\|\|Z^{-1}\|}.$$

Then, $T + \beta P \in GL(Y, X)$ and

$$\mathcal{F}'(Z)P = \lim_{\beta \to 0} \frac{1}{\beta}\left(\mathcal{F}(Z + \beta P) - \mathcal{F}(Z)\right) = -Z^{-1}PZ^{-1}.$$

Taking into account the last, it is now easy to write Newton's method as

$$Z_{n+1} = 2Z_n - Z_n T Z_n, \quad n \geq 0, \quad \text{for given } Z_0. \tag{4}$$

As we can see, algorithm (4) does not use inverse operators to appoximate T^{-1}. Notice that the quadratic order of the method is held.

2.2. Chebyshev's Method

If we now apply Chebyshev's method,

$$Z_{n+1} = Z_n - \left(I + \frac{1}{2}L_{\mathcal{F}}(Z_n)\right)[\mathcal{F}'(Z_n)]^{-1}\mathcal{F}(Z_n), \quad n \geq 0, \quad \text{for given } Z_0 \in GL(Y, X),$$

where $L_{\mathcal{F}}(Z_n) = [\mathcal{F}'(Z_n)]^{-1}\mathcal{F}''(Z_n)[\mathcal{F}'(Z_n)]^{-1}\mathcal{F}(Z_n)$, and do the same as in Newton's method, we observe that the method does not use inverse operators to approximate T^{-1} if the method is written as

$$\begin{cases} Z_0 \text{ given,} \\ \mathcal{F}(Z_n) + \mathcal{F}'(Z_n)(U_n - Z_n) = 0, \quad n \geq 0, \\ \mathcal{F}''(Z_n)(U_n - Z_n)^2 + 2\mathcal{F}'(Z_n)(Z_{n+1} - U_n) = 0. \end{cases}$$

So, as before, given $P, Q \in GL(Y, X)$, as Z^{-1} exists, if

$$0 < \beta < \frac{1}{\|Q\|\|Z^{-1}\|},$$

then $Z + \beta Q \in GL(Y, X)$ and

$$\mathcal{F}''(Z)PQ = \lim_{\beta \to 0} \frac{1}{\beta}\left(\mathcal{F}'(Z + \beta Q)P - \mathcal{F}'(Z)P\right) = Z^{-1}PZ^{-1}QZ^{-1} + Z^{-1}QZ^{-1}PZ^{-1}.$$

Taking into account the last, it is now easy to write Chebyshev's method as

$$Z_{n+1} = 3Z_n - 3Z_n T Z_n + Z_n T Z_n T Z_n, \quad n \geq 0, \quad \text{for given } Z_0. \tag{5}$$

As we can see, algorithm (5) does not use inverse operators to appoximate T^{-1}. Notice that the cubical order of the method is held.

2.3. Chebyshev–Type Methods

If we now apply the family of iterations given in (3) to approximate T^{-1},

$$Z_{n+1} = Z_n - \left(I + \frac{1}{2}L_{\mathcal{F}}(Z_n) + \alpha L_{\mathcal{F}}(Z_n)^2\right)[\mathcal{F}'(Z_n)]^{-1}\mathcal{F}(Z_n), \quad n \geq 0,$$

for given $Z_0 \in GL(Y, X)$, and do the same as in the Newton and Chebyshev methods, we see that these iterations do not use inverse operators to approximate T^{-1} if they are written as

$$\begin{cases} Z_0 \text{ given,} \\ \mathcal{F}(Z_n) + \mathcal{F}'(Z_n)(U_n - Z_n) = 0, \quad n \geq 0, \\ \mathcal{F}''(Z_n)(U_n - Z_n)^2 + \mathcal{F}'(Z_n)V_n = 0, \\ \mathcal{F}''(Z_n)(U_n - Z_n)\left((U_n - Z_n) + 2\alpha V_n\right) + 2\mathcal{F}'(Z_n)(Z_{n+1} - U_n) = 0. \end{cases}$$

Proceeding as for Chebyshev's method, we can write the iterations as

$$\begin{cases} Z_0 \text{ given,} \\ Z_{n+1} = (3 + 4\alpha)Z_n - 3(1 + 4\alpha)Z_n T Z_n \\ \qquad + (1 + 12\alpha)Z_n T Z_n T Z_n - 4\alpha Z_n T Z_n T Z_n T Z_n, \quad n \geq 0, \end{cases} \quad (6)$$

and see that algorithm (6) does not use inverse operators to approximate T^{-1}. Notice that the cubical order of the method is held.

3. CONVERGENCE ANALYSIS

In this section, we study the convergence of iterations given in (6). For this, we use a simple technique which is different from the commonly used, which are based on the majorant principle and on recurrence relations. We fist see that four is the order of convergence of (6) for $\alpha = 1/4$.

Theorem 1. *Iteration (6) has order of convergence at least four if $\alpha = 1/4$.*

Proof. If $r_n = Z_n - T^{-1}$, then

$$r_{n+1} + T^{-1} = Z_{n+1}$$
$$= T^{-1}\left((3 + 4\alpha)TZ_n - 3(1 + 4\alpha)(TZ_n)^2 + (1 + 12\alpha)(TZ_n)^3 - 4\alpha(TZ_n)^4\right).$$

Observe that, if $\alpha = 1/4$, the last expression is then reduced to

$$r_{n+1} + T^{-1} = Z_{n+1} = T^{-1}\left(I - (Tr_n)^4\right) = T^{-1} - r_n(Tr_n)^3,$$

so that $r_{n+1} = -r_n(Tr_n)^3$ and $\|r_{n+1}\| \leq \|T\|^3\|r_n\|^4$. Therefore, the order of convergence at least four is guaranteed for $\alpha = 1/4$. $\square$

Taking then into account that we have order of convergence four for method (6) with $\alpha = 1/4$, namely,

$$\begin{cases} Z_0 \text{ given,} \\ Z_{n+1} = 4Z_n - 6Z_n T Z_n + 4Z_n T Z_n T Z_n - Z_n T Z_n T Z_n T Z_n, \quad n \geq 0, \end{cases} \quad (7)$$

we establish now the convergence of this method.

Theorem 2. If $\|I - TZ_0\| < 1$, iteration (7) is convergent. In addition, if $TZ_0 = Z_0T$, then $\lim_{n\to\infty} Z_n = T^{-1}$.

Proof. If $I - TZ_n = \varepsilon_n$, then

$$\varepsilon_{n+1} = I - TZ_{n+1} = I - 4TZ_n + 6(TZ_n)^2 - 4(TZ_n)^3 + (TZ_n)^4 = \varepsilon_n^4.$$

After that, from $\|\varepsilon_0\| = \|I - TZ_0\| < 1$, it follows that $\lim_{n\to\infty}(I - TZ_n) = 0$ and, as a consequence, $TZ^* = I$, where $Z^* = \lim_{n\to\infty} Z_n$.

Next, as $TZ^* = I$, if $Z^*T = I$, we have $G^* = T^{-1}$. For this, it is enough to see that $TZ_n = Z_nT$, for $n \in \mathbb{N}$. If $n = 1$, then

$$\begin{aligned} TZ_1 &= 4TZ_0 - 6(TZ_0)^2 + 4(TZ_0)^3 - (TZ_0)^4 \\ &= Z_0\left(4 - 6(TZ_0) + 4(TZ_0)^2 - (TZ_0)^3\right)T \\ &= Z_1T. \end{aligned}$$

Following now mathematical induction on n, it is easy to complete the proof. □

Observe that $TZ^* = I$ is only satisfied if $TZ_0 \neq Z_0T$ and, as a consequence, sequence $\{Z_n\}$ converges to the right inverse of T. If we prove that T^{-1} exists, then $Z^* = T^{-1}$ without demanding the commutativeness of Z_0 and T.

4. NUMERICAL EXAMPLES

In this section, we illustrate the application of iteration (7) to approximate inverse operators.

Example 3. *In general, we do not need to invert a matrix to solve a linear system in most practical applications. It is well–known that techniques based on the decomposition of the matrix involved are much faster than the inversion of the matrix. We can find many fast methods for special kinds of systems of linear equations in the mathematical literature. But, inversion of matrices plays an important role in some specific applications, such as computer graphics and wireless communications. In this work, we present an example where the use of method (4) is made good.*

For example, if we follow a process of discretization, by using finite differences, to transform the boundary value problem given by

$$y''(t) = v\,y(t) + w\,y'(t), \qquad y(a) = A, \quad y(b) = B,$$

into a linear system, the $m \times m$ matrix

$$M = \begin{pmatrix} x & y & 0 & \cdots & 0 \\ z & x & y & \cdots & 0 \\ 0 & z & x & \cdots & 0 \\ \vdots & \vdots & \vdots & \ddots & \vdots \\ 0 & 0 & 0 & \cdots & x \end{pmatrix}$$

is involved in the process, where $x = 2 + k^2v$, $y = -1 + \frac{k}{2}w$, $z = -1 - \frac{k}{2}w$ (with $k, v, w \in \mathbb{R}$) and m is the appropriate integer used to introduce the nodes $t_i = a + ih$, with $i = 0, 1, \ldots, m+1$ and $h = \frac{1}{m+1}$.

From Table 1, we see that matrix M is badly conditioned, since $\text{cond}(M) = 8.7466$. But, if we approximate M^{-1} by method (7), starting at matrix $Z_0 = \text{diag}\{1/x, 1/x, \ldots, 1/x\}$, which satisfies condition $\|I - TZ_0\|_1 < 1$ of Theorem 2, since $\|I - MZ_0\|_1 = 0.8 < 1$, we only need three iterations, Z_3, to get the inverse matrix M^{-1} without any stability problem.

Table 1. Approximation of M^{-1} by Z_3 with $k = 1$, $v = 1/2$, $w = 2$ and $m = 32$

$\det(M)$	5.4210×10^{12}
$\text{cond}(M)$	8.9928
$\|I - MZ_0\|_1$	0.8
$\|I - MZ_3\|_1$	0

Example 4. *Consider the Fredholm integral equation of the second kind given by*

$$x(s) = \ell(s) + \lambda \int_a^b K(s,t)x(t)\,dt, \qquad (8)$$

where $-\infty < a < b < +\infty$, $\ell(s) \in \mathcal{C}[a,b]$, the kernel $K(s,t)$ is a known function in $[a,b] \times [a,b]$ and $x(s) \in \mathcal{C}[a,b]$ is the unknown function to find.

If we consider the linear operator $\mathfrak{K} : \mathcal{C}[a,b] \to \mathcal{C}[a,b]$, given by $[\mathfrak{K}x](s) = \int_a^b K(s,t)x(t)\,dt$, we can write equation (8) as

$$x(s) = \ell(s) + \lambda[\mathfrak{K}x](s),$$

so that $[(I - \lambda\mathfrak{K})x](s) = \ell(s)$. So, if operator $[I - \lambda\mathfrak{K}]^{-1}$ exists, we can find a solution of (8) by solving

$$x(s) = \left[[I - \lambda\mathfrak{K}]^{-1}\ell\right](s). \qquad (9)$$

Following the study presented in this work, if we consider $T = I - \lambda\mathfrak{K}$ and $Z_0 \in \mathcal{L}(\mathcal{C}[a,b], \mathcal{C}[a,b])$ such that $\|I - TZ_0\| < 1$, then method (7) allows constructing a sequence of linear operators $\{Z_n\}$ such that $\lim_n Z_n = Z^*$ with $TZ^* = I$. Moreover, from the existence of T^{-1}, it follows $Z^* = T^{-1}$. As a consequence, we construct the iterative method

$$x_n(s) = [Z_n(\ell)](s), \quad n \in \mathbb{N}, \quad \text{for given } Z_0 \in \mathcal{L}(\mathcal{C}[a,b], \mathcal{C}[a,b]).$$

As $\{Z_n\}$ is convergent, it is a Cauchy sequence, so that $\{x_n\}$ is also a Cauchy sequence in $\mathcal{C}[a,b]$, since

$$\|x_{n+m} - x_n\| \leq \|Z_{n+m} - Z_n\|\|\ell\|.$$

Therefore, there exists $\lim_n x_n = x^*$ and then x^* is a solution of (8).

Taking now the above, if we consider the Fredholm integral equation of the second kind

$$x(s) = (1+s)^2 + \frac{1}{2}\int_{-1}^{1} \left(st + s^2 t^2\right) x(t)\, dt, \qquad (10)$$

we can approximate the exact solution $x^*(s) = 1 + 3s + \frac{5}{3}s^2$. Fredholm integral equations of this type are given in [8].

Following the Banach lemma on invertible operators and choosing the max–norm, we observe that operator $[I - \lambda \mathfrak{K}]^{-1}$ exists if $\lambda = 1/2$, since then $\|\lambda \mathfrak{K}\| \leq 5/6 < 1$. Next, if $Z_0 = I$, we can prove by mathematical induction on n that iteration (7) is reduced to $Z_n = \sum_{i=0}^{2^{2n}-1} \mathfrak{K}^i$, for all $n \geq 0$. Then, if we choose $x_0(s) = (1+s)^2$, we obtain the first three iterates given in Table 2, where we can also see that the iterates converge to the exact solution $x^*(s)$ of integral equation (10).

Table 2. Approximation of solution $x^*(s)$ of (10) and absolute errors

n	$x_n(s)$	$\|x^*(s) - x_n(s)\|_\infty$
0	$(1+s)^2$	1.6666
1	$1.00000000 + 2.96296296\, s + 1.66133333\, s^2$	4.2370×10^{-2}
2	$1.00000000 + 2.99999993\, s + 1.66666666\, s^2$	6.9713×10^{-8}
3	$1.00000000 + 3.00000000\, s + 1.66666666\, s^2$	8.7369×10^{-31}

REFERENCES

[1] Traub, J. F. 1964. *Iterative methods for the solution of equations.* New Jersey: Prentice–Hall.

[2] Lasiecka, I. and Triggiani, R. 2000. *Control theory for partial differential equations: continuous and approximation theories, Part 1.* Cambridge: Cambridge University Press.

[3] Lasiecka, I. and Triggiani, R. 2000. *Control theory for partial differential equations: continuous and approximation theories, Part 2.* Cambridge: Cambridge University Press.

[4] Ezquerro, J. A., Gutiérrez, J. M., Hernández, M. A. and Salanova, M. A. 1999. "Chebyshev–like methods and quadratic equations". *Rev. Anal. Numér. Théor. Approx.* 28(1): 23–35.

[5] Amat, S., Busquier, S. and Gutiérrez, J. M. 2006. "An adaptative version of a fourth–order iterative method for quadratic equations". *J. Comput. Appl. Math.* 191(2): 259–268.

[6] Gander, W. 1985. "On Halley's iteration method". *Amer. Math. Monthly.* 92: 131–134.

[7] Yurko, V. A. 2000. "Inverse problems of spectral analysis for differential operators and their applications". *J. Math. Sci.* 98(3): 319–426.

[8] Wazwaz, A.-M. 2011. *Linear and nonlinear integral equations: methods and applications*. Berlin: Springer–Verlag.

In: Understanding Banach Spaces
Editor: Daniel González Sánchez

ISBN: 978-1-53616-745-0
© 2020 Nova Science Publishers, Inc.

Chapter 19

INITIAL VALUE PROBLEMS IN CLIFFORD ANALYSIS USING ASSOCIATED SPACES

David Armendáriz[1,*], *Johan Ceballos*[2,†] *and Antonio Di Teodoro*[1,‡]
[1]Departamento de Matemáticas, Colegio de Ciencias e Ingenierías
Universidad San Francisco de Quito, Quito, Ecuador
[2]Escuela de Ciencias Físicas y Matemáticas
Universidad de Las Américas, Quito, Ecuador

Abstract

In the present chapter we propose and solve an initial value problem

$$\partial_t u = F(t, x, u, \partial_{x_i} u), \quad u(0, x) = \varphi(x),$$

where the initial data belong to an associated space in a Clifford algebra. The method of associated spaces allows to construct and guaranteed the existence of a fixed point from the contractivity in a Banach space.

Keywords: matrix representations, Clifford–type algebras, initial value problems, monogenic functions, associated space, interior estimate

AMS Subject Classification: 11E88, 15A66, 16W55, 35F10, 35B45, 46B25, 47H10, 65F30

1. THE GOAL OF THE PROBLEM

Initial value problems

$$\partial_t u = F(t, x, u, \partial_{x_i} u), \quad u(0, x) = \varphi(x), \tag{1}$$

---

[*]E-mail: darmendarizp@estud.usfq.edu.ec.
[†]Corresponding Author's E-mail: johan.ceballos@udla.edu.ec.
[‡]E-mail: nditeodoro@usfq.edu.ec.

where t is the time, F is a linear first order operator in a Clifford Analysis [1, 2] and φ is a monogenic function, can be solved by applying the method of associated spaces how is constructed by W. Tutschke [3].

The concept of associated spaces generalizes the connection between holomorphic right–hand sides and holomorphic initial data: If the initial data belongs to an associated space of the right–hand side, then the initial value problem is (uniquely) solvable. The desired solution of the initial value problem can be found as fixed point of a related operator

$$U(x,t) = \varphi(x) + \int_0^t F(\tau, x, u, \partial_j u) d\tau. \qquad (2)$$

To construct fixed points of this operator, we have to estimate the integro–differential operator on the right–hand side of (2) over a Banach space. This is a little bit difficult because the integrand contains derivatives with respect to the spacelike variables x_j. One possibility to solve this problem is using the Cauchy–Pompeiu integral representation. The existence of an associated differential operator satisfying an interior estimate, implies not only the solvability of the differential equation but also the possibility to choose arbitrary solutions of the associated differential equation as initial functions.

Summarizing, one looks for an associated function space having two properties:

- The operator maps the space into itself over a Banach space, which means that we have to use the concept of associated pair (see [3–5]).

- The elements of the associated space satisfy a first order interior estimate [5–12], that is, the norms of the derivatives with respect to spacelike variables of the elements of the associated space can be estimated by the norms of the elements.

The present chapter formulates sufficient conditions on the coefficients of operator L under which L is associated to differential equations with monogenic right–hand sides. Using an algorithm to characterize the operators over a high order Clifford structures, we shall exhibit an interior estimate for the monogenic functions using the Cauchy integral formula and then give out conditions under which this initial value problem (1) is uniquely solvable in the context of Clifford algebras over an appropriate Banach space.

Scheme of the two chapters:

- Preliminaries and notation are exhibit in Section 2. Next to it, we construct a matrix representation for real Clifford type algebras focusing on the computer implementation. Moreover, in Section 3, we present the Dirac operator using these matrices without Leibniz's rule.

- In Section 4, we present a discussion about the importance of the Banach space to construct the interior estimate, also we present an interior estimate using a Cauchy–Pompeiu formula, focusing our attention in the construction of conditions to have associated pairs using the matrix approach.

- Finally, we discuss about the uniqueness of the solution of the initial value problem and we present an extension of solutions using matrix approach.

- In the next chapter, we will present explicit algorithms in MATLAB to obtain the matrix representation of Clifford Algebras. We show how to calculate the derivative of the product of functions in Clifford Algebras and a much longer example to calculate the associated spaces where the need of this matrix representation is shown.

2. PRELIMINARIES AND NOTATIONS

Let $\mathbb{R}_{0,n}$ be the 2^n–dimensional real Clifford algebra generated by the orthonormal basis $\{e_1, \ldots, e_n\}$ of the Euclidean space $\mathbb{R}^n$ (see [1, 13, 14]). The multiplication in $\mathbb{R}_{0,n}$ is associative and is governed by the following rules

$$\begin{aligned} e_j^2 &= -1, \quad j = 1, \ldots, n, \\ e_i e_j + e_j e_i &= 0, \quad 1 \leq i \neq j \leq n. \end{aligned} \quad (3)$$

A vector space basis for $\mathbb{R}_{0,m}$ is given by $\{e_A | A \subseteq \{1, 2, \ldots, n\}\}$ with $e_A = e_{\alpha_1} e_{\alpha_1} \cdots e_{\alpha_r}$, where $1 \leq \alpha_1 < \cdots < \alpha_r \leq n$, $0 \leq r \leq n$, $e_0 := 1$ is the identity element with respect to the multiplication of the Clifford algebra $\mathbb{R}_{0,n}$.

The subspace of k–vectors is defined as $\mathbb{R}_{0,n}^{(k)} = \{u = \sum_A u_A e_A, u_A \in \mathbb{R}, |A| = k\}$ leading to the decomposition $\mathbb{R}_{0,n} = \bigoplus_{k=0}^{n} \mathbb{R}_{0,n}^{(k)}$. $\mathbb{R}^{n+1}$ may be naturally embedded in $\mathbb{R}_{0,n}$ by associating to any element $(x_0, x_1, \ldots, x_n) \in \mathbb{R}^{n+1}$ the paravector:

$$x = x_0 e_0 + \tilde{x} = x_0 e_0 + \sum_{j=1}^{n} x_j e_j \in \mathbb{R}_{0,n}^{(0)} \bigoplus \mathbb{R}_{0,n}^{(k)}$$

Conjugation in $\mathbb{R}_{0,n}$ is given by

$$\overline{u} = \sum_A u_A \overline{e}_A,$$

where $\overline{e}_A = \overline{e}_{\alpha_k} \cdots \overline{e}_{\alpha_1}, \overline{e}_j = -e_j, j = 1, \ldots, n$.

One natural way to generalize the theory of holomorphic functions of a complex variable to higher dimensions, is by considering the null solutions of the so–called generalized Cauchy-Riemann operator in $\mathbb{R}^{n+1}$, given by

$$\tilde{\partial} = \sum_{j=0}^{n} e_j \partial_{x_j} = e_0 \partial_{x_0} + \sum_{j=0}^{n} e_j \partial_{x_j} = e_0 \partial_{x_0} + D,$$

where $D = \sum_{j=1}^{n} e_j \partial_{x_j}$ is the Dirac operator in $\mathbb{R}^n$. More precisely, an $\mathbb{R}_{0,n}$–valued function f defined and continuously differentiable in an open set $\Omega \subseteq \mathbb{R}^{n+1}$, is called left(resp. right)–monogenic in Ω if $\tilde{\partial} f = 0$ (resp. $f\tilde{\partial} = 0$) in Ω ([1, 2, 14]). In a similar fashion is defined monogenicity with respect to D.

Finally, we recall that the Euclidean Dirac operator D, factorizes the n–dimensional Euclidean Laplace type operator ([15]):

$$\tilde{\Delta} = DD = -\sum_{j=1}^{n} \partial_j^2 \quad (4)$$

similarly, the Cauchy–Riemann operator and it's conjugate factorize the standard Euclidean Laplace operator $\tilde{\partial}\overline{\tilde{\partial}} = \sum_{j=0}^{d} \partial_{x_j}^2 = \Delta_{n+1}$ in $\mathbb{R}^{n+1}$.

Example 1. *Using the rules given in equation (3) find a simplified expression for* $e_1 e_2 e_3 e_4 e_2 e_1 e_3$.

$$\begin{aligned}
e_1\, e_2\, e_3\, e_4\, (e_2\, e_1)\, e_3 &= (-1)^1 e_1\, e_2\, e_3\, (e_4\, e_1)\, e_2\, e_3 \\
&= (-1)^2 e_1\, e_2\, (e_3\, e_1)\, e_4\, e_2\, e_3 \\
&= (-1)^3 e_1\, (e_2\, e_1)\, e_4\, e_4\, e_2\, e_3 \\
&= (-1)^4 (e_1\, e_1)\, e_2\, e_3\, e_4\, e_2\, e_3 \\
&= (-1)^5 e_2\, e_3\, (e_4\, e_2)\, e_3 \\
&= (-1)^6 e_2\, (e_3\, e_2)\, e_4\, e_3 \\
&= (-1)^7 (e_2\, e_2)\, e_3\, e_4\, e_3 \\
&= (-1)^8 e_3\, (e_4\, e_3) \\
&= (-1)^9 (e_3\, e_3)\, e_4 \\
&= (-1)^{10} e_4 \\
&= + e_4
\end{aligned}$$

We observe that this is a recursively processes, therefore, we propose an algorithm to compute any product between elements of $\mathbb{R}_{0,n}$.

Algorithm 2 (Product between elements of $\mathbb{R}_{0,n}$).
Input : *Vector x with indices of the elements of $\mathbb{R}_{0,n}$ as entries.*
Output: *x and sign such that the product is equal to $sign * e_x$, $e_x \in \mathbb{R}_{0,n}$.*

Step 1: *Sort the vector x in ascending order, in each exchange multiply by -1 as in Definition 3.*
Step 2: *Each pair of consecutive equal indexes is replaced by -1 as in Definition 3.*

Remark 3. *To obtain the result of Example (1) using the previuos algorithm, we need to introduce as input, the subscripts of $e_1 e_2 e_3 e_4 e_2 e_1 e_3$ as a vector x=[1 2 3 4 2 1 3]. The outputs are: s, representing the sign of the result product, 1 for positive, -1 for negative, and the subscript of the result product x. The solution is $s = 1$ and $x = 4$, this implies that,* $e_1 e_2 e_3 e_4 e_2 e_1 e_3 = + e_4$

Example 4. *In $\mathbb{R}_{0,3}$ the structure relation is given by*

$$\begin{cases} e_0 = 1 \\ e_i^2 = -1 \quad \text{where} \quad i = 1, 2, 3 \\ e_i e_j + e_j e_i = 0, \quad \text{for} \quad i \neq j \quad i, j = 1, 2, 3 \end{cases}$$

the dimension is $\dim \mathbb{R}_{0,3} = 2^3 = 8$, whereas the basis is

$$\{e_0 = 1, e_1, e_2, e_3, e_{12}, e_{23}, e_{13}, e_{123}\}.$$

An element different from 0 in $\mathbb{R}_{0,3}$ is expressed as

$$\sum_{i=0}^{3} a_i e_i + a_{12}e_{12} + a_{13}e_{13} + a_{23}e_{23} + a_{123}e_{123},$$

where e_i are the vectors for $i = 0, 1, 2, 3$; e_{12}, e_{13}, e_{23} are the bi–vectors and e_{123} is the tree–vector.

Example 5. *For $n = 2$.*
Let $u = u_0 + u_1 e_1 + u_2 e_2 + u_{12} e_{12}$ be a monogenic function with values in $\mathbb{R}_{0,2}$. Then, the real–valued components u_0, u_1, u_2 and u_{12} satisfy the following system of differential equations:

$$\partial_0 u_0 - \partial_1 u_1 - \partial_2 u_2 = 0,$$
$$\partial_0 u_1 + \partial_1 u_0 + \partial_2 u_{12} = 0,$$
$$\partial_0 u_2 - \partial_1 u_{12} + \partial_2 u_0 = 0,$$
$$\partial_0 u_{12} + \partial_1 u_2 - \partial_2 u_1 = 0.$$

Indeed, the basis elements of $\mathbb{R}_{0,2}$ is $\{e_0 = 1, e_1, e_2, e_{12}\}$ over $\mathbb{R}^3$. Carrying out the multiplication

$$\tilde{\partial} u = (\partial_0 + e_1 \partial_1 + e_2 \partial_2)(u_0 + u_1 e_1 + u_2 e_2 + u_{12} e_{12})$$

and taking into account the relations

$$e_1 e_1 = e_2 e_2 = -1$$
$$e_1 e_{12} = e_1(e_1 e_2) = -e_2$$
$$e_2 e_1 = -e_1 e_2$$
$$e_2 e_{12} = e_2(-e_2 e_1) = e_1,$$

the equation $\tilde{\partial} u = 0$ leads to the above equations because the basis elements $1, e_1, e_2, e_{12}$ are linearly independent.

Example 6. *In $\mathbb{R}_{0,2}$, the function $u(x_0, x_1, x_2) = -x_0 e_1 + x_2 e_{12}$ is left–monogenic but not right–monogenic.*

Consider the function $u(x_0, x_1, x_2) = -x_0 e_1 + x_2 e_{12}$. Then we have

$$\tilde{\partial} u = (\partial_0 + e_1 \partial_1 + e_2 \partial_2)(-x_0 e_1 + x_2 e_{12}) = -e_1 + e_2 e_{12} = 0$$

and

$$u\tilde{\partial} = (-x_0 e_1 + x_2 e_{12})(\partial_0 + e_1 \partial_1 + e_2 \partial_2) = -e_1 + e_{12} e_2 = -2e_1 \neq 0.$$

Example 7. *In $\mathbb{R}_{0,2}$, the function $u(x_0, x_1, x_2) = x_1 e_1 - x_2 e_2$ is both left–monogenic and right–monogenic.*

The $\mathbb{R}_{0,2}$–valued function is $u(x_0, x_1, x_2) = x_1 e_1 - x_2 e_2$. Here we have

$$\tilde{\partial} u = (\partial_0 + e_1 \partial_1 + e_2 \partial_2)(x_1 e_1 - x_2 e_2) = 0 + e_1 e_1 + e_2(-e_2) = 0$$

and

$$u\tilde{\partial} = (x_1 e_1 - x_2 e_2)(\partial_0 + e_1 \partial_1 + e_2 \partial_2) = 0 + e_1 e_1 + (-e_2)e_2 = 0.$$

3. FUNDAMENTAL MATRICES AND CLIFFORD MATRIX REPRESENTATION

It is well known that the matrix basis for the quaternions algebras are given by ([13, 16, 17]):

$$(i) = \begin{pmatrix} 0 & -1 & 0 & 0 \\ 1 & 0 & 0 & 0 \\ 0 & 0 & 0 & -1 \\ 0 & 0 & 1 & 0 \end{pmatrix}, \qquad (j) = \begin{pmatrix} 0 & 0 & -1 & 0 \\ 0 & 0 & 0 & 1 \\ 1 & 0 & 0 & 0 \\ 0 & -1 & 0 & 0 \end{pmatrix}$$

and

$$(k) = \begin{pmatrix} 0 & 0 & 0 & -1 \\ 0 & 0 & -1 & 0 \\ 0 & 1 & 0 & 0 \\ 1 & 0 & 0 & 0 \end{pmatrix}.$$

such that $(i)^2 = (j)^2 = (k)^2 = -1$ and $(i)(j) = (k)$. This represents an advantage when handling objects. For example, any Quaternions can be written as $v = a_0 \mathbf{I}_4 + a_1(i) + a_2(j) + a_3(k)$, where a_i are constant and $\mathbf{I}_4$ is the identity matrix.

Two examples where this may be useful is when manipulating a huge amount of information or for physics applications ([13, 17–22]).

An extension of the Quaternions is the Non–conmutative and associative algebra "the Clifford algebras". Throughout the last two decades, it was found a matrix representation of Clifford algebra and different kind of constructions were found ([23–26]), with some specific properties like representations of the basis elements of Clifford algebra, chosen to satisfy some particular conditions ([16]), including condition on the invertibility of the elements ([27]).

An important part of the matrix approach is that all the information about the structure relations of the algebra ($e_j e_k + e_k e_j = -2\delta_{jk}$, δ_{jk} being the Kronecker delta) is contained in the matrices. Thus, when we operate with such matrices, it is not necessary to take into account these structure relations anymore.

Multiplying elements of a Clifford algebra is very difficult, the matrix representation provides a simple tool to multiply these objects in a simple way and from a much simpler computational point of view. The idea of using this matrix representation is to simplify the calculations and characterize any case, in any dimension of a Clifford algebra. It is not necessary the use of a product rule for valued Clifford functions, because there is no Leibniz rule, alternative formulas for the product can be obtained. However, with the use of a matrix representation it is possible to go through the Leibniz rule for real valued functions to simplify calculations.

3.1. Construction of the Fundamental Matrices

We use the matrix representation given in [28–31] to construct fundamental matrices E_{jn}, $j = 1, 2, \ldots n$, that satisfy the structural relation:

$$E_{j,n}^2 = \mathbf{I_{2^n}},$$
$$E_{i,n}E_{j,n} + E_{j,n}E_{i,n} = \mathbf{0_{2^n}}, \quad i \neq j, \tag{5}$$

for $i, j = 1, 2, \ldots, n$. Where $\mathbf{I_{2^n}}$ and $\mathbf{0_{2^n}}$ are the identity matrix and the cero matrix respectively (sometimes to clarify we use the symbol $E_{0,n}$ to refer to the identity matrix), recursively allowing the possibility to develop the computer implementation for the Clifford algebras.

Example 8. *If $n = 1$, the fundamental matrices for $\mathbb{R}_{0,1}$ are $E_{1,1} = \begin{pmatrix} 0 & -1 \\ 1 & 0 \end{pmatrix}$, and the identity matrix $\mathbf{I_2}$. This example gives a matrix representation for the complex case.*

Example 9. *If $n = 2$, the fundamental matrices for $\mathbb{R}_{0,2}$ are*

$$E_{1,2} = \begin{pmatrix} 0 & -1 & 0 & 0 \\ 1 & 0 & 0 & 0 \\ 0 & 0 & 0 & -1 \\ 0 & 0 & 1 & 0 \end{pmatrix} = \left(\begin{array}{c|c} E_{1,1} & \mathbf{0_2} \\ \hline \mathbf{0_2} & E_{1,1} \end{array}\right),$$

$$E_{2,2} = \begin{pmatrix} 0 & 0 & -1 & 0 \\ 0 & 0 & 0 & 1 \\ 1 & 0 & 0 & 0 \\ 0 & -1 & 0 & 0 \end{pmatrix} = \left(\begin{array}{c|c} \mathbf{0_2} & -\mathbf{M_2} \\ \hline \mathbf{M_2} & \mathbf{0_2} \end{array}\right),$$

where $\mathbf{M_2}^2 = \mathbf{I_2}$ and

$$E_{12,2} = \begin{pmatrix} 0 & 0 & 0 & -1 \\ 0 & 0 & -1 & 0 \\ 0 & 1 & 0 & 0 \\ 1 & 0 & 0 & 0 \end{pmatrix} = E_{1,2} \cdot E_{2,2}.$$

Remark 10. *The fundamental matrices $E_{j,n}$, $j = 1, \ldots, n$ can be obtained recursively using block matrices of the previous dimension and the matrix product, which will allow us to develop the computer algorithms to represent the Clifford algebras.*

Algorithm 11 (Recursive matrix representation algorithm for $E_{i,n}$ in $\mathbb{R}_{0,n}$).
Input : $n \in \mathbb{Z}^+$.
Output: Matrix $E_{i,n} \in \mathbb{R}^{2^n \times 2^n}$

Step 1: Compute the matrix $E_{i,i} = \left(\begin{array}{c|c} \mathbf{0_{2^{n-1}}} & -\mathbf{M_{2^{n-1}}} \\ \hline \mathbf{M_{2^{n-1}}} & \mathbf{0_{2^{n-1}}} \end{array}\right)$, where $\mathbf{M}^2_{2^{n-1}} = E_{0,n-1}$.

Step 2: Compute the matrix $E_{i,k} = \left(\begin{array}{c|c} E_{i,k-1} & \mathbf{0_{2^{n-1}}} \\ \hline \mathbf{0_{2^{n-1}}} & E_{i,k-1} \end{array}\right)$, for $k = i+1, \ldots, n$.

Example 12. *Using the rules given in equation (5), Algorithm 11 and matrix product, find a simplified expression for $E_{1,2} E_{2,2} E_{12,2} E_{1,2} \in \mathbb{R}_{0,2}$.*

$$E_{1,2} E_{2,2} E_{12,2} E_{1,2} = -\begin{pmatrix} 0 & 1 & 0 & 0 \\ -1 & 0 & 0 & 0 \\ 0 & 0 & 0 & 1 \\ 0 & 0 & -1 & 0 \end{pmatrix} = -E_{1,2}.$$

Matrices $E_{1,3}, E_{2,3}, E_{3,3}$ are given in Example 9.

Example 13. *A Clifford valued function $u : \mathbb{R}^3 \to \mathbb{R}_{0,2}$ can be written in the matrix form, using the fundamental matrices given in the example 9 for $\mathbb{R}_{0,2}$ and $u_0, u_1, u_2, u_{12} : \mathbb{R}^3 \to \mathbb{R}$, in the following form*

$$(u) = u_0 \mathbf{I_4} + u_1 E_{1,2} + u_2 E_{2,2} + u_{12} E_{12,2}$$

$$= \begin{pmatrix} u_0 & 0 & 0 & 0 \\ 0 & u_0 & 0 & 0 \\ 0 & 0 & u_0 & 0 \\ 0 & 0 & 0 & u_0 \end{pmatrix} + \begin{pmatrix} 0 & -u_1 & 0 & 0 \\ u_1 & 0 & 0 & 0 \\ 0 & 0 & 0 & -u_1 \\ 0 & 0 & u_1 & 0 \end{pmatrix} +$$

$$\begin{pmatrix} 0 & 0 & -u_2 & 0 \\ 0 & 0 & 0 & u_2 \\ u_2 & 0 & 0 & 0 \\ 0 & -u_2 & 0 & 0 \end{pmatrix} + \begin{pmatrix} 0 & 0 & 0 & -u_{12} \\ 0 & 0 & -u_{12} & 0 \\ 0 & u_{12} & 0 & 0 \\ u_{12} & 0 & 0 & 0 \end{pmatrix}$$

$$= \begin{pmatrix} u_0 & -u_1 & -u_2 & -u_{12} \\ u_1 & u_0 & -u_{12} & u_2 \\ u_2 & u_{12} & u_0 & -u_1 \\ u_{12} & -u_2 & u_1 & u_0 \end{pmatrix}.$$

Example 14. *(Product of two matrices functions) Let u and v $\mathbb{R}_{0,2}$–Clifford valued functions such that*

$$u = u_1 E_{1,2} + u_2 E_{2,2} \quad \text{and}$$
$$v = v_{12} E_{12,2},$$

where u_1, u_2, v_{12} are real valued functions.
The matrix product $u \cdot v$ is given by:

$$u \cdot v = (u_1 E_{1,2} + u_2 E_{2,2})(v_{12} E_{12,2})$$

$$= \begin{pmatrix} 0 & -u_1 & -u_2 & 0 \\ u_1 & 0 & 0 & u_2 \\ u_2 & 0 & 0 & u_1 \\ 0 & -u_2 & u_1 & 0 \end{pmatrix} \cdot \begin{pmatrix} 0 & 0 & 0 & -v_{12} \\ 0 & 0 & -v_{12} & 0 \\ 0 & v_{12} & 0 & 0 \\ v_{12} & 0 & 0 & 0 \end{pmatrix}$$

$$= \begin{pmatrix} 0 & u_2 v_{12} & u_1 v_{12} & 0 \\ u_2 v_{12} & 0 & 0 & -u_1 v_{12} \\ -u_1 v_{12} & 0 & 0 & -u_2 v_{12} \\ 0 & u_1 v_{12} & u_2 v_{12} & 0 \end{pmatrix} = -u_1 v_{12} E_{2,2} + u_2 v_{12} E_{1,2}.$$

3.2. Cauchy–Riemann and Laplace Operators in the Matrix Form

It is useful to write the Cauchy–Riemann and the Dirac operator in matrix form to do calculations in advance. In our case, is to compute associated operators. Using the fundamental matrices, we can obtain the matrix representation for both operators using the corresponding basis for $\mathbb{R}_{0,n}$.

Definition 15. *The matrix representation for the Cauchy–Riemann operator in $\mathbb{R}_{0,n}$ is defined by $\mathbb{R}^{2^n \times 2^n}$ matrix:*

$$\tilde{\partial}_{n+1} = \mathbf{I}_{2^n} \partial_0 + \sum_{i=1}^{n} E_{i,n} \partial_i = \mathbf{I}_{2^n} \partial_0 + \mathcal{D}_n$$

where, $\mathcal{D}_n$ is the Dirac operator and ∂_i represents the derivative with respect to x_i. Its conjugate is given by

$$\overline{\tilde{\partial}}_{n+1} = \mathbf{I}_{2^n} \partial_0 - \sum_{i=1}^{n} E_{i,n} \partial_i = \mathbf{I}_{2^n} \partial_0 - \mathcal{D}_n$$

Remark 16. *The Clifford algebra basis for $\mathbb{R}_{0,3}$ is*

$$\{E_{0,3}, E_{1,3}, E_{2,3}, E_{3,3}, E_{12,3}, E_{13,3}, E_{23,3}, E_{123,3}\},$$

for simplicity, we identify each basis element as a bijective function:

$$\{0, 1, 2, 3, 12, 13, 23, 123\} \to \{0, 1, 2, 3, 4, 5, 6, 7, 8\}$$

in other words,

$$\begin{cases} E_{0,3} = E_{0,3}, & E_{1,3} = E_{1,3}, \\ E_{2,3} = E_{2,3}, & E_{3,3} = E_{3,3}, \\ E_{12,3} = E_{12,3}, & E_{13,3} = E_{13,3}, \\ E_{23,3} = E_{23,3}, & E_{123,3} = E_{123,3}. \end{cases}$$

Example 17. *Consider the cases $n = 1, 2, 3$ in $\mathbb{R}_{0,n}$ respectively to build the matrix, Cauchy–Riemann and Dirac operators:*

$$\tilde{\partial}_2 = \begin{pmatrix} \partial_0 & -\partial_1 \\ \partial_1 & \partial_0 \end{pmatrix} = \mathbf{I}_{2^1} \partial_0 + E_{1,1} \partial_1, \tag{6}$$

and

$$\tilde{\partial}_3 = \begin{pmatrix} \partial_0 & -\partial_1 & -\partial_2 & 0 \\ \partial_1 & \partial_0 & 0 & \partial_2 \\ \partial_2 & 0 & \partial_0 & -\partial_1 \\ 0 & -\partial_2 & \partial_1 & \partial_0 \end{pmatrix} = \mathbf{I}_{2^2} \partial_0 + E_{1,2} \partial_1 + E_{2,2} \partial_2, \tag{7}$$

$$\tilde{\partial}_4 = \mathbf{I}_{2^3} \partial_0 + \mathcal{D}_3,$$

and the Dirac operators are, respectively:

$$\mathcal{D}_1 = \begin{pmatrix} 0 & -\partial_1 \\ \partial_1 & 0 \end{pmatrix} = E_{1,1}\partial_1,$$

$$\mathcal{D}_2 = \begin{pmatrix} 0 & -\partial_1 & -\partial_2 & 0 \\ \partial_1 & 0 & 0 & \partial_2 \\ \partial_2 & 0 & 0 & -\partial_1 \\ 0 & -\partial_2 & \partial_1 & 0 \end{pmatrix} = E_{1,2}\partial_1 + E_{2,2}\partial_2$$

Example 18. *The Cauchy–Riemann System in the matrix form*

$$\tilde{\partial}_3 u = \begin{pmatrix} \partial_0 & -\partial_1 & -\partial_2 & 0 \\ \partial_1 & \partial_0 & 0 & \partial_2 \\ \partial_2 & 0 & \partial_0 & -\partial_1 \\ 0 & -\partial_2 & \partial_1 & \partial_0 \end{pmatrix} \begin{pmatrix} u_0 & -u_1 & -u_2 & -u_{12} \\ u_1 & u_0 & -u_{12} & u_2 \\ u_2 & u_{12} & u_0 & -u_1 \\ u_{12} & -u_2 & u_1 & u_0 \end{pmatrix} = 0. \qquad (8)$$

This system is equivalent to the systems exhibited in example 5.

An important consequence related to the operator $\tilde{\partial}$ (as well $\mathcal{D}_n$) and the relations given in 5, is the factorization of the Laplace operator through the Cauchy–Riemann (Dirac) operator, see (4).

Lemma 19. *Let $\tilde{\partial}$ and $\mathcal{D}_n$ be the matrices $\mathbb{R}^{2^n \times 2^n}$ in $\mathbb{R}_{0,n}$ and $\Delta_{n+1} = \sum_{i=0}^{n} \partial_i^2 = \partial_0^2 + \sum_{i=1}^{n} \partial_i^2 = \partial_0^2 + \Delta_n$ the $(n+1)$–Laplace operator. Then:*

i $\mathcal{D}_n^2 = -I_{2^n}\Delta_n$.

ii $\tilde{\partial}_{n+1}\overline{\tilde{\partial}_{n+1}} = I_{2^n}\Delta_{n+1}$

Proof. By direct calculation, we have

i.

$$\mathcal{D}_n^2 = \left(\sum_{i=1}^{n-1} E_{i,n}\partial_i + E_{n,n}\partial_n\right)\left(\sum_{j=1}^{n-1} E_{j,n}\partial_j + E_{n,n}\partial_n\right) =$$

$$\sum_{j=1}^{n-1}\sum_{i=1}^{n-1} E_{i,n}E_{j,n}\partial_i\partial_j + \sum_{i=1}^{n-1} E_{i,n}E_{n,n}\partial_i\partial_n + \sum_{j=1}^{n-1} E_{n,n}E_{j,n}\partial_n\partial_j + E_{n,n}^2\partial_n^2$$

$$= \sum_{i=1}^{n-1} E_{i,n}^2\partial_i^2 + E_{n,n}^2\partial_n^2 = -\mathbf{I_{2^n}}\left(\sum_{i=1}^{n-1}\partial_i^2 + \partial_n^2\right)$$

$$= -\mathbf{I_{2^n}}\Delta_n$$

ii.

$$\tilde{\partial}\overline{\tilde{\partial}} = (\mathbf{I_{2^n}}\partial_0 + \mathcal{D}_n)(\mathbf{I_{2^n}}\partial_0 - \mathcal{D}_n) =$$
$$\mathbf{I_{2^n}}\partial_0^2 - \mathbf{I_{2^n}}\partial_0\mathcal{D}_n + \mathcal{D}_n\mathbf{I_{2^n}}\partial_0 + \mathbf{I_{2^n}}\mathcal{D}_n^2 = \mathbf{I_{2^n}}\Delta_{n+1}$$

In the last line we use the fact that $\mathcal{D}_n^2 = -\mathbf{I}_{2^n}\mathcal{D}_n^2$ □

Example 20. *As well known, the Leibniz's rule is not valid for the product of two Clifford valued functions. See [11, 32]. In the next examples we are going to show how to compute the product using the matrix representations of the Clifford valued functions.*

Let $u, v \in \mathbb{R}_{0,2}$ such that $u = u_1 E_{1,2} + u_2 E_{2,2}$ and $v = v_{12} E_{2,2}$, then, applying the matrix Dirac operator $\mathcal{D}_2$ (Example 17) to the product give in Example 14, we have:

$$\mathcal{D}_2(u \cdot v) =$$

$$\begin{pmatrix} 0 & -\partial_1 & -\partial_2 & 0 \\ \partial_1 & 0 & 0 & \partial_2 \\ \partial_2 & 0 & 0 & -\partial_1 \\ 0 & -\partial_2 & \partial_1 & 0 \end{pmatrix} \cdot \begin{pmatrix} 0 & -u_2 v_{12} & u_1 v_{12} & 0 \\ u_2 v_{12} & 0 & 0 & -u_1 v_{12} \\ -u_1 v_{12} & 0 & 0 & -u_2 v_{12} \\ 0 & u_1 v_{12} & u_2 v_{12} & 0 \end{pmatrix} =$$

$$\begin{pmatrix} \partial_2 k_1 - \partial_1 k_2 & 0 & 0 & \partial_1 k_1 + \partial_2 k_2 \\ 0 & \partial_2 k_1 - \partial_1 k_2 & \partial_1 k_1 + \partial_2 k_2 & 0 \\ 0 & -\partial_1 k_1 - \partial_2 k_2 & \partial_2 k_1 - \partial_1 k_2 & 0 \\ -\partial_1 k_1 - \partial_2 k_2 & 0 & 0 & \partial_2 k_1 - \partial_1 k_2 \end{pmatrix} =$$

$$(\partial_2 k_1 - \partial_1 k_2) E_{0,2} - (\partial_1 k_1 + \partial_2 k_2) E_{12,2}$$

where $k_1 = u_1 v_{12}, k_2 = u_2 v_{12}$.

Example 21. *Let u a function defined on $\mathbb{R}_{0,2}$, $\tilde{\partial}_3$ the matrix representation for the Cauchy–Riemann operator in $\mathbb{R}_{0,2}$ and $\overline{\tilde{\partial}}_3$ its conjugate, then*

$$\tilde{\partial}_3 \overline{\tilde{\partial}}_3 u = \Delta_3 u_0 E_{0,2} + \Delta_3 u_1 E_{1,2} + \Delta_3 u_2 E_{2,2} + \Delta_3 u_{12} E_{12,2},$$

where, $\Delta_3 = (\partial_0^2 + \partial_1^2 + \partial_2^2) E_{0,2}$.

4. SUFFICIENT CONDITIONS FOR ASSOCIATED PAIRS AND APPLICATIONS TO INITIAL VALUE PROBLEMS

In this section we deepen in the discussion of how two associated pairs can help us to solve initial value problem.

The solutions of equation (2) exist in a canonical set M_η in the (t, x)–space over a bounded domain Ω in the x–space in which the initial function φ is given (The parameter η will describe the height of the canonical set).

The desired fixed point will be constructed in a Banach space of functions $u = u(t, x)$ defined in M_η. To construct the Banach space, we consider an exhaustion of Ω by sub–domains $\Omega_s, 0 < s < s_0$, satisfying the following conditions:

- To each $x \neq x_0$ there exists $s(x), 0 < s(x) < s_0$, such that $x \in \partial \Omega_{s(x)}$.

- $dist(\Omega_{s'}, \partial \Omega_{s''}) \leq c_1(s'' - s')$ (*dist* is the distance).

Then $s_0 - s(x)$ can be interpreted as a measure of the distance of x from the boundary $\partial\Omega$ of Ω, and M_η can be defined by

$$M_\eta = \{(t,x) : x \in \Omega, 0 \leq t < \eta.dist(x, \partial\Omega)\},$$

where η will be fixed later. $d(t,x) = s_0 - s(x) - \frac{t}{\eta}$ can be interpreted as pseudo–distance of a point $(t,x) \in M_\eta$ from the lateral surface of M_η.

Let $\mathcal{B}(\Omega)$ be any Banach space of functions defined in Ω. Its norm will be denoted by $||\cdot||_\Omega$. The only condition imposed on $\mathcal{B}$ is the existence of a constant c_2 such that

$$\sup_\Omega |f| \leq c_2 ||f||_\Omega,$$

for the elements $f \in \mathcal{B}$. This condition is satisfied for the Banach space of $Holder$ continuous functions or with $c_2 = 1$ for the Banach space of continuous and bounded functions.

Define $\mathcal{B}_s = \mathcal{B}(\Omega_s)$, and denote the norm of $||\cdot||_s$. Consider functions $u = u(x,t)$ defined in M_η. Then $\mathcal{B}_*(M_\eta)$ contains all functions $u = u(t,x)$ having the following properties:

- $u(t,x)$ is continuous in M_η.

- $u(\tilde{t},x)$ belongs to $\mathcal{B}_{s(x)}$ for fixed $\tilde{t}$ if only $s(x) < \tilde{s}$ where $\tilde{s} = s_0 - \frac{\tilde{t}}{\eta}$.

- The norm $||u||_* = \sup_{(t,x) \in M_\eta} ||u(t,\cdot)||_{s(x)} d(t,x)$ is finite.

Then one can show that $\mathcal{B}_*(M_\eta)$ is a Banach space, see [33, 34].

To estimate the operator defined by the right–hand side of (2), one needs an estimate of $\partial_j u$ because F depends also on those derivatives. This, however, can be realized by an interior estimate of first order

$$||\partial_j||_{\mathcal{B}(s')} \leq \frac{C}{dist(s''-s')} \cdot ||u||_{s''}$$

using the exhaustion of Ω by Ω_s.

Consider the subspace $\mathcal{B}_*^G(M)$ of $\mathcal{B}_*(M_\eta)$ whose elements satisfy a linear elliptic equation $Gu = 0$ for each fixed t. And assume that the right–hand side F of the differential equation (1) satisfies the following global Lipschitz condition

$$||Fu - Fv||_s \leq K_0 ||u-v||_s + \sum_j K_j ||\partial_j u - \partial_j||_s$$

where $Fu = f(t,x,u,\partial_j u)$ and $Fv = f(t,x,v,\partial_j v)$

Consider two arbitrary elements u and v of $\mathcal{B}_*^G(M)$ and their images defined by the right hand side of (2) $U = U(t,x)$ and $V = V(t,x)$ respectively. The estimate

$$||U-V||_* \leq \eta L ||u-v||_* \qquad (9)$$

is true where $L = s_0 K_0 + \frac{4C}{c_1} \sum_j K_j$

Remark 22. *To construct such estimates, we can use Poisson and Cauchy–Pompeiu formulas, also L_p norm can be considered. See [7–10, 12]*

4.1. Interior Estimates by Using a Cauchy Integral Formula in the Supreme Norm

One way to calculate the interior estimate is through Cauchy's integral formula. For this purpose, in our case (we are going to study the IVP using monogenic functions), we need the fundamental solution of the operator $\tilde{\partial}$ and the Cauchy representation.

The function $E(x, \xi)$ defined by

$$E(x,\xi) = \frac{1}{\omega_{n+1}} \frac{1}{\varrho^{n+1}} \left((x_0 - \xi_0) - \sum_{i,j=1}^{n} e_i(x_j - \xi_j) \right),$$

for $x, \xi \in \Omega$, and $\lambda \in \Re$. This function is a fundamental solution for the operator $\tilde{\partial}$ in $\mathbb{R}_{0,n}$ (see [15, 35]), also known as the Cauchy kernel.

The Cauchy integral type formula for the monogenic function u is given by

$$u(\xi) = \int_{\partial \Omega} E(x, \xi) \cdot d\sigma \cdot u, \tag{10}$$

where $u, v \in \mathbb{R}_{0,n}$ (see [15, 35]).

Let's define the sup–norm of an $\mathbb{R}_{0,n}$–valued function $u(x) = \sum_{A \in \Gamma_n} u_A(x) e_A$ on a set $\mathcal{K}$ as the maximum of the sup–norms of its 2^n real–valued components, i.e.,

$$\|u\| = \sup_{\mathcal{K}} |u(x)| := \max_{A \in \Gamma_n} (\sup_{\mathcal{K}} |u_A(x)|)$$

Where $\Gamma_n = \{0, 1, 2, \ldots, 12, 13, \ldots, 123 \ldots n\}$. To obtain an interior estimate using the equation (10) for monogenic functions we must compute the derivative of the fundamental solution E. Using (10) with $E(x, \xi)$ instead of u, and integrating over $|x - \xi| = r$ instead of $\partial \Omega$, we obtain

$$|\partial_{\xi_s} u(\xi)| \leq \frac{K^*}{r} \sup_{|x-\xi|=r} |u(x)|, \tag{11}$$

where $K^* = \frac{M^*}{M_*}$, $M^* = (n+1)4^n \max_{B,j,A} b_{B,j,A}$ and $M_* = (n+1)2^n \max_{A,j} |e_A e_j| \cdot b_{B,j,A}$ is an upper bound for $|e_B e_j e_A|$.

Hence the limiting process $r \to d(\xi) = \inf_{x \in \partial \Omega} |x - \xi|$ leads to the interior estimate for the derivative of a monogenic function (see [12] for details).

4.2. Initial Value Problems

The estimate (9) implies contractivity of the operator, provided the height η of the canonical set is small enough. In general, the images U and V do not belong to $\mathcal{B}_*^G(M_\eta)$ although they do belong to $\mathcal{B}_*(M_\eta)$. To conclude the existence of a fixed point from the contractivity we have to know that the images U and V belong to $\mathcal{B}_*^G(M_\eta)$ as u and v do so. This is the case if the operators F and G form an associated pair.

Definition 23. *Suppose that the coefficients of G depend only on the space–like variable x, whereas the coefficients of F may depend on t. Then the pair F, G is said to be associated in case $Gu = 0$ implies $G(Fu) = 0$ for each t.*

Theorem 24. *Suppose that F, G are an associated pair, where G satisfies an interior estimate of first order and $||F\theta||_* = \sup_{M_\eta} ||\theta||_{s(x)} d(t,x)$ has a finite $*$–norm. Suppose, further, that the initial function φ satisfies the condition $Gu = 0$ and has a finite $*$–norm, too. Then the initial value problem*

$$\partial_t u = F(t, x, u, \partial_{x_i} u), \quad u(0, x) = \varphi(x)$$

Is uniquely solvable in $\mathcal{B}_*^G(M_\eta)$ provided the height of the conical M_η is small enough ($\eta L < 1$). Moreover, the solution $u = u(t,x)$ satisfies the condition $Gu = 0$ for each t.

Theorem 25. *[Clifford context] Consider the initial value problem:*

$$\partial_t u = F(t, x, u, \partial_{x_i} u), \quad u(0, x) = \varphi(x), \tag{12}$$

where t is the time, F is a linear first order operator in a Clifford Analysis and φ is a monogenic function, This problem is solvable in the conical domain

$$M_\mu = \{(t, x) : x \in \Omega, 0 \le t < \mu.\mathrm{dist}(x, \partial\Omega)\}$$

by the contraction–mapping principle provided μ is small enough. Moreover, the solution $u(t,x)$ belongs to the associated space for each t.

This theorem is valid in all the cases presented below with the interior estimate given in (11). In this chapter we focus our attention on calculating the sufficient conditions for the operators to obtain an associated pair using the matrix representation and computational tools, omitting the use of a product rule in Clifford analysis (Leibniz's type rule), and without deepening in the calculation of the interior estimates.

4.3. Complex Case

For the complex case, or equivalently for $\mathbb{R}_{0,1}$, we are interested in obtaining necessary and sufficient conditions such that the operator $\mathcal{L}$ is associated with the operator $\tilde{\partial}_2$. This means that if Ω is a bounded and simply connected domain of the Euclidean space $\mathbb{R}^2$ and $u : \Omega \to \mathbb{C}$ is a function satisfying $\tilde{\partial}_2 u = 0$ then $\tilde{\partial}_2(\mathcal{L}u) = 0$.

Consider the operator $\mathcal{L}$ applied to a continuously differentiable function u defined as follows:

$$\mathcal{L}u = \sum_{i=0}^{1} A^{(i)} u(t,x) \partial_i u + B(t,x) u, \tag{13}$$

where

$$A^{(i)}(t,x) = \sum_{k=0}^{1} a_k^{(i)}(t,x) E_{k,1},$$

$$B(t,x) = \sum_{k=0}^{1} b_k(t,x) E_{k,1},$$

$a_k, b_k \in C^1[x_0, x_1]$ and $E_{k,1}$ ($k = 0, 1$) are defined as in Example 8. Applying $\tilde{\partial}_2$ to (13), then the 2×1 vector $\tilde{\partial}_2(\mathcal{L}u)$ is expressed as:

$$\tilde{\partial}_2(\mathcal{L}u) = \begin{pmatrix} \partial_0(a_0^{(0)}\partial_0 u_0) - \partial_0(a_1^{(0)}\partial_0 u_1) + \partial_0(a_0^{(1)}\partial_1 u_0) - \partial_0(a_1^{(1)}\partial_1 u_1) \\ +\partial_0(b_0 u_0) - \partial_0(b_1 u_1) - \partial_1(a_0^{(0)}\partial_0 u_1) - \partial_1(a_1^{(0)}\partial_0 u_0) \\ -\partial_1(a_0^{(1)}\partial_1 u_1) - \partial_1(a_1^{(1)}\partial_1 u_0) - \partial_1(b_0 u_1) - \partial_1(b_1 u_0) \\ \\ \partial_0(a_0^{(0)}\partial_0 u_1) + \partial_0(a_1^{(0)}\partial_0 u_0) + \partial_0(a_0^{(1)}\partial_1 u_1) + \partial_0(a_1^{(1)}\partial_1 u_0) \\ +\partial_0(b_0 u_1) + \partial_0(b_1 u_0) + \partial_1(a_0^{(0)}\partial_0 u_0) - \partial_1(a_1^{(0)}\partial_0 u_1) \\ +\partial_1(a_0^{(1)}\partial_1 u_0) - \partial_1(a_1^{(1)}\partial_1 u_1) + \partial_1(b_0 u_0) - \partial_1(b_1 u_1) \end{pmatrix} = \mathbf{0}.$$

After applying the product rule to each term, we can decompose the resulting matrix into the sum of seven matrices as follows:

$$\tilde{\partial}_2(\mathcal{L}u) = \begin{pmatrix} \partial_0 a_0^{(0)} \partial_0 u_0 - \partial_0 a_1^{(0)} \partial_0 u_1 + \partial_0 a_0^{(1)} \partial_1 u_0 - \partial_0 a_1^{(1)} \partial_1 u_1 \\ +\partial_0 b_0 u_0 - \partial_0 b_1 u_1 - \partial_1 a_0^{(0)} \partial_0 u_1 - \partial_1 a_1^{(0)} \partial_0 u_0 - \partial_1 a_0^{(1)} \partial_1 u_1 \\ -\partial_1 a_1^{(1)} \partial_1 u_0 - \partial_1 b_0 u_1 - \partial_1 b_1 u_0 \\ \\ \partial_0 a_0^{(0)} \partial_0 u_1 + \partial_0 a_1^{(0)} \partial_0 u_0 + \partial_0 a_0^{(1)} \partial_1 u_1 + \partial_0 a_1^{(1)} \partial_1 u_0 \\ +\partial_0 b_0 u_1 + \partial_0 b_1 u_0 + \partial_1 a_0^{(0)} \partial_0 u_0 - \partial_1 a_1^{(0)} \partial_0 u_1 + \partial_1 a_0^{(1)} \partial_1 u_0 \\ -\partial_1 a_1^{(1)} \partial_1 u_1 + \partial_1 b_0 u_0 - \partial_1 b_1 u_1 \end{pmatrix}$$

$$+ b_0 \begin{pmatrix} \partial_0 u_0 - \partial_1 u_1 \\ \partial_0 u_1 + \partial_1 u_0 \end{pmatrix} + b_1 \begin{pmatrix} -\partial_0 u_1 - \partial_1 u_0 \\ \partial_0 u_0 - \partial_1 u_1 \end{pmatrix}$$

$$+ a_0^{(0)} \begin{pmatrix} \partial_0^2 u_0 - \partial_1 \partial_0 u_1 \\ \partial_0^2 u_1 + \partial_1 \partial_0 u_0 \end{pmatrix} + a_1^{(0)} \begin{pmatrix} -\partial_0^2 u_1 - \partial_1 \partial_0 u_0 \\ \partial_0^2 u_0 - \partial_1 \partial_0 u_1 \end{pmatrix}$$

$$+ a_0^{(1)} \begin{pmatrix} \partial_0 \partial_1 u_0 - \partial_1^2 u_1 \\ \partial_0 \partial_1 u_1 + \partial_1^2 u_0 \end{pmatrix} + a_1^{(1)} \begin{pmatrix} -\partial_0 \partial_1 u_1 - \partial_1^2 u_0 \\ \partial_0 \partial_1 u_0 - \partial_1^2 u_1 \end{pmatrix}.$$

The Cauchy–Riemann system $\tilde{\partial}_2 u = 0$ in the matrix form is given by:

$$\tilde{\partial}_2 u = \begin{pmatrix} \partial_0 u_0 - \partial_1 u_1 \\ \partial_1 u_0 + \partial_0 u_1 \end{pmatrix} = \mathbf{0}. \qquad (14)$$

Also note that $\partial_i \tilde{\partial}_2 u = 0$ for $i = 0, 1$.

$$\partial_0 \tilde{\partial}_2 u = \begin{pmatrix} \partial_0^2 u_0 - \partial_0 \partial_1 u_1 \\ \partial_0 \partial_1 u_0 + \partial_0^2 u_1 \end{pmatrix} = \mathbf{0},$$

$$\partial_1 \tilde{\partial}_2 u = \begin{pmatrix} \partial_1 \partial_0 u_0 - \partial_1^2 u_1 \\ \partial_1^2 u_0 + \partial_1 \partial_0 u_1 \end{pmatrix} = \mathbf{0}.$$

Thus last six matrices of $\tilde{\partial}_2(\mathcal{L}u)$ are zero by hypothesis $\tilde{\partial}_2 u = 0$. So we have that

$$\tilde{\partial}_2(\mathcal{L}u) = \begin{pmatrix} \partial_0 a_0^{(0)} \partial_0 u_0 - \partial_0 a_1^{(0)} \partial_0 u_1 + \partial_0 a_0^{(1)} \partial_1 u_0 - \partial_0 a_1^{(1)} \partial_1 u_1 \\ + \partial_0 b_0 u_0 - \partial_0 b_1 u_1 - \partial_1 a_0^{(0)} \partial_0 u_1 - \partial_1 a_1^{(0)} \partial_0 u_0 - \partial_1 a_0^{(1)} \partial_1 u_1 \\ - \partial_1 a_1^{(1)} \partial_1 u_0 - \partial_1 b_0 u_1 - \partial_1 b_1 u_0 \\ \\ \partial_0 a_0^{(0)} \partial_0 u_1 + \partial_0 a_1^{(0)} \partial_0 u_0 + \partial_0 a_0^{(1)} \partial_1 u_1 + \partial_0 a_1^{(1)} \partial_1 u_0 \\ + \partial_0 b_0 u_1 + \partial_0 b_1 u_0 + \partial_1 a_0^{(0)} \partial_0 u_0 - \partial_1 a_1^{(0)} \partial_0 u_1 + \partial_1 a_0^{(1)} \partial_1 u_0 \\ - \partial_1 a_1^{(1)} \partial_1 u_1 + \partial_1 b_0 u_0 - \partial_1 b_1 u_1 \end{pmatrix}.$$

This matrix can be decomposed into six matrices.

$$\tilde{\partial}_2(\mathcal{L}u) = \begin{pmatrix} \partial_0 a_0^{(0)} - \partial_1 a_1^{(0)} \\ \partial_0 a_1^{(0)} + \partial_1 a_0^{(0)} \end{pmatrix} \partial_0 u_0 + \begin{pmatrix} -\partial_0 a_1^{(0)} - \partial_1 a_0^{(0)} \\ \partial_0 a_0^{(0)} - \partial_1 a_1^{(0)} \end{pmatrix} \partial_0 u_1$$

$$+ \begin{pmatrix} \partial_0 a_0^{(1)} - \partial_1 a_1^{(1)} \\ \partial_0 a_1^{(1)} + \partial_1 a_0^{(1)} \end{pmatrix} \partial_1 u_0 + \begin{pmatrix} -\partial_0 a_1^{(1)} - \partial_1 a_0^{(1)} \\ \partial_0 a_0^{(1)} - \partial_1 a_1^{(1)} \end{pmatrix} \partial_1 u_1$$

$$+ \begin{pmatrix} \partial_0 b_0 - \partial_1 b_1 \\ \partial_0 b_1 + \partial_1 b_0 \end{pmatrix} u_0 + \begin{pmatrix} -\partial_0 b_1 - \partial_1 b_0 \\ \partial_0 b_0 - \partial_1 b_1 \end{pmatrix} u_1$$

Using (14), we get

$$\tilde{\partial}_2(\mathcal{L}u) = \begin{pmatrix} \partial_0 a_1^{(0)} + \partial_1 a_0^{(0)} + \partial_0 a_0^{(1)} - \partial_1 a_1^{(1)} \\ -\partial_0 a_0^{(0)} + \partial_1 a_1^{(0)} + \partial_0 a_1^{(1)} + \partial_1 a_0^{(1)} \end{pmatrix} \partial_1 u_0$$

$$+ \begin{pmatrix} \partial_0 a_0^{(0)} - \partial_1 a_1^{(0)} - \partial_0 a_1^{(1)} - \partial_1 a_0^{(1)} \\ \partial_0 a_1^{(0)} + \partial_1 a_0^{(0)} + \partial_0 a_0^{(1)} - \partial_1 a_1^{(1)} \end{pmatrix} \partial_1 u_1$$

$$+ \begin{pmatrix} \partial_0 b_0 - \partial_1 b_1 \\ \partial_0 b_1 + \partial_1 b_0 \end{pmatrix} u_0 + \begin{pmatrix} -\partial_0 b_1 - \partial_1 b_0 \\ \partial_0 b_0 - \partial_1 b_1 \end{pmatrix} u_1$$

Finally,

$$\tilde{\partial}_2(\mathcal{L}u) = \begin{pmatrix} \partial_0 a_1^{(0)} + \partial_1 a_0^{(0)} + \partial_0 a_0^{(1)} - \partial_1 a_1^{(1)} \\ -\partial_0 a_0^{(0)} + \partial_1 a_1^{(0)} + \partial_0 a_1^{(1)} + \partial_1 a_0^{(1)} \end{pmatrix} \partial_1 u$$

$$+ \begin{pmatrix} \partial_0 b_0 - \partial_1 b_1 \\ \partial_0 b_1 + \partial_1 b_0 \end{pmatrix} u \tag{15}$$

The rows of these two last matrices gives us the necessary conditions for the operator $\tilde{\partial}_2$ to be associated with $\mathcal{L}$:

$$\begin{cases} \partial_0 b_0 - \partial_1 b_1 = 0, & (16) \\ \partial_0 b_1 + \partial_1 b_0 = 0, & (17) \\ \partial_0 a_1^{(0)} + \partial_0 a_0^{(1)} + \partial_1 a_0^{(0)} - \partial_1 a_1^{(1)} = 0, & (18) \\ -\partial_0 a_0^{(0)} + \partial_0 a_1^{(1)} + \partial_1 a_1^{(0)} + \partial_1 a_0^{(1)} = 0. & (19) \end{cases}$$

Now, we can state the following theorem:

Theorem 26. *Assume that $A^{(i)}(t,x) \in C^2(\Omega, \mathbb{C})$ ($i = 0, 1$) and $B(t,x) \in C^1(\Omega, \mathbb{C})$ for each $t \in [0, T]$. Then, the operator $\mathcal{L}$ is associated with the operator $\tilde{\partial}_2$ if and only if the conditions (16), (17), (18) and (19) are satisfied.*

Proof. We assume that the operator $\mathcal{L}$ is associated with the operator $\tilde{\partial}_2$. Choose, $u \equiv 1$. Then (15) reduces to:

$$\tilde{\partial}_2(\mathcal{L}u) = \begin{pmatrix} \partial_0 b_0 - \partial_1 b_1 \\ \partial_0 b_1 + \partial_1 b_0 \end{pmatrix}$$

thus, we need that (16) and (17) have to be satisfied. Now choose, $u \equiv x_1$. Then, we have that (15) reduces to:

$$\tilde{\partial}_2(\mathcal{L}u) = \begin{pmatrix} \partial_0 a_1^{(0)} + \partial_1 a_0^{(0)} + \partial_0 a_0^{(1)} - \partial_1 a_1^{(1)} \\ -\partial_0 a_0^{(0)} + \partial_1 a_1^{(0)} + \partial_0 a_1^{(1)} + \partial_1 a_0^{(1)} \end{pmatrix}$$

for this expression to be zero, we need that (18) and (19) have to be satisfied.
The second implication is an immediate consequence. □

Example 27. *Explicit construction of the operator $\mathcal{L}u$.*
Let

$$b_0(x_0, x_1, t) = e^{x_1 - ix_0 + t} + e^{x_1 + ix_0 + t},$$
$$b_1(x_0, x_1, t) = -ie^{x_1 - ix_0 + t} + ie^{x_1 + ix_0 + t}.$$

where i is the imaginary unit that we can identify with e_1. We can choose $a_0^{(0)}(x_0, x_1, t)$ and $a_0^{(1)}(x_0, x_1, t)$ arbitrarily, let them be zero, and

$$a_1^{(0)}(x_0, x_1, t) = 2tx_1,$$
$$a_1^{(1)}(x_0, x_1, t) = t(1 - 2x_0).$$

It is straightforward to check that these functions satisfy the conditions (16),(17),(18) and (19). Thus, (13) can be written explicitly as:

$$\mathcal{L}u = (2tx_1)e_1\partial_0 u + t(1 - 2x_0)e_1\partial_1 u$$
$$+ \left[(e^{x_1 - ix_0 + t} + e^{x_1 + ix_0 + t})e_0 + (-ie^{x_1 - ix_0 + t} + ie^{x_1 + ix_0 + t})e_1\right]u.$$

4.4. The Quaternion Case or $\mathbb{R}_{0,2}$

As in the previous case, we are interested in knowing the necessary and sufficient conditions for the operator $\mathcal{L}$ to be associated with the operator $\tilde{\partial}_3$. This task has already been done previously in [36], where the author finds the conditions in a more compact form. However, the advantage of our method is the facility that matrices offer to do calculations.

Consider Ω a bounded and simply connected domain of the Euclidean space $\mathbb{R}^4$ and $u : \Omega \to \mathbb{R}_{0,2}$ is a continuously and differentiable function satisfying $\tilde{\partial}_{\mathbf{3}} u = 0$. The operator $\mathcal{L}$ for this case is given by:

$$\mathcal{L}u = \sum_{i=0}^{3} A^{(i)}(t,x)\partial_i u + B(t,x)u, \qquad (20)$$

where $A^{(i)}(t,x) = \sum_{k=0}^{3} a_k^{(i)}(t,x) E_{k,2}$ and $B(t,x) = \sum_{k=0}^{3} b_k(t,x) E_{k,2}$.
Recalling that the matrix basis was given in the Example 9, $\{\mathbf{I}_4, E_{1,2}, E_{2,2}, E_{12,2}\}$, then using $\tilde{\partial}_{\mathbf{3}} u = 0$ and $\partial_i \tilde{\partial}_{\mathbf{3}} u = 0$ for $i = 0, 1, 2, 3$. Thus, we have that:

$\mathcal{L}(\tilde{\partial}_{\mathbf{3}} u) =$

$$\begin{pmatrix} \partial_0 a_1^{(0)} + \partial_0 a_0^{(1)} + \partial_1 a_0^{(0)} - \partial_1 a_1^{(1)} + \partial_2 a_3^{(0)} - \partial_2 a_2^{(1)} - \partial_3 a_2^{(0)} - \partial_3 a_3^{(1)} \\ -\partial_0 a_0^{(0)} \partial_0 a_1^{(1)} + \partial_1 a_1^{(0)} + \partial_1 a_0^{(1)} + \partial_2 a_3^{(0)} + \partial_2 a_3^{(1)} + \partial_3 a_3^{(0)} - \partial_3 a_2^{(1)} \\ -2b_3 - \partial_0 a_3^{(0)} + \partial_0 a_2^{(1)} - \partial_1 a_2^{(0)} - \partial_1 a_3^{(1)} + \partial_2 a_0^{(0)} + \partial_2 a_1^{(1)} - \partial_3 a_0^{(0)} + \partial_3 a_1^{(1)} \\ 2b_2 + \partial_0 a_2^{(0)} + \partial_0 a_3^{(1)} - \partial_1 a_3^{(0)} + \partial_1 a_2^{(1)} + \partial_2 a_0^{(0)} - \partial_2 a_1^{(1)} + \partial_3 a_1^{(0)} + \partial_3 a_0^{(1)} \end{pmatrix} \partial_1 u$$

$$+ \begin{pmatrix} \partial_0 a_2^{(0)} + \partial_0 a_0^{(2)} - \partial_1 a_3^{(0)} - \partial_1 a_1^{(2)} + \partial_2 a_0^{(0)} - \partial_2 a_2^{(2)} + \partial_3 a_1^{(0)} - \partial_3 a_3^{(2)} \\ 2b_3 + \partial_0 a_3^{(0)} + \partial_0 a_1^{(2)} + \partial_1 a_2^{(0)} + \partial_1 a_0^{(2)} - \partial_2 a_1^{(0)} + \partial_2 a_3^{(2)} + \partial_3 a_0^{(0)} - \partial_3 a_2^{(2)} \\ -\partial_0 a_0^{(0)} + \partial_0 a_2^{(2)} + \partial_1 a_1^{(0)} - \partial_1 a_3^{(2)} + \partial_2 a_2^{(0)} + \partial_2 a_0^{(2)} + \partial_3 a_3^{(0)} + \partial_3 a_1^{(2)} \\ -2b_1 - \partial_0 a_1^{(0)} + \partial_0 a_3^{(2)} - \partial_1 a_0^{(0)} + \partial_1 a_2^{(2)} - \partial_2 a_3^{(0)} - \partial_2 a_1^{(2)} + \partial_3 a_2^{(0)} + \partial_3 a_0^{(2)} \end{pmatrix} \partial_2 u$$

$$+ \begin{pmatrix} \partial_0 a_3^{(0)} + \partial_0 a_0^{(3)} + \partial_1 a_2^{(0)} - \partial_1 a_1^{(3)} - \partial_2 a_1^{(0)} - \partial_2 a_2^{(3)} + \partial_3 a_0^{(0)} - \partial_3 a_3^{(3)} \\ -2b_2 - \partial_0 a_2^{(0)} + \partial_0 a_1^{(3)} + \partial_1 a_3^{(0)} + \partial_1 a_0^{(3)} - \partial_2 a_0^{(0)} + \partial_2 a_3^{(3)} - \partial_3 a_1^{(0)} - \partial_3 a_2^{(3)} \\ 2b_1 + \partial_0 a_1^{(0)} + \partial_0 a_2^{(3)} + \partial_1 a_0^{(0)} - \partial_1 a_3^{(3)} + \partial_2 a_3^{(0)} + \partial_2 a_0^{(3)} - \partial_3 a_2^{(0)} + \partial_3 a_1^{(3)} \\ -\partial_0 a_0^{(0)} + \partial_0 a_3^{(3)} + \partial_1 a_1^{(0)} + \partial_1 a_2^{(3)} + \partial_2 a_2^{(0)} - \partial_2 a_1^{(3)} + \partial_3 a_3^{(0)} + \partial_3 a_0^{(3)} \end{pmatrix} \partial_3 u$$

$$+ \begin{pmatrix} 0 \\ 2a_2^{(0)} + 2a_3^{(1)} \\ 2a_1^{(0)} - 2a_3^{(2)} \\ -2a_1^{(1)} + a_2^{(2)} \end{pmatrix} \partial_1 \partial_2 u + \begin{pmatrix} 0 \\ 2a_3^{(0)} - 2a_1^{(1)} \\ 2a_1^{(1)} - 2a_3^{(3)} \\ 2a_1^{(0)} + 2a_2^{(3)} \end{pmatrix} \partial_1 \partial_3 u + \begin{pmatrix} 0 \\ -2a_2^{(2)} + 2a_3^{(3)} \\ 2a_3^{(0)} + 2a_1^{(2)} \\ a_2^{(0)} - 2a_1^{(3)} \end{pmatrix} \partial_2 \partial_3 u$$

$$+ \begin{pmatrix} 0 \\ 0 \\ -2a_2^{(0)} - 2a_3^{(1)} \\ -2a_3^{(0)} + 2a_2^{(1)} \end{pmatrix} \partial_1^2 u + \begin{pmatrix} -2a_1^{(0)} + 2a_3^{(2)} \\ 0 \\ -2a_3^{(0)} - 2a_1^{(2)} \end{pmatrix} \partial_2^2 u + \begin{pmatrix} 0 \\ -2a_1^{(0)} - 2a_2^{(3)} \\ -2a_2^{(0)} + 2a_1^{(3)} \\ 0 \end{pmatrix} \partial_3^2 u$$

$$+ \begin{pmatrix} \partial_0 b_0 - \partial_1 b_1 - \partial_2 b_2 - \partial_3 b_3 \\ \partial_0 b_1 + \partial_1 b_0 + \partial_2 b_3 - \partial_3 b_2 \\ \partial_0 b_2 - \partial_1 b_3 + \partial_2 b_0 + \partial_3 b_1 \\ \partial_0 b_3 + \partial_1 b_2 - \partial_2 b_1 + \partial_3 b_0 \end{pmatrix} u \qquad (21)$$

Initial Value Problems in Clifford Analysis Using Associated Spaces

Thus, as the previous case, using the rows of these matrices, the following conditions are necessary for $\mathcal{L}$ to be associated with $\tilde{\partial}_3$:

$$\begin{cases}
\partial_0 b_0 - \partial_1 b_1 - \partial_2 b_2 - \partial_3 b_3 = 0, \\
\partial_0 b_1 + \partial_1 b_0 + \partial_2 b_3 - \partial_3 b_2 = 0, \\
\partial_0 b_2 - \partial_1 b_3 + \partial_2 b_0 + \partial_3 b_1 = 0, \\
\partial_0 b_3 + \partial_1 b_2 - \partial_2 b_1 + \partial_3 b_0 = 0, \\
\partial_0 a_1^{(0)} + \partial_0 a_0^{(1)} + \partial_1 a_0^{(0)} - \partial_1 a_1^{(1)} + \partial_2 a_3^{(0)} - \partial_2 a_2^{(1)} + \partial_3 a_2^{(0)} - \partial_3 a_3^{(1)} = 0, \\
\partial_0 a_0^{(0)} + \partial_0 a_1^{(1)} + \partial_1 a_1^{(0)} + \partial_1 a_0^{(1)} + \partial_2 a_2^{(0)} + \partial_2 a_3^{(1)} + \partial_3 a_3^{(0)} - \partial_3 a_2^{(1)} = 0, \\
-2b_3 - \partial_0 a_3^{(0)} + \partial_0 a_2^{(1)} - \partial_1 a_2^{(0)} - \partial_1 a_3^{(1)} + \partial_2 a_1^{(0)} + \partial_2 a_1^{(1)} - \partial_3 a_0^{(0)} + \partial_3 a_1^{(1)} = 0, \\
2b_2 + \partial_0 a_2^{(0)} + \partial_0 a_3^{(1)} - \partial_1 a_3^{(0)} + \partial_1 a_2^{(1)} + \partial_2 a_0^{(0)} - \partial_2 a_1^{(1)} + \partial_3 a_1^{(0)} + \partial_3 a_0^{(1)} = 0, \\
\partial_0 a_2^{(0)} + \partial_0 a_0^{(2)} - \partial_1 a_3^{(0)} - \partial_1 a_1^{(2)} + \partial_2 a_0^{(0)} - \partial_2 a_2^{(2)} + \partial_3 a_1^{(0)} - \partial_3 a_3^{(2)} = 0, \\
2b_3 + \partial_0 a_3^{(0)} + \partial_0 a_1^{(2)} + \partial_1 a_2^{(0)} + \partial_1 a_0^{(2)} - \partial_2 a_1^{(0)} + \partial_2 a_3^{(2)} + \partial_3 a_0^{(0)} - \partial_3 a_2^{(2)} = 0, \\
-\partial_0 a_0^{(0)} + \partial_0 a_2^{(2)} + \partial_1 a_1^{(0)} - \partial_1 a_3^{(2)} + \partial_2 a_2^{(0)} + \partial_2 a_0^{(2)} + \partial_3 a_3^{(0)} + \partial_3 a_1^{(2)} = 0, \\
-2b_1 - \partial_0 a_1^{(0)} + \partial_0 a_3^{(2)} - \partial_1 a_0^{(0)} + \partial_1 a_2^{(2)} - \partial_2 a_3^{(0)} - \partial_2 a_1^{(2)} + \partial_3 a_0^{(0)} + \partial_3 a_0^{(2)} = 0, \\
\partial_0 a_3^{(0)} + \partial_0 a_0^{(3)} + \partial_1 a_2^{(0)} - \partial_1 a_1^{(3)} - \partial_2 a_1^{(0)} - \partial_2 a_2^{(3)} + \partial_3 a_0^{(0)} - \partial_3 a_3^{(3)} = 0, \\
-2b_2 - \partial_0 a_2^{(0)} + \partial_0 a_1^{(3)} + \partial_1 a_3^{(0)} + \partial_1 a_0^{(3)} - \partial_2 a_0^{(0)} + \partial_2 a_3^{(3)} - \partial_3 a_1^{(0)} - \partial_3 a_2^{(3)} = 0, \\
2b_1 + \partial_0 a_1^{(0)} + \partial_0 a_2^{(3)} + \partial_1 a_0^{(0)} - \partial_1 a_3^{(3)} + \partial_2 a_3^{(0)} + \partial_2 a_0^{(3)} - \partial_3 a_2^{(0)} + \partial_3 a_1^{(3)} = 0, \\
-\partial_0 a_0^{(0)} + \partial_0 a_3^{(3)} + \partial_1 a_1^{(0)} + \partial_1 a_2^{(3)} + \partial_2 a_2^{(0)} - \partial_2 a_1^{(3)} + \partial_3 a_3^{(0)} + \partial_3 a_0^{(3)} = 0, \\
-a_2^{(0)} - a_3^{(1)} = 0, \quad -a_3^{(0)} + a_2^{(1)} = 0, \\
-a_1^{(0)} + a_3^{(2)} = 0, \quad -a_3^{(0)} - a_1^{(2)} = 0, \\
-a_1^{(0)} - a_2^{(3)} = 0, \quad -a_2^{(0)} + a_1^{(3)} = 0, \\
a_2^{(0)} + a_3^{(1)} = 0, \quad a_1^{(0)} - a_3^{(2)} = 0, \\
-a_1^{(1)} + a_2^{(2)} = 0, \quad a_3^{(0)} - a_2^{(1)} = 0, \\
a_1^{(1)} - a_3^{(3)} = 0, \quad a_1^{(0)} + a_2^{(3)} = 0, \\
-a_2^{(2)} + a_3^{(3)} = 0, \quad a_3^{(0)} + a_1^{(2)} = 0, \\
a_2^{(0)} - a_1^{(3)} = 0.
\end{cases}$$

We can state a similar theorem for this case as seen previously (Theorem 26).

Theorem 28. *Assume that $A^{(i)}(t,x) \in C^2(\Omega, \mathbb{H})$ ($i = 0, 1, 2, 3$) and $B(t,x) \in C^1(\Omega, \mathbb{H})$ for each $t \in [0, T]$. Then the operator $\mathcal{L}$ is associated with the operator $\tilde{\partial}_3$ if and only if the conditions (16), (17), (18) and (19) are satisfied.*

Proof. We assume that the operator $\mathcal{L}$ is associated with the operator $\tilde{\partial}_3$. Choose, $u \equiv 1$. Then we are left with the last matrix of (21):

$$\mathcal{L}(\tilde{\partial}_3 u) = \begin{pmatrix} \partial_0 b_0 - \partial_1 b_1 - \partial_2 b_2 - \partial_3 b_3 \\ \partial_0 b_1 + \partial_1 b_0 + \partial_2 b_3 - \partial_3 b_2 \\ \partial_0 b_2 - \partial_1 b_3 + \partial_2 b_0 + \partial_3 b_1 \\ \partial_0 b_3 + \partial_1 b_2 - \partial_2 b_1 + \partial_3 b_0 \end{pmatrix} = \mathbf{0},$$

we need these four equations to be zero. If we choose $u \equiv x_1$, $u \equiv x_2$ and $u \equiv x_3$, then the equations in the rows of the first, second and third matrices of (21) respectively need to be zero. Taking $u \equiv x_1 x_2$, $u \equiv x_1 x_3$ and $u \equiv x_2 x_3$, then the equations in the rows of the fourth, fifth and sixth matrices of (21) need to be zero.

Finally, considering $u \equiv \frac{x_1^2}{2}$, $u \equiv \frac{x_2^2}{2}$ and $u \equiv \frac{x_3^2}{2}$, then the equations of the seventh, eight and ninth matrices of (21) need to be zero.

Once again, the second implication is an immediate consequence. □

In the next chapter we will focus on showing the details of how matrix calculations are done.

REFERENCES

[1] Brackx, F., Delanghe, R. and Sommen, F. 1982. *Clifford analysis*. London: Pitman research notes in mathematics, Vol. 76.

[2] Delanghe, R., Sommen, F. and Soucek. V. 1992. "Clifford algebra and Spinor–valued functions: a function theory for the Dirac operator", *Mathematics and its Applications*. Volume 53. Springer.

[3] Tutschke, W. 1989. *Solution of initial value problems in classes of generalized analytic functions*. Teubner Leipzig and Springer Verlag.

[4] Son, L. H. and Tutschke, W. 2003. "First order differential operator associated to the Cauchy–Riemann Operator in the Plane". *Complex Variables*. 48: 797–801.

[5] Tutschke, W. 2008. "Associated spaces: a new tool of real and complex analysis", contained in *Function Spaces in Complex and Clifford Analysis*. National University Publ, Hanoi. 253–268.

[6] Son, L. H. and Hung, N. Q. 2008. "The initial value problems in Clifford and quaternion analysis". *Proceedings of the 15th ICFIDCAA 2007*, Osaka Municipal Universities Press. 3: 317–323.

[7] Yüksel, U. and Çelebi, O. 2010. "Solution of initial value problems of Cauchy–Kovalevsky type in the space of generalized monogenic functions". *Adv. Appl. Clifford Algebras*. 20(2): 427–444.

[8] Tutschke, W. and Van, N. T. 2007. "Interior estimates in the sup–norm for generalized monogenic functions satisfying a differential equation with anti–monogenic right–hand sides". *Complex variable and elliptic Equation*. 52(5): 367–375.

[9] Tutschke, W. and Yüksel, U. 1999. "Interrior Lp–estimates for functions with integral representations". *Appl. Anal.* 73: 281–294.

[10] Yüksel, U. 2010. "Solution of initial value problems with monogenic initial functions in Banach spaces with Lp–norm". *Adv. Appl. Clifford Alg.* 20: 201–209.

[11] Bolívar, Y., Lezama, L., Mármol, L. and Vanegas, J. 2015. "Associated spaces in Clifford analysis". *Adv. Appl. Clifford Alg.* 25: 539–551.

[12] Ariza, E., Di Teodoro, A. and Vanegas, J. 2019. "Interior estimates in the sup–norm for first order meta–monogenic functions". To appear *Bull CompAMa*.

[13] Lounesto, P. 2001. *Clifford algebras and spinors* (London Mathematical Society Lecture Note Series). LMS No. 286. Cambridge University Press, Second edition.

[14] Peña, D. and Sommen, F. 2012. "Vekua–type systems related to two–sided monogenic functions". *Complex Anal. Oper. Theory.* 6: 397âĂŞ-405.

[15] Balderrama, C., Di Teodoro, A. and Infante, A. 2013. "Some Integral representation for meta and anti–meta monogenic function in Clifford depending parameter algebra". *Adv. Appl. Clifford Algebras.* 23(4)4: 793–813.

[16] Hestenes, D. and Sobczyk, G. 1980. *Clifford slgebras and geometric calculus: a unified language for mathematics and physics*. Mathematics lecture note series, Addison Wesley, Reading, Mass.

[17] Hile, G. and Lounesto, P. 1990. "Matrix representation of Clifford algebras". *Linear Algebra and its Applications.* 128: 51–63.

[18] Poole, Jr. C., Farach, H. and Aharonov, Y. 1980. "A vector product formulation of special relativity and electromagnetism". *Found. of Phys.* 10: 531–553.

[19] Poole, Jr. C. and Farach, H. 1982. "Pauli Dirac matrix generators of Clifford algebras". *Found. of Phys.* 12: 719–738.

[20] Tian, Y. 1998. "Universal similarity factorization equalities over real Clifford algebras". *Adv. Appl. Clifford Algebras.* 8(2): 365–402.

[21] Ablamowicz, R. 2004. "Clifford Algebras: Applications to mathematics, physics, and engineering". Progress in mathematical physics. Birkhäuser.

[22] Brihaye, Y., Maslanka, P., Giler, S. and Kosinski, P. 1992. "Real representations of Clifford algebras". *J. Math. Phys.* 3(5): 1579–1581.

[23] Lee, D. and Song, Y. 2010. "The matrix representation of Clifford algebra". *J. Chungcheong Math. Soc.* 23: 363–368.

[24] Lee, D. and Song, Y. 2013. "Explicit matrix realization of Clifford algebras". *Adv. Appl. Clifford Algebras.* 23: 441–451.

[25] Lee, D. and Song, Y. 2013. "Matrix representations of the low order real Clifford algebras". *Adv. Appl. Clifford Algebras.* 23: 965–980.

[26] Lee, D. and Song, Y. 2015. "A Construction of matrix representation of Clifford algebras". *Adv. Appl. Clifford Algebras.* 25: 719–731.

[27] Lee, D. and Song, Y. 2012. "Applications of matrix algebra to Clifford groups". *Adv. Appl. Clifford Algebras*. 22: 391–398.

[28] Hamaker, Z. 2008. *Clifford algebra representation*. The Division of Science, Mathematics, and Computing of Bard College.

[29] Lee, D. and, Song, Y. 2010. "The matrix representation of Clifford algebra". *Journal of the Chungcheong Mathematical Society*. 23(2).

[30] Lee, D. and Song, Y. 2013. "Explicit matrix realization of Clifford algebras". *Adv. Appl. Clifford Algebras*. 23: 441–451.

[31] Song, Y. and Lee, D. 2015. "A construction of matrix representation of Clifford algebras". *Adv. Appl. Clifford Algebras*. 25: 719–731.

[32] Yüksel, U. 2013. "Necessary and sufficient conditions for associated differential operators in Quaternionic Analysis and Applications to Initial Value Problems". *Adv. Appl. Clifford Alg*. 23(4): 981âĂŞ-990.

[33] Tutschke, W. 2004. "Associated partial differential operators– applications to well- and ill- posed problems". *Abstract and applied analysis*. 373–383.

[34] Tutschke, W. 1994. "Interior L_p–estimates for functions with integral representations". *Applicable Analysis*. 73(1): 281– 294.

[35] Di Teodoro, A. and Vanegas, J. 2012. "Fundamental solutions for the first order metamonogenic operator". *Adv. Appl. Clifford Algebras*. 22(1): 49–58.

[36] Yüksel, U. 2013. "Necessary and sufficient conditions for associated differential operators in quaternionic analysis and applications to initial value problems". *Advances in Applied Clifford Algebras*. 23(4): 981–990.

In: Understanding Banach Spaces
Editor: Daniel González Sánchez

ISBN: 978-1-53616-745-0
© 2020 Nova Science Publishers, Inc.

Chapter 20

COMPUTATIONAL APPROACH OF INITIAL VALUE PROBLEMS IN CLIFFORD ANALYSIS USING ASSOCIATED SPACES

David Armendáriz[1,*], *Johan Ceballos*[2,†] *and Antonio Di Teodoro*[1,‡]
[1]Departamento de Matemáticas, Colegio de Ciencias e Ingenierías
Universidad San Francisco de Quito, Quito, Ecuador
[2]Escuela de Ciencias Físicas y Matemáticas
Universidad de Las Américas, Quito, Ecuador

Abstract

In the present chapter we continue the previous one related with the initial value problem in Clifford analysis using the method of associated spaces over a Banach space; but in this chapter we present the advantage of using a matrix representation for the basis of a Clifford algebra, when you have a high order algebra in particular in $\mathbb{R}_{0,3}$. We present the result of the product rule for the derivatives of two Clifford valued function and the calculations in $\mathbb{R}_{0,3}$, finally we present the calculation of the conditions over the coefficients in order to construct the associated space.

Keywords: matrix representations, Clifford algebras, initial value problems, monogenic functions, associated space

AMS Subject Classification: 11E88, 15A66, 35F10, 35B45, 47H10, 65F30

1. Previous

Consider the Initial value problem:

$$\partial_t u = L(t, x, u, \partial_{x_i} u), \quad u(0, x) = \varphi(x),$$

---

[*]E-mail: darmendarizp@estud.usfq.edu.ec.
[†]Corresponding Author's E-mail: johan.ceballos@udla.edu.ec.
[‡]E-mail: nditeodoro@usfq.edu.ec.

In the previous chapter, we formulated sufficient conditions on the coefficients of L operator under which L is associated to differential equations with monogenic right–hand sides, using an algorithm to characterize the operators over Clifford structures. We exhibited an interior estimate for the monogenic functions using the Cauchy integral formula and then gave out conditions under which the initial value problem is uniquely solvable in the context of Clifford algebras over an appropriate Banach space (see [1–4]).

In this chapter, we will show how the elements of a Clifford algebra can be represented and operate using a matrix form, from simple algorithms written in MATLAB. To illustrate this situation we use the Clifford algebra $\mathbb{R}_{0,3}$. Finally we will show the solution of an initial value problem in $\mathbb{R}_{0,3}$ justifying the use of the matrix representation and the computer implementation.

2. Fundamental Matrices Using MATLAB Algorithms

Recalling from the previous chapter that the fundamental matrices E_{jn}, $j = 1, 2, \ldots n$, satisfy the structural relation:

$$E_{j,n}^2 = \mathbf{I_{2^n}},$$
$$E_{i,n}E_{j,n} + E_{j,n}E_{i,n} = \mathbf{0_{2^n}}, \quad i \neq j, \tag{1}$$

for $i, j = 1, 2, \ldots, n$. Where $\mathbf{I_{2^n}}$ and $\mathbf{0_{2^n}}$ are the identity matrix and the cero matrix respectively (sometimes to clarify we use the symbol $E_{0,n}$ to refer to the identity matrix), recursively allowing the possibility to develop the computer implementation for the Clifford algebras (see [5–7]).

2.1. Algorithm to Construct Matrices

In this section we will present the algorithms that allow the building of a matrix representation for a fixed dimension Clifford algebra. We will show some examples on how to represent an element and a function with values in a Clifford algebra in a matrix form. In addition, we will show how to represent and operate the Cauchy–Riemann operator in a matrix form and finally we will show why it is not necessary to use a formula for the product of the derivatives in a Clifford algebra.

Algorithm 1. Recursive matrix representation algorithm for $E_{n,n}$ in $\mathbb{R}_{0,n}$

```
function Enn=RecursiveEnn(n)
vec=1;
for j=2:n
    vec=[vec,-vec];
end
Enn=zeros(2^n);
Enn(2^(n-1)+1:2^n,1:2^(n-1))=diag(vec);
Enn(1:2^(n-1),2^(n-1)+1:2^n)=diag(-vec);
```

Computational Approach of Initial Value Problems in Clifford Analysis ... 233

Algorithm 2. Recursive matrix representation algorithm for $E_{k,n}$ in $\mathbb{R}_{0,n}$ ($k = 1, 2, \ldots n$)
```
function  Ekn=RecursiveEkn(k,n)
if  k==0
    Ekn=eye(2^n);
else if  k>0
        Ekn=RecursiveEnn(k);
        for  i=k+1:n
            Aux=zeros(2^i);
            Aux(1:2^(i-1),1:2^(i-1))=Ekn;
            Aux(2^(i-1)+1:2^i,2^(i-1)+1:2^i)=Ekn;
            Ekn=Aux;
        end
    end
end
```

In the following example, we will illustrate how to construct the fundamental matrices. By multiplying these matrices properly, we obtain all the elements of the basis for a fixed n.

Example 1. *If $n = 3$, the fundamental matrices for $\mathbb{R}_{0,3}$ are*

$$E_{1,3} = \begin{pmatrix} 0 & -1 & 0 & 0 & 0 & 0 & 0 & 0 \\ 1 & 0 & 0 & 0 & 0 & 0 & 0 & 0 \\ 0 & 0 & 0 & -1 & 0 & 0 & 0 & 0 \\ 0 & 0 & 1 & 0 & 0 & 0 & 0 & 0 \\ 0 & 0 & 0 & 0 & 0 & -1 & 0 & 0 \\ 0 & 0 & 0 & 0 & 1 & 0 & 0 & 0 \\ 0 & 0 & 0 & 0 & 0 & 0 & 0 & -1 \\ 0 & 0 & 0 & 0 & 0 & 0 & 1 & 0 \end{pmatrix} = \left(\begin{array}{c|c} E_{1,2} & \mathbf{0}_4 \\ \hline \mathbf{0}_4 & E_{1,2} \end{array} \right),$$

$$E_{2,3} = \begin{pmatrix} 0 & 0 & -1 & 0 & 0 & 0 & 0 & 0 \\ 0 & 0 & 0 & 1 & 0 & 0 & 0 & 0 \\ 1 & 0 & 0 & 0 & 0 & 0 & 0 & 0 \\ 0 & -1 & 0 & 0 & 0 & 0 & 0 & 0 \\ 0 & 0 & 0 & 0 & 0 & 0 & -1 & 0 \\ 0 & 0 & 0 & 0 & 0 & 0 & 0 & 1 \\ 0 & 0 & 0 & 0 & 1 & 0 & 0 & 0 \\ 0 & 0 & 0 & 0 & 0 & -1 & 0 & 0 \end{pmatrix} = \left(\begin{array}{c|c} E_{2,2} & \mathbf{0}_4 \\ \hline \mathbf{0}_4 & E_{2,2} \end{array} \right),$$

$$E_{3,3} = \begin{pmatrix} 0 & 0 & 0 & 0 & -1 & 0 & 0 & 0 \\ 0 & 0 & 0 & 0 & 0 & 1 & 0 & 0 \\ 0 & 0 & 0 & 0 & 0 & 0 & 1 & 0 \\ 0 & 0 & 0 & 0 & 0 & 0 & 0 & -1 \\ 1 & 0 & 0 & 0 & 0 & 0 & 0 & 0 \\ 0 & -1 & 0 & 0 & 0 & 0 & 0 & 0 \\ 0 & 0 & -1 & 0 & 0 & 0 & 0 & 0 \\ 0 & 0 & 0 & 1 & 0 & 0 & 0 & 0 \end{pmatrix} = \left(\begin{array}{c|c} \mathbf{0}_4 & -\mathbf{M}_4 \\ \hline \mathbf{M}_4 & \mathbf{0}_4 \end{array} \right),$$

where $\mathbf{M}_4{}^2 = \mathbf{I}_4$.

2.2. Algorithm to Construct Basis

Algorithm 3. Recursive matrix representation algorithm for $E_{k,n}$ in $\mathbb{R}_{0,n}$

```
function Ekn=RecursiveMatrixRepresentationEkn(k,n)
t=baseelements(n);
if length(find(t==k))==1
    if k==n
        Ekn=RecursiveEnn(n);
    else
        digik=numendigi(k);
        lengthk=length(digik);
        Ekn=eye(2^n);
        for i=1:lengthk
            if digik(i)==n
                Aux=RecursiveEnn(n);
            else
                Aux=RecursiveEkn(digik(i),n);
            end
            Ekn=Ekn*Aux;
        end
    end
else
    Ekn='The matrix Ekn does not exist';
end
```

Example 2. *Computing the following products, we can construct the complete basis for* $\mathbb{R}_{0,3}$. $E_{12,3} = E_{1,3} \cdot E_{2,3}$, $E_{13,3} = E_{1,3} \cdot E_{3,3}$, $E_{23,3} = E_{2,3} \cdot E_{3,3}$, $E_{123,3} = E_{1,3} \cdot E_{2,3} \cdot E_{3,3}$ *and the identity* $\mathbf{I}_8$. *Then the basis of* $\mathbb{R}_{0,3}$ *is given by:*

$$\beta_{\mathbb{R}_{0,3}} = \{\mathbf{I}_8, E_{1,3}, E_{2,3}, E_{3,3}, E_{12,3}, E_{13,3}, E_{23,3}, E_{123,3}\}$$

Note that the dimension of $\beta_{\mathbb{R}_{0,3}} = 2^3 = 8$ *and* $E_{1,3}^2 = E_{2,3}^2 = E_{3,3}^2 = -\mathbf{I}_8$.

A product of the elements of the basis $\beta_{\mathbb{R}_{0,3}}$ is shown in the following example.

Example 3. *Using the rules given in* (1), *Algorithm 1 and matrix product, we find a simpli-*

fied expression for $E_{1,3} E_{2,3} E_{3,3} E_{2,3} E_{1,3} E_{3,3} \in \mathbb{R}_{0,3}$.

$$E_{1,3} E_{2,3} E_{3,3} E_{2,3} E_{1,3} E_{3,3} = -\begin{pmatrix} 1 & 0 & 0 & 0 & 0 & 0 & 0 & 0 \\ 0 & 1 & 0 & 0 & 0 & 0 & 0 & 0 \\ 0 & 0 & 1 & 0 & 0 & 0 & 0 & 0 \\ 0 & 0 & 0 & 1 & 0 & 0 & 0 & 0 \\ 0 & 0 & 0 & 0 & 1 & 0 & 0 & 0 \\ 0 & 0 & 0 & 0 & 0 & 1 & 0 & 0 \\ 0 & 0 & 0 & 0 & 0 & 0 & 1 & 0 \\ 0 & 0 & 0 & 0 & 0 & 0 & 0 & 1 \end{pmatrix} = -E_{0,3} = \mathbf{I_8}.$$

Matrices $E_{1,3}, E_{2,3}, E_{3,3}$ *are given in Example 1.*

Example 4. *[Matrix representation of a Cillford Valued function] A Clifford valued function* $u : \mathbb{R}^3 \to \mathbb{R}_{0,3}$ *can be written in the matrix form, using the basis* $\beta_{\mathbb{R}_{0,3}}$ *given in the example 2 and* $u_0, u_1, u_2, u_{12} : \mathbb{R}^3 \to \mathbb{R}$, *in the following form:*

$$u = u_0 \, \mathbf{I_8} + u_1 \, E_{1,3} + u_2 \, E_{2,3} + u_3 \, E_{3,3} + u_{12} \, E_{12,3} + u_{13} \, E_{13,3} + u_{23} \, E_{23,3} + u_{123} \, E_{123,3}$$

$$= \begin{pmatrix} u_1 & -u_1 & -u_2 & -u_{12} & -u_3 & -u_{13} & -u_{23} & u_{123} \\ u_1 & u_0 & -u_{12} & u_2 & -u_{13} & u_3 & -u_{123} & -u_{23} \\ u_2 & u_{12} & u_0 & -u_1 & -u_{23} & u_{123} & u3 & u_{13} \\ u_{12} & -u_2 & u_1 & u_0 & -u_{123} & -u_{23} & u_{13} & -u_3 \\ u_3 & u_{13} & u_{23} & -u_{123} & u_0 & -u_1 & -u_2 & -u_{12} \\ u_{13} & -u_3 & u_{123} & u_{23} & u_1 & u_0 & -u_{12} & u_2 \\ u_{23} & -u_{123} & -u_3 & -u_{13} & u_2 & u_{12} & u_0 & -u_1 \\ u_{123} & u_{23} & -u_{13} & u_3 & u_{12} & -u_2 & u_1 & u_0 \end{pmatrix}.$$

Let u and v $\mathbb{R}_{0,3}$–Clifford valued functions in the matrix form as in example 4, then the first column of the matrix product $u \cdot v$ is given by:

$$\begin{cases} u_0 v_0 - u_1 v_1 - u_2 v_2 - u_3 v_3 - u_{12} v_{12} - u_{13} v_{13} - u_{23} v_{23} + u_{123} v_{123} \\ u_0 v_1 + u_1 v_0 + u_2 v_{12} - u_{12} v_2 + u_3 v_{13} - u_{13} v_3 - u_{23} v_{123} - u_{123} v_{23} \\ u_0 v_2 + u_2 v_0 - u_1 v_{12} + u_{12} v_1 + u_3 v_{23} - u_{23} v_3 + u_{13} v_{123} + u_{123} v_{13} \\ u_1 v_2 - u_2 v_1 + u_0 v_{12} + u_{12} v_0 + u_{13} v_{23} - u_{23} v_{13} - u_3 v_{123} - u_{123} v_3 \\ u_0 v_3 + u_3 v_0 - u_1 v_{13} + u_{13} v_1 - u_2 v_{23} + u_{23} v_2 - u_{12} v_{123} - u_{123} v_{12} \\ u_1 v_3 - u_3 v_1 + u_0 v_{13} + u_{13} v_0 - u_{12} v_{23} + u_{23} v_{12} + u_2 v_{123} + u_{123} v_2 \\ u_2 v_3 - u_3 v_2 + u_0 v_{23} + u_{23} v_0 + u_{12} v_{13} - u_{13} v_{12} - u_1 v_{123} - u_{123} v_1 \\ u_3 v_{12} - u_2 v_{13} + u_{12} v_3 - u_{13} v_2 + u_1 v_{23} + u_{23} v_1 + u_0 v_{123} + u_{123} v_0 \end{cases}$$

We don't exhibit the complete matrix because the matrix product is too big, due to all the calculations, but we can conclude, that the matrix representation is useful to do calculation in a high order Clifford algebra.

Example 5. *[Product of two functions matrices] Let u and v $\mathbb{R}_{0,3}$–Clifford valued functions such that*

$$u = u_1 E_{1,3} \quad \text{and}$$
$$v = v_{123} E_{123,3},$$

where u_1 and v_{123} are real valued functions.
The matrix product $u \cdot v$ is given by:

$$u \cdot v = (u_1 E_{1,3}) \cdot (v_{123} E_{123,3})$$

$$= \begin{pmatrix} 0 & 0 & 0 & 0 & 0 & 0 & u_1 v_{123} & 0 \\ 0 & 0 & 0 & 0 & 0 & 0 & 0 & u_1 v_{123} \\ 0 & 0 & 0 & 0 & u_1 v_{123} & 0 & 0 & 0 \\ 0 & 0 & 0 & 0 & 0 & u_1 v_{123} & 0 & 0 \\ 0 & 0 & -u_1 v_{123} & 0 & 0 & 0 & 0 & 0 \\ 0 & 0 & 0 & -u_1 v_{123} & 0 & 0 & 0 & 0 \\ -u_1 v_{123} & 0 & 0 & 0 & 0 & 0 & 0 & 0 \\ 0 & -u_1 v_{123} & 0 & 0 & 0 & 0 & 0 & 0 \end{pmatrix}$$

2.3. Algorithm of the Cauchy–Riemann Operator in the Matrix Form

Recalling from the previous chapter that using the fundamental matrices, we can obtain the matrix representation of the Cauchy–Riemann in $\mathbb{R}_{0,n}$

$$\tilde{\partial}_{\mathbf{n+1}} = \mathbf{I}_{2^{\mathbf{n}}} \partial_0 + \sum_{i=1}^{n} E_{i,n} \partial_i = \mathbf{I}_{2^{\mathbf{n}}} \partial_0 + \mathcal{D}_n$$

where, $\mathcal{D}_n$ is the Dirac operator and ∂_i represents the derivative with respect to x_i. Its conjugate is given by

$$\overline{\tilde{\partial}}_{n+1} = \mathbf{I}_{2^{\mathbf{n}}} \partial_0 - \sum_{i=1}^{n} E_{i,n} \partial_i = \mathbf{I}_{2^{\mathbf{n}}} \partial_0 - \mathcal{D}_n$$

Example 6. *Consider the Clifford algebra $\mathbb{R}_{0,3}$ the matrix, Cauchy–Riemann and Dirac operators are given by:*

$$\mathcal{D}_3 = \begin{pmatrix} 0 & -\partial_1 & -\partial_2 & 0 & -\partial_3 & 0 & 0 & 0 \\ \partial_1 & 0 & 0 & \partial_2 & 0 & \partial_3 & 0 & 0 \\ \partial_2 & 0 & 0 & -\partial_1 & 0 & 0 & \partial_3 & 0 \\ 0 & -\partial_2 & \partial_1 & 0 & 0 & 0 & 0 & -\partial_3 \\ \partial_3 & 0 & 0 & 0 & 0 & -\partial_1 & -\partial_2 & 0 \\ 0 & -\partial_3 & 0 & 0 & \partial_1 & 0 & 0 & \partial_2 \\ 0 & 0 & -\partial_3 & 0 & \partial_2 & 0 & 0 & -\partial_1 \\ 0 & 0 & 0 & \partial_3 & 0 & -\partial_2 & \partial_1 & 0 \end{pmatrix}$$

$$= E_{1,3} \partial_1 + E_{2,3} \partial_2 + E_{3,3} \partial_3.$$

and the Cauchy–Riemann operator $\tilde{\partial}_4 = \mathbf{I}_8 \partial_0 + \mathcal{D}_3$.

Example 7. *The Cauchy–Riemann System in the reduced matrix form*

$$\tilde{\partial}_4 u = \begin{pmatrix} \partial_0 & -\partial_1 & -\partial_2 & 0 & -\partial_3 & 0 & 0 & 0 \\ \partial_1 & \partial_0 & 0 & \partial_2 & 0 & \partial_3 & 0 & 0 \\ \partial_2 & 0 & \partial_0 & -\partial_1 & 0 & 0 & \partial_3 & 0 \\ 0 & -\partial_2 & \partial_1 & \partial_0 & 0 & 0 & 0 & -\partial_3 \\ \partial_3 & 0 & 0 & 0 & \partial_0 & -\partial_1 & -\partial_2 & 0 \\ 0 & -\partial_3 & 0 & 0 & \partial_1 & \partial_0 & 0 & \partial_2 \\ 0 & 0 & -\partial_3 & 0 & \partial_2 & 0 & \partial_0 & -\partial_1 \\ 0 & 0 & 0 & \partial_3 & 0 & -\partial_2 & \partial_1 & \partial_0 \end{pmatrix} \begin{pmatrix} u_0 \\ u_1 \\ u_2 \\ u_3 \\ u_{12} \\ u_{13} \\ u_{23} \\ u_{123} \end{pmatrix} = 0.$$

2.4. Product Rule in a Matrix Form

As well known, the Leibniz's rule is not valid for the product of two Clifford valued functions (see [7–9]). In the next examples we are going to show how to compute the product using the matrix representations of the Clifford valued functions.

$$\tilde{\partial}_{n+1}(u \cdot v) = \mathbf{I}_{2^n} \partial_0 (u \cdot v) + \sum_{i=1}^{n} E_{i,n} \partial_i (u \cdot v) = \mathbf{I}_{2^n} \left[\partial_0(u) \cdot v + u \cdot \partial_0(v) \right] + \mathcal{D}_n(u \cdot v).$$

Example 8. *Given $u, v \in \mathbb{R}_{0,3}$ such that $u = u_1 E_{1,3}$ and $v = v_{123} E_{123,3}$, then, applying the matrix Dirac operator $\mathcal{D}_3$ (example 6) to the product give in example 5, we have:*

$$\mathcal{D}_3(u \cdot v)$$

$$= \begin{pmatrix} 0 & 0 & \partial_3 u_1 v_{123} & 0 & -\partial_2 u_1 v_{123} & 0 & 0 & -\partial_1 u_1 v_{123} \\ 0 & 0 & 0 & -\partial_3 u_1 v_{123} & 0 & \partial_2 u_1 v_{123} & \partial_1 u_1 v_{123} & 0 \\ -\partial_3 u_1 v_{123} & 0 & 0 & 0 & 0 & -\partial_1 u_1 v_{123} & \partial_2 u_1 v_{123} & 0 \\ 0 & \partial_3 u_1 v_{123} & 0 & 0 & \partial_1 u_1 v_{123} & 0 & 0 & -\partial_2 u_1 v_{123} \\ \partial_2 u_1 v_{123} & 0 & 0 & \partial_1 u_1 v_{123} & 0 & 0 & 0 & \partial_3 u_1 v_{123} \\ 0 & -\partial_2 u_1 v_{123} & -\partial_1 u_1 v_{123} & 0 & 0 & 0 & -\partial_3 u_1 v_{123} & 0 \\ 0 & \partial_1 u_1 v_{123} & -\partial_2 u_1 v_{123} & 0 & 0 & -\partial_3 u_1 v_{123} & 0 & \partial_3 u_1 v_{123} \\ -\partial_1 u_1 v_{123} & 0 & 0 & 0 & \partial_2 u_1 v_{123} & 0 & 0 & 0 \end{pmatrix}.$$

Remark 9. *Let u and v $\mathbb{R}_{0,3}$–Clifford valued functions in the matrix form as in example 4, and $\mathcal{D}_3$ the matrix Dirac operator given in example 6. The element $(1,1)$ of $\mathcal{D}_3(u \cdot v)$ is* given by:

$$-\partial_1(u_0 v_1 + u_1 v_0 + u_2 v_{12} - u_{12} v_2 + u_3 v_{13} - u_{13} v_3 - u_{23} v_{123} - u_{123} v_{23})$$
$$-\partial_2(u_0 v_2 + u_2 v_0 - u_1 v_{12} + u_{12} v_1 + u_3 v_{23} - u_{23} v_3 + u_{13} v_{123} + u_{123} v_{13})$$
$$-\partial_3(u_0 v_3 + u_3 v_0 - u_1 v_{13} + u_{13} v_1 - u_2 v_{23} + u_{23} v_2 - u_{12} v_{123} - u_{123} v_{12}).$$

Note that, $\mathcal{D}_3(u \cdot v)$ is a 8×8 matrix, and the last expression is only for one of the 64 components. Once again, we can observe the importance of having algorithms to operate in Clifford algebras.

3. SUFFICIENT CONDITIONS FOR ASSOCIATED PAIRS AND APPLICATIONS TO INITIAL VALUE PROBLEMS

In the previous chapter we presented a discussion of how two associated pairs can help us to solve initial value problem. In this section we follow the ideas exhibited previously, but we present a more complicated case in detail to illustrate the potentiality of the matrices.

3.1. Associated Spaces: The Basic Idea Using to Solve Initial Value Problem

In the previous chapter, we found solutions to the initial value problems of the form:

$$\partial_t u = F(t, x, u, \partial_{x_i} u), \quad u(0, x) = \varphi(x), \tag{2}$$

where t is time, using associated pairs on monogenic functions (or more general) that has interior estimates (see [8–10]).

Theorem 10. *The initial value problem (2), where F is a linear first order operator in a Clifford algebra and φ is a monogenic function, suppose that F and φ are an associated pair, where φ satisfies an interior estimate of first order. This problem is solvable in the conical domain*
$$M_\mu = \{(t, x) : x \in \Omega, 0 \leq t < \mu.dist(x, \partial\Omega)\}$$
by the contraction–mapping principle provided μ is small enough. Moreover, the solution $u(t, x)$ belongs to the associated space for each t.

In this section the sufficient conditions for associated operator pairs are given using a computer implementation.

3.2. Calculation of the Sufficient Conditions in $\mathbb{R}_{0,3}$

The systems involved in these calculations are too big, therefore we follow the Algorithm 14 (pseudo code is presented in section 4). Once again, we want to find conditions for the $\mathcal{L}$ operator to be associated with the $\tilde{\partial}_4$ operator. Consider Ω a bounded and simply connected domain of the Euclidean space $\mathbb{R}^8$ and $u : \Omega \to \mathbb{R}_{0,3}$ is a continuously and differentiable function satisfying $\tilde{\partial}_4 u = 0$ then $\tilde{\partial}_4(\mathcal{L}u) = 0$. In this case the $\mathcal{L}u$ operator is given by:

$$\mathcal{L}u = \sum_{i=0}^{7} A^{(i)}(t, x)\partial_i u + B(t, x)u, \tag{3}$$

where $A^{(i)}(t,x) = \sum_{k=0}^{7} a_k^{(i)}(t,x)E_{k,3}$, $B(t,x) = \sum_{k=0}^{7} b_k(t,x)E_{k,3}$ and $\tilde{\partial}_4$ is defined as:

$$\tilde{\partial}_4 = \mathbf{I}_8 \partial_0 + \sum_{i=1}^{7} \partial_i E_{i,3}. \tag{4}$$

Remark 11. *The Clifford algebra basis for $\mathbb{R}_{0,3}$ is*

$$\{E_{0,3}, E_{1,3}, E_{2,3}, E_{3,3}, E_{12,3}, E_{13,3}, E_{23,3}, E_{123,3}\},$$

for simplicity, we identify each basis element as a bijective function:

$$\{0, 1, 2, 3, 12, 13, 23, 123\} \leftrightarrow \{0, 1, 2, 3, 4, 5, 6, 7, 8\}$$

in other words,

$$\begin{cases} E_{0,3} = E_{0,3}, & E_{1,3} = E_{1,3}, \\ E_{2,3} = E_{2,3}, & E_{3,3} = E_{3,3}, \\ E_{12,3} = E_{12,3}, & E_{13,3} = E_{13,3}, \\ E_{23,3} = E_{23,3}, & E_{123,3} = E_{123,3}. \end{cases}$$

3.3. The $\mathcal{L}u$ Operator in Matrix Form

Using the basis for $\mathbb{R}_{0,3}$ and the Remark 11, we can construct a matrix representation for $A^{(i)}(t,x)$ and $B(t,x)$. For example, if $i = 1$ we have:

$$A^{(1)}(t,x) = \sum_{k=0}^{7} a_k^{(1)}(t,x)E_{k,3} =$$
$$a_0^{(1)}(t,x)\mathbf{I}_8 + a_1^{(1)}(t,x)E_{1,3} + a_2^{(1)}(t,x)E_{2,3} + a_3^{(1)}(t,x)E_{3,3} +$$
$$a_{12}^{(1)}(t,x)E_{12,3} + a_{13}^{(1)}(t,x)E_{13,3} + a_{23}^{(1)}(t,x)E_{23,3} + a_{123}^{(1)}(t,x)E_{123,3}$$

and

$$B(t,x) = \sum_{k=0}^{7} b_k(t,x)E_{k,3} =$$
$$b_0^{(1)}(t,x)\mathbf{I}_8 + b_1^{(1)}(t,x)E_{1,3} + b_2^{(1)}(t,x)E_{2,3} + b_3^{(1)}(t,x)E_{3,3} +$$
$$b_{12}^{(1)}(t,x)E_{12,3} + b_{13}^{(1)}(t,x)E_{13,3} + b_{23}^{(1)}(t,x)E_{23,3} + b_{123}^{(1)}(t,x)E_{123,3}.$$

Hence, from equation (3) and example 4

$$\begin{aligned} \mathcal{L}u &= \sum_{i=0}^{7} A^{(i)}(t,x)\partial_i u + B(t,x)u \\ &= \left(\sum_{i=0}^{7} A^{(i)}(t,x)\partial_i + B(t,x)\right)u. \end{aligned}$$

$$\mathcal{L}u = \left(\sum_{i=0}^{7} A^{(i)}(t,x)\partial_i + B(t,x)\right) \cdot$$

$$\begin{pmatrix} u_1 & -u_1 & -u_2 & -u_{12} & -u_3 & -u_{13} & -u_{23} & u_{123} \\ u_1 & u_0 & -u_{12} & u_2 & -u_{13} & u_3 & -u_{123} & -u_{23} \\ u_2 & u_{12} & u_0 & -u_1 & -u_{23} & u_{123} & u_3 & u_{13} \\ u_{12} & -u_2 & u_1 & u_0 & -u_{123} & -u_{23} & u_{13} & -u_3 \\ u_3 & u_{13} & u_{23} & -u_{123} & u_0 & -u_1 & -u_2 & -u_{12} \\ u_{13} & -u_3 & u_{123} & u_{23} & u_1 & u_0 & -u_{12} & u_2 \\ u_{23} & -u_{123} & -u_3 & -u_{13} & u_2 & u_{12} & u_0 & -u_1 \\ u_{123} & u_{23} & -u_{13} & u_3 & u_{12} & -u_2 & u_1 & u_0 \end{pmatrix}.$$

3.4. The Necessary Conditions Over the Coefficients in $\mathbb{R}_{0,3}$ for the $\tilde{\partial}_4$ Operator to be Associated with $\mathcal{L}$

If $\tilde{\partial}_4 = 0$ (example 7), then $\tilde{\partial}_4\left(\mathcal{L}u\right)u = 0$ and we obtain the following conditions on the coefficients $A^{(i)}(t,x)$ y $B(t,x)$

$$\begin{cases}
\partial_0 b_1 + \partial_1 b_0 + \partial_2 b_4 + \partial_3 b_5 = 0, \\
\partial_0 b_2 - \partial_1 b_4 + \partial_2 b_0 + \partial_3 b_6 = 0, \\
\partial_0 b_4 + \partial_1 b_2 - \partial_2 b_1 - \partial_3 b_7 = 0, \\
\partial_0 b_3 - \partial_1 b_5 - \partial_2 b_6 + \partial_3 b_0 = 0, \\
\partial_0 b_5 + \partial_1 b_3 + \partial_2 b_7 - \partial_3 b_1 = 0, \\
\partial_0 b_6 - \partial_1 b_7 + \partial_2 b_3 - \partial_3 b_2 = 0, \\
\partial_0 b_7 + \partial_1 b_6 - \partial_2 b_5 + \partial_3 b_4 = 0, \\
\partial_0 a_1^{(0)} + \partial_0 a_0^{(1)} + \partial_1 a_0^{(0)} - \partial_1 a_1^{(1)} + \partial_2 a_4^{(0)} - \partial_2 a_2^{(1)} + \partial_3 a_5^{(0)} - \partial_3 a_3^{(1)} = 0, \\
-\partial_0 a_0^{(0)} + \partial_0 a_1^{(1)} + \partial_1 a_1^{(0)} + \partial_1 a_0^{(1)} + \partial_2 a_2^{(0)} + \partial_2 a_4^{(1)} + \partial_3 a_3^{(0)} + \partial_3 a_5^{(1)} = 0, \\
2b_2 + \partial_0 a_2^{(0)} + \partial_0 a_4^{(1)} - \partial_1 a_4^{(0)} + \partial_1 a_2^{(1)} + \partial_2 a_0^{(0)} - \partial_2 a_1^{(1)} + \partial_3 a_6^{(0)} - \partial_3 a_7^{(1)} = 0, \\
-2b_5 - \partial_0 a_5^{(0)} + \partial_0 a_3^{(1)} - \partial_1 a_3^{(0)} - \partial_1 a_5^{(1)} - \partial_2 a_7^{(0)} - \partial_2 a_6^{(1)} + \partial_3 a_1^{(0)} + \partial_3 a_0^{(1)} = 0, \\
2b_3 + \partial_0 a_3^{(0)} + \partial_0 a_5^{(1)} - \partial_1 a_5^{(0)} + \partial_1 a_3^{(1)} - \partial_2 a_6^{(0)} + \partial_2 a_7^{(1)} + \partial_3 a_0^{(0)} - \partial_3 a_1^{(1)} = 0, \\
\partial_0 a_7^{(0)} + \partial_0 a_6^{(1)} + \partial_1 a_6^{(0)} - \partial_1 a_7^{(1)} - \partial_2 a_5^{(0)} + \partial_2 a_3^{(1)} + \partial_3 a_4^{(0)} - \partial_3 a_2^{(1)} = 0, \\
-\partial_0 a_6^{(0)} + \partial_0 a_7^{(1)} + \partial_1 a_7^{(0)} + \partial_1 a_6^{(1)} - \partial_2 a_3^{(0)} - \partial_2 a_5^{(1)} + \partial_3 a_2^{(0)} + \partial_3 a_4^{(1)} = 0, \\
\partial_0 a_2^{(0)} + \partial_0 a_0^{(2)} - \partial_1 a_4^{(0)} - \partial_1 a_1^{(2)} + \partial_2 a_0^{(0)} - \partial_2 a_2^{(2)} + \partial_3 a_6^{(0)} - \partial_3 a_3^{(2)} = 0, \\
2b_4 + \partial_0 a_4^{(0)} + \partial_0 a_1^{(2)} + \partial_1 a_2^{(0)} + \partial_1 a_0^{(2)} - \partial_2 a_1^{(0)} + \partial_2 a_4^{(2)} - \partial_3 a_7^{(0)} + \partial_3 a_5^{(2)} = 0, \\
-\partial_0 a_0^{(0)} + \partial_0 a_2^{(2)} + \partial_1 a_1^{(0)} - \partial_1 a_4^{(2)} + \partial_2 a_2^{(0)} + \partial_2 a_0^{(2)} + \partial_3 a_3^{(0)} + \partial_3 a_6^{(2)} = 0, \\
-2b_1 - \partial_0 a_1^{(0)} + \partial_0 a_4^{(2)} - \partial_1 a_0^{(0)} + \partial_1 a_2^{(2)} - \partial_2 a_4^{(0)} - \partial_2 a_1^{(2)} - \partial_3 a_5^{(0)} - \partial_3 a_7^{(2)} = 0, \\
-2b_6 - \partial_0 a_6^{(0)} + \partial_0 a_3^{(2)} + \partial_1 a_7^{(0)} - \partial_1 a_5^{(2)} - \partial_2 a_3^{(0)} - \partial_2 a_6^{(2)} + \partial_3 a_2^{(0)} + \partial_3 a_0^{(2)} = 0, \\
-\partial_0 a_7^{(0)} + \partial_0 a_5^{(2)} - \partial_1 a_6^{(0)} + \partial_1 a_3^{(2)} + \partial_2 a_5^{(0)} + \partial_2 a_7^{(2)} - \partial_3 a_4^{(0)} - \partial_3 a_1^{(2)} = 0, \\
2b_3 + \partial_0 a_3^{(0)} + \partial_0 a_6^{(2)} - \partial_1 a_5^{(0)} - \partial_1 a_7^{(2)} - \partial_2 a_6^{(0)} + \partial_2 a_3^{(2)} + \partial_3 a_0^{(0)} - \partial_3 a_2^{(2)} = 0, \\
\partial_0 a_5^{(0)} + \partial_0 a_7^{(2)} + \partial_1 a_3^{(0)} + \partial_1 a_6^{(2)} + \partial_2 a_7^{(0)} - \partial_2 a_5^{(2)} - \partial_3 a_1^{(0)} + \partial_3 a_4^{(2)} = 0,
\end{cases}$$

$$\begin{cases}
\partial_0 a_3^{(0)} + \partial_0 a_0^{(3)} - \partial_1 a_1^{(0)} - \partial_1 a_5^{(3)} - \partial_2 a_6^{(0)} - \partial_2 a_2^{(3)} + \partial_3 a_0^{(0)} - \partial_3 a_3^{(3)} = 0, \\
2b_5 + \partial_0 a_5^{(0)} + \partial_0 a_1^{(3)} + \partial_1 a_3^{(0)} + \partial_1 a_0^{(3)} + \partial_2 a_7^{(0)} + \partial_2 a_4^{(3)} - \partial_3 a_1^{(0)} + \partial_3 a_5^{(3)} = 0, \\
2b_6 + \partial_0 a_6^{(0)} + \partial_0 a_2^{(3)} - \partial_1 a_7^{(0)} - \partial_1 a_4^{(3)} + \partial_2 a_3^{(0)} + \partial_2 a_0^{(3)} - \partial_3 a_2^{(0)} + \partial_3 a_6^{(3)} = 0, \\
\partial_0 a_7^{(0)} + \partial_0 a_4^{(3)} + \partial_1 a_6^{(0)} + \partial_1 a_2^{(3)} - \partial_2 a_5^{(0)} - \partial_2 a_1^{(3)} + \partial_3 a_4^{(0)} - \partial_3 a_7^{(3)} = 0, \\
-\partial_0 a_0^{(0)} + \partial_0 a_3^{(3)} + \partial_1 a_1^{(0)} - \partial_1 a_5^{(3)} + \partial_2 a_2^{(0)} - \partial_2 a_6^{(3)} + \partial_3 a_3^{(0)} + \partial_3 a_0^{(3)} = 0, \\
-2b_1 - \partial_0 a_1^{(0)} + \partial_0 a_5^{(3)} - \partial_1 a_0^{(0)} + \partial_1 a_3^{(3)} - \partial_2 a_4^{(0)} + \partial_2 a_7^{(3)} - \partial_3 a_5^{(0)} - \partial_3 a_1^{(3)} = 0, \\
-2b_2 - \partial_0 a_2^{(0)} + \partial_0 a_6^{(3)} + \partial_1 a_4^{(0)} - \partial_1 a_7^{(3)} - \partial_2 a_0^{(0)} + \partial_2 a_3^{(3)} - \partial_3 a_6^{(0)} - \partial_3 a_2^{(3)} = 0, \\
-\partial_0 a_4^{(0)} + \partial_0 a_7^{(3)} - \partial_1 a_2^{(0)} + \partial_1 a_6^{(3)} + \partial_2 a_1^{(0)} - \partial_2 a_5^{(3)} + \partial_3 a_7^{(0)} + \partial_3 a_4^{(3)} = 0, \\
\partial_0 a_4^{(0)} + \partial_0 a_0^{(4)} + \partial_1 a_2^{(0)} - \partial_1 a_1^{(4)} - \partial_2 a_1^{(0)} - \partial_2 a_2^{(4)} - \partial_3 a_7^{(0)} - \partial_3 a_3^{(4)} = 0, \\
-2b_2 - \partial_0 a_2^{(0)} + \partial_0 a_1^{(4)} + \partial_1 a_4^{(0)} + \partial_1 a_0^{(4)} - \partial_2 a_0^{(0)} + \partial_2 a_4^{(4)} - \partial_3 a_6^{(0)} + \partial_3 a_5^{(4)} = 0, \\
2b_1 + \partial_0 a_1^{(0)} + \partial_0 a_2^{(4)} + \partial_1 a_0^{(0)} - \partial_1 a_4^{(4)} + \partial_2 a_4^{(0)} + \partial_2 a_0^{(4)} + \partial_3 a_5^{(0)} + \partial_3 a_5^{(4)} = 0, \\
-\partial_0 a_0^{(0)} + \partial_0 a_4^{(4)} + \partial_1 a_1^{(0)} + \partial_1 a_2^{(4)} + \partial_2 a_2^{(0)} - \partial_2 a_1^{(4)} + \partial_3 a_3^{(0)} - \partial_3 a_7^{(4)} = 0, \\
\partial_0 a_7^{(0)} + \partial_0 a_3^{(4)} + \partial_1 a_6^{(0)} - \partial_1 a_5^{(4)} - \partial_2 a_5^{(0)} - \partial_2 a_6^{(4)} + \partial_3 a_4^{(0)} + \partial_3 a_0^{(4)} = 0, \\
-2b_6 - \partial_0 a_6^{(0)} + \partial_0 a_5^{(4)} + \partial_1 a_7^{(0)} + \partial_1 a_3^{(4)} - \partial_2 a_3^{(0)} + \partial_2 a_7^{(4)} + \partial_3 a_2^{(0)} - \partial_3 a_1^{(4)} = 0, \\
2b_5 + \partial_0 a_5^{(0)} + \partial_0 a_6^{(4)} + \partial_1 a_3^{(0)} - \partial_1 a_7^{(4)} + \partial_2 a_7^{(0)} + \partial_2 a_3^{(4)} - \partial_3 a_1^{(0)} - \partial_3 a_2^{(4)} = 0, \\
-\partial_0 a_3^{(0)} + \partial_0 a_7^{(4)} + \partial_1 a_5^{(0)} + \partial_1 a_6^{(4)} + \partial_2 a_6^{(0)} - \partial_2 a_5^{(4)} - \partial_3 a_0^{(0)} + \partial_3 a_4^{(4)} = 0, \\
\partial_0 a_5^{(0)} + \partial_0 a_0^{(5)} + \partial_1 a_3^{(0)} - \partial_1 a_1^{(5)} + \partial_2 a_7^{(0)} - \partial_2 a_2^{(5)} - \partial_3 a_1^{(0)} - \partial_3 a_3^{(5)} = 0, \\
-2b_3 - \partial_0 a_3^{(0)} + \partial_0 a_1^{(5)} + \partial_1 a_5^{(0)} + \partial_1 a_0^{(5)} + \partial_2 a_6^{(0)} + \partial_2 a_4^{(5)} - \partial_3 a_0^{(0)} + \partial_3 a_5^{(5)} = 0, \\
-\partial_0 a_7^{(0)} + \partial_0 a_2^{(5)} - \partial_1 a_6^{(0)} - \partial_1 a_4^{(5)} + \partial_2 a_5^{(0)} + \partial_2 a_0^{(5)} - \partial_3 a_4^{(0)} + \partial_3 a_6^{(5)} = 0, \\
2b_6 + \partial_0 a_6^{(0)} + \partial_0 a_4^{(5)} - \partial_1 a_7^{(0)} + \partial_1 a_2^{(5)} + \partial_2 a_3^{(0)} - \partial_2 a_1^{(5)} - \partial_3 a_2^{(0)} - \partial_3 a_7^{(5)} = 0, \\
2b_1 + \partial_0 a_1^{(0)} + \partial_0 a_3^{(5)} + \partial_1 a_0^{(0)} - \partial_1 a_5^{(5)} + \partial_2 a_4^{(0)} - \partial_2 a_6^{(5)} + \partial_3 a_5^{(0)} + \partial_3 a_0^{(5)} = 0, \\
-\partial_0 a_0^{(0)} + \partial_0 a_5^{(5)} + \partial_1 a_1^{(0)} + \partial_1 a_3^{(5)} + \partial_2 a_2^{(0)} + \partial_2 a_7^{(5)} + \partial_3 a_3^{(0)} - \partial_3 a_1^{(5)} = 0, \\
-2b_4 - \partial_0 a_4^{(0)} + \partial_0 a_6^{(5)} - \partial_1 a_2^{(0)} - \partial_1 a_7^{(5)} + \partial_2 a_1^{(0)} + \partial_2 a_3^{(5)} + \partial_3 a_7^{(0)} - \partial_3 a_2^{(5)} = 0, \\
-\partial_0 a_2^{(0)} + \partial_0 a_7^{(5)} - \partial_1 a_4^{(0)} + \partial_1 a_6^{(5)} + \partial_2 a_0^{(0)} - \partial_2 a_5^{(5)} + \partial_3 a_6^{(0)} + \partial_3 a_4^{(5)} = 0, \\
\partial_0 a_6^{(0)} + \partial_0 a_0^{(6)} - \partial_1 a_7^{(0)} - \partial_1 a_1^{(6)} + \partial_2 a_3^{(0)} - \partial_2 a_2^{(6)} - \partial_3 a_2^{(0)} - \partial_3 a_3^{(6)} = 0, \\
\partial_0 a_7^{(0)} + \partial_0 a_1^{(6)} + \partial_1 a_6^{(0)} + \partial_1 a_0^{(6)} - \partial_2 a_5^{(0)} + \partial_2 a_4^{(6)} + \partial_3 a_4^{(0)} + \partial_3 a_5^{(6)} = 0, \\
-2b_3 - \partial_0 a_3^{(0)} + \partial_0 a_2^{(6)} + \partial_1 a_5^{(0)} - \partial_1 a_4^{(6)} + \partial_2 a_6^{(0)} + \partial_2 a_0^{(6)} - \partial_3 a_0^{(0)} + \partial_3 a_6^{(6)} = 0, \\
-2b_5 - \partial_0 a_5^{(0)} + \partial_0 a_4^{(6)} - \partial_1 a_3^{(0)} + \partial_1 a_2^{(6)} - \partial_2 a_7^{(0)} - \partial_2 a_1^{(6)} + \partial_3 a_1^{(0)} - \partial_3 a_7^{(6)} = 0, \\
2b_2 + \partial_0 a_2^{(0)} + \partial_0 a_3^{(6)} - \partial_1 a_4^{(0)} - \partial_1 a_5^{(6)} + \partial_2 a_0^{(0)} - \partial_2 a_6^{(6)} + \partial_3 a_6^{(0)} + \partial_3 a_0^{(6)} = 0, \\
2b_4 + \partial_0 a_4^{(0)} + \partial_0 a_5^{(6)} + \partial_1 a_2^{(0)} + \partial_1 a_3^{(6)} - \partial_2 a_1^{(0)} + \partial_2 a_7^{(6)} - \partial_3 a_7^{(0)} - \partial_3 a_1^{(6)} = 0, \\
-\partial_0 a_0^{(0)} + \partial_0 a_6^{(6)} + \partial_1 a_1^{(0)} - \partial_1 a_7^{(6)} + \partial_2 a_2^{(0)} + \partial_2 a_3^{(6)} + \partial_3 a_3^{(0)} - \partial_3 a_2^{(6)} = 0, \\
-\partial_0 a_1^{(0)} + \partial_0 a_7^{(6)} - \partial_1 a_0^{(0)} + \partial_1 a_6^{(6)} - \partial_2 a_4^{(0)} - \partial_2 a_5^{(6)} - \partial_3 a_5^{(0)} + \partial_3 a_4^{(6)} = 0,
\end{cases}$$

$$\begin{cases}
-\partial_0 a_7^{(0)} + \partial_0 a_0^{(7)} - \partial_1 a_6^{(0)} - \partial_1 a_1^{(7)} + \partial_2 a_5^{(0)} - \partial_2 a_2^{(7)} - \partial_3 a_4^{(0)} - \partial_3 a_3^{(7)} = 0, \\
\partial_0 a_6^{(0)} + \partial_0 a_1^{(7)} - \partial_1 a_7^{(0)} + \partial_1 a_0^{(7)} + \partial_2 a_3^{(0)} + \partial_2 a_4^{(7)} - \partial_3 a_2^{(0)} + \partial_3 a_5^{(7)} = 0, \\
-\partial_0 a_5^{(0)} + \partial_0 a_2^{(7)} - \partial_1 a_3^{(0)} - \partial_1 a_4^{(7)} - \partial_2 a_7^{(0)} + \partial_2 a_0^{(7)} + \partial_3 a_1^{(0)} + \partial_3 a_6^{(7)} = 0, \\
\partial_0 a_3^{(0)} + \partial_0 a_4^{(7)} - \partial_1 a_5^{(0)} + \partial_1 a_2^{(7)} - \partial_2 a_6^{(0)} - \partial_2 a_1^{(7)} + \partial_3 a_0^{(0)} - \partial_3 a_7^{(7)} = 0, \\
\partial_0 a_4^{(0)} + \partial_0 a_3^{(7)} + \partial_1 a_2^{(0)} - \partial_1 a_5^{(7)} - \partial_2 a_1^{(0)} - \partial_2 a_6^{(7)} - \partial_3 a_7^{(0)} + \partial_3 a_0^{(7)} = 0, \\
-\partial_0 a_2^{(0)} + \partial_0 a_5^{(7)} + \partial_1 a_0^{(0)} + \partial_1 a_4^{(7)} - \partial_2 a_0^{(0)} + \partial_2 a_7^{(7)} - \partial_3 a_6^{(0)} - \partial_3 a_1^{(7)} = 0, \\
\partial_0 a_1^{(0)} + \partial_0 a_6^{(7)} + \partial_1 a_0^{(0)} - \partial_1 a_7^{(7)} + \partial_2 a_4^{(0)} + \partial_2 a_3^{(7)} + \partial_3 a_5^{(0)} - \partial_3 a_2^{(7)} = 0,
\end{cases}$$

This system continues, but for simplicity we omit the rest of coefficients. The previous process is summarized in the following theorem.

Theorem 12. *Assume that $A^{(i)}(t,x) \in C^2(\Omega, \mathbb{R}_{0,3})$ ($i = 0,1,2,3$) and $B(t,x) \in C^1(\Omega, \mathbb{R}_{0,3})$ for each $t \in [0,T]$. Then the $\mathcal{L}$ operator is associated with the $\tilde{\partial}_4$ operator if and only if the conditions mentioned above are satisfied.*

The rows give us the necessary conditions for the $\tilde{\partial}_4$ operator to be associated with $\mathcal{L}$.

Remark 13. *From Theorem 12 and interior estimate, the hypotheses of Theorem 10 are satisfied.*

4. EXTENSION

The problem of finding sufficient and necessary conditions for associated differential operators may be solved if we consider higher dimensions. If we take into account the Clifford algebra $\mathbb{R}_{0,n}$ with dimension 2^n, then, the following result holds:

Consider the Clifford algebra $\mathbb{R}_{0,n}$ with matrix base $E = \{E_{\hat{p}_n(k),n} : k = 1,2,3,\ldots,2^n\}$ that satisfies:

$$\begin{aligned}
E_{0,n} &= I_{2^n}. \\
E_{i,n}E_{j,n} &= -E_{j,n}E_{i,n}, \text{ for } i = 1,2,\ldots,2^n - 1. \\
E_{i,n}^2 &= -I_{2^n}, \text{ for } i = 1,2,\ldots,2^n - 1.
\end{aligned}$$

The Cauchy–Riemann operator in a matrix form

$$\tilde{\partial}_{\mathbf{n+1}} = \sum_{i=0}^{n} E_{i,n} \partial_i. \tag{5}$$

and the operator

$$\mathcal{L} = \sum_{i=0}^{2^n-1} A^{(i)}(t,x)\partial_i + B(t,x) \tag{6}$$

where $A^{(i)} = \sum_{k=1}^{2^n} a_k^{(i)}(t,x) E_{k,n}$ and $B(t,x) = \sum_{k=1}^{2^n} b_k(t,x) E_{k,n}$.

Then, if $u : \Omega \subset \mathbb{R}^{2^n} \to \mathbb{R}_{0,n}$ is any solution of $\tilde{\partial}_{n+1}u = 0$, it is possible to find the necessary conditions for the $\mathcal{L}$ operator to be associated to $\tilde{\partial}_{n+1}$ following the next algorithm.

Algorithm 14 (Algorithm to find the conditions of the coefficients).
Input: Dimension n of the Clifford algebra $\mathbb{R}_{0,n}$
Output: Necessary and sufficient conditions for $\mathcal{L}$ to be associated with $\tilde{\partial}_{n+1}$ as defined in equations (5) and (6).
Step 1: Construct the matrix base using Algorithm 1.
Step 2: Create operators $\mathcal{L}$ and $\tilde{\partial}_{n+1}$ and function u with the matrices of **Step 1**.
Step 3: Compute the matrix product $\tilde{\partial}_{n+1}(\mathcal{L}u)$, and use the structural relations of $\mathbb{R}_{0,n}$ to simplify the result.

Sufficient conditions can be found following the previous algorithm 14:

Let $I = \{0, 1, \ldots, 2^n - 1\}$. If we choose $a_i^{(j)} = b_i = 0$ for all $i, j \in I$ then $\tilde{\partial}_{n+1}$ and $\mathcal{L}$ are automatically associated. If not, suppose $\tilde{\partial}_{n+1}u = 0$ but $\partial_0 u_i \neq 0$ for $i \in I$. Then, if $\partial_l(a_i^{(j)}\partial_j u_k) = \partial_l a_i^{(j)}\partial_j u_k + a_i^{(j)}\partial_l\partial_j u_k \neq 0$ and $\partial_i(b_j u_k) = \partial_i b_j u_k + b_j \partial_i u_k \neq 0$ for all $i, j, k, l \in I$, then the expression $\tilde{\partial}_{n+1}(\mathcal{L}u)$ is not zero, unless conditions on the $a_i^{(j)}$ and on the b_i are established.

The problem of finding necessary and sufficient conditions for two differential operators $\mathcal{L}$ and $\tilde{\partial}_{n+1}$ to be associated is solved for any $n \in \mathbb{N}$. In fact, this chapter generalizes what has been done in [10] because, with the help of the written code based on the algorithm described, we can find the conditions explicitly for any n. Moreover, we observe the following advantages of matrix representation: the Leibniz rule is not necessary, all conditions appears as matrix rows; simplification of the calculations, provided that the matrices have the information of the non–commutativeness of the algebra and allow to perform an algorithmic extension to any dimension.

REFERENCES

[1] Yüksel, U. 2010. "Solution of initial value problems with monogenic initial functions in Banach spaces with Lp–norm". *Adv. Appl. Clifford Alg.* 20: 201–209.

[2] Tutschke, W. 1989. *Solution of initial value problems in classes of generalized analytic functions.* Teubner Leipzig and Springer Verlag.

[3] Son, L. H. and Hung, N. Q. 2008. "The initial value problems in Clifford and quaternion analysis". *Proceedings of the 15th ICFIDCAA 2007, Osaka Municipal Universities Press.* 3: 317–323.

[4] Tutschke, W. 2008. "Associated spaces: a new tool of real and complex analysis", contained in Function Spaces in Complex and Clifford Analysis. *National University Publ, Hanoi.* 253–268.

[5] Lounesto, P. 2001. *Clifford algebras and spinors* (London Mathematical Society Lecture Note Series). LMS No. 286. Cambridge University Press, Second edition.

[6] Ablamowicz, R. 2004. "Clifford Algebras: Applications to mathematics, physics, and engineering". Progress in mathematical physics. Birkhäuser.

[7] Delanghe, R., Sommen, F. and Soucek. V. 1992. *Clifford algebra and Spinor–valued functions: a function theory for the Dirac operator (Mathematics and its Applications)*. Volume 53. Springer.

[8] Yüksel, U. 2013. "Necessary and sufficient conditions for associated differential operators in Quaternionic Analysis and Applications to Initial Value Problems". *Adv. appl. Clifford alg.* 23(4): 981âĂŞ-990.

[9] Bolívar, Y., Lezama, L., Mármol, L. and Vanegas, J. 2015. "Associated spaces in Clifford analysis". *Adv. Appl. Clifford Alg.* 25: 539–551.

[10] Yüksel, U. 2013. "Necessary and sufficient conditions for associated differential operators in quaternionic analysis and applications to initial value problems". *Advances in Applied Clifford Algebras.* 23(4): 981–990.

In: Understanding Banach Spaces
Editor: Daniel González Sánchez

ISBN: 978-1-53616-745-0
© 2020 Nova Science Publishers, Inc.

Chapter 21

ASYMMETRIC CONVEXITY AND SMOOTHNESS: QUANTITATIVE LYAPUNOV THEOREMS, LOZANOVSKII FACTORISATION, WALSH CHAOS AND CONVEX DUALITY

Sergey S. Ajiev[*]
University of Technology, Sydney, Australia

Abstract

The limits of the applicability of abstract methods of modern functional analysis in PDE, real analysis, optimisation and other mathematical disciplines are governed by understanding the basic properties of particular Banach spaces employed, at least, as well as we know the properties of Lebesgue and l_p spaces. Proceeding from abstract to numerous specific Banach spaces and starting with the choice of geometrically friendly equivalent norms, this chapter lays the foundation for our quasi–Euclidean approach to understanding the local geometric properties of the majority of specific parameterised families of Banach spaces in use. Being also helpful in non–local geometry of Banach spaces, these results have found applications in the approximation of uniformly continuous mappings between Banach spaces by Hölder ones, isometric and bounded extension and interpolation of the latter and the Hölder classification of Banach spaces. Quantitative and occasionally sharp study of the uniform convexity and smoothness of convex functionals on abstract and specific Banach spaces in terms of our asymmetric and double moduli, their comparison to power functions and related convexity and smoothness classes of functionals and spaces, including a reverse triangle and multi–homogeneous inequalities, explicit Hölder continuity and quantitative monotonicity of the subgradient mapping, sharp counterparts of the Fenchel–Young–Lindenstrauss duality formulae, cosine and Babylonian/Pythagorean type theorems and explicit estimates for various groups of parameterised classes of specific spaces, leads to occasionally sharp immediate quantitative applications, including Lyapunov type theorems on the range of an atomic vector–valued measure and Lozanovskii factorisation, counterparts of Parseval's identity for both asymmetric Bernoulli and abstract vector–valued Walsh chaos including high order characterisations of the uniform

---

[*]Corresponding Author's E-mail: sergey.ajiev@uts.edu.au.

convexity and smoothness of Banach spaces, a version of Pisier's martingale inequality and occasionally sharp Markov and Rademacher type and cotype constant estimates, as well as to further 2– and 3–step applications. The specific spaces ranging from auxiliary classes IG and IG_+ to general noncommutative L_p also include several types of Besov, Lizorkin–Triebel and Sobolev spaces on open subsets and their duals, their l_p–sums and "Bochnerisations" endowed with geometrically friendly norms.

Keywords: uniform convexity and smoothness, asymmetric and double moduli, reverse triangle inequality, subgradient mapping, Walsh chaos, Fenchel–Young–Lindenstrauss duality, Rademacher type and cotype, Lyapunov theorem, Lozanovskii factorisation, Pisier inequality, asymmetric and abstract Rademacher and Walsh systems, Besov, Lizorkin–Triebel, Sobolev and IG–spaces, wavelets, Schatten–von Neumann classes, noncommutative L_p, Birkhoff–Fortet orthogonality, Kadets and atomic Lyapunov constants, duality mapping, multi–homogeneous inequality, cosine and Babylonian/Pythagorean theorems, Hanner convexity and smoothness, Bynum–Drew–Gross inequality

AMS Subject Classification: 46E35, 46B09, 46B10, 52A41, 52A40, 46G10, 46B42, 47H05, 60G50, 26D20, 28B05, 54C20, 46B70, 42C40, 46N10, 46N30, 49J30, 46E30, 46L52

1. Introduction

Although our quantitative study of Banach spaces and functionals (and mappings) defined on them starts with abstract spaces and classes of spaces, the meaning of the terms "quantitative" and "abstract" themselves and the value of the abstract results are governed by the spread of their applicability to the settings of specific groups of parameterised spaces employed in the other mathematical disciplines and their applications.

The groups of the main and auxiliary parameterised Banach spaces under consideration, including various types of Lebesgue, Sobolev, Besov, Lizorkin–Triebel and noncommutative spaces, as well as their duals (for example, function spaces of negative smoothness accommodating weak solutions in PDE), l_p–sums (as in the maximal regularity problems in parabolic PDE) and "Bochnerisations" (as in stochastic PDE and Besov–regular stochastic processes), are described in Sections 2 and 3 and "processed" mostly in Section 4.4. Moreover, our results, occasionally excluding the matter of sharpness, are easily extended to the setting of the function spaces of variable smoothness (as considered by O. V. Besov in [1]), particularly including weighted spaces, and the function spaces on Riemannian manifolds and homogeneous metric measure spaces. *Short motivation of this branch of our study*: making the quantitative geometric properties of the other parameterised spaces as well–known as those of the Lebesgue spaces allows the application of abstract methods developed in functional analysis in the other mathematical disciplines and their applications.

Dense Section 4.1 dealing with the most abstract settings is dedicated to the definitions of the asymmetric and double moduli of the uniform convexity and smoothness of convex functionals on a Banach space, related background and the investigation of their properties and applications constituting our most fundamental tools: *a)* mostly sharp upper and lover estimates, particularly covering the 1st order approximation of a convex functional, that is our inequality counterpart of the cosine theorem for the Birkhoff–Fortet orthogonality (see also Section 5.1); b) sharp duality and mutual relations between our moduli, including our sharp counterparts of the Fenchel–Young–Lindenstrauss duality formulae; c) monotonicity of the ratios of the moduli and the 1st and 2nd powers of the parameter generalising the classical estimates in [2, Page 63]; d) limiting behavior with respect to the convex functionals and finite representability of Banach spaces; e) our key abstract zero order approximation and oscillation inequalities for discrete and continuous random variables; f) various abstract counterparts of certain Pafnutii L. Chebyshev's and Gilles Pisier's martingale inequalities and reverse triangle inequality.

Section 4.3 introduces still rather abstract notions of uniform log–convexity and log–smoothness of convex functionals that are specific enough to establish their high order characterisations in terms of our inequality counterparts of Parseval's identity for both asymmetric and general versions of full Walsh chaos. While the classical Lyapunov theorem on the convexity of the range of a nonatomic vector measure, established by Aleksei A. Lyapunov [3] in 1940 (see [4] for details), substantiates the "bang–bang" principle (optimality of the step–function controls) in optimal control theory, the measures in numerical applications are atomic and require the estimates for a stronger measure of nonconvexity (see Sections 4.2 and 5.2) than the midpoint nonconvexity estimated in [5].

In addition to dealing closely with all the specific groups of classes of particular parameterised spaces under consideration, Section 4.4 develops and houses our important analytic tools including the linear and nonlinear approaches to the duality of our most specific and sharp uniform convexity and smoothness, the detailed behavior of the related descriptive functions, the duality of the Hanner types of convexity and smoothness and the Bynum–Drew–Gross and our multi–homogeneous inequality (naturally appeared also [6]).

We have already dealt with the asymmetric uniform convexity of more general functionals in connection with the Hölder regular retractions and Chebyshev centre mappings [7], the explicit estimates for the Kantorovich functional in the mass transportation problem [8] and explicit deviation and concentration inequalities [8], while the greedy approximation algorithms (by a dictionary) in the modern approximation theory [9] are governed by the uniform smoothness. While Lemma 4.1, d) below is formally sharper than Lemma 5.2 in [7], Theorems 4.13 and 4.17 − 4.23 show that the endpoint explicit estimates are the same. Theorem 4.24 leads to sharper estimates for the Rademacher type and cotype constants governing the quantitative Dvoretzky theorem in function spaces in [8] than ones used there. The estimates for the kth layer of the Gaussian and Rademacher Walsh chaos with the powers of norm as functionals and their applications are presented in [10]. We establish explicit estimates for an arbitrary complete (all layers) Walsh chaos with abstract and specific functionals and, thus, higher order characterisations of the uniform convexity and smoothness in Sections 4.3 and 5.3.

The closing Section 5.4 is dedicated to the explicit and occasionally sharp estimates reflecting the quantitative monotonicity and Hölder regularity of the duality mapping and the Hölder regularity of the Lozanovskii factorisation mapping and the same properties of their inverses.

The list of existing (cited) applications is: a) global (on bounded subsets) Hölder continuity (and existence) of the Chebyshev centre mappings, retractions and metric projections onto closed convex subsets and homogeneous right inverses of closed linear endomorphisms in [7]; b) Hölder classification of infinite–dimensional unit spheres, Tsar'kov's phenomenon and commutative homogeneous Hölder group structures on the spaces under consideration in [11, 12]; c) explicit quantitative Lozanovskii factorisation (Theorem 5.6) and the duality mapping (Theorem 5.5) are used in [11, 12] to show that the uniform classification of the Banach lattices due to E. Odell and Th. Schlumprecht in [13] coincides with our Hölder classification; d) Pythagorean theorems for the Birkhoff–Fortet orthogonality (Theorem 5.2) is used in [14] to show that the bounded linear operators between various pairs of spaces under consideration with strictly changing geometry/summability coincide with either compact or Kato-Pełczyński operators; e) three–space problem for the Hölder classification of spheres, Hölder continuity of N. Kalton's nonlinear version of Pełczyński's decomposition method [11, 12]; f) explicit quantitative real interpolation and bounded extension properties of the Hölder mappings between the pairs of spaces under consideration in [15]; g) two–sided estimates for the Berdyshev, James, Lifshits, Schäffer, hyperplane projection and related constants and applications to the metric fixed point theory in [16, 6].

The numbering of the equations is used sparingly. Since the majority of references inside every logical unit are to the formulas inside the unit, equations are numbered independently inside every proof of a corollary, lemma or a theorem, or definition (if there are any numbered formulas).

2. Definitions, Designations and Basic Properties

Let $\mathbb{N}$ be the set of the natural numbers; $\mathbb{N}_0 = \mathbb{N} \cup \{0\}$; $I_n = [1, n] \cap \mathbb{N}$ for $n \in \mathbb{N}$.

Let p' be the conjugate to $p \in [1, \infty]^n$, i.e., $1/p'_i + 1/p_i = 1$ ($1 \leq i \leq n$). For $q \in \mathbb{R}$ and $n \in \mathbb{N}$, let $\overline{q} = (q, \ldots, q) \in \mathbb{R}^n$.

By means of $M(\Omega, \mu)$, we designate the space of all μ–measurable functions defined on the measure space (Ω, μ). For a Lebesgue measurable $E \subset \mathbb{R}^n$ and a finite D, let $|E|$ and $|D|$ be, correspondingly, the Lebesgue measure $\int \chi_E(x) dx$ of E and the number of the elements of D.

For $n \in \mathbb{N}$, $r \in (0, \infty]^n$, $p \in (0, \infty)$, $q \in (0, \infty]$, a (countable) index set I, and a quasi–Banach space A, let $l_q(A) := l_q(I, A)$ be the space of all sequences $\alpha = \{\alpha_k\}_{k \in I} \subset A$ with the finite quasi–norm $\|\alpha\|_{l_q} = (\sum_{k \in I} |\alpha_k|^q)^{1/q} < \infty$;

$$l_r(A) := l_{r_n}(l_{r_{n-1}} \ldots (l_{r_1}(A) \underbrace{\ldots))}_{n \text{ brackets}}.$$

For $G \subset \mathbb{R}^n$ and $f : G \to \mathbb{R}$ by means of $\overline{f} : \mathbb{R}^n \to \mathbb{R}$, we designate the function

$$\overline{f(x)} := \begin{cases} f(x), & \text{for } x \in G, \\ 0, & \text{for } x \in \mathbb{R}^n \setminus G. \end{cases}$$

For $p \in (0, \infty]^n$, let $L_p(G)$ be the space of all measurable functions $f : G \to \mathbb{R}^n$ with the finite mixed quasi–norm

$$\|f|L_p(G)\| = \left(\int_\mathbb{R} \left(\int_\mathbb{R} \cdots \left(\int_\mathbb{R} |\overline{f}|^{p_1} dx_1 \right)^{p_2/p_1} \cdots \right)^{p_n/p_{n-1}} dx_n \right)^{1/p_n},$$

where, for $p_i = \infty$, one understands $\left(\int_{\mathbb{R}} |g(x_i)|^{p_i} dx_i\right)^{1/p_i}$ as $\operatorname{ess\,sup}_{x_i \in \mathbb{R}} |g(x_i)|$. The classical quantitative geometry of these spaces had been studied for a long time (see, for example, [17]).

For an ideal space $Y = Y(\Omega)$ for a measurable space (Ω, μ) and a Banach space X, let $Y(\Omega, X)$ be the space of the Bochner–measurable functions $f : \Omega \mapsto X$ with the finite (quasi)norm
$$\|f|Y(\Omega, X)\| := \|\|f(\cdot)\|_X|Y(\Omega)\|.$$

If another measure ν, absolutely continuous with respect to μ with the density $\frac{d\nu}{d\mu} = \omega$, is used instead of μ, the corresponding space is denoted by $Y(G, \omega, X)$.

For example, $L_p(\mathbb{R}^n, l_q)$ with $p, q \in [1, \infty]$ is a Banach space of the measurable function sequences $f = \{f_k(x)\}_{k=0}^\infty$ with the finite norm $\|\|\{f_k(\cdot)\}_{k \in \mathbb{N}_0}|l_q\|\|L_p(\mathbb{R}^n)\|$.

The construction in the definition of the auxiliary $lt_{p,q}$–spaces (next) suits both the classical dyadic approach [18] and our slightly more optimal covering approach [19] to the wavelet norms (characterisations) of anisotropic function spaces.

Definition 2.1. *For $p \in (0, \infty)^n$, $q \in (0, \infty)$ and $n \in \mathbb{N}$, let $lt_{p,q} = lt_{p,q}(\mathbb{Z}^n \times J)$ be the quasi–Banach space of the sequences $\{t_{i,j}\}_{i \in \mathbb{Z}^n}^{j \in J}$ with $J \in \{\mathbb{N}_0, \mathbb{Z}\}$ endowed with the (quasi)norm*

$$\|\{t_{i,j}\}|lt_{p,q}\| := \left\|\left\{\sum_{i \in \mathbb{Z}^n} t_{i,j}\chi_{F_{i,j}}\right\}_{j \in J}\bigg| L_p(\mathbb{R}^n, l_q(J))\right\|,$$

where $\{F_{i,j}\}_{i \in \mathbb{Z}^n}^{j \in J}$ is a fixed nested family of the decompositions $\{F_{i,j}\}_{i \in \mathbb{Z}^n}$ of $\mathbb{R}^n$ into unions of congruent parallelepipeds $\{F_{i,j}\}_{i \in \mathbb{Z}^n}$ satisfying

$$\cup_{i \in \mathbb{Z}^n} F_{i,j} = \mathbb{R}^n, \ |F_{i,j} \cap F_{k,j}| = 0 \text{ for every } j \in J, \ i \neq k,$$

and either $|F_{i_0,j_0} \cap F_{i_1,j_1}| = 0$, or $F_{i_0,j_0} \cap F_{i_1,j_1} = F_{i_0,j_0}$ for every i_0, i_1 and $j_0 > j_1$. We shall assume that this system regular in the sense that the length of the kth side $l_{k,j}$ of the parallelepipeds $\{F_{i,j}\}_{i \in \mathbb{Z}^n}$ of the jth decomposition (level) satisfies

$$c_l b^{-j(\gamma_a)_k} \leq l_{k,j} \leq c_u b^{-j(\gamma_a)_k} \text{ for } k \in I_n$$

for some positive constants $b > 1$ and c_l, c_u (see Section 3.2 below regarding the anisotropy vector γ_a). Let also the 0–level parallelepipeds $\{F_{i,0}\}_{i \in \mathbb{Z}^n}$ be of the form $Q_t(z)$.

For a parallelepiped $F \in \{F_{i,j}\}_{i \in \mathbb{Z}^n}^{j \in J}$ or $F = \mathbb{R}^n$, by means of $lt_{p,q}(F)$ we designate the subspace of $lt_{p,q}(\mathbb{Z}^n \times J)$ defined by the condition: $t_{i,j} = 0$ if $F_{i,j} \not\subset F$. The symbol $lt_{p,q}$ denotes either of the spaces $lt_{p,q}(\mathbb{Z}^n \times J)$ or $lt_{p,q}(F)$.

The space $lt_{p,q}$ is isometric to a complemented subspace of $L_p(\mathbb{R}^n, l_q)$, while its dual $lt_{p,q}^*$ is isomorphic to $lt_{p',q'}$ for $p \in (1, \infty)^n$, $q \in (1, \infty)$ (see Section 2.1 below).

Remark 2.1. *There are many books and articles dedicated to the wavelet decompositions (and their predecessors) and related characterisations of the function spaces, including [20, 21, 22, 18]. The construction of $lt_{p,q}$ above is compatible not only with the known results but also with a more optimal version [19]. Thus, Besov $B_{p,q}^s(\mathbb{R}^n)_w$ and*

Lizorkin–Triebel $L_{p,q}^s(\mathbb{R}^n)_w$ sequence spaces of a related wavelet basis expansion coefficients, that we shall call Besov and Lizorkin–Triebel spaces with wavelet norms, can be defined in various constructive ways (we omit the lengthy details found in [19]; see also the end of Chapter [23]) but the following always holds. Throughout the article, we only need to know the immediate corollary of the definition(s) that $B_{p,q}^s(\mathbb{R}^n)_w$ is, in fact, isometric to $l_q(\mathbb{N}_0, l_p(\mathbb{Z}^n))$, while $L_{p,q}^s(\mathbb{R}^n)_w$ is isometric to a 1–complemented subspace of $lt_{p,q}(\mathbb{Z}^n \times \mathbb{N}, l_q(I_M))$ for certain $M \in \mathbb{N}$ and contains, in turn, an isometric and 1–complemented copy of $lt_{p,q} = lt_{p,q}(\mathbb{Z}^n \times \mathbb{N}_0)$ (see Section 2.1 for details).

For (quasi) Banach spaces X and Y, $\mathcal{C}(X, Y)$ and $\mathcal{L}(X, Y)$ are the classes of the closed and the bounded linear operators correspondingly. For $C \geq 1$, we say that a subspace Y of a (quasi) Banach space X is C–complemented (in X) if there exists a projection P onto Y satisfying $\|P|\mathcal{L}(X)\| \leq C$.

For $B \subset X$, let $\mathrm{co}(B)$ and $\overline{\mathrm{co}}(B)$ be the convex envelope and the closed convex envelope of B in X respectively. Similarly, let $\mathrm{lin}(B)$ and $\overline{\mathrm{lin}}(B)$ be the linear envelope and the closed linear envelope of B in X correspondingly. Let also

$$S_X = \{x \in X : \|x\|_X = 1\} \text{ and } B_X = \{x \in X : \|x\|_X \leq 1\}$$

be the unit sphere and the unit ball of a Banach space X, and $B(r, y) = y + rB_X$ for $r > 0$ and $y \in X$.

The bi–linear form representing the duality between a (quasi) normed space A and $A^* = \mathcal{L}(A, \mathbb{R})$ is written as

$$(A^*, A) \ni (f, x) \mapsto \langle f, x \rangle = f(x).$$

Definition 2.2. *For $r \in [1, \infty]$, a finite, or countable set I and a set of quasi–Banach spaces $\{X_i\}_{i \in I}$, let theirs l_r–sum $l_r(I, \{A_i\}_{i \in I})$ be the space of the sequences $x = \{x_i\}_{i \in I} \in \prod_{i \in I} X_i$ with the finite norm*

$$\|x|l_r(I, \{A_i\}_{i \in I})\|^r := \sum_{i \in I} \|x_i\|_{A_i}^r.$$

Definition 2.3. *Let X, Y be (quasi)Banach spaces and $\lambda \geq 1$. Then the Banach–Mazur distance $d_{BM}(X, Y)$ between them is equal to ∞ if they are not isomorphic and is, otherwise, defined by*

$$d_{BM}(X, Y) := \inf\{\|T\| \cdot \|T^{-1}\| : T : X \xrightarrow{onto} Y, \mathrm{Ker}\, T = 0\}.$$

The space X is λ–finitely represented in Y if for every finite–dimensional subspace $X_1 \subset X$,

$$\inf\{d_{BM}(X_1, Y_1) : Y_1 \text{ is a subspace of } Y\} = \lambda.$$

If λ is equal to 1, then X is simply said to be finitely represented in Y.

It is said that Y contains almost isometric copies of X (or contains X almost isometrically) if

$$\inf\{d_{BM}(X, Y_1) : Y_1 \text{ is a subspace of } Y\} = 1.$$

For example, it is well–known that, for a Banach space X, its second conjugate X^{**} is finitely represented in X, and X itself is finitely represented in c_0. While, according to Banach, $d_{BM}(l_p, L_p) = \infty$ for $p \in [1, \infty) \setminus \{2\}$ (see [24]), L_p is finitely represented in l_p for $p \in [1, \infty]$.

2.1. Independently Generated Spaces, $lt_{p,q}$ and $lt^*_{p,q}$

The purpose of this subsection is to introduce a wide class of auxiliary spaces containing not only almost all the auxiliary spaces related to the function spaces defined above but also the Lebesgue spaces with mixed norm and l_p–sums of various spaces used in counterexamples.

First let us note some important properties of the space $lt_{p,q}$ (see Definition 2.1) and its dual and explain the necessity to introduce the latter. It will be enough to deal with $lt_{p,q}(\mathbb{R}^n)$. The next lemma is very helpful despite to its simplicity.

Lemma 2.1 ([16]). *Let X be a Banach space, and $P \in \mathcal{L}(X)$ be a projector onto its subspace $Y \subset X$. Assume also that $Q_Y : X \to \tilde{X} = X/\operatorname{Ker} P$ is the quotient map. Then we have*

$$\|Q_Y x\|_{\tilde{X}} \leq \|Px\|_X \leq \|P|\mathcal{L}(X)\| \|Q_Y x\|_{\tilde{X}} \text{ for every } x \in X.$$

In particular, the dual space $Y^* = X^*/Y^\perp$ and Y are isometric to P^*X^* and $\tilde{X}$ respectively if, and only if, Y is 1–complemented in X, i.e $\|P|\mathcal{L}(X)\| = 1$.

As mentioned after Definition 2.1, the space $lt_{p,q}$ is isometric to a subspace of $L_p(\mathbb{R}^n, l_q)$ for $p \in (1,\infty)^n$, $q \in (1,\infty)$ and, as shown in Section 2.1 in [16] both $lt_{p,q}$ and $lt^*_{p',q'}$ are complemented (1–complemented if $p = \bar{q}$) subspaces of $L_p(\mathbb{R}^n, l_q)$ for $p \in (1,\infty)^n$ and $q \in (1,\infty)$, while $lt^*_{p,q}$ is isomorphic to $lt_{p',q'}$ thanks to Lemma 2.1. Therefore, the families $\{lt_{p,q}\}_{p \in (1,\infty)^n}^{q \in (1,\infty)}$ and $\{lt^*_{p,q}\}_{p \in (1,\infty)^n}^{q \in (1,\infty)}$ inherit the complex interpolation properties of the family $\{L_p(\mathbb{R}^n, l_q)\}_{p \in (1,\infty)^n}^{q \in (1,\infty)}$ in the sense of the isomorphisms: for $p_j \in (1,\infty)^n$ and $q_j \in (1,\infty)$ for $j = 0, 1$, $\theta \in (0,1)$, $1/p = (1-\theta)/p_0 + \theta/p_1$ and $1/q = (1-\theta)/q_0 + \theta/q_1$, $(lt_{p_0,q_0}, lt_{p_1,q_1})_{[\theta]} \asymp lt_{p,q}$ and $(lt^*_{p_0,q_0}, lt^*_{p_1,q_1})_{[\theta]} \asymp lt^*_{p,q}$.

Nevertheless, the spaces $lt_{p,q}$ and $lt^*_{p,q}$ contain an isometric and 1–complemented copies of l_p and l_q according to the next lemma. Its proof in [16] is purely constructive.

Lemma 2.2 ([16]). *Let $p \in [1,\infty)^n$ and $q \in [1,\infty)$. Then the spaces $lt_{p,q}(\mathbb{R}^n)$ and $lt_{p',q'}(\mathbb{R}^n)^*$ contain isometric 1–complemented copies of $l_p(\mathbb{N}^n)$, $l_q(\mathbb{N})$ and $l_p(\mathbb{Z}^n, l_q(\mathbb{N}))$, and the spaces $lt_{p,q}(F)$ (see Def. 2.1) and $lt_{p',q'}(F)^*$ contain isometric 1–complemented copies of $l_q(\mathbb{N})$, $l_p(I_m, l_q(\mathbb{N}))$ for every $m \in \mathbb{N}^n$.*

Remark 2.2. *There are several known ways of detecting copies of l_σ–spaces in a Banach space. It seems that, at least, in the setting of a regular domain G one can uncover almost isometric copies in Besov and Lizorkin–Triebel spaces with the aid of the remarkable results due to M. Levy [25], M. Mastyło [26] and V. Milman [24] (in combination with the results in [23]). Direct constructions of the isomorphic (and often complemented) copies of various sequence spaces (ex. mixed l_p and Lorentz $l_{p,r}$) in various anisotropic Besov, Sobolev and Lizorkin–Triebel spaces of functions defined on open subsets of $\mathbb{R}^n$ are presented in Chapter [23].*

Definition 2.4. *Independently generated spaces*
Let $\mathcal{S}$ be a set of ideal (quasi–Banach) spaces, such that every element $Y \in \mathcal{S}$ is either a sequence space $Y = Y(I)$ with a finite or countable I, or a space $Y = Y(\Omega)$, where (Ω, μ) is a measure space with a countably additive measure μ without atoms.

By means of the leaf growing process (step) from some $Y \in \mathcal{S}$, we shall call the substitution of Y with:

(*Type A*) either $Y(I, \{Y_i\}_{i \in I})$ for some $\{Y_i\}_{i \in I} \subset \mathcal{S}$ if $Y = Y(I)$, or
(*Type B*) $Y(\Omega, Y_0)$ for some $Y_0 \in \mathcal{S}$ if $Y = Y(\Omega)$.

Here the quasi–Banach space $Y(I, \{Y_i\}_{i \in I})$ is the linear subset of $\prod_{i \in I} Y_i$ of the elements $\{x_i\}_{i \in I}$ with the finite quasi-norm $\|\{x_i\}_{i \in I}|Y(I, \{Y_i\}_{i \in I})\| := \|\{\|x_i\|_{Y_i}\}_{i \in I}\|_Y$.

Note that a type B leaf (i.e., of the form $Y = Y(\Omega)$) can grow only one leaf of its own.

We shall also refer to either $\{Y_i\}_{i \in I}$, or Y_0 as to the leaves growing from Y, which could have been a leaf itself before the tree growing process. Let us designate by means of $IG(\mathcal{S})$ the class of all spaces obtained from an element of $\mathcal{S}$ in a finite number of the tree growing steps consisting of the tree growing processes for some or all of the current leaves.

Thus, there is a one–to–one correspondence between $IG(\mathcal{S})$ and the trees of the finite depth with the vertexes from $\mathcal{S}$, such that every vertex of the form $Y(I)$ has at most I branches and every vertex of the form $Y(\Omega)$ has at most one branch. The tree corresponding to a space $X \in IG(\mathcal{S})$ is designated by $T(X)$. The appearance of the space forming the root of $T(X)$ is counted as the first step of the tree–growing process. The minimal number of steps necessary to "grow" X is designated by $N_{\min}(X)$.

The set of all vertexes (elements of $\mathcal{S}$) of the tree corresponding to some $X \in IG(\mathcal{S})$ will be denoted by means of $\mathcal{V}(X)$.

We shall always assume that the generating set $\mathcal{S}$ of $IG(\mathcal{S})$ is minimal in the sense that there does not exist a proper subset $\mathcal{Q} \subset \mathcal{S}$, such that $\mathcal{S} \subset IG(\mathcal{Q})$.

If the set $\mathcal{S}$ includes only the spaces described by (different) numbers of parameters from $[1, \infty]$ and $X \in IG(\mathcal{S})$, we assume that $I(X)$ is the set of all the parameters of the spaces at the vertexes of the tree T corresponding to X and

$$p_{\min}(X) := \inf I(X) \text{ and } p_{\max}(X) := \sup I(X).$$

For the sake of brevity, we also assume that

$$IG := \{X \in IG(l_p, L_p, lt_{p,q}, lt^*_{p',q'}) : [p_{\min}(X), p_{\max}(X)] \subset (1, \infty)\} \text{ and}$$
$$IG_0 := IG \cap IG(l_p, L_p).$$

We say that two IG–spaces are of the same tree type if their trees are congruent and the spaces at the corresponding vertexes are both either l_p–spaces, or $L_p(\Omega)$–spaces on two measure spaces (Ω, μ_0) and (Ω, μ_1) with their (non–negative) measures μ_0 and μ_1 being absolutely continuous with respect to σ–additive and σ–finite a measure μ on Ω, or $lt_{p,q}$–spaces, or $lt^*_{p,q}$–spaces.

It is convenient to think about the parameters as parameter functions $p = p_X : P \to [1, \infty]$ defined on one and the same parameter position set $P = \mathcal{V}(X)$ (here we slightly abuse the notation in the sense that the vertexes that are classes $lt_{p,q}$ or $lt^*_{p,q}$ are doubled to cover both their parameters, or that the value of p on them is 2–dimensional with the vector operations as in Section 3.2) determined by T and the spaces at its vertexes.

Every IG space X can be interpreted as a space of functions defined on some set $\Omega = \Omega(X)$ that is obtained by (repeated) combinations of the operations of taking sums and Cartesian products from a family of measurable spaces and index sets.

We also assume that an abstract L_p is an IG space with $I(L_p) = \{p\}$.

Let us also define the class IG_+. A space X belongs to IG_+ if it is either in IG, or obtained from a space $X_- \in IG$ by means of the leaf growing process, in the course of which some leaves (or just one leaf) of X_- have grown some new leaves from the set

$$\{S_p, S_p^n, L_p(\mathcal{M}, \tau)\}_{p \in [1,\infty]}^{n \in \mathbb{N}},$$

where $(\mathcal{M}, \tau)$ is a semifinite von Neumann algebra and their simplest versions — infinite and finite–dimensional Schatten–von Neumann classes S_p, S_p^n (see Section 2.2 below). The set of the parameters p of these "last noncommutative leaves" is included into $I(X)$, and they are part of the corresponding tree $T(X)$. Let also

$$IG_{0+} = \{Y \in IG_+ : Y_- \in IG_0\}.$$

Remark 2.3. a) We shall deal with the set $\{l_p, L_p, lt_{p,q}, lt_{p',q'}^*\}$, where l_p, L_p, $lt_{p,q}$ and $lt_{p',q'}^*$ designate, respectively, the classes $\{l_p(I)\}_{p \in [1,\infty]}$, $\{L_p(\Omega)\}_{p \in [1,\infty]}$ for all the Ω with some countably additive and not purely atomic measure μ on it, $\{lt_{p,q}\}_{p,q \in [1,\infty]}$ and $\{lt_{p',q'}^*\}_{p,q \in [1,\infty]}$.

b) The subclass of l_p–spaces can formally be excluded from the definition of the class IG because l_p is isometric to $lt_{p,p} = lt_{p,p}^*$ but is left there for the sake of technical convenience. The subclass of $lt_{p,q}^*$–spaces is included to make IG closed with respect to passing to dual spaces.

c) The Lebesgue or sequence spaces with mixed norm and the l_p–sums of them are particular elements of $IG(l_p, L_p)$.

d) Two $IG(\mathcal{S})$ spaces X and Y form a compatible couple of Banach spaces if their trees are congruent and the spaces standing at the corresponding vertexes form compatible couples themselves, that is when $T(X) = T(Y)$ (meaning, particularly, the same domain of the parameter functions $P = \mathcal{V}(X) = \mathcal{V}(Y)$).

To sidestep the fact that $lt_{p,q}$ is not a 1–complemented subspace in $L_p(\mathbb{R}^n, l_q(\mathbb{N}))$, we shall need the following structural observation.

Lemma 2.3 ([16]). *Let $X \in IG_+$ ($X \in IG$). Then there exists $Y \in IG_{0+}$ ($Y \in IG_0$) with $I(X) = I(Y)$ and $T(X) \subset T(Y)$, such that X is a complemented subspace of a quotient Y/Z, where Z is a complemented subspace of Y. Moreover, if $lt_{p,q}^* \notin T(X)$, then X is a complemented subspace of Y.*

The proof of this lemma [16] relies on the fact, established by J. Bourgain, that the Fefferman–Stein inequality remains valid in the setting of UMD–valued functions.

2.2. Noncommutative Spaces

The related definitions, examples and all needed information about the noncommutative $L_p(\mathcal{M}, \tau)$, where $(\mathcal{M}, \tau)$ is a von Neumannn algebra with a normal semifinite faithful trace τ, and the more general Haagerup $L_p(\mathcal{M}) = L_p(\mathcal{M}, \phi)$, where $(\mathcal{M}, \phi)$ is a von Neumannn algebra with a normal semifinite faithful weight ϕ (that exists for every von Neumann algebra $\mathcal{M}$), are found in the open access [16] and in this section. Apart from Lemma 2.4 below on the existence of 1–complemented isometric copies of l_p in every infinite-dimensional noncommutative L_p, it is adapted from [27]. Typical examples of the noncommutative $L_p(\mathcal{M}, \tau)$ are the infinite–dimensional S_p and the finite–dimensional S_p^n

Schatten–von Neumann classes with the classical trace. The only knowledge needed for us to apply and relate Theorems 2.1 and 2.2 below is that a finite trace is semifinite, as well as that a von Neumann algebra $\mathcal{M}$ is said to be semifinite (finite) if it admits a normal semifinite (finite) faithful (n. s.(f.) f.) trace.

Examples 2.1 ([28]). *Let H_1, H_2 be Hilbert spaces and $p \in [1, \infty)$. The Schatten–von Neumann class $S_p = S_p(H_1, H_2)$ is the Banach space of all compact operators $A \in \mathcal{L}(H_1, H_2)$ with the finite norm*

$$\|A\|_{S_p} = \|A|S_p(H_1, H_2)\| := \left(tr(A^*A)^{p/2}\right)^{1/p}.$$

The class $S_\infty(H_1, H_2)$ is the space of all compact operators with the norm inherited from $\mathcal{L}(H_1, H_2)$. The trace of a projector P, $\tau(P) = tr(P) = \dim(\mathrm{Im}\, P)$ corresponds to the counting measure and

$$S_p(H_1, H_2) = L_p\left(\mathcal{L}(H_1, H_2), tr\right) \text{ and } S_p = S_p(H_1, H_1) \text{ for } \dim(H_1) = \infty.$$

This trace is normal, semifinite and faithful. When $\dim(H_1) = \dim(H_2) = n < \infty$, the elements of the Schatten–von Neumann classes can be represented by matrixes with a dense subset of the invertible matrixes, and the trace is finite. In this case we designate $S_p^n = S_p(H_1, H_2)$.

Theorem 2.1 ([29, 30]). *Let $\mathcal{M}$ be a semifinite von Neumann algebra, $p_0, p_1 \in [1, \infty]$, $\theta \in (0, 1)$ and $1/p = (1 - \theta)/p_0 + \theta/p_1$. Then*

a) $(L_{p_0}(\mathcal{M}, \tau), L_{p_1}(\mathcal{M}, \tau))_{[\theta]} = L_p(\mathcal{M}, \tau)$ (isometry);

b) $(L_{p_0}(\mathcal{M}, \tau), L_{p_1}(\mathcal{M}, \tau))_{\theta, p} = L_p(\mathcal{M}, \tau)$ (isomorphism).

The following theorem permits to reduce the study of the local properties of the Haagerup L_p–spaces to checking them for the L_p spaces of finite von Neumann algebras.

Theorem 2.2 (Haagerup, see [27]). *Let $\mathcal{M}$ be a von Neumannn algebra with a normal semifinite faithful weight ϕ, $p \in (0, \infty)$, and let $L_p(\mathcal{M}, \phi)$ be the associated Haagerup L_p-space. Then there are a $\min(p, 1)$–Banach space X, a directed family $\{(\mathcal{M}_i, \tau_i)\}_{i \in I}$ of finite von Neumann algebras and a family $\{J_i\}_{i \in I}$ of isometric embeddings $J_i : L_p(\mathcal{M}_i, \tau_i) \hookrightarrow X$ satisfying*

1) $\mathrm{Im}\, J_i \subset \mathrm{Im}\, J_{i'}$ *for all i, i' with $i \leq i'$;*

2) $\bigcup_{i \in I} \mathrm{Im}\, J_i$ *is dense in X;*

3) $L_p(\mathcal{M}, \phi)$ *is isometric to a subspace of X, complemented if $p \in [1, \infty)$.*

We finish this section with the counterpart of Lemma 2.2 for Schatten–von Neumann classes S_p and general $L_p(\mathcal{M})$, where $\mathcal{M}$ is a von–Neumann algebra with a normal semifinite faithful weight (that always exists).

Lemma 2.4 ([16]). *For $p \in [1, \infty]$, the space S_p contains 1–complemented isometric copies of S_p^n, $l_p(I_n)$ and l_p for $n \in \mathbb{N}$. Moreover, an infinite–dimensional $L_p(\mathcal{M})$ contains a 1–complemented isometric copy of l_p.*

3. CLASSES OF BESOV–NIKOL'SKII, LIZORKIN–TRIEBEL AND SOBOLEV SPACES

As mentioned in Introduction, our approach to the asymmetric uniform convexity and smoothness is general enough to cover various classes of (anisotropic) spaces of functions defined on an open subset of an Euclidean space or a Riemannian manifold, as well as their duals, viewed as spaces of negative smoothness containing weak solutions in PDE, their l_p–sums, such as, for example, the l_p–sum of the range space of the (pseudo)differential operator and the initial and boundary value spaces (the ranges of the trace operators) for a given PDE or ΨDE, and the vector–valued versions of function spaces (for example, in stochastic PDE). Moreover, the spaces of variable smoothness [1] (particularly, weighted ones) are treated in the same manner.

Instead of considering a generic space as in Definition 2.4, here we only outline the basic groups of classes of functions spaces and explain the role of equivalent geometrically friendly norms [31] and their relation to our auxiliary IG above (the basic properties are in [32, 33, 34, 35, 36, 18]).

The parameters of the spaces are in the reflexive/UMD ranges [31, 23]: $p, a \in (1, \infty)^n$ and $q, \varsigma \in (1, \infty)$. If the underlying mixed Lebesgue L_p–norm is substituted with a geometrically friendly mixed Lorentz $L_{p,\theta}$–norm, one also has $\theta \in (1, \infty)$ but, in this chapter, we do not establish explicit estimates sharper than Lemma 2.1 in [14] in this setting.

3.1. Abstract Form and Connections between Function and IG Spaces

The homogeneous (seminormed) versions of spaces are defined as follows. Given a linear topological space W, a (quasi) Banach space Y and an injective linear operator $A : W \supset D(A) \to Y$, let $z_{A,Y}$ be the completion of the linear space $\{x \in W : Ax \in Y\}$ endowed with the (quasi) norm $\|x|z_{A,Y}\| := \|Ax\|_Y$. If $\operatorname{Ker} A \neq \{0\}$ is closed, we use $W/\operatorname{Ker} A$ instead of W. We note that, according to this definition, $z_{A,Y}$ is isometric to the closure $\overline{\operatorname{Im} A}^Y$ of the image of A in Y.

The ordinary (normed) spaces are endowed with the graph norms following Sergey L. Sobolev's idea [32]. Given $\varsigma \in (0, \infty]$, Banach spaces X and Y and a closed linear operator $A \in \mathcal{C}(X, Y)$, let $Z_{A,X,Y} = Z_{A,X,Y,\varsigma}$ be the Banach linear space $D_X(A) = D(A) \cap X$ endowed with the norm

$$\|x|Z_{A,X,Y}\|^\varsigma := \|x\|_X^\varsigma + \|Ax\|_Y^\varsigma.$$

Here the term $\|Ax\|_Y$ defines the corresponding seminormed space. Particular examples of A are given in Chapter [23]. Note that the completeness of $Z_{A,X,Y}$ is equivalent to the closedness of A.

The exact operators A and the spaces X and Y that are (quotients of) IG–spaces for every spaces under consideration are found in Section 2.2 in Chapter [23]. It means that every function space under consideration and its dual, is either a subspace or a subspace of a quotient of an IG–space, and the same is true for their l_p–sums and vector–valued versions because IG–class is closed with respect to these operations.

3.2. Group Γ_1: Anisotropic Besov and Lizorkin–Triebel Spaces with Intrinsic Norms

This group includes, correspondingly, the anisotropic Besov and Lizorkin–Triebel spaces defined in terms of the (locally) averaged differences $B^s_{p,q,a}(G)$ and $L^s_{p,q,a}(G)$, in terms of the local polynomial approximations ($\mathcal{D}$–functional) $\tilde{B}^{s,A}_{p,q,a}(G)$ and $\tilde{L}^{s,A}_{p,q,a}(G)$ and the classical anisotropic Besov spaces $B^s_{p,q}(G)$ [34] defined in terms of the ordinary differences, as well as their homogeneous versions and duals (see Section 2.1 in [23]). The dependence (including equivalent norms) on a, the set of the allowed polynomial exponents $A \subset \mathbb{N}_0^n$ and G is studied in [31]. As shown in [31], the straightforward definition of $L^s_{p,q}(G)$ for a vide class of (not necessarily extension) domains G gives an equivalent norm in the corresponding Lizorkin–Triebel space [34] if the smoothness in every direction exceeds 1. As will be seen from what follows, the factor spaces of the corresponding seminormed spaces by their null spaces share their properties with their respective normed versions mentioned above.

We shall need a rather technical definition of the admissibility of the auxiliary summability parameter $a \in (1,\infty)^n$ explained in detail in [23] and refrence therein. Let $\mathbb{R}^n$ be the n–dimensional Euclidean space with the standard basis $\langle e^1,\ldots,e^n \rangle$, $x = (x_i) = (x_1,\ldots,x_n) = \sum_1^n x_i e^i = (x_i)$; $x_{\min} := \min_i x_i$ and $x_{\max} := \max_i x_i$. For instance, $\bar{1} = (1) = (1,\ldots,1) \in \mathbb{N}^n$. Recall also that p' is the conjugate to $p \in [1,\infty]^n$, i.e., $1/p'_i + 1/p_i = 1$ ($1 \leq i \leq n$). For $x, y \in \mathbb{R}^n$, $t > 0$, let $[x,y]$ be the segment in $\mathbb{R}^n$ with the ends x and y; $xy = (x_i y_i)$; $x/y = \left(\frac{x_i}{y_i}\right)$ and $z/y = z\bar{1}/y$ for $z \in \mathbb{R}$ and $y_i \neq 0$ for $1 \leq i \leq n$. Recall also the notation for the scalar product (x,y) of $x,y \in \mathbb{R}^n$: $(x,y) = \sum_{i=1}^n x_i y_i$.

In what follows, one assumes that the anisotropy vector $\gamma_a = ((\gamma_a)_1,\ldots,(\gamma_a)_n) \in (0,\infty)^n$ possesses the normalisation $|\gamma_a| = \sum_1^n (\gamma_a)_i = n$ and related to the anisotropic smoothness vector $s \in (0,\infty)^n$ and the adjusted smoothness $s_* > 0$ by the relation $s = s_*/\gamma_a$.

Definition 3.1. *We say that the parameter a is admissible or in the admissible range for $Y \in \Gamma_1$ if either*

$$Y \in \left\{ B^s_{p,q,a}(G),\ L^s_{p,q,a}(G),\ b^s_{p,q,a}(G),\ l^s_{p,q,a}(G) \right\}$$

and $s_* > (\gamma_a(1/p - 1/a))_{\max}$, or

$$Y \in \left\{ \tilde{B}^{s,A}_{p,q,a}(G),\ \tilde{L}^{s,A}_{p,q,a}(G),\ \tilde{b}^{s,A}_{p,q,a}(G),\ \tilde{l}^{s,A}_{p,q,a}(G) \right\}$$

and $s_* > (\gamma_a, 1/p - 1/a)$, or

$$Y \in \left\{ B^s_{p',q',a'}(G)^*,\ L^s_{p',q',a'}(G)^*,\ b^s_{p',q',a'}(G)^*,\ l^s_{p',q',a'}(G)^* \right\}$$

and $s_* > -(\gamma_a(1/p - 1/a))_{\max}$, or

$$Y \in \left\{ \tilde{B}^{s,A}_{p',q',a'}(G)^*,\ \tilde{L}^{s,A}_{p',q',a'}(G)^*,\ \tilde{b}^{s,A}_{p',q',a'}(G)^*,\ \tilde{l}^{s,A}_{p',q',a'}(G)^* \right\}$$

and $s_* > -(\gamma_a, 1/p - 1/a)$.

3.3. Group Γ_2: Anisotropic Sobolev Spaces

This group includes the classical Sobolev space $W_p^s(G)$, its homogeneous version $w_p^s(G)$ and their duals. The geometrically friendly norm defined in terms of the Sobolev generalised derivatives $D_i^{s_i} f$ is

$$\|f|W_p^s(G)\|^\varsigma := \|f|L_p(G)\|^\varsigma + \|f|w_p^s(G)\|^\varsigma = \|f|L_p(G)\|^\varsigma + \sum_{i=1}^n \|D_i^{s_i} f|L_p(G)\|^\varsigma,$$

where we assume the reflexive (and UMD) range [23] of $p \in (1, \infty)^n$ and $\varsigma \in (1, \infty)$ but the optimal geometric properties are brought by choosing $\varsigma \in [p_{\min}, p_{\max}]$.

3.4. Group Γ_3: Anisotropic Besov and Lizorkin–Triebel Spaces with Wavelet Norms

This group includes the anisotropic Besov $B_{p,q}^s(\mathbb{R}^n)_w$ and Lizorkin–Triebel $L_{p,q}^s(\mathbb{R}^n)_w$ spaces sufficiently described for our purposes in Remark 2.1 above and explicitly defined in [19, 23] and their duals. Their traditional counterparts [18] based on the dyadic decompositions possess the same properties.

3.5. Group Γ_4: Anisotropic Besov and Lizorkin–Triebel Spaces via Closed Operators

This group includes both the anisotropic Besov $B_{p,q}^s(G)_\mathcal{F}$ and Lizorkin–Triebel $L_{p,q}^s(G)_\mathcal{F}$ spaces, explicitly defined in Definition 2.10 in [23], and their duals. This group is very large and contains such popular classes as the spaces defined by Sergey M. Nikol'skii [33] in terms of the entire functions of exponential type (as the ranges of Fourier multipliers), the spaces defined in terms of more general Littlewood–Paley decompositions introduced by Peter I. Lizorkin ($q = 2$) and Hans Triebel [35, 18] and the Besov and Lizorkin–Triebel spaces defined in terms of a holomorphic functional calculus of a closed operator. Note that the "domain" versions of the spaces introduced by Lizorkin and Triebel are defined as restrictions and are actually subspaces of the corresponding classes defined in terms of intrinsic norms.

4. ASYMMETRIC UNIFORM CONVEXITY AND SMOOTHNESS: QUANTITATIVE PROPERTIES

4.1. Convex and Smooth Functions Defined on Abstract Banach Spaces

Since the ordinary (qualitative) convexity of a function F defined on a convex set G is equivalent to its midpoint convexity $2F\left(\frac{x+y}{2}\right) \leq F(x) + F(y)$ for $x, y \in G$, the most well–studied type of the quantitative/uniform convexity (and smoothness) is the midpoint type. The importance of the complete asymmetric uniform convexity was highlighted by S. M. Nikol'skii [33, Section 1.3.5], and Yu. N. Chekanov, Yu. E. Nesterov and A. A. Vladimirov [37] started the systematic study of the uniform convexity of the convex functions on Banach spaces evaluated by the modulus of the "double" uniform convexity $\delta(t, F) = \inf_{\lambda \in (0,1), x, y \in X, \|x-y\|=t} \omega(\lambda, x, y)$, where

$$\omega(\lambda, x, y) = \frac{\lambda F(x) + (1-\lambda)F(y) - F(\lambda x + (1-\lambda)y)}{\lambda(1-\lambda)}.$$

In this section we introduce the analogous modulus of the double uniform smoothness and investigate the properties (including the dependence on λ and the counterparts of the Fenchel–Young–Lindenstrauss duality) of the moduli of the asymmetric uniform convexity and smoothness defined, correspondingly, as $\delta_\lambda(t, F) = \inf_{x,y \in X} \omega(\lambda, x, y)$ and $\rho_\lambda(t, F) = \sup_{x,y \in X, \|x-y\|=t} \omega(\lambda, x, y)$. In particular, this leads also to the duality theory and estimates for the "double" moduli. Results of this section are extensively used in the other sections.

Throughout Section 4.1, we assume that F is a convex real–valued function with the convex domain $\mathrm{D}(F)$ in a Banach space X with the norm $\|\cdot\|_X$, and that $f_x \in \partial F(x)$ (a subgradient) for every x from the interior of $\mathrm{D}(F)$. For example, f_x is defined by $f_x(x) = \|x\|_X$ and $\|f_x\|_{X^*} = 1$ if $F = \|\cdot\|_X$. Since the majority of our estimates will be modulo the affine functions of the form $f(x) + c$, we may often assume without loss of generality that $F(0) = 0$ and $F \geq 0$ (i.e., $0 \in \partial F(0)$).

Definition 4.1. *For $1 - \nu = \mu \in (0, 1)$, the moduli of the uniform μ–convexity and uniform μ–smoothness of a function $F : X \supset \mathrm{D}(F) \to \mathbb{R}$ are the functions*

$$\delta_\mu(t, F, X) = (\mu\nu)^{-1} \inf\{\mu F(x) + \nu F(y) - F(\mu x + \nu y) : x, y \in \mathrm{D}(F), \|x - y\|_X = t\},$$
$$\rho_\mu(t, F, X) = (\mu\nu)^{-1} \sup\{\mu F(x) + \nu F(y) - F(\mu x + \nu y) : x, y \in \mathrm{D}(F), \|x - y\|_X = t\}$$

correspondingly defined, together with their upper and lower estimates, on $[0, d\,(\mathrm{D}(F)))$, where $d\,(\mathrm{D}(F))$ is the diameter of $\mathrm{D}(F)$.

Let $p, q \in [1, \infty]$ and h_c, h_s be non–negative functions defined on $(0, 1)$. For $\mu \in (0, 1)$, we say that a function F is (p, μ)–uniformly convex $((q, \mu)$–uniformly smooth$)$ if, for some $C_s, C_c > 0$, $\delta_\mu(t, F, X) \geq C_c t^p$ $(\rho_\mu(t, F, X) \leq C_s t^q)$.

We also say that a function F is (p, h_c)–uniformly convex $((q, h_s)$–uniformly smooth$)$ if

$$\delta_\mu(t, F, X) \geq h_c(\mu) t^p \ (\rho_\mu(t, F, X) \leq h_s(\mu) t^q) \ \textit{for all } \mu \in (0, 1).$$

Since $\delta_\mu(t, F, X) = \delta_\nu(t, F, X)$ and $\rho_\mu(t, F, X) = \rho_\nu(t, F, X)$, we can and always assume that $h_c(\mu) = h_c(\nu)$ and $h_s(\mu) = h_s(\nu)$. For the sake of convenience, we also assume that

$$h_c(0) = \lim_{\mu \to 0} h_c(\mu) \textit{ and } h_s(0) = \lim_{\mu \to 0} h_s(\mu)$$

if the corresponding limit exists. We also define the double moduli

$$\delta(t, F, X) := \inf_{\mu \in (0,1)} \delta_\mu(t, F, X) \textit{ and } \rho(t, F, X) := \sup_{\mu \in (0,1)} \rho_\mu(t, F, X).$$

In the case of a Hilbert space H, we have, for example,

$$\delta_\mu\left(t, \|\cdot\|_H^2, H\right) = \delta\left(t, \|\cdot\|_H^2, H\right) = \rho_\mu\left(t, \|\cdot\|_H^2, H\right) = \rho\left(t, \|\cdot\|_H^2, H\right) = t^2.$$

Remark 4.1. *a) Note that the definitions of our moduli are quite natural because they are just the best majorants and minorants of the divided differences $[0, \lambda, 1]\phi$ for $\phi(t) = F(tx + (1-t)y)$.*
b) According to the definition, we always have $0 \leq \delta_\mu(t, F, X) \leq \rho_\mu(t, F, X) \leq 2\|F|L_\infty(D(F))\|$. Noticing that, for $x', y', z \in X$ satisfying $z = \mu x + \nu y = \mu x' + \nu y'$ and $[x, y] \subset [x', y']$, one has $(F(x') - F(z))/\|x' - y'\|_X \geq (F(x) - F(z))/\|x - y\|_X$,

$(F(y') - F(z))/\|x' - y'\|_X \geq (F(y) - F(z))/\|x - y\|_X$, *and, therefore,*

$(\mu F(x') + \nu F(y') - F(z))/\|x' - y'\|_X \geq (\mu F(x) + \nu F(y) - F(z))/\|x - y\|_X$,

we see that $\rho_\mu(\cdot, F, X)$ is convex and $\delta_\mu(\cdot, F, X)/t$ is increasing on $[0, \infty)$ if $D(F) = X$.
c) It follows from Part h) of the next lemma that a function F is (p, μ_0)–uniformly convex ((q, μ_0)–uniformly smooth) for some particular $\mu \in (0, 1)$ if, and only if, it is (p, h_c)–uniformly convex ((q, h_s)–uniformly smooth) with $h_c(\mu_0) \geq C_c$ ($h_s(\mu_0) \leq C_s$). Thus, we prefer to work with (p, h_c)–uniform convexity and (q, h_s)–uniform smoothness as providing sharper information.

The following lemma contains the basic properties of the moduli defined above and the first general application, a reverse triangle inequality. Let us recall that, when F is defined on X, the Fenchel–Young transform F^* of F is defined by

$$F^*(f) = \sup_{x \in X} f(x) - F(x), \text{ and } F^*(f_x) + F(x) = f_x(x). \quad (FY)$$

Parts c) and i) of the next key lemma are the first applications added as sample "advertisement" of few other parts. The former anticipates Section 4.4 and uses the insight $q \leq 2$, particularly, following from Part g) in the setting of Section 4.4.1. Indeed, Part g) is an alternative approach, working even in our more general setting, to the beautiful asymptotic behavior of the classical moduli of uniform smoothness and convexity [2, Page 63] (see also Section 4.4.2 relating our moduli with the classical ones).

Lemma 4.1. *For a Banach space X and a convex function real–valued F defined on convex $D(F) \subset X$, let $f_z \in \partial F(z)$ for $z \in \text{int}\, D(F)$.*
a) For $x, y \in \text{int}\, D(F)$, we have

$$\delta_\mu(\|x - y\|_X, F, X) \leq (\mu\nu)^{-1} (\mu F(x) + \nu F(y) - F(\mu x + \nu y)) \leq \langle f_x - f_y, x - y \rangle \text{ and}$$
$$\delta_\mu(\|f_x - f_y\|_{X^*}, F^*, X^*) \leq (\mu\nu)^{-1} (\mu F^*(f_x) + \nu F^*(f_y) - F^*(\mu f_x + \nu f_y)) \leq$$
$$\leq \langle f_x - f_y, x - y \rangle.$$

b) For $\xi > 0$, $1 - \nu = \mu \in (0, 1)$ and $B(\xi\nu\|x - y\|_X, x) \cup B(\xi\mu\|x - y\|_X, y) \subset \text{int}\, D(F)$, we have
$$\|f_x - f_y\|_{X^*} \|x - y\|_X \leq \xi^{-1} \rho_\mu(\|x - y\|(1 + \xi), F, X) - \xi^{-1} \delta_\mu(\|x - y\|_X, F, X).$$
c) For a (q, μ)–uniformly smooth $F = \|\cdot\|_X^q$ in b) with $1 < q = 1 + 1/\xi \leq 2$ and $x, y \in S_X$ and $g_x, g_y \in S_{X^*}$ with $g_x(x) = g_y(y) = 1$, we have
$$\|g_x - g_y\|_{X^*} \leq q^{q/q'} C_s(\|x - y\|_X)^{q-1} - q'^{-1} \delta_\mu(\|x - y\|_X, F, X) \leq 2C_s\|x - y\|_X^{q-1}.$$
d) For $x \in \text{int}\, D(F)$ and $x \pm y \in D(F)$, we have

$$\max(\mu, \nu)\delta_\mu(\|y\|_X, F, X) \leq F(x + y) - F(x) - f_x(y) \leq$$
$$\leq \max(\mu, \nu)\rho_\mu(\|y\|_X/\max(\mu, \nu), F, X).$$

e) For $\mathrm{D}(F) = X$ and $t > 0$, we have

$$\rho_\mu(t, F^*, X^*) = \delta_\mu^*(t, F, X) \text{ and } \rho_\mu(t, F, X) = \delta_\mu^*(t, F^*, X^*) \text{ and, thus,}$$
$$\rho_\mu^*(t, F^*, X^*) \leq \delta_\mu(t, F, X) \text{ and } \rho_\mu^*(t, F, X) \leq \delta_\mu(t, F^*, X^*).$$

f) For $\mathrm{D}(F) = X$ and $t > 0$, we have

$$\rho(t, F^*, X^*) = \delta^*(t, F, X) \text{ and } \rho(t, F, X) = \delta^*(t, F^*, X^*) \text{ and, thus,}$$
$$\rho^*(t, F^*, X^*) \leq \delta(t, F, X) \text{ and } \rho^*(t, F, X) \leq \delta(t, F^*, X^*).$$

g) For $\mathrm{D}(F) = X$, the functions $\delta_\mu(t, F, X)/t$, $\delta(t, F, X)/t$, $\rho_\mu(t, F, X)/t$ and $\rho(t, F, X)/t$ are nondecreasing on $(0, \infty)$, while the functions $\delta(t, F, X)/t^2$ and $\rho(t, F, X)/t^2$ are, respectively, nondecreasing and nonincreasing on $(0, \infty)$.

h) For $t > 0$, $\mu, \alpha \in (0, 1)$, and $\mu + \nu = \alpha + \beta = 1$ one has the estimates:

$$\max(\mu, \nu)\delta_\mu(t, F, X) \leq \max(\alpha, \beta)\delta_\alpha(t, F, X) \text{ and}$$
$$\min(\mu, \nu)\rho_\mu(t, F, X) \geq \max(\alpha, \beta)\rho_\alpha(t, F, X) \text{ for } \min(\alpha, \beta) \leq \min(\mu, \nu);$$
$$\min(\mu, \nu)\delta_\mu(t, F, X) \leq \min(\alpha, \beta)\delta_\alpha(t, F, X) \text{ and}$$
$$\max(\mu, \nu)\rho_\mu(t, F, X) \geq \min(\alpha, \beta)\rho_\alpha(t, F, X) \text{ for } \min(\alpha, \beta) \geq \min(\mu, \nu);$$
$$c_\mu \delta_\mu(t, F, X) \leq \delta(t, F, X) \text{ and } \rho(t, F, X) \leq c_\mu \rho_\mu(t/c_\mu, F, X)$$
$$\text{for } c_\mu = 4 \min(1/4, \mu, 1 - \mu) \max(\mu, 1 - \mu).$$

i) For a normed X and every $x, y \in X$, we have a reverse triangle inequality

$$\|x\|_X + \|y\|_X - \|x + y\|_X \leq 2 \inf_{z \in [x, -y]} \|z\|_X \leq 2 \frac{\|\|y\|_X x - \|x\|_X y\|_X}{\|x\|_X + \|y\|_X}.$$

k) If Y is finitely represented in X, then one has, for every $t > 0$ and convex f,

$$\delta_\mu(t, f(\|\cdot\|_X), X) \leq \delta_\mu(t, f(\|\cdot\|_Y), Y) \text{ and}$$
$$\rho_\mu(t, f(\|\cdot\|_X), X) \geq \rho_\mu(t, f(\|\cdot\|_Y), Y).$$

l) For $t > 0$, $\mu \in (0, 1)$ and a family $\{F_i\}_{i \in I}$ with common domain $\Omega = \mathrm{D}(F_i) \subset X$, one has, for $I = \mathbb{N}$:

$$\delta_\mu\left(t, \sup_{i \in I} F_i, X\right) \geq \inf_{i \in I} \delta_\mu(t, F_i, X), \rho_\mu\left(t, \inf_{i \in I} F_i, X\right) \leq \sup_{i \in I} \rho_\mu(t, F_i, X);$$
$$\delta_\mu\left(t, \limsup_{i \to \infty} F_i, X\right) \geq \liminf_{i \to \infty} \delta_\mu(t, F_i, X), \rho_\mu\left(t, \liminf_{i \to \infty} F_i, X\right) \leq \limsup_{i \to \infty} \rho_\mu(t, F_i, X).$$

Remark 4.2. a) The relation $x \in \partial F^*(f_x)$ shows that the inequalities in a) and b) remain true if we substitute x, y, F and X with f_x, f_y, F^* and X^* correspondingly.

b) A stronger analog of a) with $X = L_p$, $F(x) = \|x\|_{L_p}^p$ and $\mu = 1/2$ is valid in L_p with $p \geq 2$ thanks to the Chernykh inequality [38]:

$$2^{2-p}|a - b|^p \leq |a|^p + |b|^p - ab(|a|^{p-2} + |b|^{p-2}).$$

Asymmetric Convexity and Smoothness

c) Pichugov [39] noticed and utilised the case of Part b) with $F(x) = \|x\|_X$:

$$\|x\|_X + \|y\|_X - f_x(y) - f_y(x) = \langle f_x - f_y, x - y \rangle \leq 2\|x - y\|_X.$$

d) Concerning the dependence on μ, the estimate $\delta_{1/2}(t, F, X) \leq 2\delta(t, F, X)$ established in [37] and Part h) demonstrate that it is enough to work with the middle point convexity and smoothness looking for rough estimates.

e) Part h) also shows that the moduli depend continuously on μ.

f) Part h) implies, in particular, that, for $\mu \in (0, 1)$, every μ–uniformly convex (smooth) space is midpoint–uniformly convex (smooth) implying the reflexivity of these spaces thanks to the classical results due to D. P. Milman and V. L. Shmul'yan.

g) Part i) is a reverse triangle inequality.

h) Part l) shows that the class of (p, h_c)–uniformly convex functions is closed with respect to the operations of taking "sup" and "lim sup", while the class of (q, h_s)–uniformly smooth functions is closed with respect to the operations of taking "inf" and "lim inf".

Proof of Lemma 4.1. We start by noticing the following general convexity estimates

$$\mu\nu\delta_\mu(\|x-y\|_X, F, X) - (\mu F(x-\nu z) - F(\mu x + \nu y) + \nu F(y + \mu z)) \leq$$
$$\leq \mu(F(x) - F(x - \nu z)) + \nu(F(y) - F(y + \mu z)) \leq \mu\nu\langle f_x - f_y, z\rangle \leq$$
$$\leq \xi^{-1}\mu(F(x + \nu\xi z) - F(x)) + \xi^{-1}\nu(F(y - \mu\xi z) - F(y)) \leq$$
$$\leq \xi^{-1}\mu\nu\rho_\mu(\|x - y + \xi z\|_X, F, X) - \xi^{-1}\mu\nu\delta_\mu(\|x - y\|_X, F, X). \quad (1)$$

Now we obtain a) and b) by means of choosing, respectively, $z = x - y$ and $z \in \{z_k\}_{k=1}^\infty$ with

$$\|z_k\|_X = \|x - y\|_X \text{ and } \lim_k (f_x - f_y)(z_k) = \|f_x - f_y\|_{X^*}\|z_k\|_X \quad (2)$$

and noticing that $x \in \partial F^*(f_x)$ for $x \in X$.

To see that c) is a typical application of b), let us note that, for $F(x) = \|x\|_x^q$, one has $f_x = q\|x\|_X^{q-1} g_x$, and the maximum $h(q) = q'^{q/q'}$ of $(1 + \xi)^q/\xi$ on $(0, +\infty)$ is achieved at $\xi = q'/q$. In turn, the maximum 2 of $h(q) = (1 + 1/(q-1))^{q-1}$ on $(1, 2]$ is achieved at $q = 2$.

Eventually, we establish d), relying on the other convexity estimates

$$t^{-1}\mu\nu\delta_\mu(\|y\|_X, F, X) \leq t^{-1}(tF(x+y) - F(x+ty) + (1-t)F(x)) =$$
$$= F(x+y) - F(x) + t^{-1}(F(x) - F(x+ty)) \leq F(x+y) -$$
$$- F(x) - f_x(y) \leq F(x+y) - F(x) + s^{-1}(F(x-sy) - F(x)) \leq$$
$$\leq (1 + s^{-1})\left(\frac{s}{s+1}F(x+y) - F(x) + \frac{1}{s+1}F(x-sy)\right) \leq$$
$$\leq \mu\nu(1 + s^{-1})\rho(\|y\|_X(1+s), F, X), \quad (3)$$

where we take $t = \min(\mu, \nu) = s/(s+1)$.

Let us start the proof of Part e) with the inequalities

$$\rho_\mu(t, F^*, X^*) \leq \delta_\mu^*(t, F, X) \text{ and } \rho_\mu(t, F, X) \leq \delta_\mu^*(t, F^*, X^*). \quad (4)$$

For every $f, g \in X^*$ and $\varepsilon > 0$, there are $x, y \in X$ with the properties

$$F^*(f) \leq f(x) - F(x) + \varepsilon \text{ and } F^*(g) \leq g(x) - F(y) + \varepsilon.$$

Hence, one has the chain of inequalities:

$$\mu F^*(f) + \nu F^*(g) - F^*(\mu f + \nu g) \leq \varepsilon + \mu f(x) + \nu g(y) - \langle \mu f + \nu g, \mu x + \nu y \rangle -$$
$$- (\mu F(x) + \nu F(y) - F(\mu x + \nu y)) \leq \mu\nu\langle f - g, x - y \rangle - \mu\nu\delta_\mu(\|x - y\|_X, F, X) \leq$$
$$\leq \mu\nu\delta_\mu^*(\|x - y\|_X, F, X). \quad (5)$$

In a simpler manner, for every $x, y \in X$, there are $f \in \partial F(x)$ and $g \in \partial F(y)$. Thus, due to (FY), we have

$$\mu F(x) + \nu F(y) - F(\mu x + \nu y) \leq \mu f(x) + \nu g(y) - \langle \mu f + \nu g, \mu x + \nu y \rangle -$$
$$- (\mu F^*(f) + \nu F^*(g) - F^*(\mu f + \nu g)) \leq \mu\nu\langle f - g, x - y \rangle - \mu\nu\delta_\mu(\|f - g\|_{X^*}, F^*, X^*) \leq$$
$$\leq \mu\nu\delta_\mu^*(\|f - g\|_{X^*}, F^*, X^*). \quad (6)$$

Taking the supremum over all f, g with $\|f - g\|_{X^*} = t$ in (5) and over all x, y with $\|x - y\|_X = t$ in (6), we establish (4).

To approach the opposite inequalities, let us say that a pair $(x, y) \in X \times X$ is μ–described by the pair $(z, w) \in X \times X$ with $x - y = w$ and $\mu x + \nu y = z$.

First, for an arbitrary $\varepsilon > 0$, we find $s > 0$ and, then, $x, y \in X$, satisfying

$$\delta_\mu^*(t, F, X) \leq \varepsilon + ts - \delta_\mu(s, F, X), \text{ and}$$

$$\mu F(x) + \nu F(y) - F(\mu x + \nu y) - \mu\nu\varepsilon \leq \mu\nu\delta_\mu(s, F, X). \quad (7)$$

Now one chooses $u, v \in X^*$ with $u \in \partial F(\mu x + \nu y)$ and $\|v\|_{X^*} = t$, such that

$$F^*(u) = u(\mu x + \nu y) - F(\mu x + \nu y) \text{ and } v(x - y) = \|v\|_{X^*}\|x - y\|_X. \quad (8)$$

For the pair (f, g) μ–described by (u, v), the relations (7) and (8) imply

$$\delta_\mu^*(t, F, X) \leq 2\varepsilon + v(x - y) - (\mu\nu)^{-1}(\mu F(x) + \nu F(y) - F(\mu x + \nu y)) =$$
$$2\varepsilon + (\mu\nu)^{-1}\left(\mu(f(x) - F(x)) + \nu(g(y) - F(y)) - (u(\mu x + \nu y) - F(\mu x + \nu y))\right) \leq$$
$$\leq \rho_\mu(t, F^*, X^*) + 2\varepsilon. \quad (9)$$

In a similar manner, for an arbitrary $\varepsilon > 0$, we find $s > 0$, $f, g \in X^*$, and, then, $(z, w) \in X \times X$ describing (x, y), so that $\|w\|_X = t$,

$$\delta_\mu^*(t, F^*, X^*) \leq \varepsilon + ts - \delta_\mu(s, F^*, X^*),$$

$$\mu F^*(f) + \nu F^*(g) - F^*(\mu f + \nu g) - \mu\nu\varepsilon \leq \mu\nu\delta_\mu(s, F^*, X^*),$$

$$F^*(\mu f + \nu g) \leq \mu\nu\varepsilon + \langle \mu f + \nu g, z \rangle - F(z) \text{ and } \|w\|_X\|f - g\|_{X^*} \leq \langle f - g, w \rangle + \varepsilon. \quad (10)$$

the estimates (10) imply the counterpart of (9),

$$\delta_\mu^*(t, F^*, X^*) \leq 3\varepsilon + \langle f - g, z \rangle - (\mu\nu)^{-1}(\mu F^*(f) + \nu F^*(g) - F^*(\mu f + \nu g)) \leq$$

Asymmetric Convexity and Smoothness

$$4\varepsilon + (\mu\nu)^{-1}(\mu F(x) + \nu F(y) - F(\mu x + \nu y)) \leq \rho_\mu(t, F, X) + 4\varepsilon. \tag{11}$$

The second part of e) follows from the first thanks to the fact: $\phi^{**}(t) \leq \phi(t)$ for $t > 0$.
In the case of f), let us observe the implication

$$\delta_\mu(\cdot, F, X) \geq \delta(\cdot, F, X) \Rightarrow \delta_\mu^*(\cdot, F, X) \leq \delta^*(\cdot, F, X). \tag{12}$$

At the same time, for $t, \varepsilon > 0$, one can find $s > 0$ and $\mu \in (0, 1)$, satisfying

$$\delta^*(t, F, X) \leq \varepsilon + ts - \delta(s, F, X) \leq 2\varepsilon + ts - \delta_\mu(s, F, X) \leq \delta_\mu^*(s, F, X) + 2\varepsilon. \tag{13}$$

Now (12, 13) and Part d) imply the first identity of f). The proof of the second identity is very similar and, thus, omitted.

The second part of Part f) follows from the first as in the case of Part e).

To establish the first half of Part g), let us consider the auxiliary functional: for $t \in \mathbb{R}_+$, $\mu \in (0, 1)$, $z \in X$ and $w \in S_X$,

$$\phi_{\mu,z,w}(t, F) := (\mu\nu)^{-1}(\mu F(z - t\nu w) + \nu F(z + t\mu w) - F(z)),$$

and note that it is a convex function of t on $\mathbb{R}_+$ (as the sum of such functions). Therefore, $\phi_{\mu,z,w}(t, F)/t$ is nondecreasing, and this property is inherited by

$$\delta_\mu(t, F, X) = \inf_{(z,w) \in X \times S_X} \phi_{\mu,z,w}(t, F), \quad \rho_\mu(t, F, X) = \sup_{(z,w) \in X \times S_X} \phi_{\mu,z,w}(t, F),$$

$$\delta(t, F, X) = \inf_{\mu \in (0,1)} \delta_\mu(t, F, X) \text{ and } \rho(t, F, X) = \sup_{\mu \in (0,1)} \rho_\mu(t, F, X).$$

Since Chekanov, Nesterov and Vladimirov [37] have established that $\delta(t, F, X)/t^2$ is nondecreasing, we need to prove the second part of g) only. But it is also true, since Part d) provides us with the key relation (for $c > 1$)

$$c^{-2}\rho(ct, F, X) = c^{-2}\delta^*(ct, F^*, X^*) = (c^{-2}\delta(c\cdot, F^*, X^*))^*(t) \leq \delta^*(t, F^*, X^*) =$$
$$= \rho(t, F, X).$$

Since the proofs of the first 4 inequalities of Part h) follow the same steps, we consider only the first. Assuming $\mu \geq \nu$, $\alpha \geq \beta$, and $z = \mu x + \nu y$, we use the uniform convexity, and, then, the convexity of F to observe the estimates

$$F(\alpha x + \beta y) \leq \left(1 - \frac{\beta}{\nu}\right)F(x) + \frac{\beta}{\nu}F(z) \leq$$
$$\leq \left(1 - \frac{\beta}{\nu}\right)F(x) + \frac{\beta}{\nu}(\mu F(x) + \nu F(y) - \mu\nu\delta_\mu(\|x - y\|_X, F, X)) \leq$$
$$\leq \alpha F(x) + \beta F(y) - \mu\beta\delta_\mu(\|x - y\|_X, F, X),$$

followed by the first inequality of Part h). The fifth inequality is equivalent to $c_\mu\delta_\mu(t, F, X) \leq \delta_\eta(t, F, X)$ for $\eta \in (0, 1)$ and follows from the previous ones when $\eta \in (0, \mu] \cup [1-\mu, 1)$ and from $\mu(1-\mu)\delta_\mu(t, F, X) \leq \eta(1-\eta)\delta_\eta(t, F, X)$ for $\eta \in [\mu, 1-\mu]$. The duality in f) finishes the proof of h).

Part i) is a consequence of a) with $F(x) = \|x\|_X$ and x', y' defined by $x = \mu x'$ and $y = \nu y'$. We just need to take $\mu/\nu = \|x\|_X / \|y\|_X$:

$$\|x\|_X + \|y\|_X - \|x + y\|_X \leq \inf_\mu \mu\nu\langle f_x - f_y, x' - y'\rangle \leq 2\min_{\mu \in [0,1]}\|\nu x - \mu y\|_X \leq$$
$$\leq 2\frac{|\|y\|_X x - \|x\|_X y|_X}{\|x\|_X + \|y\|_X}.$$

Part k) follows immediately from the definition of the finite representability (Definition 2.3).

The last two inequalities in Part l) follow from the first two by taking limits, while the latter follows immediately from the definition of the moduli. $\square$

Next corollary is a quantitative Banach–space version of an inequality due to Pafnutiy L. Chebyshev.

Corollary 4.1. *For $\xi > 0$, let $\psi_{\mu,\xi}(t) = \xi^{-1}\rho_\mu(t(1+\xi), F, X) - \xi^{-1}\delta_\mu(t, F, X)$, and I be finite or countable.*
a) For a finite or countable index set I, let $\{x_i\}_{i\in I} \subset D(F)$ and $\{1 - \nu_{i,j} = \mu_{i,j}\}_{i,j \in I} \subset (0,1)$ with $\bigcup_{i,j\in I} B(\xi\nu_{i,j}\|x_i - x_j\|_X, x_i) \cup B(\xi\mu_{i,j}\|x_i - x_j\|_X, x_j) \subset \operatorname{int}D(F)$, and $\{\alpha_i\}_{i\in I} \subset \mathbb{R}_+$ with $\sum_{i\in I}\alpha_i = 1$. Then we have

$$2^{-1}\sum_{i,j\in I}\alpha_i\alpha_j\delta_{\mu_{i,j}}(\|x_i - x_j\|_X, F, X) \leq \sum_{i\in I}\alpha_i f_{x_i}(x_i) - \left(\sum_{i\in I}\alpha_i f_{x_i}\right)\left(\sum_{i\in I}\alpha_i x_i\right) \leq$$
$$\leq 2^{-1}\sum_{i,j\in I}\alpha_i\alpha_j\psi_{\mu_{i,j},\xi}(\|x_i - x_j\|_X).$$

b) For independent identically distributed X–valued stochastic variables ξ and η on a probability measure space (Ω, μ) and a convex function F defined on X, assume that $f_\xi \in \partial F(\xi)$ is measurable on (Ω, μ). Then we have $2^{-1}\mathrm{E}\delta_\mu(\|\xi - \eta\|_X, F, X) \leq \mathrm{E}f_\xi(\xi) - \langle \mathrm{E}f_\xi, \mathrm{E}\xi\rangle \leq 2^{-1}\mathrm{E}\psi_\theta(\xi - \eta)$.

Proof of Corollary 4.1. According to Parts a) and b) of Lemma 4.1, one has, for $i, j \in I$,

$$\delta_{\mu_{i,j}}(\|x_i - x_j\|_X, F, X) \leq (f_{x_i} - f_{x_j})(x_i - x_j) \leq \psi_{\mu_{i,j},\xi}(\|x_i - x_j\|_X). \tag{1}$$

Summing up over $I \times I$ with the weight $\alpha_i\alpha_j$, we establish Part a). Part b) follows in the same manner. $\square$

Let X be a Banach space, (Ω, σ, p) a probability space, $\{\sigma_i\}_{i\in\mathbb{N}_0}$ a nondecreasing sequence of σ–subalgebras of σ, and ξ_i a X–valued Bochner σ_i–measurable stochastic variable for every $i \in \mathbb{N}_0$. Let us recall that the sequence $\{\xi_i\}_{i\in\mathbb{N}} \subset L_1(\Omega, X)$ is a martingale difference sequence (with respect to $\{\sigma_n\}_{i\in\mathbb{N}}$) if $\mathrm{E}(\xi_{i+1}|\sigma_i) = 0$. In what follows we adapt a simplified designation for the mathematical expectation of a stochastic variable ξ on (Ω, μ) with respect to a (sub)algebra σ: $\mathrm{E}_\sigma\xi := \mathrm{E}(\xi|\sigma)$.

Part c) of the next lemma corresponds to Pisier's martingale inequality [40] when $F(x) = \|x\|_X^p$ in a p–uniformly convex or smooth space X.

Lemma 4.2. *a) For a finite or countable I and $\{\alpha_i\}_{i \in I} \subset \mathbb{R}_+$ with $\sum_{i \in I} \alpha_i = 1$, let $\mathcal{X} = \{x_i\}_{i \in I} \subset \mathrm{D}(F)$ with $x_c = \sum_{i \in I} \alpha_i x_i$ satisfy $x_c \in \mathrm{int}\,\mathrm{D}(F)$ and $2x_c - \mathcal{X} \subset \mathrm{D}(F)$. Then we have*

$$\max(\mu, \nu) \sum_{i \in I} \alpha_i \delta_\mu (\|x_i - x_c\|_X, F, X) \le \sum_{i \in I} \alpha_i F(x_i) - F\left(\sum_{i \in I} \alpha_i x_i\right) \le$$
$$\le \max(\mu, \nu) \sum_{i \in I} \alpha_i \rho_\mu (\|x_i - x_c\|_X / \max(\mu, \nu), F, X).$$

b) For a convex F and a X–valued stochastic variable ξ defined on a probability measure space (Ω, μ) with subalgebra σ satisfying $\xi(\Omega) \subset \mathrm{D}(F)$, $\mathrm{E}_\sigma \xi \in \mathrm{int}\,\mathrm{D}(F)$ and $2\mathrm{E}_\sigma \xi - \xi(\Omega) \subset \mathrm{D}(F)$, we have

$$\max(\mu, \nu) \mathrm{E}_\sigma \delta_\mu (\|\xi - \mathrm{E}_\sigma \xi\|_X, F, X) \le \mathrm{E}_\sigma F(\xi) - F(\mathrm{E}_\sigma \xi) \le$$
$$\le \max(\mu, \nu) \mathrm{E}_\sigma \rho_\mu (\|\xi - \mathrm{E}_\sigma \xi\|_X / \max(\mu, \nu), F, X) \; \sigma\text{-a.e.}$$

c) For an X–valued martingale difference sequence $\{\xi_i\}_{i \in \mathbb{N}_0}$ and a convex function F defined on X, we have, for every $n \in \mathbb{N}$,

$$\max(\mu, \nu) \sum_{i=1}^n \mathrm{E}\delta_\mu (\|\xi_i\|_X, F, X) \le \mathrm{E}F\left(\sum_{i=0}^n \xi_i\right) - \mathrm{E}F(\xi_0) \le$$
$$\le \max(\mu, \nu) \sum_{i=1}^n \mathrm{E}\rho_\mu (\|\xi_i\|_X / \max(\mu, \nu), F, X).$$

d) For a finite or countable I and $\{\alpha_i\}_{i \in I} \subset \mathbb{R}_+$ with $\sum_{i \in I} \alpha_i = 1$, let $\mathcal{X} = \{x_i\}_{i \in I} \subset \mathrm{D}(F)$ with $x_c = \sum_{i \in I} \alpha_i x_i$ satisfy $\mathcal{X} - x_c \subset \mathrm{int}\,\mathrm{D}(F)$ and $(\mathcal{X} - \mathcal{X}) \bigcup (\mathcal{X} + \mathcal{X} - 2x_c) \subset \mathrm{D}(F)$. Then we have

$$\sum_{i \in I} \alpha_i \left(\max(\mu, \nu) \delta_\mu(\|x_i - x_c\|_X, F, X) + F(x_i - x_c)\right) \le \sum_{i,j \in I} \alpha_i \alpha_j F(x_i - x_j) \le$$
$$\le \sum_{i \in I} \alpha_i \left(\max(\mu, \nu) \rho_\mu(\|x_i - x_c\|_X / \max(\mu, \nu), F, X) + F(x_i - x_c)\right).$$

e) For a convex F and independent identically distributed X–valued stochastic variables ξ and η on a probability measure space (Ω, μ) satisfying $\xi(\Omega) - \mathrm{E}\xi \subset \mathrm{int}\,\mathrm{D}(F)$ and $(\xi(\Omega) - \xi(\Omega)) \bigcup (\xi(\Omega) + \xi(\Omega) - 2\mathrm{E}\xi) \subset \mathrm{D}(F)$, we have

$$\max(\mu, \nu) \mathrm{E}\delta_\mu(\|\eta - \mathrm{E}\eta\|_X, F, X) \le \mathrm{E}F(\xi - \eta) - \mathrm{E}F(\xi - \mathrm{E}\eta) \le$$
$$\le \max(\mu, \nu) \mathrm{E}\rho_\mu(\|\eta - \mathrm{E}\eta\|_X / \max(\mu, \nu), F, X).$$

f) For arbitrary independent X–valued stochastic variables ξ and η defined on a probability measure space (Ω, μ) and a convex function F defined on X, we have

$$\max(\mu, \nu) \mathrm{E}\delta_\mu(\|\xi - \mathrm{E}\xi\|_X, F, X) \le \mathrm{E}F(\xi - \eta) - \mathrm{E}F(\mathrm{E}\xi - \eta) \le$$
$$\le \max(\mu, \nu) \mathrm{E}\rho_\mu(\|\xi - \mathrm{E}\xi\|_X / \max(\mu, \nu), F, X).$$

Proof of Lemma 4.2. Part a) is a particular case of Part b). According to Part d) of Lemma 4.1, one has, μ–a. e.,

$$\max(\mu,\nu)\delta_\mu\left(\|\xi(\omega) - \mathrm{E}_\sigma\xi\|_X, F, X\right) \leq F(\xi(\omega)) - F(\mathrm{E}_\sigma\xi) - f_{\mathrm{E}_\sigma\xi(\omega)}(\xi(\omega) - \mathrm{E}_\sigma\xi) \leq$$
$$\leq \max(\mu,\nu)\rho_\mu\left(\|\xi(\omega) - \mathrm{E}_\sigma\xi\|_X/\max(\mu,\nu), F, X\right).$$

We finish the proof of Part b) by taking the conditional mathematical expectation E_σ.

To establish c), we note that, according to b), one has, for every $1 \leq k \leq n$,

$$\max(\mu,\nu)\mathrm{E}_{\sigma_{k-1}}\delta_\mu\left(\|\xi_k\|_X, F, X\right) \leq \mathrm{E}_{\sigma_{k-1}}F\left(\sum_{i=0}^{k}\xi_i\right) - \mathrm{E}_{\sigma_{k-1}}F\left(\sum_{i=0}^{k-1}\xi_i\right) \leq$$
$$\leq \max(\mu,\nu)\mathrm{E}_{\sigma_{k-1}}\rho_\mu\left(\|\xi_i\|_X/\max(\mu,\nu), F, X\right).$$

Taking the full mathematical expectations and summing these inequalities up for all $1 \leq k \leq n$, we obtain c).

Part d) is a particular case of Part e). To establish the latter, let us assume that the subalgebra σ is generated by ξ. Hence we know, according to Part b) with $\xi - \eta$ instead of ξ, that, σ–a.e.,

$$\max(\mu,\nu)\mathrm{E}\delta_\mu\left(\|\eta - \mathrm{E}\eta\|_X, F, X\right) \leq \mathrm{E}_\sigma F(\xi - \eta) - F(\xi - \mathrm{E}\eta) \leq$$
$$\leq \max(\mu,\nu)\mathrm{E}\rho_\mu\left(\|\eta - \mathrm{E}\eta\|_X/\max(\mu,\nu), F, X\right).$$

Taking the full mathematical expectation finishes the proof of Part e).

Part f) is established in exactly the same way as Part e) with σ generated by η:

$$\max(\mu,\nu)\mathrm{E}\delta_\mu(\|\xi - \mathrm{E}\xi\|_X, F, X) \leq \mathrm{E}_\sigma F(\xi - \eta) - F(\mathrm{E}\xi - \eta) \leq$$
$$\leq \max(\mu,\nu)\mathrm{E}\rho_\mu(\|\xi - \mathrm{E}\xi\|_X/\max(\mu,\nu), F, X).$$

Taking the full mathematical expectation changes only the middle part of this inequality. $\square$

For $i \in \mathbb{N}$ and $1 - \nu_i = \mu_i \in (0,1)$, assume that (D_i, μ_i) is a copy of $D = \{0,1\}$ endowed with the probability measure p_i defined by $p_i(\{1\}) = \mu_i$. For $n \in \mathbb{N}$ and $\varepsilon \in D^n$, let D^n be endowed with the corresponding product probability measure $p_{d^n} := \prod_{i=1}^n p_i$, and b_i be the function on D^n defined by

$$b_i(\varepsilon) = b(\varepsilon_i, \mu_i), \text{ where } b(\delta, \mu) := \begin{cases} \mu & \text{for } \delta = 0, \\ \mu - 1 & \text{for } \delta = 1. \end{cases} \quad (A)$$

Lemma 4.3. *For a convex F defined on X, $n \in \mathbb{N}$ and $\{x_i\}_{i=0}^n \subset X$, we have:*

$$a) \sum_{i=1}^n \mu_i\nu_i\delta_{\mu_i}(\|x_i\|_X, F, X) \leq \int_{D^n} F\left(x_0 + \sum_{i=1}^n b_i x_i\right) dp_{d^n} - F(x_0) \leq$$
$$\leq \sum_{i=1}^n \mu_i\nu_i\rho_{\mu_i}(\|x_i\|_X, F, X).$$

b) *Particularly, for all $\mu_i = 1/2$, one has*

$$\sum_{i=1}^{n} \delta_{1/2}(2\|x_i\|_X, F, X) \leq 2^{-n} \sum_{\varepsilon \in D^n} F((2\varepsilon - \bar{1}, \vec{x})) - F(0) \leq \sum_{i=1}^{n} \rho_{1/2}(2\|x_i\|_X, F, X),$$

where $(\varepsilon, \vec{x}) = \sum_{i=1}^{n} \varepsilon_i x_i$.

Proof of Lemma 4.3. We just need to sum the particular cases of the corresponding reformulations of the definitions of δ_μ and ρ_μ: for $1 \leq i \leq n$,

$$\mu_i \nu_i \delta_{\mu_i}(\|x_i\|_X, F, X) \leq \int_{D^n} F\left(x_0 + \sum_{k=1}^{i} b_k x_k\right) dp_{d^n} - \int_{D^n} F\left(x_0 + \sum_{k=1}^{i-1} b_k x_k\right) dp_{d^n} \leq$$
$$\leq \mu_i \nu_i \rho_{\mu_i}(\|x_i\|_X, F, X).$$

$\square$

4.2. Quantitative Lyapunov Theorems

The second part of the celebrated Lyapunov theorem asserts that the image of an atomless countably–additive measure with the values in a finite–dimensional space is compact and convex. There are many research directions, as well as alternative proofs and generalisations, branching from this result. We postpone the further historical remarks and references till Section 5.2, where, after the investigation of the uniform smoothness for a wide range of spaces in Section 4, we compare the related Kadets constant with the atomic Lyapunov constants computing the latter for IG_+–spaces and discuss the question of sharpness.

We estimate (from above) the (strong) measure of the nonconvexity of the image of a vector–valued measure with bounded variation taking values in a Banach space admitting a suitable uniformly smooth convex functions defined on it. The estimates for measures with the values in an arbitrary finite-dimensional normed space are established as well, utilising Lemma 2.1.1 from [41] in the style of the Kantorovich theory.

Definition 4.2. *For a bounded subset A of a Banach space X its measure of nonconvexity $\lambda(A)$ and Chebyshev radius $r(A)$ are defined by*

$$r(A) = \inf_{x \in X} \sup_{y \in A} \|x - y\|_X \quad \text{and} \quad \lambda(A) = \sup_{x \in \mathrm{co}(A)} \inf_{y \in A} \|x - y\|_X.$$

Remark 4.3. *There is a very simple but important and representative model of the (purely) atomic measure ν_m defined on the algebra σ_m of the subsets $\mathcal{P}(I_m)$ of $I_m = [1, m] \cap \mathbb{N}$ by the relations*

$$\nu_m(\{i\}) = e_i \text{ for } 1 \leq i \leq m,$$

where $\{e_i\}_{i=1}^m$ is the Euclidean basis in $\mathbb{R}^m$. In this case, the image of the measure $\nu_m(\sigma_m) = \{0, 1\}^m$ is the vertexes of the cube $[0, 1]^m$, while its convex envelope is the cube itself.

Considering now the linear map $T : \mathbb{R}^m \to X$ defined by the relations

$$Te_i = x_i \in X \text{ for } 1 \leq i \leq m,$$

we obtain another measure $\nu_T = T\nu_m$ with the vertexes $T(\{0,1\}^m)$ of the (possibly degenerated) parallelepiped $T([0,1]^m)$ being the image $\nu_T(\sigma_m)$ and, thanks to the linearity, with
$$\mathrm{co}(\nu_T(\sigma_n)) = T([0,1]^n).$$

We shall deal with an algebra σ of subsets of Ω (with $\Omega \in \sigma$) endowed with a countably additive measure μ taking values in a Banach space X. Let us recall that an atom for μ is a set $A \in \sigma$ with $\mu(A) \neq 0$, such that, for every subset $B \subset A$, $B \in \sigma$, one has either $\mu(B) = 0$ or $\mu(B) = \mu(A)$. The degree of atomicity and the variation of the measure μ are defined (see [5]) by

$$\mathrm{at}(\mu) := \sup\{\|\mu(A)\|_X : A \text{ is an atom for } \mu\}.$$

For a convex function F with $\mu(\sigma) \subset \mathrm{D}(F) = \mathrm{co}(\mathrm{D}(F))$, we also define the F-variation of μ by

$$\mathrm{Var}_F(\mu) := \sup\left\{\sum_{i=1}^n F(\mu(A_i)) : n \in \mathbb{N}, A_i \in \sigma, \mu(A_i \cap A_j) = 0 \text{ for } i \neq j\right\}.$$

For $F(x) = \|x\|_X$, we obtain the ordinary variation

$$\mathrm{Var}(\mu) := \sup\left\{\sum_{i=1}^n \|\mu(A_i)\|_X : n \in \mathbb{N}, A_i \in \sigma, \mu(A_i \cap A_j) = 0 \text{ for } i \neq j\right\}.$$

Corollary 3.3 shows the correctness of the following definition.

Definition 4.3. *For a finite-dimensional Banach space X, we define its atomic Lyapunov constant $C_{AL} = C_{AL}(X)$ as the infimum of the constants C such that the following inequality holds for every measure space (Ω, σ, ν) with the atomic countably-additive X-valued measure ν:*

$$\lambda(\nu(\sigma)) \leq C \mathrm{at}(\nu).$$

Definition 4.4. *For a convex function F on a Banach space X, we designate by means of F_{rl} its maximal radial minorant defined by*

$$F_{rl}(t) := \inf_{\|x\|_X = t} F(x) \text{ for } t > 0.$$

Similarly, we designate by means of F_{ru} its minimal radial majorant defined by

$$F_{ru}(t) := \sup_{\|x\|_X = t} F(x) \text{ for } t > 0.$$

Lemma 4.4. *Let F be a convex function on a Banach space X with $F(0) = 0$. Then, both $F_{rl}(t)/t$ and $F_{ru}(t)/t$ are non-decreasing, and $F_{ru}(t)$ is convex on $[0, \infty)$, while F_{rl} is right continuous (upper semicontinuous).*

Asymmetric Convexity and Smoothness

Proof of Lemma 4.4. As in the proof of Part g) of Lemma 4.1, we consider the auxiliary families of convex functions

$$\{\phi_x\}_{x \in S_X}, \text{ where } \phi_x(t) = F(tx) \text{ for } x \in S_X.$$

Since

$$F_{rl}(t) = \inf_{x \in S_X} \phi_x(t) \text{ and } F_{ru}(t) = \sup_{x \in S_X} \phi_x(t),$$

their convexity means that $F_{ru}(t)$ is convex and $\phi_x(t)/t$ and, hence, $F_{rl}(t)/t$ and $F_{ru}(t)/t$ are non–decreasing on $[0, \infty)$ because $F(0) = 0$. $\square$

Theorem 4.1. *Let X be a Banach space and F be a nonnegative convex function on X, such that $F_{rl}(z) = 0$ implies $z = 0$, and F_{rl} is (left) continuous. Assume also that (Ω, σ, ν) is a measure space with the countably–additive X–valued measure ν defined on σ (with $\Omega \in \sigma$). Then we have*

a) $\lambda(\nu(\sigma)) \le F_{rl}^{-1}\left(\min\left(\text{Var}_{h_\rho(\cdot, F, X)}(\nu), \dfrac{h_\rho(\text{at}(\nu), F, X)\text{Var}(\nu)}{\text{at}(\nu)}\right)\right),$

where $h_\rho(t, F, X) := \sup_{\mu \in (0,1)} \mu(1-\mu)\rho_\mu(t, F, X)$;

b) $\lambda(\nu(\sigma)) \le F_{rl}^{-1}\left(\min\left(2^{-2}\text{Var}_{\rho(\cdot, F, X)}(\nu), \dfrac{\rho(\text{at}(\nu), F, X)\text{Var}(\nu)}{4\text{at}(\nu)}\right)\right).$

Proof of Theorem 4.1. Under the conditions of the theorem F_{rl} is increasing and right continuous thanks to Lemma 4.4. This means that F_{rl}^{-1} exists (and continuous). Let us note that Part b) follows from Part a) because

$$h_\rho(t, F, X) \le \max_{\mu \in (0,1)} \mu(1-\mu)\rho(t, F, X) = 2^{-2}\rho(t, F, X)$$

As in the proof of Part g) of Lemma 4.1 and Lemma 4.4, we see that $h_\rho(t, F, X)$ is convex and $h_\rho(t, F, X)/t$ is non-decreasing. Thus, for any partition $\{A_i\}_{i=1}^m$ of Ω, we have the estimate

$$\sum_{i=1}^m h_\rho(\|\nu(A_i)\|_X) \le \max_{i=1}^m \dfrac{h_\rho(\max_{i=1}^m \|\nu(A_i)\|_X, F, X)}{\max_{i=1}^m \|\nu(A_i)\|_X} \text{Var}(\nu) \qquad (1)$$

showing that the whole Part a) almost follows from the estimate in Part a) involving the h_ρ-variation of ν. Thus, our main task is to estimate the measure of nonconvexity $\lambda(\nu(\sigma))$ in terms of the left–hand side of (1).

Every point $x \in \text{co}(\nu(\sigma))$ has the form

$$x = \sum_{j=1}^l \alpha_j \nu(B_j) \text{ for some } \{B_j\}_{j=1}^l \subset \sigma, \{\alpha_j\}_{j=1}^l \subset [0, 1], \sum_{j=1}^l \alpha_j = 1. \qquad (2)$$

Considering various intersections of the elements of the family $\{B_j\}_{j=1}^l$, we find another family $\{A_i\}_{i=1}^m$ such that $A_i \cap A_j = \emptyset$ and every B_j belongs to the algebra generated by the unions of $\{A_i\}_{i=1}^m$. Continuing the subdivision process, if necessary, we may further assume that

$$\|\nu(A_i)\|_X \leq \mathrm{at}(\mu) \text{ (or } \|\nu(A_i)\|_X \leq \mathrm{at}(\mu) + \varepsilon/2^i) \text{ for } 1 \leq i \leq m. \tag{3}$$

It is easily seen that the finite subalgebra of σ generated by the sets $\{A_i\}_{i=1}^m$ has the same image as the algebra $\nu_T = T\nu_m$ described in Remark 4.3 with $x_i = \nu(A_i)$ for $1 \leq i \leq m$, and that $x \in \mathrm{co}(\nu_T) = T([0,1]^m)$. In other word, there are $\{\mu_i\}_{i=1}^m \subset (0,1]$ satisfying

$$x = \sum_{i=1}^m \mu_i \nu(A_i). \tag{4}$$

Noticing the remarkable relation

$$\sum_{i=1}^m b_i(s)\nu(A_i) = x - \nu\left(\sum_{i=1}^m s_i A_i\right) \tag{5}$$

involving the asymmetric system $\{b_i\}_{i=1}^m$ defined in (A) before Lemma 4.3, we use the inclusion $T(\{0,1\}^m) \subset \nu(\sigma)$ and apply Lemma 4.3 with the probability measure on $D^m = \{0,1\}^m$ defined in (P) before it to establish the chain of inequalities

$$F_{rl}\left(\inf_{y \in \nu(\sigma)} \|x-y\|_X\right) \leq F_{rl}\left(\min_{s \in D^m} \|x - Ts\|_X\right) \leq$$

$$\leq \int_{D^m} F\left(\sum_{i=1}^m b_i(s)\nu(A_i)\right) dp_{d^m} \leq \sum_{i=1}^m h_\rho(\|\nu(A_i)\|_X, F, X). \tag{6}$$

Since the right–hand side of (6) is the left–hand side of (1), and F_{rl} is lower semicontinuous, we finish the proof of the theorem by taking supremum over all $x \in \mathrm{co}(\nu(\sigma))$. □

Let us recall that $|A|$ denotes the number of elements in a finite set A, or a counting measure.

Lemma 4.5 ([42, 41]). *Let $K \subset \mathbb{R}^n$ be a polyhedron given by the system*

$$\begin{cases} f_i(x) = a_i, & 1 \leq i \leq p, \\ g_j(x) \leq b_j, & 1 \leq j \leq q, \end{cases}$$

where $\{f_i\}_{i=1}^p \cup \{g_j\}_{j=1}^q \subset \mathbb{R}^{n*}$ and $\{a_i\}_{i=1}^p \cup \{b_j\}_{j=1}^q \subset \mathbb{R}$. Assume also that x_0 is a vertex (extreme point) of K and $A = \{j : g_j(x_0) = b_j\}$. Then one has $|A| \geq n - p$.

Theorem 4.2. *Assuming that X is an n–dimensional space under the conditions of Theorem 4.1, we have*

$$a) \ \lambda(\nu(\sigma)) \leq F_{rl}^{-1}(nh_\rho(\mathrm{at}(\nu), F, X)),$$

where $h_\rho(t, F, X) := \sup_{\mu \in (0,1)} \mu(1-\mu)\rho_\mu(t, F, X)$;

$$b) \ \lambda(\nu(\sigma)) \leq F_{rl}^{-1}\left(\frac{n}{4}\rho(\mathrm{at}(\nu), F, X)\right).$$

Proof of Theorem 4.2. Let us start by estimating of the right–hand side of (6) from the proof of Theorem 4.1:

$$\sum_{i=1}^{m} h_\rho(\|\nu(A_i)\|_X, F, X) \leq m h_\rho(\mathrm{at}(\nu), F, X) \leq \frac{m}{4}\rho(\mathrm{at}(\nu), F, X). \qquad (1)$$

Therefore, our strategy is to show that m in this formula can be chosen to satisfy $m \leq n$. We accomplish this task by repeating the first step from the proof of Lemma 2.1.2 from [42, 41] with the aid of Lemma 4.5.

Namely, to optimise the representation of x in formula (4) from the proof of Theorem 4.1, we apply Lemma 4.5 to the system (in the variables $\mu = (\mu_1, \ldots, \mu_m) \in \mathbb{R}^m$)

$$\begin{cases} x = \sum_{i=1}^{m} \mu_i x_i, & n \text{ equations,} \\ \mu \in [0,1]^m, & 2m \text{ inequalities} \end{cases}$$

and obtain another vector $\mu' \in [0,1]^m$ satisfying

$$|\{j : \mu'_j \in \{0,1\}\}| \geq m - n.$$

This means that one can represent x in the form

$$x = \sum_{i=1}^{m} \mu'_i \nu(A_i) = \nu(A_0) + \sum_{j=1}^{k} \mu'_{i_j} \nu(A_{i_j}), \text{ where}$$

$$\{\mu'_{i_j}\}_{j=1}^{k} \subset (0,1), \quad A_0 = \sum_{\mu'_i \in \{0,1\}} \mu'_i A_i \text{ and } k \leq n. \qquad (2)$$

Renumbering $\{\mu'_{i_j}\}_{j=1}^{k}$ and $\{A_{i_j}\}_{j=1}^{k}$, we obtain the following counterpart of formula (6) from the proof of Theorem 4.1:

$$F_{rl}\left(\inf_{y \in \nu(\sigma)} \|x - y\|_X\right) \leq F_{rl}\left(\min_{s \in D^k} \|x - (\nu(A_0) + Ts)\|_X\right) \leq$$

$$\leq \int_{D^k} F\left(\sum_{i=1}^{k} b_i(s)\nu(A_i)\right) dp_{d^k} \leq \sum_{i=1}^{k} h_\rho(\|\nu(A_i)\|_X, F, X). \qquad (3)$$

We finish the proof by applying (1) and taking supremum over all $x \in \mathrm{co}(\nu(\sigma))$. □

The following corollary generalises and strengthens the main result of [43] to the setting of an arbitrary finite–dimensional Banach space and strengthens the first theorem from [5].

Corollary 4.2. *For an n–dimensional normed space X, let (Ω, σ, ν) be a measure space with an atomic countably–additive X–valued measure ν. Then, the following estimate holds:*

$$\lambda(\nu(\sigma)) \leq \frac{n}{2}\mathrm{at}(\nu) \text{ (i.e., } C_{AL} \leq n/2).$$

Proof of Corollary 4.2. Choosing $F(x) = \|x\|_X$ under the conditions of Theorem 4.2, we obtain

$$F_{rl}(t) = t \text{ and } \rho(t, \|\cdot\|_X, X) \leq 2t, \tag{1}$$

where the second estimate is sharp and follows from

$$\mu\|x\|_X + (1-\mu)\|y\|_X - \|\mu x + (1-\mu)y\|_X \leq 2\mu(1-\mu)\|x-y\|_X \tag{2}$$

becoming an equality in the case $\mu x + (1-\mu)y = 0$. Therefor, the corollary follows from Part *b*) of Theorem 4.2. $\square$

The next corollary is a simple consequence of the definition and triangle inequality.

Corollary 4.3. *For an n–dimensional normed space X, let (Ω, σ, ν) be a measure space with an atomic countably–additive X–valued measure ν. Then, the following estimate holds:*

$$r(\nu(\sigma)) \geq \frac{1}{2} \sup_{A,B \in \sigma} \|\nu(A) - \nu(B)\|_X \geq \frac{1}{2} \sup_{A \in \sigma} \|\nu(A)\|_X.$$

4.3. Log–Convexity and Log–Smoothness and High Order Characterisations

In this section we establish interesting generalisations of the inequalities from Lemmas 4.2 and 4.3 for a wide subclasses of log–convex and log–smooth functions on a Banach space.

Definition 4.5. *Let X be a Banach space, and let $F : X \to \mathbb{R}$ be convex with $F(0) = 0$.*

For a nonnegative function $h_c = h_c(\cdot, F, X)$ defined on $(0, 1)$, we say that the function F is log–h_c–convex on X if one has

$$\delta_\mu(\|x\|_X, F, X) \geq h_c(\mu) F(x) \text{ for every } \mu \in (0,1),\ x \in X,$$

or, that is equivalent to

$$\delta_\mu(t, F, X) \geq h_c(\mu) F_{ru}(t) \text{ for every } \mu \in (0,1),\ t \in [0, \infty).$$

For a nonnegative function $h_s = h_s(\cdot, F, X)$ defined on $(0, 1)$, we say that the function F is log–h_s–smooth on X if one has

$$\rho_\mu(\|x\|_X, F, X) \leq h_s(\mu) F(x) \text{ for every } \mu \in (0,1),\ x \in X,$$

or, that is equivalent to

$$\rho_\mu(t, F, X) \leq h_s(\mu) F_{rl}(t) \text{ for every } \mu \in (0,1),\ t \in [0, \infty).$$

Without loss of generality, we shall always assume that $h_c(\mu) = h_c(1-\mu)$ and $h_s(\mu) = h_s(1-\mu)$.

Whenever one or both of the limits $\lim_{\mu \to 0+} h_c(\mu)$ and $\lim_{\mu \to 0+} h_s(\mu)$ exist, we assume that

$$h_c(0) = h_c(1) = \lim_{\mu \to 0+} h_c(\mu) \text{ and } h_s(0) = h_s(1) = \lim_{\mu \to 0+} h_s(\mu).$$

Asymmetric Convexity and Smoothness 273

It is convenient to use a variant of the Fourier–Walsh formalism.

Definition 4.6. *For $i \in \mathbb{N}$ and $1 - \nu_i = \mu_i \in (0,1)$, assume that (D_i, μ_i) is a copy of $D = \{0,1\}$ endowed with the probability measure p_i defined by $p_i(\{1\}) = \mu_i$.*

For $n \in \mathbb{N}$ and $\varepsilon \in D^n$, let D^n be endowed with the corresponding product probability measure $p_{d^n} := \prod_{i=1}^n p_i$, and b_i be the function on D^n defined by ((A) in Section 4.1)

$$b_i(\varepsilon) = b(\varepsilon_i, \mu_i), \text{ where } b(\delta, \mu) := \begin{cases} \mu & \text{for } \delta = 0, \\ \mu - 1 & \text{for } \delta = 1. \end{cases}$$

Let us also recall the set $I_n = \mathbb{N} \cap [1, n]$ and the algebra σ_n of all its subsets with $|\sigma_n| = 2^n$ discussed in Remark 4.3 in Section 4.2. In the space $L_\infty(D^n) = \mathbb{R}^{2^n}$, we consider complete orthogonal systems of functions $\{b_A\}_{A \subset I_n}$ and $\{\tilde{b}_A\}_{A \subset I_n}$ defined by $b_\varnothing = \tilde{b}_\varnothing = 1$ and

$$b_A := \prod_{i \in A} b_i \text{ and } \tilde{b}_A := \|b_A | L_2(D^n)\|^{-2} b_A \text{ for } A \neq \varnothing.$$

We also define the related asymmetric Fourier transform $\mathcal{F}_a : L_p(D^n, X) \to l_q(\sigma_n, X)$ for any $p, q \in (0, \infty]$ and its inverse by

$$\mathcal{F}_a : f \mapsto \hat{f} \text{ with } \hat{f}(A) = \int_{D^n} f \tilde{b}_A \, dp_{d^n} \text{ and}$$

$$\mathcal{F}_a^{-1} : \hat{f} \mapsto f \text{ with } f = \sum_{A \in \sigma_n} \hat{f}(A) b_A.$$

Let us note that $\mathcal{F}_a$ is the same for different n in the sense that the restriction of $\mathcal{F}_a$ defined on $L_p(D^n, X)$ coincides with $\mathcal{F}_a$ defined on $L_p(D^m, X)$ for any $m \leq n$.

Let also $\mathcal{F}_{ai}$ be the asymmetric Fourier transform in the variable ε_i for $\varepsilon \in D^n$:

$$\mathcal{F}_{ai} f(\varepsilon_1, \ldots, \varepsilon_{i-1}, A, \varepsilon_{i+1}, \ldots, \varepsilon_n) = \mathcal{F}_a g(A) \text{ for } g(\varepsilon_i) = f(\varepsilon)$$

with ε_j fixed for $j \neq i$ and $A \in \sigma_1 = \{\varnothing, \{1\}\}$.

Remark 4.4. *a) Let us note an interesting method of constructing a new log–h_c–convex (log–h_s–smooth) function F from a family $\{F_\omega\}_{\omega \in \Omega}$ of such functions, where (Ω, σ, ν) is a measure space. We assume that, for any $x \in X$, the function $f_x(\omega) = F_\omega(x)$ belongs to a function space $\Psi(\Omega)$ of real-valued measurable functions, and that $g : \Omega \to \mathbb{R}_+$ is a non-negative measurable function on Ω from the dual space $\Psi(Q)^*$. Let also $F_\omega(x)$ be log–$h_{c\omega}$–convex (log–$h_{s\omega}$–smooth) for ν–almost every ω in Ω. Then, the function $F(x) := \int_\Omega g(\omega) f_x(\omega) d\nu$ is log–h'_c–convex (log–h'_s–smooth) for $h'_c = \text{ess inf}_{\omega \in \Omega} h_{c\omega}$ ($h'_s = \text{ess sup}_{\omega \in \Omega} h_{s\omega}$).*

b) It is useful to note that, for $A \in \sigma_n$, $\|b_A | L_1(D^n)\| = 2^{|A|} \|b_A | L_2(D^n)\|^2 = 2^{|A|} \prod_{i \in A} \mu_i \nu_i$.

c) The basic function $b(\mu, \varepsilon) : (0,1) \times \{0,1\} \to \mathbb{R}$ standing behind the definition of b_A has an important symmetry between μ and $\nu = 1 - \mu$: $b(\nu, \varepsilon) = -b(\mu, 1 - \varepsilon)$.

Theorem 4.3. *Let X be a Banach space, F be a convex function defined on X with $F(0) = 0$, and $n \in \mathbb{N}$. Then, the following criteria hold:*
a) *F is log–h_c–convex if, and only if, for every $\mu \in (0,1)^n$ and $f \in L_\infty(D^n, X)$, we have*

$$\int_{D^n} F(f) dp_{d^n} \geq \sum_{A \in \sigma_n} \prod_{i \in A} \mu_i \nu_i h_c(\mu_i) F(\hat{f}(A));$$

b) *F is log–h_s–smooth if, and only if, for every $\mu \in (0,1)^n$ and $f \in L_\infty(D^n, X)$, we have*

$$\int_{D^n} F(f) dp_{d^n} \leq \sum_{A \in \sigma_n} \prod_{i \in A} \mu_i \nu_i h_s(\mu_i) F(\hat{f}(A)).$$

Proof of Theorem 4.3. We shall prove only Part a) because the proof of b) is completely symmetric, i.e., it is enough to reverse all the inequalities and substitute h_c with h_s (and the related terms) in the former to obtain the latter.

To establish the "if"–implication, we consider $f(\varepsilon) = g(\varepsilon_n)$ for $\varepsilon \in D^n$ with $g(1) = x$, $g(0) = y$ and, therefore, $\mathcal{F}_{an} f(\ldots, \varnothing) = \mu_n x + \nu_n y$ and $\mathcal{F}_{an} f(\ldots, \{1\}) = y - x$ to observe that the inequality in a) takes the form

$$\mu_n F(x) + \nu_n F(y) \geq F(\mu_n x + \nu_n y) + \mu_n \nu_n h_c(\mu_n) F(y - x) \text{ for any } x, y \in X, \quad (1)$$

that is equivalent to log–h_c–convexity of F since μ_n is arbitrary.

Note that we have also proved the "only if"–implication in the case $n = 1$. To finish the proof by induction, let us assume that the "only if"–implication holds for $n = m - 1$. Assume that $\bar{\varepsilon} = (\varepsilon_1, \ldots, \varepsilon_{m-1})$. Given $f \in L_\infty(D^m, X)$, we recall that (1) implies

$$\int_{D_m} F(f(\bar{\varepsilon}, \cdot)) \geq F(\mathcal{F}_{am} f(\bar{\varepsilon}, \varnothing)) + \mu_n \nu_n h_c(\mu_n) F(\mathcal{F}_{am} f(\bar{\varepsilon}, \{1\})) \text{ for any } \bar{\varepsilon} \in D^{m-1},$$
(2)

while the "only if"–implication for $n = m - 1$ means that, for $g = g(\bar{\varepsilon}) \in \{\mathcal{F}_{am} f(\bar{\varepsilon}, \varnothing), \mathcal{F}_{am} f(\bar{\varepsilon}, \{1\})\}$, we have

$$\int_{D^{m-1}} F(g) dp_{d^{m-1}} \geq \sum_{A \in \sigma_{m-1}} \prod_{i \in A} \mu_i \nu_i h_c(\mu_i) F(\hat{g}(A)). \quad (3)$$

The proof is finished by integrating both parts of (2) over D^{m-1} and inserting (3). □

In the next theorem, a scalar–valued function F is called *positively p–homogeneous* for some $p \in [1, \infty)$ if $F(tx) = |t|^p F(x)$ for $t \in \mathbb{R}$. It is a generalisation of the previous theorem to the case of an arbitrary system of mutually independent stochastic variables. Recall that $I_n = \mathbb{N} \cap [1, n]$ and $\sigma_n = \mathcal{P}(I_n)$ for $n \in \mathbb{N}$.

Theorem 4.4. *Let X be a Banach space, $n \in \mathbb{N}$, $p, q \in [1, \infty)$, F be a convex function defined on X with $F(0) = 0$, and $\{\xi_i\}_{i=1}^n$ be a set of mutually independent scalar–valued stochastic variables on a measure space (Ω, Ξ, ν) with $E\xi_i = 0$ for $i \in I_n$. Assume also that $\{x_A\}_{A \in \sigma_n} \subset X$, $\xi_A = \prod_{i \in A} \xi_i$, $\xi_\varnothing = 1$ for $A \in \sigma_n$ and*

$$C_c = C_c(F, X) := \sup_{\mu \in [0, 1/2]} (1 - \mu) h_c(\mu) \text{ and } C_s = C_s(F, X, q) := \inf_{\mu \in [0, 1/2]} (1 - \mu)^{1-q} h_s(\mu).$$

Then, it follows that

a) the log–h_c–convexity and positive p–homogeneity of F implies

$$\mathrm{E}F\left(\sum_{A\in\sigma_n} x_A\xi_A\right) \geq \sum_{A\in\sigma_n} C_c^{|A|} F(x_A) \prod_{i\in A} \mathrm{E}|\xi_i|^p$$

and, if $\{\xi_i\}_{i=1}^n$ are also symmetric,

$$\mathrm{E}F\left(\sum_{A\in\sigma_n} x_A\xi_A\right) \geq \sum_{A\in\sigma_n} \left(2^{p-2}h_c(1/2)\right)^{|A|} F(x_A) \prod_{i\in A} \mathrm{E}|\xi_i|^p;$$

b) the log–h_s–smoothness and positive q–homogeneity of F implies

$$\mathrm{E}F\left(\sum_{A\in\sigma_n} x_A\xi_A\right) \leq \sum_{A\in\sigma_n} C_s^{|A|} F(x_A) \prod_{i\in A} \mathrm{E}|\xi_i|^q$$

and, if $\{\xi_i\}_{i=1}^n$ are also symmetric,

$$\mathrm{E}F\left(\sum_{A\in\sigma_n} x_A\xi_A\right) \leq \sum_{A\in\sigma_n} \left(2^{q-2}h_s(1/2)\right)^{|A|} F(x_A) \prod_{i\in A} \mathrm{E}|\xi_i|^q.$$

Remark 4.5. *Similarly to the previous theorem, this theorem can be turned into criteria for log–convexity and log–smoothness with the very possible loss in the sharpness of the constants involved.*

Proof of Theorem 4.4. This theorem follows from Part b) of Lemma 4.2 by means of mathematical induction by n. In the symmetric case, instead of Lemma 4.2, we use the relations $\mathrm{E}F(z-\xi) = \mathrm{E}F(z+\xi)$ and

$$\mathrm{E}2^{-2}\delta_{1/2}(2\|\xi\|_X, F, X) \leq \mathrm{E}F(z+\xi) - F(z) \leq \mathrm{E}2^{-2}\rho_{1/2}(2\|\xi\|_X, F, X)$$

for $z \in X$. We shall prove only Part b) because the proof of a) is completely symmetric, i.e., it is enough to reverse all the inequalities and substitute h_s with h_c (and the related terms) in the former to obtain the latter.

For $n = 1$, Part b) is just the right–hand side inequality of Lemma 4.2, b) with $\xi = x_\varnothing + x_{\{1\}}\xi_{\{1\}}$ combined with the positive q–homogeneity of F. Assuming Part b) to hold for $n-1$, we use the independence of ξ_n on the sigma–algebra Ξ_{n-1} generated by $\{\xi_i\}_{i=1}^{n-1}$, positive q–homogeneity of F and the second inequality from Part b) of Lemma 4.2 to obtain the estimates

$$\mathrm{E}_{\Xi_{n-1}}F\left(\sum_{A\in\sigma_n} x_A\xi_A\right) - F\left(\sum_{B\in\sigma_{n-1}} x_B\xi_B\right) \leq C_s\mathrm{E}|\xi_n|^q F\left(\sum_{n\in A\in\sigma_n} x_A\xi_{A\setminus\{n\}}\right) \quad \Xi_{n-1}-a.e. \tag{1}$$

We now finish the proof by taking the full mathematical expectation E of both sides of (1) and using the inductive hypothesis twice:

$$\mathrm{E}F\left(\sum_{A\in\sigma_n} x_A\xi_A\right) \leq \sum_{B\in\sigma_{n-1}} C_s^{|B|} F(x_B) \prod_{i\in B} \mathrm{E}|\xi_i|^q + \sum_{n\in A\in\sigma_n} C_s^{|A|} F(x_A) \prod_{i\in A} \mathrm{E}|\xi_i|^q.$$

□

4.4. Asymmetric Uniform and Hanner Types of Convexity and Smoothness

Here we return to the question of the uniform convexity and smoothness of the functions $F_q(x) = \|x\|_X^q$ for $q > 1$. The uniform convexity and smoothness are understood in a wider sense then the midpoint ones. Bynum–Drew–Gross, Hanner and our multihomogeneous inequalities are investigated and used at some stages.

The pathway from Part g) of Lemma 4.1 to the condition $2 \in [q,p] \subset [1,\infty]$, its alternatives and historical remarks are discussed in front of Lemma 4.1.

4.4.1. Definitions and Abstract Results

Definition 4.7. *Let X be a Banach space and $2 \in [q,p] \subset [1,\infty]$.*

We say that the space X is (p, h_c)–uniformly convex if the function $F_p(x) = \|x\|_X^p$ is log–h_c–uniformly convex, that is if, for every $x, y \in X$ and $\mu = 1 - \nu \in (0, 1)$, we have

$$\mu\|x\|_X^p + \nu\|y\|_X^p \geq \|\mu x + \nu y\|_X^p + h_c(\mu)\mu\nu\|y - x\|_X^p.$$

We say that the space X is (q, h_s)–uniformly smooth if the function $F_q(x) = \|x\|_X^q$ is log–h_s–uniformly smooth, that is if, for every $x, y \in X$ and $\mu = 1 - \nu \in (0, 1)$, we have

$$\mu\|x\|_X^q + \nu\|y\|_X^q \leq \|\mu x + \nu y\|_X^q + h_s(\mu)\mu\nu\|y - x\|_X^q.$$

Having in mind the non–improvable estimates

$$\max\left(\|\mu x + \nu y\|_X, \|x - y\|_X/2\right) \leq \max(\|x\|_X, \|y\|_X) \text{ and}$$

$$\mu\|x\|_X + \nu\|y\|_X \leq \|\mu x + \nu y\|_X + 2\mu\nu\|x - y\|_X,$$

valid for every Banach space X, we shall refer to them as to $(\infty, 1)$–uniform convexity and $(1, 1)$–uniform smoothness respectively for the sake of the convenience explained in the proof of Theorem 4.17, a) and the preservation of the duality.

Choosing x and y with $\mu x + \nu y = 0$, we see that h_c (h_s) must satisfy

$$h_c(\mu) \leq \mu^{p-1} + \nu^{p-1} \leq 1 \ (h_s(\mu) \geq \mu^{q-1} + \nu^{q-1} \geq 1),$$

if X is (q, h_c)–uniformly convex (smooth). Better estimates based on the duality in Lemma 4.1 are estblished in Lemma 4.6, and the fact of being better is our multi–homogeneous inequality Theorem 4.13, d). The function $\mu^{q-1} + \nu^{q-1}$ appears also if one uses Pisier's martingale approach (see [44]).

Definition 4.8. *For $q \in (1, 2]$, we say that a Banach space X is Hanner q–smooth if one has*

$$(\|x\|_X + \|y\|_X)^q + |\|x\|_X - \|y\|_X|^q \leq \|x + y\|_X^q + \|x - y\|_X^q \text{ for every } x, y \in X.$$

For $q \in [2, \infty)$, we say that the space X is Hanner q–convex if one has

$$(\|x\|_X + \|y\|_X)^q + |\|x\|_X - \|y\|_X|^q \geq \|x + y\|_X^q + \|x - y\|_X^q \text{ for every } x, y \in X.$$

According to Corollary 4.4, the properties defined in these definitions are inherited by subspaces, quotients and almost isometrically finitely–represented spaces.

Remark 4.6. *a) Every Banach space X is Hanner ∞–convex and 1–smooth:*

$$\|x\|_X + \|y\|_X + |\|x\|_X - \|y\|_X| = 2\max(\|x\|_X, \|y\|_X) \leq \|x+y\|_X + \|x-y\|_X;$$

$$\max(\|x+y\|_X, \|x-y\|_X) \leq \max(\|x\|_X + \|y\|_X, |\|x\|_X - \|y\|_X|).$$

There exists no Banach space X that would be either Hanner p–convex for $p \in [1,2)$ or Hanner q–smooth for $q \in (2,\infty]$ because it is impossible for $X = \mathbb{R}$, and, thus, any 1–dimensional subspace.
b) The implicit constant 1 in Definition 4.8 is important because we always have a constant provided by

$$2^{1/q}\max(\|x\|_X, \|y\|_X) \leq ((\|x\|_X + \|y\|_X)^q + |\|x\|_X - \|y\|_X|^q)^{1/q} \leq$$
$$\leq 2\max(\|x\|_X, \|y\|_X).$$

c) Part h) of Lemma 4.1 implies that $h_c > 0$ on $(0,1)$ if $h_c(\mu_0) > 0$ for some $\mu_0 \in (0,1)$, and that $h_s < \infty$ on $(0,1)$ if $h_s(\mu_0) < \infty$ for some $\mu_0 \in (0,1)$.
d) According to Part h) of Remark 4.2, the functions $\sup_{y \in A} \|x-y\|_X^p + \phi(x)$ for a bounded $A \subset X$ and $\limsup_{k \to \infty} \|x - y_k\|_X^p + \phi(x)$ for a bounded $\{y_k\}_{k=1}^\infty \subset X$ are (p, h_c)–uniformly convex if X is (p, h_c)–uniformly convex and $\phi : X \to \mathbb{R}$ is convex, and the functions $\inf_{y \in A} \|x-y\|_X^q + g(x)$ for $A \subset X$ and $\liminf_{k \to \infty} \|x - y_k\|_X^q + g(x)$ for a bounded $\{y_k\}_{k=1}^\infty \subset X$ are (q, h_s)–uniformly smooth if X is (q, h_s)–uniformly smooth and $g : X \to \mathbb{R}$ is affine.
e) Hanner [45] has shown that both $\mathbb{C}$ and L_p is Hanner p–smooth for $p \in [1,2]$ and Hanner p–convex for $p \in [2, \infty]$.
f) Note that the Hanner 2–convexity is equivalent to the Hanner 2–smoothness thanks to the linear substitution of the variables in the inequalities

$$2\|x\|_X^2 + 2\|x\|_X^2 \geq \|x+y\|_X^2 + \|x-y\|_X^2 \text{ and } 2\|x\|_X^2 + 2\|x\|_X^2 \leq \|x+y\|_X^2 + \|x-y\|_X^2,$$

and, thus, equivalent to the parallelogram identity. Hence, the only Hanner 2–convex and/or Hanner 2–smooth spaces are the inner product spaces.

Similarly to the classical uniform convexity and smoothness (see Definition 4.11), there is considerable duality in our setting. We provide two different proofs relying on linear and nonlinear duality arguments.

Theorem 4.5. *Let X be a Banach space and $p, q \in [1, \infty]$. Then:*
a) the space X is (p, h_c)–uniformly convex if and only if its dual X^* is $(p', h_c^{1-p'})$–uniformly smooth;
b) the space X is (q, h_s)–uniformly smooth if and only if its dual X^* is $(q', h_s^{1-q'})$–uniformly convex.

The first proof of Theorem 4.5. For every linear space Y, we define a pair of mutually inverse operators A_μ and A_μ^{-1} transforming $Y \times Y$ according to the formulas

$$A_\mu : (x,y) \longmapsto (\mu x + \nu y, y - x) \text{ and } A_\mu^{-1} : (x,y) \longmapsto (x - \nu y, x + \mu y).$$

Let $Y_{X,q,\mu}$ and $Z_{X,q,\mu,h}$ be the space $X \times X$, equipped, correspondingly, with the norms

$$\|(x,y)|Y_{X,q,\mu}\| := (\mu\|x\|_X^q + \nu\|y\|_X^q)^{1/q}, \quad \|(x,y)|Z_{X,q,\mu,h}\| := (\|x\|_X^q + h(\mu)\mu\nu\|y\|_X^q)^{1/q}.$$

Considering the duality relations

$$(Y_{X,q,\mu})^* = Y_{X^*,q',\mu} \text{ with } \langle (f,g), (x,y) \rangle := \mu f(x) + \nu g(y) \text{ and}$$

$$(Z_{X,q,\mu,h})^* = Z_{X^*,q',\mu,h^{1-q'}} \text{ with } \langle (f,g), (x,y) \rangle := f(x) + \mu\nu g(y),$$

we see that the theorem follows from the relations $A_\mu^* = A_\mu^{-1}$, $A_\mu^{-1*} = A_\mu$,

$$\|A_\mu|\mathcal{L}(Y_{X,p,\mu}, Z_{X,p,\mu,h_c})\| = \|A_\mu^*|\mathcal{L}(Z_{X^*,p',\mu,h_c^{1-p'}}, Y_{X^*,p',\mu})\| \text{ and}$$

$$\|A_\mu^{-1}|\mathcal{L}(Z_{X,q_s,\mu,h}, Y_{X,q,\mu})\| = \|A_\mu^{-1*}|\mathcal{L}(Y_{X^*,q',\mu}, Z_{X^*,q',\mu,h_s^{1-q'}})\|.$$

$\square$

The second proof of Theorem 4.5. Having in mind the relations

$$(s^q/q)^*(t) = t^{q'}/q', \quad (\|x\|_X^q/q)^* = \|y\|_{X^*}^{q'}/q', \quad (c_1 f(c_2 s))^* = c_1 f^*(t/(c_1 c_2))$$

and Lemma 4.1, f), we observe that the conditions of the theorem, written as

a) $\delta_\mu(s, \|\cdot\|_X^p/p, X) \geq h_c(\mu) s^p/p$ and $\rho_\mu(s, \|\cdot\|_X^q/q, X) \leq h_s(\mu) s^q/q$,

imply respectively (and independently)

$$\rho_\mu(t, \|\cdot\|_{X^*}^{p'}/p', X^*) \leq h_c^{1-p'}(\mu) t^{p'}/p' \text{ and } \delta_\mu(t, \|\cdot\|_{X^*}^{q'}/q', X^*) \geq h_s^{1-q'}(\mu) t^{q'}/q'.$$

$\square$

With the aid of Lemma 4.1, we can also establish the duality for (p, μ)–uniform convexity and (q, μ)–uniform smoothness.

To study the asymptotic power convexity and smoothness of Banach spaces, it can be useful to define the coefficients measuring their relative behavior with respect to the best possible midpoint one.

Definition 4.9. *For $\mu \in (0,1)$ and a Banach space X, let the relative asymptotic power convexity and smoothness constants be defined, respectively, by*

$$\mathrm{AC}(\mu, X) := \limsup_{p \to \infty} \frac{\delta_\mu(t, \|\cdot\|_X^p, X)}{2^{2-p} t^p} \text{ and } \mathrm{AS}(\mu, X) := \liminf_{p \to 1} \frac{\rho_\mu(t, \|\cdot\|_X^p, X)}{2^{2-p} t^p}.$$

Lemma 4.6. *Let X be an arbitrary normed space, $t > 0$, $2 \in [q,p] \subset (1,\infty)$ and $1 - \nu = \mu \in (0,1)$. Then we have*

a) $\delta_\mu(t, \|\cdot\|_X^p, X) \leq \left(\mu^{1/(p-1)} + \nu^{1/(p-1)}\right)^{1-p} t^p \leq t^p$;

b) $\rho_\mu(t, \|\cdot\|_X^q, X) \geq \left(\mu^{(q-1)} + \nu^{(q-1)}\right) t^q \geq t^q$;

c) $\mathrm{AC}(\mu, X) \leq (2\sqrt{\mu\nu})^{-1}$;

d) $\mathrm{AS}(\mu, X) \geq 1$.

Proof of Lemma 4.6. Taking $x = \nu w$ and $y = -\mu w$ for a $w \in X$ with $\|w\|_X = t$, one obtains $b)$ and a weaker version of $a)$ (see the multi–homogeneous inequality, Theorem 4.13, $d)$):

$$\delta_\mu(t, \|\cdot\|_X^p, X) \leq (\mu^{p-1} + \nu^{p-1})t^p \leq t^p \text{ and } \rho_\mu(t, \|\cdot\|_X^q, X) \geq (\mu^{q-1} + \nu^{q-1})t^q \geq t^q,$$

which is also valid for q', p' and X^* instead of p, q and X correspondingly. Applying Theorem 4.5, $a)$, we obtain $a)$ by duality. The duality consideration does not provide any improvement in the case of the uniform smoothness, as seen from the inequality of Theorem 4.13, $d)$.

Parts $c)$ and $d)$ follows from $a)$ and $b)$ respectively. □

Definition 4.10. *For a Banach space X and $q \in (1, \infty)$, let $S_{X,q}$ be the space $X \times X$ endowed with the quasi–norm (norm thanks to Theorem 4.6)*

$$\|(x,y)|S_{X,q}\| := 2^{-1}\left((\|x+y\|_X + \|x-y\|_X)^q + |\|x+y\|_X - \|x-y\|_X|^q\right)^{1/q}.$$

The next theorem demonstrates the perfect duality between $S_{X,q}$ and $S_{X^*,q'}$ (cf. the proof of Lemma 6 in [46]) implying the validity of the triangle inequality.

Theorem 4.6.

a) For every $x, y \in X$ and $f, g \in X^*$, one has

$$\langle (f,g), (x,y) \rangle = |f(x) + g(y)| \leq \|(f,g)|S_{X^*,q'}\| \cdot \|(x,y)|S_{X,q}\|.$$

b) For every $x, y \in X$, there are $f, g \in X^*$ satisfying

$$\|(f,g)|S_{X^*,q'}\| = 1 \text{ and } f(x) + g(y) = \|(x,y)|S_{X,q}\|.$$

c) For every $f, g \in X^*$ and $c \in (0,1)$, there are $x, y \in X$ satisfying

$$\|(x,y)|S_{X,q}\| = 1 \text{ and } f(x) + g(y) \geq c\|(f,g)|S_{X^*,q'}\|.$$

Moreover, both $S_{X,q}$ and $S_{X^*,q'}$ are Banach spaces.

Proof of Theorem 4.6. $a)$ Using the Hölder inequality, we obtain the estimates

$$4f(x) + 4g(y) = 2\langle f+g, x+y \rangle + 2\langle f-g, x-y \rangle \leq 2\|f+g\|_{X^*} \cdot \|x+y\|_X +$$
$$+ 2\|f-g\|_{X^*} \cdot \|x-y\|_X = (\|f+g\|_{X^*} + \|f-g\|_{X^*})(\|x+y\|_X + \|x-y\|_X) +$$
$$+ |\|f+g\|_{X^*} - \|f-g\|_{X^*}| \cdot |\|x+y\|_X - \|x-y\|_X| \leq 4\|(x,y)|S_{X,q}\| \cdot \|(f,g)|S_{X^*,q'}\|.$$

To establish b), we employ the Hahn–Banach theorem twice to find $a > 0$, $b \in [-a, a]$ and $u, v \in X^*$, satisfying $a^{q'} + |b|^{q'} = 1$, $\|u\|_{X^*} = \|v\|_{X^*} = 1$, $u(x+y) = \|x+y\|_X$, $v(x-y) = \|x-y\|_X$ and, hence,

$$2\|(x,y)|S_{X,q}\| = \langle (a+b)u, x+y \rangle + \langle (a-b)v, x-y \rangle = 2f(x) + 2g(y), \text{ where}$$
$$2f = (a+b)u + (a-b)v, \quad 2g = (a+b)u - (a-b)v,$$
$$\|f+g\|_{X^*} + \|f-g\|_{X^*} = 2a, \quad \|f+g\|_{X^*} - \|f-g\|_{X^*} = 2b.$$

These relations imply Part b). The proof of c) is symmetric except for the choice of $e, h \in X$ satisfying $f(e) + g(e) \geq c\|f+g\|_{X^*}$ and $f(h) - g(h) \geq c\|f-g\|_{X^*}$ and followed by the choice
$$2x = (a+b)e + (a-b)h \text{ and } 2y = (a+b)e - (a-b)h.$$

The space $S_{X,q}$ is Banach because

$$\|(x,y)|S_{X,q}\| = \max\left\{ f(x) + g(y) : \|(f,g)|S_{X^*,q'}\| = 1 \right\}$$

according to Parts a) and b). Similar argument for $S_{X^*,q'}$ finishes the proof. $\square$

Lemma 4.7. *Let X, Y be (quasi)Banach spaces, M, N be their (closed) subspaces respectively with inherited (quasi)norms, and $A \in \mathcal{L}(X, Y)$ with $AM \subset N$. Then one has $\|A_{|M}|\mathcal{L}(M,N)\| \leq \|A|\mathcal{L}(X,Y)\|$ and $\|\tilde{A}|\mathcal{L}(X/M, Y/N)\| \leq \|A|\mathcal{L}(X,Y)\|$, where $\tilde{A}$ is induced by A and correctly defined.*

Proof of Lemma 4.7. The first estimate and the existence of $\tilde{A}$ follows from the inclusion $AM \subset N$, while the second one is implied by $\inf_{z \in N} \|y - z\|_Y \leq \inf_{z \in AM} \|y - z\|_Y$. $\square$

Using either the duality relations for subspaces and quotients, or the last simple lemma, one establishes the following useful corollary.

Corollary 4.4. *Let X be a Banach space, Y be its subspace, and a normed space Z be finitely represented in X. If X is either (q, h)–uniformly convex, or (q, h)–uniformly smooth, or Hanner q–convex, or Hanner q–smooth, then all the spaces Y, X/Y and Z possess the same property.*

Corollary 4.5 (Lemma 6 in [46]). *Let X be a Banach space and $q \in (1, \infty)$. Then:*
a) the space X is Hanner q–convex if, and only if, its dual X^* is Hanner q'–smooth;
b) the space X is Hanner q–smooth if, and only if, its dual X^* is Hanner q'–convex.

Proof of Corollary 4.5. Repeating word by word the proof of the classical duality relation (for Banach spaces Y, Z) $\|A|\mathcal{L}(Y,Z)\| = \|A^*|\mathcal{L}(Z^*, Y^*)\|$ and utilising Theorem 4.6, we obtain the desirable relations $\|I|\mathcal{L}(X_q, S_{X,q})\| = \|I|\mathcal{L}(S_{X^*,q'}, X_{q'}^*)\|$ and $\|I|\mathcal{L}(S_{X,q}, X_q)\| = \|I|\mathcal{L}(X_{q'}, S_{X^*,q'})\|$, where $\|(x,y)|X_q\| = (\|x\|_X^q + \|y\|_X^q)^{1/q}$. $\square$

The non–weighted case $u_0 = v_0 = u_1 = v_1 = 1$ of Part b) of the next theorem is Theorem 5.1.2 from [47], its proof also works for Part a) (see also [48, 49]). This general setting, established with the aid of Besov's trick described by Lizorkin in [50], is taken from [16].

Theorem 4.7 ([16]). *a) For an index set I, let $\{\bar{A}_i\}_{i \in I}$ and $\{\bar{B}_i\}_{i \in I}$ be systems of compatible couples of Banach spaces. Assume that $p_j, q_j \in [1, \infty]$ for $j = 0, 1$, $\theta \in (0, 1)$, $1/p_\theta = (1-\theta)/p_0 + \theta/p_1$ and $1/q_\theta = (1-\theta)/q_0 + \theta/q_1$. Let also T be a bounded linear operator from $l_{p_j}(I, \{A_{ji}\}_{i \in I})$ into $l_{q_j}(K, \{B_{ji}\}_{i \in I})$ with the norm M_j for $j = 0, 1$ and finite, or countable I and K, where we use, correspondingly, $c_0(I, \{A_{ji}\}_{i \in I})$ and $c_0(K, \{B_{ji}\}_{i \in I})$ instead of $l_{p_j}(I, \{A_{ji}\}_{i \in I})$ if $p_j = \infty$ and $l_{q_j}(K, \{B_{ji}\}_{i \in I})$ if $q_j = \infty$. Then T is also bounded from*

$$l_{p_\theta}(I, \{\bar{A}_{i[\theta]}\}_{i \in I}) = (l_{p_0}(I, \{A_{0i}\}_{i \in I}), l_{p_1}(I, \{A_{1i}\}_{i \in I}))_{[\theta]} \text{ (isometry) into}$$

$$l_{q_\theta}(K, \{\bar{B}_{i[\theta]}\}_{i \in I}) = (l_{q_0}(K, \{B_{0i}\}_{i \in I}), l_{q_1}(K, \{B_{1i}\}_{i \in I}))_{[\theta]} \text{ (isometry) with the norm}$$

$$\|T|\mathcal{L}(l_{p_\theta}(I, \{\bar{A}_{i[\theta]}\}_{i \in I}), l_{q_\theta}(K, \{\bar{B}_{i[\theta]}\}_{i \in I}))\| \leq M_0^{1-\theta} M_1^\theta.$$

b) Let (A_0, A_1) and (B_0, B_1) be compatible couples of Banach spaces, and let also (Ω, μ) and (G, ν) be measure spaces. Assume that u_j and v_j for $j = 0, 1$ are μ–measurable and ν measurable weights on Ω and G correspondingly. Let $p_j, q_j \in [1, \infty]$ for $j = 0, 1$, $\theta \in (0, 1)$, $1/p_\theta = (1-\theta)/p_0 + \theta/p_1$, $1/q_\theta = (1-\theta)/q_0 + \theta/q_1$, $u_\theta = u_0^{\frac{p_1 - p_\theta}{p_1 - p_0}} u_1^{\frac{p_0 - p_\theta}{p_0 - p_1}}$ and $v_\theta = v_0^{\frac{q_1 - q_\theta}{q_1 - q_0}} v_1^{\frac{q_0 - q_\theta}{q_0 - q_1}}$. Assume also that a linear operator T is bounded from $L_{p_j}(\Omega, A_j)$ into $L_{q_j}(G, B_j)$ with the norm M_j for $j = 0, 1$, where the closures $L_\infty^0(\Omega, u_j, A_j)$ and/or $L_\infty^0(\Omega, v_j, B_j)$ of the simple functions are used instead of $L_{p_j}(\Omega, u_j, A_j)$ and $L_{q_j}(G, v_j, B_j)$ correspondingly if $p_j = \infty$ and/or $q_j = \infty$ for $j = 0, 1$. Then T is also bounded from

$$L_{p_\theta}(\Omega, u_\theta, \bar{A}_\theta) = (L_{p_0}(\Omega, u_0, A_0), L_{p_1}(\Omega, u_1, A_1))_{[\theta]} \text{ (isometry) into}$$

$$L_{q_\theta}(G, v_\theta, \bar{B}_\theta) = (L_{q_0}(G, v_0, B_0), L_{q_1}(G, v_1, B_1))_{[\theta]} \text{ (isometry) with the norm}$$

$$\|T|\mathcal{L}(L_{p_\theta}(\Omega, u_\theta, \bar{A}_\theta), L_{q_\theta}(G, v_\theta, \bar{B}_\theta))\| \leq M_0^{1-\theta} M_1^\theta.$$

Multiple iterations of the Banach space versions of both parts of Theorem 4.7 with the aid of Theorem 4.14 and Lemma 2.3 leads to the complex interpolation result for IG–spaces described in the following corollary.

Corollary 4.6. *Let (A_0, A_1) be a compatible pair (see Remark 2.6, d)) of IG_+-spaces $(A_0, A_1 \in IG_+)$ with the parameter functions p_0 and p_1 taking values in $[1, \infty]$ and $\theta \in (0, 1)$. Assume also that $1/p_\theta = (1-\theta)/p_0 + \theta/p_1$, and that the space $A_\theta \in IG_+$ defined by the parameter function p_θ is the space with the same tree type as both A_0 and A_1. Then we have the isomorphism $(A_0, A_1)_{[\theta]} \asymp A_\theta$ that becomes the isometry $(A_0, A_1)_{[\theta]} = A_\theta$ if $(A_0, A_1 \in IG_{0+})$.*

4.4.2. Classical Moduli of Uniform Convexity and Smoothness

In this section, we find, often sharp, estimates for the classical moduli of uniform smoothness and uniform convexity. The equivalence between the classical uniform convexity and some homogeneous variants is discussed also in [51]. Let us recall equivalent definitions of the classical moduli of uniform convexity and uniform smoothness (see [52, 2]).

Definition 4.11 (Classical moduli of uniform smoothness and convexity). *Let X be a Banach space. The classical modulus of uniform convexity of X is the function $\delta_X : (0, 2] \longmapsto \mathbb{R}_+$ defined by*

$$\delta_X(\varepsilon) := \inf \left\{ \frac{\|x + y\|_X}{2} - 1 : \|x\|_X = \|y\|_X = 1, \|x - y\|_X \geq \varepsilon \right\}.$$

The classical modulus of uniform smoothness of X is the convex function $\rho_X : (0, \infty) \longmapsto \mathbb{R}_+$ defined by

$$\rho_X(t) := \sup \left\{ \frac{\|x + y\|_X + \|x - y\|_X}{2} - 1 : \|x\|_X = 1, \|y\|_X \leq t \right\}.$$

The next lemma transforms Theorem 4.24 into rather good, and occasionally sharp, estimates for the classical moduli of uniform smoothness and uniform convexity.

Theorem 4.8. *For a Banach space Y, let X be a Banach space that is either a subspace, or a quotient of Y, or finitely represented in Y. Then we have the estimates:*

a) $\delta_X(\varepsilon) \geq h_c(1/2)\varepsilon^p/4p$ if Y is (p, h_c)–uniformly convex;

b) $\rho_X(t) \leq h_s(1/2)(2t)^q/4q$ if Y is (q, h_s)–uniformly smooth.

Proof of Theorem 4.8. Thanks to Corollary 4.4, the general case is reduced to $X = Y$. We use only the Hölder inequality and the trivial inequality $(1+x)^\alpha \leq 1+\alpha x$ for $\alpha \in (0, 1)$ and $x \in (-1, \infty)$ (along with the definitions of the (p, h_c)–uniform convexity and the (q, h_s)–uniform smoothness). Taking supremum over the admissible x and y in the left–hand sides of the following estimates, we obtain, respectively, $1 - \delta_X(\varepsilon)$ and $1 + \rho_X(t)$:

$$\left\| \frac{x + y}{2} \right\|_X \leq \left(\frac{\|x\|_X + \|y\|_X}{2} - 2^{-2}h_c(1/2)\|x - y\|_X^p \right)^{1/p} = (1 - h_c(1/2)\varepsilon^p/4)^{1/p} \leq$$
$$\leq 1 - h_c(1/2)\varepsilon^p/4p;$$

$$\frac{\|x + y\|_X + \|x - y\|_X}{2} \leq \left(\frac{\|x + y\|_X^q + \|x - y\|_X^q}{2} \right)^{1/q} \leq$$
$$\leq \left(\|x\|_X^q + 2^{-2}h_s(1/2)\|2y\|_X^q \right)^{1/q} = (1 + h_s(1/2)(2t)^q/4)^{1/q} \leq 1 + h_s(1/2)(2t)^q/4q. \quad \square$$

4.4.3. Analytic Tools and Bynum–Drew–Gross and Multi–Homogeneous Inequalities

The next theorem reveals the behavior of our properties with respect to the complex interpolation.

Theorem 4.9. *Let (A_0, A_1) be a compatible pair of Banach spaces, $q_j \in [1, \infty]$, and h_j be a positive function defined on $(0, 1)$ with $h_j(1 - t) = h_j(t)$ for $j = 0, 1$, and $\theta \in (0, 1)$, $1/q_\theta = (1 - \theta)/q_0 + \theta/q_1$, and $h_\theta = h_0^{\frac{q_1 - q_\theta}{q_1 - q_0}} h_1^{\frac{q_0 - q_\theta}{q_0 - q_1}}$. Then:*

a) the space $(A_0, A_1)_{[\theta]}$ is (q_θ, h_θ)–uniformly convex if A_j is (q_j, h_j)–uniformly convex for $j = 0, 1$;

b) the space $(A_0, A_1)_{[\theta]}$ is (q_θ, h_θ)–uniformly smooth if A_j is (q_j, h_j)–uniformly smooth for $j = 0, 1$.

Proof of Theorem 4.9. As we have seen in the proof of Theorem 4.5, to establish $a)$ and $b)$ means to establish the boundedness of, correspondingly, the operators A_μ and A_μ^{-1} between the spaces $Y_{\bar{A}_{[\theta]},q_0,\mu}$ and $Z_{\bar{A}_{[\theta]},q_0,\mu,h_\theta}$. But the boundedness of these operators is obtained by means of the complex interpolation and Theorem 5.6.3 from [47] (see Part $a)$ of Theorem 4.7) implying the identities $(Y_{A_0,q_0,\mu}, Y_{A_1,q_1,\mu})_{[\theta]} = Y_{\bar{A}_{[\theta]},q_0,\mu}$ and $(Z_{A_0,q_0,\mu,h_0}, Z_{A_1,q_1,\mu,h_1})_{[\theta]} = Z_{\bar{A}_{[\theta]},q_0,\mu,h_\theta}$. □

The next theorem deals with the limiting cases of the previous one. It is particularly interesting in the case $A_0 = A_1 = X$.

Theorem 4.10. *Let (A_0, A_1) be a compatible pair of Banach spaces, $p \in [1,2]$ and $q \in [2,\infty]$, and h be a positive function defined on $(0,1)$ with $h(1-t) = h(t)$.*
a) If A_1 is (q,h)–uniformly convex, then $(A_0, A_1)_{[q/r]}$ is $(r, 2^{q-r}h)$–uniformly convex for $r \in [q,\infty)$;
b) If A_1 is (p,h)–uniformly smooth, then $(A_0, A_1)_{[p'/r']}$ is $(r, 2^{\frac{p-r}{p-1}} h^{\frac{r-1}{p-1}})$–uniformly smooth for $r \in (1,p]$.

Proof of Theorem 4.10. As one has seen in Definition 3.3, the extra factor 2 is present if $q_1 = 1$ or $q_1 = \infty$. Remembering the agrement concerning the $(1,1)$–uniform smoothness, we obtain Part $b)$ as a particular case of Theorem 4.7, $b)$. To take into consideration the problem with the pseudo–weight $1/2$ for $q_1 = \infty$, we introduce the additional scaling by considering the operator B_μ defined by $B_\mu : (x,y) \longmapsto (\mu x + \nu y, (y-x)/2)$. Then we follow the same steps as in Theorem 4.9, $a)$ with $h'_j = 2^{q_j} h_j$ instead of h_j for $j = 0, 1$ and B_μ instead of A_μ. □

With the aid of the identity $(X, X)_{[\theta]} = X$, we obtain an important corollary of the previous theorems about the connectivity of the sets of the exponents of the uniform and Hanner convexity and smoothness.

Corollary 4.7. *Let X be a Banach space, $p \in [1,2]$ and $q \in [2,\infty]$, and h be a positive function defined on $(0,1)$ with $h(1-t) = h(t)$.*
a) If X is (q,h)–uniformly convex, then it is $(r, 2^{q-r}h)$–uniformly convex for $r \in [q,\infty)$;
b) If X is (p,h)–uniformly smooth, then it is $(r, 2^{\frac{p-r}{p-1}} h^{\frac{r-1}{p-1}})$–uniformly smooth for $r \in (1,p]$.

Let us recall the Bynum–Drew–Gross and Hanner inequalities. Part $a)$ is further sharpened in Theorem 6.3, $a), d)$ in [14] in terms of Milman's moduli for Lebesgue spaces.

Lemma 4.8 ([53, 45]).
a)[Bynum–Drew–Gross] For $a,b \in \mathbb{R}$, one has

$$\left(\left(\frac{a+b}{2}\right)^2 + (q-1)\left(\frac{a-b}{2}\right)^2\right)^{1/2} \leq \left(\frac{|a|^q + |b|^q}{2}\right)^{1/q} \quad \text{for } q \in [1,2];$$

$$\left(\left(\frac{a+b}{2}\right)^2 + (q-1)\left(\frac{a-b}{2}\right)^2\right)^{1/2} \geq \left(\frac{|a|^q + |b|^q}{2}\right)^{1/q} \quad \text{for } q \in [2,\infty).$$

b)[*Hanner*] For $f, g \in L_q$, one has

$$(\|f\|_{L_q} + \|g\|_{L_q})^q + |\|f\|_{L_q} - \|g\|_{L_q}|^q \leq \|f+g\|_{L_q}^q + \|f-g\|_{L_q}^q \text{ for } q \in [1,2];$$
$$(\|f\|_{L_q} + \|g\|_{L_q})^q + |\|f\|_{L_q} - \|g\|_{L_q}|^q \geq \|f+g\|_{L_q}^q + \|f-g\|_{L_q}^q \text{ for } q \in [2,\infty).$$

The following theorem reveals the relations between the Hanner q–convexity and smoothness and $(2, q-1)$–uniform convexity and smoothness with the aid of Lim and Smarzewski's trick.

Theorem 4.11. *Let X be a Banach space, $q \in (1, \infty)$, $\mu \in (0, 1)$ and $h > 0$. Assume also that A_μ, $Y_{X,q,\mu}$ and $Z_{X,q,\mu,h}$ are as in the proof of Theorem 4.5. Then*
a) the space X is $(2, q-1)$–uniformly convex if it is Hanner q–smooth;
b) the space X is $(2, q-1)$–uniformly smooth if it is Hanner q–convex;
c) $\|A_\mu| \mathcal{L}(Y_{X,2,\mu}, Z_{X,2,\mu,h})\| \leq 1$ if $\|A_\mu| \mathcal{L}(Y_{X,2,1/2}, Z_{X,2,1/2,h})\| \leq 1$;
d) $\|A_\mu^{-1}| \mathcal{L}(Z_{X,2,\mu,h}, Y_{X,2,\mu})\| \leq 1$ if $\|A_\mu^{-1}| \mathcal{L}(Z_{X,2,1/2,h}, Y_{X,2,1/2})\| \leq 1$.

Remark 4.7. *As shown in [54] the constant $q-1$ is sharp in the case of a Lebesgue space X in both a) and b). The examples leading to the sharpness in the spaces under consideration are presented in the proof of Theorem 4.15.*

Proof of Theorem 4.11. To establish the restricted version of a),

$$\|(x+y)/2\|_X^2 + (q-1)\|(x-y)/2\|_X^2 \leq \|x\|_X^2/2 + \|y\|_X^2/2 \text{ for } x, y \in X, \quad (1)$$

one uses the approach from [53] in the form of Lemma 4.8, a). Indeed, we set $a = \|x+y\|_X + \|x-y\|_X$ and $b = |\|x+y\|_X - \|x-y\|_X|$ and use the Hanner q–smoothness and Hölder inequality to obtain

$$\|2(x+y)\|_X^2 + (q-1)\|2(x-y)\|_X^2 \leq 2^{2/q'}(a^q + b^q)^{2/q} \leq$$
$$\leq 2^{2/q'}(\|2x\|_X^q + \|2y\|_X^q)^{2/q} \leq 2(\|2x\|_X^2 + \|2y\|_X^2).$$

Part a) follows from its restricted form (1) with the aid of Part c) that is the "trick" found in [54].

To establish c), let us notice that the set V of all ν, such that

$$\|\mu x + \nu y\|_X^2 + (q-1)\mu\nu\|x-y\|_X^2 \leq \mu\|x\|_X^2 + \nu\|y\|_X^2 \quad (2)$$

holds for all x and y, possesses the properties:

$$1/2 \in V \subset [0,1], V \text{ is closed}, V = 1 - V \text{ and } 2^{-1}V \subset V \text{ (even } V \cdot V \subset V).$$

The observation that the orbit of $1/2$ with respect to the transformation semigroup of $[0, 1]$ generated by $t \to 1-t$ and $t \to t/2$ is dense in $[0, 1]$ finishes the proof of c).

There are two ways, leading to b): we can either repeat the approach used to prove a) in the symmetric form utilising the second inequality of Lemma 4.8, a), or to use the duality (Corollary 4.5, a)), then Part a) with X^* and q', and, eventually, to use the second duality argument — Theorem 4.5, b). Similarly there are two ways to prove d): by means of either the symmetric approach, or the duality in Theorem 4.5. □

In the next two theorems, for $1 - \nu = \mu \in (0, 1/2]$ and $p \in (1, \infty) \setminus \{2\}$, the parameter $z_p(\mu)$ is the unique solution of the equation $h_p(z) = \nu z^{p-1} - \mu - (\nu z - \mu)^{p-1} = 0$ on the interval $[\mu/\nu, 1]$ and $z_2(\mu) = 1$.

While the Hanner inequality (Lemma 4.8, b)) and Theorem 4.11 describe the uniform convexity of the Lebesgue spaces L_p for $p \in (1, 2]$, the case $p > 2$ has required a completely different approach and been settled by Chekanov, Nesterov and Vladimirov [37] for $\mu = 1/2$ and Lim [55, 56] for an arbitrary μ.

Theorem 4.12 ([55, 56]). *For $p > 2$ and $1 - \nu = \mu \in (0, 1/2]$, let*

$$\psi_p(\mu) := \frac{1 + z_p(\mu)^{p-1}}{(1 + z_p(\mu))^{p-1}}.$$

Then,
a) $k(\mu) \leq \mu\nu\psi_p(\mu)$, if and only if one has, for every $x, y \in L_p(E, \mathbb{C})$,

$$\|\mu x + \nu y\|_{L_p}^p + k(\mu)\|x - y\|_{L_p}^p \leq \mu\|x\|_{L_p}^p + \nu\|y\|_{L_p}^p;$$

b) the function $z_p(\mu)$ ($\psi_p(\mu)$) is strictly increasing (decreasing) in μ, $z_p(1/2) = 1$ ($\psi_p(1/2) = 2^{2-p}$), and $z_p(0+)$ is the unique solution to the equation

$$(p-2)z^{p-1} + (p-1)z^{p-2} = 1 \text{ on } [0,1], \text{ hence } 2^{2-p} \leq \psi_p(\mu) \leq \psi_p(0+) =: c_p;$$

c) it is also true that $\max(c_p(\mu^{p-1} + \nu^{p-1}), (\mu\nu)^{p/2-1}) \leq \psi_p(\mu)$, and the constant c_p is the best possible in this inequality.

In the end of [56], Lim has also stated that all the inequalities in the proof, and, thus, in the statement of Theorem 4.12 can be converted in the case $p \in (1, 2]$. Let us note that the dual version of a) can also be derived from a) itself and Theorems 4.5 and 4.13, c).

We add to this line of observations stressing on the dependence on p, rather than on μ. Namely, let us assume $\mu \in (0, 1/2]$ fixed and designate

$$z(p, \mu) := z_p(\mu) \text{ and } \phi(p, t) = \frac{1 + t^{p-1}}{(1 + t)^{p-1}}.$$

The observation $\psi_p(1/2) = 2^{2-p}$ means that Part c) of the next theorem describes also the mutual asymptotic behavior of $\psi_p(\mu)$ for different μ. Its Part d) is an interesting multi–homogeneous inequality appearing also in an equivalent form in the lower estimates for the Chebyshev radii ratio constants and other constants dominating them in [6].

Theorem 4.13. *a) The function $z_p(\mu)$ is (strictly) increasing in $\mu \in (0, 1/2)$ for $p \in (1, \infty)$ and*

$$\max(\mu/\nu, (\mu/\nu)^{1/p-1}, 1) < z_p(\mu) < 1;$$

as a function of $t \in [0, 1]$, $\phi(p, t)$ is (strictly) increasing for $p \in (1, 2)$ and decreasing for $p \in (2, \infty)$; while $\phi(p, t)$ is decreasing in p, and $2^p\phi(p, t)$ is increasing in p.
b) There exist limits $\lim_{p \to \infty} \frac{\psi_p(\mu)}{2^{2-p}}$ and $\lim_{p \to 1} \psi_p(\mu)$, and we have, for $p > 2$,

$$\left(\frac{\mu}{\nu}\right)^{\frac{1}{p-1}} \leq z(p) \leq z_0^{\frac{1}{p-1}}, \quad \frac{1 + z_0}{(1 + z_0^{\frac{1}{p-1}})^{p-1}} \leq \psi_p(\mu) \leq \left(\mu^{\frac{1}{p-1}} + \nu^{\frac{1}{p-1}}\right)^{1-p} \leq \mu^{p-1} + \nu^{p-1},$$

and
$$\frac{\sqrt{z_0(\mu)} + 1/\sqrt{z_0(\mu)}}{2} \leq \lim_{p \to \infty} \frac{\psi_p(\mu)}{2^{2-p}} \leq \frac{1}{2\sqrt{\mu\nu}},$$

where $z_0 = z_0(\mu) \in [(\mu/\nu)^{1/(p-1)}, 1)$ is the only root of the equation

$$h(z) = \nu z \ln z - (\nu z - \mu) \ln(\nu z - \mu) = 0.$$

And, for $p \in (1, 2]$, we have

$$\frac{\mu}{\nu} \leq z(p) \leq z_0, \quad \left(\mu^{\frac{1}{p-1}} + \nu^{\frac{1}{p-1}}\right)^{1-p} \leq \mu^{p-1} + \nu^{p-1} \leq \psi_p(\mu) \leq \frac{1 + z_0^{p-1}}{(1+z_0)^{p-1}}, \text{ and}$$

$$\lim_{p \to 1} \psi_p(\mu) = \lim_{p \to 1} 2^{2-p} = 2,$$

where $z_0 = z_0(\mu) \in [(\mu/\nu)^{p-1}, 1)$ is the only root of the equation

$$h(z) = \nu z \ln z - (\nu z - \mu) \ln(\nu z - \mu) = 0.$$

c) $\psi_p(0+) = (p-1)\left(\dfrac{z_p(0+)}{1 + z_p(0+)}\right)^{p-2}$, $(z_p(\mu))^{p-1} = z_{p'}(\mu)$ and $\psi_{p'}(\mu) = (\psi_p(\mu))^{1-p'}$.

d) For $a, b \geq 0$ and $r > 0$, we have multi–homogeneous inequality

$$\left(\frac{a+b}{a^r + b^r}\right)^{1/r} \leq \frac{a^{1/r} + b^{1/r}}{a+b},$$

where the equality is equivalent to $ab(a-b)(r-1) = 0$.

Proof of Theorem 4.13. The fact that the function $z(p, \mu)$ is increasing in μ on $(0, 1/2)$ for $p > 2$ is established in Theorem 4.12, b). To consider $p \in (1, 2)$, we notice that the implicit function approach still works in this case. Namely, for $\mu = 1 - \nu \in [0, 1/2)$, let

$$g(\mu, y) = \nu y^{p-1} - \mu - (\nu y - \mu)^{p-1}.$$

Then one has, for $p \in (1, 2)$ and $\mu \in (0, 1/2)$,

$$g'_y = \nu(p-1)(y^{p-2} - (\nu y - \mu)^{p-2}) < 0; \quad g''_{\mu\mu} = (p-1)(2-p)(1+y)^2(\nu y - \mu)^{p-3} > 0,$$

meaning that g is decreasing in y on $[\mu/\lambda, 1)$ and convex and, therefore, increasing in μ ($g(0, y) = 0$). Hence, the implicit function $y(\mu)$ defined by $g(\mu, y(\mu)) = 0$ is increasing.

The behavior of $\phi(p, t)$ is well–described by its derivatives

$$\phi'_t = \frac{(p-1)(t^{p-2} - 1)}{(1+t)^p}; \quad \phi'_p = (1+t)^{1-p}\left(t^{p-1} \ln \frac{t}{1+t} - \ln(1+t)\right) < 0.$$

The rest of Part a) follows from the observations:

$$(2^{p-1}\phi)'_p(p, t) = \left(\frac{2}{1+t}\right)^{p-1} q(t), \text{ where } q(t) = \ln \frac{2}{1+t} + t^{p-1} \ln \frac{2t}{1+t}, \tag{1}$$

$$q(1) = 0, \text{ and } q'(t) = \frac{t^{p-2}-1}{1+t} + (p-1)t^{p-2}\ln\frac{2t}{1+t} < 0 \text{ for } t \in (0,1), p > 2. \quad (2)$$

In what follows, we make the change of the variable $r = p - 1$. We deduce the case $p \in (1, 2]$ of Part b) from the case $p > 2$ and the duality identities (involving p and p') in Part c). To approach Part b) with $p > 2$, we investigate the behavior of the function

$$u(r, t) = (\nu t^r - \mu)^{1/r} - (\nu t - \mu) \text{ for } r > 1 \text{ and } t \in \left[\left(\frac{\mu}{\nu}\right)^{1/r}, 1\right].$$

Since its derivatives are

$$u'_r(r, t) = r^{-2}(\nu t^r - \mu)^{1/r-1}h(t^r) \text{ and } u'_t(r, t) = \nu \left(\frac{t^r}{\nu t^r - \mu}\right)^{1-1/r} - \nu > 0, \text{ and}$$

$h'(s) > 0$ for $s \in (\mu/\lambda, 1)$, we see that $u(r, t)$ is strictly increasing in r in the domain $t > z_0^{1/r}$ and strictly decreasing in the domain $t < z_0^{1/r}$, making $r = \ln z_0 / \ln t$ the global maximum for a fixed t. The last problem is to understand, which domain the solution $t(r)$ of the equation $u(t(r), r) = 0$ is in.

If $t(r) > z_0^{1/r}$ for some $r > 1$, then the implicit derivative $t' = -u'_r/u'_t < 0$ and, thus, there is $r_0 > r$, such that

$$t(r_0) = z_0^{1/r_0} \text{ and } u(t(r_0), r_0) = (\nu z_0 - \mu)^{1/r_0} - (\nu z_0^{1/r_0} - \mu) = 0 = u(t(r_0), 1). \quad (3)$$

But this is impossible, since r_0 is the strict global maximum of $u(r, z_0^{1/r_0})$. Hence, we have the desirable inclusion of b): $t(r) \in \left[(\mu/\nu)^{1/r}, z_0^{1/r}\right]$. Using the monotonicity of $\phi(p, t)$ in t (see a)), Lemma 4.6 and taking the limit, one recovers the rest of b) with the aid of d).

The first identity in c) holds because the solution of $(p-2)z^{p-1} + (p-1)z^{p-2} = 1$ satisfies

$$1 + z^{p-1} = (p-1)(1+z)z^{p-2} \text{ and } \phi(p, z) = (p-1)\left(\frac{z}{1+z}\right)^{p-2}.$$

The second identity is implied by $t(1/r) = (t(r))^{1/r}$ and $\phi(p, t) = (\phi(p', t^{p-1}))^{1-p}$.

The inequality d) is symmetric with respect to the transformations

$$a \leftrightarrow b, (a, b) \leftrightarrow (\lambda a, \lambda b), \text{ and } r \leftrightarrow 1/r.$$

Therefore, we can set $r = a \geq 1$, $b \leq a \neq 0$ and $x = b/a$ to observe that it is equivalent to

$$f(x) \geq f(x^{1/r}) \text{ on } [0, 1] \text{ for } f(x) = g\left(\frac{1}{1+x}\right) = \frac{1+x^r}{(1+x)^r} \text{ with } g(y) = y^r + (1-y)^r.$$

Since f is decreasing on $[0, 1]$ because g is increasing on $[1/2, 1]$, we only need to note that $x \leq x^{1/r}$ to finish the proof of the inequality d) together with the equality conditions. □

We also need the following quantitative version of the result due to M. S. Baouendi and G. Goulaouic [57, 35] in terms of the Banach-Mazur distance (see Definition 2.3).

Theorem 4.14. ([57, 35]) *For $p \in [1, \infty]$, $\theta \in (0, 1)$, let (A_0, A_1) be a compatible couple of Banach spaces, and B be a complemented subspace of $A_0 + A_2$, whose projector $P \in \mathcal{L}(A_0) \cap \mathcal{L}(A_1)$. Then $(B_0, B_1) = (A_0 \cap B, A_1 \cap B)$ is also compatible, and we have*

a) $d_{BM}\left((B_0, B_1)_{\theta,p}, (A_0, A_1)_{\theta,p} \cap B\right) \leq \|P|\mathcal{L}(A_0)\|^{1-\theta} \|P|\mathcal{L}(A_1)\|^\theta$;

b) $d_{BM}\left((B_0, B_1)_{[\theta]}, (A_0, A_1)_{[\theta]} \cap B\right) \leq \|P|\mathcal{L}(A_0)\|^{1-\theta} \|P|\mathcal{L}(A_1)\|^\theta$.

While the lower estimates for $\|x|(B_0, B_1)_{\theta,p}\|$ and $\|x|(B_0, B_1)_{[\theta]}\|$ are provided by the definitions of the interpolation functors, the upper estimates follow from the exactness of these functors:
$\max\left(\|P|\mathcal{L}\left((A_0, A_1)_{\theta,p}\right)\|, \|P|\mathcal{L}\left((A_0, A_1)_{[\theta]}\right)\|\right) \leq \|P|\mathcal{L}(A_0)\|^{1-\theta} \|P|\mathcal{L}(A_1)\|^\theta$.

4.4.4. Independently Generated IG_+ and Noncommutative L_p Spaces

Theorem 4.15. *For $r, q \in [1, \infty]$ with $\{p_{\min}(X), p_{\max}(X), 2\} \subset [r, q]$, let X be either a subspace or a quotient, or finitely represented in $Y \in IG_+$. Then, for $f, g \in X$, one has*

a) $\left(\left\|\frac{f+g}{2}\right\|_X^2 + (r-1)\left\|\frac{f-g}{2}\right\|_X^2\right)^{1/2} \leq \left(\frac{\|f\|_X^r + \|g\|_X^r}{2}\right)^{1/r}$ *if $p_{\max}(Y) \leq 2$;*

b) $\left(\left\|\frac{f+g}{2}\right\|_X^2 + (q-1)\left\|\frac{f-g}{2}\right\|_X^2\right)^{1/2} \geq \left(\frac{\|f\|_X^q + \|g\|_X^q}{2}\right)^{1/q}$ *if $p_{\min}(Y) \geq 2$.*

The inequalities themselves and the bounds on the parameters r, q are sharp for $X = Y$.

Proof of Theorem 4.15. Thanks to Corollary 4.4, the assertion of the theorem for subspaces and quotients follows from the one for $X = Y \in IG_+$. Moreover, due to Lemma 2.3, we further assume, without loss of generality, that $X = Y \in IG_{0+}$. Thanks to the Hölder inequality, one also assumes that $r = p_{\min}(X)$ and $q = p_{\max}(Y)$. We prove *a)* because *b)* follows from *a)* either by duality, or by the argument symmetrically repeating the proof of *a)*.

We say that a Banach space Z has the property $BDG(r-1)$ if the inequality in Part *a)* holds for this space. Using the Minkowski inequality twice, we see that the Bochner–Lebesgue space $L_p(Z)$ has $BDG(r-1)$ if $p \in [r, 2]$ and Z has $BDG(r-1)$. The same argument shows that $l_p(I, \{Z_i\}_{i \in I})$ has $BDG(r-1)$ if $p \in [r, 2]$ and every Z_i has $BDG(r-1)$. Moreover, a noncommutative $L_p(\mathcal{M}, \tau)$, where $\mathcal{M}$ is a von Neumann algebra with a normal semifinite faithful weight τ, has $BDG(r-1)$ if $p \in [r, 2]$ according to Theorem 4.16. The original Bynum–Drew–Gross inequality (Lemma 4.8) means that $\mathbb{R}$ has $BDG(r-1)$ for every $r \in [1, 2]$.

After this preparation we see that the induction by the depth of the tree $T(X)$ permits us to finish the proof. Namely, we are using the "tree–decaying" process. We can assume that every last leaf (of the tree-growing process generating X) is either $\mathbb{R}$ or a noncommutative $L_p(\mathcal{M}, \tau)$ with $p \in [r, 2]$ and, thus, has $BDG(r-1)$. Then we notice that all the spaces that are the "leaves" combined with the immediate "branches" they have "grown" from have $BDG(r-1)$ too. We continue adding immediate branches until reaching the whole space X that is, therefore, has $BDG(r-1)$.

The sharpness of the bounds $p_{\min}$ and $p_{\max}$ for the parameters r and q respectively, as well as the constant 1 in both inequalities, follows from the properties of the sharpness of the corresponding parameters r and q for the (two–dimensional) spaces $l_{p_{\min}}(I_2)$ and $l_{p_{\max}}(I_2)$. The latter spaces are either isometric to subspaces of X, or finitely represented in it by a sequence of isometric copies of $l_{p_k}(I_2)$ for a monotone sequence $\{p_k\}_{k\in\mathbb{N}} \subset I(X)$ converging to $p_{\min}(X)$ or $p_{\max}(X)$. One also uses Lemma 2.4 to find the copies of the 2–dimensional l_p–spaces in the end–point leaves that can be noncommutative spaces and Lemma 2.2 in the cases of $lt_{p,q}$ or $lt^*_{p,q}$.

To demonstrate the sharpness of the bound $p_{\min}(X)$ or $p_{\max}(X)$ for r or q respectively, one takes $f = (1+t, 1-t)$ and $g = (1-t, 1+t)$ for some $t \to 0$. We finish the proof by the observation that the inequalities are sharp for $X = \mathbb{R}$. For example, one can take $f = (1+t, 0)$ and $g = (1-t, 0)$. Indeed, the nature of the both examples is reduced to the fact that both $(1 + (s-1)t^2)^{1/2}$ and $2^{-1/s}((1+t)^s + |1-t|^s)^{1/s}$ ($s \in \{r, p_{\min}(X), p_{\max}(X)\}$) have the same behavior when $t \to 0$. $\square$

The succeeding theorem is an immediate corollary of a deep result due to K. Ball, E. Carlen and E. Lieb [46, 27] (Theorem 1 in [46]). For $X = Y = S_p^n$, $r = \min(p, 2)$ and $q = \max(p, 2)$, it is presented in [46], along with the comments regarding its further extension to the general case (see also [27]). The last step of extending from the setting of semifinite von Neumann algebras can be performed with the aid of Theorem 2.3.

Theorem 4.16. *For $r, q \in [1, \infty]$ and $p \in (1, \infty)$ with $\{p, 2\} \subset [r, q]$, let X be either a subspace or a quotient of $Y \in \{L_p(\mathcal{M}, \tau), S_p\}$, or finitely represented in Y, where $(\mathcal{M}, \tau)$ is a von Neumannn algebra with a normal semifinite faithful weight τ. Then, for $f, g \in X$, one has*

a) $\left(\left\| \dfrac{f+g}{2} \right\|_X^2 + (r-1) \left\| \dfrac{f-g}{2} \right\|_X^2 \right)^{1/2} \leq \left(\dfrac{\|f\|_X^r + \|g\|_X^r}{2} \right)^{1/r}$;

b) $\left(\left\| \dfrac{f+g}{2} \right\|_X^2 + (q-1) \left\| \dfrac{f-g}{2} \right\|_X^2 \right)^{1/2} \geq \left(\dfrac{\|f\|_X^q + \|g\|_X^q}{2} \right)^{1/q}$.

The inequalities themselves and the bounds on the parameters p, r, q are sharp for $X = Y$.

Remark 4.8. *a) This result was used by Carlen and Lieb in [58] to confirm the Gross conjecture on the optimal hypercontractivity bounds for the fermion oscillator semigroup playing an important role in Mathematical Physics.*
b) Since the statement of this theorem is weaker than the corresponding Hanner convexity and smoothness (see Theorem 4.11), let us note that the validity of Hanner's inequalities for noncommutative spaces was established by K. Ball, E. Carlen and E. Lieb in [46] both in the special case of positive semidefinite $f + g$ and $f - g$ and in the reduced general case $p = r \in [1, 4/3]$ for the Hanner smoothness and $p = q \geq 4$ for the Hanner convexity (see Theorem 2 in [46]).

Proof of Theorem 4.16. Since b) is treated in a symmetric fashion with respect to a), we consider only the latter. The sharpness is established as in the proof of Theorem 4.15. Thanks to Theorem 2.3, we can assume that $(\mathcal{M}, \tau)$ is semifinite.

The limiting case $r = p$ is known as mentioned above, while the treatment of the case $p \in (r, 2]$ is very similar to the proof of Theorem 4.9. We use the notation from the proof of Theorem 4.5. Taking $A_0 = L_r(\mathcal{M}, \tau)$ and $A_1 = L_2(\mathcal{M}, \tau)$, we interpolate the relations

$$\|A_{1/2}|\mathcal{L}(Y_{A_0,r,1/2}, Z_{A_0,2,1/2,r-1})\| \leq 1 \text{ and } \|A_{1/2}|\mathcal{L}(Y_{A_1,r,1/2}, Z_{A_1,2,1/2,r-1})\| \leq 1 \quad (1)$$

with the aid of Theorem 4.7 for $1/p = (1 - \theta)/r + \theta/2$ to obtain $\|A_{1/2}|\mathcal{L}(Y_{A_{[\theta]},r,1/2}, Z_{A_{[\theta]},2,1/2,r-1})\| \leq 1$ with $A_{[\theta]} = L_p(\mathcal{M}, \tau)$ (Theorem 2.2). The second estimate in (1) holds because A_1 is Hilbert and, thus, isometric to some $l_2(I)$ with countable or uncountable I, while $l_2(I)$ is an IG-space treated in Theorem 4.15. □

To formulate the first main theorem of this section determining the (p, h_c)–uniform convexity and (q, h_s)–uniform smoothness of IG–spaces, we define two functions. For $s, t \in (1, \infty)$ and $\mu \in [0, 1/2]$, let

$$\omega_c(\mu, s, t) = \begin{cases} (s-1)2^{2-t} & \text{for } s \leq 2, \\ \psi_s(\mu)2^{s-t} & \text{for } s \geq 2; \end{cases}$$

$$\omega_s(\mu, s, t) = \begin{cases} (\psi_s(\mu))^{\frac{t-1}{s-1}} 2^{\frac{s-t}{s-1}} & \text{for } s \leq 2, \\ (s-1)^{t-1}2^{2-t} & \text{for } s \geq 2, \end{cases}$$

where

$$\psi_s(\mu) = \frac{1 + (z_s(\mu))^{s-1}}{(1 + z_s(\mu))^{s-1}} \text{ and}$$

$$z_s(\mu) = \begin{cases} \text{positive root of } \nu z^{s-1} - \mu = (\nu z - \mu)^{s-1} & \text{for } s \neq 2 \text{ and } \mu \neq 0, \\ \text{positive root of } (s-2)z^{s-1} + (s-1)z^{s-2} = 1 & \text{for } s \neq 2 \text{ and } \mu = 0, \\ 1 & \text{for } s = 2. \end{cases}$$

Theorem 4.17. *Let $1 - \nu = \mu \in [0, 1]$, and let X be either a subspace or a quotient of a space $Y \in IG$, or finitely represented in Y with $[p_{\min}(Y), p_{\max}(Y)] \cup \{2\} \subset [r, q] \subset (1, \infty)$. Then, for $f, g \in X$, we have*

a) $\|\mu f + \nu g\|_X^q + \mu\nu\omega_c(\min(\mu, \nu), p_{\min}(Y), q)\|f - g\|_X^q \leq \mu\|f\|_X^q + \nu\|g\|_X^q$;

b) $\mu\|f\|_X^r + \nu\|g\|_X^r \leq \|\mu f + \nu g\|_X^r + \mu\nu\omega_s(\min(\mu, \nu), p_{\max}(Y), r)\|f - g\|_X^r$;

c) *the bound $q \geq \max(p_{\max}(Y), 2)$ in a) is sharp if $l_{\max(p_{\max}(Y),2)}$ is finitely represented in X, and the bound $r \leq \min(p_{\min}(Y), 2)$ is sharp if $l_{\min(p_{\min}(Y),2)}$ is finitely represented in X; particularly, these are the cases when $X = Y \in IG$ and the spaces at the vertexes of $T(Y)$ that either possess the parameters $p_{\max}(Y)$ and $p_{\min}(Y)$, or possess the parameters converging to $p_{\max}(Y)$ and $p_{\min}(Y)$ are infinite-dimensional;*

d) *the constants $\omega_c(1/2, p_{\min}(Y), q) = 2^{2-q}$ and $\omega_s(1/2, p_{\max}(Y), r) = 2^{2-r}$ are sharp if $p_{\min}(Y) \geq 2$ and $p_{\max}(Y) \leq 2$ correspondingly;*

e) *the constant $\omega_c(\mu, p_{\min}(Y), 2) = p_{\min}(Y) - 1$ is sharp uniformly by μ if $l_{p_{\max}(Y)}(I_2)$ is finitely represented in X and $p_{\max}(Y) \leq 2$, and the constant $\omega_s(\mu, p_{\max}(Y), 2) = p_{\max}(Y) - 1$ is sharp uniformly by μ if $l_{p_{\min}(Y)}(I_2)$ is finitely represented in X and $p_{\min}(X) \geq 2$; this is, particularly the case when $X = Y \in IG$.*

Remark 4.9. a) *The finite representability of $l_{p_{\min}(Y)}$ and/or $l_{p_{\max}(Y)}$ in $Y \in IG$ is relatively easy to check. The "worst" cases may look like $l_{p_0}(\mathbb{N}, \{l_{p_i}(I_{n_i})\}_{i\in\mathbb{N}})$ for some unbounded $\{n_i\}_{i\in\mathbb{N}} \subset \mathbb{N}$ with $\inf_{i\in\mathbb{N}_0} p_i \notin \{p_i\}_{i\in\mathbb{N}_0}$ and/or $\sup_{i\in\mathbb{N}_0} p_i \notin \{p_i\}_{i\in\mathbb{N}_0}$.*
b) *The sharpness of the range for r is important, for example, for estimating the atomic Lyapunov and Kadets constants in Corollary 5.1 and Theorem 5.4 and for the variety of results in Sections 6 and 7.*

Proof of Theorem 4.17. As in the proof of Theorem 4.15, we notice that Corollary 4.4 and Lemma 2.3 reduce the proof of a) and b) to the case $X = Y \in IG_0$. Part b) is proved either in the completely symmetric way with respect to the proof of Part a), or with the aid of the duality in the form of Theorem 4.5. In the latter case, b) follows from a) and Theorem 4.13, c).

In the proof, we shall deal only with the IG_0 spaces that possess the same tree type $T = T(X)$ as the space X itself but different values of the parameters of the spaces at the vertexes of T. As in Definition 2.4, the parameter functions $p : P \to [1, \infty]$ defined on one and the same parameter position set $P = \mathcal{V}(X)$ determined by T and the spaces at its vertexes. Thus, $I(X) = p(P)$. The IG_0 space of the same tree type T with the parameter function p will be designated by X_p. We also adopt the notations introduced in Theorem 4.13.

Let us establish Part a). One assumes also that $\chi_{\lambda,+}$ and $\chi_{\lambda,-}$ are the indicator functions of the sets

$$P_{\lambda,+} = \{\varpi \in P : p(\varpi) > \lambda\} \text{ and } P_{\lambda,-} = P \setminus P_{\lambda,+}.$$

For a scalar $\gamma > 0$, let also $\bar{\gamma}$ be the constant parameter function defined by $\bar{\gamma}(P) = \{\gamma\}$.

Given $p \in I(X)$, there are two opportunities: $p_{\min}(X) \leq 2$ and $p_{\min}(X) > 2$. In the first case, we have either $2 \in [p_{\min}(X), p_{\max}(X))$, or $p_{\max}(X) \leq 2$. We construct the parameter function $q_2 := \chi_{2,-}p + \chi_{2,+}\bar{2}$ and observe that X_{q_2} is $(2, p_{\min}(X) - 1)$–uniform convex due to Part a) of Theorem 4.15 combined with Part c) of Theorem 4.11.

It is the last step, if $p_{\max}(X) \leq 2$. Otherwise, we find $\eta \in (0, 1)$ and r_2 with $r_2(P) \in (1, \infty]$ with the properties

$$1/p = (1-\eta)/q_2 + \eta/r_2, \ \chi_{2,-}p = \chi_{2,-}r_2 \text{ and } 1/q = (1-\eta)/2. \tag{1}$$

Thanks to the $(\infty, 1)$–uniform convexity of L_{r_2}, we now apply Theorem 4.9, a) and Corollary 4.6 ($IG \subset IG_+$) to finish the proof of a) in the case $p_{\min}(X) \leq 2$.

If $p_{\min}(X) \geq 2$, we skip the first step and use Theorem 4.12, a) to establish the $(p_{\min}(X), \psi_{p_{\min}(X)})$–uniform convexity of X_{q_2}, where q_2 is the constant parameter function $q_2(P) = \{p_{\min}(X)\}$. In the same manner as in (1), one chooses $\eta \in (0, 1)$ and $r_2 \in (1, \infty)$, satisfying

$$1/p = (1-\eta)/q_2 + \eta/r_2 \text{ and } 1/q = (1-\eta)/p_{\min}(X).$$

The proof is finished by the application of Theorem 4.9, a) and Corollary 4.6 as it is done in the first case.

The sharpness of the bounds on the exponents q and r in c) follows from the sharpness of the maximal type and minimal cotype exponents (see Definition 4.12) of the spaces $l_{\min(p_{\min}(X),2)}$ and $l_{\max(p_{\max}(X),2)}$ respectively because these spaces are finitely represented in $X = Y$ and, thus, inherit all its type and cotype exponents (see [52, 28, 2]). The type and cotype exponents of the space $X = Y$ itself is given in Theorem 4.20 below.

In Part d), the constants $\omega_c(\mu, p_{\min}(Y), q)$ and $\omega_s(\mu, p_{\max}(Y), r)$ are sharp because they coincide with the best constants provided by, respectively, Parts a) and b) of Lemma 4.6 for an arbitrary Banach space, starting with the real line.

In Part e), the constants $\omega_c(\mu, p_{\min}(X), 2) = p_{\min}(X) - 1$ and $\omega_s(\mu, p_{\max}(X), 2)$ are sharp for every μ because they coincide with the corresponding constants for the two–dimensional spaces $l_{p_{\min}(X)}(I_2)$ and $l_{p_{\max}(X)}(I_2)$ that are sharp thanks to the choice $x = (1 + t\nu, 1 - t\nu)$, $y = (1 - t\mu, 1 + t\mu)$ followed by taking the limit $t \to 0$ (also cf. [54]). Indeed, if $p_{\min}(X) \in I(X)$ ($p_{\max}(X) \in I(X)$), then $l_{p_{\min}(X)}(I_2)$ ($l_{p_{\max}(X)}(I_2)$) is a subspaces of X. Otherwise, there exists a monotone sequence $\{p_k\}_{k \in \mathbb{N}} \subset I(X)$ converging to $p_{\min}(X)$ ($p_{\max}(X)$), and, therefore, a sequence of subspaces of X isometric to $l_{p_k}(I_2)$ with the sharp $(2, p_k - 1)$–uniform convexity (smoothness) constants $p_k - 1$ converging to $p_{\min}(X) - 1$ ($p_{\max}(X) - 1$) providing the sharpness of the latter constant. □

Before presenting the next theorem describing the (q, h_c)–uniform convexity and (r, h_s)–uniform smoothness properties of noncommutative Lebesgue spaces (ex. Schatten–von Neumann classes), we recall that $\omega_c(1/2, s, t) = (\min(s, 2) - 1) 2^{2-t}$ and $\omega_s(1/2, s, t) = (\max(s, 2) - 1)^{t-1} 2^{2-t}$ for $s, t \in [1, \infty]$. The midpoint case $\mu = \nu = 1/2$ of the next theorem was considered by Dixmier [59] in 1953, Simon [60] in 1987 and Ball, Carlen and Lieb [46] in 1994. Namely (for $\mu = 1/2$), the cases of both Part a) with $p = q \in [2, \infty]$ and Part b) with $p = r \in [1, 2]$ were treated in [59, 60], while the cases of both Part a) with $p = r \in [1, 2]$ and Part b) with $p = q \in [2, \infty]$ were considered in [46].

Theorem 4.18. *Let $1 - \nu = \mu \in [0, 1]$, $r, q \in [1, \infty]$ and $p \in (1, \infty)$, and let X be either a subspace or a quotient of $Y \in \{L_p(\mathcal{M}, \tau), S_p\}$, or finitely represented in Y, where $(\mathcal{M}, \tau)$ is a von Neumannn algebra with a normal semifinite faithful weight τ. Assume also that $\{p, 2\} \subset [r, q]$. Then, for $f, g \in X$, we have*

a) $\|\mu f + \nu g\|_X^q + \mu\nu\omega_c(1/2, p, q)\|f - g\|_X^q \leq \mu\|f\|_X^q + \nu\|g\|_X^q$;

b) $\mu\|f\|_X^r + \nu\|g\|_X^r \leq \|\mu f + \nu g\|_X^r + \mu\nu\omega_s(1/2, p, r)\|f - g\|_X^r$;

c) the bound $q \geq \max(p, 2)$ in a) is sharp if $l_{\max(p, 2)}$ is finitely represented in X, and the bound $r \leq \min(p, 2)$ is sharp if $l_{\min(p, 2)}$ is finitely represented in X;

d) for the space X, the constants $\omega_c(1/2, p, q) = 2^{2-q}$ and $\omega_s(1/2, p, r) = 2^{2-r}$ are sharp if $p \geq 2$ and $p \leq 2$ correspondingly;

e) the constants $\omega_c(1/2, p, 2) = p - 1$ and $\omega_s(1/2, p, 2) = p - 1$ are sharp uniformly by μ if $l_p(I_2)$ is finitely represented in X and, correspondingly, $p \in [1, 2]$ and $p \in [2, \infty]$; this is, particularly, the case when $X = Y$.

Remark 4.10. *It is natural to investigate the (q, h_c)–uniform convexity and (r, h_s)–uniform smoothness properties of noncommutative Lebesgue spaces in a more precise manner (with h_c and h_s depending on μ) in the Dixmier–Simon ranges of the parameter p and compare them with the corresponding properties of the (commutative) Lebesgues spaces and l_p.*

Proof of Theorem 4.18. Regarding Parts a) and b), we notice (as in some proofs above) that X inherits the corresponding convexity and smoothness properties of Y and, thus, assume that $X = Y$.

Our proof of Part a) with $p \in [2, \infty]$ and Part b) with $p \in [1, 2]$ will follow closely the proof of Corollary 4.8 relying on the complex interpolation method. The conditions on q and r permit us to represent $X = (L_2(\mathcal{M}, \tau), Z)_{[\theta]}$ for some $Z = L_{p_0}(\mathcal{M}, \tau)$ with $p_0 \in [1, \infty]$ and with either $(1 - \theta)/2 = 1/q$ in the former case, or $(1 - \theta)/2 + \theta = 1/r$ in the latter case. Then, noticing that Z (as any Banach space) is $(\infty, 1)$–uniformly convex and $(1, 1)$–uniformly smooth (see Definition 4.7), while L_2 is both $(2, 1)$–uniformly smooth and $(2, 1)$–uniformly convex as a Hilbert space, we apply Theorem 4.10 to establish both Part a) with $p \in [2, \infty]$ and Part b) with $p \in [1, 2]$.

The sharpness of the constants in d) is implied by their sharpness for $\mu = \nu = 1/2$ established in Part d) of Theorem 4.17 (i.e., they are the best possible for an arbitrary Banach space according to Lemma 4.6).

Theorem 4.16 combined with Theorem 4.11, c), d) immediately implies the $(2, p-1)$–uniform convexity in a) with $p \in [1, \infty]$ and the $(2, p-1)$–uniform smoothness in b) with $p \in [2, \infty]$. Eventually we finish the proof of Parts a) and b) by interpolating these properties with the help of Corollary 4.7.

The proof of c) and e) repeats the proof of Parts c) and e) of Theorem 4.17. We only note that Y definitely contains an isometric copy of the (commutative) space $l_p(I_2)$ for every exponent $p \in I(Y)$ (1–dimensional leaves do not count as leaves). □

Theorem 4.19. Let $1 - \nu = \mu \in [0, 1]$, and let X be either a subspace or a quotient of a space $Y \in IG_+$, or finitely represented in Y with $[p_{\min}(Y), p_{\max}(Y)] \cup \{2\} \subset [r, q] \subset (1, \infty)$. Then, for $f, g \in X$, we have

a) $\|\mu f + \nu g\|_X^q + \mu \nu \omega_c(1/2, p_{\min}(Y), q)\|f - g\|_X^q \leq \mu\|f\|_X^q + \nu\|g\|_X^q$;

b) $\mu\|f\|_X^r + \nu\|g\|_X^r \leq \|\mu f + \nu g\|_X^r + \mu \nu \omega_s(1/2, p_{\max}(Y), r)\|f - g\|_X^r$;

c) the bound $q \geq \max(p_{\max}(Y), 2)$ in a) is sharp if $l_{\max(p_{\max}(Y), 2)}$ is finitely represented in X, and the bound $r \leq \min(p_{\min}(Y), 2)$ is sharp if $l_{\min(p_{\min}(Y), 2)}$ is finitely represented in X;

d) the constants $\omega_c(1/2, p_{\min}(Y), q) = 2^{2-q}$ and $\omega_s(1/2, p_{\max}(Y), r) = 2^{2-r}$ are sharp uniformly by μ if $p_{\min}(Y) \geq 2$ and $p_{\max}(Y) \leq 2$ correspondingly;

e) the constant $\omega_c(\mu, p_{\min}(Y), 2) = p_{\min}(Y) - 1$ is sharp uniformly by μ if $l_{p_{\max}(Y)}(I_2)$ is finitely represented in X and $p_{\max}(Y) \leq 2$, and the constant $\omega_s(\mu, p_{\max}(Y), 2) = p_{\max}(Y) - 1$ is sharp uniformly by μ if $l_{p_{\min}(Y)}(I_2)$ is finitely represented in X and $p_{\min}(X) \geq 2$; this is, particularly the case when $X = Y \in IG$.

Proof of Theorem 4.19. The proof of this theorem almost coincides with the proof of Theorems 4.17. More precisely, instead of the sharp $(p_{\min}(X), \psi_{p_{\min}(X)})$–uniform convexity of X_{q_2} provided by Theorem 4.12, a), where q_2 is the constant parameter function $q_2(P) = \{p_{\min}(X)\}$, we have to employ the rough $(p_{\min}(X), 2^{2-p_{\min}(X)})$–uniform convexity of X_{q_2} provided by Theorem 4.10, a) as the result of the identity $[X_{\bar{2}}, X_{\bar{1}}]_{[2/p_{\min}(x)]} = X_{q_2}$ given by Corollary 4.6. Here the spaces $X_{\bar{2}}$ and $X_{\bar{1}}$ are defined as X_{q_2} but with the constant values $\{2\}$ and $\{1\}$ respectively.

Keeping in mind Lemma 2.4, one establishes the sharpness in c) and e) exactly as in the proof of Theorem 4.17. The uniform sharpness of the constants in d) (as constants and not as functions) is implied by the sharpness of them for $\mu = 1/2$ that is obtained exactly as in the proof of Theorem 4.17. □

Remark 4.11. *To illustrate the reason for the sharpness of some constants in Theorems 4.17 − 4.19 and Corollary 4.8, let us notice, in particular, that the space $l_p(A)$ with $A = \prod_{i=1}^n A_i$ and $p \in (1, \infty)^n$ for $\cup_{i=1}^n A_i \subset \mathbb{N}$, we necessarily have $|A_i| > 1$. Otherwise, it would be just the space $l_{\tilde{p}}\left(\prod_{j \in I_n}^{j \neq i} A_i\right)$ with the corresponding $\tilde{p} \in (1, \infty)^{n-1}$.*

4.4.5. Setting of Lebesgue, Sobolev, Besov and Lizorkin–Triebel Spaces

The interpretation of the function spaces as subspaces of some IG–spaces in Section 3 and Theorem 4.17, a), b) imply Theorem 4.23 and Parts a) of Theorems 4.20 − 4.22. The scheme of the proof of Parts $c - e$) of Theorem 4.17 assisted by the finite representability of l_σ–spaces [23] (alternative proofs in Sections 3,4.4 and 6 there) yield the rest of these theorems. Let us recall that the functions ω_c and ω_s are defined in front of Theorem 4.17, and the admissibility of the auxiliary summability parameter a is found in Definition 3.1.

Theorem 4.20. *Let open $G \subset \mathbb{R}^n$, $p, a \in (1, \infty)^n$, $q, \varsigma \in (1, \infty)$, $s \in (0, \infty)$ and $\{p_{\min}, p_{\max}, q, 2, a_{\min}, a_{\max}\} \subset [r_s, r_c] \subset (1, \infty)$. Assume also that*

$$Y \in \left\{ B^s_{p,q,a}(G), \tilde{B}^{s,A}_{p,q,a}(G), L^s_{p,q,a}(G), \tilde{L}^{s,A}_{p,q,a}(G), b^s_{p,q,a}(G), \tilde{b}^{s,A}_{p,q,a}(G), l^s_{p,q,a}(G), \tilde{l}^{s,A}_{p,q,a}(G), \right.$$
$$B^s_{p',q',a'}(G)^*, \tilde{B}^{s,A}_{p',q',a'}(G)^*, L^s_{p',q',a'}(G)^*, \tilde{L}^{s,A}_{p',q',a'}(G)^*, b^s_{p',q',a'}(G)^*, \tilde{b}^{s,A}_{p',q',a'}(G)^*,$$
$$\left. l^s_{p',q',a'}(G)^*, \tilde{l}^{s,A}_{p',q',a'}(G)^* \right\},$$

and X is either a subspace, or a quotient, or finitely represented in Y. Then
a) the space X is (r_c, h_c)–uniformly convex and (r_s, h_s)–uniformly smooth with $h_c(\mu) = \omega_c(\mu, \min(p_{\min}, q, a_{\min}), r_c)$ and $h_s(\mu) = \omega_s(\mu, \max(p_{\max}, q, a_{\max}), r_s)$ for $\mu \in [0, 1]$.
b) If a is admissible [23, 11] for Y, the condition $\{p_{\min}, p_{\max}, q, 2\} \subset [r_s, r_c]$ is sharp.
c) In Part a) the constants $h_c(1/2) = \omega_c(1/2, \min(p_{\min}, q, a_{\min}), r_c) = 2^{2-r_c}$ and $h_s(1/2) = \omega_s(1/2, \max(p_{\max}, q, a_{\max}), r_s) = 2^{2-r_s}$ are sharp if $\min(p_{\min}, q, a_{\min}) = \min(p_{\min}, q) \geq 2$ and $\max(p_{\max}, q, a_{\max}) = \max(p_{\max}, q) \leq 2$ correspondingly.
d) For $X = Y$ with admissible a, the constant $\omega_c(\mu, \min(p_{\min}, q, a_{\min}), 2) = \min(p_{\min}, q, a_{\min}) - 1$ is sharp uniformly by μ if $\min(p_{\min}, q, a_{\min}) = \min(p_{\min}, q)$ and $\max(p_{\max}, q, a_{\max}) \leq 2$, and the constant $\omega_s(\mu, \max(p_{\max}, q, a_{\max}), 2) = \max(p_{\max}, q, a_{\max}) - 1$ is sharp uniformly by μ if $\max(p_{\max}, q, a_{\max}) = \max(p_{\max}, q)$ and $\min(p_{\min}, q, a_{\min}) \geq 2$.

Theorem 4.21. *Let $Y \in \left\{ L_p(G), W^s_p(G), W^s_{p'}(G)^* \right\}$ for open $G \subset \mathbb{R}^n$, $p \in (1, \infty)^n$, $\varsigma \in (1, \infty)$, $s \in \mathbb{N}_0^n$ and $\{p_{\min}, p_{\max}, 2\} \subset [r_s, r_c] \subset (1, \infty)$. Assume also that X is either a subspace, or a quotient, or almost isometrically finitely represented in Y. Then*
a) the space X is (r_c, h_c)–uniformly convex and (r_s, h_s)–uniformly smooth with $h_c(\mu) = \omega_c(\mu, p_{\min}, r_c)$ and $h_s(\mu) = \omega_s(\mu, p_{\max}, r_s)$ for $\mu \in [0, 1]$.
b) The condition $\{p_{\min}, p_{\max}, 2\} \subset [r_s, r_c]$ is sharp.
c) The constants $\omega_c(1/2, p_{\min}, r_c) = 2^{2-r_c}$ and $\omega_s(1/2, p_{\max}, r_s) = 2^{2-r_s}$ are sharp if $p_{\min} \geq 2$ and $p_{\max} \leq 2$ correspondingly.
d) For $X = Y$ the constants $\omega_c(\mu, p_{\min}, 2) = p_{\min} - 1$ and $\omega_s(\mu, p_{\max}, 2) = p_{\max} - 1$ are sharp uniformly by μ if $p_{\max} \leq 2$ and $p_{\min} \geq 2$ correspondingly.

Theorem 4.22. *Let* $Y \in \left\{B^s_{p,q}(G), B^s_{p,q}(\mathbb{R}^n)_w, L^s_{p,q}(\mathbb{R}^n)_w, B^s_{p',q'}(\mathbb{R}^n)^*_w, L^s_{p',q'}(\mathbb{R}^n)^*_w\right\}$ *for an open* $G \subset \mathbb{R}^n$, $p \in (1,\infty)^n$, $q, \varsigma \in (1,\infty)$, $s \in (0,\infty)$ *and* $\{p_{\min}, p_{\max}, q, 2\} \subset [r_s, r_c] \subset (1,\infty)$. *Assume also that* X *is either a subspace, or a quotient, or finitely represented in* Y. *Then*

a) *the space* X *is* (r_c, h_c)*–uniformly convex and* (r_s, h_s)*–uniformly smooth with* $h_c(\mu) = \omega_c(\mu, \min(p_{\min}, q), r_c)$ *and* $h_s(\mu) = \omega_s(\mu, \max(p_{\max}, q), r_s)$ *for* $\mu \in [0,1]$.
b) *The condition* $\{p_{\min}, p_{\max}, q, 2\} \subset [r_s, r_c]$ *is sharp.*
c) *The constants* $\omega_c(1/2, \min(p_{\min}, q), r_c) = 2^{2-r_c}$ *and* $\omega_s(1/2, \max(p_{\max}, q), r_s) = 2^{2-r_s}$ *are sharp if* $\min(p_{\min}, q) \geq 2$ *and* $\max(p_{\max}, q) \leq 2$ *correspondingly.*
d) *For* $X = Y$ *the constants* $\omega_c(\mu, \min(p_{\min}, q), 2) = \min(p_{\min}, q) - 1$ *and* $\omega_s(\mu, \max(p_{\max}, q), 2) = \max(p_{\max}, q) - 1$ *are sharp uniformly by* μ *if* $\max(p_{\max}, q) \leq 2$ *and* $\min(p_{\min}, q) \geq 2$ *correspondingly.*

Theorem 4.23. *Let* $Y \in \left\{B^s_{p,q,\mathcal{F}}(G), L^s_{p,q,\mathcal{F}}(G), B^s_{p',q',\mathcal{F}}(G)^*, L^s_{p',q',\mathcal{F}}(G)^*\right\}$ *for* $G \subset \mathbb{R}^n$, $p \in (1,\infty)^n$, $q, \varsigma \in (1,\infty)$, $s \in \mathbb{R}$, $\{p_{\min}, p_{\max}, q, 2\} \subset [r_s, r_c] \subset (1,\infty)$. *Assume also that* X *is either a subspace, or a quotient, or finitely represented in* Y. *Then* X *is* (r_c, h_c)*–uniformly convex and* (r_s, h_s)*–uniformly smooth with* $h_c(\mu) = \omega_c(\mu, \min(p_{\min}, q), r_c)$ *and* $h_s(\mu) = \omega_s(\mu, \max(p_{\max}, q), r_s)$ *for* $\mu \in [0,1]$.

4.5. Midpoint Uniform Convexity and Smoothness and Rademacher Constants

In this section we establish sharper estimates for the parameters $h_s(1/2)$ and $h_c(1/2)$ implying also the estimates for the Rademacher type and cotype constants entering various important explicit estimates of the local geometry of Banach spaces (see [61, 62, 63]). For example, they were required for the explicit estimates describing quantitative Dvoretzky's theorems for finite–dimensional subspaces of Banach spaces (see [8] for the setting of function and IG–spaces). Thanks to the point of view at function spaces presented and employed in Sections 3 and 4.4.5, we deduce the corresponding estimates for various types of function spaces (and noncommutative L_p) from Theorem 4.24.

We recall the definitions of the Rademacher type and cotype (constants), B–convexity and C–convexity. Let $\{r_i\}_{i\in\mathbb{N}}$ be the family of the Rademacher functions on $[0,1]$.

Definition 4.12. *Let* X *be a Banach space and* $p, q \in [1,\infty]$. *If, for some constant* $T > 0$ *and every finite* $I \subset \mathbb{N}$ *and* $\{x_i\}_{i\in I} \subset X$, *one has*

$$\left\| \sum_{i\in I} r_i x_i \,\Big|\, L_q([0,1], X) \right\| \leq T \left\| \{x_i\}_{i\in I} \,\Big|\, l_p(I, X) \right\|, \qquad (RT)$$

X *is said to possess the Rademacher type* p.
If, for some constant $C > 0$ *and every finite* $I \subset \mathbb{N}$ *and* $\{x_i\}_{i\in I} \subset X$, *one has*

$$\left\| \{x_i\}_{i\in I} \,\Big|\, l_q(I, X) \right\| \leq C \left\| \sum_{i\in I} r_i x_i \,\Big|\, L_p([0,1], X) \right\|, \qquad (RC)$$

X *is said to possess the Rademacher cotype* q. *The best constants* T *and* C *in* (RT) *and* (RC) *respectively are the Rademacher type* $T_{p,q} = T_{p,q}(X)$ *and cotype* $C_{p,q} = C_{p,q}(X)$ *constants. Let* $I_T(X)$ *and* $I_C(X)$ *be the sets (closed or semiclosed intervals) of those* p *and*

q respectively, such that X possesses the Rademacher type p and the Rademacher cotype q. A Banach space X is said to be B–convex (C–convex) if the interval $I_T(X)$ ($I_C(X)$) is nontrivial, i.e., contains, at least, two different points of $[1, 2]$ ($[2, \infty]$).

Note that every Banach space is of Rademacher type 1 and of Rademacher cotype ∞ and that $T_{p,p}(X), C_{q,q}(X) \geq 1$ for $p, q \in [1, \infty]$.

The next lemma is an abstract form of the Clarkson inequalities and their relation with the uniform convexity and smoothness. The study of the Clarkson classes of Banach spaces leads to (occasionally sharp) quantitative Hahn–Banach theorems [16].

Lemma 4.9 (Abstract Clarkson inequality). *For $\theta \in (0, 1)$, let Y and H be a Banach and a Hilbert spaces and $X = [Y, H]_{[\theta]}$. Then X is (q, h_c)–uniformly convex for $q \in [2/\theta, \infty]$ with $h_c > 0$ on $(0, 1)$ and $h_c(1/2) = 2^{2-q}$ and (r, h_s)–uniformly smooth for $q \in [1, (2/\theta)']$ with $h_s(1/2) = 2^{2-r}$, and we have*

$$\left(\frac{\|x\|_X^{2/\theta} + \|y\|_X^{2/\theta}}{2}\right)^{\theta/2} \leq \left(\left\|\frac{x+y}{2}\right\|_X^{(2/\theta)'} + \left\|\frac{x-y}{2}\right\|_X^{(2/\theta)'}\right)^{1/(2/\theta)'} \quad \text{and}$$

$$\left(\left\|\frac{x+y}{2}\right\|_X^{2/\theta} + \left\|\frac{x-y}{2}\right\|_X^{2/\theta}\right)^{\theta/2} \leq \left(\frac{\|x\|_X^{(2/\theta)'} + \|y\|_X^{(2/\theta)'}}{2}\right)^{1/(2/\theta)'} \quad \text{for } x, y \in X.$$

Proof of Lemma 4.9. Thanks to the Hölder inequality, embedding $l_{(2/\theta)'} \subset l_{2/\theta}$ and Remark 4.6, c), we see that the inequalities imply the smoothness and convexity claimed. Moreover, the inequalities are equivalent. We obtain the second inequality by applying the complex interpolation to the pairs $(L_\infty(D, Y), L_2(D, H))$ and $(l_1(I_2, Y), l_2(I_2, H))$ (see Theorem 4.7). □

In the next theorem we sharpen the uniform midpoint convexity and smoothness constants and, thus, also estimate Rademacher type and cotype constants.

Theorem 4.24. *Let X be either a subspace or a quotient of a space $Y \in IG_+$, or finitely represented in Y with $[p_{\min}(Y), p_{\max}(Y)] \cup \{2\} \subset [r, q] \subset (1, \infty)$. Then X is (q, h_c)–uniformly convex and (r, h_s)–uniformly smooth with*

a) $h_c(1/2) = 2^{2-q} \left(\min(p_{\min}(Y), 2) - 1\right)^{\frac{(p_{\min}(Y)'-q)_+}{p_{\min}(Y)'-2}}$;

b) $h_s(1/2) = 2^{2-r} \left(\max(p_{\max}(Y), 2) - 1\right)^{\frac{(r-p_{\max}(Y)')_+}{2-p_{\max}(Y)'}}$;

c) $C_{q,q}(X) \leq \left(\min(p_{\min}(Y), 2) - 1\right)^{-\frac{(p_{\min}(Y)'/q-1)_+}{p_{\min}(Y)'-2}}$, *particularly*, $C_{q,q}(X) = 1$ *when* $1/p_{\min}(Y) + 1/p_{\max}(Y) \leq 1$;

d) $T_{r,r}(X) \leq \left(\max(p_{\max}(Y), 2) - 1\right)^{\frac{(1-p_{\max}(Y)'/r)_+}{2-p_{\max}(Y)'}}$, *particularly*, $T_{r,r}(X) = 1$ *when* $1/p_{\min}(Y) + 1/p_{\max}(Y) \geq 1$.

Proof of Theorem 4.24. As in the proof of Theorem 4.15, we notice that Corollary 4.4 and Lemma 2.3 reduce the proof of a) and b) to the case $X = Y \in IG_{0+}$. Part b) is proved either in the completely symmetric way with respect to the proof of Part a), or with the aid of the duality in the form of Theorem 4.5. In the latter case, b) follows from a) and

Theorem 4.13, c). Upper estimates for the Rademacher constants in c) and d) follow from the symmetric variable setting of Corollary 5.4 (in Section 5.3) applied to a function f with $\hat{f}(A) = 0$ for $|A| \neq 1$. Since the Rademacher type and cotype constants are always not less than 1, we have the sharpness of the value 1.

As earlier we shall deal only with the IG spaces that possess the same tree type $T = T(X)$ as the space X itself but different values of the parameters of the spaces at the vertexes of T and adopt the notations introduced in the proof of Theorem 4.17.

There is a mild difference with the proof of Theorem 4.17: we just use $p_{\min}(Y)'$ instead of ∞ in the proof of a) if $p_{\max}(Y) < p_{\min}(Y)'$, and we make this change at the last step of the proof of Theorem 4.17 (establishing better estimates).

Moreover, if $p_{\min}(X) \geq 2$, Corollary 4.8 gives the required estimate already. Otherwise, there are two choices: either $p_{\min}(Y)' \leq p_{\max}(Y)$, or $p_{\min}(Y)' > p_{\max}(Y)$. In the former case we have $q \geq \max(p_{\min}(Y)', p_{\max}(Y))$ and can find an IG-space Z satisfying $Y = [X_{\bar{2}}, Z]_{[\theta]}$ and, thus, apply Lemma 4.9 to establish the desirable estimate. Here $X_{\bar{2}}$ is a Hilbert space that is IG–space with the same tree as Y but with the constant parameter function $\bar{2}$ (i.e., all the parameters are equal to 2), while Y has the same tree and the parameters in the interval $[1, \infty]$.

Note that we have just covered also the case of $q \geq p'_{\min}(Y) > p_{\max}(Y)$. For $p'_{\min}(Y) > q > p_{\max}(Y)$ the proof of Theorem 4.17 provides us with the $(2, p_{\min}(X)-1)$– uniform convexity of the space X_{q_2}. Note that $X = X_{q_2}$ and this is the last step if $p_{\max}(Y) \leq 2$. Otherwise, there exists an IG space X_{r_3} with the same tree and the exponent function r_3 with values in $[p_{\min}(Y), p'_{\min}(Y)]$ with the properties

$$1/p = (1-\eta)/q_2 + \eta/r_3, \quad \chi_{2,-p} = \chi_{2,-r_2} \text{ and } 1/q = (1-\eta)/2 + \eta/p'_{\min}(Y). \quad (1)$$

The above argument using Lemma 4.9 shows that X_{r_3} is $(p_{\min}(Y)', h_3)$–uniformly convex with $h_3(1/2) = 2^{2-p_{\min}(Y)'}$. As in the proof of Theorem 4.17, we now apply Theorem 4.9, a) and Corollary 4.6 to finish the proof of the theorem. □

5. Quasi–Euclidean Approach and Some Applications

5.1. Counterparts of Pythagorean Theorem for Birkhoff–Fortet Orthogonality

Birkhoff and Fortet considered a nonsymmetric extension of the orthogonality in a pre–Hilbert (inner/scalar product) space relying on its extremal distance property.

Definition 5.1 (Birkhoff–Fortet orthogonality). *For a Banach space X and $x, y \in X$. The element x is orthogonal to y, that is $x \perp y$, if $\|x\|_X \leq \|x + \lambda y\|_X$ for every $\lambda \in \mathbb{R}$.*

The definition shows that it is actually the orthogonality $x \perp H$ where H is a hyperplane containing y, while the hyperplanes are in one–to–one correspondence with the elements of the unit sphere of X^*.

This notion was implicitly used in the following result of V. L. Shmul'yan. Let us recall that a Banach space is rotund if its unit ball is strictly convex meaning that every point of the unit sphere is an extreme point of the unit ball. A Banach space is smooth if its norm is Gâteaux–differentiable. Namely, Shmul'yan has noticed that X is strictly convex if, and only if, for every hyperplane $H \subset X$ there exists no more than one $x \in S_X$ satisfying $x \perp H$, and that X is smooth if, and only if, for every $x \in X$ there exists only one hyperplane H satisfying $x \perp H$. The existence of the latter hyperplane is just a form of the Hahn–Banach theorem.

The qualitative study of the Birkhoff–Fortet orthogonality was finished by deep results due to R. C. James including his celebrated characterisation of reflexivity.

We summarise all these achievements in the next theorem. Its Part a) is due to Hahn and Banach, Parts b) and c) belong to Shmul'yan, and the rest is established by James.

Theorem 5.1. *Let X be a Banach space. Then*
a) for every $x \in X$, there exists a hyperplane $H \subset X$ satisfying $x \perp H$;
b) X is smooth if X^* is rotund;
c) X is rotund if X^* is smooth;
d) X is reflexive if, and only if, for every hyperplane $H \subset X$, there exists $x \in X$ satisfying $x \perp H$;
e) the space X with $\dim(X) > 2$ is pre–Hilbert if, and only if, for every $x \in X$, there exists a hyperplane $H \subset X$ satisfying $H \perp x$,
f) and also if, and only if, for every hyperplane $H \subset X$, there exists $x \in X$ satisfying $H \perp x$.

Let us present a counterpart of the Pythagorean theorem for the Birkhoff–Fortet orthogonality. Parts a) and b) of the next theorem are the counterparts of the Pythagorean theorem for the Birkhoff–Fortet orthogonality in p–uniformly convex and q–uniformly smooth Banach spaces correspondingly. We shall call them as the upper and the lower Pythagorean theorems.

Let us notice that the triangle inequality can be viewed as an extremal case of the upper Pythagorean theorems.

Theorem 5.2. *Let X be a Banach space, $x, y \in X$, $f_x \in X^*$ and $p, q \in [1, \infty)$. Assume also that $f_x(y) = 0$ and $f_x(x) = \|x\|_X$. Then,*
a) if X is (p, h_c)–uniformly convex, we have

$$\|x\|_X^p + \sup_{\mu \in (0, 1/2]} (1 - \mu) h_c(\mu) \|y\|_X^p \leq \|x + y\|_X^p;$$

b) if X is (q, h_s)–uniformly smooth, we have

$$\|x + y\|_X^q \leq \|x\|_X^q + \inf_{\mu \in (0, 1/2]} (1 - \mu)^{1-q} h_s(\mu) \|y\|_X^q.$$

Remark 5.1. *According to the same point of view, Lemma 5.5 in [7] is a counterpart of the cosine theorem extending the Pythagorean theorem. Part d) of Lemma 4.1 is its stronger version and a further extension to the case of more general convex functions then the powers $\|\cdot\|^p$.*

Proof of Theorem 5.2. Considering Lemma 4.1, d) with $F(x) = \|x\|_X^p$ and $F(x) = \|x\|_X^p$, we obtain in the cases of Parts a) and b), respectively,

$$\|x\|_X^p + \sup_{\mu \in (0,1)} \max(\mu, \nu) \delta_\mu(\|y\|_X, \|\cdot\|_X^p, X) \leq \|x + y\|_X^p \text{ and} \tag{1}$$

$$\|x+y\|_X^q \leq \|x\|_X^q + \inf_{\mu \in (0,1)} \max(\mu,\nu)\rho_\mu(\|y\|_X/\max(\mu,\nu), \|\cdot\|_X^q, X). \quad (2)$$

These estimates, together with the definitions imply the theorem. $\square$

5.2. Kadets and Atomic Lyapunov Constants for (q, h_s)–Uniformly Smooth Spaces

In this section we continue to "grow" the research direction initiated by J. Elton and Th. P. Hill [43] and V. M. Kadets [5] and devoted to the evaluation of the nonconvexity of the image of a vector measure with atoms in terms of their size. The authors of the former article dealt with the case of the vector measures with the values in finite–dimensional Euclidean spaces but considered a stronger measure of nonconvexity. Their approach has essentially relied on the compactness of the image of the measure, while the countable additivity, along with James' criterion, gives us only the weak compactness of the range for the measures with values in infinite–dimensional spaces.

V. M. Kadets has treated the cases of the measures taking values in either abstract finite–dimensional normed space, or a B–convex (infinite–dimensional) Banach space and considered a weaker midpoint measure of nonconvexity. His estimates (Theorem 5.3 below) imply the Lyapunov theorem for an atomless measures with bounded variation and the values in a B–convex space. We show that the exact value of the Kadets constant coincides with the lower estimate found by him [5] in 1990 in the case of the n–dimensional Hilbert space.

In this section we estimate the measures of nonconvexity of the ranges and the atomic Lyapunov constants of Banach–valued measures in terms of either their variation and atomicity, or their dimension and compare these results with the corresponding estimates of the mid–point measure of nonconvexity and the Kadets constants. Our generalisation of Kadets' quantitative version [5] of the Lyapunov theorem on the convexity of the range of a measure will be applied to the measures of bounded variation with values in mixed–norm Lebesgue, Besov and Lizorkin–Triebel spaces.

The notions of the atom, atomicity, measure of nonconvexity, variation and the atomic constant of an X–valued measure are defined in Section 4.2.

V. M. Kadets [5] has defined a measure of the nonconvexity of a subset $F \subset X$ as

$$\mathcal{C}(F) := \sup_{x,y \in F} \inf_{z \in F} \left\| z - \frac{x+y}{2} \right\|_X.$$

He has also established the following deep result extending the Lyapunov theorem that can be applied to the ranges of measures of bounded variation with the values in a wide class of B–convex spaces that is wider than the class of the (q, h_s)–uniformly smooth spaces. It seems natural that imposing more restrictive conditions on a Banach space one establishes twice sharper (or, occasionally, sharp) estimates, even in the case of the Hilbert–valued measures.

Theorem 5.3 ([5]). *For every countably additive X–valued measure (ν, σ), one has*

$$a)\ \mathcal{C}(\nu(\sigma)) \leq \sup_{A \in \Sigma} \inf \left\{ \min_{\varepsilon \in D_n} \left\| \sum_{i=1}^n \varepsilon_i \nu(A_i) \right\|_X : \cup_{i=1}^n A_i = A,\ A_i \cap A_j = \varnothing \right\}.$$

b) *Thus, one also has* $C(\nu(\sigma)) \leq T_{p,p}(X)(\operatorname{at}(\mu))^{1/p'}(\operatorname{Var}(\nu))^{1/p}$ *if X has the Rademacher type $p > 1$.*

To formulate another result due to V. M. Kadets, we need to define the Kadets constant.

Definition 5.2. *For a finite–dimensional Banach space X we define its Kadets constant $C_K = C_K(X)$ as the infimum of the constants C such that the following inequality holds for every measure space (Ω, σ, ν) with the atomic countably–additive X–valued measure ν: $C(\mu(\sigma)) \leq C\operatorname{at}(\nu)$.*

In 1990 V. M. Kadets [5] has established the estimate $n^{1/2}/2 \leq C_K(H) \leq n^{1/2}$ for an n–dimensional Hilbert space H. More generally, he showed that $C_K(X) \leq T_{p,p}(X)n^{1/p}$ if X is an n–dimensional Banach space of the Rademacher type $p > 1$ with the constant $T_{p,p}(X)$ (see Definition 4.12). Corollary 5.2 shows, in particular, that the lower bound is its exact value, that is $C_K = C_{AL}(H) = n^{1/2}/2$ if $\dim(X) = n$, because $C_K(X) \leq C_{AL}(X)$.

Corollary 5.1. *For $q \in (1, 2]$, let X be a (q, h_s)–uniformly smooth Banach space, and let (Ω, σ, ν) be an X–valued countably additive measure on σ with bounded variation. Then we have*

$$\lambda(\nu(\sigma)) \leq \left(\sup_{\mu \in (0,1)} \mu(1-\mu)h_s(\mu)\right)^{1/q} (\operatorname{Var}(\nu))^{1/q} (\operatorname{at}(\nu))^{1/q'}.$$

Moreover, if $\dim(X) = n$, we also have

$$C_K(X) \leq C_{AL}(X) \leq \left(n \sup_{\mu \in (0,1)} \mu(1-\mu)h_s(\mu)\right)^{1/q}.$$

Remark 5.2. *a) Let us note that pursuing Kadets' approach we could notice that the (q, h_s)-uniform smoothness implies the estimate $T_{q,q}(X) \leq (2^{q-2}h_s(1/2))^{1/q}$ (see Corollary 5.4 below and the proof of Theorem 4.20 above), while using Theorem 4.13, a), we actually have the twice better estimate*

$$\left(\sup_{\mu \in (0,1)} \mu(1-\mu)h_s(\mu)\right)^{1/q} = 2^{-1}(2^{q-2}h_s(1/2))^{1/q}.$$

b) Note that both C_{AL} and C_K do not depend on q.
c) Explicit values $h_s(1/2) = \max_{\mu \in (0,1)} h_s(\mu)$ for particular spaces are provided in Section 4.4.4.

Proof of Corollary 5.1. Applying Parts *a)* of Theorems 4.1 and 4.2, we only need to note that, for $F(x) = \|x\|_X^q$, one has $\rho_\mu(t, x, X) \leq h_s(\mu)t^q$ and $F_{rl} = t^q$ if X is (q, h_s)–uniformly smooth. $\square$

Combining Corollary 5.1 with the (q, h_s)–uniform smoothness established in Theorem 4.19 (see also the comments preceding Theorem 4.18), we obtain the following theorem.

Theorem 5.4. *Let X be either a subspace or a quotient of, or finitely represented in $Y \in IG_+$ with $[p_{\min}(Y), p_{\max}(Y)] \subset (1, \infty)$, and let (Ω, σ, ν) be an X-valued countably additive measure on σ with bounded variation. Then we have*

$$(2\lambda(\nu(\sigma)))^{\min(p_{\min}(Y),2)} \leq (\max(p_{\max}(Y),2)-1)^{\min(p_{\min}(Y),2)-1} \operatorname{Var}(\nu) \times$$
$$\times (\operatorname{at}(\nu))^{\min(p_{\min}(Y),2)-1} \text{ and}$$
$$(2C_{AL}(X))^{\min(p_{\min}(Y),2)} \leq \dim(X) \, (\max(p_{\max}(Y),2)-1)^{\min(p_{\min}(Y),2)-1}.$$

Corollary 5.2. *For an arbitrary Hilbert space H of the dimension n, we have $C_K = C_{AL}(H) = n^{1/2}/2$. One also has $C_K(l_p(I_n)) = C_{AL}(l_p(I_n)) = n^{1/p}/2$ for $p \in (1, 2]$.*

Proof of Corollary 5.2. It is only left to establish the lower estimates for the Kadets constants. The model example of the measure ν_n from Remark 4.3 shows that these bounds are sharp because the middle $\nu_n(I_n)/2$ of the unit cube $[0,1]^n = \operatorname{co}(\nu_n(\sigma_n))$ is at the distances $n^{1/p}/2$ for $p \in (1,2]$ from all its vertexes $\{0,1\}^n = \nu_n(\sigma_n)$, and the atomicity of ν_n is 1. □

5.3. Counterparts of Parseval's Identity for Abstract and Asymmetric Walsh Systems

Using the notation of Section 4.2 in this section, we discuss the application of Theorems 4.3 and 4.4 to the setting of (q, h_c)–uniformly convex and (r, h_s)–uniformly smooth Banach spaces from different points of view, pointing out the connection with the atomic Lyapunov contants discussed above. Our two asymmetric models can also be interpreted as the models of complete chaos (see [10])) up to an arbitrary integer order for both the simplest and the most general variants of Walsh systems (X–bases).

Theorems 4.3 and 4.4 imply the following two corollaries (respectively).

Corollary 5.3. *Let X be a Banach space, $2 \in [r, q] \subset [1, \infty]$, and $n \in \mathbb{N}$. Then the following criteria hold:*
a) X is (q, h_c)–uniformly convex if, and only if, for every $\mu \in (0,1)^n$ and $f \in L_\infty(D^n, X)$, we have

$$\int_{D^n} \|f\|_X^q dp_{d^n} \geq \sum_{A \in \sigma_n} \prod_{i \in A} \mu_i \nu_i h_c(\mu_i) \left\| \hat{f}(A) \right\|_X^q ;$$

b) X is (r, h_s)–uniformly smooth if, and only if, for every $\mu \in (0,1)^n$ and $f \in L_\infty(D^n, X)$, we have

$$\int_{D^n} \|f\|_X^r dp_{d^n} \leq \sum_{A \in \sigma_n} \prod_{i \in A} \mu_i \nu_i h_s(\mu_i) \left\| \hat{f}(A) \right\|_X^r .$$

Remark 5.3. *a) Note that Part b) implies the following inequality with the same expression that we used to estimate the atomic Lyapunov constant in Corollary 5.1:*

$$\int_{D^n} \|f\|_X^r dp_{d^n} \leq \sum_{A \in \sigma_n} \left(\sup_{\mu \in (0,1)} \mu(1-\mu) h_s(\mu) \right)^{|A|} \left\| \hat{f}(A) \right\|_X^r .$$

b) *Moreover, Part b) also implies the approximation result: for* $m \in I_n$ *(including $\varnothing$), we have (for a (r, h_s)–uniformly smooth X)*

$$\int_{D^n} \left\| f - \sum_{A \in \sigma_n}^{|A| \leq m} b_A \hat{f}(A) \right\|_X^r dp_{d^n} \leq \sum_{A \in \sigma_n}^{|A| > m} \prod_{i \in A} \mu_i \nu_i h_s(\mu_i) \left\| \hat{f}(A) \right\|_X^r. \tag{1}$$

Thus, taking $m = 1$, we obtain the estimate for the approximation of a polyhedron F defined by f (with repeating values–vertexes if F has less then 2^n vertexes) by the translated zonotope $\sum_{A \in \sigma_n}^{|A| \leq 1} b_A \hat{f}(A)$. Whenever h_s is well–described explicitly, we can minimise the right–hand side of (1) by μ.

c) *The values of (the corresponding chaos of the mth order in terms of [10]) $\sum_{A \in \sigma_n}^{|A|=m} b_A \hat{f}(A)$ can be interpreted as the mth order generalisation of a zonotope.*

The next immediate corollary to Theorem 4.4 (with $F \in \{\|\cdot\|_X^q, \|\cdot\|_X^r\}$) covers the case of an abstract Walsh system.

Corollary 5.4. *For $2 \in [p, q] \subset [1, \infty]$ and $n \in \mathbb{N}$, let X be a (p, h_c)–uniformly convex and (q, h_s)–uniformly smooth Banach space, and let $\{\xi_i\}_{i=1}^n$ be a set of mutually independent scalar–valued stochastic variables on a measure space (Ω, Ξ, p) with $\mathrm{E}\xi_i = 0$ for $i \in I_n$. Assume also that $\{x_A\}_{A \in \sigma_n} \subset X$, $\xi_A = \prod_{i \in A} \xi_i$, $\xi_\varnothing = 1$ for $A \in \sigma_n$ and*

$$C_c = C_c(F, X) := \sup_{\mu \in [0, 1/2]} (1 - \mu) h_c(\mu) \text{ and } C_s = C_s(F, X, q) := \inf_{\mu \in [0, 1/2]} (1 - \mu)^{1-q} h_s(\mu).$$

Then it follows that

a) $\displaystyle \mathrm{E} \left\| \sum_{A \in \sigma_n} x_A \xi_A \right\|_X^q \geq \sum_{A \in \sigma_n} C_c^{|A|} \|x_A\|_X^q \prod_{i \in A} \mathrm{E}|\xi_i|^q,$

and, if $\{\xi_i\}_{i=1}^n$ are also symmetric,

$$\mathrm{E} \left\| \sum_{A \in \sigma_n} x_A \xi_A \right\|_X^q \geq \sum_{A \in \sigma_n} \left(2^{q-2} h_c(1/2)\right)^{|A|} \|x_A\|_X^q \prod_{i \in A} \mathrm{E}|\xi_i|^q,$$

b) $\displaystyle \mathrm{E} \left\| \sum_{A \in \sigma_n} x_A \xi_A \right\|_X^r \leq \sum_{A \in \sigma_n} C_s^{|A|} \|x_A\|_X^r \prod_{i \in A} \mathrm{E}|\xi_i|^r.$

and, if $\{\xi_i\}_{i=1}^n$ are also symmetric,

$$\mathrm{E} \left\| \sum_{A \in \sigma_n} x_A \xi_A \right\|_X^r \leq \sum_{A \in \sigma_n} \left(2^{r-2} h_s(1/2)\right)^{|A|} \|x_A\|_X^r \prod_{i \in A} \mathrm{E}|\xi_i|^r.$$

5.4. Duality Mapping and Lozanovskii Factorisation

In this section we establish the Hölder regularity of the duality mapping and the Lozanovskii factorisation mapping in the settings of (p, h_c)–uniformly convex and (q, h_s)–uniformly smooth spaces.

Definition 5.3. *Assume that (M_0, d_0) and (M_1, d_1) are metric spaces, and $\alpha \in (0, 1]$.*
Let us denote by $H^\alpha(M_0, M_1)$ the family of all Hölder mappings $f : X \to Y$ satisfying:

$$\|f | H^\alpha(M_0, M_1)\| := \sup\{d_1(f(x), f(y))/d_0(x, y)^\alpha : x, y \in M_0 \text{ and } x \neq y\} < \infty.$$

The family $H^1(M_0, M_1)$ is know as the class of Lipschitz mappings.

In our setting, the metric spaces are subsets of Banach spaces endowed with the inherited metric.

Definition 5.4. *Let X and Y be (quasi) Banach spaces and $\alpha, \beta \in (0, 1]$. We say that the unit spheres S_X and S_Y are (α, β)–Hölder homeomorphic, or that X and Y are homogeneously (α, β)–Hölder homeomorphic, and write $X \xleftrightarrow{(\alpha,\beta)} Y$ if there exists a homeomorphism $\phi : S_X \leftrightarrow S_Y$ satisfying $\phi \in H^\alpha(S_X, Y)$ and $\phi^{-1} \in H^\beta(S_Y, X)$.*

5.4.1. Duality Mapping: Quantitative Monotonicity and Hölder Regularity

Let us recall that the duality mapping $J_X : S_X \to S_{X^*}$ is correctly defined by $\langle J_X x, x \rangle = 1$ in the case of a smooth X thanks to the Hahn–Banach theorem. It has its natural inverse $J_X^{-1} = J_{X^*}$ if X is also strictly convex and reflexive.

The following lemma, describing the monotonicity of the duality mapping in a quantitative manner, is a particular case of Part $a)$ of Lemma 4.1 combined with the quantitative duality of our notions of smoothness and convexity established in Theorem 4.5 and Remark 4.2, $a)$. Let us also recall that the monotonicity of $(1 + x)^{1/x}$ on $(0, +\infty)$ implies that the maximum 2 of $h(q) = q'^{q/q'} = (1 + 1/(q-1))^{q-1}$ is achieved at $q = 2$ on $(1, 2]$.

Lemma 5.1. *For $2 \in [q, p] \subset (1, \infty)$, let X be a (p, h_c)–uniformly convex and (q, h_s)–uniformly smooth Banach space and $x, y \in S_X$ with $f_x, f_y \in S_{X^*}$ satisfying $\langle f_x, x \rangle = \langle f_y, y \rangle = 1$. Then we have*

$$\sup_{\mu \in (0,1)} h_c(\mu) \|x - y\|^p / p \leq \langle f_x - f_y, x - y \rangle;$$

$$\left(\inf_{\mu \in (0,1)} h_s(\mu) \right)^{1-q'} \|f_x - f_y\|^{q'} / q' \leq \langle f_x - f_y, x - y \rangle;$$

$$\|f_x - f_y\|_{X^*} \leq q'^{q/q'} \inf_{\mu \in (0,1)} h_s(\mu) \|x - y\|_X^{q-1} \leq 2 \inf_{\mu \in (0,1)} h_s(\mu) \|x - y\|_X^{q-1}.$$

Moreover, the corresponding relation holds if X is either (p, h_c)–uniformly convex, or (q, h_s)–uniformly smooth.

We reformulate an immediate corollary to this lemma (and the definition of x^*) in terms of the following explicit estimates for the Hölder noms of the duality mapping. Theorems 4.17 − 4.19, combined with the point of view at the function spaces given in Section 3, provide explicit expressions for the quantities in brackets in terms of the parameters of IG_+−spaces and various classes of Besov, Lizorkin–Triebel and Sobolev spaces (and their duals, subspaces etc.).

Theorem 5.5. *For $2 \in [q,p] \subset (1,\infty)$, let X be a (p,h_c)-uniformly convex and (q,h_s)-uniformly smooth Banach space and $J_X : S_X \to S_{X^*}$ the duality mapping. Then we have $X \overset{(q-1,p'-1)}{\longleftrightarrow} X^*$ with $J_X^{-1} = J_{X^*}$,*

$$\left\| J_X | H^{q-1}(S_X, S_{X^*}) \right\| \le q'^{q/q'} \inf_{\mu \in (0,1)} h_s(\mu) \text{ and}$$

$$\left\| J_X^{-1} | H^{(p-1)^{-1}}(S_{X^*}, S_X) \right\| \le p^{p'/p} \left(\sup_{\mu \in (0,1)} h_c(\mu) \right)^{(1-p)^{-1}}.$$

One has a related uniform bound $\max\left(q'^{q/q'}, p^{p'/p} \right) \le 2$ *achieved at $p = q = 2$.*

Remark 5.4. *This abstract result provides sharp exponents of the Hölder regularity for $X \in \{L_p, l_p\}$ with $p \in (1, \infty)$. Indeed, the duality map is just the Mazur map m_{p-1} covered in [64]. This leads to the sharpness for $X \in IG$ (see the proofs of Theorems 4.17 − 4.19).*

5.4.2. Lozanovskii Factorisation: Hölder Regularity

With the aid of Theorem 5.5 and Lemma 5.1, we establish, in an explicit quantitative manner, the Hölder–regular version of the Lozanovskii factorisation mapping [65, 66] (see [64, 67, 13, 68, 11] for more references and related applications).

The following theorem contains strengthened versions of Lemma 9.5 and Corollary 9.6 from [64] where the same mapping was shown to be a uniform homeomorphism. Let us also notice that function $r(p) = (2p)^{1/p}$ decreases on $[2, +\infty)$, thus, achieving its maximum 2 at $p = 2$.

Theorem 5.6. *For $2 \in [q,p] \subset (1,\infty)$, let X be a (p,h_c)-uniformly convex and (q,h_s)-uniformly smooth Banach space that is also a lattice. Then there exist Lozanovskii mappings $L_X : X \to L_1(E,\mu)$ and $L_{X^*} : X^* \to L_1(E,\mu)$ for some measure space (E,μ) that are correctly defined and realise the equivalences*

$$X \overset{(q-1,1/p)}{\longleftrightarrow} L_1(E,\mu) \overset{(1/q',p'-1)}{\longleftrightarrow} X^*.$$

Moreover, we also have the estimates

$$\left\| L_X | H^{q-1}(S_X, S_{L_1(E,\mu)}) \right\| \le 2^{2-q} + q'^{q/q'} \inf_{\mu \in (0,1)} h_s(\mu),$$

$$\left\| L_X^{-1} | H^{1/p}(S_{L_1(E,\mu)}, S_X) \right\| \le (2p)^{1/p} \left(\sup_{\mu \in (0,1)} h_c(\mu) \right)^{-1/p}, \text{ and}$$

$$\left\| L_{X^*}^{-1} | H^{1/q'}(S_{L_1(E,\mu)}, S_{X^*}) \right\| \le (2q')^{1/q'} \left(\inf_{\mu \in (0,1)} h_s(\mu) \right)^{1/q}.$$

One has a related uniform bound $\max\left(q'^{q/q'}, (2q')^{1/q'}, (2p)^{1/p}\right) \leq 2$ *simultaneously achieved at* $p = q = 2$.

Remark 5.5. *The exponents of the Hölder regularity provided for L_X^{-1} are sharp for $X \in \{L_p, l_p\}$ with $p \geq 2$ since L_X^{-1} becomes a Mazur mapping covered in [64].*

Proof of Theorem 5.6. According to the representation theorem for the lattices (see [2, 64]), X and X^* are (linearly) isometric to some Banach function or sequence spaces on a measure space (Ω, μ) with $\langle h, x \rangle = \int_\Omega h(t)x(t)d\mu(t)$ for $h \in X^*$ and $x \in X$. It is known (for example, see the proof of Theorem 9.7 and Section 9.6 in [64]) that the mapping

$$L_X : X \longrightarrow L_1(\Omega, \mu) : x \longmapsto |x(t)|J_X x(t) \qquad (L)$$

is the uniform homeomorphism of the unit spheres but we only need to know that it is a homeomorphism of dense subsets (such as the union of the finite–dimensional subspaces spanned by simple functions). The "onto" property for the restrictions onto the finite–dimensional subspaces follows from the continuity (particularly, the Hölder estimates) and the Brouwer theorem (the argument from [69, 64]). Thus, we only establish the estimates for the Hölder norms assuming that $\|\cdot\|_1$ is the $L_1(E, \mu)$–norm.

Let $x, y \in S_X$, $f = L_X x = |x|J_X x \in S_{L_1(E,\mu)}$, $g = L_X y = |y|J_X y \in S_{L_1(E,\mu)}$ and $\|f - g\|_1 = \varepsilon$. Assume that $G = \{t \in E : x(t)y(t) > 0\}$ and

$$h = \chi_G \frac{x}{|x|} \min(|f|, |g|). \qquad (1)$$

Comparing $|f| + |g|$ with $|f - g|$ on G and $E \setminus G$, we see that $\|h\|_1 = 1 - \varepsilon/2$. For $\lambda > 2$, let

$$B_\lambda = \left\{ t \in G : \frac{x(t)}{y(t)} + \frac{y(t)}{x(t)} \geq \lambda \right\} \text{ and } D_\lambda = G \setminus B_\lambda. \qquad (2)$$

The lattice properties also suggest that $|x|, |y| \in S_X$ and $|J_X x|, |J_X y| \in S_{X^*}$ meaning $\langle |J_X x|, |y| \rangle \leq 1$, $\langle |J_X x|, |y| \rangle \leq 1$, and

$$2 \geq \langle |J_X x|, |y| \rangle + \langle |J_X y|, |x| \rangle \geq \left\| h\left(\frac{x}{y} + \frac{y}{x}\right) \right\|_1 \geq 2\|h\|_1 + (\lambda - 2)\|\chi_{B_\lambda} h\|_1. \qquad (3)$$

Therefore, comparing with the norm of h, we see that

$$\|\chi_{B_\lambda} h\|_1 \leq \varepsilon/(\lambda - 2) \text{ and } \|\chi_{D_\lambda} h\|_1 \geq 1 - \varepsilon/2 - \varepsilon/(\lambda - 2). \qquad (4)$$

Now remembering the definition of D_λ, we deduce from (4) that

$$\langle |J_X x|, |y| \rangle + \langle |J_X y|, |x| \rangle \geq \langle \chi_{D_\lambda} J_X x, y \rangle + \langle \chi_{D_\lambda} J_X y, x \rangle \geq$$
$$\geq \left\| \chi_{D_\lambda} h\left(\frac{x}{y} + \frac{y}{x}\right) \right\|_1 \geq 2\|\chi_{D_\lambda} h\|_1 \geq 2 - \varepsilon\frac{\lambda}{\lambda - 2}. \qquad (5)$$

Combined with the first inequality in (3), this gives us

$$|\langle \chi_{\bar{D}_\lambda} J_X x, y \rangle| + |\langle \chi_{\bar{D}_\lambda} J_X y, x \rangle| \leq \varepsilon \frac{\lambda}{\lambda - 2}, \qquad (6)$$

where $\bar{D}_\lambda = E \setminus D_\lambda$. Now (5) and (6) naturally imply the key estimate

$$\langle J_X x, y \rangle + \langle J_X y, x \rangle \geq 2 - 2\varepsilon \frac{\lambda}{\lambda - 2}, \tag{7}$$

where we take the limit $\lambda \to \infty$ and obtain

$$\langle J_X x - J_X y, x - y \rangle \leq 2\varepsilon. \tag{8}$$

Combining (8) with Lemma 5.1, we arrive at the desirable estimates for the Hölder norms of L_X^{-1} and $L_{X^*}^{-1}$. To finish the proof, we deduce the first estimate of the theorem from Theorem 5.5, the triangle inequality and the representation

$$L_X x - L_X y = (|x| - |y|) J_X x + |y| (J_X x - J_X y) :$$

$$\|L_X x - L_X y\|_1 \leq \|x - y\|_X + \|J_X x - J_X y\|_{X^*} \leq$$
$$\leq \left(2^{2-q} + q'^{q/q'} \inf_{\mu \in (0,1)} h_s(\mu) \right) \|x - y\|_X^{q-1}. \quad \Box$$

Acknowledgments

I am grateful to Professor Oleg V. Besov for the most essential part of my knowledge related to Besov, Lizorkin–Triebel and other function spaces and appreciates his encouragement and support over the years.

At different times and under different circumstances, the work related to this chapter had been partially supported by the the Australian Research Council (project numbers DP0881037, DP120100097 and DP150102758), the Russian Fund for Basic Research (project numbers 08–01–00443, 11–01–00444–a and 14–01–00684) and the EIS Research Grant held by James McCoy at the Institute for Mathematics and its Applications at the University of Wollongong.

It is also my pleasure to thank the President and Scientific and Technical Editor(s) of Nova Science Publishers, Inc. for the opportunity and related collaboration.

REFERENCES

[1] Besov, Oleg Vladimirovich. 2005. "Interpolation, embedding, and extension of spaces of functions of variable smoothness." *Tr. Matem. Inst. Steklova*. 248: 52–63. Engl. transl. *Proc. Steklov Math. Inst.* 248(1): 47–58.

[2] Lindenstrauss, Joram and Tzafriri, Lior. 1979. *Classical banach spaces. Volume II.* (In *Ergebnisse, v. 97.*) Berlin-Heidelberg-New York: Springer Verlag. P. 1–243+i–x.

[3] Lyapunov, Alekseiy Andreevich (Liapounoff). 1940. "On completely additive vector functions. Part I." *Izvestiya Akademii Nauk SSSR (USSR)* 4: 465–478.

[4] Bogachev, Vladimir Igorevich. 2007. *Measure theory. Volume II.* Berlin–New York: Springer Verlag.

[5] Kadets, Vladimir Mikhailovich. 1992. "Remark on the Lyapunov theorem on vector measures." *Func. Anal. Appl.* 25(4): 295–297.

[6] Ajiev, S. S. "Amir–Franchetti problems, quantitative K-convexity test and Jacobi convexity hierarchy." *Manuscript* 1–35.

[7] Ajiev, S. S. 2009. "On Chebyshev centres, retractions, metric projections and homogeneous inverses for Besov, Lizorkin-Triebel Sobolev and other Banach spaces." *East J. Approx.* 15(3): 375-428.

[8] Ajiev, S. S. 2010. "On concentration, deviation and Dvoretzky's theorem for Besov, Lizorkin–Triebel and other spaces." *Compl. Var. Ellipt. Eq.* (Special issue dedicated to Gérard Bourdaud.) 55(8–10): 693–726.

[9] Temlyakov, Vladimir. 2011. *Greedy Approximation.* (In *Cambridge Monographs on Applied and Computational Mathematics, V. 20.*) Cambridge–Mexico City: Cambridge University Press. P. 1–418+i–xiv.

[10] Ledoux, Michel and Talagrand, Michel. 1991. *Probability in Banach Spaces: Isoperimetry and Processes.* Berlin–Heidelberg-New York: Springer Verlag. P. 1–482+i–xii.

[11] Ajiev, S. S. 2014. "Hölder analysis and geometry on Banach spaces: homogeneous homeomorphisms and commutative group structures, approximation and Tzar'kov's phenomenon. Part I." *Eurasian Math. J.* 5(1): 7–60.

[12] Ajiev, S. S. 2014. "Hölder analysis and geometry on Banach spaces: homogeneous homeomorphisms and commutative group structures, approximation and Tzar'kov's phenomenon. Part II." *Eurasian Math. J.* 5(2): 6–50.

[13] Odell, Elton and Schlumprecht, Thomas. 1994. "The distortion problem." *Acta Math.* 173: 259–281.

[14] Ajiev, S. S. "Nonlocal geometry of function and other Banach spaces: constants, moduli and compact and Kato-Pełczyński operators." *Manuscript* 1–51.

[15] Ajiev, S. S. 2019. "Interpolation and bounded extension of Hölder–Lipschitz mappings between function, noncommutative and other Banach spaces." In *Understanding Banach Spaces*, edited by Daniel González. Long Island: Nova Science Publishers, Inc.

[16] Ajiev, S. S. 2013. "Quantitative Hahn–Banach theorems and isometric extensions for wavelet and other Banach spaces." *Axioms* 2: 224-270. (Open access: doi:10.3390/axioms2020224.)

[17] Maleev, R. P. and Trojanski, S. L. 1971/2. "Moduli of convexity and of smoothness for the spaces L_{pq}." *Annuaire Univ. Sofia Fac. Math.* 66: 331–339.

[18] Triebel, Hans. 2006. *Theory of Function Spaces III.* (In *Monographs in Mathematics, V. 100.*) Basel–Boston–Berlin: Birkhäuser Verlag. P. 1–426+i–xii.

[19] Ajiev, S. S. "Optimal non–dual local non–stationary multiresolution analysis for anisotropic function spaces." *Manuscript* 1–51.

[20] Berkolaiko, M. Z. and Novikov, I. Ya. 1994. "Unconditional bases in spaces of functions of anisotropic smoothness." *Proc. Steklov Math. Inst.* 204(4): 27–41.

[21] Cohen, A., Daubechies, Ingrid, and Feauveau, J.-C. 1992. "Biorthogonal bases of compactly supported wavelets." *Comm. Pure Appl. Math.* 45: 485–560.

[22] Daubechies, Ingrid. 1992. *Ten lectures on wavelets*. Philadelphia: Soc. Industr. Appl. Math. P. 1–357+i–xix.

[23] Ajiev, S. S. 2019. "Copies of sequence spaces and basic properties of function spaces." In *Understanding Banach Spaces*, edited by Daniel González. Long Island: Nova Science Publishers, Inc.

[24] Milman, Vitaliy Davidovich. 1969. "Geometric theory of Banach spaces. II. Geometry of the unit ball." *Uspehi. Mat. Nauk* 26(6(162)): 73–149.

[25] Levy, Mireille. 1980. "L'espace d'interpolation réel $(A_0, A_1)_{\theta,p}$ contient l_p." In *Initiation Seminar on Analysis, 19th year: 1979/80. Publ. Math. Univ. Pierre et Marie Curie* 41(3):9.

[26] Mastylo, M. 1991. "Banach spaces via sublinear operators." *Math. Japonica* 36(1): 85–92.

[27] Pisier, Gilles and Xu, Q. 2001. "Non–Commutative L_p-Spaces." In *Handbook of the geometry of Banach spaces, V. II*, 1459–1517. Amsterdam–Tokyo: Elsevier.

[28] Diestel, Joseph, Jarchow, Hans, and Tonge, Andrew. 1995. *Absolutely summing operators*. Cambridge: Cambridge University Press. P. 1–474+i–xv.

[29] Ovchinnikov, V. I. 1970. "Symmetric spaces of measurable operators." *Soviet. Math. Dokl.* 11: 448–451.

[30] Ovchinnikov, V. I. 1971. Symmetric spaces of measurable operators. *Trudy Inst. Math. VGU* 3: 88–107.

[31] Ajiev, S. S. 1999. "Characterizations of $B^s_{p,q}(G), L^s_{p,q}(G), W^s_p(G)$ and other function spaces. Applications." *Proc. Steklov Math. Inst.* 227(4): 1–36.

[32] Sobolev, Sergey L'vovich. 1950. *Some applications of functional analysis in mathematical physics*. Leningrad: Izd. Leningr. Gos. Univ. 1950. P. 1–255. Third edition. Moscow: "Nauka". 1988. P. 1–334. English transl. by H. H. McFaden. (In *Translations of Mathematical Monographs, V. 90.*) Providence, Rhode Island: American Mathematical Society. 1991. P. 1–286+i–viii.

[33] Nikol'skii, Sergey Mikhailovich. 1969. *Approximation of function of several variables and embedding theorems*. Moscow: Nauka. Engl. transl. Berlin–Heidelberg–New York: Springer–Verlag. 1975.

[34] Besov, Oleg Vladimirovich, Il'in, Valentin Petrovich, and Nikol'skiy, Sergey Mikhailovich. 1996. *Integral representations of functions and embedding theorems*. Moscow: Nauka–Fizmatgiz. (1st Ed. Moscow: Nauka. 1975.)

[35] Triebel, Hans. 1995. *Interpolation Theory, Function Spaces, Differential Operators*. Amsterdam–Oxford: North–Holland Publishing Company. P. 1–528.

[36] Burenkov, Viktor Ivanovich. 1998. *Sobolev Spaces on Domains*. (In *Teubner Texts in Mathematics, V. 137*), Stuttgart: B. G. Teubner Verlagsgesellschaft GmbH.

[37] Chekanov, Yu. N., Nesterov, Yu. E., and Vladimirov, A. A. 1978. "On uniformly convex functionals." *Vestnik Mosk. Univ., Ser. Vych. Mat. Kibern.* 3: 12–23.

[38] Pichugov, S. A. 1988. "The Jung constant of L_p." *Math. Notes* 43(5): 604–614.

[39] Pichugov, S. A. 1991. "On the separability of sets by a hyperplane in L_p." *Analysis Math.* 17: 21–33.

[40] Pisier, Gilles. 1975. "Martingales with values in uniformly convex spaces." *Israel J. Math.* 20(3–4): 326-350.

[41] Kadets, Mikhail I. and Kadets, Vladimir Mikhailovich. 1997. *Series in Banach Spaces: Conditional and Unconditional Convergence*. (In *Operator Theory, V. 94*.) Basel-Boston-Berlin: Birkhäuser Verlag. P. 1–156+i–viii.

[42] Kadets, Mikhail I. and Kadets, Vladimir Mikhailovich. 1988. *Rearrangements of Series in Banach Spaces*. Tartu: Tartu Gos. University. English transl. (In *Translation of Math. Monographs, V. 86*.) Providence, R.I.: Amer. Math. Soc. 1991.

[43] Elton, J. and Hill, Th. P. 1987. "A generalization of Lyapunov's convexity theorem to measures with atoms." *Proc. Amer. Math. Soc.* 99(2): 297–304.

[44] Prus, B. and Smarzewski, R. 1987. "Strongly unique best approximations and centres in uniformly convex spaces." *J. Math. Anal. Appl.* 121: 10–21.

[45] Hanner, Otto. 1955. "On the uniform convexity of L_p and l_p." *Arkiv för Math.* 3(19): 239–244.

[46] Ball, K., Carlen, E. A., and E. H. Lieb. 1994. "Sharp uniform convexity and smoothness inequalities for trace norms." *Invent. Math.* V. 115. No. 3. 1994. P. 463–482.

[47] Bergh, J. and Löfström, J. 1976. *Interpolation spaces. An introduction*. (In *Die Grundlehren der mathematischen Wissenschaften, B. 223*.) Berlin–Heidelberg–New York–Tokyo: Springer–Verlag.

[48] Cobos, Fernando, Peetre, Jaak, and Persson, Lars Eric. 1998. "On the connection between real and complex interpolation of quasi–Banach spaces." *Bull. Sci. Math.* 122: 17–37.

[49] Cwickel, Michael, Milman, Mario, and Sagher, Yoram. 1986. "Complex interpolation of some quasi-Banach spaces." *J. Func. Anal.* 65: 339–347.

[50] Lizorkin, Pyotr Ivanovich. 1976. "Interpolation of L_p spaces with weight." *Tr. Matem. Inst. Steklova.* 140: 201–211, 289.

[51] Deville, R. M., Godefroy, Gilles, and Zizler, Vaclav. 1993. *Smoothness and renormings in Banach spaces.* (In *Pitman Mon. Surv. Pure Appl. Math., V. 64.*) Harlow: Longman Scientific & Technical. P. 1–376+i–xii.

[52] Beauzamy, B. 1985. *Introduction to Banach spaces and their geometry.* (In *North-Holland Mathematics Studies, V. 68.*) Amsterdam: North-Holland Publishing Co. P. 1–338+i–xv.

[53] Bynum, W. L. and Drew, J. H. 1972. "A weak parallelogram law for l_p." *Amer. Math. Monthly* 118:1012–1015.

[54] Lim, T.-Ch. and Smarzewski, R. 1991. "On an L_p inequality." *Glasnik Mat.* 26(46): 79–81.

[55] Lim, T.-Ch. 1983. "Fixed point theorems for uniformly Lipschitzian mappings in L_p spaces." *Nonlinear Analysis, Theory, Methods and Applications* 7(5): 555–563.

[56] Lim, T.-Ch. 1991. "On some L_p inequalities in best approximation theory." *J. Math. Anal. Appl.* 154: 23–528.

[57] Baouendi, M. S. and Goulaouic, G. 1967. "Commutation de l'intersection et des founcteurs d'interpolation." *Compt. Rend. Acad. Sci. Paris* 265: 313–315.

[58] Carlen, E. A. and Lieb, E. H. 1993. "Optimal hypercontractivity for Fermi fields and related non-commutative intergration inequalities." *Comm. Math. Phys.* 155: 27–46.

[59] Dixmier, J. 1953. "Formes linéaires sur un anneau d'opérateurs." *Bull. Soc. Math. France* 81: 222–245.

[60] Simon, B. 1979. *Trace ideals and their applications.* Cambridge: Cambridge University Press.

[61] Milman, Vitaliy Davidovich and Schechtman, Gideon. 1986. "Asymptotic theory of finite dimensional normed spaces" (with an appendix by M. Gromov). (In *Lect. Notes Math., V. 1200.*) Berlin–Heidelberg–New York: Springer Verlag. P. 1–156+i–viii.

[62] Tomczak-Jaegermann, Nicole. 1989. *Banach–Mazur Distances and Finite dimensional Operator Ideals.* Harlow: Longman Scientific & Technical. P. 1–395+i–xii.

[63] Pisier, Gilles. 1986. "Probabilistic Methods in the Geometry of Banach Spaces." (In *Probability and Analysis, Varenna, 1985.*) *Lect. Notes Math. V. 1206*, 167–241.

[64] Benyamini, Y. and Lindenstrauss, J. 2000. *Geometric nonlinear functional analysis. Volume 1.* (In *Colloquium Publications, V. 48.*) Providence, Rhode Island: American Mathematical Society. P. 1–489+i–xii.

[65] Lozanovskii, G. Ja. 1969. "On some Banach lattices." *Siberian Math. J.* 10: 584–599.

[66] Lozanovskii, G. Ja. 1972. "On some Banach lattices III." *Siberian Math. J.* 13: 1304–1313.

[67] Chaatit, F. 1995. "On the uniform homeomorphisms of the unit spheres of certain Banach lattices." *Pacific J. Math.* 168: 11–31.

[68] Gillespie, T. A. 1981. "Factorization in Banach function spaces." *Indag. Math.* 43: 287–300.

[69] Bollobas, B., Leader, I., and Radcliffe, A. J. 1989. "Reverse Kleitman inequalities." *Proc. London Math. Soc.* 58: 153–168.

In: Understanding Banach Spaces
Editor: Daniel González Sánchez
ISBN: 978-1-53616-745-0
© 2020 Nova Science Publishers, Inc.

Chapter 22

COPIES OF SEQUENCE SPACES AND BASIC PROPERTIES OF ANISOTROPIC FUNCTION SPACES

Sergey S. Ajiev[*]
University of Technology, Sydney, Australia

Abstract

Employing both harmonic/real and functional analysis tools, we study the existence of isomorphic, almost isometric and complemented copies of a number of basic sequence spaces and their finite–dimensional subspaces in various classes of spaces of functions defined on an open subset of an Euclidean space and extensively employed in various branches of mathematics (for example, PDE). The spaces under consideration include the classes of anisotropic Besov, Lizorkin–Triebel and Sobolev spaces whose norms are defined in terms of averaged differences, local approximations by polynomials and wavelet decompositions with underlying mixed Lebesgue, Lorentz, or general ideal norms. Among special cases are anisotropic $BMO(G)$ and $VMO(G)$ and their positively smooth counterparts. Our main functional analysis tools are three theorems due to James, Krivine, Maurey and Pisier, lattice convexity and concavity, upper and lower estimates and interplay between them leading to the Rademacher type and cotype sets of the spaces under consideration, among which are simple examples of spaces with semi–closed type or cotype sets complementing the classical constructions based on the Tzirel'son space. The presence of the properties α, UMD, stability, separability and superreflexivity is also studied with applications to semi–embeddings. Established results have already allowed to establish sharpness in a number of questions beyond the scope of the article.

Keywords: Besov, Lizorkin–Triebel and Sobolev spaces, wavelets, stability, UMD, complemented subspaces, Rademacher type and cotype, maximum principle, semi-embeddings, Krein–Milman and Radon–Nikodým properties, property α, superreflexivity, separability

AMS Subject Classification: 42B35, 46E35, 42C40, 46A45, 46B20, 46B03, 46B07, 46E30

---

[*]Corresponding Author's E-mail: sergey.ajiev@uts.edu.au.

1. INTRODUCTION

Many local and nonlocal properties of Banach spaces, such as Markov and Rademacher type and cotype sets and estimates for various constants and moduli (for example, [1, 2, 3]), are inherited by all their subspaces (or the spaces that are λ–finitely represented in them; see Definition 3.1), while others are inherited only by complemented, or even 1–complemented subspaces (for example, see [4, 5, 6, 7, 8, 9, 1, 10, 11]). Therefore, the existence of isomorphic and almost isometric (or λ–finitely represented) copies of such well–studied spaces as l_p becomes a valuable universal tool for establishing the sharpness of various related results.

As typical examples in the setting of the anisotropic Sobolev spaces of functions defined on an open subset of an Euclidean space, let us consider two classical properties: the Rademacher cotype (see Definition 4.1) governing, in particular, Orlich type theorems (see also [12, 13, 14]) and the quantitative uniform convexity reflecting the geometric quality of the norm of a Banach space. The former property is isomorphic (i.e., preserved by isomorphic spaces, or equivalent norms), while the latter is geometric (i.e., peculiar to the concrete choice of an equivalent norm, when such choice is possible). For $s \in \mathbb{N}^n$, $p \in [1, \infty]^n$ and an open $G \subset \mathbb{R}^n$, let $X = W_p^s(G)$ be the Sobolev space of the measurable functions f defined on G endowed with the finite norm

$$\|f|W_p^s(G)\|^2 := \|f|L_p(G)\|^2 + \sum_{i=1}^{n} \|D_i^{s_i} f|L_p(G)\|^2. \qquad (S)$$

In fact, we can substitute the exponent 2 with any $\varsigma \in [p_{\min}, p_{\max}]$ with the same outcome for not only the matter of Rademacher cotype (and type) but also Markov type and cotype, moduli of uniform convexity and smoothness, Milman's moduli and many other local and asymptotic properties and constants. We have to introduce $\varsigma \in (1, \infty)$, while the traditional choice is $\varsigma = 1$, because it solves the problem of finding a geometrically friendly norm. S. L. Sobolev himself has originally defined the Sobolev space exactly with $\varsigma = 2$, thus, anticipating its remarkable geometric properties in comparison with the traditional choice $\varsigma = 1$. According to the point of view in Section 2.2, X is a subspace of the l_2-sum $Y = l_2(I_{n+1}, L_p(G))$, where $I_{n+1} = [1, n+1] \cap \mathbb{N}$. Therefore, with the aid of Theorems 4.3, a) and 4.1, we note that the Rademacher cotype set $I_C(Y) = [\max(p_{\max}, 2), \infty]$ (i.e., the set of all cotypes of Y), where $p_{\max} = \max_i p_i$, is a subset of the Rademacher cotype set $I_C(X)$ of X. Note that l_2 is finitely represented in every Banach space due to the Dvoretzky theorem (see also Remark 6.1), and all $\{l_{p_i}\}_{i=1}^{n}$ are λ–finitely represented in X for certain $\lambda \geq 1$ because every l_{p_i} is a π_λ-space (see Remark 3.1 and [15]), and X contains a λ–isomorphic copy of $l_{p_i}(I_m)$ for every integer m (see Theorem 6.2, or Section 3). Therefore, the cotype sets of X and $Z = l_{\max(p_{\max},2)}$ are related by $I_C(X) \subset I_C(Z) = I_C(Y)$, and, thus, $I_C(X) = I_C(Y) = I_C(Z) = [\max(p_{\max}, 2), \infty]$.

As the second particular example, let us observe that the original Sobolev norm (S) is one of the best equivalent norms of the space $X = W_p^s(G)$ from the point of view of the asymmetric uniform convexity [2]. For the sake of simplicity, we assume that $p \in (1, 2]^n$ (see [2] for the general case $p \in (1, \infty)$). Indeed, since X is a subspace of Y, it inherits the asymmetric uniform convexity of Y established in Theorem 4.17 in [2] in the form

$$\|\mu f + \nu g\|_X^2 + \mu\nu(p_{\min} - 1)\|f - g\|_X^2 \leq \mu\|f\|_X^2 + \nu\|g\|_X^2 \qquad (AUC)$$

for every $f, g \in X$ and $\mu = 1 - \nu \in (0, 1)$. While we already know that $W = l_{p_{\min}}$ is λ–finitely represented in X with the given norm and, thus, also with an arbitrary equivalent norm as well, the James–Krivine theorem (see Theorem 3.1) implies that W is actually finitely represented in X endowed with an arbitrary equivalent norm. Therefore, $W = l_{p_{\min}}$ and its two–dimensional subspace $l_{p_{\min}}(I_2)$ possess the same asymmetric uniform convexity (satisfies (AUC)). Moreover, choosing $x = (1 + t\nu, 1 - t\nu)$ and $y = (1 - t\mu, 1 + t\mu)$ and considering the limit $t \to 0$ (cf. [16]), we notice that, for every $\mu = 1 - \nu \in (0, 1)$, the constant $p_{\min} - 1$ is the best possible for $l_{p_{\min}}(I_2)$ and, thus, also for both W and X. Eventually the finite representability of W (and its two–dimensional subspace $l_{p_{\min}}(I_2)$) in X endowed with an arbitrary equivalent norm means that the given norm (the original Sobolev norm) is one of the geometrically friendly norms (every $\varsigma \in [p_{\min}, 2]$ is suitable [2]) achieving the best constant $p_{\min} - 1$ in (AUC).

Section 2 not only contains the basic notation and the definitions of almost all function spaces dealt with but also reveals the crucial relation between the function spaces and auxiliary ideal space containing them as subspaces. The spaces under consideration include various classes of anisotropic Sobolev, Besov and Lizorkin–Triebel spaces of functions on an open subset G of an Euclidean space defined in terms of either averaged differences, local polynomial approximations, wavelet decompositions, or a family of closed operators with underlying Lorentz or mixed Lebesgue (quasi) norms. Typical examples of a suitable family of operators are the Fourier multipliers of smooth Littlewood–Paley decompositions and certain operators defined in terms of holomorphic functional calculus (see Remarks 2.4 and 2.5). The precise (but lengthy and technical) definition of the anisotropic function spaces endowed with the wavelet norms based on the optimal wavelet decompositions from [17] are postponed till Section 7 but the classical "dyadic" approach presented in [18] can also be employed (with essentially the same outcome). General related information on function spaces can be found in [19, 20, 21, 22, 23, 24, 25, 18, 26].

In Sections 3, 4.4, 5 and 6, we identify the most representative subspaces of the function spaces under consideration extending and strengthening the corresponding results from [27, 4, 5] using mostly real analysis tools (in §3 and 6), as well as the James–Krivine theorem (in §3), the Maurey–Pisier theorem (in §4.4) and the James–Krivine–Maurey theorem (in §5). One adopts the approach leading to the non–complementability of $VMO(G)$ in $BMO(G)$ and the analogous subspaces of Nikol'skii (i.e., Besov with $q = \infty$) and Lizorkin–Triebel spaces (also with $q = \infty$) in [4] following G. M. Fichtenholtz and L.V. Kantorovich's result [28] on the non–complementability of $C([0, 1])$ in $L_\infty([0, 1])$ and their upgrades to the corresponding Phillips type theorems in [5]. While many results hold in the setting of an arbitrary open $G \subset \mathbb{R}^n$, the condition of almost boundedness and the quadrangle condition are introduced (see §2) and occasionally imposed. To avoid (minimise) the restrictions on G, we develop and apply an approach based on our wavelet decompositions in Section 6 only.

Sections 4 and 5 are mainly dedicated to immediate applications. In §4, we establish the Rademacher type and cotype exponent sets of the spaces under consideration employing both the existence of the representative subspaces of function spaces isomorphic to the copies of auxiliary sequence spaces and the lattice methods (see [29]) based on the study of the upper and lower estimates and lattice convexity and concavity, as well as the abstract p–convexity result from [30] used to describe the p–convex envelopes of the Hardy–Lorentz spaces (the p–convex envelopes of Hardy spaces were described by A. B. Aleksandrov [31] using different approach). Examples of the spaces with half–open type and cotype exponent intervals, that are noticeably simpler than the modifications of the Tsirel'son space (see [32]), are found among both simple auxiliary ideal and Lizorkin–Triebel spaces. Alternative approaches can be found in [33, 2]. The latter source contains also explicit estimates for the Rademacher type and cotype constants for many spaces under consideration.

While the main (auxiliary) role of §5 is to establish the stability of the spaces of interest to apply the James–Krivine–Maurey theorem upgrading the λ–representation of l_σ in the spaces under consideration to the almost isometric representability used extensively in the following sections, we also identify the presence of a number of other properties, including α, non–separability, UMD, superreflexivity, Radon–Nikodým and Krein–Milman properties, and derive immediate applications to semi–embeddings.

The most technical (but conceptually simple) estimates are placed in Appendix.

The numbering of the equations is used sparingly. Since the majority of references inside every logical unit are to the formulas inside the unit, equations are numbered independently inside every proof of a corollary, lemma and theorem, or a definition (if there are any numbered formulas). The name of the corresponding logical unit does not accompany the number of the formula in the references from inside this unit.

2. DEFINITIONS, DESIGNATIONS AND BASIC PROPERTIES

Let $\mathbb{N}$ be the set of the natural numbers; $\mathbb{N}_0 = \mathbb{N} \cup \{0\}$; $I_n = [1, n] \cap \mathbb{N}$ for $n \in \mathbb{N}$ and $I_m = \prod_{i=1}^n I_{m_i}$ for $m \in \mathbb{N}^n$; for $\alpha, \beta \in \mathbb{N}_0^n$, assume that $\alpha \leq \beta$ means the partial order relation generated by the coordinate order relations; $\max(\alpha, \beta) = \min\{\gamma : \gamma \geq \alpha, \gamma \geq \beta\}$; $\mathbb{R}^n$ is the n–dimensional Euclidean space with the standard basis $\langle e^1, \ldots, e^n \rangle$, $x = (x_1, \ldots, x_n) = \sum_1^n x_i e^i = (x_i)$; $x_{\min} := \min_i x_i$ and $x_{\max} := \max_i x_i$.

Let p' be the conjugate to $p \in [1, \infty]^n$, i.e. $1/p_i' + 1/p_i = 1$ for $i \in I_n$. For a scalar $p \in \mathbb{R}$, let $\bar{p} = p\bar{1} = (p, \ldots, p) \in \mathbb{R}^n$.

For $A \subset \mathbb{N}_0^n$, let $|A|$ designate the number of the elements of A, for $\alpha \in \mathbb{N}_0^n$,

$$\hat{A}_\alpha = \{\beta : \beta \in \mathbb{N}_0^n, \beta \leq \alpha\}; \quad \hat{A} = \bigcup_{\alpha \in A} \hat{A}_\alpha.$$

In what follows, the anisotropy vector $\gamma_a = ((\gamma_a)_1, \ldots, (\gamma_a)_n) \in (0, \infty)^n$ with $|\gamma_a| = \sum_1^n (\gamma_a)_i = n$ is assumed to be fixed.

For $x, y \in \mathbb{R}^n$, $t > 0$, let $[x, y]$ be the segment in $\mathbb{R}^n$ with the ends x and y; $xy = (x_i y_i)$, $t^y = (t^{y_i})$; $x/y = \left(\frac{x_i}{y_i}\right)$ for $y_i \neq 0$, and $t/\gamma_a = \left(\frac{t}{(\gamma_a)_i}\right)$. Assuming $|x|_{\gamma_a} = \max_{1 \leq i \leq n} |x_i|^{1/(\gamma_a)_i}$, we have $|x + y|_{\gamma_a} \leq c_{\gamma_a} (|x|_{\gamma_a} + |y|_{\gamma_a})$. For $E \subset \mathbb{R}^n$ and $b \in \mathbb{R}^n$, let $|E|_{\gamma_a} = \inf_{x \in E} |x|_{\gamma_a}$, $x \pm bE = \{y : y = x \pm bz, z \in E\}$ and $G_t = \{x : x \in G, |x - \partial G|_{\gamma_a} > t\}$ for an open $G \subset \mathbb{R}^n$. We say that a domain $G \subset \mathbb{R}^n$ is almost bounded if G_t is bounded for some $t > 0$. We also say that G satisfies the quadrangle condition if G contains a translation of one of the 2^n coordinate quadrangles $L_{\bar{\varepsilon}} \subset \mathbb{R}^n$ defined by the conditions $\varepsilon_i x_i \geq 0$ for $i \in I_n$ and indexed by $\{\varepsilon_i\}_{i \in I_n} \in \{-1, 1\}^n$.

Definition 2.1. *For $m \in \mathbb{N}$, $h > 0$, $\gamma \in \mathbb{R}$, $x \in \mathbb{R}^n$, assume that $\Delta_i^m(h) f(x)$ is the difference of the function f of the mth order with the step h in the direction of e^i at the* *point x, and*

$$\Delta_i^m(h, E)f(x) = \begin{cases} \Delta_i^m(h)f(x) & \text{for } [x, x+mhe^i] \in E, \\ 0 & \text{for } [x, x+mhe^i] \notin E, \end{cases}$$

$$\delta_{i,a}^m(t, x, f, E)_\gamma = \left(\int_{-1}^{1} |\Delta_i^m(t^{(\gamma_a)_i}u, E)f(x+\gamma t^{(\gamma_a)_i}u)|^a \, du \right)^{1/a}.$$

Sometimes, in the absence of ambiguity, we use a shorter form $\delta_{i,a}^m(t, x, f)$ instead of $\delta_{i,a}^m(t, x, f, E)_\gamma$.

By means of $M(\Omega, \mu)$, we designate the space of all μ–measurable functions defined on the measure space (Ω, μ). For a Lebesgue measurable $E \subset \mathbb{R}^n$ and a finite D, let $|E|$ and $|D|$ be, correspondingly, the Lebesgue measure $\int \chi_E(x)dx$ of E and the number of the elements of D. For $f \in M(\Omega, \mu)$, one has $\sigma(f, t) := |\{x : |f(x)| > t\}|$, and f^* denotes the nonincreasing right–continuous rearrangement of f. As to quasi–additivity, we shall refer to the following property of $*$–operation and σ–functional: for $f, g \in M(\Omega, \mu), t, \tau \in \mathbb{R}_+$,

$$(f+g)^*(t+\tau) \leq f^*(t) + g^*(\tau), \quad \sigma(f+g, t+\tau) \leq \sigma(f, t) + \sigma(g, \tau).$$

If ϕ is an integrable function on a Lebesgue–measurable set $E \subset \mathbb{R}^n$, and $|E|$ is the Lebesgue measure, then $\phi_E = |E|^{-1} \int_E \phi \, dx$. Assume $Q_0 = [-1, 1]^n$. For $v \in \mathbb{R}_+$ and $x \in \mathbb{R}^n$, we say that $Q_v(x) = x + v^{\gamma_a}Q_0$ is the parallelepiped of the γ_a–radius v with the centre x; χ_E is the characteristic function of E, and $\Theta = \chi_{(0,\infty)} : \mathbb{R} \to \{0, 1\}$ is Heaviside's Θ–function.

For $n \in \mathbb{N}, r \in (0, \infty]^n, p \in (0, \infty), q \in (0, \infty]$, a (countable) index set I, and a quasi–Banach space A, let $l_q(A) := l_q(I, A)$ be the space of all sequences $\alpha = \{\alpha_k\}_{k \in I} \subset A$ with the finite quasi–norm $\|\alpha\|_{l_q} = (\sum_{k \in I} |\alpha_k|^q)^{1/q} < \infty$; assume also that $c_0(A)$ is the subspace of $l_\infty(A)$ determined by $\lim_k \alpha_k = 0$;

$$l_r(A) := l_{r_n}(l_{r_{n-1}} \ldots (l_{r_1}(A)\underbrace{) \ldots)}_{n \text{ brackets}}.$$

Let also $l_{p,q}(A)$ be the space of all sequences $\alpha = \{\alpha_k\}_{k \in \mathbb{N}_0} \subset A$ with the finite quasi–norm

$$\|\alpha|l_{p,q}(A)\| := \left(\beta_0^{*q} + \sum_{i=1}^{\infty} \beta_i^{*q}(i^{q/p} - (i-1)^{q/p}) \right)^{1/q}, \text{ where }$$

$\{\beta_i^*\}_{i=0}^{\infty}$ is a non–increasing rearrangement of the sequence $\beta = \{\|\alpha_i|A\|\}_{i=0}^{\infty}$. For $m \in \mathbb{N}$ and either $l = l_r(A)$, or $l \in \{l_q(A), l_{p,q}(A), c_0(A)\}$, by means of l^m, we designate the subspace of l satisfying $\alpha_i = 0$ for either $i_{\max} > m$, or $i > m$ correspondingly.

Let $E \subset \mathbb{R}^n$ be a Lebesgue measurable set (of σ–finite measure). For $p \in (0, \infty) \setminus \{1\}$ and $q \in (0, \infty]$, let $L_{p,q}(E)$ be the (quasi)normed space of all functions $f \in M(E, dx)$ with the finite (quasi)norm

$$\|f|L_{p,q}(E)\| := \left(\int_{\mathbb{R}_+} \left(t^{1/p} f^*(t) \right)^q dt/t \right)^{1/q} \text{ for } q < \infty, \text{ or}$$

$$\|f|L_{p,\infty}(E)\| := \|t^{1/p} f^*|L_\infty(\mathbb{R}_+)\|.$$

For $G \subset \mathbb{R}^n$ and $f : G \to \mathbb{R}$ by means of $\overline{f} : \mathbb{R}^n \to \mathbb{R}$, we designate the function

$$\overline{f(x)} := \begin{cases} f(x), & \text{for } x \in G, \\ 0, & \text{for } x \in \mathbb{R}^n \setminus G. \end{cases}$$

For $p \in (0,\infty]^n$, let $L_p(G)$ be the space of all measurable functions $f : G \to \mathbb{R}^n$ with the finite mixed quasi–norm

$$\|f|L_p(G)\| = \left(\int_\mathbb{R} \left(\int_\mathbb{R} \cdots \left(\int_\mathbb{R} |\overline{f}|^{p_1} dx_1 \right)^{p_2/p_1} \cdots \right)^{p_n/p_{n-1}} dx_n \right)^{1/p_n},$$

where, for $p_i = \infty$, one understands $\left(\int_\mathbb{R} |g(x_i)|^{p_i} dx_i \right)^{1/p_i}$ as $\operatorname{ess\,sup}_{x_i \in \mathbb{R}} |g(x_i)|$. The classical quantitative geometry of these spaces had been studied for a long time (see, for example, [34]).

Similarly, for $p, q \in (0,\infty]^n$, let $L_{p,q}(G)$ be the quasi–normed space of the (measurable) functions on G with

$$\|f|L_{p,q}(G)\| = \underbrace{\|\ldots\|}_{n\ times}\overline{f}|L_{p_1,q_1}(\mathbb{R}, dx_1)\| \ldots |L_{p_n,q_n}(\mathbb{R}, dx_n)\|.$$

Definition 2.2. *For $p \in (0,\infty]^n$ and an open $G \subset \mathbb{R}^n$, we say that an ideal space $Y = Y(G)$ is L_p–similar, or, respectively, strongly L_p–similar, if $\|\chi_{Q_t(z)}|Y(G)\| \asymp \|\chi_{Q_t(z)}|L_p(G)\|$ for every $Q_t(z) \subset G$, or $\|\chi_E|Y(G)\| \asymp \|\chi_E|L_p(G)\|$ for every measurable subset $E \subset G$.*

An ideal sequence space $Y = Y(I)$ with finite or countable I is regular if its natural (coordinate) basic sequence is an (orthogonal) basis in Y. Thus, l_p and $lt_{p,q}$ (Definition 2.3) for $p \in [1,\infty)^n$ and $q \in [1,\infty)$ and c_0 are regular, while l_∞ is not.

For an ideal space $Y = Y(\Omega)$ for a measurable space (Ω, μ) and a Banach space X, let $Y(\Omega, X)$ be the space of the Bochner–measurable functions $f : \Omega \mapsto X$ with the finite (quasi)norm

$$\|f|Y(\Omega, X)\| := \|\|f(\cdot)\|_X|Y(\Omega)\|.$$

If another measure ν, absolutely continuous with respect to μ with the density $\frac{d\nu}{d\mu} = \omega$, is used instead of μ, the corresponding space is denoted by $Y(G, \omega, X)$.

For example, $L_p(\mathbb{R}^n, l_q)$ with $p, q \in [1,\infty]$ is a Banach space of the measurable function sequences $\tilde{f} = \{f_k(x)\}_{k=0}^\infty$ with the finite norm $\|\|\{f_k(\cdot)\}_{k \in \mathbb{N}_0}|l_q\||L_p(\mathbb{R}^n)\|$.

For $s \in \mathbb{N}^n$, $\varsigma \in [1,\infty)$, $p \in [1,\infty]^n$ and an open $G \subset \mathbb{R}^n$, let $W_p^s(G) = W_p^s(G)_\varsigma$ be the Banach space of the measurable functions f defined on G possessing the Sobolev generalised derivatives $D_i^{s_i} f$ and the finite norm

$$\|f|W_p^s(G)\|^\varsigma := \|f|L_p(G)\|^\varsigma + \|f|w_p^s(G)\|^\varsigma = \|f|L_p(G)\|^\varsigma + \sum_{i=1}^n \|D_i^{s_i} f|L_p(G)\|^\varsigma.$$

Definition 2.3. *For $p \in (0,\infty)^n$, $q \in (0,\infty)$ and $n \in \mathbb{N}$, let $lt_{p,q} = lt_{p,q}(\mathbb{Z}^n \times J)$ be the quasi–Banach space of the sequences $\{t_{i,j}\}_{i\in\mathbb{Z}^n}^{j\in J}$ with $J \in \{\mathbb{N}_0, \mathbb{Z}\}$ endowed with the (quasi)norm*

$$\|\{t_{i,j}\}|lt_{p,q}\| := \left\| \left\{ \sum_{i\in\mathbb{Z}^n} t_{i,j}\chi_{F_{i,j}} \right\}_{j\in J} \bigg| L_p(\mathbb{R}^n, l_q(J)) \right\|,$$

where $\{F_{i,j}\}_{i\in\mathbb{Z}^n}^{j\in J}$ is a fixed nested family of the decompositions $\{F_{i,j}\}_{i\in\mathbb{Z}^n}$ of $\mathbb{R}^n$ into unions of congruent parallelepipeds $\{F_{i,j}\}_{i\in\mathbb{Z}^n}$ satisfying

$$\cup_{i\in\mathbb{Z}^n} F_{i,j} = \mathbb{R}^n, \ |F_{i,j} \cap F_{k,j}| = 0 \text{ for every } j \in J, \ i \neq k,$$

and either $|F_{i_0,j_0} \cap F_{i_1,j_1}| = 0$, or $F_{i_0,j_0} \cap F_{i_1,j_1} = F_{i_0,j_0}$ for every i_0, i_1 and $j_0 > j_1$. We shall assume that this system regular in the sense that the length of the kth side $l_{k,j}$ of the parallelepipeds $\{F_{i,j}\}_{i\in\mathbb{Z}^n}$ of the jth decomposition (level) satisfies

$$c_l b^{-j(\lambda_a)_k} \leq l_{k,j} \leq c_u b^{-j(\lambda_a)_k} \text{ for } k \in I_n$$

for some positive constants $b > 1$ and c_l, c_u. Let also the 0-level parallelepipeds $\{F_{i,0}\}_{i\in\mathbb{Z}^n}$ be of the form $Q_t(z)$.

For a parallelepiped $F \in \{F_{i,j}\}_{i\in\mathbb{Z}^n}^{j\in J}$ or $F = \mathbb{R}^n$, by means of $lt_{p,q}(F)$, we designate the subspace of $lt_{p,q}(\mathbb{Z}^n \times J)$ defined by the condition: $t_{i,j} = 0$ if $F_{i,j} \not\subset F$.

The symbol $lt_{p,q}$ denotes either of the spaces $lt_{p,q}(\mathbb{Z}^n \times J)$ or $lt_{p,q}(F)$.

Thanks to the Fefferman–Stein inequality, the space $lt_{p,q}$ is isometric to a complemented subspace of $L_p(\mathbb{R}^n, l_q)$, while its dual $lt_{p,q}^*$ is isomorphic to $lt_{p',q'}$ for $p \in (1,\infty)^n$, $q \in (1,\infty)$ (see [1]).

Remark 2.1. *The study of the absolute majority of geometric, isomorphic and structural properties of anisotropic Besov $B_{p,q}^s(\mathbb{R}^n)_w$ and Lizorkin–Triebel $L_{p,q}^s(\mathbb{R}^n)_w$ spaces (see Definition 7.1) endowed with wavelet norms (making them sequence spaces) is reduced to the study of the same properties of $l_q(\mathbb{N}_0, l_p(\mathbb{Z}^n))$ and $lt_{p,q} = lt_{p,q}(\mathbb{Z}^n \times \mathbb{N}_0)$ respectively (see Remark 7.1).*

We say that a subspace Y of a Banach space X is C–complemented (in X) if there exists a projection P onto Y satisfying $\|P|\mathcal{L}(X)\| \leq C$.

For $B \subset X$, let $\mathrm{lin}(B)$ and $\overline{\mathrm{lin}}(B)$ be the linear envelope and the closed linear envelope of B in X correspondingly.

The bi–linear form representing the duality between a (quasi)normed space A and its dual A^* is written as

$$(A^*, A) \ni (f, x) \mapsto \langle f, x \rangle = f(x).$$

For a subspace E of a (quasi) Banach space A, let also $A_{|E}$ denote E endowed with $\|\cdot|A\|$. For a subset Q of a (quasi) Banach space X, int Q denotes its interior.

For an operator T from X into Y, let $D(T)$, $\mathrm{Ker}\,T$ and $\mathrm{Im}\,T$ be its domain, kernel and image correspondingly, and $\mathcal{C}(X, Y)$ and $\mathcal{L}(X, Y)$ be the spaces of closed and bounded operators respectively.

A quasi–normed space A is said to be p–convex for some $p \in (0, 1]$, if

$$\left\|\sum_{i \in \mathbb{N}} a_i\right\|_A \leq \|\{\|a_i\|_Y\}_{i \in \mathbb{N}}\|_{l_p(\mathbb{N})}.$$

We also define the range $\alpha_c = \alpha_c(A)$ of the p–convexity of A as

$$\alpha_c(A) = \{p : A \text{ is } p\text{-convex}\}.$$

Definition 2.4. *For $r \in [1, \infty]$, a finite, or countable set I and a set of quasi–Banach spaces $\{X_i\}_{i \in I}$, let their l_r–sum $l_r(I, \{A_i\}_{i \in I})$ be the space of the sequences $x = \{x_i\}_{i \in I} \in \prod_{i \in I} X_i$ with the finite norm*

$$\|x|l_r(I, \{A_i\}_{i \in I})\|^r := \sum_{i \in I} \|x_i\|_{A_i}^r.$$

For $\gamma_a \in (0, \infty)^n$ and $s \in [0, \infty)$, let $A_s^* := \{\alpha : \alpha \in \mathbb{N}_0^n, (\alpha, \gamma_a) \leq s\}$.

For $z \in \mathbb{R}^n$ and $v > 0$, assume that $\tau_z f(x) = f(x - z)$ and $\sigma_v f(x) = f(v^{-\gamma_a} x)$. For a Banach space X and $A \subset \mathbb{N}_0^n$, $|A| < +\infty$, let $\mathcal{P}_A(X)$ be the space of the polynomials of the form $\sum_{\alpha \in A} c_\alpha x^\alpha$ with $\{c_\alpha\}_{\alpha \in A} \subset X$, $\mathcal{P}_A = \mathcal{P}_A(\mathbb{R})$. For $a \in [1, \infty]^n$, let $p_A \in \mathcal{L}(L_a(Q_0), \mathcal{P}_A)$ be a certain projector onto $\mathcal{P}_A$. To insure the translation invariance of $\mathcal{P}_A(X)$, we shall always assume that $A = \hat{A}$.

While some results can be extended to the case of the X–valued spaces, we consider the scalar–valued case for simplicity. In this case (as well as when $\dim X < \infty$) there always exists the best (or 0–best) L_a approximation by $\mathcal{P}_A$ that is also used in the definition of the $\mathcal{D}$–functional below.

Definition 2.5. *For $a \in [1, \infty]^n$, let $p_{A,v,z} = \tau_z \circ \sigma_v \circ p_A \circ \sigma_v^{-1} \circ \tau_z^{-1}$.*

For $\varepsilon > 0$, $a \in (0, \infty]^n$, let $\pi_{A,v,z} : L_a(Q_v(z)) \to \mathcal{P}_A$ be an operator of the best L_a–approximation satisfying

$$\|f - \pi_{A,v,z} f | L_a(Q_v(z))\| = \min_{g \in \mathcal{P}_A} \|f - g | L_a(Q_v(z))\|, \ f \in L_a(Q_v(z)).$$

For $f \in L_{a,loc}(G)$, $v > 0$ and $a \in (0, \infty]^n$, we define the $\mathcal{D}$–functionals

$$\mathcal{D}_a(v, x, f, G, A) := \begin{cases} \|f|L_a(Q_v(x))/\mathcal{P}_A\| v^{-(\gamma_a, 1/a)}, & \text{if } Q_v(x) \subset G, \\ 0 & \text{otherwise.} \end{cases}$$

$$\mathcal{D}_a(v, x, f, G, p_A) := \begin{cases} \|f - p_{A,v,x}|L_a(Q_v(x))\| v^{-(\gamma_a, 1/a)}, & \text{if } Q_v(x) \subset G, \\ 0 & \text{otherwise.} \end{cases}$$

Remark 2.2. *Let us note that*

$$\|f - \pi_{A,v,x} f|L_a(Q_v(x))\| \asymp \|f - p_{A,v,x}|L_a(Q_v(x))\| \asymp \|f|L_a(Q_v(x))/\mathcal{P}_A\|$$

uniformly in v and x when they all are well–defined. It is interesting to note that, while switching from one $\mathcal{D}$–functional to another provides only an equivalent norm, the geometric properties under consideration will depend on the parameters only, remaining identical. The same is true regarding switching to another projector in the definition of the second $\mathcal{D}$–functional.

2.1. Classes of Spaces of Besov and Lizorkin–Triebel Types

Above we have already considered Sobolev spaces and Besov and Lizorkin–Triebel spaces endowed with the wavelet norms. Let us define several more classes of spaces of Besov and Lizorkin–Triebel types. In these definitions, we use a parameter $\varsigma \in (0, \infty]$, which is essential in the study of the geometric properties of function spaces but not the topological (isomorphic) ones (we have equivalent (quasi)norms for different $\varsigma \in (0, \infty]$). It will normally be omitted for the sake of simplicity. If its presence and value should be emphasised, we say that the space under consideration is endowed with the ς–product norm, or, just, the ς–norm, and/or add ς as a subindex. For a seminormed (homogeneous) space $x(G)$ of functions defined on $G \subset \mathbb{R}^n$ and an ideal space $Y(G)$, we assume that its intersection $x(G) \cap Y(G)$ is endowed with the ς–norm too:

$$\|f|x(G) \cap Y(G)\|^\varsigma := \|f|Y(G)\|^\varsigma + \|f|x(G)\|^\varsigma.$$

Moreover, we shall always assume the parameter ς to be equal to one of the other parameters or its components of $x(G)$ and $Y(G)$ except for the smoothness and anisotropy components.

We start with the spaces defined in terms of the averaged (shifted) axis–directional differences. While the study of these norms and their equivalence with the other norms was one of the primary tasks of, for example, [35] (including the setting of arbitrary open subsets G), we shall refrain from the usage of the results of this type here in order to cover the sets of the parameters not covered by the equivalence results, and because the geometric constants depend on the specific equivalent norm chosen.

Let Pr_i be the orthogonal projector on the ith axis in $\mathbb{R}^n$, and, for any $y \in (I - Pr_i)(G)$,

$$In_i(y) := (I - Pr_i)^{-1}(y) \cap G = \{x \in G : x = y + te_i, t \in \mathbb{R}\}.$$

Definition 2.6. *For an ideal space $Y = Y(G)$, $p \in (0, \infty]^n$, $q \in (0, \infty]$, $r > 0$, $b > 1$, $s \in (0, \infty)$, $s/\gamma_a < m \in \mathbb{N}_0^n$, $a \in (0, \infty]^n$, $\gamma \geq 0$, and an open set $G \subset \mathbb{R}^n$, by means of $b^s_{Y,q,a}(G)$, we designate the (quasi) semi–normed space of the measurable functions $f \in L_{a_i,loc}(In_i(y), dx_i)$ for a.e. $y \in (I - Pr_i)(G)$ for $i \in I_n$ with the finite (quasi) semi–norm*

$$\|f|b^s_{Y,q,a}(G)\|^\varsigma := \sum_{i=1}^n \left(\int_0^\infty \left\| \delta^{m_i}_{i,a_i}(t, \cdot, f, G_{rt})_\gamma | Y(G) \right\|^q t^{-qs} \frac{dt}{t} \right)^{\varsigma/q} \tag{1}$$

or, that is equivalent, with the finite (quasi) semi–norm

$$\left(\sum_{i=1}^n \left\| \left\{ b^{-sj} \left\| \delta^{m_i}_{i,a_i}(b^{-j}, \cdot, G_{rb^{-j}}, f)_\gamma \right| Y(G) \right\| \right\}_{j \in \mathbb{Z}} \Big| l_q \right\|^\varsigma \right)^{1/\varsigma}. \tag{2}$$

By means of $\overset{\circ}{b}^s_{Y,\infty,a}(G)$, we designate the subspace of $b^s_{Y,\infty,a}(G)$ of the functions satisfying the condition $\lim_{t \to 0} \left\| \delta^{m_i}_{i,a_i}(t, f, G_{rt})_\gamma | Y(G) \right\| t^{-s} = 0$.

$$B^s_{Y,q,a}(G) := b^s_{Y,q,a}(G) \cap Y(G) \text{ and } \overset{\circ}{B}^s_{Y,\infty,a}(G) = \overset{\circ}{b}^s_{Y,\infty,a}(G) \cap Y(G);$$

We designate the completions of these spaces by means of the same symbols whenever they appear not to be complete.

Note that $b^s_{p,q,a}(G) := b^s_{L_p,q,a}(G)$, and, for $\gamma = 0$, one has $B^s_{p,q,1}(G) = B^s_{p,q}(G)$ [23].

Definition 2.7. *For an ideal space $Y = Y(G)$, $p \in (0,\infty]^n$, $q \in (0,\infty]$, $r > 0$, $b > 1$, $s \in (0,\infty)$, $s/\gamma_a < m \in \mathbb{N}_0^n$, $a \in (0,\infty]^n$, $\gamma \in \mathbb{R}$, and an open set $G \subset \mathbb{R}^n$, by means of $l^s_{p,q,a}(G)$, we designate the (quasi) semi–normed space of the measurable functions $f \in L_{a_i,loc}(In_i(y), dx_i)$ for a.e. $y \in (I - Pr_i)(G)$ for $i \in I_n$ with the finite (quasi) semi–norm*

$$\|f|l^s_{Y,q,a}(G)\|^{\varsigma} := \sum_{i=1}^{n} \left\| \left(\int_0^\infty \left(\delta^{m_i}_{i,a_i}(t,\cdot,G_{rt},f)_\gamma \right)^q t^{-qs} \frac{dt}{t} \right)^{1/q} \Big| Y(G) \right\|^{\varsigma} \quad (3)$$

or, that is equivalent, with the finite (quasi) semi–norm

$$\left(\sum_{i=1}^{n} \left\| \left\| \{ b^{-sj} \delta^{m_i}_{i,a_i}(b^{-j},\cdot,G_{rb^{-j}},f)_\gamma \}_{j\in\mathbb{Z}} \Big| l_q \right\| |Y(G)| \right\|^{\varsigma} \right)^{1/\varsigma}. \quad (4)$$

By means of $\overset{\circ}{l}{}^s_{Y,\infty,a}(G)$, we designate the subspace of $l^s_{Y,\infty,a}(G)$ of the functions satisfying the condition

$$\lim_{\tau \to 0} \left\| \sup_{t \in (0,\tau)} \delta^{m_i}_{i,a_i}(t,\cdot,f,G_{rt})_\gamma t^{-s} \Big| Y(G) \right\| = 0,$$

or, that is equivalent for $Y = L_p$ with $p \in (0,\infty)^n$ due to the Lebesgue and Levi theorems, satisfying $\lim_{t\to 0} \delta^{m_i}_{i,a_i}(t,x,f,G_{rt})_\gamma t^{-s} = 0$ for a.e. x.

$$L^s_{Y,q,a}(G) := l^s_{Y,q,a}(G) \cap Y(G) \text{ and } \overset{\circ}{L}{}^s_{Y,\infty,a}(G) = \overset{\circ}{l}{}^s_{Y,\infty,a}(G) \cap Y(G).$$

We designate the completions of these spaces by means of the same symbols whenever they appear not to be complete.

Note that $l^s_{p,q,a}(G) := l^s_{L_p,q,a}(G)$, and, for $\gamma = 0$, one has $L^s_{p,q}(G) = L^s_{p,q,1}(G)$ [23].

Let us define the anisotropic local approximation spaces of Besov and Lizorkin–Triebel type in terms of the $\mathcal{D}$–functional as follows.

Definition 2.8. *For $p \in (0,\infty]^n$, $q \in (0,\infty]$, $a \in (0,\infty]^n$, $s \in (0,\infty)$, $D = \hat{D} \subset \mathbb{N}_0^n$, $|D| < \infty$ and an ideal space $Y = Y(G)$ by means of $\widetilde{b}^{s,D}_{Y,q,a}(G)$ and $\widetilde{l}^{s,D}_{Y,q,a}(G)$, we designate, correspondingly, the anisotropic (quasi) semi–normed space of the functions $f \in L_{a,loc}(G)$ with the finite (quasi) semi–norm*

$$\left\| f | \widetilde{b}^{s,D}_{Y,q,a}(G) \right\| := \left(\int_0^\infty \| t^{-s} \mathcal{D}_a(t,\cdot,f,G,D) | Y(G) \|^q \frac{dt}{t} \right)^{\frac{1}{q}}, \quad \widetilde{b}^{s,D}_{p,q,u}(G) := \widetilde{b}^{s,D}_{L_p,q,u}(G),$$

and

$$\left\| f | \widetilde{l}^{s,D}_{Y,q,a}(G) \right\| := \left\| \left(\int_0^\infty (t^{-s} \mathcal{D}_a(t,\cdot,f,G,D))^q \frac{dt}{t} \right)^{\frac{1}{q}} \Big| Y(G) \right\|, \quad \widetilde{l}^{s,D}_{p,q,u}(G) := \widetilde{l}^{s,D}_{L_p,q,u}(G).$$

Assume also that

$$\widetilde{B}^{s,D}_{Y,q,a}(G) = \widetilde{b}^{s,D}_{Y,q,a}(G) \cap Y(G) \text{ and } \widetilde{L}^{s,D}_{Y,q,a}(G) = \widetilde{l}^{s,D}_{Y,q,a}(G) \cap Y(G).$$

By means of $\mathring{\widetilde{b}}_{Y,\infty,a}^{s,D}(G)$, we designate the subspace of $\widetilde{b}_{Y,\infty,a}^{s,D}(G)$ of the functions f satisfying the condition
$$\lim_{t\to 0}\|\mathcal{D}_a(t,\cdot,f,G,D)|Y(G)\|\,t^{-s}=0.$$

By means of $\mathring{\widetilde{l}}_{Y,\infty,a}^{s,D}(G)$, we designate the subspace of $\widetilde{l}_{Y,\infty,a}^{s,D}(G)$ of the functions f satisfying the condition
$$\lim_{\tau\to 0}\left\|\sup_{t\in(0,\tau)}\delta_{i,a_i}^{m_i}(t,\cdot,f,G_{rt})_\gamma t^{-s}\Big|Y(G)\right\|=0,$$

or, that is equivalent for $Y=L_p$ with $p\in(0,\infty)^n$ due to the Lebesgue and Levi theorems, satisfying
$$\lim_{t\to 0}\mathcal{D}_a(t,\cdot,f,G,D)t^{-s}=0\text{ for a.e. }x.$$

We designate the completions of these spaces by means of the same symbols whenever they appear not to be complete.

Definition 2.9. *Under the conditions of Definitions 2.6 − 2.8, we say that $b_{Y,q,a}^s(G)$ ($B_{Y,q,a}^s(G)$), $l_{Y,q,a}^s(G)$ ($L_{Y,q,a}^s(G)$), $\widetilde{b}_{Y,q,a}^{s,D}(G)$ ($\widetilde{B}_{Y,q,a}^{s,D}(G)$), or $\widetilde{l}_{Y,q,a}^{s,D}(G)$ ($\widetilde{L}_{Y,q,a}^{s,D}(G)$) is compatible with its underlying space $Y(G)$ if, for some $C>0$ and every $t>0$, one has, correspondingly,*

$$\left(\int_t^\infty\left\|\tau^{-s}\delta_{i,a_i}^{m_i}(\tau,\cdot,G_{r\tau},f)_\gamma\Big|Y(G)\right\|^q d\tau/\tau\right)^{1/q}\le Ct^{-s}\|f|Y(G)\|,$$

$$\left\|\left(\int_t^\infty\left(\tau^{-s}\delta_{i,a_i}^{m_i}(\tau,\cdot,G_{r\tau},f)_\gamma\right)^q d\tau/\tau\right)^{1/q}\Big|Y(G)\right\|\le Ct^{-s}\|f|Y(G)\|,$$

$$\left(\int_t^\infty\bigl\|\tau^{-s}\mathcal{D}_a(\tau,\cdot,f,G,D)\bigm|Y(G)\bigr\|^q d\tau/\tau\right)^{1/q}\le Ct^{-s}\|f|Y(G)\|,\text{ or}$$

$$\left\|\left(\int_t^\infty(\tau^{-s}\mathcal{D}_a(\tau,\cdot,f,G,D))^q d\tau/\tau\right)^{1/q}\Big|Y(G)\right\|\le Ct^{-s}\|f|Y(G)\|.$$

Generalised Minkowski inequality implies that we, particularly, have the compatibility for $s>0$ and $Y=L_p$ with $a_{\max}\le p_{\min}$ in the case of any Besov, or Lizorkin–Triebel space under the consideration.

We assume all the nonhomogeneous spaces under consideration to be compatible with their underlying ideal spaces.

Remark 2.3. *It is important to note that, in the presence of the compatibility, we obtain equivalent norms in the nonhomogeneous Besov and Lizorkin–Triebel spaces defined above by substituting the integration $\int_0^\infty$ in their seminorms with the integration $\int_0^h$ for any fixed $h>0$. This fact is used in §3. At the same time, the geometric properties will remain the same for every $h\in(0,\infty]$, depending on the parameters of $Y(G)$, q and a only (as in the previous remark).*

Let us note that $BMO^{\gamma_a}(G) = \tilde{b}^{0,A_0^*}_{\infty,\infty,1}(G)$ and $VMO^{\gamma_a}(G) = \overset{\circ}{\tilde{b}}{}^{0,A_0^*}_{\infty,\infty,1}(G)$.

We also define variants of abstract classes of Besov and Lizorkin–Triebel spaces containing the spaces defined in terms of both a smooth Littlewood–Paley decomposition and a holomorphic functional calculus.

Definition 2.10. *Let $G \subset \mathbb{R}^n$, $p \in [1,\infty]^n$, $q,\varsigma \in [1,\infty]$, $s \in \mathbb{R}$, $b > 1$ and $\mathcal{F} = \{F_k\}_{k \in \mathbb{N}_0} \subset \mathcal{C}(L_p(G))$ be a system of closed operators satisfying*

$$f \in L_p(G) \text{ and } F_k f = 0 \text{ for } k \in \mathbb{N}_0 \Rightarrow f = 0.$$

By means of $B^s_{p,q,\mathcal{F}}(G)$ and $L^s_{p,q,\mathcal{F}}(G)$, we designate, correspondingly, the Banach spaces of functions defined on G with the finite norms

$$\|f\| B^s_{p,q,\mathcal{F}}(G)\|^\varsigma = \|f\| L_p(G)\|^\varsigma + \left(\sum_{k \in \mathbb{N}_0} b^{ksq} \|F_k f\| L_p(G)\|^q\right)^{\varsigma/q},$$

$$\|f\| L^s_{p,q,\mathcal{F}}(G)\|^\varsigma = \|f\| L_p(G)\|^\varsigma + \left\|\left(\sum_{k \in \mathbb{N}_0} b^{ksq} |F_k f(\cdot)|^q\right)^{1/q}\bigg| L_p(G)\right\|^\varsigma.$$

We designate the completions of these spaces by means of the same symbols whenever they appear not to be complete.

Remark 2.4. *Under the conditions of the last definition, let also $\Omega \subset \mathbb{C}$ be open with $b^{-k}\Omega \subset \Omega$ for $k \in \mathbb{N}_0$, $g \in H_\infty(\Omega)$ (bounded holomorphic function on Ω), and $A \in \mathcal{C}(L_p(G))$ admits the bounded $H_\infty(\Omega)$ functional calculus*

$$H_\infty(\Omega) \ni h \mapsto h(A) \in \mathcal{L}(L_p(G)) \text{ with } \|f(A)|\mathcal{L}(L_p(G))\| \leq \|f|H_\infty(\Omega)\|.$$

Assuming that $F_k = g(b^{-k}A)$, we obtain the Besov and Lizorkin–Triebel spaces $B^s_{p,q,\mathcal{F}}(G)$ and $L^s_{p,q,\mathcal{F}}(G)$ defined in terms of the bounded H_∞–calculus.

The following definition is suggested by Theorems 3.2 and 3.3 and Part $a)$ of Remark 2.5.

Definition 2.11. *We say that the parameter a is admissible or in the admissible range for*

$$Y \in \left\{B^s_{p,q,a}(G), \tilde{B}^{s,A}_{p,q,a}(G), L^s_{p,q,a}(G), \tilde{L}^{s,A}_{p,q,a}(G), b^s_{p,q,a}(G), \tilde{b}^{s,A}_{p,q,a}(G), l^s_{p,q,a}(G), \tilde{l}^{s,A}_{p,q,a}(G)\right\}$$

if either

$$Y \in \left\{B^s_{p,q,a}(G), L^s_{p,q,a}(G), b^s_{p,q,a}(G), l^s_{p,q,a}(G)\right\}, \ s > (\gamma_a(1/p - 1/a))_{\max}, \text{ or}$$

$$Y \in \left\{\tilde{B}^{s,A}_{p,q,a}(G), \tilde{L}^{s,A}_{p,q,a}(G), \tilde{b}^{s,A}_{p,q,a}(G), \tilde{l}^{s,A}_{p,q,a}(G)\right\}, \ s > (\gamma_a, 1/p - 1/a).$$

Remark 2.5. *a) The admissibility of a for Y is closely related to the cases when we do not need to take the completion in Definitions 2.6 – 2.8.*

b) The Besov spaces $B^s_{p,q}(G)$ and Triebel–Lizorkin spaces $F^s_{p,q}(G)$ (see [24]) defined in terms of a smooth Littlewood–Paley decomposition are particular cases of $B^s_{p,q,\mathcal{F}}(G)$ and $L^s_{p,q,\mathcal{F}}(G)$ and their subspaces.

2.2. Function Spaces as Subspaces of Auxiliary Spaces

In this subsection we show that the second (if not the first) major idea standing behind the introduction of the Sobolev spaces in [19], besides the generalised functions and generalised derivatives, still makes a lot of sense for various Besov and Lizorkin–Triebel spaces.

It is possible to classify all the function spaces of Besov, Lizorkin–Triebel and Sobolev types defined above into two categories: homogeneous (semi–normed) spaces and normed spaces.

Given a linear topological space W, a (quasi) Banach space Y and an injective linear operator $A : W \supset D(A) \to Y$, let $z_{A,Y}$ be the completion of the linear space $\{x \in W : Ax \in Y\}$ endowed with the (quasi) norm $\|x|z_{A,Y}\| := \|Ax\|_Y$. If $\operatorname{Ker} A \neq \{0\}$ is closed, we use $W/\operatorname{Ker} A$ instead of W. We note that, according to this definition, $z_{A,Y}$ is isometric to the closure $\overline{\operatorname{Im} A}^Y$ of the image of A in Y.

Given $\varsigma \in (0, \infty]$, (quasi) Banach spaces X and Y and a closed linear operator $A \in \mathcal{C}(X, Y)$, let $Z_{A,X,Y} = Z_{A,X,Y,\varsigma}$ be the (quasi) Banach linear space $D_X(A) = D(A) \cap X$ endowed with the (quasi) norm

$$\|x|Z_{A,X,Y}\|^\varsigma := \|x\|_X^\varsigma + \|Ax\|_Y^\varsigma.$$

Note that the completeness of $Z_{A,X,Y}$ is equivalent to the closedness of A.

In what follows we shall often deduce some properties of the spaces $z_{A,Y}$ and $Z_{A,X,Y}$ from the corresponding properties of the spaces Y and the l_ς–sum $l_\varsigma(I_2, \{X, Y\})$ respectively.

The following auxiliary operators introduced in [4] provide the particular examples of A corresponding to the spaces

$$B_{p,q,a}^s(G), \tilde{B}_{p,q,a}^{s,A}(G), L_{p,q,a}^s(G), \tilde{L}_{p,q,a}^{s,A}(G), b_{p,q,a}^s(G), \tilde{b}_{p,q,a}^{s,A}(G), l_{p,q,a}^s(G), \tilde{l}_{p,q,a}^{s,A}(G),$$
$$B_{p,q,\mathcal{F}}^s(G), L_{p,q,\mathcal{F}}^s(G).$$

Definition 2.12. *For $s \geq 0$, $a \in (0, \infty]^n$, $A \in \mathbb{N}_0^n$, $|A| < +\infty$, let*

$$\Upsilon_{t,z,A} = \begin{cases} t^{-s} Q_{A,1,0} \circ \sigma_t^{-1} \circ \tau_z^{-1} & \text{for } Q_t(z) \subset G, \\ 0 & \text{for } Q_t(z) \not\subset G, \end{cases}$$

where $Q_{A,1,0} : L_a(Q_1(0)) \to L_a(Q_1(0))/\mathcal{P}_A$ is the quotient map.

$$\Xi_{i,t,z} = t^{-s} \tau_{\gamma u}^{-1} \circ \Delta_i^{m_i}(u, G_{kt})) \circ \sigma_t^{-1} \circ \tau_z^{-1}.$$

Let us designate, by means of $\Upsilon_{\tilde{B}}$ and $\Upsilon_{\tilde{L}}$, the operators

$$\Upsilon_{\tilde{B}} : \tilde{b}_{p,q,a}^{s,A}(G) \longrightarrow L_{*q}\left(\mathbb{R}_+, L_p\left(G, L_a(Q_1(0))/\mathcal{P}_A\right)\right),$$

$$\Upsilon_{\tilde{L}} : \tilde{l}_{p,q,a}^{s,A}(G) \longrightarrow L_p\left(G, L_{*q}\left(\mathbb{R}_+, L_a(Q_1(0))/\mathcal{P}_A\right)\right),$$

where

$$\Upsilon_{\tilde{L}} = \Upsilon_{\tilde{B}} = \{\Upsilon_{t,z,A}\}_{z \in G}^{t \in \mathbb{R}_+}.$$

By means of Ξ_B and Ξ_L, we designate the operators

$$\Xi_B : b^s_{p,q,a}(G) \longrightarrow \prod_{i=1}^n L_{*q}\left(\mathbb{R}_+, L_p\left(G, L_{a_i}([-1,1])\right)\right),$$

$$\Xi_L : l^s_{p,q,a}(G) \longrightarrow \prod_{i=1}^n L_p\left(G, L_{*q}\left(\mathbb{R}_+, L_{a_i}([-1,1])\right)\right),$$

where
$$\Xi_L = \Xi_B = \{\Xi_{i,t,z}\}_{z \in G,\ t \in \mathbb{R}_+}^{i \in I_n}.$$

In the case of $B^s_{p,q,\mathcal{F}}(G)$ and $L^s_{p,q,\mathcal{F}}(G)$, the corresponding closed operator A is defined by

$$A : f \mapsto \{F_k f\}_{k \in \mathbb{N}_0}.$$

3. Copies of Sequence Spaces and λ–Finite Representability

It is peculiar to our approach that we start with the investigation of the existence of the isomorphic copies of elementary sequence spaces (not only l_p) in the spaces under the consideration. For example, the Maurey–Pisier theorem uses the (pre–acquired) knowledge of the Rademacher type and cotype sets to establish the existence of the copies of $l_p(I_n)$, while we use this existence to establish the Rademacher type and cotype ranges of our spaces and, afterwards, to upgrade our knowledge of the Minkowski spectrum (i.e., the set of σ, such that l_σ is finitely represented in the space under consideration) with the aid of the Maurey–Pisier theorem in §4.4. The existence of the copies of $l_p(\mathbb{N})$ is treated similarly (see §5).

For the sake of convenience (in the view of the extensive usage), in this section and in Appendix (where some proofs are) we use the following abridged designation of sequence spaces: for $m \in \mathbb{N}, p \in (0, \infty]^n, r \in (0, \infty], \theta \in (1, \infty]$ and $X \in \{l_q, c_0\}$, let $l^m_p = l_p(I^n_m)$, $l^m_r = l_r(I_m)$, $l^m_p(X) = l_p(I^n_m, X)$, $l^m_r(X) = l_r(I_m, X)$, $l^m_{r,\theta} = l_{r,\theta}(I_m)$, $l^m_{r,\theta}(X) = l_{r,\theta}(I_m, X)$.

Definition 3.1. *Let X, Y be (quasi)Banach spaces and $\lambda \geq 1$. Then the Banach–Mazur distance $d_{BM}(X,Y)$ between them is equal to ∞ if they are not isomorphic and is, otherwise, defined by*

$$d_{BM}(X,Y) := \inf\{\|T\| \cdot \|T^{-1}\| : T : X \xrightarrow{onto} Y,\ \operatorname{Ker} T = 0\}.$$

The space X is λ–finitely represented in Y if for every finite–dimensional subspace $X_1 \subset X$,
$$\inf\{d_{BM}(X_1, Y_1) : Y_1 \text{ is a subspace of } Y\} = \lambda.$$

If λ is equal to 1, then X is simply said to be finitely represented in Y.

It is said that Y contains almost isometric copies of X (or contains X almost isometrically) if
$$\inf\{d_{BM}(X, Y_1) : Y_1 \text{ is a subspace of } Y\} = 1.$$

For example, it is well–known that, for a Banach space X, its second conjugate X^{**} is finitely represented in X, and X itself is finitely represented in c_0. While, according to Banach, $d_{BM}(l_p, L_p) = \infty$ for $p \in [1, \infty) \setminus \{2\}$ (see [36]), L_p is finitely represented in l_p for $p \in [1, \infty]$.

Remark 3.1. *If X is a π_λ–space [15], i.e., there exists a monotone (by inclusion) and dense sequence (net) of uniformly complemented finite–dimensional subspaces, in order to establish the λ–finite representability of X in Y, it is sufficient to check the conditions of the previous definition for all X_1 from this sequence (net) only. We note that the spaces $l_q(l_{p,p_1}), c_0(l_{p,p_1}), l_{p,p_1}(l_q)$ and $l_{p,p_1}(l_q)$ are π_λ–spaces.*

The next classical theorem was proved by R. C. James for c_0 and l_1 and by J. L. Krivine for $l_p, p \in (1, \infty)$ (see also Theorem 5.2).

Theorem 3.1. [James, Krivine] *Let X be either c_0 or l_p for $p \in [1, \infty)$, Y a Banach space, and $\lambda \geq 1$. Then the λ–finite representability of X in Y implies the finite representability of X in Y.*

Now it is left to prove our main λ–finite representability results extensively used later.

Let us give the definitions of the mappings dealt with in the theorems of this section. As far as Λ_B is concerned, assume that $\phi \in C_0^\infty$ with supp $\phi \subset Q_1(0)$, $s \in \mathbb{R}$, $p \in (0, \infty)^n$ and $U_i \cap U_j = \emptyset$ for $i \neq j$,

$$U := \bigcup_{i \in \mathbb{N}_0} U_i \subset G_{\varkappa \delta_0 M}, \text{ where } U_k := (Mb^{-k})^{\gamma_a} \operatorname{supp} \phi + Q_{Mb^{-k}\varkappa_0\delta_0}(z_k)$$

for $\varkappa_0 := \max_i(|\gamma + m_i|^{1/(\gamma_a)_i}, 1) > 1$ and some $\delta_0 \in (0,1)$, $M > 0$, $\{z_k\}_{k \in \mathbb{N}_0} \subset G$ and related function space parameters $\gamma > 0$, $\varkappa \geq 0$ and $m \in \mathbb{N}^n$ (see §2.1). Then $\Lambda_B = \Lambda_B^{s,p,\phi}$ denotes the mapping of the space l_∞ of all bounded sequences into C_0^∞ defined by $\Lambda_B(\lambda) = \sum_{i=0}^\infty \lambda_k \phi_k$, where $\phi_k = b^{-k(s-(\gamma_a, 1/p))}(\tau_{z_k} \circ \sigma_{Mb^{-k}})\phi$.

To define the mapping $\Lambda_L = \Lambda_L^{s,\psi}$, we take advantage of the construction in the proof of the following lemma from [4]. It is an anisotropic variant of the Calderón–Zygmund decomposition and presented below. Special parabolic anisotropic variant of Part a) of the lemma was considered in [37], while Part a) in its generality was proved in [23].

Lemma 3.1 ([4]). *For $\delta \in \left(0, 2^{1/(\gamma_a)_{\max}} - 1\right)$, let $u \in L_1(\mathbb{R}^n)$, $t > 0$ and*

$$0 < \xi \leq \min_{i \in I_n}(2(1+\delta)^{-(\gamma_a)_i} - 1)^{(\gamma_a)_i^{-1}}.$$

Then there is a family $\{Q_k\}_{k \in \mathbb{N}}$ of the closed parallelepipeds possessing pairwise disjoint interiors and satisfying $Q_{a_k}(x^{(k)}) \subset Q_k \subset Q_{(1+\delta)a_k}(x^{(k)})$ for every k and some $\{a_k\}_{k \in \mathbb{N}} \subset (0, \infty)$, $\{x^{(k)}\}_{k \in \mathbb{N}} \subset \mathbb{R}^n$ and:
a) $t \leq |Q_k|^{-1} \int_{Q_k} |u| dx \leq \varkappa^{|\gamma_a|} t$, $|u(x)| \leq t$ for a.e. $x \in \mathbb{R}^n \setminus \cup_k Q_k$, where $\varkappa = \varkappa(\delta, \gamma_a)$;
b) for some $\delta > 0$, the family $\{Q_k\}_{k \in \mathbb{N}}$ possesses also the property:
if $Q_{(1+\delta)^{-1}b}(z) \subset Q_k \subset Q_b(y)$, then $Q_{b\xi}(y) \subset Q_k$.

For example, under the additional condition $\delta \geq (3/2)^{(\gamma_a)_{\max}} - 1$, one simply takes

$$0 < \xi \leq (2(1+\delta)^{-(\gamma_a)_{\min}} - 1)^{(\gamma_a)_{\min}^{-1}}.$$

Proof of Lemma 3.1. As in the proof of Lemma on Covering from [23], let us start with a subdivision of $\mathbb{R}^n$ into a net of the (anisotropic) non–overlapping cubes $Q_{l_0}(y)$, where l_0 is chosen to satisfy $|Q_{l_0}(0)| > t^{-1}\|u|L_1(\mathbb{R}^n)\|$. Let

$$l \geq \max_{1 \leq i \leq n} \left(1 - (1+\delta)^{-(\gamma_a)_i}\right)^{-(\gamma_a)_i^{-1}}. \tag{1}$$

At the kth step, we divide the ribs parallel to the ith coordinate axis of every parallelepiped obtained at the $k - 1$th step into s_{ki} equal segments, where $s_{1i} = [l^{(\gamma_a)_i}]$, $s_{ki} = \left[\frac{l^{k(\gamma_a)_i}}{\prod_{j=1}^{k-1} s_{ji}}\right]$. This process results in the net of parallelepipeds of the kth level. Thus, for the half–lengths of the ribs of the parallelepipeds obtained, one has the estimates analogous to those from [23]:

$$(l^{-k}l_0)^{(\gamma_a)_i} \leq \frac{l_0^{(\gamma_a)_i}}{\prod_{j=1}^{k} s_{ji}} \leq \frac{l_0^{(\gamma_a)_i}}{\prod_{j=1}^{k-1} s_{ji}(\frac{l^{k(\gamma_a)_i}}{\prod_{j=1}^{k-1} s_{ji}} - 1)} \leq \frac{l_0^{(\gamma_a)_i}}{l^{k(\gamma_a)_i} - l^{(k-1)(\gamma_a)_i}} \leq$$

$$\leq (l^{-k}l_0)^{(\gamma_a)_i}(1+\delta)^{(\gamma_a)_i}.$$

Here one uses (1), $\prod_{j=1}^{k} s_{ji} \leq l^{k(\gamma_a)_i}$ and a corollary of (1):

$$l^{-(\gamma_a)_i} + (1+\delta)^{-(\gamma_a)_i} \leq 1 \text{ for } i \in I_n.$$

Let $\varkappa = l(1+\delta)$. Now the literal repetition of the related considerations from [23] proves a). Namely, at the kth step, we divide only those parallelepipeds Q, obtained as the result of the division at the $k - 1$th step, that satisfy $|Q|^{-1}\int_Q |u|dx \leq t$. All the "stopping–time" parallelepipeds, whose division process has ceased at some step, are exactly our selection $\{Q_k\}_{k\in\mathbb{N}}$. If Q is selected at the kth step as a result of the division of $F \supset Q$ (that was not selected at the $k - 1$th step), then we have

$$t < |Q|^{-1}\int_Q |u|dx \leq \frac{|F|}{|Q|}|F|^{-1}\int_F |u|dx \leq t\varkappa^{|\gamma_a|}.$$

Studying the coordinate projections of $Q_k, Q_b(y), Q_{b\xi}(y)$ satisfying the conditions of b) shows that $Q_{b\xi}(y) \subset Q_k$, if δ and ξ satisfy the relation

$$\min_i (2(1+\delta)^{-(\gamma_a)_i} - 1)^{(\gamma_a)_i^{-1}} \geq \xi.$$

$\square$

Now we choose a parallelepiped $F_0^0 := Q_M(z_0) \subset G_{M\varkappa\delta_0}$ and assume $l_0 = M > 0$, and that $l = b > 1$ is large enough. Let $\{F_i^k\}_i$ be the family of the parallelepipeds obtained at the kth step in the proof of Lemma 3.1 contained in F_0^0. The half–length l_{kj} of the ribs of $\{F_i^k\}_i$ parallel to the jth coordinate axis is contained in the interval $[(Mb^{-k})^{(\gamma_a)_j}, (\lambda_0 Mb^{-k})^{(\gamma_a)_j}]$. In addition, one has $F_i^k \supset Q_{Mb^{-k}}(O_i^k)$, where O_i^k is the centre of F_i^k. Let $\psi \in C_0^\infty$, supp $\psi = Q_1(0)$, and $\psi_k = b^{-ks}\sum_i \psi((\cdot - O_i^k)/l_k)$ for $k \in \mathbb{N}_0$, where $l_k = (l_{k1}, \ldots, l_{kn})$. Eventually $\Lambda_L = \Lambda_L^{s,\psi}$ is defined by $\Lambda_L : \lambda \mapsto \sum_{k=0}^\infty \lambda_k \psi_k$.

Definition 3.2. Assume that, for $\Lambda \in \{\Lambda_B, \Lambda_L\}$, $\Lambda : l_\infty \supset D(\Lambda) \to Y$, where Y is a (quasi) semi–normed space, the space ZY is the image Im Λ endowed with the (quasi) semi–norm of Y.

Theorem 3.2. Let $G \subset \mathbb{R}^n$ be an open set, $a, p \in (0, \infty]^n$, $q \in (0, \infty]$, $A_s^* \subset A \subset \mathbb{N}_0^n$, $|A| < +\infty$, and $Y = Y(G)$ be an ideal space with a translation–invariant (quasi)norm and $r \in \alpha_c(Y)$. Then, there are ϕ and ψ determining Λ_B and Λ_L, such that the following holds.
If Y is L_p–similar, and either $s > 0$, or $Y = L_\infty(G)$ and $s \geq 0$, then:
a) for $s > (\gamma_a(1/p - 1/a))_{\max}$, Λ_B induces the isomorphisms

$$\Lambda_B : l_q \longleftrightarrow X \in \{ZB^s_{Y,q,a}, zb^s_{Y,q,a}\} \text{ and } \Lambda_B : c_0 \longleftrightarrow X \in \left\{\mathring{ZB}^s_{Y,\infty,a}, \mathring{zb}^s_{Y,\infty,a}\right\},$$

and, in addition, $zb^s_{Y,\infty,a} \cap \mathring{b}^s_{Y,\infty,a}(G) = \mathring{zb}^s_{Y,\infty,a}$;
b) for $s > (\gamma_a, 1/p - 1/a)$, Λ_B induces the isomorphisms:

$$\Lambda_B : l_q \longleftrightarrow X \in \left\{\widetilde{ZB}^{s,A}_{Y,q,a}, \widetilde{zb}^{s,A}_{Y,q,a}\right\} \text{ and } \Lambda_B : c_0 \longleftrightarrow X \in \left\{\mathring{\widetilde{ZB}}^{s,A}_{Y,\infty,a}, \mathring{\widetilde{zb}}^{s,A}_{Y,\infty,a}\right\},$$

and, in addition, $\widetilde{zb}^{s,A}_{Y,\infty,a} \cap \mathring{\widetilde{b}}^{s,A}_{Y,\infty,a}(G) = \mathring{\widetilde{zb}}^{s,A}_{Y,\infty,a}$.
If $s > 0$, then:
c) for $s > (\gamma_a(1/p - 1/a))_{\max}$, Λ_L induces the isomorphisms

$$\Lambda_L : l_q \longleftrightarrow X \in \{ZL^s_{Y,q,a}, zl^s_{Y,q,a}\} \text{ and } \Lambda_L : c_0 \longleftrightarrow X \in \left\{\mathring{ZL}^s_{Y,\infty,a}, \mathring{zl}^s_{Y,\infty,a}\right\},$$

and, in addition, $zl^s_{Y,\infty,a} \cap \mathring{l}^s_{Y,\infty,a}(G) = \mathring{zl}^s_{Y,\infty,a}$;
d) for $s > (\gamma_a, 1/p - 1/a)$, Λ_L induces the isomorphisms

$$\Lambda_L : l_q \longleftrightarrow X \in \left\{\widetilde{ZL}^{s,A}_{Y,q,a}, \widetilde{zl}^{s,A}_{Y,q,a}\right\} \text{ and } \Lambda_L : c_0 \longleftrightarrow X \in \left\{\mathring{\widetilde{ZL}}^{s,A}_{Y,\infty,a}, \mathring{\widetilde{zl}}^{s,A}_{Y,\infty,a}\right\},$$

and, in addition, $\widetilde{zl}^{s,A}_{Y,\infty,a} \cap \mathring{\widetilde{l}}^{s,A}_{Y,\infty}(G) = \mathring{\widetilde{zl}}^{s,A}_{Y,\infty,a}$.

The lengthy and relatively technical proof of this theorem is placed in Appendix.

Remark 3.2. a) *Of course, the statement of Theorem 3.2 means, in addition, that the spaces in brackets are isomorphic to each other.*
b) *H. Triebel [27] established the presence of the isomorphic copies of l_p (where p is a*

scalar integrability parameter and $G = \mathbb{R}^n$) in the case of the spaces defined in terms of the Littlewood–Paley decompositions (see Remark 2.5, b)). The results of this type are also known as the localization property. The next theorem extends these results to wider classes of anisotropic function spaces with mixed norm.

c) Theorem 6 from [35], describing the degenerated ($A_s^* \setminus A \neq \emptyset$) spaces as sums of the non–degenerated spaces of lower dimensions, shows that the condition $A_s^* \subset A$ can be omitted in the theorems of this section.

Theorem 3.3. a) Let $G \subset \mathbb{R}^n$ be an open set, $a, p \in (0, \infty]^n$, $q \in (0, \infty]$, $s > 0$, $A_s^* \subset A \subset \mathbb{N}_0^n$, $|A| < +\infty$, and let $X(G) \in$

$$\in \left\{ b_{p,q,a}^s(G), \tilde{b}_{p,q,a}^{s,A}(G), l_{p,q,a}^s(G), \tilde{l}_{p,q,a}^{s,A}(G), B_{p,q,a}^s(G), \tilde{B}_{p,q,a}^{s,A}(G), L_{p,q,a}^s(G), \tilde{L}_{p,q,a}^{s,A}(G) \right\}$$

be a space compatible with $L_p(G)$. Then, there is a constant $C_0 > 0$, such that for every $m \in \mathbb{N}$ and $i \in I_n$, there is an injective linear operator $\Lambda_{X,m,i} : l_{p_i}^m \to X$ satisfying
$$\max\left(\|\Lambda_{X,m,i}|\mathcal{L}(l_{p_i}^m, X(G))\|, \|\Lambda_{X,m,i}^{-1}|\mathcal{L}(X(G)_{|\mathrm{Im}\Lambda_{X,m,i}}, l_{p_i}^m)\| \right) < C_0.$$

b) We can take $m = \infty$, i.e., substitute the space $l_{p_i}^m$ with l_{p_i}, if either G satisfies the quadrangle condition, or p is a constant vector ($p_i = p_j$ for $i, j \in I_n$) and G is not almost bounded.

Proof of Theorem 3.3. Let us start with a). We provide the proof for the case of the spaces defined in terms of differences. For the other spaces the proof is practically the same.

Given integer m and $1 \leq i \leq n$, one can choose parameters $\delta, \delta_2, \delta_1 > 0$ with $Q_{\varkappa_0\delta}(0) + Q_{\delta_2}(0) \subset Q_{\delta_1}$, a function $\phi \in C_0^\infty$ with $\operatorname{supp} \phi \subset Q_1(0)$ and cubes $\{Q_j^{m,i}\}_{j=0}^{m-1} = \{Q_{r_1\delta_1}(z_j)\}_{j=0}^{m-1}$ with $z_j := z_0 + 2j(r_1\delta_1)^{(\gamma_a)_i} e_i$ satisfying $\cup_{j=1}^{m-1} Q_j^{m,i} \subset Q_t(z)$. We also define $\phi_j(x) := (r_1\delta_2)^{s-(\gamma_a, 1/p)} \phi\left(\frac{x-z_j}{(r_1\delta_2)^{\gamma_a}}\right)$, and the mapping $\Lambda_{m,i} = \Lambda_{X,m,i} : y = \{y_j\}_{j=0}^{m-1} \mapsto \sum_{j=0}^{m-1} y_j \phi_j$. Now to prove the theorem, it is sufficient to establish the following estimates with the constants not depending on m:

$$\|y|l_{p_i}^m\| \leq C \left(\int_0^{\delta r_1} \|\tau^{-s} \delta_{i,a_i}^{m_i}(\tau, \cdot, \Lambda_{m,i}(y))|L_p(Q_t(z))\|^q d\tau/\tau \right)^{1/q}, \tag{1}$$

$$\|y|l_{p_i}^m\| \leq C \left\| \left(\int_0^{\delta r_1} (\tau^{-s} \delta_{i,a_i}^{m_i}(\tau, \cdot, \Lambda_{m,i}(y)))^q d\tau/\tau \right)^{1/q} \middle| L_p(Q_t(z)) \right\|, \tag{2}$$

$$\|\Lambda_{m,i}(y)|B_{p,q,a}^s(Q_t(z))\| \leq C\|y|l_{p_i}^m\|, \tag{3}$$

$$\|\Lambda_{m,i}(y)|L_{p,q,a}^s(Q_t(z))\| \leq C\|y|l_{p_i}^m\|. \tag{4}$$

Because ϕ is not a polynomial, (1) and (2) follow, respectively, from the simple identities:

$$\|y|l_{p_i}^m\| \cdot \left(\int_0^1 \|\tau^{-s}\delta_{i,a_i}^{m_i}(\tau,\cdot,\phi)|L_p(\mathbb{R}^n)\|^q d\tau/\tau \right)^{1/q} =$$

$$= \left(\int_0^{\delta r_1} \|\tau^{-s}\delta_{i,a_i}^{m_i}(\tau,\cdot,\Lambda_{m,i}(y))|L_p(Q_t(z))\|^q d\tau/\tau \right)^{1/q}, \tag{5}$$

$$\|y|l_{p_i}^m\| \cdot \left\| \left(\int_0^1 (\tau^{-s} \delta_{i,a_i}^{m_i}(\tau, \cdot, \phi))^q d\tau/\tau \right)^{1/q} \Big| L_p(\mathbb{R}^n) \right\| =$$

$$= \left\| \left(\int_0^{\delta r_1} (\tau^{-s} \delta_{i,a_i}^{m_i}(\tau, \cdot, \Lambda_{m,i}(y)))^q d\tau/\tau \right)^{1/q} \Big| L_p(Q_t(z)) \right\|. \tag{6}$$

The compatibility assumption means that

$$\left(\int_b^\infty \|\tau^{-s} \delta_{i,a_i}^{m_i}(\tau, \cdot, f) | L_p(\mathbb{R}^n)\|^q d\tau/\tau \right)^{1/q} \le Cb^{-s}\|f|L_p(\mathbb{R}^n)\|, \tag{7}$$

$$\left\| \left(\int_b^\infty (\tau^{-s} \delta_{i,a_i}^{m_i}(\tau, \cdot, f))^q d\tau/\tau \right)^{1/q} \Big| L_p(\mathbb{R}^n) \right\| \le Cb^{-s}\|f|L_p(\mathbb{R}^n)\|. \tag{8}$$

The proof is finished by the observation that $(5, 7)$ and $(6, 8)$ imply (3) and (4) respectively because

$$\|\Lambda_{m,i}(y)|L_p(Q_t(z))\| = (r_1\delta_2)^s \|\phi|L_p(\mathbb{R}^n)\| \cdot \|y|l_{p_i}^m\|. \tag{9}$$

In Part $b)$, it is enough to consider only G containing the positive quadrangle (cone) and set $m = \infty$ for a non–constant p. Otherwise, the absence of the almost boundedness of G allows us to choose shifts z_i and proceed as in the proof of Theorem 3.4, $b)$. $\square$

Theorem 3.4. *a) Let $G \subset \mathbb{R}^n$ be an open set, $a, p \in [1, \infty)$, $q, \theta \in (0, \infty]$, $s > 0$ and*

$$X(G) \in \left\{ b_{L_{p,\theta},q,a}^s(G), \tilde{b}_{L_{p,\theta},q,a}^{s,A}(G), l_{L_{p,\theta},q,a}^s(G), \tilde{l}_{L_{p,\theta},q,a}^{s,A}(G), \right.$$
$$\left. B_{L_{p,\theta},q,a}^s(G), \tilde{B}_{L_{p,\theta},q,a}^{s,A}(G), L_{L_{p,\theta},q,a}^s(G), \tilde{L}_{L_{p,\theta},q,a}^{s,A}(G) \right\}.$$

Then, there is a constant $C_0 > 0$, such that for every $m \in \mathbb{N}$, there is an injective operator $\Lambda_{X,m} : l_{p,\theta}^m \to X(G)$ satisfying

$$\max\left(\|\Lambda_{X,m}|\mathcal{L}(l_{p,\theta}^m, X(G))\|, \|\Lambda_{X,m}^{-1}|\mathcal{L}(X(G)_{|\operatorname{Im}\Lambda_{X,m}}, l_{p,\theta}^m)\|\right) < C_0.$$

b) We can take $m = \infty$, i.e., substitute the space $l_{p,\theta}^m$ with $l_{p,\theta}$, if G is not almost bounded.

Proof of Theorem 3.4. a) This time we deal only with the spaces of Besov and Lizorkin–Triebel type defined in terms of $\mathcal{D}$–functional. The case of the other spaces is treated in the same way.

Moreover, let us use the construction from the proof of Theorem 3.3 and take $\Lambda_m = \Lambda_{X,m} := \Lambda_{X,m,1}$. The rest of this proof is very similar to the previous one too. Thus, we concentrate mainly on the alterations required. Assume, in addition, that the function ϕ is non-negative and $|E_\varepsilon(\phi)| > 0$, where $E_\varepsilon(\phi) = \{x : \phi(x) > \varepsilon\}$ and $\varepsilon > 0$.

One needs to establish the following estimates with the constants not depending on m:

$$\|y|l_{p,\theta}^m\| \le C \left(\int_0^{\delta r_1} \|\tau^{-s}\mathcal{D}_a(\tau, \cdot, \Lambda_m(y), Q_t(z), A)|L_{p,\theta}(Q_t(z))\|^q d\tau/\tau \right)^{1/q}, \tag{1}$$

$$\|y|l_{p,\theta}^m\| \le C \left\| \left(\int_0^{\delta r_1} (\tau^{-s}\mathcal{D}_a(\tau, \cdot, \Lambda_m(y), Q_t(z), A))^q d\tau/\tau \right)^{1/q} \Big| L_{p,\theta}(Q_t(z)) \right\|, \tag{2}$$

$$\|\Lambda_m(y)|\tilde{B}^{s,A}_{L_{p,\theta},q,a}(Q_t(z))\| \leq C\|y|l^m_{p,\theta}\|, \qquad (3)$$

$$\|\Lambda_m(y)|\tilde{L}^{s,A}_{L_{p,\theta},q,a}(Q_t(z))\| \leq C\|y|l^m_{p,\theta}\|. \qquad (4)$$

The estimates (1) and (2) follow, respectively, from the relations:

$$C_1(\phi)\|y|l^n_{p,\theta}\| \leq \left(\int_0^{\delta r_1} \|\tau^{-s}\mathcal{D}_a(\tau,\cdot,\Lambda_m(y),Q_t(z),A)|L_{p,\theta}(Q_t(z))\|^q d\tau/\tau\right)^{1/q} \leq$$
$$\leq C_2(\phi)\|y|l^m_{p,\theta}\|, \quad (5)$$

$$C_1(\phi)\|y|l^n_{p,\theta}\| \leq \left\|\left(\int_0^{\delta r_1}(\tau^{-s}\mathcal{D}_a(\tau,\cdot,\Lambda_m(y),Q_t(z),A))^q d\tau/\tau\right)^{1/q}\bigg| L_{p,\theta}(Q_t(z))\right\| \leq$$
$$\leq C_2(\phi)\|y|l^m_{p,\theta}\|. \quad (6)$$

These relations, in turn, can be established with the aid of the fact that, for $\{E_j\}_{j=0}^{m-1}$ with $|E_i \cap E_j| = 0$ if $i \neq j$, one has

$$C_3 \min_{j=0}^{m-1}|E_j|^{1/p}\|y|l^m_{p,\theta}\| \leq \left\|\sum_{j=0}^{m-1} y_j\chi_{E_j}|L_{p,\theta}(\mathbb{R}^n)\right\| \leq C_4 \max_{j=0}^{m-1}|E_j|^{1/p}\|y|l^m_{p,\theta}\|. \qquad (7)$$

For example, in the case of (6), one can apply (7) to the groups of characteristic functions

$$\{\chi_{E_\varepsilon(g_j)}\}_{j=0}^{m-1} \text{ and } \{\chi_{Q_j^{m,1}}\}_{j=0}^{m-1},$$

$$\text{where } g_j = \chi_{Q_j^{m,1}}(\cdot)\left(\int_0^{\delta r_1}(\tau^{-s}\mathcal{D}_a(\tau,\cdot,\Lambda_m(y),Q_t(z),A))^q d\tau/\tau\right)^{1/q},$$

and use the estimates

$$\varepsilon\chi_{E_\varepsilon(g_j)} \leq g_j \leq \|g_j|L_\infty(\mathbb{R}^n)\|\chi_{Q_j^{m,1}} \text{ for } 0 \leq j \leq m-1.$$

Note that $|E_\varepsilon(g_j)| = |E_\varepsilon(g_1)| \neq 0$ for some $\varepsilon > 0$ and every j because ϕ is not a polynomial.

The compatibility assumption means that

$$\left(\int_b^\infty \|\tau^{-s}\mathcal{D}_a(\tau,\cdot,f,\mathbb{R}^n,A)|L_{p,\theta}(\mathbb{R}^n)\|^q d\tau/\tau\right)^{1/q} \leq Cb^{-s}\|f|L_{p,\theta}(\mathbb{R}^n)\|, \qquad (8)$$

$$\left\|\left(\int_b^\infty (\tau^{-s}\mathcal{D}_a(\tau,\cdot,f,\mathbb{R}^n,A))^q d\tau/\tau\right)^{1/q}\bigg| L_{p,\theta}(\mathbb{R}^n)\right\| \leq Cb^{-s}\|f|L_{p,\theta}(\mathbb{R}^n)\|. \qquad (9)$$

The proof is finished by the observation that (5, 8) and (6, 9) imply (3) and (4) respectively because of another implication of (7):

$$C_5(\phi)(r_1\delta_2)^s\|y|l^m_{p,\theta}\| \leq \|\Lambda_m(y)|L_p(Q_t(z))\| \leq C_6(\phi)(r_1\delta_2)^s\|y|l^m_{p,\theta}\|. \qquad (10)$$

Part b) follows immediately from the definition of the almost boundedness and the proof of a). Indeed, the absence of the almost boundedness means that there exists a properly separated sequence $\{z_j\}_{j\in\mathbb{N}_0}$ (not necessarily on a line) defining $\{\phi_j\}_{j\in\mathbb{N}_0}$ in the proof of Theorem 3.3, such that the estimates in the proof of a) hold for

$$\Lambda_\infty : y = \{y_j\}_{j\in\mathbb{N}_0} \longmapsto \sum_{j=0}^{\infty} y_j \phi_j, \quad l_{p,\theta} \longrightarrow X(G).$$

□

After this preparation, we combine the proofs of the previous three theorems to achieve the following improvements (the proofs of the next two theorems are in Appendix). Let us recall that $\bar{p} = (p, p, \ldots, p)$ when p is a constant.

Theorem 3.5. *Let $G \subset \mathbb{R}^n$ be an open set, $a, p \in (0, \infty]^n$, $q \in (0, \infty]$, $s \in \mathbb{R}$, $A_s^* \subset A \subset \mathbb{N}_0^n$, $|A| < +\infty$. Then, there is a constant $C_0 > 0$, such that for every $m \in \mathbb{N}$, there are injective linear operators $\Lambda_{B,m} : Y_B \to X_B(G)$ and $\Lambda_{L,m} : Y_L \to X_L(G)$ satisfying*

$$\max\left(\|\Lambda_{B,m}|\mathcal{L}(Y_B, X_B(G))\|, \|\Lambda_{B,m}^{-1}|\mathcal{L}(X_B(G)_{|\mathrm{Im}\,\Lambda_{B,m}}, Y_B)\|\right) < C_0$$

$$\max\left(\|\Lambda_{L,m}|\mathcal{L}(Y_B, X_L(G))\|, \|\Lambda_{L,m}^{-1}|\mathcal{L}(X_L(G)_{|\mathrm{Im}\,\Lambda_{L,m}}, Y_L)\|\right) < C_0$$

in the following cases.
If either $s > 0$, or $p = \infty$ and $s \geq 0$, the cases are:
a) $s > (\gamma_a(1/p - 1/a))_{\max}$ and either $X_B(G) \in \{b_{p,q,a}^s(G), B_{p,q,a}^s(G)\}$ and $Y_B = l_q(l_p^m)$, or $X_B(G) \in \{\mathring{b}_{p,\infty,a}^s(G), \mathring{B}_{p,\infty,a}^s(G)\}$ and $Y_B = c_0(l_p^m)$;
b) $s > (\gamma_a, 1/p - 1/a)$ and either $X_B(G) \in \{\tilde{b}_{p,q,a}^{s,A}(G), \tilde{B}_{p,q,a}^{s,A}(G)\}$ and $Y_B = l_q(l_p^m)$, or $X_B(G) \in \{\mathring{\tilde{b}}_{p,\infty,a}^{s,A}(G), \mathring{\tilde{B}}_{p,\infty,a}^{s,A}(G)\}$ and $Y_B = c_0(l_p^m)$.
If $s > 0$, the cases are:
c) $s > (\gamma_a(1/p - 1/a))_{\max}$ and either $X_L(G) \in \{l_{p,q,a}^s(G), L_{p,q,a}^s(G)\}$ and $Y_L = l_p^m(l_q)$, or $X_L(G) \in \{\mathring{l}_{p,\infty,a}^s(G), \mathring{L}_{p,\infty,a}^s(G)\}$ and $Y_L = l_p^m(c_0)$;
d) $s > (\gamma_a, 1/p - 1/a)$ and either $X_L(G) \in \{\tilde{l}_{p,q,a}^{s,A}(G), \tilde{L}_{p,q,a}^{s,A}(G)\}$ and $Y_L = l_p^m(l_q)$, or $X_L(G) \in \{\mathring{\tilde{l}}_{p,\infty,a}^{s,A}(G), \mathring{\tilde{L}}_{p,\infty,a}^{s,A}(G)\}$ and $Y_L = l_p^m(c_0)$.
e) One can take $m = \infty$ (i.e., substitute l_p^m with l_p) in a) – d) if G satisfies the quadrangle condition.
f) One can take $m = \infty$ (i.e., substitute l_p^m with l_p) in a) – d) if G is not almost bounded and p is constant vector ($p_i = p_j$ for $i, j \in I_n$).

Theorem 3.6. *Let $G \subset \mathbb{R}^n$ be an open set, $a \in (0, \infty]^n$, $\theta, q \in (0, \infty]$, $p \in (0, \infty)$, $s > 0$, $A_s^* \subset A \subset \mathbb{N}_0^n$, $|A| < +\infty$. Then, there is a constant $C_0 > 0$, such that for every $m \in \mathbb{N}$,* there are injective linear operators $\Lambda_{B,m} : Y_B \to X_B(G)$ and $\Lambda_{L,m} : Y_L \to X_L(G)$ satisfying

$$\max\left(\|\Lambda_{B,m}|\mathcal{L}(Y_B, X_B(G))\|, \|\Lambda_{B,m}^{-1}|\mathcal{L}(X_B(G)_{|\operatorname{Im}\Lambda_{B,m}}, Y_B)\|\right) < C_0$$

$$\max\left(\|\Lambda_{L,m}|\mathcal{L}(Y_B, X_L(G))\|, \|\Lambda_{L,m}^{-1}|\mathcal{L}(X_L(G)_{|\operatorname{Im}\Lambda_{L,m}}, Y_L)\|\right) < C_0$$

in the following cases:

a) $s > (\gamma_a(1/\bar{p} - 1/a))_{\max}$ and either $X_B(G) \in \left\{b^s_{L_{p,\theta},q,a}(G), B^s_{L_{p,\theta},q,a}(G)\right\}$ and $Y_B = l_q(l^m_{p,\theta})$, or

$X_B(G) \in \left\{\mathring{b}^s_{L_{p,\theta},\infty,a}(G), \mathring{B}^s_{L_{p,\theta},\infty,a}(G)\right\}$ and $Y_B = c_0(l^m_{p,\theta})$;

b) $s > (\gamma_a, 1/\bar{p} - 1/a)$ and either $X_B(G) \in \left\{\tilde{b}^{s,A}_{L_{p,\theta},q,a}(G), \tilde{B}^{s,A}_{L_{p,\theta},q,a}(G)\right\}$ and $Y_B = l_q(l^m_{p,\theta})$, or

$X_B(G) \in \left\{\mathring{\tilde{b}}^{s,A}_{L_{p,\theta},\infty,a}(G), \mathring{\tilde{B}}^{s,A}_{L_{p,\theta},\infty,a}(G)\right\}$ and $Y_B = c_0(l^m_{p,\theta})$;

c) $s > (\gamma_a(1/\bar{p} - 1/a))_{\max}$ and either $X_L(G) \in \left\{l^s_{L_{p,\theta},q,a}(G), L^s_{L_{p,\theta},q,a}(G)\right\}$ and $Y_L = l^m_{p,\theta}(l_q)$, or

$X_L(G) \in \left\{\mathring{l}^s_{L_{p,\theta},\infty,a}(G), \mathring{L}^s_{L_{p,\theta},\infty,a}(G)\right\}$ and $Y_L = l^m_{p,\theta}(c_0)$;

d) $s > (\gamma_a, 1/\bar{p} - 1/a)$ and either $X_L(G) \in \left\{\tilde{l}^{s,A}_{L_{p,\theta},q,a}(G), \tilde{L}^{s,A}_{L_{p,\theta},q,a}(G)\right\}$ and $Y_L = l^m_{p,\theta}(l_q)$, or

$X_L(G) \in \left\{\mathring{\tilde{l}}^{s,A}_{L_{p,\theta},\infty,a}(G), \mathring{\tilde{L}}^{s,A}_{L_{p,\theta},\infty,a}(G)\right\}$ and $Y_L = l^m_{p,\theta}(c_0)$.

e) One can take $m = \infty$ (i.e., substitute $l^m_{p,\theta}$ with $l_{p,\theta}$) in a) – d) if G is not almost bounded.

Let us also note the following observation, which is established by means of a slight modification of the proof of Theorem 3.2, a), b).

Theorem 3.7. *Let $G \subset \mathbb{R}^n$ be an open set, $a \in (0, \infty]^n$, $p \in (0, \infty]$, $q \in (0, \infty]$, $A^*_s \subset A \subset \mathbb{N}^n_0$, $|A| < +\infty$. Then, there are ϕ defining Λ_B such that, for either $s > 0$, or $Y = L_\infty(G)$ and $s \geq 0$,*

a) *if $s > (\gamma_a(1/p - 1/a))_{\max}$, Λ_B induces the isomorphisms*

$$\Lambda_B : l_p \longleftrightarrow X \in \left\{ZL^s_{p,q,a}, zl^s_{p,q,a}\right\};$$

b) *if $s > (\gamma_a, 1/p - 1/a)$, Λ_B induces the isomorphisms:*

$$\Lambda_B : l_p \longleftrightarrow X \in \left\{\widetilde{ZL}^{s,A}_{p,q,a}, \widetilde{zl}^{s,A}_{p,q,a}\right\}.$$

Combining Remark 3.1 and Theorem 3.1 with either Theorems 3.1 – 3.3 or Theorems 3.4 – 3.6, we obtain the key corollary of this section summarizing the results of the section to simplify cross–referencing below.

Corollary 3.1. *Assume that the parameter a is admissible for the spaces dealt with in a) – f).*

a) Under the conditions of Theorems 3.1 − 3.3, the spaces c_0, l_q and $\{l_{p_i}\}_{i\in I_n}$ are finitely represented in the corresponding function spaces endowed with any equivalent norm.

b) Under the conditions of Theorem 3.5, the spaces $l_p(l_q)$, $l_p(c_0)$, $l_q(l_p)$ and $c_0(l_p)$ are λ–finitely represented in the corresponding function spaces endowed with any equivalent (quasi)norm.

c) Under the conditions of Theorems 3.4 and 3.6, the spaces $l_{p,\theta}$, $l_q(l_{p,\theta})$, $c_0(l_{p,\theta})$, $l_{p,\theta}(l_q)$ and $l_{p,\theta}(c_0)$ are λ–finitely represented in the corresponding function spaces endowed with any equivalent (quasi)norm.

d) Under the conditions of Theorems 3.2 and 3.7, the spaces c_0, l_q and l_p are λ–represented in the corresponding function spaces endowed with any equivalent (quasi)norm.

e) Under the conditions of Theorems 3.3 and 3.5, the spaces $\{l_{p_i}\}_{i\in I_n}$, $l_p(l_q)$, $l_p(c_0)$, $l_q(l_p)$ and $c_0(l_p)$ are λ–represented in the corresponding function spaces endowed with any equivalent (quasi)norm if G satisfies the quadrangle condition.

f) Under the conditions of Theorems 3.3, 3.4 and 3.6, the spaces l_p with a scalar p, $l_{p,\theta}$, $l_q(l_{p,\theta})$, $c_0(l_{p,\theta})$, $l_{p,\theta}(l_q)$ and $l_{p,\theta}(c_0)$ are λ–represented in the corresponding function spaces endowed with any equivalent (quasi)norm if G is not almost bounded.

Remark 3.3. *a) In §5 (Corollary 5.2), we show, in particular, that the Banach–Mazur distance between certain sequence spaces and their isomorphic copies can be made arbitrary close to 1.*

b) In §4, we extend the range of the sequence spaces finitely representable in the function spaces under the consideration with the aid of the Maurey–Pisier theorem.

c) The comparison of Theorems 3.3 and 3.7 shows that our approach to the λ–representability of l_p with a scalar p in the spaces of Besov and Lizorkin–Triebel types requires to impose a restriction on G.

d) There are several other known ways of detecting copies of l_σ–spaces in a Banach space. One can uncover almost isometric copies of l_σ for some σ in Besov and Lizorkin–Triebel spaces with the aid of the remarkable results due to M. Levy [38], M. Mastylo [39] and V. Milman [36], at least in the setting of a sufficiently regular domain G.

4. RADEMACHER TYPE AND COTYPE FOR ABSTRACT AND CONCRETE SPACES: PROPERTIES, RELATED NOTIONS AND APPLICATIONS

This section is devoted to the classical Rademacher type and cotype and related notions of lattice convexity, concavity and upper and lower estimates studied for a particularly wide range of function spaces including the spaces of Besov and Lizorkin–Triebel type with the underlying Lorentz (quasi)norm used instead of the traditional Lebesgue norm (i.e., $Y = L_{p,\theta}$ in Definitions 2.6, 2.7 and 2.8). In the course of this activity we provide simple examples of ideal and even Lizorkin–Triebel spaces whose type and cotype intervals are not closed (i.e., semi–open intervals not achieving the supremum of the type set and the infimum of the cotype set). The traditional examples constructed as the q–convexifications of the Tsirel'son space and their duals are discussed in [32] (page 305).

Utilizing the Maurey–Pisier theorem, we complement the results of §3 on the representability of the l_p–sequence spaces by extending the ranges of the exponents, particularly, for the spaces defined with the aid of the Lorentz (quasi)norms.

In addition, we provide an alternative approach to the convexity and concavity properties of Lorentz spaces implying the convexity properties of Hardy–Lorentz spaces. Estimates for the Rademacher type and cotype constants for many spaces under consideration (endowed with geometrically friendly equivalent norms) are established in §4.5 of [2] as a by–product of the (p, h_c)–uniform convexity and (q, h_s)–uniform smoothness estimates.

4.1. Related Definitions and Basic Relations

To shorten the proofs of the main theorems in Section 4, we recall the definitions and basic relations regarding Rademacher type, cotype, B–convexity and C–convexity. Let $\{r_i\}_{i \in \mathbb{N}}$ be the family of the Rademacher functions on $[0, 1]$.

Definition 4.1. *Let X be a Banach space and $p, q \in [1, \infty]$. If, for some constant $T > 0$ and every finite $I \subset \mathbb{N}$ and $\{x_i\}_{i \in I} \subset X$, one has*

$$\left\| \sum_{i \in I} r_i x_i \,\middle|\, L_q([0,1], X) \right\| \leq T \left\| \{x_i\}_{i \in I} \,\middle|\, l_p(I, X) \right\|, \qquad (RT)$$

it is said that X possesses the Rademacher type p.

If, for some constant $C > 0$ and every finite $I \subset \mathbb{N}$ and $\{x_i\}_{i \in I} \subset X$, one has

$$\left\| \{x_i\}_{i \in I} \,\middle|\, l_q(I, X) \right\| \leq C \left\| \sum_{i \in I} r_i x_i \,\middle|\, L_p([0,1], X) \right\|, \qquad (RC)$$

it is said that X possesses the Rademacher cotype q.

The best constants T and C in (RT) and (RC) respectively are the Rademacher type $T_{p,q} = T_{p,q}(X)$ and cotype $C_{p,q} = C_{p,q}(X)$ constants.

Let $I_T(X)$ and $I_C(X)$ be the sets (closed or semiclosed intervals) of those p and q respectively, such that X possesses the type p and the cotype q.

A Banach space X is said to be B–convex (C–convex) if the interval $I_T(X)$ ($I_C(X)$) is nontrivial, i.e., contains, at least, two different points of $[1, 2]$ ($[2, \infty]$).

Note that every Banach space is of Rademacher type 1 and of Rademacher cotype ∞ (with the natural interpretation of the latter). It is also clear that $T_{p,p}(X) \geq 1$ and $C_{q,q}(X) \geq 1$.

Kahane's generalisantion of the Khinchine inequality (see Theorem 4.2) shows the correctness of the definition of the Rademacher type and cotype.

In the next theorem we gathered the basic properties of the Rademacher type and cotype for easy interim referencing. They can be found for example, in [29, 15, 32].

Assume that (E, μ) is a countably–additive (atomic) measure space.

Theorem 4.1. *For $\lambda \geq 1$ and Banach spaces X, Y and Z, let X be λ–finitely represented in Y, and subspaces $X_1 \subset X$ and $Z_1 \subset Z$. Assume also that Z is B–convex and $p \in [1, \infty]$. Then one has:*
a) $I_T(X) \subset [1, 2]$ and $I_C(X) \subset [2, \infty]$ are convex;
b) $I_T(X) \supset I_T(Y)$, $I_C(X) \supset I_C(Y)$, and $I_T(X) \subset I_T(X/X_1)$;
c) $p \in I_T(Y)$ implies $p' \in I_C(Y^*)$ and $C_{p',p'}(Y^*) \leq T_{p,p}(Y)$ for $p \in [1, 2]$;
d) $I_C(Z) \subset I_C(Z/Z_1)$;

e) $p \in I_T(Z) \Leftrightarrow p' \in I_C(Z^*)$, $p \in I_C(Z) \Leftrightarrow p' \in I_T(Z^*)$;
f) $I_T(L_p(E, X)) = [1, \min(p, 2)] \cap I_T(X)$ and $I_C(L_p(E, X)) = [\max(p, 2), \infty] \cap I_C(X)$;
g) X is B–convex if, and only if, X^* is B–convex;
h) X is B–convex if, and only if, l_1 is not λ–represented in X for any $\lambda > 0$;
i) X is C–convex if, and only if, c_0 is not λ–represented in X for any $\lambda > 0$.

Part a) of the next theorem is the classical Khinchin inequality, while b) and c) are its celebrated generalizations due to Kahane and Maurey.

Theorem 4.2 (Khinchin, Kahane, Maurey). *Let X be a Banach space, Y be an s–concave lattice, and $p, q \in (0, \infty)$. Then there are (finite closed) intervals $K_0(p), K_1(p, q), K_2(p)$ satisfying $1 \in K_0(p) \cap K_1(p, q) \cap K_2(p)$ and $K_0(p) \cup K_1(p, q) \cup K_2(p) \subset (0, \infty)$, such that, for every finite $I \subset \mathbb{N}$, one has*
a) $\|\sum_{i \in I} r_i \alpha_i | L_p([0, 1])\| / \|\{\alpha_i\}_{i \in I} | l_2^{|I|}\| \in K_0(p)$ *for every* $\{\alpha_i\}_{i \in I} \subset \mathbb{R}$;
b) $\|\sum_{i \in I} r_i x_i | L_p([0, 1], X)\| / \|\sum_{i \in I} r_i x_i | L_q([0, 1], X)\| \in K_1(p, q)$ *for every* $\{x_i\}_{i \in I} \subset X$, *and*
c) $\|\sum_{i \in I} r_i f_i | L_p([0, 1], Y)\| / \left\| |\{f_i\}_{i \in I} | l_2^{|I|} | X \right\| \in K_2(p)$ *for every* $\{f_i\}_{i \in I} \subset Y$.
It is assumed that $0/0 = 1$.

4.2. Ideal Spaces

Let us recall that measurable functions $\{f_i\}_{i \in I}$ defined on E are disjoint if there are measurable subsets $\{E_i\}_{i \in I}$ satisfying $|E_i \cap E_j| = 0$ for $i \neq j$ and $f_i \chi_{E_i} = f_i$ for every $i, j \in I$.

Definition 4.2. *For $p, q \in (0, \infty]$, an ideal space Y is said to be satisfying the lower p–estimate or the upper p–estimate, or to be lattice (p, q)–convex, or lattice (p, q)–concave if, for every $\{f_i\}_{i \in I} \subset Y$ with $|I| < \infty$ and some $C_1, C_2, C_3 > 0$, one has, correspondingly,*
$(\sum_{i \in I} \|f_i | Y\|^p)^{1/p} \leq C_1 \|\sum_{i \in I} f_i | Y\|$ *if $\{f_i\}_{i \in I}$ are disjoint or*
$\|\sum_{i \in I} f_i | Y\| \leq C_2 (\sum_{i \in I} \|f_i | Y\|^p)^{1/p}$ *if $\{f_i\}_{i \in I}$ are disjoint, or*
$\left\| (\sum_{i \in I} |f_i|^q)^{1/q} | Y \right\| \leq C_3 (\sum_{i \in I} \|f_i | Y\|^p)^{1/p}$, *or*
$(\sum_{i \in I} \|f_i | Y\|^p)^{1/p} \leq C_4 \left\| (\sum_{i \in I} |f_i|^q)^{1/q} | Y \right\|$.
When $p = q$, the second parameter is usually omitted. The corresponding ranges of the parameters p for which one has the lower p–estimate, upper p–estimate, (p, p)–convexity, or (p, p)–concavity will be designated by $I_{le}(Y), I_{ue}(Y), I_{conv}(Y)$, or $I_{conc}(Y)$ respectively.

The subset of $I_{conv}(Y)$ ($I_{conc}(Y)$) of the parameters, such that the corresponding inequality holds with the constant $C_3 = 1$ ($C_4 = 1$) will be designated by $I_{conv}^1(Y)$ ($I_{conc}^1(Y)$).

We say that a set of the parameters defined above is nontrivial if it intersects with $(1, \infty)$.

For example, a quasi–Banach space A, that is also an ideal space, is $(p, 1)$–convex for $p \in (0, 1]$ if, and only if $p \in \alpha_c(A)$. The latter notation is used to avoid the ambiguity between the lattice p–convexity (that is, (p, p)–convexity) and the p–convexity of quasi–normed spaces (see §2).

Remark 4.1. a) *For every ideal Y, one has $I_{ue}(Y) \supset I_{conv}(Y)$, $I_{le}(Y) \supset I_{conc}(Y)$, $1 \in I_{ue}(Y)$, and $\infty \in I_{le}(Y)$.*
b) *Both the (p,q)–convexity and (q,p)–concavity are possible for $p \leq q$ only.*
c) *The conditions $1 \in I_{le}(Y)$ and $\infty \in I_{ue}(Y)$ are very restrictive. For example, Kakutani [29] showed that a lattice with the lower 1–estimate is (order) isometric to some $L_1(E, \mu)$, while Benyamini [29] established that a separable lattice satisfying upper ∞–estimate is isomorphic to some $C(K)$.*

The Minkowski inequality and its dualisation generate the examples from Parts a) and b) of the next theorem. Parts b), c) and the related type–cotype results are established by Creekmore [40] and Novikov [41]. To prove Part c), Creekmore [40] has provided a direct proof of the upper and lower estimates in the case $q < r$ and used the duality.

Theorem 4.3. *Let $p \in (0, \infty]^n$, $q \in (0, \infty)$, $r \in (0, \infty]$ and $q \neq r$. Then one has:*
a) $I_{conv}(L_p) = I_{ue}(L_p) = (0, p_{\min}]$ and $I_{conc}(L_p) = I_{le}(L_p) = [p_{\max}, \infty]$;
b) $I_{conv}(L_{q,r}) = (0, q) \cap (0, r]$ and $I_{conc}(L_{q,r}) = (q, \infty] \cap [r, \infty)$;
c) $I_{ue}(L_{q,r}) = (0, \min(q, r)]$ and $I_{le}(L_{q,r}) = [\max(q, r), \infty]$

Assume that (E, μ) is a countably–additive (possibly atomic) measure space. Part a) of the next theorem is established by Maurey and Pisier (see [42]), while Parts $a) - e)$ and $h)$ can be found, for instance, in [29]. It is noticed, particularly, in [32] that Parts $f)$ and $g)$ follow from the Khinchin–Maurey inequality (Theorem 4.2, c)). In turn, Parts $f)$ and $g)$ and Remark 4.1, a) imply h) and i).

Theorem 4.4. *Let Y be an ideal space, $p \in [1, \infty]$ and $q, r \in (1, \infty)$. Then one has:*
a) $q \in I_{ue}(Y)$ *implies* $(1, q) \subset I_{conv}(Y)$, *and* $q \in I_{le}(Y)$ *implies* $(q, \infty) \subset I_{conc}(Y)$;
b) $p \in I_{conv}(Y) \Leftrightarrow p' \in I_{conc}(Y^*)$ *and* $p \in I_{conc}(Y) \Leftrightarrow p' \in I_{conv}(Y^*)$;
c) $q \in I_{ue}(Y) \Leftrightarrow q' \in I_{le}(Y^*)$ *and* $q \in I_{le}(Y) \Leftrightarrow q' \in I_{ue}(Y^*)$;
d) Y *is (q, r)–convex if and only if Y^* is (q', r')–concave;*
e) Y *is (q, r)–concave if and only if Y^* is (q', r')–convex;*
f) $p \in I_T(Y)$ *if and only if Y is $(p, 2)$–convex and s–concave for some $s < \infty$;*
g) $p \in I_C(Y)$ *if and only if Y is $(p, 2)$–concave;*
h) $I_T(Y) \subset I_{ue}(Y)$ *and* $I_{conc}(Y) \cap [2, \infty] \subset I_C(Y) \subset I_{le}(Y)$;
i) $I_{conv}(Y) \cap [1, 2] \subset I_T(Y)$ *if Y is lattice s–concave for some $s < \infty$.*

Remark 4.2. *Thanks to Parts a), h) and i) of the last theorem, the sets $I_{conv}(Y) \cap [1, 2]$, $I_C(Y)$ and $I_{le}(Y) \cap [1, 2]$, and the sets $I_{conc}(Y) \cap [2, \infty]$, $I_C(Y)$ and $I_{le}(Y) \cap [2, \infty]$ coincide modulo one point (and even more precise information may be inferred).*

Let us observe that Creekmore's key result mentioned around Theorem 4.3 can be improved with the aid of the next result from [30] dealing with the notion of p–convexity of quasi–Banach spaces (Abelian groups).

Lemma 4.1. *For $p \in (0, 1)$ and $q \in (0, \infty]$, the Lorentz space $L_{p,q}$ and the Hardy–Lorentz space $H_{p,q}$ are $\min(p, q)$–convex as quasi–Banach spaces (see §2).*

Namely, we obtain its corollary.

Corollary 4.1. *For $p \in (0, \infty)$ and $q \in (0, \infty]$, the Lorentz space $L_{p,q}$ is:*
a) *lattice (r, s)–convex for $r \in (0, \min(p, q)]$ and $s \in (p, \infty]$;*
b) *lattice (r, s)–concave for $r \in [\max(p, q), \infty)$ and $s \in (0, p)$.*

Remark 4.3. *Corollary 4.1 immediately implies Parts c) and b) of Theorem 4.3 and the description of the Rademacher type and cotype sets with the exception of the point 2.*

Proof of Corollary 4.1. With the aid of the appropriate Mazur transforms $Z_\sigma : f \to \mathrm{sign}(f)|f|^\sigma$, we see that Part $a)$ is equivalent to the Lorentz-space case of Lemma 4.1 with the parameters p/s and q/s instead of p and q there ($\sigma := s$), and that Part $b)$ follows from its particular case with $p, q, r, s \in (1, \infty)$ ($\sigma < \min(p, q, r, s)$). It is only left to use the duality in the form of Part $d)$ of Theorem 4.4 and the relation

$$L_{p,q}^* = L_{p',q'} \text{ for } p \in (1, \infty) \text{ and } q \in [1, \infty). \quad \square$$

The next lemma is a key ingredient of our approach. It provides simple examples of spaces with open type and cotype sets. Let us notice that the consideration of the spaces of functions is essential.

Lemma 4.2. *Let $Y_1 = Y_1(E_1)$ and $Y_2 = Y_2(E_2)$ be ideal spaces. Then we have*
a) $I_{ue}(Y_1(E_1, Y_2)) \subset I_{ue}(Y_2)$ *and* $I_{conv}(Y_1(E_1, Y_2)) \subset I_{conv}(Y_2)$;
b) $I_{le}(Y_1(E_1, Y_2)) \subset I_{le}(Y_2)$ *and* $I_{conc}(Y_1(E_1, Y_2)) \subset I_{conc}(Y_2)$.
Moreover, assuming that, for some $p \in (0, \infty)$ and every $n \in \mathbb{N}$, Y_2 contains an order isomorphic copy X_n of $l_p(I_n)$, and $\sup_{n \in \mathbb{N}} \|T_n\| \cdot \|T_n^{-1}\| := C_I < \infty$, where T_n is the corresponding order isomorphism, we also have the properties:
c) *the upper p–estimate for $Y_1(E_1, Y_2)$ implies the lattice p–convexity of Y_1 ($I_{ue}(Y_1(E_1, Y_2)) \subset I_{conv}(Y_1)$);*
d) *the lower p–estimate for $Y_1(E_1, Y_2)$ implies the lattice p–concavity of Y_1 ($I_{le}(Y_1(E_1, Y_2)) \subset I_{conc}(Y_1)$).*

Proof of Lemma 4.2. Let $\{g_i\}_{i=1}^n \subset Y_2(E_2)$ be disjoint, and let $h \in Y_1(E_1)$ be nontrivial. Then their products $\{g_i h\}_{i=1}^n \subset Y_1(E_1, Y_2)$ (or better to write $\{g_i \otimes h\}_{i=1}^n \subset Y_1(E_1, Y_2)$) are disjoint too and

$$\|g_i \otimes h | Y_1(E_1, Y_2)\| = \|h_i | Y_2(E_2)\| \cdot \|h | Y_1(E_1)\|.$$

This observation imply a) and b).

We fix some $m \in \mathbb{N}$. Let $\{e_i\}_{i \in \mathbb{N}} \subset Y$ be the natural basis of l_p, $\{f_i\}_{i=1}^m \subset Y_1$, and the set $\{h_i\}_{i=1}^m$ be defined by $h_i := f_i \otimes T_m e_i$. Then $\{T_m e_i\}_{i=1}^m$ are disjoint in Y_2, and, therefore, $\{h_i\}_{i=1}^m$ are disjoint in $Y_1(E_1, Y_2)$ too. Since we are speaking about spaces of functions, one has

$$\left(\sum_{i=1}^m |f_i(x)|^p\right)^{1/p} \asymp \left\|\sum_{i=1}^m T e_i f_i(x)\right\|_{Y_2} \quad \text{a.e. on } E_1. \qquad (1)$$

Now Parts c) and d) follow from (1) and the upper and lower estimates correspondingly:

$$\left\|\sum_{i=1}^m h_i \bigg| Y_1(E_1, Y_2)\right\| \le C \left(\sum_{i=1}^m \|T_m e_i\|_{Y_2}^p \|f_i\|_{Y_1}^p\right)^{1/p} \quad \text{and}$$

$$C\left\|\sum_{i=1}^{m}h_i|Y_1(E_1,Y_2)\right\| \geq \left(\sum_{i=1}^{m}\|T_m e_i\|_{Y_2}^p\|f_i\|_{Y_1}^p\right)^{1/p}.$$

Note also that $\{\|T_m e_i\|_{Y_2}\}_{i=1}^{M} \subset [\|T_m^{-1}\|^{-1}, \|T_m\|]$. □

Another ingredient is the next lemma. It deals with some features typical for the mixed–norm spaces.

For a family of ideal spaces $\{Y_i\}_{i=1}^n$ of functions defined on $\{E_i\}_{i=1}^n$, let $L_{Y_1,\ldots,Y_n} = L_{Y_1,\ldots,Y_n}(E)$ with $E = E_1 \times \cdots \times E_n$ be the ideal space of functions with the finite (quasi)norm

$$\|f|L_{Y_1,\ldots,Y_n}\| := \underbrace{\|\ldots\|}_{n\ norms} f|Y_1(E_1)\|\ldots|Y_n(E_n)\|.$$

The next simple but useful lemma sharpens the ideas behind the proof of Part a) of Theorem 4.3.

Lemma 4.3. *Let A be a Banach space, $Y = Y(\Omega)$ an ideal space and $p, q \in (0,\infty]$. Then one has:*
a) $\bigcap_{i=1}^n I_{conv}(Y_i) \subset I_{conv}(L_{Y_1,\ldots,Y_n})$;
b) $\bigcap_{i=1}^n I_{conc}(Y_i) \subset I_{conc}(L_{Y_1,\ldots,Y_n})$;
c) $I_{ue}(Y_1) \cap \bigcap_{i=2}^n I_{conv}(Y_i) \subset I_{ue}(L_{Y_1,\ldots,Y_n})$;
d) $I_{le}(Y_1) \cap \bigcap_{i=2}^n I_{conc}(Y_i) \subset I_{le}(L_{Y_1,\ldots,Y_n})$;
e) If Y_1 is lattice (p,q)–convex, then $L_{Y_1,\ldots,Y_n}$ is lattice (r,q)-convex for $r \in (0,p] \cap \bigcap_{i=2}^n I_{conv}(Y_i)$;
f) If Y_1 is lattice (p,q)–concave, then $L_{Y_1,\ldots,Y_n}$ is lattice (r,q)-concave for $r \in [p,\infty] \cap \bigcap_{i=2}^n I_{conc}(Y_i)$;
g) $I_T(A) \cap I_{conv}(Y) \subset I_T(Y(\Omega,A))$ and $I_C(A) \cap I_{conc}(Y) \subset I_C(Y(\Omega,A))$.

Proof of Lemma 4.3. It is sufficient to apply the definitions in the by–coordinate manner. □

Theorem 4.5. *Let $\{E_i\}_{i=1}^3$ be measurable and $a \in (0,\infty]^n$, $q, r \in (0,\infty]$, $p \in (0,\infty) \setminus \{r,1\}$. Assume also that*

$$Y_B := L_q(L_{p,r}(L_a)) = L_q(E_3, L_{p,r}(E_2, L_a(E_1)))\ and$$
$$Y_L := L_{p,r}(L_q(L_a)) = L_{p,r}(E_3, L_q(E_2, L_a(E_1))).$$

Then one has:
a) $I_{conv}(Y_B) = I_{conv}(Y_L) = (0, \min(q, r, a_{\min})] \cap (0, p]$;
b) $I_{conc}(Y_B) = I_{conc}(Y_L) = [\max(q, r, a_{\max}), \infty] \cap (p, \infty]$;
*c) $I_{ue}(Y_B) = (0, \min(p, q, r, a_{\min})] \setminus (\{p\} \cap \{a_{\min}\})$ and
$I_{ue}(Y_L) = (0, \min(p, q, r, a_{\min})] \setminus (\{p\} \cap \{\min(q, a_{\min})\})$;*
*d) $I_{le}(Y_B) = [\max(p, q, r, a_{\max}), \infty] \setminus (\{p\} \cap \{a_{\max}\})$ and
$I_{le}(Y_L) = [\max(p, q, r, a_{\max}), \infty] \setminus (\{p\} \cap \{\max(q, a_{\max})\})$;*
*e) $I_T(Y_B) = [1, \min(2, p, q, r, a_{\min})] \setminus (\{p\} \cap \{2, a_{\min}\})$ and
$I_T(Y_L) = [1, \min(2, p, q, r, a_{\min})] \setminus (\{p\} \cap \{2, \min(q, a_{\min})\})$;*
*f) $I_C(Y_B) = [\max(2, p, q, r, a_{\max}), \infty] \setminus (\{p\} \cap \{2, a_{\max}\})$ and
$I_C(Y_L) = [\max(2, p, q, r, a_{\max}), \infty] \setminus (\{p\} \cap \{2, \max(q, a_{\max})\})$.*

Proof of Theorem 4.5. As far as the identities in Parts $a) - d)$ are concerned, we obtain one inclusion in each part as an immediate consequence of Theorem 4.3 and Lemma 4.3 and, then, we sharpen these inclusions using Lemma 4.2 in the contrapositive way, establishing the opposite inclusions.

Parts $e)$ and $f)$ are derived from the ones already proved with the aid of Parts $h)$ and $i)$ of Theorem 4.4. The cases of the Rademacher type 2 in $e)$ and the Rademacher cotype 2 in $f)$ require extra attention according to Parts $f)$ and $g)$ of Theorem 4.4. $\square$

In a similar but simpler way, one obtains the following theorem.

Theorem 4.6. *Let* $q, r \in (0, \infty]$, $p \in (0, \infty) \setminus \{r, 1\}$. *Then one has:*
a) $I_{conv}(l_q(l_{p,r})) = I_{conv}(l_{p,r}(l_q)) = (0, \min(q, r)] \cap (0, p)$;
b) $I_{conc}(l_q(l_{p,r})) = I_{conc}(l_{p,r}(l_q)) = [\max(q, r), \infty] \cap (p, \infty]$;
c) $I_{ue}(l_q(l_{p,r})) = (0, \min(p, q, r)]$ *and*
$I_{ue}(l_{p,r}(l_q)) = (0, \min(p, q, r)] \setminus (\{p\} \cap \{q\})$;
d) $I_{le}(l_q(l_{p,r})) = [\max(p, q, r), \infty]$ *and*
$I_{le}(l_{p,r}(l_q)) = [\max(p, q, r), \infty] \setminus (\{p\} \cap \{q\})$;
e) $I_T(l_q(l_{p,r})) = [1, \min(2, p, q, r)] \setminus (\{p\} \cap \{2\})$ *and*
$I_T(l_{p,r}(l_q)) = [1, \min(2, p, q, r)] \setminus (\{p\} \cap \{2, q\})$;
f) $I_C(l_q(l_{p,r})) = [\max(2, p, q, r), \infty] \setminus (\{p\} \cap \{2\})$ *and*
$I_C(l_{p,r}(l_q)) = [\max(2, p, q, r), \infty] \setminus (\{p\} \cap \{2, q\})$.

Remark 4.4. a) According to Parts $e)$ and $f)$ of Theorem 4.6, the spaces $l_{p,q}(l_p)$ with $p < q$ and $p > q$, correspondingly, are examples of spaces without maximal type and minimal cotype (like A_p–weights). Note that these examples are more elementary than the traditional modifications of the Tzirel'son space mentioned in the beginning of the section. More spaces manifesting the same phenomenon are given in Theorem 4.5 and can further be constructed in the same manner.
b) In each identity of Theorem 4.5, the parameter a may be chosen in the right or left vicinity of p or q, so that the right–hand side of this identity will coincide with the right–hand side of the corresponding part of Theorem 4.6.

Proof of Theorem 4.6. As before, Parts $e)$ and $f)$ are derived from the rest of the theorem with the aid of Parts $h)$ and $i)$ of Theorem 4.4 and the treatment of the type/cotype 2 presence with the aid of Parts $f)$ and $g)$ of Theorem 4.4.

As in the proof of Theorem 4.5, we establish the $\supset$–inclusion in every other part using Theorem 4.3 and Lemma 4.3, while the $\subset$–inclusion is deduced with the aid of Lemma 4.2. $\square$

4.3. Rademacher Type and Cotype Sets of Anisotropic Function Spaces

The next two theorems are the main theorems of this section. We establish them with the aid of Theorems 4.5, 4.6 and 4.3, $a)$ and results in Section 3. Let us recall that, thanks to the Hardy inequality, a Lorentz space $L_{p,\theta}(E)$ with $p \in (1, \infty)$ and $\theta \in [1, \infty]$ can be equipped with an equivalent norm

$$\|f\|_{**} = \left(\int_0^\infty \left(f^{**}(\tau) \tau^{1/p} \right)^\theta d\tau/\tau \right)^{1/\theta}, \text{ where } f^{**}(t) = \frac{1}{t} \int_0^t f^*(\tau) d\tau$$

and, thus, $(f+g)^{**} \leq f^{**} + g^{**}$. Therefore, we are dealing with Banach spaces in the next theorem.

Theorem 4.7. *Let $G \subset \mathbb{R}^n$ be an open set, $a \in (0,\infty]^n$, $\theta, q \in [1,\infty]$, $\theta \neq p$, $p \in (1,\infty)$, $s > 0$, $A_s^* \subset A \subset \mathbb{N}_0^n$, $|A| < +\infty$. Assume also that a satisfies the conditions of Theorem 3.6. Then we have:*

a) $[1, \min(2, p, q, \theta, a_{\min})] \setminus (\{p\} \cap \{a_{\min}, 2\}) \subset I_T(X_B) \subset$
$$\subset [1, \min(2, p, q, \theta)] \setminus (\{p\} \cap \{2\}) \text{ and}$$

$$[\max(2, p, q, \theta, a_{\max}), \infty] \setminus (\{p\} \cap \{a_{\max}, 2\}) \subset I_C(X_B) \subset$$
$$\subset [\max(2, p, q, \theta), \infty] \setminus (\{p\} \cap \{2\})$$
for $X_B \in \{b^s_{L_{p,\theta},q,a}(G), B^s_{L_{p,\theta},q,a}(G), \tilde{b}^{s,A}_{L_{p,\theta},q,a}(G), \tilde{B}^{s,A}_{L_{p,\theta},q,a}(G)\};$

b) $[1, \min(2, p, q, \theta, a_{\min})] \setminus (\{p\} \cap \{q, a_{\min}, 2\}) \subset I_T(X_L) \subset$
$$\subset [1, \min(2, p, q, \theta)] \setminus (\{p\} \cap \{q, 2\}) \text{ and}$$

$$[\max(2, p, q, \theta, a_{\max}), \infty] \setminus (\{p\} \cap \{q, a_{\max}, 2\}) \subset I_C(X_L) \subset$$
$$\subset [\max(2, p, q, \theta), \infty] \setminus (\{p\} \cap \{q, 2\})$$
for $X_L \in \{l^s_{L_{p,\theta},q,a}(G), L^s_{L_{p,\theta},q,a}(G), \tilde{l}^{s,A}_{L_{p,\theta},q,a}(G), \tilde{L}^{s,A}_{L_{p,\theta},q,a}(G)\}.$

Moreover, we have the identities in a) and b) if $a \in (\min(2, p, q, \theta), \max(2, p, q, \theta))^n$.

Remark 4.5. a) *The type and cotype sets of the dual spaces, subspaces and quotients of Besov and Lizorkin–Triebel type spaces from Theorem 4.7 can be derived from Theorems 4.7 and 4.5 with the aid of Parts b), d) and e) of Theorem 4.1 and Parts c) of Corollary 3.1.*
b) *The case $p = \theta$ is a special case of the next theorem.*

Proof of Theorem 4.7. Let us start with the homogeneous variants of the corresponding function spaces.

The proof relies on the principle of "two policemen" following from Theorem 4.1, b): if a Banach space Y_1 is λ–finitely represented in a Banach space X, while X is λ–finitely represented in a Banach space Y_2, and the type and cotype sets of Y_1 and Y_2 coincide, then X, Y_1 and Y_2 have identical type and cotype sets. For every function space X under consideration, the corresponding Y_1 is provided by Part c) of Corollary 3.1 with its type and cotype sets identified in Theorem 4.6, while the corresponding Y_2 is represented by one of the spaces dealt with in Theorem 4.5, or its quotient according to Section 2.2.

When X is defined in terms of a local approximation by polynomials, one also takes advantage of the following simple but handy observation. For an arbitrary ideal space $Z(E)$ and $A = \widehat{A}$ with $|A| < \infty$, we have $Z(E, L_a/\mathcal{P}_A) = Z(E, L_a)/Z(E, \mathcal{P}_A)$, where $Z(E, \mathcal{P}_A)$ is complemented in $Z(E, L_a)$ because the $|A|$–dimensional $\mathcal{P}_A$ is complemented in $L_a = L_a(Q_1(0))$.

Since the non–homogeneous counterpart $\widehat{X}$ of the corresponding homogeneous function space X is an l_p–sum of X and $L_{p,\theta}(G)$, we analyse the type and cotype sets of $\widehat{X}$ substituting Y_1 and Y_2 in the above considerations with their l_p–sums with $L_{p,\theta}(G)$. Theorem 4.3 combined with Parts a), h) and i) of Theorem 4.4 shows that the type and cotype sets of $L_{p,\theta}(G)$ are larger than those of X. Therefore, the application of Parts a) and b) of Theorem 4.1 leads to the identities $I_T(\widetilde{X}) = I_T(X)$ and $I_T(\widetilde{X}) = I_T(X)$. □

The next theorem identifies the Rademacher type and cotype sets for the majority of function spaces dealt with in various applications (for example, in PDE). For the sake of brevity, we divide all the "ordinary" classes of Besov and Lizorkin–Triebel spaces under consideration into four groups. Let G be an open subset of $\mathbb{R}^n$ for some $n \in \mathbb{N}$, and $J = [1, \infty]$. The groups are:

$$\Gamma_1 = \left\{ B^s_{p,q,a}(G), \tilde{B}^{s,A}_{p,q,a}(G), L^s_{p,q,a}(G), \tilde{L}^{s,A}_{p,q,a}(G), b^s_{p,q,a}(G), \tilde{b}^{s,A}_{p,q,a}(G), l^s_{p,q,a}(G), \tilde{l}^{s,A}_{p,q,a}(G) \right\},$$

where $p, a \in J^n$, $q, \varsigma \in J$, $s \in (0, \infty)$ and $A \subset \mathbb{N}_0^n$ with $|A| < \infty$;

$$\Gamma_2 = \left\{ W^s_p(G) : p \in J^n, s \in \mathbb{N}_0^n \right\};$$

$$\Gamma_3 = \left\{ B^s_{p,q}(\mathbb{R}^n)_w, L^s_{p,q}(\mathbb{R}^n)_w : p \in J^n, q \in J, s \in (0, \infty) \right\};$$

$$\Gamma_4 = \left\{ B^s_{p,q,\mathcal{F}}(G), L^s_{p,q,\mathcal{F}}(G) : p \in J^n, q, \varsigma \in J, s \in \mathbb{R} \right\}.$$

Theorem 4.8. *Let $X_i \in \Gamma_i$ for $i \in \{1, 2, 3, 4\}$. Assume also that the parameter a of X_1 is admissible. Then we have:*

$$[1, \min(2, p_{\min}, q, a_{\min})] \subset I_T(X_1) \subset [1, \min(2, p_{\min}, q)];$$

$$[1, \max(2, p_{\max}, q, a_{\max})] \subset I_C(X_1) \subset [\max(2, p_{\max}, q), \infty];$$

$$I_T(X_2) = [1, \min(2, p_{\min})] \text{ and } I_C(X_2) = [\max(2, p_{\max}), \infty];$$

$$I_T(X_3) = [1, \min(2, p_{\min}, q)] \text{ and } I_C(X_3) = [\max(2, p_{\max}, q), \infty];$$

$$I_T(X_4) \supset [1, \min(2, p_{\min}, q)] \text{ and } I_C(X_4) \supset [\max(2, p_{\max}, q), \infty].$$

Moreover, we have the identities for X_1 if $a \in (\min(2, p_{\min}, q), \max(2, p_{\max}, q))^n$.

Remark 4.6. *a) The type and cotype sets of the dual spaces, subspaces and quotients of Besov and Lizorkin–Triebel type spaces from Theorem 4.8 can be derived from Theorems 4.8 and 4.5 with the aid of Parts b), d) and e) of Theorem 4.1 and Part a) of Corollary 3.1. b) A combination of results in Sections 4.4.5 and 4.5 in [2] provides, with the aid of the observations in Section 2.2 above, explicit estimates for the type and cotype constants of the spaces dealt with in Theorem 4.8 (and their l_σ–sums, duals and quotients) when they are endowed with geometrically friendly norms. Theorem 4.8 stands on the top of the "food chain" of our qualitative approach that is tangibly shorter and addresses the sharpness of the exponent sets as well. The main reason for this seems to be that the notions studied in this section are isomorphic (preserved by the isomorphic spaces), while the uniform convexity and smoothness, quantified in [2], is geometric, depending on the nature of the particular equivalent norm (more examples of similar phenomena are to appear).*

Proof of Theorem 4.8. The proof is essentially the same as that of the preceding theorem and even simpler. Therefore, we point out the differences.

In the case of X_1, one needs to use Part $e)$ of Corollary 3.1 instead of Part $c)$ to treat Y_1 and Part $f)$ of Theorem 4.1 (including the discrete version corresponding to a purely atomic measure) instead of Theorems 4.5 and 4.5 to treat Y_2.

As long as $X_2 = W_p^s(G)$ is concerned, the corresponding Y_2 can be chosen as the l_{p_n}–sum of isometric copies of $L_p(G)$ (see Subsection 2.2 and the definition of Sobolev spaces), while the search for Y_1 can be reduced to the considered case when X_1 is a Lizorkin–Triebel spaces with $q = 2$. Indeed, there is a (closed) parallelepiped $Q_v(z) \subset G$. According to [23], there is a bounded linear extension operator from $W_p^s(Q_v(z))$ into X_1, meaning that $W_p^s(Q_v(z))$ is isomorphic to the range/image of this extension operator. Furthermore, it is shown in [23] that $W_p^s(Q_v(z))$ is isomorphic to $L_{p,2,\bar{1}}^s(Q_v(z))$, while the latter spaces is isomorphic to $L_{p,2,a}^s(Q_v(z))$ with some admissible $a \in (\min(2, p_{\min}, q), \max(2, p_{\max}, q))^n$ according to [35] (the direct isomorphism between the latter space and $W_p^s(Q_v(z))$ is given in [35] as well). Eventually Part $e)$ of Corollary 3.1 provides an appropriate choice of $Y_1 = l_p(l_2)$ that is finitely λ–represented in $L_{p,2,a}^s(Q_v(z))$ and, therefore, also in X_2. Part $f)$ of Theorem 4.1 describes the type and cotype sets of this Y_1.

The case of a Besov space Y_3 is covered by Part $f)$ of Theorem 4.1 because it is isometric to $l_q(\mathbb{N}_0, l_p(\mathbb{Z}^n))$ (see Remark 7.1). When X_3 is a Lizorkin–Triebel space, Remark 7.1 provides us with $Y_1 = lt_{p,q}(\mathbb{Z}^n \times \mathbb{N}_0)$ and $Y_2 = lt_{p,q}(\mathbb{Z}^n \times \mathbb{N}, l_q(I_M))$ that possess identical type and cotype sets because they both are isometric to subspaces of $L_p(G, l_q(\mathbb{N}))$ and contain an isometric copy of $l_p(\mathbb{Z}^n, l_q(\mathbb{N}))$ due to Lemma 2.2 in [1]. In turn, the latter two spaces have identical type and cotype sets thanks to Part $f)$ of Theorem 4.1.

For X_4 we only need to find Y_2 that is clearly suggested by Definition 2.10 and Section 2.2. Namely, we take $\varsigma = p_n$ if X_4 is a Lizorkin–Triebel space and $\varsigma = q$ if X_4 is a Besov space. In both cases, the corresponding choice of Y_2 is treated with the aid of Part $f)$ of Theorem 4.1. □

Now we use the Maurey–Pisier theorem to extend the ranges of l_σ–spaces finitely representable in the spaces that are in Γ_i for $i \in I_3$ (established in §3).

4.4. Finite Representability of l_σ–Spaces

This subsection is dedicated to the Maurey–Pisier theorem that further extends our knowledge about the Minkowski spectrum of an infinite–dimensional Banach space X, that is the set of σ, such that $l_\sigma(I_n)$ is λ–represented in X for $n \in \mathbb{N}$ and some $\lambda \geq 1$. This deep and celebrated result was established by B. Maurey and G. Pisier [42] in 1976 and was further analysed by many authors (see, for example, [43, 44, 45]).

Theorem 4.9 ([42, 43, 44, 45]). *Let X be an infinite–dimensional Banach space, $p(X) := \sup I_T(X)$, $q(X) := \inf I_C(X)$, and $\sigma \in [p(X), 2] \cup \{q(X)\}$. Then the space l_σ is finitely represented in X.*

Combining Theorems 4.7, 4.8 and 4.9, we establish the following results on the finite representability of sequence spaces.

Theorem 4.10. *Under the conditions of Theorem 4.7, the space l_σ is finitely represented in every function space mentioned there and the duals of those spaces with the adjoint parameters p', q' and a' if $\sigma \in [\min(2, p, q, \theta), 2] \cup \{\max(2, p, q, \theta)\}$ and $a \in (\min(2, p, q, \theta), \max(2, p, q, \theta))^n$ (the spaces can be endowed with arbitrary equivalent norms).*

Theorem 4.11. *Let a Banach space X_i be from the group Γ_i for $i \in \{1, 2, 3\}$ with*

$$a \in (\min(2, p_{\min}, q), \max(2, p_{\max}, q))^n \text{ (for } i = 1\text{).}$$

Assume also that the parameter a of X_1 is admissible. Then the space l_σ is finitely represented in X_i for the following ranges of σ:

$$\sigma \in [\max(2, p_{\max}, q), 2] \cup \{\max(2, p_{\max}, q)\} \text{ for } i = 1, 3 \text{ and}$$
$$\sigma \in [\min(2, p_{\min}), 2] \cup \{\max(2, p_{\max})\} \text{ for } i = 2.$$

5. PROPERTIES: α, REFLEXIVITY, SEPARABILITY, STABILITY AND UMD. APPLICATIONS

Let us recall the definitions, basic properties and basic relations among stable, UMD, reflexive and superreflexive spaces and those with the Krein–Milman and Radon–Nykodým properties.

In this section we shall also deal with the mixed norm Lorentz spaces $L_{p,\theta}$ for $p, \theta \in [1, \infty]^n$ with $p_i = 1$ if $\theta_i = 1$ defined in the same way as the mixed norm L_p–spaces.

Definition 5.1. *Let $p \in (1, \infty)$. A Banach space X is said to be a UMD space if there is a constant $\beta_p = \beta_p(X)$, such that, for every X–valued martingale difference sequence $\{\xi_i\}_{i=1}^\infty$ on any probability space (Ω, σ, ν) and every sequence $\{\varepsilon_i\}_{i=1}^\infty$ of numbers 1 and -1, one has*

$$\sup_{i \in \mathbb{N}} \left\| \sum_{j=1}^i \varepsilon_j \xi_j \bigg| L_p(\Omega, X) \right\| \leq \beta_p \sup_{i \in \mathbb{N}} \left\| \sum_{j=1}^i \xi_i \bigg| L_p(\Omega, X) \right\|.$$

A Banach space X is said to be ζ–convex if there is a bi–convex function $\zeta : X \times X \to \mathbb{R}$ satisfying

$$\zeta(0, 0) > 0 \text{ and } \zeta(x, y) \leq \|x + y\| \text{ for } \|x\| = \|y\| = 1.$$

A Banach space X is said to be an HT space if the Hilbert transform is a bounded operator in the Bochner–Lebesgue space $L_p(\mathbb{R}, X)$.

A Banach space X is said to possess the property α if there exist a parameter $p > 0$ and a constant $\alpha = \alpha(p) > 0$ satisfying

$$\left\| \sum_{i,j=1}^n c_{i,j} r_i r'_j x_{i,j} \bigg| L_p(\Omega \times \Omega', X) \right\| \leq \alpha \max_{i,j=1}^n |c_{i,j}| \left\| \sum_{i,j=1}^n r_i r'_j x_{i,j} \bigg| L_p(\Omega \times \Omega', X) \right\|$$

for every $n \in \mathbb{N}$, $\{x_{i,j}\}_{i,j=1}^n \subset X$, $\{c_{i,j}\}_{i,j=1}^n \subset \mathbb{C}^{n^2}$, and independent Rademacher stochastic variables $\{r_i\}_{i=1}^n$ and $\{r'_i\}_{i=1}^n$ defined on Ω and Ω' respectively.

The properties HT and UMD, as well as ζ–convexity, were introduced by Burkholder [46, 47], while the property α is introduced by Pisier [48]. It is shown by Bourgain [49] and Burkholder [46, 47] that the classes of UMD, HT and ζ–convex spaces coincide and do not depend on p. Thanks to Kahane's generalization (Theorem 4.2, b)) of the Khinchine inequality and Minkowski one, the presence of α does not depend on p. Using the Lyons–Peetre interpolation formula from [50] (see Lemma 5.1 below), Cobos [51] proved that Lorentz–Zygmund spaces and their noncommutative analogs are UMD spaces.

Definition 5.2. *A Banach space X is said to be stable if, for every couple of non–trivial ultrafilters $\mathfrak{A}$ and $\mathfrak{B}$ on $\mathbb{N}$ and every two bounded sequences $\{x_i\}_{i\in\mathbb{N}} \subset X$ and $\{y_i\}_{i\in\mathbb{N}} \subset X$, one has*
$$\lim_{i,\mathfrak{A}} \lim_{j,\mathfrak{B}} \|x_i + y_j\| = \lim_{j,\mathfrak{B}} \lim_{i,\mathfrak{A}} \|x_i + y_j\|.$$

A Banach space X is said to be superreflexive if it is reflexive and every Banach space Y, that is finitely represented in X, is reflexive too.

It is known, for example, that c_0 is not stable, while $\{l_p\}_{p\in[1,\infty)}$ are.

Definition 5.3. *A Banach space X is said to possess the Krein–Milman property if its every closed convex bounded subset can be represented as the convex hull of its extreme points.*

A Banach space X is said to possess the Radon–Nikodým property if the Radon–Nikodým theorem holds for the X-valued measures. That is, for every probability space (Ω, σ, ν) and every countably additive X–valued measure μ on Ω, which is absolutely continuous with respect to ν, there is a Bochner–ν–measurable X–valued derivative $\frac{d\mu}{d\nu}$.

The relations between these properties are best seen from the following results of M. G. Krein, D. P. Milman and Huff, Morris [52, 53]. A Banach space X possesses the Radon–Nikodým (Krein–Milman) property if, and only if, every bounded closed (convex) subset of X has an extreme point. For dual spaces both properties coincide.

It is known that the UMD spaces are superreflexive (see Aldous [54]), and all the reflexive spaces have the Radon–Nikodým property (see Bessaga, Pełczyński [55]). However, these inclusions are strict.

All the above properties are inherited by subspaces.

Part b) of the next lemma was established by J. L. Krivine and B. Maurey [56, 57]. The case $p = q$ of c) is due to D. Burkholder, while the general case is established by F. Cobos [51]. Parts a) and d) follow from the definitions. In e), we use the definitions related to ideal spaces (or, more generally, Banach lattices) from §4.2. We prove e) and outline the proof of c) for comparison.

Lemma 5.1. *Let (E, σ, μ) be a countably additive measure space.*
a) If X is a UMD space, and X and Y are isomorphic, then the dual X^* and Y are UMD spaces.
b) Let X be stable and $p \in [1, \infty)$. Then $L_p(E, X)$ is stable.
c) Let X be a UMD space and $p, q \in (1, \infty)$. Then $L_{p,q}(E, X)$ is a UMD space.
d) Let X_1 and X_2 be UMD (stable) spaces and $p \in (1, \infty)$. Then the l_p-sum $l_p(I_2, \{X_i\}_{i\in I_2})$ is a UMD (stable) space;
e) Assume that X has the property α, and $Y = Y(E)$ is a quasi-Banach lattice with $I_{le}(Y) \cap (1, \infty) \neq \emptyset$. Then $Y(E, X)$ has the property α.

Remark 5.1. *Thanks to Parts b), c) and d) of the lemma, it is enough to analyse only the mixed norm spaces to detect the stability and UMD property for various types of Besov and Lizorkin–Triebel spaces and and their duals.*

Proof of Lemma 5.1, c), e). As it was noticed by Burkholder, $L_p(E, X)$ is a UMD (HT) space if X is. To consider the case $p \neq q$, one follows the general idea of [51].

For a compatible couple (A_0, A_1), $\theta \in (0, 1)$ and $q \in [1, \infty)$, Lions and Peetre [50] established that the spaces $(L_q(\mathbb{R}, A_0), L_q(\mathbb{R}, A_1))_{\theta,q}$ and $L_q(\mathbb{R}, (A_0, A_1)_{\theta,q})$ are isomorphic. Thus, $(A_0, A_1)_{\theta,q}$ is a UMD space, when A_0 and A_1 are UMD spaces.

To finish the proof of c), we choose $A_i := L_{p_i}(E, X)$ for $i = 0, 1$ in the Lions–Peetre result and take advantage of the isomorphism (see [58])

$$(L_{p_0}(E, X), L_{p_1}(E, X))_{\theta,q} \asymp L_{p_\theta,q}(E, X) \text{ for } 1/p_\theta = (1 - \theta)/p_1 + \theta/p_1.$$

To establish Part e), we just need to parody the proof of the least trivial part of the Khinchin–Maurey inequality (Theorem 4.2, c)) from [32], having in mind the comments following Definition 5.1 and the Hölder and Khinchin–Kahane inequalities. Indeed, Theorem 4.4, a) (due to Maurey and Pisier) says that

$$I_{le}(Y) \cap (1, \infty) \neq \emptyset \Leftrightarrow I_{conc}(Y) \cap (1, \infty) \neq \emptyset,$$

while Y is r–convex for some $r \in (0, 1)$ according to the Aoki–Rolewicz theorem. Hence, for every $m \in \mathbb{N}$, $q \in I_{conc}(Y) \cap (1, \infty)$ and $\{f_{i,j}\}_{i,j=1}^m \subset Y(E, X)$, one has

$$\left\| \sum_{i,j=1}^n c_{i,j} r_i r'_j f_{i,j} | L_q(\Omega \times \Omega', Y(E, X)) \right\| \leq$$

$$\leq C \left\| \sum_{i,j=1}^n c_{i,j} r_i r'_j x_{i,j} | Y(E, L_q(\Omega \times \Omega', X)) \right\| \leq$$

$$\leq C \max_{i,j=1}^n |c_{i,j}| \left\| \sum_{i,j=1}^n c_{i,j} r_i r'_j x_{i,j} | Y(E, L_q(\Omega \times \Omega', X)) \right\| \leq$$

$$\leq C \max_{i,j=1}^n |c_{i,j}| \left\| \sum_{i,j=1}^n c_{i,j} r_i r'_j x_{i,j} | Y(E, L_r(\Omega \times \Omega', X)) \right\| \leq$$

$$\leq C \max_{i,j=1}^n |c_{i,j}| \left\| \sum_{i,j=1}^n c_{i,j} r_i r'_j x_{i,j} | L_r(\Omega \times \Omega', Y(E, X)) \right\| \leq$$

$$\leq C \max_{i,j=1}^n |c_{i,j}| \left\| \sum_{i,j=1}^n c_{i,j} r_i r'_j x_{i,j} | L_q(\Omega \times \Omega', Y(E, X)) \right\|.$$

□

Lemma 5.2. *a) If X is a UMD space, and Y is its subspace, then the quotient X/Y is a UMD space.*

b) If X is stable space, and Y is its subspace of finite dimension, then the quotient space X/Y is stable. Moreover, for every two bounded sequences $\{x_i\}_{i\in\mathbb{N}} \subset X$ and $\{y_i\}_{i\in\mathbb{N}} \subset X$ and the quotient mapping $Q: X \to X/Y$, one has

$$\lim_{i,\mathfrak{A}}\lim_{j,\mathfrak{B}} \|Qx_i + Qy_j\|_{X/Y} = \inf_{\substack{f\in Y \\ z_i\to 0}} \inf_{\{z_i\}\subset Y} \lim_{i,\mathfrak{A}} \inf_{\substack{\{t_j\}\subset Y \\ t_j\to 0}} \lim_{j,\mathfrak{B}} \|x_i + y_j + z_i + t_j + f\|_X =$$

$$= \inf_{f\in Y} \lim_{i,\mathfrak{A}} \lim_{j,\mathfrak{B}} \|x_i + y_j + f\|_X.$$

Proof of Lemma 5.2. Thanks to Lemma 5.1, a), Part a) follows from the relation $(X/Y)^* = Y^\perp$.

The first half of Part b) follows from the second. To prove this identity, we modify the proof of Lemma 1 from [59] due to Bastero and Raynaud.

For a positive sequence $\varepsilon_{i,j}$ decreasing to 0 by both i and j, let us choose $\{f_{i,j}\}_{i,j\in\mathbb{N}}$ to satisfy

$$\|Qx_i + Qy_j\|_{X/Y} \leq \|x_i + y_j + f_{i,j}\|_X \leq \|Qx_i + Qy_j\|_{X/Y} + \varepsilon_{i,j}. \tag{1}$$

Since Y is locally compact, $\{f_{i,j}\}$ is precompact, and, for every $i\in\mathbb{N}$, there exist

$$f_i = \lim_{j,\mathfrak{B}} f_{i,j}, \quad f = \lim_{i,\mathfrak{A}} f_i, \quad h_{i,j} := f_{i,j} - f_j, \quad r_i := f_i - f.$$

Using (1), one obtains

$$\lim_{i,\mathfrak{A}}\lim_{j,\mathfrak{B}} \|Qx_i + Qy_j\|_{X/Y} = \lim_{i,\mathfrak{A}}\lim_{j,\mathfrak{B}} \|x_i + y_j + f + r_i + h_{i,j}\|_X \geq$$

$$\geq \inf_{\substack{f\in Y \\ z_i\to 0}} \inf_{\{z_i\}\subset Y} \lim_{i,\mathfrak{A}} \inf_{\substack{\{t_j\}\subset Y \\ t_j\to 0}} \lim_{j,\mathfrak{B}} \|x_i + y_j + z_i + t_j + f\|_X. \tag{2}$$

So, the first equality of a) follows from (2) and $\|Qx_i + Qy_j\|_{X/Y} \leq \|x_i + y_j + z_i + t_j + f\|_X$. The second equality follows from the conditions $\lim_{j,\mathfrak{B}} t_j = 0$, $\lim_{i,\mathfrak{A}} z_i = 0$ and the fact that the norm is a 1–Lipshitz functional on X. $\square$

The proof of the next theorem is the observation that the proof of the reflexivity and separability of Besov and Lizorkin–Triebel spaces from [35] provides, in fact, their stability, UMD property and superreflexivity. In [1], we establish the superreflexivity of even wider ranges of spaces than those dealt with in the next theorem by means of computing Dol'nikov–Pichugov, Jacobi and self–Jung constants in terms of the parameters of spaces.

Theorem 5.1. *Let $G \subset \mathbb{R}^n$ be an open set, $Y = Y(G)$ be an ideal space with $I_{le}(Y) \cap (1,\infty) \neq \emptyset$, $a, p, \theta \in [1,\infty)^n$ with $\theta_i = 1$ if $p_i = 1$, $q \in [1,\infty)$, $A \subset \mathbb{N}^n$ be finite, and*

$$X \in \Big\{ B^s_{Y,q,a}(G), L^s_{Y,q,a}(G), \tilde{B}^s_{Y,q,a}(G), \tilde{L}^s_{Y,q,a}(G), b^s_{Y,q,a}(G),$$

$$l^s_{Y,q,a}(G), \tilde{b}^s_{Y,q,a}(G), \tilde{l}^s_{Y,q,a}(G), B^s_{p,q,\mathcal{F}}(G), L^s_{p,q,\mathcal{F}}(G) \Big\}$$

Then:
a) X is stable for $Y = L_p(G)$;
b) X and X^* endowed with an arbitrary equivalent norm are UMD and superreflexive spaces for $p, \theta, a \in (1, \infty)^n$, $q \in (1, \infty)$ and $Y = L_{p,\theta}$.
c) X possesses the property α. In particular, this is the case for $Y = L_{p,\theta}$ with $p, \theta, a \in (1, \infty)^n$.

Proof of Theorem 5.1. The operators $\Upsilon_{\tilde{B}}, \Upsilon_{\tilde{L}}, \Xi_B, \Xi_L$ and A from Definition 2.12 define, respectively, the isomorphisms between the spaces $\tilde{b}_{Y,q,a}^{s,A}(G), \tilde{l}_{Y,q,a}^{s,A}(G), b_{Y,q,a}^{s}(G), l_{Y,q,a}^{s}(G)$ and the corresponding subspaces of the spaces

$$L_{*q}\left(\mathbb{R}_+, Y\left(G, L_a(Q_1(0))/\mathcal{P}_A\right)\right), Y\left(G, L_{*q}\left(\mathbb{R}_+, L_a(Q_1(0))/\mathcal{P}_A\right)\right),$$
$$l_\varsigma\left(I_n, \{L_{*q}\left(\mathbb{R}_+, Y\left(G, L_{a_i}([-1,1])\right)\right)\}_{i=1}^n\right), l_\varsigma\left(I_n, \{Y\left(G, L_{*q}\left(\mathbb{R}_+, L_{a_i}([-1,1])\right)\right)\}_{i=1}^n\right),$$

while the rest of the spaces are subspaces of their l_ς–sums with $Y(G)$ (see §2.2 for more details). The stability and the UMD property of the latter follow from Parts a) – d) of Lemma 5.1 but, in the case of the spaces defined in terms of the local approximation by polynomials, we apply Lemma 5.2 to the spaces $X = L_a(Q_1(0))$ and $Y = \mathcal{P}_A$ with $\dim \mathcal{P}_A = |A|$ first.

Part c) follows from Lemma 5.1, e) assisted by Lemma 4.3. □

The next corollary also follows from either the reflexivity [35], or the separability [4] of the corresponding function spaces.

Corollary 5.1. *The function spaces in Part b) of the last theorem possesses the Radon–Nikodým and Krein–Milman properties.*

Proof of Corollary 5.1. The reflexive spaces possess the Radon–Nikodým property, implying the Krein–Milman one (see [54, 55]). □

Remark 5.2. *The absence of the UMD property follows from the absence of reflexivity implied by Theorems 3.2, 3.3 and 3.5 and treated in [4, 5], while the absence of stability is discussed in Corollary 5.3, a).*

Now we need a well–known infinite–dimensional counterpart of Theorem 3.1. The next classical theorem was proved by R. C. James [60] for c_0 and l_1 and by J. L. Krivine and B. Maurey [56] for l_p, $p \in [1, \infty)$.

Theorem 5.2. [James, Krivine, Maurey] *Let X be either c_0 or l_p for $p \in [1, \infty)$, Y be a Banach space, and $\lambda \geq 1$. Assume also that $d_Y(X) = \inf\{d(X, Y_1) : Y_1$ is a subspace of $Y\} < \infty$. Then $d_Y(X) = 1$ if X is*
a) either c_0 or l_1, or
b) $p > 1$, $X = l_p$, and Y is stable.

Let us recall that we say that X is almost isometrically represented in Y if $d_Y(X) = 1$.
The following corollary is as important in our investigation of the nonlocal properties, as Theorem 3.1 for the study of local properties.

Corollary 5.2. *Let a satisfy the admissability condition. Then we have the following.*
a) Under the conditions of Theorems 3.2 and 3.7, the spaces c_0, l_q and l_p with scalar p are almost isometrically represented in the corresponding function spaces.
b) Under the conditions of Theorem 3.3, the spaces $\{l_{p_i}\}_{i \in I_n}$ are almost isometrically represented in the corresponding function spaces if G satisfies the quadrangle condition.
c) Under the conditions of Theorem 3.3, the space l_p with a scalar p is almost isometrically represented in the corresponding function spaces if G is not almost bounded.

Proof of Corollary 5.2. The combinations of Theorem 5.2 and Parts $d) - f)$ of Corollary 3.1 imply Parts $a) - c)$ correspondingly. □

The next corollary is an extension of the corresponding theorem concerning the nonseparability and nonreflexivity of some Besov and Lizorkin–Triebel spaces from [4, 5].

Corollary 5.3. *a) Let X be isomorphic to any of the spaces dealt with in either Theorem 3.2 with $q = \infty$ or Theorem 3.3, b) with $p_{\max} = \infty$, or Theorem 3.7 with $p = \infty$, and let $\overset{\circ}{X}$ be its corresponding subspace. Then X and $\overset{\circ}{X}$ are not stable, and X is nonseparable.*
b) Let X be isomorphic to any of the spaces dealt with in either Theorem 3.2 with $q \in \{1, \infty\}$ or Theorem 3.3, b) with some $p_i \in \{1, \infty\}$, or Theorem 3.7 with $p \in \{1, \infty\}$. Then X is not reflexive, superreflexive, or a UMD space.

Proof of Corollary 5.3. Both parts are established in one and the same way. Were X stable, then, for every $\varepsilon > 0$, X would contain a (stable) subspace X_ε with $d(X_1, c_0) \leq 1 + \varepsilon$ according to Corollary 5.2 and, thus, implying the stability and reflexivity of c_0. But c_0 is not stable. (In fact, the proof of the latter result shows that any space that is isomorphic to c_0 is not stable. Thus, we can use only Corollary 3.1 to rule out the stability.) □

The idea of §2.2 combined with Corollary 3.1 also leads to the following result from [4, 5].

Theorem 5.3. *Let $G \subset \mathbb{R}^n$ be an open set, $s \geq 0$, $A \subset \mathbb{N}_0^n$, $|A| < +\infty$, $a \in [1, \infty]^n$. Then the following spaces are separable:*
a) $\overset{\circ}{B}_{p,\infty}^s(G)$, $\overset{\circ}{B}_{p,\infty,a}^s(G)$, $\overset{\circ}{\widetilde{B}}_{p,\infty,a}^{s,A}(G)$, $\overset{\circ}{b}_{p,\infty}^s(G)$, $\overset{\circ}{b}_{p,\infty,a}^s(G)$, $\overset{\circ}{\widetilde{b}}_{p,\infty,a}^{s,A}(G)$, $\overset{\circ}{L}_{p,\infty}^s(G)$, $\overset{\circ}{L}_{p,\infty,a}^s(G)$, $\overset{\circ}{\widetilde{L}}_{p,\infty,a}^{s,A}(G)$, $\overset{\circ}{l}_{p,\infty}^s(G)$, $\overset{\circ}{l}_{p,\infty,a}^s(G)$, $\overset{\circ}{\widetilde{l}}_{p,\infty,a}^{s,A}(G)$ for $s \geq 0$, $p, a \in [1, \infty)^n$;
b) The spaces $B_{p,\infty}^s(G)$, $B_{p,\infty,a}^s(G)$, $\widetilde{B}_{p,\infty,a}^{s,A}(G)$, $b_{p,\infty}^s(G)$, $b_{p,\infty,a}^s(G)$ and $\widetilde{b}_{p,\infty,a}^{s,A}(G)$ are not separable   for $A \supset A_s^*$ and for either $s > 0$, $p \in [1, \infty]^n$, or $s \geq 0$, $p = \infty$.

5.1. Application to Semi–Embeddings

Let us recall that a semi–embedding of a Banach space X into a Banach space Y is an injective operator $A \in \mathcal{L}(X, Y)$, for which the image $A(B_X)$ of the closed unit ball B_X is closed in Y.

The related abstract theorem is due to Fonf [61].

Theorem 5.4 ([61]). *If a Banach space X contains an infinite–dimensional subspace isomorphic to a separable conjugate Banach space, then there is a Banach space Y and a non–isomorphic semi–embedding of X into Y.*

Thus, we establish one more application of the theorems from §3.

Theorem 5.5. *Let X be isomorphic to any of the spaces dealt with in either Theorem 3.2 with $q < \infty$ or Theorem 3.3, b) with $p_{\min} < \infty$, or Theorem 3.7 with $p < \infty$. Then there is a Banach space Y and a non–isomorphic semi–embedding of X into Y.*

6. COMPLEMENTED COPIES

Occasionally one needs to know not only that a Banach space X contains a subspace X_n with $d_{BM}(l_p(I_n, X_n)) < \lambda$ for $n \in \mathbb{N}$ but also that every X_n is, in addition, complemented in X, and there is a uniformly bounded family of projectors onto $\{X_n\}_{n\in\mathbb{N}}$. For example, this knowledge is crucial for establishing the sharpness of the results on the bounded extension of Hölder–Lipschitz mappings between function spaces [10], as well as the detection of the presence of Tsar'kov's phenomenon [8] leading to the sharpness of the regularity exponents in the Hölder classification of infinite–dimensional spheres [9]. The most natural approach is to span every X_n with appropriate members of a suitable basic sequence in X. Unfortunately, the existence of wavelet bases in anisotropic Besov and Lizorkin–Triebel spaces of functions defined on open $G \subset \mathbb{R}^n$ is associated with very strong restrictions on G, such as, for example, $G = \mathbb{R}^n$, or G is an extension domain. In this section, we sidestep this obstacle establishing the desirable result for the various anisotropic Besov, Sobolev and Lizorkin–Triebel spaces under consideration in the setting of an arbitrary open $G \subset \mathbb{R}^n$. For various geometric problems, one even needs the existence of 1–complemented copies of sequence spaces (see, for example, [1]).

Remark 6.1. *Let us recall that, for $p \in (1, \infty)$, $l_p(I_{2^n})$ contains a C_0–isomorphic and C_2–complemented copy of $l_2(I_n)$ for every $n \in \mathbb{N}$ thanks to the Hölder and Khinchin inequalities and the following lemma.*

Lemma 6.1 ([1]). *Let X be a Banach space, and $P \in \mathcal{L}(X)$ be a projector onto its subspace $Y \subset X$. Assume also that $Q_Y : X \to \tilde{X} = X/\mathrm{Ker}\, P$ is the quotient map. Then we have*
$$\|Q_Y x\|_{\tilde{X}} \leq \|Px\|_X \leq \|P|\mathcal{L}(X)\|\|Q_Y x\|_{\tilde{X}} \text{ for every } x \in X.$$
In particular, the dual space $Y^* = X^*/Y^\perp$ and Y are isometric to P^*X^* and $\tilde{X}$ if, and only if, Y is 1–complemented in X, i.e $\|P|\mathcal{L}(X)\| = 1$.

Definition 6.1. *For $\lambda \geq 1$, we say that a quasi–Banach space X is λ–isomorphic to a quasi–Banach space Y if there exists an isomorphism $T : X \leftrightarrow Y$ with*
$$\|T|\mathcal{L}(X,Y)\|\|T^{-1}|\mathcal{L}(Y,X)\| \leq \lambda.$$
For $\beta \geq 1$, we say that a subspace Z of X is β–complemented in X if there exists a projector P onto Z with $\|P|\mathcal{L}(X)\| \leq \beta$.

To establish the next two theorems one could work (without tangible changes in the proof) with the dyadic wavelet basis given in [18] but we prefer to employ our optimal (and more general) counterpart constructed in [17].

Theorem 6.1. Let an open $G \subset \mathbb{R}^n$, $p, a \in (1, \infty)^n$, $q, \varsigma \in (1, \infty)$, $s \in (0, \infty)$ and
$$r \in \{p_{\min}, p_{\max}, q, 2\}.$$

Assume also that

$$Y \in \Big\{ B^s_{p,q,a}(G), \tilde{B}^{s,A}_{p,q,a}(G), L^s_{p,q,a}(G), \tilde{L}^{s,A}_{p,q,a}(G), b^s_{p,q,a}(G), \tilde{b}^{s,A}_{p,q,a}(G), l^s_{p,q,a}(G), \tilde{l}^{s,A}_{p,q,a}(G),$$
$$B^s_{p',q',a'}(G)^*, \tilde{B}^{s,A}_{p',q',a'}(G)^*, L^s_{p',q',a'}(G)^*, \tilde{L}^{s,A}_{p',q',a'}(G)^*, b^s_{p',q',a'}(G)^*, \tilde{b}^{s,A}_{p',q',a'}(G)^*,$$
$$l^s_{p',q',a'}(G)^*, \tilde{l}^{s,A}_{p',q',a'}(G)^* \Big\}$$

and that a is admissible for Y if it is not defined as dual, while a' is admissible for Y^* if $Y = Y^{**}$ is defined as dual. Then there are constants $C_0, C_1 > 0$, such that Y contains an C_0–isomorphic and C_1–complemented copy of $l_r(I_m)$ for every $m \in \mathbb{N}$.

Proof of Theorem 6.1. Since every projector P in a Banach space Y defines the projector $P^* : f \mapsto f \circ P$ and $l_p^* = l_{p'}$ for $p \in [1, \infty)$, it is sufficient to consider only the non–dual spaces. We also consider only non–homogeneous spaces because the homogeneous spaces are treated identically.

We assume that $Y_w = B^s_{p,q}(\mathbb{R}^n)_w$ if Y is a Besov space and $Y_w = L^s_{p,q}(\mathbb{R}^n)_w$ otherwise. Let us fix anisotropic cubes $Q_v(z) \subset Q_{v_0}(z) \subset G$ and designate by $W_Q \subset W_Y = \bigcup_{j \in \mathbb{N}_0, k \in \mathbb{Z}^n} \{(j,k)\} \times [0, a_{j+1}) \cap \mathbb{N}$ the triplets (l, j, k) numbering those elements of the local (i.e., compactly supported) wavelet basis $\{b_j^{\gamma_a/p} \psi_{l,j}(b_j x - k)\}_{j \in \mathbb{N}_0, k \in \mathbb{Z}}^{1 \le l < a_{j+1}}$ in Theorem 7.1 that satisfy the inclusion $\operatorname{supp} \psi_{l,j}(b_j x - k) \cup \operatorname{supp} \psi^*_{l,j}(b_j \cdot - k) \subset Q_v(z)$. Assume now that

$$Y_Q = \left\{ \sum_{(l,j,k) \in W_Q} \lambda_{l,j,k} \psi_{l,j,k} : \lambda \in Y_w, \lambda_{0,k} = \lambda_{l,j,k} = 0 \text{ for } (l,j,k) \notin W_Q \right\}.$$

Thanks to Theorem 7.1, we know that Y_Q is a subspace of $Y(\mathbb{R}^n)$ and, thus, also $Y = Y(G)$. Thanks to the same theorem, we also observe that Y_Q (endowed with the Y–norm) contains a C'_0–isomorphic and C'_1–complemented copy of either an l_q–sum $l_q \left(\mathbb{N}, \{l_p(I^n_{m_d})\}_{d \in \mathbb{N}} \right)$ of the finite–dimensional l_p–spaces with $\lim_{d \to \infty} m_d = \infty$ if Y is a Besov space, or $lt_{p,q}(F)$ for some rectangle $F \subset Q_v(z)$ (see Definitions 2.3 and 7.1 and Remark 7.1) if Y is a Lizorkin–Triebel space. Dealing with $lt_{p,q}(F)$ with the aid of Lemma 6.2 if Y is a Lizorkin–Triebel space, we see that Y_Q, as a (closed) subspace of Y, contains a $\tilde{C}_0$–isomorphic and $\tilde{C}_1$–complemented copy of $l_r(I_m)$ for every $m \in \mathbb{N}$.

Now it is left to construct the projector from Y onto Y_Q. There exists $g \in C_0^\infty$, such that $\operatorname{supp} g \subset Q_{v_1}(z)$ for some $v_1 \in (v, v_0)$ and $g(x) = 1$ for $x \in Q_v(z)$. A particular case of the Hölder inequalities for various Besov and Lizorkin–Triebel spaces from [35] provides the estimate

$$\|gf|Y(Q_{v_0}(z))\| \le C \|f|Y(Q_{v_0}(z))\| \text{ for } f \in Y(Q_{v_0}(z)). \tag{1}$$

Let M_g be the corresponding operator of the point–wise multiplication by g, $R : Y \longrightarrow Y(Q_{v_0}(z))$ the restriction operator onto $Q_{v_0}(z)$ and

$$Pw : f \mapsto \sum_{(l,j,k) \in W_Q} \langle f, b_j^{\gamma_a/p'} \psi^*_{l,j}(b_j \cdot - k) \rangle b_j^{\gamma_a/p} \psi_{i,j}(b_j x - k).$$

While R is a norm 1 operator, $P_w \in \mathcal{L}(Y(\mathbb{R}^n))$ thanks to Theorem 7.1 and, thus, also $P_w \in \mathcal{L}(Y)$. Combining their boundedness with (1), we finish the proof by observing that $P = P_w \circ M_g \circ R$ is a projector onto Y_Q we are looking for. $\square$

Lemma 6.2 ([1]). *Let $p \in [1,\infty)^n$ and $q \in [1,\infty)$. Then the spaces $lt_{p,q}(\mathbb{R}^n)$ and $lt_{p',q'}(\mathbb{R}^n)^*$ contain isometric 1–complemented copies of $l_p(\mathbb{N}^n)$, $l_q(\mathbb{N})$ and $l_p(\mathbb{Z}^n, l_q(\mathbb{N}))$, and the spaces $lt_{p,q}(F)$ and $lt_{p',q'}(F)^*$ contain isometric 1–complemented copies of $l_q(\mathbb{N})$, $l_p(I_m, l_q(\mathbb{N}))$ for every $m \in \mathbb{N}^n$.*

Theorem 6.2. *For an open $G \subset \mathbb{R}^n$, $p \in (1,\infty)^n$, $\varsigma \in (1,\infty)$, $s \in \mathbb{N}_0^n$, let*

$$Y \in \{W_p^s(G), W_{p'}^s(G)^*\} \text{ and } r \in \{p_{\min}, p_{\max}, 2\}.$$

Then there are constants $C_0, C_1 > 0$, such that Y contains an C_0–isomorphic and C_1–complemented copy of $l_r(I_m)$ for every $m \in \mathbb{N}$.

Proof of Theorem 6.2. As in the proof of Theorem 6.1, we reduce the theorem to the case $Y = W_p^s(G)$. In the notations of the proof of Theorem 6.1, the restriction R onto $Q_{v_0}(z)$ is still a norm 1 operator. It is also shown in [62, 23] that the $W_p(Q_{v_0}(z))$–norm is equivalent to $L_{p,2,1}^s(Q_{v_0}(z))$, and in [35] that the $W_p(Q_{v_0}(z))$–norm is equivalent to the $L_{p,2,a}^s(Q_{v_0}(z))$–norm (and also to the $\tilde{L}_{p,2,a}^{s,A}(G)$–norm if $A \supset A_s^*$) for an admissible a. Hence, we can use the operators M_g, P_w and subspace Y_Q from the proof of Theorem 6.1 for $Y = L_{p,2,a}^s(Q_{v_0}(z))$ (or $\tilde{L}_{p,2,a}^{s,A}(G)$) to construct the bounded projector $P = P_w \circ M_g \circ R$ from Y onto $Y_Q \subset Y$. The existence of uniformly isomorphic and uniformly complemented copies of $l_r(I_m)$ in Y_Q is established in the proof of Theorem 6.1. This observation finishes the proof. $\square$

The next theorem is trivial in the case of Besov spaces (see Remark 7.1). With the aid of Lemma 6.2, it follows from Remark 7.1 (see also Definition 7.1).

Theorem 6.3. *Let $Y \in \left\{B_{p,q}^s(\mathbb{R}^n)_w, L_{p,q}^s(\mathbb{R}^n)_w, B_{p',q'}^s(\mathbb{R}^n)_w^*, L_{p',q'}^s(\mathbb{R}^n)_w^*\right\}$ for $p \in (1,\infty)^n$, $\varsigma \in (1,\infty)$, $s \in \mathbb{R}$ and $r \in \{p_{\min}, p_{\max}, q, 2\}$. Then Y contains an isometric 1–complemented copy of $l_r(I_m)$ for every $m \in \mathbb{N}$.*

7. WAVELET NORMS AND CHARACTERIZATIONS OF BESOV AND LIZORKIN–TRIEBEL SPACES

Here we introduce more multi–index notations in addition to those in Section 2. For $n \in \mathbb{N}$, $m \in \mathbb{N}^n$ and $i \in I_n$, let $\{V_j^{m_i}\}_{j \in \mathbb{Z}}$ be the local spline multiresolution analysis with the re–scaling factors $\{a_{i,j}\}_{j \in \mathbb{Z}}$ provided by the application of Theorem 4.5 in [17] with $m = m_i$ and $a_k = s_{k,i}$, where $s_{k,i}$ is defined in the proof of Lemma 3.1. Thus, we also have complete scaling factors $\{b_{i,j}\}_{j \in \mathbb{N}_0}^{i \in I_n}$, associated scaling functions $\{\phi_i\}_{i \in I_n}$, generators $\{\phi_i^*\}_{i \in I_n}$ of their bi–orthogonal bases, wavelet layer generators $\{\psi_{i,l,j}\}_{j \in \mathbb{Z}}^{1 \leq l < a_{i,j+1}}$ for $i \in I_n$, and the generators $\{\psi_{i,l,j}^*\}_{j \in \mathbb{Z}}^{1 \leq l < a_{i,j+1}}$ for $i \in I_n$ of the basis that is bi–orthogonal to the wavelet bases. They are all provided by Theorem 4.5 in [17]. Let us denote $J_a = \prod_{i \in I_n}[0, a_i] \cap$

$\mathbb{Z}^n \setminus \{\overline{0}\}$ for $a \in \mathbb{N}^n$ and $\overline{0} = (0, 0, \ldots, 0)$. Assume also that $V_j^m = \bigotimes_{i \in I_n} V_j^{m_i}$, $\phi(x) = \psi_0(x) = \prod_{i \in I_n} \phi_i(x_i)$, $\phi^*(x) = \psi_0^*(x) = \prod_{i \in I_n} \phi_i^*(x_i)$ and $\psi_{l,j}(x) = \prod_{i \in I_n} \psi_{i,l_i,j}(x_i)$, $\psi_{l,j}^*(x) = \prod_{i \in I_n} \psi_{i,l_i,j}^*(x_i)$ for $l \in J_{a_{j+1}}$, where $\psi_{i,0,j} := \phi_i$ and $\psi_{i,0,j}^* := \phi_i^*$ for $i \in I_n$. Let us also recall that, in multiindex notation, $\overline{1} = (1, 1, \ldots, 1)$ and, for $\alpha, \beta \in (0, \infty)^n$, one has $\alpha/\beta = (\alpha_1/\beta_1, \ldots, \alpha_n/\beta_n)$ and $\alpha^\beta = \prod_{i \in I_n} \alpha_i^{\beta_i}$. For example, $b^{\overline{1}} = \prod_{i \in I_n} b_i$ for $b \in \mathbb{R}^n$.

Definition 7.1 ([17]). *For $j \in \mathbb{N}_0$, let $\{F_{k,j}\}_{k \in \mathbb{Z}^n}$ be the system of parallelepipeds obtained at the jth step in the proof of Lemma 3.1 and, thus, possessing the properties described in Definition 2.3. Assume also that $p \in [1, \infty]^n$, $q \in [1, \infty]$ and $s \in \mathbb{R}$. The space $B_{p,q}^s(\mathbb{R}^n)_w$ is the Banach space of all sequences $\lambda = \{\lambda_{0,k}, \lambda_{l,j,k}\}_{j \in \mathbb{N}_0, k \in \mathbb{Z}^n}^{l \in J_{a_{j+1}}}$ with the finite norm*

$$\|\lambda|B_{p,q}^s(\mathbb{R}^n)_w\| := \left(\|\{\lambda_{0,k}\}|l_p(\mathbb{Z}^n)\|^q + \sum_{j \in \mathbb{N}_0} b_j^{q(s\overline{1}/n - \gamma_a/p)} \sum_{l \in J_{a_{j+1}}} \|\{\lambda_{l,j,k}\}|l_p(\mathbb{Z}^n)\|^q \right)^{1/q}.$$

The space $L_{p,q}^s(\mathbb{R}^n)_w$ is the Banach space of all $\lambda = \{\lambda_{0,k}, \lambda_{l,j,k}\}_{j \in \mathbb{N}_0, k \in \mathbb{Z}^n}^{l \in J_{a_{j+1}}}$ with the finite norm

$$\|\lambda|L_{p,q}^s(\mathbb{R}^n)_w\| := \left\| \left(\sum_{k \in \mathbb{Z}^n} |\lambda_{0,k}|^q \chi_{F_{k,0}} + \sum_{k \in \mathbb{Z}^n} \sum_{j \in \mathbb{N}_0} b_j^{sq\overline{1}/n} \chi_{F_{k,j}} \sum_{l \in J_{a_{j+1}}} |\lambda_{l,j,k}|^q \right)^{1/q} \bigg| L_p(\mathbb{R}^n) \right\|.$$

Remark 7.1. Definition 7.1 shows that the spaces $B_{p,q}^s(\mathbb{R}^n)_w$ and $L_{p,q}^s(\mathbb{R}^n)_w$ are isometric to $B_{p,q}^0(\mathbb{R}^n)_w$ and $L_{p,q}^0(\mathbb{R}^n)_w$ correspondingly. This observation reflects the lifting property of function spaces. The isomorphisms $B_{p,q}^0(\mathbb{R}^n)_w^* \asymp B_{p',q'}^0(\mathbb{R}^n)_w$ and $L_{p,q}^0(\mathbb{R}^n)_w^* \asymp L_{p',q'}^0(\mathbb{R}^n)_w$ explain why the function spaces of negative smoothness are defined as duals. Moreover, $B_{p,q}^0(\mathbb{R}^n)_w$ is $l_q(\mathbb{N}_0, l_p(\mathbb{Z}^n))$, while $L_{p,q}^0(\mathbb{R}^n)_w$ is a 1–complemented subspace of $lt_{p,q}(\mathbb{Z}^n \times \mathbb{N}, l_q(I_M))$ and contains, in turn, an isometric and 1–complemented copy of $lt_{p,q} = lt_{p,q}(\mathbb{Z}^n \times \mathbb{N}_\mathbb{C})$, where $M = \max_{j \in \mathbb{N}} a_j^{\overline{1}}$. Many quantitative properties of $lt_{p,q}(\mathbb{Z}^n \times \mathbb{N}, l_q(I_M))$ and $lt_{p,q}$ are the same because they depend on p and q and the latter is a 1–complemented copy of the former. Therefore, these quantitative properties of $B_{p,q}^s(\mathbb{R}^n)_w$ and $L_{p,q}^s(\mathbb{R}^n)_w$ and, correspondingly, $l_q(\mathbb{N}_0, l_p(\mathbb{Z}^n))$ and $lt_{p,q}(\mathbb{Z}^n \times \mathbb{N}, l_q(I_M))$ are the same as well.

Theorem 7.1 ([17]). *For $n \in \mathbb{N}$, $p, a \in [1, \infty]^n$, $q \in [1, \infty]$ and $s \in (0, \infty)$, let*

$$Y_B \in \left\{ B_{p,q,a}^s(\mathbb{R}^n), \tilde{B}_{p,q,a}^{s,A}(\mathbb{R}^n) \right\} \text{ and } Y_L \in \left\{ L_{p,q,a}^s(\mathbb{R}^n), \tilde{L}_{p,q,a}^{s,A}(\mathbb{R}^n) \right\}.$$

Let also a be in the admissible range for $Y \in \{Y_B, Y_L\}$ and $s/\gamma_a < m - \overline{1}$. Then $\{V_j^m\}_{j \in \mathbb{N}_0}$ described above is a local multiresolution analysis in $Y \in \{Y_B, Y_L\}$ with associated wavelets $\{\psi_0, \psi_{l,j}\}_{j \in \mathbb{N}_0}^{l \in J_{a_{j+1}}}$ generating the unconditional wavelet basis

$$\{\psi_0(x - k), \psi_{l,j}(b_j x - k)\}_{j \in \mathbb{N}_0, k \in \mathbb{Z}^n}^{l \in J_{a_{j+1}}} \text{ in } Y.$$

The system $\{\psi_0^*, \psi_{l,j}^*\}_{j\in\mathbb{N}_0}^{l\in J_{a_{j+1}}}$ generates the bi–orthogonal unconditional basis

$$\{\psi_0^*(x-k), b_j^{\bar{1}}\psi_{l,j}^*(b_jx-k)\}_{j\in\mathbb{N}_0, k\in\mathbb{Z}^n}^{l\in J_{a_{j+1}}} \text{ in } Y^*.$$

Moreover, the mapping J defined by the relations

$$\lambda_{0,k} = \int_{\mathbb{R}^n} f(x)\overline{\psi_0^*(x-k)}dx \text{ and } \lambda_{l,j,k} = b_j^{\bar{1}}\int_{\mathbb{R}^n} f(x)\overline{\psi_{l,j}^*(b_jx-k)}dx$$

for $j \in \mathbb{N}_0$, $l \in J_{a_{j+1}}$ and $k \in \mathbb{Z}^n$ is an isomorphism between Y_B and $B_{p,q}^s(\mathbb{R}^n)_w$ and, for $p \in (1,\infty)^n$ and $q \in (1,\infty]$, also between Y_L and $L_{p,q}^s(\mathbb{R}^n)_w$.

Theorem 7.2 ([17]). *For $n \in \mathbb{N}$, $p \in (1,\infty)^n$, $r \in [1,\infty]^n$, $a \in [1,\infty]^n$, $q \in [1,\infty]$ and $s \in (0,\infty)$, let*

$$Y_B \in \left\{B_{L_{p,r},q,a}^s(\mathbb{R}^n), \tilde{B}_{L_{p,r},q,a}^{s,A}(\mathbb{R}^n)\right\} \text{ and } Y_L \in \left\{L_{L_{p,r},q,a}^s(\mathbb{R}^n), \tilde{L}_{L_{p,r},q,a}^{s,A}(\mathbb{R}^n)\right\}.$$

Let also a be in the admissible range for $Y \in \{Y_B, Y_L\}$ and $s/\gamma_a < m - \bar{1}$. Then $\{V_j^m\}_{j\in\mathbb{N}_0}$ described above is a local multiresolution analysis in $Y \in \{Y_B, Y_L\}$ with associated wavelets $\{\psi_0, \psi_{l,j}\}_{j\in\mathbb{N}_0}^{l\in J_{a_{j+1}}}$ generating the unconditional wavelet basis

$$\{\psi_0(x-k), \psi_{l,j}(b_jx-k)\}_{j\in\mathbb{N}_0, k\in\mathbb{Z}^n}^{l\in J_{a_{j+1}}} \text{ in } Y.$$

The system $\{\psi_0^*, \psi_{l,j}^*\}_{j\in\mathbb{N}_0}^{l\in J_{a_{j+1}}}$ generates the bi–orthogonal unconditional basis

$$\{\psi_0^*(x-k), b_j^{\bar{1}}\psi_{l,j}^*(b_jx-k)\}_{j\in\mathbb{N}_0, k\in\mathbb{Z}^n}^{l\in J_{a_{j+1}}} \text{ in } Y^*.$$

Moreover, the mapping J defined by the relations

$$\lambda_{0,k} = \int_{\mathbb{R}^n} f(x)\overline{\psi_0^*(x-k)}dx \text{ and } \lambda_{l,j,k} = b_j^{\bar{1}}\int_{\mathbb{R}^n} f(x)\overline{\psi_{l,j}^*(b_jx-k)}dx$$

for $j \in \mathbb{N}_0$, $l \in J_{a_{j+1}}$ and $k \in \mathbb{Z}^n$ is an isomorphism between Y_B and $B_{L_{p,r},q}^s(\mathbb{R}^n)_w$ and between Y_L and $L_{L_{p,r},q}^s(\mathbb{R}^n)_w$.

Theorems 5.1 and 5.2 are also highly regular forms of atomic decompositions allowing to deal with the duals of function spaces in a constructive manner.

APPENDIX: PROOFS OF RESULTS FROM §3.1

Proof of Theorem 3.2. In addition to $r \in \alpha_c(Y)$, let $a_0 \in \alpha_c(Y) \cap (0,q] \cap (0,a_{\min}]$. Without loss of generality, we may assume $M = 1$ and $h_0 \le \delta_0$ and choose $b > 1$ to be large enough later. For the sake of simplicity, for $i, l, k \in \mathbb{N}_0$, $k < l$ and $i \in I_n$, we designate $\chi_g = \chi_{\text{supp } g}$, $\chi_l = \chi_{Q_{\varkappa_1 b^{-l}(z_l)}}$ with

$$\varkappa_1 = \max_i((\delta_0^{(\gamma_a)_i}|\gamma + m_i| + 1)^{1/(\gamma_a)_i}, (1+\delta_0^{(\gamma_a)_i})^{1/(\gamma_a)_i}) > 1 \text{ and } \chi_{k,l,i} = \chi_{F_{k,l,i}} \text{ with}$$

$$F_{k,l,i} = z_l + (b^{-l(\gamma_a)_1}, \ldots, b^{-l(\gamma_a)_{i-1}}, (\varkappa_1 b^{-k})^{(\gamma_a)_i}, b^{-l(\gamma_a)_{i+1}}, \ldots, b^{-l(\gamma_a)_n})Q_1(0).$$

Let also $\chi(x) = \chi_{Q_M(z_0)}$ and $\tilde{\chi}(x) = \chi_{Q_{\varkappa_1 M}(z_0)}$. It is more convenient to use the discrete (by t) variants of quasi–semi–norms under consideration. Moreover, thanks to the compatibility of the spaces under consideration with their underlying ideal spaces, we can treat only the range $t \in (0, \delta_0]$. It is sufficient to establish the boundedness of Λ and Λ^{-1} with $\Lambda \in \{\Lambda_L, \Lambda_B\}$ in the mappings

$$ze_{Y,q,a}^s \xrightarrow{\Lambda_L^{-1}} l_q \xrightarrow{\Lambda_L} ZE_{Y,q,a}^s, \tilde{z}e_{Y,q,a}^{s,A} \xrightarrow{\Lambda_L^{-1}} l_q \xrightarrow{\Lambda_L} \widetilde{ZE}_{Y,q,a}^{s,A},$$ where $E \in \{B, L\}$, and the inclusions

$$\Lambda(c_0) \subset X \in \left\{ \overset{\circ}{E}_{p,\infty}^s(G), \overset{\circ}{E}_{p,\infty,a}^s(G), \overset{\circ}{\tilde{E}}_{p,\infty,a}^{s,A}(G) \right\}.$$

To see how the latter task follows from the former one, let us note that

$$\lim_{k \to 0} \|(0, \ldots, 0, \lambda_k, \ldots) | l_\infty \| = 0 \text{ for } \lambda \in c_0 \tag{1}$$

and $\Lambda((\lambda_1, \ldots, \lambda_{k-1}, 0, \ldots)) \in C_0^\infty \subset X$.

We shall extensively use the inequalities

$$l_q \subset l_1 \text{ and } \|\lambda - \mu|l_q\|^q \geq |\|\lambda|l_q\|^q - \|\mu|l_q\|^q| \text{ for } q \in (0, 1] \text{ and } \lambda, \mu \in l_q. \tag{2}$$

Starting with a) and assuming that $\lambda = \{\lambda_i\} \in l_q$, one has

$$\|\Lambda_B(\lambda)|Y(G)\| \leq \left(\sum_{i=0}^\infty |\lambda_i|^r \|\phi_i|Y(G)\|^r \right)^{1/r} \leq C\|\lambda|l_\infty\| \|\phi|L_\infty\| \left(\sum_{i=0}^\infty b^{-ris} \right)^{1/r} \leq$$
$$\leq C\|\lambda|l_q\| \tag{3}$$

for $s > 0$ thanks to the r–convexity and L_p–similarity of $Y(G)$. For $Y = L_\infty$, one has

$$\|\Lambda_B(\lambda)|L_\infty(G)\| \leq \|\lambda|l_\infty\| \|\phi|L_\infty(G)\| \leq C\|\lambda|l_q\|. \tag{4}$$

Concerning the spaces defined in terms of differences in a), the Lagrange theorem, the Minkowski inequality, the a_0–convexity of $Y(G, L_{a_i})$ and the L_p–similarity of $Y(G)$ imply, for $i \in I_n$,

$$b^{a_0 ks} \|\delta_{i,a_i}^{m_i}(\delta_0 b^{-k}, \cdot, \Lambda_B(\lambda), G_{\varkappa \delta_0 b^{-k}})|Y(G)\|^{a_0} \leq$$
$$\leq \sum_{l=0}^k \left(b^{ks}|\lambda_l| \|\delta_{i,a_i}^{m_i}(\delta_0 b^{-k}, \cdot, \phi_l, G_{\varkappa \delta_0 b^{-k}})|Y(G)\| \right)^{a_0} +$$
$$+ \sum_{l=k+1}^\infty \left(b^{ks}|\lambda_l| \|\delta_{i,a_i}^{m_i}(\delta_0 b^{-k}, \cdot, \phi_l, G_{\varkappa \delta_0 b^{-k}})|Y(G)\| \right)^{a_0} \leq$$

$$\leq C \sum_{l=0}^{k} \left(b^{-k(m_i(\gamma_a)_i - s)} \| D_i^{m_i} \phi_l | L_\infty(G) \| \| \lambda_l | \chi_l | Y(G) \| \right)^{a_0} +$$

$$+ C \sum_{l=k+1}^{\infty} \left(b^{ks + (k-l)(\gamma_a)_i / a_i} \| \phi_l | L_\infty(G) \| \| \lambda_l | \chi_{k,l,i} | Y(G) \| \right)^{a_0} \leq$$

$$\leq C \| \phi | W_\infty^m(Q_1(0)) \|^{a_0} \left(\sum_{l=0}^{k} \left(b^{(l-k)(m_i(\gamma_a)_i - s)} |\lambda_l| \right)^{a_0} + \right.$$

$$\left. + \sum_{l=k+1}^{\infty} \left(b^{(k-l)(s + (\gamma_a)_i(1/a_i - 1/p_i))} |\lambda_l| \right)^{a_0} \right), \quad (5)$$

where, for $k < l$, we have used the inequality

$$\delta_{i,a_i}^{m_i} (\delta_0 b^{-k}, x, \phi_l, G_{xb^{-k}}) \leq C 2^{m_i} b^{(k-l)(\gamma_a)_i / a_i} \| \phi_l | L_\infty \| \chi_{k,l,i}(x). \quad (6)$$

Applying now the Hardy inequality, we obtain, thanks to the estimate $\min_i (m_i(\gamma_a)_i - s) > 0$, that

$$\| \Lambda_B(\lambda) | b_{Y,q,a}^s(G) \| \leq C \| \phi | W_\infty^m(Q_1(0)) \| \cdot \| \lambda | l_q \|, \text{ i.e., } \Lambda_B \in \mathcal{L}(l_q, b_{Y,q,a}^s(G)). \quad (7)$$

The boundedness (and existence) of Λ_B^{-1} from $z b_{Y,q,a}^s(G)$ into l_q follows from the estimate

$$c(\phi) |\lambda_k| \leq \| \delta_{i,a_i}^{m_i}(b^{-k}\delta_0, \cdot, \Lambda_B(\lambda), Q_{b^{-k}(z_k)}) | Y(G) \| b^{ks} \leq$$
$$\leq \| \delta_{i,a_i}^{m_i}(b^{-k}\delta_0, \cdot, \Lambda_B(\lambda), G_{cb^{-k}}) | Y(G) \| b^{ks} \quad (8)$$

relying on the L_p–similarity of Y and the function ϕ not being a polynomial locally and followed by

$$c(\phi) \| \lambda | l_q \| \leq \| \Lambda_B(\lambda) | b_{Y,q,a}^s(G) \|. \quad (9)$$

To deal with the space $\widetilde{b}_{Y,q,a}^{s,A}(G)$, let us choose $d > s$ with $A_d^* \subset A$ and observe that the Hölder inequality imply the estimate

$$\mathcal{D}_a(t, w, g, G, A) \leq C t^d \| g | \widetilde{b}_{\infty,\infty,\infty}^{d,A_d^*}(G) \| \text{ for every } g \text{ and open } G \supset Q_t(w). \quad (10)$$

Similarly to the case of $b_{Y,q,a}^s(G)$, the estimate (10), the Minkowski inequality, the

a_0–convexity of $Y(G, L_a)$ and the L_p–similarity of $Y(G)$ imply

$$b^{a_0ks}\|\mathcal{D}_a(\delta_0 b^{-k}, \cdot, \Lambda_B(\lambda), G, A)|Y(G)\|^{a_0} \leq$$

$$\leq \sum_{l=0}^{k}\left(b^{ks}|\lambda_l|\|\mathcal{D}_a(\delta_0 b^{-k}, \cdot, \phi_l, G, A)|Y(G)\|\right)^{a_0} +$$

$$+ \sum_{l=k+1}^{\infty}\left(b^{ks}|\lambda_l|\|\mathcal{D}_a(\delta_0 b^{-k}, \cdot, \phi_l, G, A)|Y(G)\|\right)^{a_0} \leq$$

$$\leq C\sum_{l=0}^{k}\left(b^{-k(d-s)}\|\phi_l|\widetilde{b}^{d,A_d^*}_{\infty,\infty,\infty}(Q_{\varkappa_1 b^{-k}}(z_k))\|\,|\lambda_l|\|\chi_l|Y(G)\|\right)^{a_0} +$$

$$+ C\sum_{l=k+1}^{\infty}\left(b^{ks+(k-l)(\gamma_a,1/a)}\|\phi_l|L_\infty(G)\|\,|\lambda_l|\|\chi_k|Y(G)\|\right)^{a_0} \leq$$

$$\leq C\|\phi|\widetilde{B}^{d,A_d^*}_{\infty,\infty,\infty}(Q_{\varkappa_1}(0))\|^{a_0}\left(\sum_{l=0}^{k}\left(b^{(l-k)(d-s)}|\lambda_l|\right)^{a_0} +\right.$$

$$\left.+ \sum_{l=k+1}^{\infty}\left(b^{(k-l)(s+(\gamma_a,1/a-1/p))}|\lambda_l|\right)^{a_0}\right), \quad (11)$$

where, for $k < l$, we have used the inequality

$$\mathcal{D}_a(\delta_0 b^{-k}, x, \phi_l, G, A) \leq C\|\phi_l|L_\infty\|b^{(k-l)(\gamma_a,1/a)}. \quad (12)$$

Applying now the Hardy inequality, we obtain, because of the inequalities $d > s > 0$, that

$$\|\Lambda_B(\lambda)|\widetilde{b}^{s,A}_{Y,q,a}(G)\| \leq C\|\phi|B^{d,A_d^*}_{\infty,\infty,\infty}(Q_{\varkappa_1}(0))\| \cdot \|\lambda|l_q\|, \text{ i.e., } \Lambda_B \in \mathcal{L}(l_q, \widetilde{b}^{s,A}_{Y,q,a}(G)). \quad (13)$$

The proof of the inclusion $\Lambda^{-1} \in \mathcal{L}(\widetilde{b}^{s,A}_{Y,q,a}(G), l_q)$ is completely analogous to $(8, 9)$. Namely, the construction of Λ_B provides the estimate

$$c(\phi)|\lambda_k| \leq \|\mathcal{D}_a(b^{-k}\delta_0, \cdot, \lambda_k\phi_k, Q_{b^{-k}}(z_k))|Y(G)\|b^{ks} \leq$$
$$\leq \|\mathcal{D}_a(b^{-k}\delta_0, \cdot, \Lambda_B(\lambda), G_{cb^{-k}}, A)|Y(G)\|b^{ks} \quad (14)$$

relying on the L_p–similarity of Y and the function ϕ not being a polynomial locally and implying

$$c(\phi)\|\lambda|l_q\| \leq \|\Lambda_B(\lambda)|\widetilde{b}^{s,A}_{Y,q,a}(G)\|. \quad (15)$$

Let us consider Part b). Since Y is an ideal space, one has

$$\|\Lambda_L(\lambda)|Y(G)\| \leq \|\lambda|l_\infty\|\left(\sum_{k=0}^{\infty}b^{-ks}\right)\|\psi|L_\infty\| \cdot \|\chi_0|Y(G)\| \leq C\|\lambda|l_q\|. \quad (16)$$

As before, with the aid of either the Lagrange theorem in the case of the spaces defined in terms of differences, or just (10) otherwise, and using the triangle inequality and the

a_0–convexity of L_{a_i}, we obtain, for $i \in I_n$ and $k \in \mathbb{N}_0$,

$$\left(b^{ks}\delta^{m_i}_{i,a_i}(b^{-k}\delta_0, x, \Lambda(\lambda), G_{\delta_0 \varkappa b^{-k}})\right)^{a_0} \leq \sum_{l=0}^{k}\left(b^{ks}|\lambda_l|\delta^{m_i}_{i,a_i}(b^{-k}\delta_0, x, \phi_l, G_{\delta_0 \varkappa b^{-k}})\right)^{a_0} +$$

$$+ \sum_{l=k}^{\infty}\left(b^{ks}|\lambda_l|\delta^{m_i}_{i,a_i}(b^{-k}\delta_0, x, \phi_l, G_{\delta_0 \varkappa b^{-k}})\right)^{a_0} \leq C\|\psi|W^m_\infty(Q_1(0))\|^{a_0}\tilde{\chi}(x)\times$$

$$\times \left(\sum_{l=0}^{k}\left(|\lambda_l|b^{(l-k)(m_i(\gamma_a)_i - s)}\right)^{a_0} + \sum_{l=k+1}^{\infty}\left(|\lambda_l|b^{(k-l)s}\right)^{a_0}\right), \text{ or } \quad (17)$$

$$\left(b^{ks}\mathcal{D}_a(b^{-k}\delta_0, x, \Lambda(\lambda), G, A)\right)^{a_0} \leq \sum_{l=0}^{k}\left(b^{ks}|\lambda_l|\mathcal{D}_a(b^{-k}\delta_0, x, \psi_l, A)\right)^{a_0} +$$

$$+ \sum_{l=k+1}^{\infty}\left(b^{ks}|\lambda_l|\mathcal{D}_a(b^{-k}\delta_0, x, \psi_l, G, A)\right)^{a_0} \leq C\|\psi|\widetilde{B}^{d,A^*_d}_{\infty,\infty,\infty}(Q_{\varkappa_1}(0))\|^{a_0}\tilde{\chi}(x)\times$$

$$\times \left(\sum_{l=0}^{k}\left(b^{(l-k)(d-s)}|\lambda_l|\right)^{a_0} + \sum_{l=k+1}^{\infty}\left(b^{(k-l)s}|\lambda_l|\right)^{a_0}\right), \quad (18)$$

where $d > s$ and $A^*_s = A^*_d$. Thus, in both cases, the lattice property of $Y(G)$ and the Hardy inequality provide the boundedness properties of Λ_L looked for.

To establish the openness properties of Λ_L, assume for the moment that we have found the basic function ψ spoken about in the following assertion.

Assertion. There is a function ψ and a constant $\xi > 0$ satisfying, for $i \in I_n$,

$$a) \quad \inf_{\substack{k \in \mathbb{N}_0 \\ x \in Q_1(z_0)}} \delta^{m_{i_0}}_{i_0,a_{i_0}}(b^{-k}\delta_0, x, b^{ks}\psi_k, G_{\varkappa \delta_0 b^{-k}})_\gamma \geq \xi;$$

$$b) \quad \inf_{\substack{k \in \mathbb{N}_0 \\ x \in Q_1(z_0)}} \mathcal{D}_a(b^{-k}\delta_0, x, b^{ks}\psi_k, G_{\varkappa \delta_0 b^{-k}}, A) \geq \xi.$$

We use (2), the definition of Λ_L in §3 and the considerations as above to derive the key pointwise estimates

$$\left(b^{ks}\delta^{m_{i_0}}_{i_0,a_{i_0}}(b^{-k}\delta_0, x, \Lambda_L(\lambda), G_{\varkappa \delta_0 b^{-k}})_\gamma\right)^{a_0}\chi(x) \geq$$

$$\geq \left|\left(|\lambda_k|b^{sk}\delta^{m_{i_0}}_{i_0,a_{i_0}}(b^{-k}\delta_0, x, \psi_k, G_{\varkappa \delta_0 b^{-k}})_\gamma\right)^{a_0} - \beta_k(x)\right|\chi(x) \text{ for } k \in \mathbb{N}_0, \text{ where } \quad (19)$$

$$\beta_k(x) = \left(b^{sk}\delta^{m_{i_0}}_{i_0,a_{i_0}}(b^{-k}\delta_0, x, \Lambda_L(\lambda) - \lambda_k\psi_k, G_{\varkappa \delta_0 b^{-k}})_\gamma\right)^{a_0} \leq$$

$$\leq \sum_{l=0}^{k-1}\left(b^{ks}|\lambda_l|\delta^{m_{i_0}}_{i_0,a_{i_0}}(b^{-k}\delta_0, x, \phi_l, G_{\delta_0 \varkappa b^{-k}})\right)^{a_0} +$$

$$+ \sum_{l=k+1}^{\infty} \left(b^{ks} |\lambda_l| \delta_{i_0,a_{i_0}}^{m_{i_0}} (b^{-k}\delta_0, x, \phi_l, G_{\delta_0 \varkappa b^{-k}}) \right)^{a_0} \leq$$

$$\leq C \tilde{\chi}(x) \left(\sum_{l=0}^{k-1} \left(\|\psi|w_\infty^m(Q_1(0))\| |\lambda_l| b^{(l-k)(m_{i_0}(\gamma_a)_{i_0}-s)} \right)^{a_0} + \right.$$

$$\left. + \sum_{l=k+1}^{\infty} \left(\|\psi|L_\infty(Q_1(0))\| 2^{m_{i_0}} |\lambda_l| b^{(k-l)s} \right)^{a_0} \right). \quad (20)$$

In a very similar way, we also deal with the estimates involving the $\mathcal{D}$–functional:

$$\left(b^{sk} \mathcal{D}_a(b^{-k}\delta_0, x, \Lambda_L(\lambda), G, A) \right)^{a_0} \chi(x) \geq$$

$$\geq \left| \left(|\lambda_k| b^{sk} \mathcal{D}_a(b^{-k}\delta_0, x, \psi_k, G, A) \right)^{a_0} - \alpha_k(x) \right| \chi(x), \text{ where} \quad (21)$$

$$\alpha_k(x) = \left(b^{sk} \mathcal{D}_a(b^{-k}\delta_0, x, \Lambda_L(\lambda) - \lambda_k \psi_k, G, A) \right)^{a_0} \leq$$

$$\leq \sum_{l=0}^{k-1} \left(b^{ks} |\lambda_l| \mathcal{D}_a(b^{-k}\delta_0, x, \psi_l, G, A) \right)^{a_0} + \sum_{l=k+1}^{\infty} \left(b^{ks} |\lambda_l| \mathcal{D}_a(b^{-k}\delta_0, x, \psi_l, G, A) \right)^{a_0} \leq$$

$$\leq C \|\psi|\widetilde{B}_{\infty,\infty,\infty}^{d,A_d^*}(Q_{\varkappa_1}(0))\|^{a_0} \tilde{\chi}(x) \left(\sum_{l=0}^{k-1} \left(b^{(l-k)(d-s)} |\lambda_l| \right)^{a_0} + \right.$$

$$\left. + \sum_{l=k+1}^{\infty} \left(b^{(k-l)s} |\lambda_l| \right)^{a_0} \right), \text{ where } d > s \text{ and } A_d^* = A_s^*. \quad (22)$$

In both cases, the Hardy inequality implies

$$\left\| \{\beta_k(x)\}_{k\in\mathbb{N}_0} | l_{q/a_0} \right\| \leq C_1 b^{-k a_0 \epsilon_1} \|\psi|W_\infty^m(Q_1(0))\|^{a_0} \|\lambda|l_q\|^{a_0}$$
$$\text{for } \epsilon_1 = \min(s, m_{i_0}(\gamma_a)_{i_0} - s); \quad (23)$$

$$\left\| \{\alpha_k(x)\}_{k\in\mathbb{N}_0} | l_{q/a_0} \right\| \leq C_2 b^{-k a_0 \epsilon_2} \|\psi|\widetilde{B}_{\infty,\infty,\infty}^{d,A_d^*}(Q_{\varkappa_1}(0))\|^{a_0} \|\lambda|l_q\|^{a_0}$$
$$\text{for } \epsilon_2 = \min(s, d-s). \quad (24)$$

Whence, using (2) another time, Assertion and (19, 20, 23, 24), we see that

$$\left\| \left\{ b^{sk} \mathcal{D}_a(b^{-k}\delta_0, x, \Lambda_L(\lambda), G_{\varkappa \delta_0 b^{-k}}, A) \right\}_k \middle| l_q \right\| \geq$$
$$\geq \chi(x) \|\lambda|l_q\| \left(\xi^{a_0} - \left(C_1 b^{-s\epsilon_1} \|\psi|W_\infty^m(Q_1(0))\| \right)^{a_0} \right)^{1/a_0}; \quad (25)$$

$$\left\| \left\{ b^{sk} \mathcal{D}_a(b^{-k}\delta_0, x, \Lambda_L(\lambda), G_{\varkappa \delta_0 b^{-k}}, A) \right\}_k \middle| l_q \right\| \geq$$
$$\geq \chi(x) \|\lambda|l_q\| \left(\xi^{a_0} - \left(C_2 b^{-s\epsilon_2} \|\psi|\widetilde{B}_{\infty,\infty,\infty}^{d,A_d^*}(Q_{\varkappa_1}(0))\| \right)^{a_0} \right)^{1/a_0}. \quad (26)$$

These estimates imply the openness of Λ_L looked for if we choose b to be large enough. This finishes the proof of the theorem. $\square$

Proof of Assertion. Let us choose a function $\phi \in C_0^\infty$ with supp $\phi = [-1, 1]$ and its m_{i_0}th derivative $\phi^{(M_{i_0})}$ possessing isolated zeros only and, therefore, not being a polynomial of the degree $m_{i_0} - 1$ on any sufficiently short subinterval of its support. Thus, we see that, for $\psi(x) = \prod_{j=1}^n \phi(x_j)$, the statement of the assertion is true with $\xi = 0$ thanks to the Lagrange theorem in $a)$, and because an arbitrary cube $Q_{\delta_0}(x)$ contains some subcube $Q_{\tau\delta_0}(y)$ with $\tau \leq 2^{-1/(\gamma_a)_{\min}}$ on which ψ is not in $\mathcal{P}_A$.

To finish the proof, one can use the fact that $Q_{b-k\delta_0}(x)$, $x \in Q_1(z_0)$ intersects no more than N parallelepipeds F_i^k for some constant $N \in \mathbb{N}$ to reduce Assertion to the following extremal problems.

In the case of Part $a)$, show the existence of the minimum of the functional

$$H_1(g) = \delta_{i_0, a_{i_0}}^{m_{i_0}}(\delta_0, 0, g, \mathbb{R}^n)_\gamma$$

on the set of functions $g(x) = \sum_{i=1}^N \psi((x - O_i)/l_i)$, where $l_i = (l_{i1}, \ldots, l_{in})$ is the vector of the semi–lengths of the ribs of a set of parallelepipeds $\{F_i\}_{i=1}^N$ covering $Q_{\delta_0}(0)$ without intersections, whose centres $\{O_i\}_{i=1}^N$ are in $Q_{d_0}(0)$ for some large $d_0 > 0$, and whose ribs are satisfying $c_0 l_i \in [1, \lambda_0]^n$ for some $c_0 > 0$.

In the case of Part $b)$, show the existence of the minimum of the functional

$$H_2(f) = \mathcal{D}_a(\delta_0, 0, f, \mathbb{R}^n, \emptyset)$$

on the set of functions $f = g - \pi$, where g is as above, and $\pi \in \mathcal{P}_A$. Note that $\|\pi|L_\infty(Q_1(0))\| \leq C \sup_g H_2(g) < \infty$.

Both $H_1(g)$ and $H_2(f)$ are continuous functions of a finite number of parameters defined on a compact set. Therefore, the minimums of both functionals are positive. $\square$

Proof of Theorem 3.5. First of all, let us note that the cases involving the space c_0 follow from their counterparts involving l_∞ as it is done in the proof of Theorem 3.2.

The notations undefined in this proof coincide with those used in the proof of Theorem 3.2. Given $m \in \mathbb{N}$ and $\mathbb{N}_{0,m}^n = \mathbb{N}_0^n \setminus [0, m)^n$, let us choose a_0 as in the proof of Theorem 3.2 and a net of cubes $\{Q_M(z_j)\}_{j \in \mathbb{N}_{0,m}^n}$ with $z_j = z_0 + 2jM^{\gamma_a}$ satisfying $\cup_j Q_M(z_j) \subset Q_M(0) \subset G$. Assume also that $\tilde{\Lambda}_B$ and $\tilde{\Lambda}_L$ are the operators Λ_B and Λ_L from the construction of Theorem 3.2. In this respect, let $\chi_j := \chi_{Q_M(z_j)}$ and $\tilde{\chi}_j := \chi_{Q_{\varkappa_1 M}(z_j)}$ for $j \in \mathbb{N}_{0,m}^n$. In addition, let

$$\lambda = \{\lambda_{j,k}\}_{j \in \mathbb{N}_{0,m}^n}^{k \in \mathbb{N}_0} = \sum_{j \in \mathbb{N}_{0,m}^n}^{k \in \mathbb{N}_0} e_{j,k} \lambda_{j,k}, \quad \lambda_k = \{\lambda_{j,k}\}_{j \in \mathbb{N}_{0,m}^n} = \sum_{j \in \mathbb{N}_{0,m}^n} e_{j,k} \lambda_{j,k},$$

$$\lambda_j = \{\lambda_{j,k}\}_{k \in \mathbb{N}_0} = \sum_{k \in \mathbb{N}_0} e_{j,k} \lambda_{j,k}, \quad (1)$$

where $\{e_{j,k}\}$ is the natural coordinate basic sequence in the corresponding sequence space. Now we take

$$\Lambda_B(\lambda) = \sum_j \Lambda_B^j(\lambda_j), \text{ where } \Lambda_B^j := \tau_{z_j} \circ \tilde{\Lambda}_B;$$

$$\Lambda_L(\lambda) = \sum_j \Lambda_L^j(\lambda_j), \text{ where } \Lambda_L^j := \tau_{z_j} \circ \tilde{\Lambda}_L. \qquad (2)$$

The motivations used in the formulas (3) and (9) from the proofs of Theorems 3.2 and 3.3 correspondingly provide

$$\|\Lambda_B(\lambda)|Y(G)\| \le \left(\sum_{k=0}^{\infty} \|\Lambda_B(\lambda_k)|L_p(G)\|^{p_{\min}} \right)^{1/p_{\min}} \le$$

$$\le C\|\lambda|l_\infty(l_p^m)\| \cdot \|\phi|L_\infty\| \left(\sum_{i=0}^{\infty} \left(Mb^{-k} \right)^{p_{\min}s} \right)^{1/p_{\min}} \le CM^s \|\lambda|l_q(l_p^m)\| \qquad (3)$$

Combining the considerations used to derive the estimates $(5, 11)$ from the proof of Theorem 3.2 and $(3, 4)$ from the proof of Theorem 3.3, we obtain, for $i \in I_n$,

$$b^{a_0 k s} \|\delta_{i,a_i}^{m_i}(\delta_0 b^{-k}, \cdot, \Lambda_B(\lambda), G_{\varkappa \delta_0 b^{-k}})|L_p(G)\|^{a_0} \le C \|\phi|W_\infty^m(Q_1(0))\|^{a_0} \times$$

$$\times \left(\sum_{l=0}^{k} \left(b^{(l-k)(m_i(\gamma_a)_i - s)} \|\lambda_l|l_p^m\| \right)^{a_0} + \sum_{l=k+1}^{\infty} \left(b^{(k-l)(s+(\gamma_a)_i(1/a_i - 1/p_i))} \|\lambda_l|l_p^m\| \right)^{a_0} \right), \qquad (4)$$

$$b^{a_0 k s} \|\mathcal{D}_a(\delta_0 b^{-k}, \cdot, \Lambda_B(\lambda), G, A)|L_p(G)\|^{a_0} \le C\|\phi|\widetilde{B}_{\infty,\infty,\infty}^{d, A_d^*}(Q_{\varkappa_1}(0))\|^{a_0} \times$$

$$\times \left(\sum_{l=0}^{k} \left(b^{(l-k)(d-s)} \|\lambda_l|l_p^m\| \right)^{a_0} + \sum_{l=k+1}^{\infty} \left(b^{(k-l)(s+(\gamma_a, 1/a-1/p))} \|\lambda_l|l_p^m\| \right)^{a_0} \right). \qquad (5)$$

Now having in mind the motivation of the estimate (7) from the proof of Theorem 3.2 and utilizing the Hardy inequality, we obtain the boundedness of the operator Λ_B in a) and b).

The motivations of $(8, 14)$ and (5) from the proofs of Theorems 3.2 and 3.3 respectively provide us with the estimates

$$c(\phi)\|\lambda_k|l_p^m\| \le \|\delta_{i,a_i}^{m_i}(b^{-k}\delta_0, \cdot, \Lambda_B(\lambda), G_{cb^{-k}})|L_p(G)\| b^{ks}, \qquad (6)$$

$$c(\phi)\|\lambda_k|l_p^m\| \le \|\mathcal{D}_a(b^{-k}\delta_0, \cdot, \Lambda_B(\lambda), G_{cb^{-k}}, A)|L_p(G)\| b^{ks} \qquad (7)$$

implying the openness of the operator Λ_B in a) and b).

The consideration of c) and d) is simpler because we can use essentially the same pointwise estimates as those obtained in the proof of Theorem 3.2. Namely, we can immediately write the counterparts of the formulas $(17, 18)$ and $(25, 26)$ as follows: for $j \in \mathbb{N}_{0,m}^n$,

$$\left(b^{ks} \delta_{i,a_i}^{m_i}(b^{-k}\delta_0, x, \Lambda_L^j(\lambda_j), G_{\delta_0 \varkappa b^{-k}}) \right)^{a_0} \le C\|\psi|W_\infty^m(Q_1(0))\|^{a_0} \tilde{\chi}_j(x) \times$$

$$\times \left(\sum_{l=0}^{k} \left(|\lambda_{j,l}| b^{(l-k)(m_i(\gamma_a)_i - s)} \right)^{a_0} + \sum_{l=k+1}^{\infty} \left(|\lambda_{j,l}| b^{(k-l)s} \right)^{a_0} \right); \qquad (8)$$

$$\left(b^{ks}\mathcal{D}_a(b^{-k}\delta_0, x, \Lambda_L^j(\lambda_j), G, A)\right)^{a_0} \leq C\|\psi|\widetilde{B}_{\infty,\infty,\infty}^{d,A_d^*}(Q_{\varkappa_1}(0))\|^{a_0}\tilde{\chi}_j(x) \times$$

$$\times \left(\sum_{l=0}^{k}\left(b^{(l-k)(d-s)}|\lambda_{j,l}|\right)^{a_0} + \sum_{l=k+1}^{\infty}\left(b^{(k-l)s}|\lambda_{j,l}|\right)^{a_0}\right); \quad (9)$$

$$\|\{b^{sk}\mathcal{D}_a(b^{-k}\delta_0, x, \Lambda_L^j(\lambda_j), G_{\varkappa\delta_0 b^{-k}}, A)\}_k|l_q\| \geq$$
$$\geq \sum_{j\in\mathbb{N}_{0,m}^n}\chi_j(x)\|\lambda_j|l_q\|\left(\xi^{a_0} - (C_1 b^{-s\epsilon_1}\|\psi|W_\infty^m(Q_1(0))\|)^{a_0}\right)^{1/a_0}; \quad (10)$$

$$\|\{b^{sk}\mathcal{D}_a(b^{-k}\delta_0, x, \Lambda_L^j(\lambda_j), G_{\varkappa\delta_0 b^{-k}}, A)\}_k|l_q\| \geq$$
$$\geq \sum_{j\in\mathbb{N}_{0,m}^n}\chi_j(x)\|\lambda_j|l_q\|\left(\xi^{a_0} - (C_2 b^{-s\epsilon_2}\|\psi|\widetilde{B}_{\infty,\infty,\infty}^{d,A_d^*}(Q_{\varkappa_1}(0))\|)^{a_0}\right)^{1/a_0}. \quad (11)$$

The assertions of c) and d) follow from these estimates with the aid of the Hardy inequality, the localisation property of L_p (the supports of $\{\Lambda_L^j(\lambda_j)\}_{j\in\mathbb{N}_{0,m}^n}$ are disjoint), and the analog of the estimate (16) from the proof of Theorem 3.2:

$$\|\Lambda_L(\lambda)|Y(G)\| \leq \left(\sum_{k=0}^{\infty}\left(Mb^{-k}\right)^s\right)\|\psi|L_\infty(\mathbb{R}^n)\| \cdot \left\|\sum_{j\in\mathbb{N}_{0,m}^n}\|\lambda_j|l_\infty\|\chi_j\right|Y(G)\right\| \leq$$
$$\leq CM^s\|\lambda|l_p^m(l_q)\|. \quad (12)$$

It is enough to establish e) for G containing the positive quadrangle (cone) for a non–constant p. In this case we just set $m = \infty$ in the above construction. The proof of f) differs only in choosing the shifts $\{z_j\}$ as in the proof of Theorem 3.4, b). $\square$

Proof of Theorem 3.6. It is almost literal repetition of that of Theorem 3.5. One can take the same operators Λ_B and λ_L acting on the subspace of all λ with $\lambda_j = 0$ for $j \notin R \subset \mathbb{N}_{0,m}^n$, $|R| = m$, and substitute the spaces l_p^m and L_p and the usage of the proof of Theorem 3.3 with the spaces $l_{p,\theta}$ and $L_{p,\theta}$ and the usage of the proof of Theorem 3.4 respectively. Fact (7) from the proof of Theorem 3.4 plays here an essential role as well. The proof of e) differs only in choosing the shifts $\{z_j\}_{j\in\mathbb{N}_0}$ as in the proof of Theorem 3.4, b). $\square$

ACKNOWLEDGMENTS

I am grateful to Professor Oleg V. Besov for the most essential part of my knowledge related to Besov, Lizorkin–Triebel and other function spaces and appreciates his encouragement and support over the years.

At different times and under different circumstances, the work related to this chapter had been partially supported by the the Australian Research Council (project numbers DP0881037, DP120100097 and DP150102758), the Russian Fund for Basic Research (project numbers 08–01–00443, 11–01–00444-a and 14–01–00684) and the EIS Research

Grant held by James McCoy at the Institute for Mathematics and its Applications at the University of Wollongong.

It is also my pleasure to thank the President and Scientific and Technical Editor(s) of Nova Science Publishers, Inc. for the opportunity and related collaboration.

REFERENCES

[1] Ajiev, S. S. 2013. "Quantitative Hahn–Banach theorems and isometric extensions for wavelet and other Banach spaces." *Axioms* 2: 224–270. (Open access: doi:10.3390/axioms2020224.)

[2] Ajiev, S. S. 2019. "Asymmetric convexity and smoothness: quantitative Lyapunov theorem and Lozanovskii factorisation, Walsh chaos and convex duality." In *Understanding Banach Spaces*, edited by Daniel González. Long Island: Nova Science Publishers, Inc.

[3] Ajiev, S. S. "Nonlocal geometry of function and other Banach spaces: constants, moduli and compact and Kato–Pełczyński operators." *Manuscript* 1-51.

[4] Ajiev, S. S. 2000. *On noncomplementable subspaces and noncompact embeddings of certain function spaces.* Moscow: VINITI. (No. 1600–BOO. P. 1–36).

[5] Ajiev, S. S. 2001. "Phillips-type theorems for Nikol'skii and certain other spaces." *Proc. Steklov Math. Inst.* 232(1): 27–38.

[6] Ajiev, S. S. 2009. "On Chebyshev centres, retractions, metric projections and homogeneous inverses for Besov, Lizorkin–Triebel Sobolev and other Banach spaces." *East J. Approx.* 15(3): 375-428.

[7] Ajiev, S. S. 2010. "On concentration, deviation and Dvoretzky's theorem for Besov, Lizorkin–Triebel and other spaces." *Compl. Var. Ellipt. Eq.* (Special issue dedicated to Gérard Bourdaud.) 55(8-10): 693–726.

[8] Ajiev, S. S. 2014. "Hölder analysis and geometry on Banach spaces: homogeneous homeomorphisms and commutative group structures, approximation and Tzar'kov's phenomenon. Part II." *Eurasian Math. J.* 5(2): 6–50.

[9] Ajiev, S. S. 2014. "Hölder analysis and geometry on Banach spaces: homogeneous homeomorphisms and commutative group structures, approximation and Tzar'kov's phenomenon. Part I." *Eurasian Math. J.* 5(1): 7–60.

[10] Ajiev, S. S. 2019. "Interpolation and bounded extension of Hölder–Lipschitz mappings between function, noncommutative and other Banach spaces." In *Understanding Banach Spaces*, edited by Daniel González. Long Island: Nova Science Publishers, Inc.

[11] Ajiev, S. S. "Amir–Franchetti problems, quantitative K–convexity test and Jacobi convexity hierarchy." *Manuscript* 1–35.

[12] Kadets, Mikhail I. and Kadets, Vladimir Mikhailovich. 1988. *Rearrangements of Series in Banach Spaces.* Tartu: Tartu Gos. University. English transl. (In *Translation of Math. Monographs, V. 86.*) Providence, R.I.: Amer. Math. Soc. 1991.

[13] Tomczak-Jaegermann, Nicole. 1989. *Banach–Mazur Distances and Finite dimensional Operator Ideals.* Harlow: Longman Scientific & Technical. P. 1–395+i–xii.

[14] Kadets, Mikhail I. and Kadets, Vladimir Mikhailovich. 1997. *Series in Banach Spaces: Conditional and Unconditional Convergence.* (In *Operator Theory, V. 94.*) Basel–Boston–Berlin: Birkhäuser Verlag. P. 1–156+i–viii.

[15] Wojtaszczyk, Przemyslav. 1991. *Banach Spaces For Analysts.* Cambridge–Sydney: Cambr. Univ. Press.

[16] Lim, T.–Ch. and Smarzewski, R. 1991. "On an L_p inequality." *Glasnik Mat.* 26(46): 79–81.

[17] Ajiev, S. S. "Optimal non–dual local non–stationary multiresolution analysis for anisotropic function spaces." *Manuscript* 1–51.

[18] Triebel, Hans. 2006. *Theory of Function Spaces III.* (In *Monographs in Mathematics, V. 100.*) Basel–Boston–Berlin: Birkhäuser Verlag. P. 1–426+i–xii.

[19] Sobolev, Sergey L'vovich. 1950. *Some applications of functional analysis in mathematical physics.* Leningrad: Izd. Leningr. Gos. Univ. 1950. P. 1–255. Third edition. Moscow: "Nauka". 1988. P. 1–334. English transl. by H. H. McFaden. (In *Translations of Mathematical Monographs, V. 90.*) Providence, Rhode Island: American Mathematical Society. 1991. P. 1–286+i–viii.

[20] Nikol'skii, Sergey Mikhailovich. 1969. *Approximation of function of several variables and embedding theorems.* Moscow: Nauka. Engl. transl. Berlin–Heidelberg–New York: Springer Verlag. 1975.

[21] Peetre, Jaak. 1969. "On the theory of $L_{p,\lambda}$ spaces." *J. Funct. Anal.* 4: 71–87.

[22] Il'in, Valentin Petrovich. 1972. "Function spaces $\mathcal{L}_{r,p,\theta}^{\lambda,a;b_s}(G)$." *Zap. Nauchn. Sem. LOMI. Acad. Sci. USSR* 23: 33–40.

[23] Besov, Oleg Vladimirovich, Il'in, Valentin Petrovich, and Nikol'skiy, Sergey Mikhailovich. 1996. *Integral representations of functions and embedding theorems.* Moscow: Nauka–Fizmatgiz. (1st Ed. Moscow: Nauka. 1975.)

[24] Triebel, Hans. 1995. *Interpolation Theory, Function Spaces, Differential Operators.* Amsterdam–Oxford.: North–Holland Publishing Company. P. 1–528.

[25] DeVore, Ronald A. and Sharpley, Robert C. 1984. "Maximal functions measuring smoothness." *Mem. Amer. Math. Soc.* 293.

[26] Burenkov, Viktor Ivanovich. 1998. *Sobolev Spaces on Domains.* (In *Teubner Texts in Mathematics, V. 137*), Stuttgart: B. G. Teubner Verlagsgesellschaft GmbH.

[27] Triebel, Hans. 1994. "A localization property for $B^s_{p,q}$ and $F^s_{p,q}$ spaces." *Studia Math.* 109(2): 183–195.

[28] Fichtenholtz, Grigoriy Mikhailovich and Kantorovich, Leonid Vital'evich. 1934. "Sur les opérations linéaires dans l'espace des fonctions bornées." *Stud. Math.* 5: 69–98.

[29] Lindenstrauss, Joram and Tzafriri, Lior. 1979. *Classical banach spaces. Volume II.* (In *Ergebnisse, v. 97.*) Berlin–Heidelberg–New York: Springer Verlag. P. 1–243+i–x.

[30] Ajiev, S. S. 2003. "On the boundedness of singular integral operators from certain classes I." *Proc. Steklov Math. Inst.* 243: 11–38.

[31] Aleksandrov, A. B. 1981. "Essays on nonlocally convex Hardy classses." In *Complex analysis and spectral theory: Seminar, Leningrad, 1979/80.*, 1–89. (*Lect. Notes Math. V. 864.*) Berlin–Heidelberg-New York: Springer–Verlag.

[32] Diestel, Joseph, Jarchow, Hans, and Tonge, Andrew. 1995. *Absolutely summing operators.* Cambridge: Cambridge University Press. P. 1–474+i–xv.

[33] Mastylo, M. 1992. "Type and cotype of some Banach spaces." *Internat. J. Math. Math. Sci.* 15(2): 235–240.

[34] Maleev, R. P. and Trojanski, S. L. 1971/2. "Moduli of convexity and of smoothness for the spaces L_{pq}." *Annuaire Univ. Sofia Fac. Math.* 66: 331–339.

[35] Ajiev, S. S. 1999. "Characterizations of $B^s_{p,q}(G), L^s_{p,q}(G), W^s_p(G)$ and other function spaces. Applications." *Proc. Steklov Math. Inst.* 227(4): 1–36.

[36] Milman, Vitaliy Davidovich. 1969. "Geometric theory of Banach spaces. II. Geometry of the unit ball." *Uspehi. Mat. Nauk* 26(6(162)): 73–149.

[37] Jones, F. 1964. "A class of singular integrals." *Amer. J. Math.* 80(2): 441–462.

[38] Levy, Mireille. 1980. "L'espace d'interpolation réel $(A_0, A_1)_{\theta,p}$ contient l_p." In *Initiation Seminar on Analysis, 19th year: 1979/80. Publ. Math. Univ. Pierre et Marie Curie* 41(3): 9.

[39] Mastylo, M. 1991. "Banach spaces via sublinear operators." *Math. Japonica* 36(1): 85–92.

[40] Creekmore, J. 1981. "Type and cotype in Lorentz $L_{p,q}$ spaces." *Nederl. Akad. Wetensch. Indag. Math.* 43(2): 145–152.

[41] Novikov, S. Ya. 1980. "Cotype and type of Lorentz function saces." *Mat. Zametki* 32(2): 213–221.

[42] Maurey, Bernhard and Pisier, Gilles. 1976. "Séries de variables aléatoires vectorielles indépendantes et géométrie des espaces de Banach." *Stud. Math.* 58: 45–90.

[43] Milman, Vitaliy Davidovich and Schechtman, Gideon. 1986. "Asymptotic theory of finite dimensional normed spaces" (with an appendix by M. Gromov). (In *Lect. Notes Math., V. 1200.*) Berlin–Heidelberg–New York: Springer Verlag. P. 1–156+i–viii.

[44] Ledoux, Michel and Talagrand, Michel. 1991. *Probability in Banach Spaces: Isoperimetry and Processes*. Berlin–Heidelberg–New York: Springer Verlag. P. 1–482+i–xii.

[45] Pietsch, Albrecht and Wenzel, Jörg. 1998. *Orthonormal systems and Banach space geometry*. (In '*Encyclopedia of Mathematics, V. 13*.) Cambridge–New York–Melbourne: Cambridge Univ. Press. P. 1–553+i–ix.

[46] Burkholder, Donald L. 1981. "A geometric characterization of Banach spaces in which martingale difference sequences are unconditional." *Ann. Prob.* 9: 997–1011.

[47] Burkholder, Donald L. 1983. "A geometric condition that implies the existence of certain singular integrals of Banach–space–valued functions." In *Conference on Harmonic Analysis in Honour of Antoni Zygmund, Chicago, 1981*. V. I, 270–286. Belmont: Wadsworth.

[48] Pisier, Gilles. 1978. "Some results on Banach spaces without unconditional structure." *Compositio Math.* 37: 3-19.

[49] Bourgain, Jean. 1983. "Some remarks on Banach spaces in which martingale difference sequences are unconditional." *Ark. Mat.* 22: 163–168.

[50] Lions, J.L. and Peetre, Jaak. 1964. "Sur une classed'espaces d'interpolation." *Inst. Hautes Etudes Sci. Publ. Mat.* 19: 5–68.

[51] Cobos, Fernando. 1986. "Some spaces in which martingale difference sequences are unconditional." *Bull. Polish Acad. Sci.* 34(11–12): 695–703.

[52] Huff, R. E. and Morris, P. 1975. "Dual spaces with the Krein–Milman property have the Radon–Nikodým property." *Proc. Amer. Math. Soc.* 49: 104–108.

[53] Huff, R. E. and Morris, P. 1976. "Geometric characterizations of the Radon–Nikodým property in Banach spaces." *Studia Math.* 56: 157–164.

[54] Aldous, D. J. 1979. "Unconditional bases and martingales in $L_p(F)$." *Math. Proc. Cambridge Phil. Soc.* 85: 117-123.

[55] Bessaga, C. and Pełczyński, Aleksandr. 1966. "On extreme points in separable conjugate spaces." *Israel J. Math.* 4: 262–264.

[56] Krivine, J. L. and Maurey, Bernhard. 1981. "Espaces de Banach stables." *Israel J. Math.* 39: 273–295.

[57] Guerre–Delabrière, Silvie. 1992. *Classical sequences in Banach spaces*. (In *Mon. Textb. Pure Appl. Math., V. 166*.) New York–Basel–Hong Kong: Marcel Dekker, Inc. P. 1–207+i–xv.

[58] Bergh, J. and Löfström, J. 1976. *Interpolation spaces. An introduction*. (In *Die Grundlehren der mathematischen Wissenschaften, B. 223*.) Berlin–Heidelberg–New York–Tokyo: Springer–Verlag.

[59] Bastero, J. and Raynaud, Y. 1989. "Quotients and interpolation spaces of stable Banach spaces." *Studia Math.* XCII: 223-239.

[60] James, Robert C. 1964. "Weak compactness and reflexivity." *Israel J. Math.* 2: 101–119.

[61] Fonf, V. P. 1986. "Semi–embeddings and G_δ–embeddings of Banach spaces." *Mat. Zametki* 39(4): 550–561.

[62] Besov, Oleg Vladimirovich. 1990. "On the Sobolev–Liouville and Lizorkin–Triebel spaces on a domain." *Tr. Matem. Inst. Steklova* 192:20–34.

In: Understanding Banach Spaces
Editor: Daniel González Sánchez

ISBN: 978-1-53616-745-0
© 2020 Nova Science Publishers, Inc.

Chapter 23

INTERPOLATION AND BOUNDED EXTENSION OF HÖLDER–LIPSCHITZ MAPPINGS BETWEEN FUNCTION, NON–COMMUTATIVE AND OTHER BANACH SPACES

Sergey S. Ajiev[*]
University of Technology, Sydney, Australia

Abstract

We expose the mostly precise relation between the geometric properties of pairs of abstract and concrete Banach and, occasionally, metric spaces and such isomorphic properties as the approximation, real interpolation and bounded extension of the Hölder–Lipschitz mappings between them, their subspaces and quotients in a quantitative manner, sharpening the uniform upper bounds for the extension operators even in the known setting of the pairs of Lebesgue spaces, particularly, by the factor $\sqrt{13/24}$ in relation to the conjecture formulated by Naor, Peres, Schramm and Sheffield in 2006. The pairs of spaces are drawn (independently) from ten parameterised groups of Banach spaces including auxiliary IG–classes, noncommutative L_p–spaces and various types of anisotropic Besov and Lizorkin–Triebel spaces of functions on open subsets with underlying mixed Lebesgue or Lorentz (quasi)norms, their duals and l_p–sums covering the majority of such spaces considered in analysis and PDE. The geometric properties, quantified in terms of asymmetric and abstract uniform convexity and smoothness, our generalised counterpart of Pisier's martingale inequality and the presence of certain (complemented) subspaces, lead to the identification of the Markov type and cotype sets with explicit estimates of the related constants, allowing an explicit quantitative and mostly sharp adaptation of K. Ball's extension scheme.

Keywords: Hölder–Lipschitz mappings, bounded extension, real interpolation, Besov, Lizorkin–Triebel, Sobolev, Lorentz and IG–spaces, Markov chains, Rademacher and Markov type and cotype, asymmetric uniform convexity and smoothness, wavelets, Schatten–von Neumann classes, noncommutative L_p, complemented subspaces

AMS Subject Classification: 54C20, 46E35, 46T20, 46B20, 60J10, 42C40, 42C40, 46E30, 46L52, 46B03, 52A40, 26D20, 60J20, 60J22, 46B09, 46B07

---

[*]Corresponding Author's E-mail: sergey.ajiev@uts.edu.au.

1. INTRODUCTION

The extendability of a function, initially defined on a subset of an Euclidean space, to the whole space with the preservation, or controlled increase, of a certain quantitative measure of its continuity is a traditional question of real/harmonic analysis utilised, for example, in the theory of function spaces and interpolation theory (see, for example, [1, 2, 3, 4]). In this classical setting of the mappings between the pairs of finite–dimensional spaces, the extendability is known to depend on the geometry of the initial domain set G (for example, on the boundary of an open $G \subset \mathbb{R}^n$) because all the spaces of the same dimension are isomorphic, thus, sharing the same isomorphic geometry. Moreover, the study of the mappings $\phi : \mathbb{R}^n \to \mathbb{R}^m$ is reduced to the study of their scalar–valued components $\phi_i : \mathbb{R}^n \to \mathbb{R}$.

While the study of the distortion–oscillation problem [5] shows the dependence of the behavior of a uniformly continuous $\psi : S_X \to \mathbb{R}$, where S_X is the unit sphere of an infinite–dimensional Banach space X, on the geometry of X, the quantitative aspect of the uniform continuity of ψ does not seem to be relevant in this infinite–dimensional setting: for example, V. D. Milman [6] noticed that every uniformly continuous ψ is oscillation stable if, and only if, every Lipschitz ψ is oscillation stable. On one hand, Tsar'kov's phenomenon [7, 8, 9, 10, 11] confirms that the mappings $\psi : S_X \to Y$ for another Banach Y discern the quantitative measures of the uniform continuity of ψ even when X and Y are Lebesgue spaces, thus, highlighting the importance of the quantitative study of the Hölder mappings ψ (i.e., explicitly quantitative uniform smoothness). On the other hand, it is shown in [10, 11] that the uniform classification of the infinite–dimensional spheres of Banach lattices established in [5] coincides with the Hölder one. Thus, seconding the importance of the quantitative study, this fact also rises the question on the role of the presence of the local unconditional structure in X and Y in booth its qualitative and quantitative aspects. To address these natural questions, we study the extendability of the Hölder mappings between the pairs of Banach spaces (X can be a metric space) drawn from ten groups of parameterised spaces including both very general IG–classes of Banach lattices and the noncommutative L_p and IG_+–spaces lacking the local unconditional structure (l.u.st.), as well as several types of anisotropic Sobolev, Besov and Lizorkin–Triebel spaces of functions on an open $G \subset \mathbb{R}^n$ with questionable l.u.st. (as shown in [10, 11], O. V. Besov's results [2] imply that that these spaces possess l.u.st. when G belongs to the class of the domains satisfying the flexible λ–horn condition including, in particular, non–extension domains).

For a Banach or metric space X and its subset D, a Banach space Y (that is assumed reflexive for simplicity) and $\alpha \in (0, 1]$, let $H^\alpha(D, Y)$ be the class of all α–Hölder mappings $\phi : D \to Y$, and let $\operatorname{Ext}_D : H^\alpha(D, Y) \longrightarrow H^\alpha(X, Y)$ be the linear extension operator with $\| \operatorname{Ext}_D | \mathcal{L}(H^\alpha(D, Y), H^\alpha(X, Y)) \| \leq d_\alpha$. If such Ext_D exists for every $D \subset X$, the pair (X, Y) is said to possess the (d_α, α)–extension property that is bounded if $d_\alpha \in [1, \infty)$ and isometric if $d_\alpha = 1$. The sets of all such exponents α are denoted by $S_b(X, Y)$ and $S_=(X, Y)$ respectively.

While the (isomorphic) geometries of X and Y represented by the summability parameters are known to influence the uniform approximation of the uniformly continuous mappings by the Hölder ones, that is Tsar'kov's phenomenon discovered by I. G. Tsar'kov [7, 8, 9] for the pairs of Lebesgue spaces and revealed in [10, 11] for the other pairs under consideration, it is shown in Section 4.3 that the real interpolation of Hölder mappings shares this dependence because of exactly the same reason providing our key sharpness tool in the study of the bounded extendability problem.

The study of the isometric extension property for the uniformly continuous, Lipschitz and Hölder mappings between the pairs of (infinite–dimensional) Hilbert and Lebesgue spaces was completely solved by Minty [12], Wells, Williams and Hayden [13, 14, 15], while the sets $S_=(X, Y)$ for the pairs (X, Y) of spaces independently drawn from the Γ–groups $\bigcup_{i=0^5} \Gamma_i$ (see Section 2.5) including the classes of auxiliary IG_+, noncommutative L_p and various types of anisotropic Sobolev, Besov and Lizorkin–Triebel spaces of functions on open subsets with underlying mixed Lebesgue norm and their duals (spaces of negative smoothness accommodating weak solutions to PDE) and l_p–sums were described in [16], where the connection with the quantitative Hahn–Banach theorems via the Jacobi classes of Banach spaces was also revealed.

K. Ball [17] had introduced the notions of Markov type of a metric space and Markov cotype of a Banach space and designed the second known scheme, after the Frechet extension [18, 19] (quantified in [10, 11, Section 10.2] in the setting of the Γ–groups), allowing to approach the bounded extendability problem for the Hölder–Lipschitz mappings between abstract Banach spaces (and from a metric space into a Banach space). While he has also treated certain pairs of Lebesgue spaces, the full solution in the setting of the pairs of Lebesgue spaces had required further insights made by A. Naor, Y. Peres, O. Schramm and S. Sheffield [20, 21]. A very good source of the information on related topics is [22]. Namely, K. Ball [17] applied his abstract result on the existence of bounded extensions of the Lipschitz mappings from a Markov type 2 space into a Markov cotype 2 space to the pairs of Lebesgue spaces (L_2, L_q) with $q \in [2, \infty)$. More precisely, he showed that the pair possesses the $(d_{p,q}, 1)$–extension property with the constant $d_{p,q} = 6(q-1)^{-1/2}$. But he had also explicitly and quantitatively related the uniform convexity and Markov cotype and shown the Markov type of L_p for $p \in (1, 2]$ to be p. A. Naor [20] mentioned that Ball's scheme works for the Hölder mappings ($\alpha \in (0, 1]$) and found out that $S_b(l_p, L_q) = p/\max(q, 2)$ for $p \in (1, 2]$ and $q \in (1, \infty)$, interpreting Hölder mappings as Lipschitz (see Section 4.2.3 for more details). A. Naor, Y. Peres, O. Schramm and S. Sheffield [21] showed that L_p has the Markov type 2 for $p \in (2, \infty)$ and estimated the Markov type constant using a representation of a Markov chain as a sum of a backward and a forward martingales. Thus, they had completed the identification of $S_b(L_p, L_q)$ for all pairs $p, q \in (1, \infty)$. Since the pair (L_p, L_q) with $2 \in [q, p]$ was shown to have the $(d_{p,q}, 1)$–extension property with $d_{p,q} = 24\sqrt{\frac{p-1}{q-1}}$, they had also conjectured that the constant 24 here can be reduced to 1. The following corollary proved in Section 4.4.2 (that is also a model case for the pairs (X, Y) under consideration) provides the estimate $d_{p,q} = 2\sqrt{78}\sqrt{\frac{p-1}{q-1}}$ not only for the pairs of commutative spaces, but also for the pairs of Schatten–von Neumann classes (and non–commutative L_p), or, even, mixed pairs with the same conditions on p and q. The constant 6 above comes from Ball's scheme meaning that the justification of the conjecture requires to improve (or sidestep) Ball's scheme itself as well. The succeeding corollary is the closest example of our results, including mixed norm and, more generally, IG, IG_+ and other Banach spaces.

Corollary 1.1. *For $\sigma \in \{q, p\} \subset (1, \infty)$, let $Y_\sigma \in \{l_\sigma, S_\sigma, L_\sigma, L_\sigma(\mathcal{M}, \tau_\sigma)\}$, where $(\mathcal{M}, \tau)$*

is a von Neumannn algebra with a normal semifinite faithful weight τ_σ. Assume also that E is a metric space, and X_σ is either a subspace or a quotient of Y_σ, or finitely represented in Y_σ for $\sigma \in \{p,q\}$. Then one has:

a) if $2 \in [q,p]$, the pair (X_p, X_q) has the $(d_1, 1)$–extension property with $d_1 = 2\sqrt{78}\sqrt{\frac{p-1}{q-1}}$;

b) $S_b(Y_p, Y_q) = (0, \min(p,2)/\max(q,2)] \subset S_b(X_p, X_q)$ and, if $2 \in [q,p]$, $S_b(X_p, X_q) = (0, 1]$;

c) $S_b(E, X_\sigma) \supset (0, 1/\max(\sigma, 2)]$;

d) $S_b(Y_1, Y_\sigma) = (0, 1/\max(\sigma, 2)]$ if $\dim(Y_1) = \dim(Y_\sigma) = \infty$.

The algorithm providing mostly precise exponent sets $S_b(X,Y)$ for the pairs of spaces under consideration and explicit values of d_α is described in Section 4.4 along with such numerical components as the Markov type and cotype exponent sets and constants and model examples. The comparison with the isometric extension sets $S_=(X,Y)$ reveals the presence of the *phase transition phenomenon*, discovered by A. Naor [20] for the pairs (X,Y) of Lebesgue spaces, in the setting of our pairs as well.

Apart from such abstract classes of Banach spaces as uniformly convex and smooth spaces, homogeneously and asymmetrically quantified [23, 24] in Section 3.1, our pairs (X,Y) of spaces under consideration (X can be a metric space) can also be arbitrarily and independently chosen from ten groups (Section 2.1) of parameterised classes of specific Banach spaces including not only noncommutative L_p, auxiliary IG–classes of ideal spaces and IG_+ (Sections 2.1 and 2.2), indicating no influence of the local unconditional structure, but also various classes of anisotropic Besov and Lizokin–Triebel spaces (Sections 2.3 and 2.5) of positive and negative (duals, weak PDE solutions) smoothness defined in terms of differences, local polynomial approximations, wavelet decompositions and systems of closed operators (for example, Fourier multipliers or holomorphic functional calculus) with underlying mixed Lebesgue or Lorentz (quasi) norms. The l_p–sums of the spaces under consideration (for example, the initial and boundary value spaces in PDE combined with the range space of the differential operator) can be treated in exactly the same fashion. While our function spaces are often set on an arbitrary open subset of an Euclidean space in order to take advantage of the existence of isomorphic copies of l_p–spaces [25], the abstract and universal nature of the asymmetric uniform convexity and smoothness presented in Section 3.1 and extensively used afterwards (constituting the base of our quasi–Euclidean approach [23, 26]) make the algorithm mentioned above equally applicable to the function spaces set on subsets of Riemannian manifolds and metric (homogeneous) spaces and function spaces of variable smoothness [27] (particularly, weighted ones) with the only possible complications or exceptions related to the matter of sharpness that is based on the finite representability of l_p–spaces and the λ–finite representability of $l_q(l_{p,\theta})$ and $l_{p,\theta}(l_q)$ treated in [25] and the existence of uniformly complemented and isomorphic copies of finite–dimensional l_p–spaces attended in Sections 2.1, 2.2 and 2.4. General information on function spaces is in [28, 29, 2, 30, 3].

Section 3 contains key tools for our quantitative and sharp adaptation of Ball's extension scheme in Sections 4.2 and 4.3: asymmetric uniform convexity and smoothness leading to abstract uniform convexity and smoothness and the Markov chain counterpart of the Pisier inequality for martingales in Lemma 3.1 and Theorems 4.2 and 4.6. The detailed description of the extension problem for Hölder–Lipschitz mappings is in the beginning of Section 4. Markov type and cotype exponents and constants are defined and studied in Sections 4.2 and 4.4.1, where our adaptation of Ball's extension scheme is presented.

The numbering of the equations is used sparingly. Since the majority of references inside every logical unit are to the formulas inside the unit, equations are numbered independently inside every proof of a corollary, lemma or a theorem, or definition (if there are any numbered formulas).

2. BASIC DEFINITIONS, SPACES AND COMPLEMENTED SUBSPACES

Let $\mathbb{N}$ be the set of the natural numbers; $\mathbb{N}_0 = \mathbb{N} \cup \{0\}$; $I_n = [1, n] \cap \mathbb{N}$ for $n \in \mathbb{N}$.

Let p' be the conjugate to $p \in [1, \infty]^n$, i.e., $1/p'_i + 1/p_i = 1$ ($1 \leq i \leq n$). For $q \in \mathbb{R}$ and $n \in \mathbb{N}$, let $\bar{q} = (q, \ldots, q) \in \mathbb{R}^n$. One assumes $\gamma_a = ((\gamma_a)_1, \ldots, (\gamma_a)_n) \in (0, \infty)^n$ and $|\gamma_a| = \sum_1^n (\gamma_a)_i = n$ to be fixed.

For $x, y \in \mathbb{R}^n$, $t > 0$, let $[x, y]$ be the segment in $\mathbb{R}^n$ with the ends x and y; $xy = (x_i y_i)$, $t^y = (t^{y_i})$; $x/y = \left(\frac{x_i}{y_i}\right)$ for $y_i \neq 0$, and $t/\gamma_a = \left(\frac{t}{(\gamma_a)_i}\right)$. Assuming $|x|_{\gamma_a} = \max_{1 \leq i \leq n} |x_i|^{1/(\gamma_a)_i}$, we have $|x + y|_{\gamma_a} \leq c_{\gamma_a}(|x|_{\gamma_a} + |y|_{\gamma_a})$. For $E \subset \mathbb{R}^n$ and $b \in \mathbb{R}^n$, let $|E|_{\gamma_a} = \inf_{x \in E} |x|_{\gamma_a}$, $x \pm bE = \{y : y = x \pm bz, z \in E\}$ and $G_t = \{x : x \in G, |x - \partial G|_{\gamma_a} > t\}$ for an open $G \subset \mathbb{R}^n$.

By means of $M(\Omega, \mu)$, we designate the space of all μ–measurable functions defined on the measure space (Ω, μ). For a Lebesgue measurable $E \subset \mathbb{R}^n$ and a finite D, let $|E|$ and $|D|$ be, correspondingly, the Lebesgue measure $\int \chi_E(x) dx$ of E and the number of the elements of D.

For $n \in \mathbb{N}$, $r \in (0, \infty]^n$, $p \in (0, \infty)$, $q \in (0, \infty]$, a (countable) index set I, and a quasi–Banach space A, let $l_q(A) := l_q(I, A)$ be the space of all sequences $\alpha = \{\alpha_k\}_{k \in I} \subset A$ with the finite quasi–norm $\|\alpha\|_{l_q} = (\sum_{k \in I} |\alpha_k|^q)^{1/q} < \infty$;

$$l_r(A) := l_{r_n}(l_{r_{n-1}} \ldots (l_{r_1}(A) \ldots)) \underbrace{}_{n \text{ brackets}} .$$

For $G \subset \mathbb{R}^n$ and $f : G \to \mathbb{R}$ by means of $\overline{f} : \mathbb{R}^n \to \mathbb{R}$, we designate the function

$$\overline{f(x)} := \begin{cases} f(x), & \text{for } x \in G, \\ 0, & \text{for } x \in \mathbb{R}^n \setminus G. \end{cases}$$

For $p \in (0, \infty]^n$, let $L_p(G)$ be the space of all measurable $f : G \to \mathbb{R}^n$ with the mixed (quasi)norm

$$\|f|L_p(G)\| = \left(\int_\mathbb{R} \left(\int_\mathbb{R} \cdots \left(\int_\mathbb{R} |\overline{f}|^{p_1} dx_1 \right)^{p_2/p_1} \cdots \right)^{p_n/p_{n-1}} dx_n \right)^{1/p_n},$$

where, for $p_i = \infty$, one understands $\left(\int_\mathbb{R} |g(x_i)|^{p_i} dx_i \right)^{1/p_i}$ as $\operatorname{ess\,sup}_{x_i \in \mathbb{R}} |g(x_i)|$. The classical quantitative geometry of these spaces had been studied for a long time (see, for example, [31]).

For an ideal space $Y = Y(\Omega)$ for a measurable space (Ω, μ) and a Banach space X, let $Y(\Omega, X)$ be the space of the Bochner–measurable functions $f : \Omega \mapsto X$ with the finite (quasi)norm

$$\|f|Y(\Omega, X)\| := \|\|f(\cdot)\|_X|Y(\Omega)\|.$$

If another measure ν, absolutely continuous with respect to μ with the density $\frac{d\nu}{d\mu} = \omega$, is used instead of μ, the corresponding space is denoted by $Y(G, \omega, X)$.

For example, $L_p(\mathbb{R}^n, l_q)$ with $p, q \in [1, \infty]$ is a Banach space of the measurable function sequences $f = \{f_k(x)\}_{k=0}^\infty$ with the finite norm $\|\|\{f_k(\cdot)\}_{k \in \mathbb{N}_0}|l_q\||L_p(\mathbb{R}^n)\|$.

The construction in the definition of the auxiliary $lt_{p,q}$–spaces (next) suits both the classical dyadic approach [4] and our slightly more optimal covering approach [32] to the wavelet norms (characterisations) of anisotropic function spaces.

Definition 2.1. *For $p \in (0, \infty)^n$, $q \in (0, \infty)$ and $n \in \mathbb{N}$, let $lt_{p,q} = lt_{p,q}(\mathbb{Z}^n \times J)$ be the quasi–Banach space of the sequences $\{t_{i,j}\}_{i \in \mathbb{Z}^n}^{j \in J}$ with $J \in \{\mathbb{N}_0, \mathbb{Z}\}$ endowed with the (quasi)norm*

$$\|\{t_{i,j}\}|lt_{p,q}\| := \left\|\left\{\sum_{i \in \mathbb{Z}^n} t_{i,j}\chi_{F_{i,j}}\right\}_{j \in J}\middle| L_p(\mathbb{R}^n, l_q(J))\right\|,$$

where $\{F_{i,j}\}_{i \in \mathbb{Z}^n}^{j \in J}$ is a fixed nested family of the decompositions $\{F_{i,j}\}_{i \in \mathbb{Z}^n}$ of $\mathbb{R}^n$ into unions of congruent parallelepipeds $\{F_{i,j}\}_{i \in \mathbb{Z}^n}$ satisfying

$$\cup_{i \in \mathbb{Z}^n} F_{i,j} = \mathbb{R}^n, \; |F_{i,j} \cap F_{k,j}| = 0 \text{ for every } j \in J, i \neq k,$$

and either $|F_{i_0,j_0} \cap F_{i_1,j_1}| = 0$, or $F_{i_0,j_0} \cap F_{i_1,j_1} = F_{i_0,j_0}$ for every i_0, i_1 and $j_0 > j_1$. We shall assume that this system regular in the sense that the length of the kth side $l_{k,j}$ of the parallelepipeds $\{F_{i,j}\}_{i \in \mathbb{Z}^n}$ of the jth decomposition (level) satisfies

$$c_l b^{-j(\lambda_a)_k} \leq l_{k,j} \leq c_u b^{-j(\lambda_a)_k} \text{ for } k \in I_n$$

for some positive constants $b > 1$ and c_l, c_u. Let also the 0–level parallelepipeds $\{F_{i,0}\}_{i \in \mathbb{Z}^n}$ be of the form $Q_t(z)$.

For a parallelepiped $F \in \{F_{i,j}\}_{i \in \mathbb{Z}^n}^{j \in J}$ or $F = \mathbb{R}^n$, by means of $lt_{p,q}(F)$ we designate the subspace of $lt_{p,q}(\mathbb{Z}^n \times J)$ defined by the condition: $t_{i,j} = 0$ if $F_{i,j} \not\subset F$.

The symbol $lt_{p,q}$ denotes either of the spaces $lt_{p,q}(\mathbb{Z}^n \times J)$ or $lt_{p,q}(F)$.

The space $lt_{p,q}$ is isometric to a complemented subspace of $L_p(\mathbb{R}^n, l_q)$, while its dual $lt_{p,q}^*$ is isomorphic to $lt_{p',q'}$ for $p \in (1, \infty)^n$, $q \in (1, \infty)$ (see Section 2.1 below).

Remark 2.1. *There are many books and articles dedicated to the wavelet decompositions (and their predecessors) and related characterisations of the function spaces, including [33, 34, 35, 4]. The construction of $lt_{p,q}$ above is compatible not only with the known results but also with a more optimal version [32]. Thus, Besov $B_{p,q}^s(\mathbb{R}^n)_w$ and Lizorkin–Triebel $L_{p,q}^s(\mathbb{R}^n)_w$ sequence spaces, that we shall call Besov and Lizorkin–Triebel spaces with wavelet norms, can be defined in various constructive ways (we omit the lengthy details) but the following always holds. Throughout the article, we only need to know the immediate* corollary of the definition(s) that $B^s_{p,q}(\mathbb{R}^n)_w$ is, in fact, isometric to $l_q(\mathbb{N}_0, l_p(\mathbb{Z}^n))$, while $L^s_{p,q}(\mathbb{R}^n)_w$ is isometric to a 1–complemented subspace of $lt_{p,q}(\mathbb{Z}^n \times \mathbb{N}, l_q(I_M))$ for certain $M \in \mathbb{N}$ and contains, in turn, an isometric and 1–complemented copy of $lt_{p,q} = lt_{p,q}(\mathbb{Z}^n \times \mathbb{N}_0)$.

We say that a subspace Y of a Banach space X is C–complemented (in X) if there exists a projection P onto Y satisfying $\|P|\mathcal{L}(X)\| \le C$. For $B \subset X$, let $\mathrm{co}(B)$ and $\overline{\mathrm{co}}(B)$ be the convex envelope and the closed convex envelope of B in X correspondingly. Similarly, let $\mathrm{lin}(B)$ and $\overline{\mathrm{lin}}(B)$ be the linear envelope and the closed linear envelope of B in X correspondingly. For a subspace E of a (quasi)Banach space A, let also $A_{|E}$ denote E endowed with $\|\cdot |A\|$.

Definition 2.2. *For $r \in [1, \infty]$, a finite, or countable set I and a set of quasi–Banach spaces $\{X_i\}_{i\in I}$, let their l_r–sum $l_r(I, \{A_i\}_{i\in I})$ be the space of the sequences $x = \{x_i\}_{i\in I} \in \prod_{i\in I} X_i$ with the finite norm*

$$\|x|l_r(I, \{A_i\}_{i\in I})\|^r := \sum_{i\in I} \|x_i\|^r_{A_i}.$$

Definition 2.3. *Let X, Y be (quasi)Banach spaces and $\lambda \ge 1$. Then the Banach–Mazur distance $d_{BM}(X, Y)$ between them is equal to ∞ if they are not isomorphic and is, otherwise, defined by*

$$d_{BM}(X, Y) := \inf\{\|T\| \cdot \|T^{-1}\| : T : X \xrightarrow{onto} Y,\ \mathrm{Ker}\, T = 0\}.$$

The space X is λ–finitely represented in Y if for every finite–dimensional subspace $X_1 \subset X$,

$$\inf\{d_{BM}(X_1, Y_1) : Y_1 \text{ is a subspace of } Y\} = \lambda.$$

If λ is equal to 1, then X is simply said to be finitely represented in Y.

It is said that Y contains almost isometric copies of X (or contains X almost isometrically) if

$$\inf\{d_{BM}(X, Y_1) : Y_1 \text{ is a subspace of } Y\} = 1.$$

For example, it is well–known that, for a Banach space X, its second conjugate X^{**} is finitely represented in X, and X itself is finitely represented in c_0. While, according to Banach, $d_{BM}(l_p, L_p) = \infty$ for $p \in [1, \infty) \setminus \{2\}$ (see [6]), and L_p is finitely represented in l_p for $p \in [1, \infty]$.

We recall the definitions of the Rademacher type and cotype (constants), B–convexity and C–convexity. Let $\{r_i\}_{i\in\mathbb{N}}$ be the family of the Rademacher functions on $[0, 1]$.

Definition 2.4. *Let X be a Banach space and $p, q \in [1, \infty]$. If, for some constant $T > 0$ and every finite $I \subset \mathbb{N}$ and $\{x_i\}_{i\in I} \subset X$, one has*

$$\left\|\sum_{i\in I} r_i x_i\, \Big|\, L_q([0,1], X)\right\| \le T \left\|\{x_i\}_{i\in I}\, |\, l_p(I, X)\right\|, \tag{RT}$$

X is said to possess the Rademacher type p.

If, for some constant $C > 0$ and every finite $I \subset \mathbb{N}$ and $\{x_i\}_{i \in I} \subset X$, one has

$$\left\| \{x_i\}_{i \in I} \big| l_q(I, X) \right\| \leq C \left\| \sum_{i \in I} r_i x_i \big| L_p([0,1], X) \right\|, \qquad (RC)$$

X is said to possess the Rademacher cotype q. The best constants T and C in (RT) and (RC) respectively are the Rademacher type $T_{p,q} = T_{p,q}(X)$ and cotype $C_{p,q} = C_{p,q}(X)$ constants. Let $I_T(X)$ and $I_C(X)$ be the sets (closed or semiclosed intervals) of those p and q respectively, such that X possesses the Rademacher type p and the Rademacher cotype q. A Banach space X is said to be B–convex (C–convex) if the interval $I_T(X)$ ($I_C(X)$) is nontrivial, i.e., contains, at least, two different points of $[1, 2]$ ($[2, \infty]$).

For example, $I_T(l_p) = [1, \min(p, 2)]$ and $I_C(l_p) = [\max(p, 2), \infty]$ (see [36, 37]). Note that every Banach space is of Rademacher type 1 and of Rademacher cotype ∞ and that $T_{p,p}(X), C_{q,q}(X) \geq 1$ for $p, q \in [1, \infty]$.

2.1. Independently Generated Spaces, $lt_{p,q}$ and $lt^*_{p,q}$

The purpose of this subsection is to introduce a wide class of auxiliary spaces containing not only almost all the auxiliary spaces related to the function spaces defined above but also the Lebesgue spaces with mixed norm and l_p–sums of various spaces used in counterexamples.

First let us note some important properties of the space $lt_{p,q}$ (see Definition 2.1 above) and its dual and explain the necessity to introduce the latter. It will be enough to deal with $lt_{p,q}(\mathbb{R}^n)$. The next lemma is very helpful despite to its simplicity.

Lemma 2.1 ([16]). *Let X be a Banach space, and $P \in \mathcal{L}(X)$ be a projector onto its subspace $Y \subset X$. Assume also that $Q_Y : X \to \tilde{X} = X/\operatorname{Ker} P$ is the quotient map. Then we have*

$$\|Q_Y x\|_{\tilde{X}} \leq \|Px\|_X \leq \||P|\mathcal{L}(X)\| \|Q_Y x\|_{\tilde{X}} \text{ for every } x \in X.$$

In particular, the dual space $Y^* = X^*/Y^\perp$ and Y are isometric to $P^* X^*$ and $\tilde{X}$ respectively if, and only if, Y is 1–complemented in X, i.e $\||P|\mathcal{L}(X)\| = 1$.

As mentioned after Definition 2.1, the space $lt_{p,q}$ is isometric to a subspace of $L_p(\mathbb{R}^n, l_q)$ for $p \in (1, \infty)^n$, $q \in (1, \infty)$. The corresponding projection operator is defined by the series of conditional expectation operators $\{E_{\Sigma_j}\}_{j \in J}$ for the subalgebras Σ_j of the Lebesgue measurable subsets of $\mathbb{R}^n$ generated by $\{F_{i,j}\}_{i \in \mathbb{Z}^n}$ (see Definition 2.1). We shall always assume that the corresponding nested family of decompositions of $\mathbb{R}^n$ is regular. Note that the operator E_{Σ_j} is defined by the kernel

$$K_j(x, y) := \sum_{i \in \mathbb{Z}^n} |F_{i,j}|^{-1} \chi_{F_{i,j}}(x) \chi_{F_{i,j}}(y).$$

Thus, the image of the mapping $E = \{E_{\Sigma_j}\}_{j \in J} : \{f_j\}_{j \in J} \mapsto \{E_j f_j\}_{j \in J}$ is complemented in $L_p(\mathbb{R}^n, l_q)$ thanks to the Fefferman-Stein inequality for mixed norm spaces (see Lemma 4 in [38] and remarks therein)

$$\|\{Mf_j\}_{j \in J} | L_p(\mathbb{R}^n, l_q)\| \leq C(p, q, n) \|\{f_j\}_{j \in J} | L_p(\mathbb{R}^n, l_q)\|$$

and the pointwise estimate
$$|\mathrm{E}_{\Sigma_j} f_j| \le CM f_j \text{ a.e.},$$
where M is the anisotropic Hardy–Littlewood maximal function. This observation, along with the Hahn–Banach theorem and Hölder inequalities for sequences and integrals, implies the estimates
$$\|f|lt_{p,q}^*\| \le \|\mathrm{E}^* \mathrm{Ext}_{HB} f|lt_{p',q'}\| \le C(p,q,n)\|f|lt_{p,q}^*\|$$
explaining the isomorphism of $lt_{p,q}^*$ and $lt_{p',q'}$ for $p \in (1,\infty)^n$, $q \in (1,\infty)$. And according to Lemma 2.1, the space $lt_{p,q}^*$ is not necessarily isometric to $lt_{p',q'}$ unless $p = \bar{q} = (q, \ldots, q)$.

Nevertheless, the spaces $lt_{p,q}$ and $lt_{p,q}^*$ contain an isometric and 1–complemented copies of l_p and l_q according to the next lemma. Its proof [16] is purely constructive.

Lemma 2.2 ([16]). *Let $p \in [1,\infty)^n$ and $q \in [1,\infty)$. Then the spaces $lt_{p,q}(\mathbb{R}^n)$ and $lt_{p',q'}(\mathbb{R}^n)^*$ contain isometric 1–complemented copies of $l_p(\mathbb{N}^n)$, $l_q(\mathbb{N})$ and $l_p(\mathbb{Z}^n, l_q(\mathbb{N}))$, and the spaces $lt_{p,q}(F)$ and $lt_{p',q'}(F)^*$ contain isometric 1–complemented copies of $l_q(\mathbb{N})$, $l_p(I_m, l_q(\mathbb{N}))$ for every $m \in \mathbb{N}^n$.*

Remark 2.2. *There are several known ways of detecting copies of l_σ–spaces in a Banach space. It seems that, at least, in the setting of a regular domain G one can uncover almost isometric copies in Besov and Lizorkin–Triebel spaces with the aid of the remarkable results due to M. Levy [39], M. Mastyło [40] and V. Milman [6] (in combination with the results in [25]). Direct constructions of the isomorphic (and often complemented) copies of various sequence spaces (ex. mixed l_p and Lorentz $l_{p,r}$) in various anisotropic Besov, Sobolev and Lizorkin–Triebel spaces of functions defined on open subsets of $\mathbb{R}^n$ are presented in [25].*

Definition 2.5. *Independently generated spaces*
Let $\mathcal{S}$ be a set of ideal (quasi–Banach) spaces, such that every element $Y \in \mathcal{S}$ is either a sequence space $Y = Y(I)$ with a finite or countable I, or a space $Y = Y(\Omega)$, where (Ω, μ) is a measure space with a countably additive measure μ without atoms.

By means of the leaf growing process (step) from some $Y \in \mathcal{S}$, we shall call the substitution of Y with:
(Type A) either $Y(I, \{Y_i\}_{i \in I})$ for some $\{Y_i\}_{i \in I} \subset \mathcal{S}$ if $Y = Y(I)$, or
(Type B) $Y(\Omega, Y_0)$ for some $Y_0 \in \mathcal{S}$ if $Y = Y(\Omega)$.

Here the quasi–Banach space $Y(I, \{Y_i\}_{i \in I})$ is the linear subset of $\prod_{i \in I} Y_i$ of the elements $\{x_i\}_{i \in I}$ with the finite quasi–norm $\|\{x_i\}_{i \in I}|Y(I, \{Y_i\}_{i \in I})\| := \|\{\|x_i\|_{Y_i}\}_{i \in I}\|_Y$. Note that a type B leaf (i.e., of the form $Y = Y(\Omega)$) can grow only one leaf of its own. We shall also refer to either $\{Y_i\}_{i \in I}$, or Y_0 as to the leaves growing from Y, which could have been a leaf itself before the tree growing process.

Let us designate by means of $IG(\mathcal{S})$ the class of all spaces obtained from an element of $\mathcal{S}$ in a finite number of the tree growing steps consisting of the tree growing processes for some or all of the current leaves.

If the set $\mathcal{S}$ includes only the spaces described by (different) numbers of parameters from $[1, \infty]$ and $X \in IG(\mathcal{S})$, we assume that $I(X)$ is the set of all the parameters of the spaces at the vertexes of the tree T corresponding to X and
$$p_{\min}(X) := \inf I(X) \text{ and } p_{\max}(X) := \sup I(X).$$

For the sake of brevity, we also assume that

$$IG := \{X \in IG(l_p, L_p, lt_{p,q}, lt^*_{p',q'}) : [p_{\min}(X), p_{\max}(X)] \subset (1, \infty)\} \text{ and}$$
$$IG_0 := IG \cap IG(l_p, L_p).$$

We also assume that an abstract L_p is an IG space $X_{\bar{p}}$ with $I(L_p) = \{p\}$ (i.e., $\bar{p} : P \to \{p\}$).

Let us also define the class IG_+. A space X belongs to IG_+ if it is either in IG, or obtained from a space $X_- \in IG$ by means of the leaf growing process, in the course of which some leaves (or just one leaf) of X_- have grown some new leaves from the set

$$\{S_p, S_p^n, L_p(\mathcal{M}, \tau)\}_{p \in [1,\infty]}^{n \in \mathbb{N}},$$

where $(\mathcal{M}, \tau)$ is a semifinite von Neumann algebra (see Section 2.2). The set of the parameters p of these "last non–commutative leaves" is included into $I(X)$ and they are part of the corresponding tree $T(X)$. Let also

$$IG_{0+} = \{Y \in IG_+ : Y_- \in IG_0\}.$$

Remark 2.3. a) *We shall deal with the set $\{l_p, L_p, lt_{p,q}, lt^*_{p',q'}\}$, where l_p, L_p, $lt_{p,q}$ and $lt^*_{p',q'}$ designate, respectively, the classes $\{l_p(I)\}_{p \in [1,\infty]}$, $\{L_p(\Omega)\}_{p \in [1,\infty]}$ for all the Ω with some countably additive and not purely atomic measure μ on it, $\{lt_{p,q}\}_{p,q \in [1,\infty]}$ and $\{lt^*_{p',q'}\}_{p,q \in [1,\infty]}$.*

b) *The subclass of l_p–spaces can formally be excluded from the definition of the class IG because l_p is isometric to $lt_{p,p} = lt^*_{p,p}$ but is left there for the sake of technical convenience. The subclass of $lt^*_{p,q}$–spaces is included to make IG closed with respect to passing to dual spaces.*

c) *The Lebesgue or sequence spaces with mixed norm and the l_p–sums of them are particular elements of $IG(l_p, L_p)$.*

2.2. Non–Commutative Spaces

The related definitions, examples and all needed information about the noncommutative $L_p(\mathcal{M}, \tau)$, where $(\mathcal{M}, \tau)$ is a von Neumannn algebra with a normal semifinite faithful trace τ, and the more general Haagerup $L_p(\mathcal{M}) = L_p(\mathcal{M}, \phi)$, where $(\mathcal{M}, \phi)$ is a von Neumannn algebra with a normal semifinite faithful weight ϕ (that exists for every von Neumann algebra $\mathcal{M}$), are found in the open access [16] and in this section. Apart from Lemma 2.3 below on the existence of 1–complemented isometric copies of l_p in every infinite–dimensional noncommutative L_p, it is adapted from [41]. Typical examples of the noncommutative $L_p(\mathcal{M}, \tau)$ are the infinite–dimensional S_p and the finite–dimensional S_p^n Schatten–von Neumann classes with the classical trace.

We need the following counterpart [16] of Lemma 2.2 for Schatten–von Neumann classes and general $L_p(\mathcal{M})$, where $\mathcal{M}$ is a von–Neumann algebra with a normal semifinite faithful weight (that always exists).

Lemma 2.3 ([16]). *For $p \in [1, \infty]$, the space S_p contains 1–complemented isometric copies of S_p^n, $l_p(I_n)$ and l_p for $n \in \mathbb{N}$. Moreover, an infinite–dimensional $L_p(\mathcal{M})$ contains a 1–complemented isometric copy of l_p.*

2.3. Sobolev, Besov and Lizorkin–Triebel Spaces

In this section, we describe how to approach the variety of the classes of anisotropic Besov, Lizorkin–Triebel and Sobolev spaces of functions defined on an arbitrary open subset $G \subset \mathbb{R}^n$ with underlying mixed Lebesgue norm. It is done with the aid of equivalent geometrically friendly norms. One of our classes includes both the spaces defined in terms of an H^∞ functional calculus and the spaces defined in terms of the Fourier multipliers (see Remark 5.1). In a similar fashion these classes can also be defined on open subsets of, for example, Riemannian manifolds leading to the same results. Moreover, our approach is as easily applicable to the l_p–sums and "Bochnerisations" of function spaces representing, for example, the initial and/or boundary value problems in PDE (see the definitions of IG and IG_+–spaces in [16]). We are also considering the duals of anisotropic function spaces of positive smoothness because they are traditionally interpreted as anisotropic function spaces of negative smoothness and play important role in quasi–linear PDE.

While many of the function spaces under consideration are known to be isomorphic (equivalent norms) for a vide range of the domains of functions and their parameters (see [42] for the norms with the additional improving parameter a), we endow every space with a geometrically friendly norm and deal with it directly. We start by describing the idea of the geometrically friendly norms via introducing the first inessential parameter ς in the simplest setting of classical Sobolev and Besov spaces, while the necessity of the second parameter a is revealed relying on the example of anisotropic Lizorkin–Triebel spaces (endowed with intrinsic norms). The technical convenience of the latter inessential parameter is in simplifying the proofs of a number of results (see, for example, the Hölder inequalities for function spaces and related results in [42]) and allowing to deal with "rough" atoms [43, 44], molecules and wavelets but also crucial in the study of the geometric (sensitive to the choice of an equivalent norm), metric, local and non–local properties of function spaces, that often do not depend on the smoothness or the domain the elements of the function space are defined on.

The basic references regarding the key properties of function spaces (including structural aspects) are [28, 29, 45, 46, 2, 30, 47, 35, 4, 3], where [29, 2, 4, 3] include also the setting of anisotropic spaces in its classical diagonal form (see also remarks below). The development of a more general spiral anisotropy originates from [48] and discussed in Remark 2.4, b).

For $s \in \mathbb{N}^n$, $\varsigma \in (1, \infty)$, $p \in [1, \infty]^n$ and an open $G \subset \mathbb{R}^n$, let the anisotropic Sobolev space $W_p^s(G) = W_p^s(G)_\varsigma$ be the Banach space of the measurable functions f defined on G possessing the Sobolev generalized derivatives $D_i^{s_i} f$ and the finite norm

$$\|f|W_p^s(G)\|^\varsigma := \|f|L_p(G)\|^\varsigma + \|f|w_p^s(G)\|^\varsigma = \|f|L_p(G)\|^\varsigma + \sum_{i=1}^n \|D_i^{s_i} f|L_p(G)\|^\varsigma.$$

Note that the traditional norm in the Sobolev space corresponds to $\varsigma = 1$ but interpreting $W_p^s(G)$ as a subspace of $l_1(I_{n+1}, L_p(G))$ either does not provide proper geometric properties of the latter space (that are inherited by the Sobolev space) for a variety of geometric properties including, for example, uniform convexity and smoothness, uniform normal structure, quantitative Lyapunov and Hahn–Banach theorems, quantitative Lozanovskiy factorisation, counterparts of Parseval's identity for asymmetric and abstract vector–valued

Walsh chaos, a version of Pisier's martingale inequality and its Markov chain counterpart, Markov type and cotype, isometric and bounded extensions of Hölder mappings between function spaces, the global Hölder regularity of both the metric projections on closed convex subsets and the Chebyshev centre mapping and exact values and estimates for various geometric constants, or still require finding a better equivalent norm that is the case with the A–convexity, Rademacher type and cotype, stability, α and UMD properties, superreflexivity and other more complex phenomena [42, 49, 24, 50, 16, 23, 25, 51, 26].

Definition 2.6. *For $m \in \mathbb{N}$, $h > 0$, $\gamma \in \mathbb{R}$, $x \in \mathbb{R}^n$, assume that $\Delta_i^m(h)f(x)$ is the difference of the function f of the mth order with the step h in the direction of e^i at the point x, and*

$$\Delta_i^m(h, E)f(x) = \begin{cases} \Delta_i^m(h)f(x) & \text{for } [x, x + mhe^i] \in E, \\ 0 & \text{for } [x, x + mhe^i] \notin E, \end{cases}$$

$$\delta_{i,a}^m(t, x, f, E)_\gamma = \left(\int_{-1}^{1} |\Delta_i^m(t^{(\gamma_a)_i}u, E)f(x + \gamma t^{(\gamma_a)_i}u)|^a \, du \right)^{1/a}.$$

Similarly to $L_p(G)$, for $p, q \in (0, \infty]^n$, let $L_{p,q}(G)$ be the quasi–normed space of the (measurable) functions on G with

$$\|f|L_{p,q}(G)\| = \underbrace{\|\dots\|\overline{f}|L_{p_1,q_1}(\mathbb{R}, dx_1)\|\dots|L_{p_n,q_n}(\mathbb{R}, dx_n)\|}_{n\ times}.$$

Let us consider a geometrically friendly norm in the classical Besov space on an open subset [2].

For $p \in [1, \infty]^n$, $q \in [1, \infty]$, $s \in (0, \infty)^n$, $s < m \in \mathbb{N}^n$, $\kappa > 0$ and an open set $G \subset \mathbb{R}^n$, by means of $B_{p,q}^s(G)$, we designate the Besov space of the measurable functions $f \in L_{1,loc}(G)$ with the finite norm

$$\|f|B_{p,q}^s(G)\|^\varsigma := \|f|L_p(G)\|^\varsigma + \sum_{i=1}^n \left(\int_0^\infty \|\Delta_i^m(t, G_{\kappa t})f|L_p(G)\|^q \, t^{-qs} \, \frac{dt}{t} \right)^{\varsigma/q}. \quad (B)$$

The proper intrinsic definition of the anisotropic Lizorkin–Triebel spaces requires the employment of the averaged differences as was understood by Oleg V. Besov [2] ($a = \bar{1}$ in Definition 2.8 below), while the ordinary differences can be used to define an equivalent norm (that is by changing the order of taking the L_p and L_q norms in the definition of the classical Besov space above) for $s \in (1, \infty)^n$ as we have shown in [42].

For $A \subset \mathbb{N}_0^n$, let $|A|$ designate the number of the elements of A; for $\alpha \in \mathbb{N}_0^n$,

$$\hat{A}_\alpha = \{\beta : \beta \in \mathbb{N}_0^n, \beta \leq \alpha\}; \quad \hat{A} = \bigcup_{\alpha \in A} \hat{A}_\alpha.$$

For $\gamma_a \in (0, \infty)^n$ and $s \in [0, \infty)$, let $A_s^* := \{\alpha : \alpha \in \mathbb{N}_0^n, (\alpha, \gamma_a) \leq s\}$.

For $z \in \mathbb{R}^n$ and $v > 0$, assume that $\tau_z f(x) = f(x - z)$ and $\sigma_v f(x) = f(v^{-\gamma_a}x)$. For $A \subset \mathbb{N}_0^n$, $|A| < +\infty$, let $\mathcal{P}_A$ be the space of the polynomials of the form $\sum_{\alpha \in A} c_\alpha x^\alpha$ with

$\{c_\alpha\}_{\alpha \in A} \subset \mathbb{R}$. To insure the translation invariance of $\mathcal{P}_A$, we shall always assume that $A = \hat{A} \supset A_s^*$.

For a Banach space X and its subspace Y, their quotient or factor space is denoted by X/Y.

Definition 2.7. *For $a \in [1,\infty]^n$, $v > 0$ and $z \in \mathbb{R}^n$, let $Q_v(z) = z + v^\lambda[-1,1]^n$ and $p_{A,v,z} = \tau_z \circ \sigma_v \circ p_A \circ \sigma_v^{-1} \circ \tau_z^{-1}$.*

For $\varepsilon > 0$, $a \in [1,\infty]^n$, let $\pi_{A,v,z} : L_a(Q_v(z)) \to \mathcal{P}_A$ be an operator of the best L_a-approximation satisfying

$$\|f - \pi_{A,v,z}f|L_a(Q_v(z))\| = \min_{g \in \mathcal{P}_A} \|f - g|L_a(Q_v(z))\|, \ f \in L_a(Q_v(z)).$$

For $f \in L_{a,loc}(G)$, $v > 0$ and $a \in (0,\infty]^n$, we define the $\mathcal{D}$–functionals

$$\mathcal{D}_a(v, x, f, G, A) := \begin{cases} \|f|L_a(Q_v(x))/\mathcal{P}_A\|v^{-(\gamma_a, 1/a)}, & \text{if } Q_v(x) \subset G, \\ 0 & \text{otherwise.} \end{cases}$$

The examples of possible G include (for different purposes) open subsets of $\mathbb{R}^n$, $\mathbb{R}^n$ itself, anisotropic cubes $Q_r(v)$, balls, domains with smooth boundary [30, 4], star–shaped domains and those with the cone condition [28, 2], Lipschitz domains (equivalent) [47, 4], domains satisfying the λ–horn, flexible λ–horn and related conditions [2]. One says that $G \subset \mathbb{R}^n$ is an extension domain for a function space X if there exists a bounded linear extension operator $E_G : X(G) \longrightarrow X(\mathbb{R}^n)$ that is the right inverse to the restriction operator $R_G : X(\mathbb{R}^n) \longrightarrow X(G), f \longmapsto \chi_G f$.

In the definitions below we shall always *assume* that the norm of the normed space $X(G) = L_p(G) \cap x(G)$ is

$$\|f|X(G)\| = (\|f|L_p(G)\|^\varsigma + \|f|x(G)\|^\varsigma)^{1/\varsigma}.$$

We start with the definitions of anisotropic Besov and Lizorkin–Triebel spaces defined in terms of the averaged (shifted) axis–directional differences.

Let Pr_i be the orthogonal projector on the ith axis in $\mathbb{R}^n$, and, for $y \in (I - Pr_i)(G)$,

$$In_i(y) := (I - Pr_i)^{-1}(y) \cap G = \{x \in G : x = y + te_i, t \in \mathbb{R}\}.$$

Definition 2.8. *For $p \in [1,\infty]^n$, $q \in [1,\infty]$, $r > 0$, $s \in (0,\infty)$, $s/\gamma_a < m \in \mathbb{N}_0^n$, $a \in [1,\infty]^n$, $\gamma \geq 0$, and an open set $G \subset \mathbb{R}^n$, by means of $b_{p,q,a}^s(G)$, we designate the semi–normed space of the measurable functions $f \in L_{a_i,loc}(In_i(y), dx_i)$ for a.e. $y \in (I - Pr_i)(G)$ with the finite semi-norm*

$$\|f|b_{p,q,a}^s(G)\|^\varsigma := \sum_{i=1}^n \left(\int_0^\infty \|\delta_{i,a_i}^{m_i}(t,\cdot,f,G_{rt})_\gamma|L_p(G)\|^q t^{-qs} \frac{dt}{t} \right)^{\varsigma/q}; \quad (1)$$

$$B_{p,q,a}^s(G) := b_{p,q,a}^s(G) \cap L_p(G).$$

Note that, for $\gamma = 0$, one has $B_{p,q,1}^s(G) = B_{p,q}^s(G)$ (isomorphism) [2].

Definition 2.9. For $p \in [1,\infty]^n$, $q \in [1,\infty]$, $r > 0$, $s \in (0,\infty)$, $s/\gamma_a < m \in \mathbb{N}_0^n$, $a \in [1,\infty]^n$, $\gamma \in \mathbb{R}$, and an open set $G \subset \mathbb{R}^n$, by means of $l_{p,q,a}^s(G)$, we designate the semi–normed space of the measurable functions $f \in L_{a_i,loc}(In_i(y), dx_i)$ for a.e. $y \in (I - Pr_i)(G)$ with the finite semi–norm

$$\|f|l_{p,q,a}^s(G)\|^\varsigma := \sum_{i=1}^n \left\| \left(\int_0^\infty \left(\delta_{i,a_i}^{m_i}(t,\cdot,G_{rt},f)_\gamma \right)^q t^{-qs} \frac{dt}{t} \right)^{1/q} \bigg| L_p(G) \right\|^\varsigma ; \quad (2)$$

$$L_{p,q,a}^s(G) := l_{p,q,a}^s(G) \cap L_p(G).$$

Note that, for $\gamma = 0$, one has $L_{p,q}^s(G) \stackrel{def}{:=} L_{p,q,1}^s(G)$ [2].

Let us define the anisotropic local approximation spaces of Besov and Lizorkin–Triebel type in terms of the $\mathcal{D}$–functional as follows.

Definition 2.10. For $p \in (0,\infty]^n$, $q \in (0,\infty]$, $a \in [1,\infty]^n$, $s \in (0,\infty)$ and $D = \hat{D} \subset \mathbb{N}_0^n$ with $|D| < \infty$ by means of $\widetilde{b}_{p,q,a}^{s,D}(G)$ and $\widetilde{l}_{p,q,a}^{s,D}(G)$, we designate, correspondingly, the anisotropic semi–normed space of the functions $f \in L_{a,loc}(G)$ with the finite semi–norm

$$\|f|\widetilde{b}_{p,q,a}^{s,D}(G)\| := \left(\int_0^\infty \|t^{-s}\mathcal{D}_a(t,\cdot,f,G,D)|L_p(G)\|^q \frac{dt}{t} \right)^{\frac{1}{q}},$$

$$\|f|\widetilde{l}_{p,q,a}^{s,D}(G)\| := \left\| \left(\int_0^\infty (t^{-s}\mathcal{D}_a(t,\cdot,f,G,D))^q \frac{dt}{t} \right)^{\frac{1}{q}} \bigg| L_p(G) \right\|.$$

Assume also that

$$\widetilde{B}_{p,q,a}^{s,D}(G) = \widetilde{b}_{p,q,a}^{s,D}(G) \cap L_p(G) \text{ and } \widetilde{L}_{p,q,a}^{s,D}(G) = \widetilde{l}_{p,q,a}^{s,D}(G) \cap L_p(G).$$

It useful to note that $BMO^{\gamma_a}(G) = \widetilde{b}_{\infty,\infty,1}^{0,A_0^*}(G)$, and, according to the particular case $p = \infty$ of a result in [42] for an open $G \subset \mathbb{R}^n$ satisfying the flexible λ–horn condition, $B_{\infty,\infty}^s(G)$ and $\overset{\circ}{B}_{\infty,\infty,a}^s(G)$ are the Hölder space and so–called "little" Hölder space correspondingly often considered in the setting of continuous maximal regularity in PDE.

Next we define a very large class of function spaces that shares quite a few geometric properties with the above classes of function spaces even in its full generality. Namely, let us define the following *abstract classes of Besov and Lizorkin-Triebel spaces*. For a Banach space X, $\mathcal{C}(X)$ is the class of all closed operators $A : X \supset D(A) \longrightarrow X$.

Definition 2.11. *Let* $G \subset \mathbb{R}^n$, $p \in [1,\infty]^n$, $q,\varsigma \in [1,\infty]$, $s \in \mathbb{R}$, $b > 1$, *and let* $\mathcal{F} = \{F_k\}_{k \in \mathbb{N}_0} \subset \mathcal{C}(L_p(G))$ *be a system of closed operators satisfying*

$$f \in L_p(G) \text{ and } F_k f = 0 \text{ for } k \in \mathbb{N}_0 \Rightarrow f = 0.$$

By means of $B_{p,q,\mathcal{F}}^s(G)$ *and* $L_{p,q,\mathcal{F}}^s(G)$, *we designate, correspondingly, the Banach spaces of functions defined on G with the finite norms*

$$\|f|B_{p,q,\mathcal{F}}^s(G)\|^\varsigma = \|f|L_p(G)\|^\varsigma + \left(\sum_{k \in \mathbb{N}_0} b^{ksq} \|F_k f|L_p(G)\|^q \right)^{\varsigma/q},$$

$$\|f|L^s_{p,q,\mathcal{F}}(G)\|^\varsigma = \|f|L_p(G)\|^\varsigma + \left\|\left(\sum_{k\in\mathbb{N}_0} b^{ksq}|F_k f(\cdot)|^q\right)^{1/q} \bigg| L_p(G)\right\|^\varsigma.$$

Remark 2.4. a) *Sergey M. Nikol'skii [29] has defined function spaces employing the anisotropic Fourier multipliers* $F_k : f \longmapsto \phi(b^{k\lambda} \cdot) * f$, *where* ϕ *and, thus,* $F_k f$ *are functions of exponential type. These spaces appeared to be global approximation spaces by the entire functions of exponential type, or the "meeting place" for the theorems of S. N. Bernstein and D. Jackson type.*

b) *The Besov spaces* $B^s_{p,q}(\mathbb{R}^n)$ *and Lizorkin–Triebel spaces* $F^s_{p,q}(\mathbb{R}^n)$ *(defined by Peter I. Lizorkin for* $q = 2$ *and by Hans Triebel for the other* q*; see [30]) defined in terms of various more general smooth Fourier multipliers (Littlewood–Paley decompositions), as well as their restrictions* $B^s_{p,q}(G)$ *and* $F^s_{p,q}(G)$ *to an open* $G \subset \mathbb{R}^n$ *are some of the most popular model cases of* $B^s_{p,q,\mathcal{F}}(G)$ *and* $L^s_{p,q,\mathcal{F}}(G)$ *and their subspaces. This setting includes the case of the spiral/parabolic anisotropy due to Calderón and Torchinski [48]. This remarkably well–developed framework does not cover, for example, Besov and Lizorkin–Triebel spaces defined on a non–extension domain, requiring the consideration of intrinsic norms as in [2] or above.*

c) *Under the conditions of the last definition, let also* $\Omega \subset \mathbb{C}$ *be open with* $b^{-k}\Omega \subset \Omega$ *for* $k \in \mathbb{N}_0$, $g \in H_\infty(\Omega)$ *(bounded holomorphic function on* Ω*), and* $A \in \mathcal{C}(L_p(G))$ *admits the bounded* $H_\infty(\Omega)$ *functional calculus* $H_\infty(\Omega) \ni h \mapsto h(A) \in \mathcal{L}(L_p(G))$ *with* $\|f(A)|\mathcal{L}(L_p(G))\| \le C\|f|H_\infty(\Omega)\|$. *Assuming that* $F_k = g(b^{-k}A)$, *we obtain the Besov and Lizorkin–Triebel spaces* $B^s_{p,q,\mathcal{F}}(G)$ *and* $L^s_{p,q,\mathcal{F}}(G)$ *defined in terms of the bounded* H_∞*–calculus.*

d) *The isomorphisms and structural properties of Besov and Lizorkin–Triebel spaces with different a and D, as well as various equivalent norms of Sobolev spaces and Hölder inequalities for function spaces were investigated in [42].*

We assume that ς and the components of a are in the convex envelope of the other summability parameters (meaning either the components of p in the case of Sobolev spaces $W^s_p(G)$, or both q and the components of p in the case of Besov and Lizorkin–Triebel spaces) and do not coincide with $p_{\min}$ and $p_{\max}$ in the setting of the underlying Lorentz quasinorms. Moreover, if the mixed Lebesgue norm $L_p(G)$ in the definition of the Sobolev space $W^s_p(G)$, or either Besov or Lizorkin–Triebel space $X^s_{p,q}(G)$ is substituted with the mixed Lorentz quasinorm $L_{p,\theta}(G)$, we denote the resulting space as $W^s_{L_{p,\theta}}(G)$, or $X^s_{L_{p,\theta},q}(G)$ respectively.

2.4. More on Uniformly Complemented and Isomorphic Copies of $l_p(I_n)$

Theorem 2.1 ([25]). *Let* $G \subset \mathbb{R}^n$, $p, a \in (1, \infty)^n$, $q, \varsigma \in (1, \infty)$, $s \in (0, \infty)$ *and* $r \in \{p_{\min}, p_{\max}, q, 2\}$. *Assume also that* $Y \in \left\{ B^s_{p,q}(G), B^s_{p',q'}(G)^* \right\}$ *or*

$$Y \in \Big\{ B^s_{p,q,a}(G), \tilde{B}^{s,A}_{p,q,a}(G), L^s_{p,q,a}(G), \tilde{L}^{s,A}_{p,q,a}(G), b^s_{p,q,a}(G), \tilde{b}^{s,A}_{p,q,a}(G), l^s_{p,q,a}(G), \tilde{l}^{s,A}_{p,q,a}(G),$$
$$B^s_{p',q',a'}(G)^*, , \tilde{B}^{s,A}_{p',q',a'}(G)^*, L^s_{p',q',a'}(G)^*, \tilde{L}^{s,A}_{p',q',a'}(G)^*, b^s_{p',q',a'}(G)^*, \tilde{b}^{s,A}_{p',q',a'}(G)^*,$$
$$l^s_{p',q',a'}(G)^*, \tilde{l}^{s,A}_{p',q',a'}(G)^* \Big\},$$

and a is in admissible range for Y. Then there are constants $C_0, C_1 > 0$, such that Y contains an C_0–isomorphic and C_1–complemented copy of $l_r(I_m)$ for every $m \in \mathbb{N}$.

Theorem 2.2 ([25]). *Let* $Y \in \{W_p^s(G), W_{p'}^s(G)^*\}$ *for* $G \subset \mathbb{R}^n$, $p \in (1, \infty)^n$, $\varsigma \in (1, \infty)$, $s \in \mathbb{N}_0^n$ *and* $r \in \{p_{\min}, p_{\max}, 2\}$. *Then there are constants* $C_0, C_1 > 0$, *such that* Y *contains an* C_0–*isomorphic and* C_1–*complemented copy of* $l_r(I_m)$ *for every* $m \in \mathbb{N}$.

Theorem 2.3 ([25]). *Let* $Y \in \{B_{p,q}^s(\mathbb{R}^n)_w, L_{p,q}^s(\mathbb{R}^n)_w, B_{p',q'}^s(\mathbb{R}^n)_w^*, L_{p',q'}^s(\mathbb{R}^n)_w^*\}$ *for* $p \in (1, \infty)^n$, $\varsigma \in (1, \infty)$, $s \in (0, \infty)$ *and* $r \in \{p_{\min}, p_{\max}, q, 2\}$. *Then* Y *contains an isometric 1–complemented copy of* $l_r(I_m)$ *for every* $m \in \mathbb{N}$.

Remark 2.5. *Let us recall that, for* $p \in (1, \infty)$, $l_p(I_{2^n})$ *contains a* C_0–*isomorphic and* C_2–*complemented copy of* $l_2(I_n)$ *for every* $n \in \mathbb{N}$ *thanks to the Hölder and Khinchin inequalities and Lemma* 2.1.

2.5. Γ–Groups of Spaces under Consideration

For the sake of convenience and brevity we divide the spaces that we will consider most often into the following 6 numbered groups of spaces. Let J be a convex subset of $[1, \infty]$, and let G be an open subset of $\mathbb{R}^n$ for some $n \in \mathbb{N}$.

$$\Gamma_0(J) = \{Y \in IG_+ : [p_{\min}(Y), p_{\max}(Y)] \subset J\}.$$

$$\Gamma_1(J) = \Big\{ B_{p,q,a}^s(G), \tilde{B}_{p,q,a}^{s,A}(G), L_{p,q,a}^s(G), \tilde{L}_{p,q,a}^{s,A}(G), b_{p,q,a}^s(G), \tilde{b}_{p,q,a}^{s,A}(G), l_{p,q,a}^s(G),$$
$$\tilde{l}_{p,q,a}^{s,A}(G), B_{p',q',a'}^s(G)^*, \tilde{B}_{p',q',a'}^{s,A}(G)^*, L_{p',q',a'}^s(G)^*, \tilde{L}_{p',q',a'}^{s,A}(G)^*, b_{p',q',a'}^s(G)^*,$$
$$\tilde{b}_{p',q',a'}^{s,A}(G)^*, l_{p',q',a'}^s(G)^*, \tilde{l}_{p',q',a'}^{s,A}(G)^* \Big\},$$

where $p, a \in J^n$, $q, \varsigma \in J$, $s \in (0, \infty)$ and $A \subset \mathbb{N}_0^n$ with $|A| < \infty$.

$$\Gamma_2(J) = \{W_p^s(G), W_{p'}^s(G)^* : p \in J^n, s \in \mathbb{N}_0^n\}.$$

$$\Gamma_3(J) = \{B_{p,q}^s(\mathbb{R}^n)_w, L_{p,q}^s(\mathbb{R}^n)_w, B_{p',q'}^s(\mathbb{R}^n)_w^*, L_{p',q'}^s(\mathbb{R}^n)_w^* : p \in J^n, q \in J, s \in (0, \infty)\}.$$

$$\Gamma_4(J) = \{B_{p,q,\mathcal{F}}^s(G), L_{p,q,\mathcal{F}}^s(G), B_{p',q',\mathcal{F}}^s(G)^*, L_{p',q',\mathcal{F}}^s(G)^* : p \in J^n, q, \varsigma \in J, s \in \mathbb{R}\}.$$

$$\Gamma_5(J) = \{L_p(\mathcal{M}, \tau), S_p : p \in J\},$$

where $(\mathcal{M}, \tau)$ is a von Neumannn algebra with a normal semifinite faithful weight τ.

For the sake of further convenience we also assume that $G \subset \mathbb{R}^n$ if it is not stated otherwise, and that

$$\Gamma_0 = IG_+ \text{ and } \Gamma_i := \Gamma_i((1, \infty)) \text{ for } i \in I_5.$$

For $i \in I_4$, let also $\Gamma_i^L(J)$ be the class of the spaces obtained from $\Gamma_i(J)$ by substituting the mixed Lebesgue L_p–norm in their definitions with a mixed Lorentz $L_{p,\theta}$–norm with $p, \theta \in J^n$ and $q \in J$ with the additional restriction $p \in (1, \infty)^n$.

3. KEY TOOLS AND OBSERVATIONS

In this section, the basic geometry of the spaces under consideration is quantified in terms of the asymmetric uniform convexity and smoothness leading, eventually, to our main tools: abstract uniform convexity and smoothness and our Markov chain counterpart of the Pisier martingale inequality.

3.1. Asymmetric Uniform Convexity and Smoothness

Definition 3.1 ([23]). Let X be a Banach space and $2 \in [q, p] \subset [1, \infty]$.

We say that the space X is (p, h_c)–uniformly convex if for every $x, y \in X$ and $\mu = 1 - \nu \in (0,1)$, we have $\mu \|x\|_X^p + \nu \|y\|_X^p \geq \|\mu x + \nu y\|_X^p + h_c(\mu)\mu\nu\|y - x\|_X^p$.

We say that the space X is (q, h_s)–uniformly smooth if for every $x, y \in X$ and $\mu = 1 - \nu \in (0,1)$, we have $\mu \|x\|_X^q + \nu \|y\|_X^q \leq \|\mu x + \nu y\|_X^q + h_s(\mu)\mu\nu\|y - x\|_X^q$.

Having in mind the non–improvable estimates

$$\max(\|\mu x + \nu y\|_X, \|x - y\|_X/2) \leq \max(\|x\|_X, \|y\|_X) \text{ and}$$

$$\mu\|x\|_X + \nu\|y\|_X \leq \|\mu x + \nu y\|_X + 2\mu\nu\|x - y\|_X,$$

valid for every Banach space X, we shall refer to them as to $(\infty, 1)$–uniform convexity and $(1,1)$–uniform smoothness respectively for the sake of convenience.

We will also assume that $h_c(0) = \lim_{\mu \to 0} h_c(\mu)$ and $h_s(0) = \lim_{\mu \to 0} h_s(\mu)$ if the corresponding limits exist.

Remark 3.1. a) Choosing x and y with $\mu x + \nu y = 0$, we see that h_c (h_s) must satisfy

$$h_c(\mu) \leq \left(\mu^{1/(p-1)} + \nu^{1/(p-1)}\right)^{1-p} \leq \mu^{p-1} + \nu^{p-1} \leq 1 \text{ and } \left(h_s(\mu) \geq \mu^{q-1} + \nu^{q-1} \geq 1\right)$$

if X is (q, h_c)–uniformly convex (smooth), where the internal term in the first relation is due to the convex or linear duality in Theorem 4.5 in [23] and also reflecting the multi–homogeneous inequality Theorem 4.13, d) in [23]. The function $\mu^{q-1} + \nu^{q-1}$ appears also if one uses Pisier's martingale approach (see [52]).

b) Part a) also shows that, whenever $h_c(1/2) = 2^{2-p}$ or $\left(h_s(1/2) = 2^{2-q}\right)$, this midpoint (i.e., $\mu = 1/2$) value is sharp. The latter is the case for Lebesque, non–commutative L_p and various spaces considered in the next subsection.

c) Related notions, generalisations and applications, such as the moduli of the uniform μ–convexity and μ–smoothness of a convex function F (not necessarily a power of the norm), the corresponding Fenchel–Young–Lindenstrauss duality, classes of log–uniformly convex and smooth functions F and their high–order characterisations (Walsh chaos estimates), are studied in [23].

To formulate Theorems 3.2 − −3.4 determining the (p, h_c)–uniform convexity and (q, h_s)–uniform smoothness of IG–spaces, we define two functions. For $s, t \in (1, \infty)$ and $\mu \in [0, 1/2]$, let

$$\omega_c(\mu, s, t) = \begin{cases} (s-1)2^{2-t} & \text{for } s \leq 2, \\ \psi_s(\mu)2^{s-t} & \text{for } s \geq 2; \end{cases}$$

$$\omega_s(\mu, s, t) = \begin{cases} (\psi_s(\mu))^{\frac{t-1}{s-1}} 2^{\frac{s-t}{s-1}} & \text{for} \quad s \leq 2, \\ (s-1)^{t-1} 2^{2-t} & \text{for} \quad s \geq 2, \end{cases}$$

where

$$\psi_s(\mu) = \frac{1 + (z_s(\mu))^{s-1}}{(1 + z_s(\mu))^{s-1}} \quad \text{and}$$

$$z_s(\mu) = \begin{cases} \text{positive root of } \nu z^{s-1} - \mu = (\nu z - \mu)^{s-1} & \text{for} \quad s \neq 2 \text{ and } \mu \neq 0, \\ \text{positive root of } (s-2)z^{s-1} + (s-1)z^{s-2} = 1 & \text{for} \quad s \neq 2 \text{ and } \mu = 0, \\ 1 & \text{for} \quad s = 2. \end{cases}$$

3.1.1. Independently Generated IG_+ and Non–Commutative L_p Spaces

The sharpness of Theorems 3.1 – 3.3 is discussed in [23] and not needed for our purposes.

Theorem 3.1 ([23])**.** *Let* $1 - \nu = \mu \in [0, 1]$, *and let* X *be either a subspace or a quotient of a space* $Y \in IG$, *or finitely represented in* Y *with* $[p_{\min}(Y), p_{\max}(Y)] \cup \{2\} \subset [r, q] \subset (1, \infty)$. *Then, for* $f, g \in X$, *we have*

a) $\|\mu f + \nu g\|_X^q + \mu \nu \omega_c(\min(\mu, \nu), p_{\min}(Y), q) \|f - g\|_X^q \leq \mu \|f\|_X^q + \nu \|g\|_X^q;$

b) $\mu \|f\|_X^r + \nu \|g\|_X^r \leq \|\mu f + \nu g\|_X^r + \mu \nu \omega_s(\min(\mu, \nu), p_{\max}(Y), r) \|f - g\|_X^r.$

Remark 3.2. *The finite representability of* $l_{p_{\min}(Y)}$ *and/or* $l_{p_{\max}(Y)}$ *in* $Y \in IG$ *is relatively easy to check. The "worst" cases may look like* $l_{p_0}(\mathbb{N}, \{l_{p_i}(I_{n_i})\}_{i \in \mathbb{N}})$ *for some unbounded* $\{n_i\}_{i \in \mathbb{N}} \subset \mathbb{N}$ *with* $\inf_{i \in \mathbb{N}_0} p_i \notin \{p_i\}_{i \in \mathbb{N}_0}$ *and/or* $\sup_{i \in \mathbb{N}_0} p_i \notin \{p_i\}_{i \in \mathbb{N}_0}$.

Before presenting the next theorem describing the (q, h_c)–uniform convexity and (r, h_s)–uniform smoothness properties of non–commutative Lebesgue spaces (ex. Schatten–von Neumann classes), we recall that, for $s, t \in [1, \infty]$, one has

$$\omega_c(1/2, s, t) = (\min(s, 2) - 1)\, 2^{2-t};$$

$$\omega_s(1/2, s, t) = (\max(s, 2) - 1)^{t-1}\, 2^{2-t}.$$

The midpoint case $\mu = \nu = 1/2$ of the next theorem was considered by Dixmier [53] in 1953, Simon [54] in 1987 and Ball, Carlen and Lieb [55] in 1994. Namely (for $\mu = 1/2$), the cases of both Part a) with $p = q \in [2, \infty]$ and Part b) with $p = r \in [1, 2]$ were treated in [53, 54], while the cases of both Part a) with $p = r \in [1, 2]$ and Part b) with $p = q \in [2, \infty]$ were considered in [55].

Theorem 3.2 ([23])**.** *Let* $1 - \nu = \mu \in [0, 1]$, $r, q \in [1, \infty]$ *and* $p \in (1, \infty)$, *and let* X *be either a subspace or a quotient of* $Y \in \{L_p(\mathcal{M}, \tau), S_p\}$, *or finitely represented in* Y, *where* $(\mathcal{M}, \tau)$ *is a von Neumannn algebra with a normal semifinite faithful weight* τ. *Assume also that* $\{p, 2\} \subset [r, q]$. *Then, for* $f, g \in X$, *we have*

a) $\|\mu f + \nu g\|_X^q + \mu \nu \omega_c(1/2, p, q) \|f - g\|_X^q \leq \mu \|f\|_X^q + \nu \|g\|_X^q;$

b) $\mu \|f\|_X^r + \nu \|g\|_X^r \leq \|\mu f + \nu g\|_X^r + \mu \nu \omega_s(1/2, p, r) \|f - g\|_X^r;$

c) the bound $q \geq \max(p,2)$ in a) is sharp if $l_{\max(p,2)}$ is finitely represented in X, and the bound $r \leq \min(p,2)$ is sharp if $l_{\min(p,2)}$ is finitely represented in X;

d) for the space X, the constants $\omega_c(1/2, p, q) = 2^{2-q}$ and $\omega_s(1/2, p, r) = 2^{2-r}$ are sharp if $p \geq 2$ and $p \leq 2$ correspondingly;

e) the constants $\omega_c(1/2, p, 2) = p - 1$ and $\omega_s(1/2, p, 2) = p - 1$ are sharp uniformly by μ if $l_p(I_2)$ is finitely represented in X and, correspondingly, $p \in [1, 2]$ and $p \in [2, \infty]$; this is, particularly, the case when $X = Y$.

Remark 3.3. *It is natural to investigate the (q, h_c)–uniform convexity and (r, h_s)–uniform smoothness properties of non–commutative Lebesgue spaces in a more precise manner (with h_c and h_s depending on μ) in the Dixmier–Simon ranges of the parameter p and compare them with the corresponding properties of the (commutative) Lebesgues spaces and l_p.*

Theorem 3.3 ([23])**.** *Let $1 - \nu = \mu \in [0, 1]$, and let X be either a subspace or a quotient of a space $Y \in IG_+$, or finitely represented in Y with $[p_{\min}(Y), p_{\max}(Y)] \cup \{2\} \subset [r, q] \subset (1, \infty)$. Then, for $f, g \in X$, we have*

a) $\|\mu f + \nu g\|_X^q + \mu\nu\omega_c(1/2, p_{\min}(Y), q)\|f - g\|_X^q \leq \mu\|f\|_X^q + \nu\|g\|_X^q;$

b) $\mu\|f\|_X^r + \nu\|g\|_X^r \leq \|\mu f + \nu g\|_X^r + \mu\nu\omega_s(1/2, p_{\max}(Y), r)\|f - g\|_X^r.$

3.1.2. Sobolev, Besov and Lizorkin–Triebel Spaces

Let us recall the agreement that the components of a and ς are in the convex envelope of the other parameters. We need the following extract (omitting the sharpness) from Theorems $4.20 - 4.23$ in [23].

Theorem 3.4 ([23])**.** *For open $G \subset \mathbb{R}^n$, $s \in (0, \infty)$ and $\{p_{\min}, p_{\max}, 2\} \subset [r_s, r_c] \subset (1, \infty)$, let*

$$X_{p,q} \in \Gamma_1 \cup \Gamma_3 \cup \Gamma_4 \cup \{B^s_{p,q}(G), B^s_{p',q'}(G)^*\}_{p\in(1,\infty)^n}^{q\in(1,\infty)} \text{ and } Y_p \in \Gamma_2 \cup \{L_p\}_{p\in(1,\infty)^n},$$

and let also X (Y) be either a subspace, or a quotient, or finitely represented in $X_{p,q}$ (Y_p). Then

a) *if $q \in [r_s, r_c]$, the space X is (r_c, h_c)–uniformly convex and (r_s, h_s)–uniformly smooth with $h_c(\mu) = \omega_c(\mu, \min(p_{\min}, q), r_c)$ and $h_s(\mu) = \omega_s(\mu, \max(p_{\max}, q), r_s)$ for $\mu \in [0, 1]$;*

b) *the space Y is (r_c, h_c)–uniformly convex and (r_s, h_s)–uniformly smooth with $h_c(\mu) = \omega_c(\mu, p_{\min}, r_c)$ and $h_s(\mu) = \omega_s(\mu, p_{\max}, r_s)$ for $\mu \in [0, 1]$. Moreover, the ranges of r_s and r_s are exact if $X = X_{p,q} \notin \Gamma_4$ and $Y = Y_p$.*

With the aid of the results on the lattice convexity and concavity in Section 4.2 in [25] and the following Theorem 3.5 from [26] relying, in turn, on $1.d.8$, $1.d.4, iii)$ and $1.f.1$ in [37] and Lemma $4.1, h)$ in [23], we obtain the existence of equivalent asymmetrically uniformly convex and uniformly smooth norms on Sobolev, Besov and Lizorkin-Triebel spaces with underlying mixed Lorentz norms in Theorem 3.6.

Theorem 3.5 ([26]). *For $2 \in [q,p] \subset (1,\infty)$ and a Banach lattice X, let $q \in I_{conv}(X)$ and $p \in I_{conc}(X)$. Then there exist an equivalent norm $\|\cdot\| \asymp \|\cdot\|_X$ and constants $C_s = C_c(q,p)$ and $C_c = C_c(q,p) > 0$, such that X endowed with $\|\cdot\|$ and the same order is not only both lattice q–convex and p–concave with the constant 1 but also (p, C_s)–uniformly convex and (q, C_s)–uniformly smooth.*

Theorem 3.6. *For an open $G \subset \mathbb{R}^n$, $s \in (0,\infty)$, $p, \theta \in (1,\infty)^n$, $q \in (1,\infty)$, $[p_{\min}, p_{\max}] \subset (r_s, r_c)$ and $\{\theta_{\min}, \theta_{\max}, 2\} \subset [r_s, r_c] \subset (1,\infty)$, let*

$$X_{L_{p,\theta,q}} \in \Gamma_1^L \cup \Gamma_3^L \cup \Gamma_4^L \cup \left\{ B^s_{L_{p,\theta,q}}(G), B^s_{L_{p',\theta',q'}}(G)^* \right\} \text{ and } Y_{L_{p,\theta}} \in \Gamma_2^L \cup \{L_{p,\theta}\},$$

and let also X (Y) be either a subspace, or a quotient, or finitely represented in $X_{L_{p,\theta,q}}$ ($Y_{L_{p,\theta}}$). Then there exist an equivalent norm on X (Y) and constants $C_c, C_s > 0$, such that
a) if $q \in [r_s, r_c]$, the space X is (r_c, h_c)-uniformly convex and (r_s, h_s)–uniformly smooth with $h_c(\mu) = C_c$ and $h_s(\mu) = C_s$ for $\mu \in [0,1]$;
b) the space Y is (r_c, h_c)-uniformly convex and (r_s, h_s)-uniformly smooth with $h_c(\mu) = C_c$ and $h_s(\mu) = C_s$ for $\mu \in [0,1]$.

Unfortunately it is not clear whether the preceding theorem gives the exact ranges of the (simultaneous) asymmetric uniform convexity and smoothness after a renorming but, allowing separate renormings for either asymmetric uniform convexity or asymmetric uniform smoothness, we obtain the exact ranges occasionally including $p_{\min}$ and $p_{\max}$ and coinciding, correspondingly, with the exact ranges of either the Rademacher cotype or Rademacher type (see [25]). Since the precise formulation in the setting of the mixed Lorentz quasinorm is rather lengthy, we consider the case of the ordinary (non–mixed) Lorentz quasinorm. Note that the spaces of the same Besov and Lizorkin–Triebel type with the same parameters can have *different ranges*. Note that the *exactness of the ranges* in Theorems 3.4 and 3.7 still takes place for many specific subclasses of Γ_4 and Γ_4^L following from the exactness for Γ_1 and Γ_1^L but not in the abstract setting.

Theorem 3.7. *For an open $G \subset \mathbb{R}^n$, $s \in (0,\infty)$ and $p, \theta, q \in (1,\infty)$, let*

$$X_{L_{p,\theta,q}} \in \Gamma_1^L \cup \Gamma_3^L \cup \Gamma_4^L \cup \left\{ B^s_{L_{p,\theta,q}}(G), B^s_{L_{p',\theta',q'}}(G)^* \right\},$$

and let also X be either a subspace, or a quotient, or finitely represented in $X_{L_{p,\theta,q}}$. Then we have:
a) the existence of an equivalent norm on $X_{L_{p,\theta,q}}$ and a constant $C_c > 0$, such that the space X is (r_c, h_c)-uniformly convex with $h_c(\mu) = C_c$ for $\mu \in [0,1]$ if either $X_{L_{p,\theta,q}}$ is a Besov type space and $r_c \in [\max(2,p,q,\theta), \infty) \setminus (\{p\} \cap \{2\})$, or $X_{L_{p,\theta,q}}$ is a Lizorkin–Triebel type space and $r_c \in [\max(2,p,q,\theta), \infty) \setminus (\{p\} \cap \{q,2\})$;
b) the existence of an equivalent norm on $X_{L_{p,\theta,q}}$ and a constant $C_s > 0$, such that the space X is (r_s, h_s)–uniformly smooth with $h_s(\mu) = C_s$ for $\mu \in [0,1]$ if either $X_{L_{p,\theta,q}}$ is a Besov type space and $r_s \in (1, \min(2,p,q,\theta)] \setminus (\{p\} \cap \{2\})$, or $X_{L_{p,\theta,q}}$ is a Lizorkin–Triebel type space and $r_s \in (1, \min(2,p,q,\theta)] \setminus (\{p\} \cap \{q,2\})$.
Moreover, the ranges of r_s and r_c in a) and b) are exact if $X = X_{L_{p,\theta,q}} \notin \Gamma_4^L$.

Proof of Theorem 3.7. As discussed (in full detail) in Section 2.2 in [25] and mentioned above, every $X_{L_{p,\theta,q}}$ is a subspace or a quotient of a subspace of an IG–space that is a reflexive function lattice (ideal space). Therefore, according to pages 100 and 101 in [37], the exponents of the classical power uniform convexity and power uniform smoothness (after a proper and, possibly, different renormings $\|\cdot\|_c$ and $\|\cdot\|_s$) coincide, correspondingly, with the Rademacher cotype and type sets found in Section 4.2 in [25]. Pisier's martingale technique [56] allows to transform the classical power type uniform convexity and smoothness into the homogeneous midpoint uniform convexity and smoothness of the same power types: for $x, y \in X_{L_{p,\theta,q}}$,

$$\frac{\|x\|_c^{r_c} + \|y\|_c^{r_c}}{2} \geq \left\|\frac{x+y}{2}\right\|_c^{r_c} + C_c \frac{\|x-y\|_c^{r_c}}{4}, \quad \frac{\|x\|_s^{r_s} + \|y\|_s^{r_s}}{2} \leq \left\|\frac{x+y}{2}\right\|_s^{r_s} + C_s \frac{\|x-y\|_s^{r_s}}{4}.$$

Eventually, Lemma 4.1, *h*) from [23] implies the (r_c, C_c)–uniform convexity of $\|\cdot\|_c$ and the (r_s, C_s)–uniform smoothness of $\|\cdot\|_s$ with the same constant functions $h_c = C_c$ and $h_s = C_s$ on $[0,1]$ finishing the proof of *a*) and *b*). The ranges of r_s and r_c are exact if $X = X_{L_{p,\theta,q}}$ because they coincide with the Rademacher type and cotype sets of X found in Theorem 4.7 in [25] due to the λ–finite representability of $l_q(\mathbb{N}, l_{p,\theta}(\mathbb{N}))$ or $l_{p,\theta}(\mathbb{N}, l_q(\mathbb{N}))$ in X [25, Section 3]. $\square$

3.2. Markov Chains and Abstract Uniform Convexity and Smoothness

This subsection explains the sense in which the asymmetric uniform convexity and smoothness imply not only the abstract uniform convexity and smoothness but also our ordinary (Lemma 3.1, *c*)) and Markov chain (Theorem 4.2 below) counterparts of the Pisier martingale inequality [56]. We need to specify the class of the Markov chains we are dealing with.

Let us recall that a Markov process with discrete time $t \in \mathbb{N}$ and finite or countable state space S is a sequence of stochastic variables $\{\xi_t\}_{t \in \mathbb{N}}$ satisfying

$$p(\xi_{t_0} = s_0 | \xi_{t_1} = s_2, \ldots, \xi_{t_n} = s_n) = p(\xi_{t_0} = s_0 | \xi_{t_1} = s_2)$$

for every $n \in \mathbb{N}$ and increasing $\{t_i\}_{i=0}^n \subset \mathbb{N}$. It is *stationary* if all $\{\xi_t\}_{t \in \mathbb{N}}$ are identically distributed, i.e., $p(\xi_t = s) = \nu_s > 0$ for every $s \in S$ and $t \in \mathbb{N}$ with $\sum_{s \in S} \nu_s = 1$. The Markov process $\{\xi_t\}_{t \in \mathbb{N}}$ with discrete time and state is *reversible* if

$$p(\xi_{t_1} = s_1, \xi_{t_2} = s_2) = p(\xi_{t_2} = s_1, \xi_{t_1} = s_2) \text{ for } t_1 \neq t_2.$$

Let $T_\xi(t_1, t_2)$ be the transition matrix $(p(\xi_{t_2} = s_2 | \xi_{t_1} = s_1))_{s_1, s_2 \in S}$ that is symmetric (and, thus, double–stochastic) for a reversible Markov process. For the sake of convenience, we assume that $T_\xi(t) = T_\xi(t, t+1)$ for $t \in \mathbb{N}$. Therefore, the properties of the Markov process $\{\xi_t\}_{t \in \mathbb{N}}$ are expressed in terms of the matrix multiplication as

$$T_\xi(t_1, t_2) = \prod_{t=t_1}^{t_2-1} T_\xi(t) \text{ for } t_1 < t_2; \quad \bar{\nu} T_\xi(t) = \bar{\nu} \text{ and } \nu_{s_1}(T_\xi(t))_{s_1, s_2} = \nu_{s_2}(T_\xi(t))_{s_2, s_1}$$

for $t \in \mathbb{N}$ correspondingly, where $\bar{\nu}$ is the row of $\{\nu_s\}_{s \in S}$.

Parts *a*) + *b*), *c*), and *d*) of the following lemma are, respectively, immediate and straightforward corollaries to Parts *b*), *c*), and *f*) of Lemma 4.2 from [23] applied to the class of the convex functions $F(x) = \|x\|_X^s$ for $s \in (1, \infty)$. Parts *a)* and *b)* simply state that the asymmetric uniform convexity and smoothness imply abstract uniform convexity and smoothness correspondingly. Note that the properties of the functions and the statements of the theorems in Sections 3.1.1 and 3.1.2 show that the constants in Parts *a)* and *b)* are either the same or sharper than in their counterparts in Lemmas 4.1 in [21] and 3.1 in [17]. Part *c)* is our counterpart of the Pisier martingale inequality. It will be crucial for estimating the Markov type and cotype constants in what follows.

Lemma 3.1. *For $2 \in [q, p] \subset (1, \infty)$, let X be a (p, h_c)–uniformly convex and (q, h_s)–uniformly smooth Banach space. Then the following holds.*
a) For an arbitrary X–valued stochastic variable ξ defined on a probability measure space (Ω, μ), we have

$$\mathrm{E}\|\xi\|_X^q - \|\mathrm{E}\xi\|_X^q \leq \inf_{\mu \in [0, 1/2]} (1-\mu)^{1-q} h_s(\mu) \mathrm{E}\|\xi - \mathrm{E}\xi\|_X^q.$$

b) For an arbitrary X–valued stochastic variable ξ defined on a probability measure space (Ω, μ), we have

$$\sup_{\mu \in [0, 1/2]} (1-\mu) h_c(\mu) \mathrm{E}\|\xi - \mathrm{E}\xi\|_X^p \leq \mathrm{E}\|\xi\|_X^p - \|\mathrm{E}\xi\|_X^p.$$

c) For an X–valued martingale difference sequence $\{\xi_i\}_{i \in \mathbb{N}_0}$ defined on a probability measure space (Ω, μ) and related filtration, we have, for every $n \in \mathbb{N}$,

$$\mathrm{E}\|\sum_{i=0}^n \xi_i\|_X^q - \mathrm{E}\|\xi_0\|_X^q \leq \inf_{\mu \in [0, 1/2]} (1-\mu)^{1-q} h_s(\mu) \sum_{i=1}^n \mathrm{E}\|\xi_i\|_X^q.$$

d) For arbitrary independent X–valued stochastic variables ξ and η defined on a probability measure space (Ω, μ), we have

$$\mathrm{E}\|\xi - \eta\|_X^q - \mathrm{E}\|\mathrm{E}\xi - \eta\|_X^q \leq \inf_{\mu \in [0, 1/2]} (1-\mu)^{1-q} h_s(\mu) \mathrm{E}\|\xi - \mathrm{E}\xi\|_X^q.$$

4. HÖLDER–LIPSCHITZ MAPPINGS: APPROXIMATION, EXTENSION AND INTERPOLATION

This section is dedicated to the extension problem for the Hölder–Lipschitz mappings from a subset of a metric or Banach space into a Banach space, and to the real interpolation of the spaces of these mappings and a related approximation problem. We start with the treatment of these problems in the setting of the pairs of abstract spaces in the first three subsections and finish with the treatment of various pairs of our concrete spaces.

Definition 4.1. *Assume that X is a metric space, $\hat{x} \in X$, and Y is a Banach space, and $\alpha, d > 0$. Let $H^\alpha(X, Y)$ be the Banach space of all Y–valued continuous functions f defined on X with $f(\hat{x}) = 0$ and the finite norm:*

$$\|f|H^\alpha(X, Y)\| := \sup\{\|f(x) - f(y)\|_Y / d_X(x, y) : x, y \in X \text{ and } x \neq y\}.$$

We say that the pair (X, Y) possesses (d, α)-extension property if, for every subset $F \subset X$ (with the induced metric) and every $f \in H^\alpha(F, Y)$, there is an extension $\tilde{f} \in H^\alpha(X, Y)$ satisfying

$$\tilde{f}(x) = f(x) \text{ for } x \in F \text{ and } \|\tilde{f}|H^\alpha(X, Y)\| \leq d\|f|H^\alpha(F, Y)\|.$$

Let $S_b(X, Y) \subset (0, \infty)$ be the set of all α, such that the pair (X, Y) possesses (d, α)-extension property for some $d < \infty$.

We say that the pair (X, Y) possesses convex (d, α)-extension property if it possesses the (d, α)-extension property, and there exists a corresponding extension $\tilde{f} \in H^\alpha(X, Y)$ of $f \in H^\alpha(F, Y)$ satisfying $\tilde{f}(X) \subset \overline{\text{co}} f(F)$.

Let also $S_=(X, Y) \subset (0, \infty)$ be the set of all α, such that the pair (X, Y) possesses $(1, \alpha)$-extension property, while $S_{=,c}(X, Y) \subset (0, \infty)$ be the set of all α, such that the pair (X, Y) possesses convex $(1, \alpha)$-extension property.

The discrepancy between an arbitrary pair of $\{S_b(X, Y), S_=(X, Y), S_{=,c}(X, Y)\}$ is called the phase transition phenomenon for the pair (ex. [20]).

We shall always assume $\alpha \in (0, 1]$ because, if X is metrically convex (see [18]; ex. X is a convex subset of a linear space) $H^\alpha(X, Y) = \{0\}$ for $\alpha > 1$ due to the triangle inequality.

Instead of enforcing $f(\hat{x}) = 0$ to define $H^\alpha(X, Y)$, we could have factorized the corresponding semi-normed space by the subspace of constant mappings (i.e., the null-space).

For the applications of the results on bounded extension it is very useful to observe that, if a pair (X, Y) has a (d, α)-extension property, and X is C_0-Lipschitz homeomorphic (or C_0-isomorphic if X is Banach) to X_0, while Y is C_1-isomorphic to Y_0, then the pair (X_0, Y_0) possesses the $(dC_0^\alpha C_1, \alpha)$-extension property, where X_0 can be a quasi-metric or a quasi-Banach space and Y_0 can be a quasi-Banach space.

4.1. Isometric Extensions

The isometric extension problem addresses the question of the existence of $(1, \alpha)$-extension property for various pairs of spaces. The general (point-by-point extension) scheme for solving the isometric extension problem, that is analogous to the proof of the Hahn–Banach theorem, was created by Minty in [12] who had treated the setting of the pairs of Hilbert spaces. Detailed presentation and further development of the scheme was undertaken by Wells, Williams and Hayden in [13, 14], who had also found the exponent sets:

$$S_{=,c}(X_p, X_q) = S_=(X_p, X_q) = \left(0, \frac{\min(p, p')}{\max(q, q')}\right]$$

for $X_\sigma \in \{l_\sigma, L_\sigma\}$, $\sigma \in \{p, q\} \subset (1, \infty)$; (WW)

$$S_{=,c}(M, Y_q) = S_=(M, Y_q) = \left(0, \frac{1}{\max(q, q')}\right]$$

for $Y_q \in \{l_q, L_q\}$, $q \in (1, \infty)$ and a metric space M.

In [16], we have identified $S_=(X,Y)$ and $S_{=,c}(X,Y)$ when $Y \in IG_+$ and X is either a metric space, or an IG_+-space and described the algorithm of dealing with X and Y that can, in addition, be also drawn from various classes of anisotropic Besov, Lizorkin–Triebel and Sobolev spaces.

4.2. Bounded Extensions

The bounded extension problem addresses the question of the existence of the (d,α)-extension property for a finite d for various pairs (X,Y) of spaces (i.e., the description of $S_b(X,Y)$ in its qualitative aspect) and the evaluation (computation) of the corresponding value of d (in its quantitative aspect). The general (multi–point extension) scheme for solving the bounded extension problem for $\alpha = 1$ in its qualitative and quantitative aspects by means of reducing it to the identification of the Markov type and cotype sets (introduced by K. Ball) and evaluating the related constants was created by K. Ball in his pioneering work [17] in 1992. In particular, he has also provided an upper estimate for the Markov cotype constant in terms of uniform convexity constants showing that the uniform convexity of the power p implies the Markov cotype p for $p \in [2,\infty)$. Ball's original extension theorem covers the case of the extension of the Lipschitz mappings from a Markov type 2 metric space into a Markov cotype 2 reflexive Banach space. In 2001, A. Naor [20] noticed that Ball's original theorem is naturally extended to the case of the Lipschitz mappings from a Markov type p metric space into a Markov cotype p reflexive Banach space. Moreover, he observed that $p/2$–Hölder mappings can be treated even using Ball's original extension theorem by means of noticing that the Markov type p property of a metric space (X,d) is equivalent to the Markov type 2 property of $(X,d^{p/2})$ for $p \in (0,2]$. By further adding some abstract stability results and sharpness counterexamples and using the isometric extension results from [13, 15, 14], Naor identified the ranges of the bounded extendability exponents $S_b(L_p,L_q)$ for the Hölder–Lipschitz mappings between the pairs of Lebesgue spaces except for the case $p \in (2,\infty)$, revealing also the presence of the phase transition phenomenon. In 2001, A. Naor, Y. Peres, O. Schramm and S. Sheffield [21] estimated Markov type 2 constants for the Lebesgue L_p–spaces with $p \in (2,\infty)$ and a number of metric spaces, thus, finishing the description of $S_b(L_p,L_q)$ for all pairs of the Lebesgue spaces and established a nonlinear version of Maurey's extension theorem. In 2006, Mendel and Naor discovered various applications of Ball's extension method including its intimate relation with a generalization of the Johnson–Lindenstrauss theorem. All these authors (except for K. Ball) were primarily interested in the applications to various metric spaces dealt with in theoretical computer science.

In this section we extend Ball's extension theorem to the "off–diagonal" setting of mappings from the Markov type p metric spaces into Markov cotype q Banach spaces with explicit bounds for the extension constant d, establish the connections between the bounded extendability of Hölder mappings and real interpolation of these mappings, investigate the properties of the Markov type and cotype for Banach spaces and apply all this results to the pairs (X,Y) of spaces drawn from the classes of metric, IG, IG_+ and non–commutative L_p–spaces.

Let us define the notions of the Markov type and Markov cotype (see Section 3.2 for some definitions regarding Markov processes).

Definition 4.2 ([21])**.** *Let (X, d) be a metric space and $p \in (0, \infty]$. The space X is said to possess the Markov type p with a constant C_{MT} if, for every $n \in \mathbb{N}$, every stationary reversible Markov chain (Markov process on $I_n \cup \{0\}$) with a state set S, and every $f : S \to X$, one has the estimate*

$$(\mathrm{E}d\,(f(\xi_n), f(\xi_0))^p)^{1/p} \leq C_{MT}\,(n\mathrm{E}d\,(f(\xi_1), f(\xi_0))^p)^{1/p}.$$

The best constant C_{MT} is designated by $C_{MTp}(X)$. Let $I_{MT}(X)$ denote the set of all Markov type exponents of X.

Remark 4.1. *a) Note that we can consider only the chains with strictly positive stationary distributions.*
b) K. Ball [17] showed that every metric space (X, d) has the Markov type 1 with the constant 1. Since d^α with $\alpha \in (0, 1)$ is still a metric on X, every metric space is also of type α. Thus, we focus on the Markov type exponents in $(1, \infty)$ and related constants.
c) Markov type properties (the type and constant) are inherited by the subsets of a metric space.
d) In fact, Theorem 1.6 in [17] shows that the definition of Markov type in [17] (Definition 1.6 in [17]) is equivalent to, at least, a formally less restrictive counterpart of Definition 4.2 where only the stationary reversible Markov chains with symmetric transition matrixes are allowed. Thus, if X is of Markov type p with a constant C_{MT} according to Definition 4.2, it is also of Markov type p with not worse constant according to Ball's original definition.
e) There are other notions of type and cotype than Markov or Rademacher ones (ex. [57]).

To define the Markov cotype, we slightly modify the original definition of Ball (written the language of matrixes) by substituting the exponent 2 with q.

Definition 4.3 ([17])**.** *Let X be a normed space and $q \in [1, \infty]$. The space X is said to possess the Markov cotype p with a constant C_{MC} if, for every $n \in \mathbb{N}$, $\beta \in (0, 1)$, symmetric (double)stochastic $n \times n$ matrix A and sequence $\{x_i\}_{i=1}^n \subset X$, one has the estimate*

$$\left(\beta \sum_{i,j=1}^n a_{i,j} \left\|\sum_{k=1}^n c_{i,k} x_k - \sum_{l=1}^n c_{j,l} x_l\right\|_X^q\right)^{1/q} \leq C_{MC} \left((1-\beta) \sum_{i,j=1}^n c_{i,j} \|x_i - x_j\|_X^q\right)^{1/q},$$

where $C = (1-\alpha)(I-\alpha A)^{-1}$ and $\{a_{i,j}\}$ and $\{c_{i,j}\}$ are the entries of A and C respectively.

For the sake of convenience, we say that X possesses the Markov cotype ∞ with the constant 1 if, instead, one has

$$\max_{i,j=1}^n \left\|\sum_{k=1}^n c_{i,k} x_k - \sum_{l=1}^n c_{j,l} x_l\right\|_X \leq \max_{i,j=1}^n \|x_i - x_j\|_X.$$

The best constant C_{MC} is designated by $C_{MCq}(X)$. Let $I_{MC}(X)$ denote the set of all Markov cotype exponents of X.

Remark 4.2. *a) Since $\sum_{k=1}^n c_{i,k} x_k - \sum_{l=1}^n c_{j,l} x_l = \sum_{k=1}^n \sum_{l=1}^n c_{i,k} c_{j,l} (x_k - x_l)$, the triangle inequality implies that every Banach space has the Markov cotype ∞ with the* *constant* 1.

b) *Let us note that, if a Banach space X is finitely represented in Y possessing the Markov type p with the constant C_{MT} and the Markov cotype q with the constant C_{MC}, then X has the same Markov type and cotype with the same constants.*

c) *Note that the Markov cotype can also be correctly defined for the convex subsets of a Banach space.*

4.2.1. Markov Type

While this section is dedicated to the Markov type sets and constants for both model and auxiliary independently generated IG_+ and noncommutative L_p–spaces, the anisotropic Sobolev, Besov and Lizorkin-Triebel spaces with underlying mixed Lebesgue and Lorentz norms and their duals from $\Gamma_i \cup \Gamma_i^L$ for $i \in I_4$ are dealt with in Section 4.4.1.

With the aid of the spectral theorem for symmetric matrixes, K. Ball showed the Markov type 2 of every Hilbert space and, then, used the isometric embedding of the Lebesgue L_p and l_p for $p \in (0, 2]$ endowed with the metric $\|f-g\|_p^{p/2}$ into a Hilbert space (Theorem 4.10 in [14]; see also Lemma 4.1, b) below) to demonstrate the Markov type p of these L_p and l_p (Theorem 1.6 in [17]). Naor [20] found a direct way to expose the Markov type p of L_p for $p \in (0, 2]$ using an integral representation in the style of §3 in [14]. Let us note that Theorem 5.11 in [14] implies the Markov type q with the constant 1 for L_p for every q satisfying $\emptyset \neq [q, p] \subset (0, 2]$. Naor, Peres, Schramm and Sheffield [21] designed an interesting approach to computing the Markov type 2 of the Lebesgues spaces L_p for $p \in (2, \infty)$ based on Pisier's inequality for martingales and a decomposition of a Markov chain as a sum of a backward and forward martingales, and estimated the Markov type 2 constant. They also computed the Markov types of various metric spaces because of their primary interest in them. We shall consider only linear spaces and use a different approach based on our generalization of Pisier's inequality for martingales to Markov processes providing explicit estimates for related constants, improving known estimates and showing the sharpness of the constant 1.

In this section we describe the Markov type sets for various spaces and estimate the corresponding constants from both sides.

We provide two proofs of the following auxiliary result.

Theorem 4.1. *The subset of matrixes with positive entries is dense in the set of the transition matrixes of the stationary reversible Markov chains with an arbitrary finite number of states (we assume that the stationary distributions have only positive entries).*

The first proof of Theorem 4.1. Indeed, let A be an $n \times n$ stochastic matrix with the stationary distribution $(w_1, w_2, \ldots, w_n)$ and $a_{i,j} = a_{j,i} = 0$ for some i and j. The matrix B, whose entries are 0 except for $b_{i,i} = b_{j,i} = w_i/(w_i + w_j)$, $b_{i,j} = b_{j,j} = w_j/(w_i + w_j)$ is a stochastic transition matrix of a stationary reversible Markov chain with the same stationary distribution, as well as the matrixes of the segment $I = \{(1-t)A + tB : t \in [0, 1]\}$. The matrixes in the interior of I have positive entries at (i, i), (i, j), (j, i) and (j, j). The claim follows by the induction on the number of 0 entries of A. □

The second proof of Theorem 4.1. For the second proof let us consider $B = (1, 1, \ldots, 1)^T(w_1, w_2, \ldots, w_n)$ and $A_t = (1-t)A + tB$. Clearly, A_t with $t \in (0, 1)$

is a transition matrix for some stationary reversible Markov chain $\xi(t)$ with the the same stationary distribution and with strictly positive entries satisfying $\lim_{t\to 0} A_t = A$. □

The following key theorem of this subsection relates the (r, h_s)–uniform smoothness to the Markov type r in a quantitative manner. It is a counterpart of Theorem 2.3 in [21] proved there with the aid of representing the stationary reversible Markov chain as a sum of forward and backward martingales (Lemma 4.3 in [21]) inspired by the continuous setting of such decomposition in [58]. We reduce the multiplicative "loss" in the Markov type constant by means of developing a different martingale decomposition in Doob's style for an auxiliary process and using uniform smoothness instead of ordinary convexity with the aid of the specially developed tools (from [23]) in Lemma 3.1 above. In particular, this approach permits us to treat the cases of odd and even n simultaneously.

Theorem 4.2. *For $r \in [1,2]$, let X be a (r, h_s)–uniformly smooth Banach space and $\{\xi_i\}_{i\in\mathbb{Z}}$ a stationary reversible Markov chain with a countable or finite state set S and $f : S \longrightarrow X$. Then, for every $n \in \mathbb{N}$ and $C_s = \inf_{\mu\in(0,1/2]}(1-\mu)^{1-q}h_s(\mu)$, we have*

$$\mathrm{E}\|f(\xi_n) - f(\xi_0)\|_X^r \leq 2^{r+1}\left(C_s n + h_s(1/2)/4\right)\mathrm{E}\|f(\xi_1) - f(\xi_0)\|_X^r \leq$$
$$\leq 2^{r+1}\left(1 + h_s(1/2)/12C_s\right)C_s n \mathrm{E}\|f(\xi_1) - f(\xi_0)\|_X^r.$$

That is, X possesses the Markov type r with the constant

$$C_{MTr} \leq C_{MT} = 2\left(2\left(1 + h_s(1/2)/12C_s\right)\right)^{1/r}\left(\inf_{\mu\in(0,1/2]}(1-\mu)^{1-q}h_s(\mu)\right)^{1/q}.$$

Moreover, in the cases of all particular groups of classes of Banach spaces under consideration, one has

$$C_{MT} \leq 2\left(13h_s(1/2)/6\right)^{1/r}.$$

Proof of Theorem 4.2. For the sake of simplicity, we denote the conditional expectation with respect to the sigma–algebra σ_j generated by ξ_j as E_{σ_j}. Let us recall that Doob's decomposition of a discrete submartingale into a martingale and predictable process is based on the manipulation with the martingale difference sequence of the form $M_i^+ = f(\xi_i) - \mathrm{E}_{\sigma_{i-1}}f(\xi_i)$. Since the Markov property withstands the change in the direction of time, we can also employ the backward martingale difference sequence $M_i^- = f(\xi_i) - \mathrm{E}_{\sigma_{i+1}}f(\xi_i)$, while the reversibility of the Markov chain allows us to relate these difference sequences with the aid of the identities $\mathrm{E}_{\sigma_i}f(\xi_{i-1}) = \mathrm{E}_{\sigma_i}f(\xi_{i+1})$. These observations justify our martingale difference representation

$$f(\xi_n) + \mathrm{E}_{\sigma_n}f(\xi_{n-1}) - f(\xi_0) - \mathrm{E}_{\sigma_0}f(\xi_1) =$$
$$= \sum_{i=1}^{n}\left(f(\xi_i) - \mathrm{E}_{\sigma_{i-1}}f(\xi_i)\right) - \sum_{i=0}^{n-1}\left(f(\xi_i) - \mathrm{E}_{\sigma_{i+1}}f(\xi_i)\right). \quad (1)$$

Note also that the convexity of the functional $\mathrm{E}\|\cdot\|_X^r$ implies the estimates

$$\mathrm{E}\|M_i^+\|_X^r \leq 2^{r-1}\left(\mathrm{E}\|f(x_i) - f(x_{i-1})\|_X^r + \mathrm{E}\left\|\mathrm{E}_{\sigma_{i-1}}(f(x_i) - f(x_{i-1}))\right\|_X^r\right) \leq$$
$$\leq 2^r\mathrm{E}\|f(x_i) - f(x_{i-1})\|_X^r = 2^r\mathrm{E}\|f(x_1) - f(x_0)\|_X^r \text{ and} \quad (2)$$

$$\mathrm{E}\|M_i^-\|_X^r \leq 2^{r-1}\left(\mathrm{E}\|f(x_i) - f(x_{i+1})\|_X^r + \mathrm{E}\left\|\mathrm{E}_{\sigma_{i-1}}(f(x_i) - f(x_{i+1}))\right\|_X^r\right) \leq$$
$$\leq 2^r \mathrm{E}\|f(x_i) - f(x_{i+1})\|_X^r = 2^r \mathrm{E}\|f(x_1) - f(x_0)\|_X^r. \quad (3)$$

Utilising these relations together with the midpoint ($\mu = 1/2$) (r, h_s)–uniform smoothness of X for $x = f(\xi_n) - f(\xi_0)$ and $y = \mathrm{E}_{\sigma_n} f(\xi_{n-1}) - \mathrm{E}_{\sigma_0} f(\xi_1)$, Part c) of Lemma 3.1 and the convexity of $\mathrm{E}\|\cdot\|_X^r$ as above, we establish the key estimates:

$$\frac{\mathrm{E}\|f(\xi_n) - f(\xi_0)\|_X^r + \mathrm{E}\|\mathrm{E}_{\sigma_n} f(\xi_{n-1}) - \mathrm{E}_{\sigma_0} f(\xi_1)\|_X^r}{2} \leq$$
$$\leq \mathrm{E}\left\|\frac{f(\xi_n) + \mathrm{E}_{\sigma_n} f(\xi_{n-1}) - f(\xi_0) - \mathrm{E}_{\sigma_0} f(\xi_1)}{2}\right\|_X^r +$$
$$+ 2^{r-2} h_s(1/2) \mathrm{E}\left\|\frac{f(\xi_n) - \mathrm{E}_{\sigma_n} f(\xi_{n-1}) - f(\xi_0) + \mathrm{E}_{\sigma_0} f(\xi_1)}{2}\right\|_X^r \leq$$
$$\leq C_s 2^{-1}\left(\sum_{i=1}^n \mathrm{E}\|M_i^+\|_X^r + \sum_{i=0}^{n-1} \mathrm{E}\|M_i^-\|_X^r\right) +$$
$$+ 2^{r-3} h_s(1/2)\left(\mathrm{E}\|\mathrm{E}_{\sigma_n}(f(\xi_n) - f(\xi_{n-1}))\|_X^r + \mathrm{E}\|\mathrm{E}_{\sigma_0}(f(\xi_0) - f(\xi_1))\|_X^r\right) \leq$$
$$\leq \left(C_s n 2^r + 2^{r-2} h_s(1/2)\right) \mathrm{E}\|f(x_1) - f(x_0)\|_X^r. \quad (4)$$

Eventually we ignore the second summand in the first numerator and multiply all terms by 2. To finish the proof with the desirable estimates

$$2^{r+1}(C_s n + h_s(1/2)/4) = 2^{r+1} C_s n(1 + h_s(1/2)/(4 C_s n)) \leq$$
$$\leq 2^{r+1} C_s n(1 + h_s(1/2)/(12 C_s)) \leq 2^r h_s(1/2) \frac{13}{6}$$

in the view of the relation $C_s \leq h_s(1/2)$ provided by Part a) of Theorem 4.13 in [23] for all particular spaces X under consideration in this chapter (note also that h_s and h_c are constant functions in Theorem 3.7 above), we only need to check the statement of the theorem for $n = 2$ to assume that $n \geq 3$. Indeed, keeping in mind the symmetric equivalent definition of the Markov property as the independence of the future and the past conditioned by now, the reversibility of the Markov chain and $C_s \geq 1$ implied by Remark 3.1, a), we use Part d) of Lemma 3.1 with E_{σ_1} instead of E and, then take the full expectation of the both sides of the result to arrive at

$$\mathrm{E}\|f(\xi_2) - f(\xi_0)\|_X^r \leq \mathrm{E}\|\mathrm{E}_{\sigma_1} f(\xi_2) - f(\xi_0)\|_X^r + C_s \mathrm{E}\|f(\xi_2) - \mathrm{E}_{\sigma_1} f(\xi_2)\|_X^r =$$
$$= \mathrm{E}\|f(\xi_0) - \mathrm{E}_{\sigma_1} f(\xi_0)\|_X^r + C_s \mathrm{E}\|f(\xi_2) - \mathrm{E}_{\sigma_1} f(\xi_2)\|_X^r \leq 2^{r+1} C_s \mathrm{E}\|f(\xi_1) - f(x_0)\|_X^r.$$

$\square$

The fact that the Markov type p is followed by the classical Rademacher type p is a deep observation due to K. Ball [17]. The straightforward generalisation of Lemma 2.1 and Theorem 2.2 from [17] from the case $p = 2$ to $p \in (1, 2]$ leads to the next (quantitative) theorem that we use to study the matter of sharpness of the sets of the parameters of the Markov type for the Banach spaces under the consideration.

Theorem 4.3. *For $p \in (1,2]$, let X be a Banach space possessing the Markov type p with a constant $C_{MTp} = C_{MTp}(X)$. Then it also has the Rademacher type p with the constant $T_{p,p}(X)$ satisfying*

$$T_{p,p}(X) \leq 2^{2-1/p} C_{MTp}(X).$$

The next lemma will help us to investigate the sharpness of the Markov type constants. The usage of the spectral theory for matrixes to establish the Markov type of a Hilbert space with the constant 1, as well as the usage of the isometric embeddings to establish the Markov type p of a Lebesgue L_p and l_p with the constant 1, is not new (see Proposition 1.4 in [17]) but we utilise these tools to show the sharpness as well.

Lemma 4.1. *a) Every Hilbert space has the Markov type r with the sharp constant 1 for $r \in (0,2]$.*
b) For $p \in (0,2]$, the space $X_{\bar{p}} \in IG$ (i.e., L_p with, possibly, atomic measure) has the Markov type r with the sharp constant 1 for $r \in (0, p]$.
c) If an infinite–dimensional Banach space X has a Markov type $r \in (0, 2]$ with the constant 1, then this constant is sharp.

Proof of Lemma 4.1. Since every Hilbert space is finitely represented in l_2, we assume that $H = l_2$. Let ξ be a stationary reversible Markov chain with an $n \times n$ transition matrix A and the stationary distribution $(w_1, w_2, \ldots, w_n)$, and let $x = (x_1, x_2, \ldots, x_n) \in H^n$ be an element of $\tilde{H} = l_2(I_n, \{w_i\}_{i=1}^n, l_2)$ (weighted vector–valued sequence space) for some $n \in \mathbb{N}$. We continue to use the designation A for its extension $A \otimes I_{l_2}$ to $l_2(I_n, \{w_i\}_{i=1}^n, l_2)$. One needs to justify the following inequality and identify its sharpness: for $m \in \mathbb{N}$, one has

$$\sum_{i,j=1}^n w_i a_{i,j}^m \|x_i - x_j\|_{l_2}^2 \leq m \sum_{i,j=1}^n w_i a_{i,j} \|x_i - x_j\|_{l_2}^2, \tag{1}$$

where $A = \{a_{i,j}\}$ and $A^m = \{a_{i,j}^m\}$. The relation (1) can be rewritten as

$$\langle x, (I - A^m)x \rangle_{\tilde{H}} \leq m \langle x, (I - A)x \rangle_{\tilde{H}}, \tag{2}$$

where A is self–adjoint because ξ is reversible. The interpolation inequality (Schur test)

$$\|A|\mathcal{L}(l_2(I_n, \{w_i\}_{i=1}^n, l_2))\|^2 \leq \|A|\mathcal{L}(l_1(I_n, \{w_i\}_{i=1}^n, l_2))\| \|A|\mathcal{L}(l_\infty(I_n, \{w_i\}_{i=1}^n, l_2))\| \leq 1 \tag{3}$$

localizes the spectrum $\sigma(A) \subset [-1, 1]$. Thanks to the spectral theorem [59] (functional calculus) for A, (2) is equivalent to the classical binomial inequality

$$(1+x)^m \geq 1 + mx \text{ for } x \in [-2, 0].$$

To show the sharpness, we note that $l_2(I_n, \{w_i\}_{i=1}^n, l_2) = l_2(\mathbb{N}, l_2(I_m, \{w_i\}_{i=1}^n))$ and interpret (2) as the sum of (more scalar) inequalities

$$\langle y_k, (I - A^m)y_k \rangle_{l_2(w)} \leq m \langle y_k, (I - A)y_k \rangle_{l_2(w)} \text{ for } k \in \mathbb{N}, \tag{4}$$

where y_k is the vector of the kth coordinates of the components of x and $l_2(w) = l_2(I_m, \{w_i\}_{i=1}^n)$. If y_k' is the vector of the coordinates of y_k in the orthonormal basis of the eigenvectors of A and $\{\lambda_l\}_{l=1}^n$ are the corresponding eigenvalues, (4) takes the form

$$\sum_{l=1}^n g(\lambda_l)|y_{k,l}'|^2 \geq 0 \text{ for } g(\lambda) = \lambda^m - m\lambda + m - 1. \tag{5}$$

Thus, the equality in $(1-5)$ is equivalent to $\sigma(A) = \{1\}$.

Let us choose $B = (1, 1, \ldots, 1)^T (w_1, w_2, \ldots, w_n)$ and $A_t = (1-t)I + tB$. Clearly, A_t is a transition matrix of some stationary reversible Markov chain $\xi(t)$ with the same stationary distribution. The $l_2(w)$–orthonormal sets of the eigenvectors of A_t and B coincide and consist of $\bar{1} = (1, 1, \ldots, 1)$ and an arbitrary $l_2(w)$–orthonormal basis in the hyperplane defined by $\{z : (\bar{1}, z)_{l_2(w)} = 0\}$. Since $\sigma(A_t) = \{1-t, 1\}$, we see that the left hand side of (5) tends to zero implying the sharpness of (1) (of the implicit constant 1) for every fixed m, n and a non–trivial sequence $\{x_i\}_{i=1}^n$. This proves a) for $r = 2$. When $r < 2$, l_2 endowed with the metric $\|\cdot\|_{l_2}^{r/2}$ is isometrically embedded into itself (see Theorem 5.11 in [14]): for every non–trivial $\{z_i\}_{i=1}^n \subset l_2$, there exist a non–trivial $\{x_i\}_{i=1}^n \subset l_2$ satisfying $\|x_i - x_j\|_{l_2} = \|z_i - z_j\|_{l_2}^{r/2}$ for every $i \neq j$ and, thus, (1) finishes the proof of a) (including the sharpness).

To approach b), let us note that $X_{\bar{2}} \in IG$ is a Hilbert space possessing the Markov type r with the constant 1 for $r \in (0, 2]$ thanks to a), while $X_{\bar{p}} \in IG$ is an $L_p(\Omega)$ space. Thanks to Theorem 5.11 in [14], $X_{\bar{p}} \in IG$ endowed with the metric $\|\cdot\|_p^{r/2}$ is isometrically embedded into L_2 for $r \in (0, p]$ and, thus, it has the Markov type r with the constant 1 (and this is also the argument in [17]).

For some $n \in \mathbb{N}$, $X_{\bar{p}}$ contains an isometric copy of $l_p(I_n)$. Now we prove the sharpness in b) in exactly the same manner as we handled the case $r < 2$ in a). Indeed, according to Theorem 5.11 in [14], $l_p(I_n)$ endowed with the metric $\|\cdot\|_p^{r/2}$ is isometrically embedded into l_2 for $0 < r < p \leq 2$.

According to Dvoretzky's theorem, every infinite–dimensional Banach space contains l_2 almost isometrically and the constant 1 is sharp for $l_2(I_n)$ for every $n \in \mathbb{N}$. Thus, this observation shows that Part c) follows from a). □

Now we are in a position to describe the Markov type sets and to find or estimate the related constants. The next theorem is a Markov type counterpart of Theorem 3.1. Let us recall that the function ω_s is defined in Section 3.2 just before and after Theorem 3.1.

Theorem 4.4. *Let X be either a subspace or a quotient of a space $Y \in IG_+$, or finitely represented in Y with $[p_{\min}(Y), p_{\max}(Y)] \cup \{2\} \subset [r, \infty) \subset (1, \infty)$. Then*
a) *X has the Markov type r with $C_{MT} = 2(13/6)^{1/r} (\max(p_{\max}(Y), 2) - 1)^{1/r'} 2^{2/r-1}$, and, if $Y \in IG$, with the constant*

$$2 \left(2 \left(\omega_s(0, p_{\max}(Y), r) + \omega_s(1/2, p_{\max}(Y), r)/12 \right) \right)^{1/r} \leq$$
$$\leq 2(13/6)^{1/r} (\max(p_{\max}(Y), 2) - 1)^{1/r'} 2^{2/r-1};$$

b) *the bound $r \leq \min(p_{\min}(X), 2)$ is sharp if $l_{\min(p_{\min}(X),2)}$ is finitely represented in X; particularly, this is the case when $X = Y \in IG_+$ and the spaces at the vertexes of $T(X)$ that either have the parameter $p_{\min}(X)$, or have the parameters converging to $p_{\min}(X)$ are infinite–dimensional.*

Proof of Theorem 4.4. The description of the Markov type set and the estimates for the constants are implied by the combination of Theorem 4.2 with either Theorem 3.3 (IG_+), or Theorem 3.1 (IG).

Since Markov type implies Rademacher type with the same exponent thanks to Theorem 4.3, the sharpness of the bounds for the exponent r in $b)$ follows from the sharpness of the type set of the space $l_{\min(p_{\min}(X),2)}$. Indeed, $l_{\min(p_{\min}(X),2)}$ inherits every Rademacher type exponent σ from every Banach space it is finitely represented in, while its own maximal Rademacher type is $\min(p_{\min}(X), 2)$. □

Theorem 4.5. *For $\{p, 2\} \subset [r, \infty) \subset [1, \infty)$, let X be either a subspace or a quotient of $Y \in \{L_p(\mathcal{M}, \tau), S_p\}$, or finitely represented in Y, where $(\mathcal{M}, \tau)$ is a von Neumannn algebra with a normal semifinite faithful weight τ. Then*
a) X has the Markov type r with $C_{MT} = 2(13/6)^{1/r} (\max(p, 2) - 1)^{1/r'} 2^{2/r-1}$;
b) the bound $r \leq \min(p, 2)$ is sharp if $l_{\min(p,2)}$ is finitely represented in X; particularly, this is the case when $X = Y$ and $\dim X = \infty$.

Proof of Theorem 4.5. Part $b)$ of Remark 4.1 handles the case $p = 1$. The rest of $a)$ follows from Theorem 4.2 with the aid of Theorem 3.2.

Since Markov type implies Rademacher type with the same exponent thanks to Theorem 4.3, the sharpness of the bounds for the exponent r in $b)$ follows from the sharpness of the type set of the space $l_{\min(p,2)}$ (as in the proof of the previous theorem). Indeed, $l_{\min(p,2)}$ is finitely represented in $X = Y$ if $\dim X = \infty$ thanks to Lemma 3.2 (see [36] on the finite representability of l_σ). □

4.2.2. Markov Cotype

While this section is dedicated to the Markov cotype sets and constants for both model and auxiliary independently generated IG_+ and noncommutative L_p-spaces, the anisotropic Sobolev, Besov and Lizorkin–Triebel spaces with underlying mixed Lebesgue and Lorentz norms and their duals from $\Gamma_i \cup \Gamma_i^L$ for $i \in I_4$ are dealt with in Section 4.4.1.

K. Ball (see Theorem 4.1 in [17]) identified the Markov cotype of the Lebesgue spaces L_p for $p \in (1, 2]$ with the aid of his version of Theorem 4.6 below. Naor [20] with Peres, Schramm and Sheffield [21] noted that his approach gives the Markov cotype p of the L_p-spaces for $p \in (2, \infty)$.

In this section we describe the Markov cotype sets for various spaces and estimate the corresponding constants from both sides.

The next theorem is a slight generalization of Theorem 4.1 from [17] with the same proof and the exponent 2 substituted with q. In our terminology, it says that abstract uniform convexity implies Markov cotype with the same exponent.

Theorem 4.6. *Let X be a Banach space and $q \in [2, \infty)$. Then X has Markov cotype q with the constant $2C_q^{-1/q}$ if, for every X-valued stochastic variable ξ, one has*

$$\|\mathrm{E}\xi\|_X^q + C_q \mathrm{E}\,\|\xi - \mathrm{E}\xi\|_X^q \leq \mathrm{E}\,\|\xi\|_X^q$$

for some positive constant C_q.

This theorem leads to the following counterpart of Theorem 4.2.

Theorem 4.7. *For $q \in [2, \infty)$, let X be a (q, h_c)–uniformly convex Banach space. Then it possesses the Markov cotype q with the constant*

$$C_{q,h_c} = 2 \left(\sup_{\mu \in (0, 1/2]} (1 - \mu) h_c(\mu) \right)^{-1/q}.$$

Proof of Theorem 4.7. It immediately follows from Lemma 3.1, b) combined with Theorem 4.6. □

The fact that the Markov cotype q is followed by the classical Rademacher cotype q is also a deep result of K. Ball [17]. The straightforward generalization of Lemma 2.1 and Theorem 2.2 from [17] from the case $q = 2$ to $q \in [2, \infty)$ combined with the duality relations of the Rademacher type and cotype leads to the next (quantitative) theorem (the cotype counterpart of Theorem 4.3) that we use to study the matter of sharpness of the sets of the parameters of the Markov cotype for the Banach spaces under the consideration.

Theorem 4.8. *For $q \in [2, \infty)$, let X be a Banach space possessing the Markov cotype q with a constant $C_{MCq} = C_{MCq}(X)$. Then it also has the Rademacher cotype q with the constant $C_{q,q}(X)$ satisfying*

$$C_{q,q}(X) \leq 2^{2-1/q'} C_{MCq}(X).$$

To establish the Markov cotypes and to estimate the related constants we need the next lemma that is a (slightly weaker) cotype counterpart of Lemma 4.1.

Lemma 4.2. *a) Every Hilbert space has the Markov cotype 2 with the constant 1 and the constant 1 is sharp.*
b) For $p \in [2, \infty]$, the space $X_{\bar{p}} \in IG$ (i.e., L_p with, possibly, atomic measure) has the Markov cotype r with the constant $2\omega_c(0, p, r)^{-1/r}$ for $r \in [p, \infty]$.
c) If an infinite–dimensional Banach space X has a Markov cotype $p \in [2, \infty)$ with the constant 1, then this constant is sharp.

Proof of Lemma 4.2. We will follow the scheme of the proof of Lemma 4.1. Since every Hilbert space is finitely represented in l_2, we assume that $H = l_2$. Let ξ be a stationary reversible Markov chain with an $n \times n$ symmetric transition matrix A and the uniform stationary distribution, and let $x = (x_1, x_2, \ldots, x_n) \in H^n$ be an element of $\tilde{H} = l_2(I_n, l_2)$ (vector–valued sequence space) for some $n \in \mathbb{N}$. We continue to use the designation A for its extension $A \otimes I_{l_2}$ to $l_2(I_n, l_2)$. One needs to justify the following inequality and identify its sharpness: for $m \in \mathbb{N}$, one has, for $\beta \in (0, 1)$ and $C = C(A) = (1 - \beta)(I - \beta A)^{-1}$,

$$\beta \sum_{i,j=1}^{n} a_{i,j} \left\| \sum_{k=1}^{n} c_{i,k} x_k - \sum_{l=1}^{n} c_{j,l} x_l \right\|_{l_2}^2 \leq (1 - \beta) \sum_{i,j=1}^{n} c_{i,j} \|x_i - x_j\|_{l_2}^2, \qquad (1)$$

where $A = \{a_{i,j}\}$ and $C = \{c_{i,j}\}$. The relation (1) can be rewritten as

$$\langle x, (C - C^2) x \rangle_{\tilde{H}} \leq \langle x, (I - C) x \rangle_{\tilde{H}}, \text{ or } \|(I - C)x\|_{\tilde{H}}^2 \geq 0. \qquad (2)$$

The interpolation inequality (Schur test) described in the proof of Lemma 4.1 implies

$$\|A|\mathcal{L}(l_2(I_n, l_2))\| \leq 1 \text{ and, thus, } \sigma(A) \subset [-1, 1]. \tag{3}$$

To show the sharpness of (1 – 2), let us choose $B = (1, 1, \ldots, 1)^T(1, 1, \ldots, 1)$ and $A_t = (1-t)I + tB$. Clearly, A_t is a transition matrix of some stationary reversible Markov chain $\xi(t)$ with the same stationary distribution, $\lim_{t\to 0} A_t = I$ and, therefore, $\lim_{t\to 0} C_t = I$ for $C_t = C(A_t)$. The latter holds thanks to the analyticity of $C(A)$. Thus, (1 – 2) are sharp for every $\{x_i\}_{i=1}^n$ thanks to

$$\lim_{t\to 0} \|(I - C_t)x\|_{\tilde{H}} = 0. \tag{4}$$

It proves a).

The approach to b) is straightforward: we use Theorem 4.7 and the simplest case (abstract $L_p \in IG$) of Theorem 3.1.

According to Dvoretzky's theorem, every infinite–dimensional Banach space contains l_2 almost isometrically and the constant 1 is sharp for $l_2(I_n)$ for every $n \in \mathbb{N}$. Thus, this observation shows that Part c) follows from a). □

Now we are in a position to describe the Markov cotype sets and to find or estimate the related constants. The next theorem is a Markov cotype counterpart of Theorem 3.1 (and Theorem 4.4).

Theorem 4.9. *Let X be either a subspace or a quotient of a space $Y \in IG_+$, or finitely represented in Y with $[p_{\min}(Y), p_{\max}(Y)] \cup \{2\} \subset (1, q] \subset (1, \infty)$. Then*
a) X possesses the Markov cotype q with $C_{MC} = 2^{2/q'}(\min(p_{\min}(Y), 2) - 1)^{-1/q}$, and, if $Y \in IG$, with the constant $2\omega_c(0, p_{\min}(Y), q)^{-1/q}$;
b) the bound $q \geq \max(p_{\max}(X), 2)$ is sharp if $l_{\max(p_{\max}(X),2)}$ is finitely represented in X; particularly, this is the case when $X = Y \in IG_+$ and the spaces at the vertexes of $T(X)$ that either have the parameter $p_{\max}(X)$, or have the parameters converging to $p_{\max}(X)$ are infinite-dimensional.

Proof of Theorem 4.9. The description of the Markov type set and the estimates for the constants are implied by the combination of Theorem 4.7 combined with either Theorem 3.3 (IG_+), or Theorem 3.1 (IG).

Since Markov cotype implies Rademacher type with the same exponent thanks to Theorem 4.8, the sharpness of the bounds for the exponent q in b) follows from the sharpness of the cotype set of the space $l_{\max(p_{\max}(X),2)}$. Indeed, $l_{\min(p_{\min}(X),2)}$ inherits every Rademacher type exponent σ from every Banach space it is finitely represented in, while its own maximal Rademacher type is $\min(p_{\min}(X), 2)$. □

The following theorem shows, relatively surprisingly, that the Markov cotype properties of the commutative and non-commutative L_p–spaces coincide (within our approach).

Theorem 4.10. *For $\{p, 2\} \subset (1, q] \subset (1, \infty]$, let X be either a subspace or a quotient of $Y \in \{L_p(\mathcal{M}, \tau), S_p\}$, or finitely represented in Y, where $(\mathcal{M}, \tau)$ is a von Neumannn algebra with a normal semifinite faithful weight τ. Then*

a) the space X possesses the Markov cotype q with the constant $2^{2/q'}(\min(p, 2) - 1)^{-1/q}$;
b) the bound $q \geq \max(p, 2)$ is sharp if $l_{\max(p,2)}$ is finitely represented in X; particularly, this is the case when $X = Y$ and $\dim X = \infty$.

Proof of Theorem 4.10. Part a) of Remark 4.2 takes care of the case $p_{\max}(X) = \infty$. The rest of a) follows from Theorem 4.7 with the aid of Theorem 3.2.

Since Markov cotype implies Rademacher type with the same exponent thanks to Theorem 4.8, the sharpness of the bounds for the exponent q in b) follows from the sharpness of the cotype set of the space $l_{\max(p,2)}$ (as in the proof of the previous theorem). Indeed, $l_{\min(p,2)}$ is finitely represented in $X = Y$ if $\dim X = \infty$ thanks to Lemma 3.2 (see [36] regarding the finite representability of l_σ). □

4.2.3. Adaptation of Ball's Scheme

We start this section with the adaptation of Ball's approach to the bounded extension problem for Hölder–Lipschitz mappings. K. Ball treated the extension of Lipschitz mappings from a Markov type 2 metric space into a Markov cotype 2 Banach space but Naor [20] mentioned that it works also for some Hölder mappings and different values of Markov types and cotypes because $d(x, y)^\alpha$ is a metric if $d(x, y)$ is and $\alpha \in (0, 1]$, and because L_p with $p \in (0, 2]$ endowed with the metric $\|x - y\|^{p/2}$ is isometrically embedded into L_2 thanks to Theorem 5.11 in [14].

The next two lemmas and a theorem are the natural slight modifications of, respectively, Lemma 1.1, Lemma 1.2 and Theorem 1.7 from [17] with essentially the same proofs. We present them to trace the numerical constants and the potential for their improvement.

Lemma 4.3. Let (X, d) be a metric space and Y be a normed space, $Z \subset X$, and let $f \in H^\alpha(Z, Y)$ for some $\alpha = p/q \in (0, 1]$ with $\emptyset \neq [p, q] \subset [1, \infty)$. Then, there is an extension $\tilde{f} \in H^\alpha(X, Y^{**})$ of f with $\|\tilde{f}|H^\alpha(X, Y^{**})\| \leq K$ if, and only if, for every $n \in \mathbb{N}$, $n \times n$ symmetric matrix H with the entries $\{h_{i,j}\}_{i,j=1}^n$ satisfying $h_{i,j} \geq 0$ and sequence $s = \{x_i\}_{i=1}^n$ in X, there is a map

$$g = g_{H,s} : \{x_i\}_{i=1}^n \longrightarrow Y^{**} \text{ satisfying } g(x) = f(x) \text{ for } x \in \{x_i\}_{i=1}^n \cap Z \text{ and}$$

$$\sum_{i,j=1}^n h_{i,j} \|g(x_i) - g(x_j)\|_{Y^{**}}^q \leq K^q \sum_{i,j=1}^n h_{i,j} d(x_i, x_j)^p.$$

Let us recall that an $n \times m$ matrix $B = \{b_{i,j}\}$ is stochastic if $b_{i,j} \geq 0$ for every i and j and $B\bar{1} = \bar{1}$ for the constant vector $\bar{1} = (1, \ldots, 1)$.

Lemma 4.4. Let (X, d) be a metric space and Y be a normed space, $Z \subset X$, and let $f \in H^\alpha(Z, Y)$ for some $\alpha = p/q \in (0, 1]$ with $\emptyset \neq [p, q] \subset [1, \infty)$. Then, there is an extension $\tilde{f} \in H^\alpha(X, Y^{**})$ of f with $\|\tilde{f}|H^\alpha(X, Y^{**})\| \leq K$ if, and only if, for every $m, n \in \mathbb{N}$, $n \times n$ symmetric stochastic matrix $A = \{a_{i,j}\}$, $n \times m$ stochastic matrix $B = \{b_{i,j}\}$, $\beta \in (0, 1)$

and sequences $\{x_i\}_{i=1}^n$ in X and $\{z_k\}_{k=1}^m$ in Z, there are points $\{y_i\}_{i=1}^n \subset Y^{**}$ satisfying

$$\beta \sum_{i,j=1}^n a_{i,j}\|y_i - y_j\|_{Y^{**}}^q + 2(1-\beta) \sum_{i=1}^n \sum_{k=1}^m b_{i,k}\|y_i - f(z_k)\|_{Y^{**}}^q \leq$$

$$\leq K^q \left(\beta \sum_{i,j=1}^n a_{i,j} d(x_i, x_j)^p + 2(1-\beta) \sum_{i=1}^n \sum_{k=1}^m b_{i,k} d(x_i, z_k)^p \right).$$

To adapt Theorem 1.7 from [17] to our setting, we use the fact that Y^{**} shares the Markov cotype properties of a Banach space Y thanks to its finite representability in Y.

Theorem 4.11. *For $\alpha = p/q \in (0,1]$ with $p, q \in [1, \infty)$, let (X, d) be a metric space possessing the Markov type p with a constant C_{MT} and Y be a Banach space possessing the Markov cotype q with a constant C_{MC}, $Z \subset X$, and also let $f \in H^\alpha(Z, Y)$. Then, there is an extension $\tilde{f} \in H^\alpha(X, Y^{**})$ of f with*

$$\|\tilde{f}|H^\alpha(X, Y^{**})\| \leq 3^{\frac{p-1}{q}} (C_{MC}^q + 2)^{\frac{1}{q}} C_{MT}^\alpha \|f|H^\alpha(Z, Y)\| \leq (3C_{MT})^\alpha C_{MC} \|f|H^\alpha(Z, Y)\|.$$

4.3. Approximation, Real Interpolation and Sharpness

In this section we reveal the intimate connection between the extensions of Hölder–Lipschitz mappings and the real interpolation of continuous and Hölder–Lipschitz mappings. In our setting, the latter comprises the direct and inverse theorems of the approximation of the elements of a Hölder space by elements of a smoother Hölder–Lipschitz space. We also investigate the sharpness of the exponents with the aid of a remarkable lemma due to I. G. Tsar'kov. One uses the notation from [1].

The space of continuous mappings from X into Y is correctly defined.

Definition 4.4. *Let X and Y be metric and Banach spaces correspondingly. Then $C(X, Y)$ is the Banach space of all continuous mappings from X into Y with the norm*

$$\|f|C(X, Y)\| = \sup_{x \in X} \|f(x)\|_Y.$$

Definition 4.5. *Metric spaces X and Y are C–Lipschitz homeomorphic if there exists an invertible homeomorphism $\phi \in H^1(X, Y)$ satisfying*

$$\|\phi|H^1(X, Y)\| \cdot \|\phi^{-1}|H^1(Y, X)\| \leq C.$$

Theorem 4.12. *Let X and Y be metric and Banach spaces correspondingly and $0 < \beta < \alpha \leq 1$ and $d \in [1, \infty)$. Assume also that the pair (X, Y) possess the (d, α)–extension property. Then we have*

$$(C(X, Y), H^\alpha(X, Y))_{\beta/\alpha, \infty} = H^\beta(X, Y) \text{ (isomorphism) with}$$

$$\left\| I \Big| \mathcal{L}\left(H^\beta(X,Y), (C(X,Y), H^\alpha(X,Y))_{\beta/\alpha,\infty}\right) \right\| \leq d^{\frac{\beta}{\alpha}}(d+1)^{1-\frac{\beta}{\alpha}} \frac{\alpha}{\beta}\left(\frac{\alpha}{\beta}-1\right)^{\frac{\beta}{\alpha}-1} \text{ and}$$

$$\left\| I \Big| \mathcal{L}\left((C(X,Y), H^\alpha(X,Y))_{\beta/\alpha,\infty}, H^\beta(X,Y)\right) \right\| \leq \frac{\alpha}{\beta}\left(\frac{\alpha}{\beta}-1\right)^{\frac{\beta}{\alpha}-1}$$

Proof of Theorem 4.12. Assume that $f \in H^\beta(X, Y)$. Given $\delta > 0$, there exists a maximal (by inclusion) δ–separated subset E_δ of X ($d(x, y) \geq \delta$ for $x, y \in E_\delta$) thanks to the Zorn lemma. Clearly, the restriction $\bar{f}_\delta$ of f onto E_δ satisfies

$$\|\bar{f}_\delta | H^\alpha(E_\delta, Y)\| \leq \|f | H^\beta(X, Y)\| \delta^{\beta-\alpha}. \tag{1}$$

The (d, α)–extension property provides an extension f_δ of $\bar{f}_\delta$ with

$$\|f_\delta | H^\alpha(X, Y)\| \leq d\|\bar{f}_\delta | H^\alpha(X, Y)\| \leq \|f | H^\beta(X, Y)\| d\delta^{\beta-\alpha}. \tag{2}$$

The maximality of E_δ in X means that, for every $x \in X$, there exists $y \in E_\delta$ satisfying $d(x, y) \leq \delta$. Therefore, one has, due to the triangle inequality,

$$\|f(x) - f_\delta(x)\|_Y \leq \|f(x) - f(y)\|_Y + \|f_\delta(x) - f_\delta(y)\|_Y \leq \|f | H^\beta(X, Y)\| \delta^\beta (1 + d). \tag{3}$$

combining (2) and (3), one finds the upper estimate for the K-functional of the pair $(C(X, Y), H^\alpha(X, Y))$: since

$$\|f - f_\delta | C(X, Y)\| + t\|f_\delta | H^\alpha(X, Y)\| \leq \|f | H^\beta(X, Y)\| \left(\delta^\beta (d + 1) + dt\delta^{\beta-\alpha} \right),$$

we choose $\delta = (\alpha/\beta - 1)^{1/\alpha} \left(\frac{dt}{d+1} \right)^{1/\alpha}$ to obtain

$$K(t, f, C(X, Y), H^\alpha(X, Y)) \leq (td)^{\frac{\beta}{\alpha}} (d+1)^{1-\frac{\beta}{\alpha}} \frac{\alpha}{\beta} \left(\frac{\alpha}{\beta} - 1 \right)^{\frac{\beta}{\alpha} - 1} \|f | H^\beta(X, Y)\|. \tag{4}$$

Since t is arbitrary, this is equivalent to the embedding

$$H^\beta(X, Y) \subset (C(X, Y), H^\alpha(X, Y))_{\beta/\alpha, \infty}.$$

Now, assuming that $f \in (C(X, Y), H^\alpha(X, Y))_{\beta/\alpha, \infty}$ with the norm C, for every $t > 0$ and $\varepsilon > 0$, one has a decomposition $f = g_t + h_t$ satisfying

$$\|g_t | C(X, Y)\| + t\|f_t | H^\alpha(X, Y)\| \leq (C + \varepsilon) t^{\beta/\alpha}. \tag{5}$$

Thus, given $x, y \in X$, we choose $t = (\alpha/\beta - 1)(d(x, y))^\alpha$ to deduce from (5) that

$$\|f(x) - f(y)\|_Y \leq (C + \varepsilon) \left(t^{\beta/\alpha} + d(x, y)^\alpha t^{\beta/\alpha - 1} \right) =$$

$$= d(x, y)^\beta (C + \varepsilon) \frac{\alpha}{\beta} \left(\frac{\alpha}{\beta} - 1 \right)^{\frac{\beta}{\alpha} - 1}. \tag{6}$$

Since $x, y \in X$ and $\varepsilon > 0$ are arbitrary, the last estimate finishes the proof implying the embedding

$$(C(X, Y), H^\alpha(X, Y))_{\beta/\alpha, \infty} \subset H^\beta(X, Y).$$

$\square$

In order to see that for various pairs of concrete spaces, the (d, α)–extension property in Theorem 4.12 cannot be substituted with a (d, λ)–extension property for some $\lambda \in (0, \alpha)$, we need the next lemma due to I. G. Tsar'kov [7, 8, 9] (see also the presentation of his results in Chapter 2 in [18]).

In this subsection it is convenient to denote the unit ball of X as $B(X)$. The Mazur mapping $M_{p,q} : L_p \to L_q$ is defined by $M_{p,q} : f \mapsto |f|^{p/q}\mathrm{sign}(f)$ and satisfies $M_{p,q} \in H^{\min(1,p/q)}(B(L_p), L_q) \subset H^\beta(B(L_p), L_q)$ for $p, q \in [1, \infty)$ and $\beta \in (0, \min(1, p/q))$.

Lemma 4.5 ([7, 8, 9, 18]). *For some $1 \leq p < q < \infty$ and $\alpha > p/q$ and every $n \in 2\mathbb{N}$, let ψ_n be an element of $H^\alpha(B(l_p(I_n)), l_q(I_n))$ satisfying*

$$C = \sup_{n \in 2\mathbb{N}} \|\psi_n | H^\alpha(B(l_p(I_n)), l_q(I_n))\| < \infty.$$

Then one has

$$\|M_{p,q} - \psi_n | C(B(l_p(I_n)), l_q(I_n))\| \geq (1 - C(n/2)^{1/q - \alpha/p})/2.$$

The next theorem shows that the real interpolation properties of the scale $\{H^\alpha(X, Y)\}_{\alpha \in (0,1]}$ depend, surprisingly, on the quantitative topology of X and Y.

Theorem 4.13. *For $1 \leq p \leq q < \infty$, $\alpha \in (0, 1]$, $\beta \in (0, \alpha) \cap (0, p/q]$ and $d, C_0, C_1, C_2 \geq 1$, let X be a metric space containing C_0–Lipschitz homeomorphic copy of the unit ball $B(l_p(I_n))$ of $l_p(I_n)$ for every $n \in \mathbb{N}$, and let Y be a Banach space containing a C_2–complemented and C_1–isomorphic copy of $l_q(I_n)$ for every $n \in \mathbb{N}$. Assume also that the pair (X, Y) possesses the (d, β)–extension property, and we have the embedding*

$$H^\beta(X, Y) \subset (C(X, Y), H^\alpha(X, Y))_{\beta/\alpha, \infty}.$$

Then we have $\alpha \leq p/q$.

Proof of Theorem 4.13. Let $\alpha > p/q$. For $n \in \mathbb{N}$, let $\phi_n : B(l_p(I_n)) \to X$, $T_n : l_q(I_n) \to Y$ and $P_n : Y \to \mathrm{Im}\, T_n$ be a corresponding C_0–homeomorphism, a C_1–isomorphism and a projector with $\|P_n | \mathcal{L}(Y)\| \leq C_2$. Choosing $\bar{f}_n = T_n \circ M_{p,q} \circ \phi_n^{-1} \in H^\beta(\mathrm{Im}\, \phi_n, Y)$, we utilize the (d, β)–extension property of the pair (X, Y) to extend it to $f_n \in H^\beta(X, Y)$ satisfying

$$\|f_n | H^\beta(X, Y)\| \leq d \|\phi_n^{-1} | H^1(\mathrm{Im}\, \phi_n, B(l_p(I_n)))\|^\beta \|T_n | \mathcal{L}(l_q(I_n), Y)\| \times$$
$$\times \|M_{p,q} | H^\beta(B(l_p(I_n)), l_q(I_n))\|. \quad (1)$$

The embedding means that, for some $C_3 > 0$ (i.e., the norm of the embedding) and for every $t > 0$, there exists $g_n \in H^\alpha(X, Y)$ satisfying

$$\|f_n - g_n | C(X, Y)\| + t \|g_n | H^\alpha(X, Y)\| \leq 2C_3 \|f_n | H^\beta(X, Y)\| t^{\beta/\alpha}. \quad (2)$$

In other words, for every $\varepsilon' > 0$, there exists $g_n \in H^\alpha(X, Y)$ satisfying

$$\|f_n - g_n | C(X, Y)\| \leq \varepsilon' \text{ and } \|g_n | H^\alpha(X, Y)\| \leq \left(2C_3 \|f_n | H^\beta(X, Y)\|\right)^{\alpha/\beta} \varepsilon'^{1-\alpha/\beta}. \quad (3)$$

Observing that $T_n^{-1} \circ P_n \circ f_n \circ \phi_n = M_{p,q}$, we define $\psi_n = T_n^{-1} \circ P_n \circ g_n \circ \phi_n$ and use (1) and (3) to infer the estimates

$$\|M_{p,q} - \psi_n | C(B(l_p(I_n)), l_q(I_n))\| \leq \varepsilon' \|T_n^{-1}|\mathcal{L}(\mathrm{Im}\, T_n, l_q(I_n))\| C_2 = \varepsilon \text{ and}$$
$$\|\psi_n|H^\alpha(B(l_p(I_n)), l_q(I_n))\| \leq$$
$$\leq (2dC_1 C_2 C_3 \|M_{p,q}|H^\beta(B(l_p(I_n)), l_q(I_n))\|)^{\alpha/\beta} C_0^\alpha \varepsilon^{1-\alpha/\beta}. \quad (4)$$

Noting that $\|M_{p,q}|H^\beta(B(l_p(I_n)), l_q(I_n))\| \leq \|M_{p,q}|H^\beta(B(l_p), l_q)\|$ for $n \in \mathbb{N}$, we choose $\varepsilon = 1/3$ in (4) to observe the contradiction with Lemma 4.5 providing, for sufficiently large n, the estimate

$$\|M_{p,q} - \psi_n | C(B(l_p(I_n)), l_q(I_n))\| > 1/3.$$

$\square$

Now we are in a position to establish our main tool for investigating the sharpness of the sets of α for which a pair (X, Y) possesses the (d, α)–extension property.

Theorem 4.14. *Under the conditions $1 \leq p \leq q < \infty$, $\alpha \in (0, 1]$, $\beta \in (0, \alpha) \cap (0, p/q]$ and $d_0, d_1, C_0, C_1, C_2 \geq 1$, let X be a metric space containing C_0–Lipschitz homeomorphic copy of the unit ball $B(l_p(I_n))$ of $l_p(I_n)$ for every $n \in \mathbb{N}$, and let Y be a Banach space containing a C_2–complemented and C_1–isomorphic copy of $l_q(I_n)$ for every $n \in \mathbb{N}$. Assume also that the pair (X, Y) possesses both the (d_0, β)–extension and the (d_1, α)–extension properties. Then we have $\alpha \leq p/q$.*

Proof of Theorem 4.14. Relying on Theorem 4.12, one sees that

$$(C(X, Y), H^\alpha(X, Y))_{\beta/\alpha, \infty} = H^\beta(X, Y) \text{ (isomorphism)}. \quad (1)$$

Thus, we can apply Theorem 4.13 leaving us with: $\alpha \leq p/q$. $\square$

The next theorem describes the scheme of establishing the majority of results in the next section dedicated to various pairs of concrete spaces. Let us recall that a subset Z (with the inherited metric) of a metric space (X, d) is a C_R–Lipschitz retract of X if there exists a Lipschitz $\phi : X \to Z$ with $\phi(z) = z$ for $z \in Z$ and $\|\phi|H^1(X, Z)\| \leq C_R$.

Theorem 4.15. *For $1 \leq p \leq q < \infty$, $\beta = p/q$, $\alpha \in (0, 1]$ and $\lambda_0, \lambda_1, \lambda_2 \geq 1$, let X be a Banach space possessing the Markov type p with the constant C_{MT}, and let Y be a Banach space possessing the Markov cotype q with the constant C_{MC}. Assume also that Y is a C_R–Lipschitz retract of Y^{**}, and that l_p is λ_0–represented in X, and that, for every $n \in \mathbb{N}$, there exists a subspace $Y_n \subset Y$ and a projector P_n onto Y_n satisfying*

$$d_{BM}(l_q(I_n), Y_n) \leq \lambda_1 \text{ and } \|P_n|\mathcal{L}(Y)\| \leq \lambda_2.$$

Let also Z be a subset of X with non–empty interior (endowed with the inherited metric). Then $S_b(Z, Y) \subset (0, p/q]$, and, if $\beta = p/q \in S_b(Z, Y)$, we also have

$$d_\beta \leq C_R (3C_{MT})^\beta C_{MC}.$$

If $Y = Y^{**}$, one takes $C_R = 1$.

Proof of Theorem 4.15. There exists a ball $B \subset Z$, and $S_b(Z,Y) \subset S_b(B,Y)$. In the "if"–direction, the statement of the theorem, including the estimate for d_β, follows from Theorem 4.11 thanks to the observation that B inherits the Markov type properties of X, according to Remark 4.1, c).

To establish the opposite inclusion $S_b(Z,Y) \subset (0, p/q]$, we note that Z contains a ball to apply Theorem 4.14 with X being some shift and dilation (if necessary) of Z. □

4.4. Pairs of Concrete Spaces

In this section, we identify the Markov type and cotype sets, describe the algorithm for the investigation of the bounded extension property for the pairs of the spaces under consideration paying special attention to the matter of sharpness and provide simple model examples of its implementation. Indeed, Theorems 4.11 and 4.15 reduce the bounded extension problem for a pair (X, Y) of Banach spaces to the identification of the Markov type set of X, the Markov cotype set of Y, the (local) spectrum of the l_r–spaces λ–finitely represented in X and the local complemented spectrum of the l_r–spaces in Y. While the representable values of the latter two sets are described in Lemmas 2.2 and 2.3, Theorems 2.1 − 2.3 and Remark 2.5 with the aid of Lemma 2.1, the results in Sections 4.2.1 and 4.2.2 reduce the problem of identifying the former sets to the investigation of the asymmetric (r_s, h_s)–uniform smoothness and the Rademacher type set of X and the asymmetric (r_c, h_c)–uniform convexity and the Rademacher cotype set of Y.

4.4.1. Markov Type and Cotype Sets and Constants of Sobolev, Besov and Lizorkin–Triebel Spaces

Theorem 4.16. *For open $G \subset \mathbb{R}^n$, $s \in (0, \infty)$ and $\{p_{\min}, p_{\max}, 2\} \subset [r_{MT}, r_{MC}] \subset (1, \infty)$, let*

$$X_{p,q} \in \Gamma_1 \cup \Gamma_3 \cup \Gamma_4 \cup \{B_{p,q}^s(G), B_{p',q'}^s(G)^*\}_{p \in (1,\infty)^n}^{q \in (1,\infty)} \text{ and } Y_p \in \Gamma_2 \cup \{L_p\}_{p \in (1,\infty)^n},$$

and let also X (Y) be either a subspace, or a quotient, or finitely represented in $X_{p,q}$ (Y_p). Then
a) if $q \in [r_{MT}, r_{MC}]$, the space X has the Markov cotype r_{MC} and Markov type r_{MT}, correspondingly, with the constants $2\left(\omega_c(0, \min(p_{\min}, q), r_{MC})\right)^{-1/r_{MC}}$ and

$$2\left(2\left(\omega_s(0, \max(p_{\max}, q), r_{MT}) + \omega_s(1/2, \max(p_{\max}, q), r_{MT})/12\right)\right)^{1/r_{MT}} \leq$$
$$\leq 2\left(13\omega_s(1/2, \max(p_{\max}, q), r_{MT})/6\right)^{1/r_{MT}};$$

b) Y possesses the Markov cotype r_{MC} and Markov type r_{MT}, respectively, with the constants $2\left(\omega_c(0, p_{\min}, r_{MC})\right)^{-1/r_{MC}}$ and

$$2\left(2\left(\omega_s(0, p_{\max}, r_{MT}) + \omega_s(1/2, p_{\max}, r_{MT})/12\right)\right)^{1/r_{MT}} \leq$$
$$\leq 2\left(13\omega_s(1/2, p_{\max}, r_{MT})/6\right)^{1/r_{MT}}.$$

The ranges of the Markov type and cotype exponents are exact if $X = X_{p,q} \notin \Gamma_4$ and $Y = Y_p$.

Proof of Theorem 4.16. Parts *a)* and *b)* follow from the combination of Theorems 4.7 and 4.2 with Theorem 3.4 and the monotonicity of ω_c and ω_s in μ established in Theorem 4.13 in [23], while the exactness of the ranges of the Markov type and cotype exponents is the consequence of Theorems 4.3 and 4.8 respectively assisted by the exactness of the Rademacher type and cotype exponents established in [25] (one can also use Lemmas 2.2 and 2.3, Theorems 2.1 − 2.3 and Dvoretzky's theorem). □

Theorem 4.17. *For an open* $G \subset \mathbb{R}^n$, $s \in (0, \infty)$ *and* $p, \theta, q \in (1, \infty)$, *let*

$$X_{L_{p,\theta,q}} \in \Gamma_1^L \cup \Gamma_3^L \cup \Gamma_4^L \cup \left\{ B^s_{L_{p,\theta,q}}(G), B^s_{L_{p',\theta',q'}}(G)^* \right\} \text{ and } Y_{L_{p,\theta}} \in \Gamma_2^L \cup \{L_{p,\theta}\},$$

and let also X be either a subspace, or a quotient, or finitely represented in $X_{L_{p,\theta,q}}$. *Then the following holds:*

a) the space X possesses the Markov cotype r_{MC} if either $X_{L_{p,\theta,q}}$ is a Besov type space and $r_{MC} \in [\max(2, p, q, \theta), \infty) \setminus (\{p\} \cap \{2\})$, or $X_{L_{p,\theta,q}}$ is a Lizorkin–Triebel type space and $r_{MC} \in [\max(2, p, q, \theta), \infty) \setminus (\{p\} \cap \{q, 2\})$;

b) the space X possesses the Markov type r_{MT} if either $X_{L_{p,\theta,q}}$ is a Besov type space and $r_{MT} \in (0, \min(2, p, q, \theta)] \setminus (\{p\} \cap \{2\})$, or $X_{L_{p,\theta,q}}$ is a Lizorkin–Triebel type space and $r_{MT} \in (0, \min(2, p, q, \theta)] \setminus (\{p\} \cap \{q, 2\})$;

c) $I_{MT}(Y_{L_{p,\theta}}) = (0, \min(2, p, \theta)] \setminus (\{p\} \cap \{2\})$ and $I_{MC}(Y_{L_{p,\theta}}) = [\max(2, p, \theta), \infty) \setminus (\{p\} \cap \{2\})$. Moreover, the ranges of r_{MT} and r_{MC} in a) and b) are exact if $X = X_{L_{p,\theta,q}} \notin \Gamma_4^L$.

Proof of Theorem 4.17. Parts *a)* and *b)* follow from the combination of Theorems 4.7 and 4.2 with Theorem 3.7, while the exactness of the ranges of the Markov type and cotype exponents is the consequence of Theorems 4.3 and 4.8 respectively assisted by the exactness of the Rademacher type and cotype exponents established in [25] (see the proof of the exactness in Theorem 3.7 above for the details). The treatment of $Y_{L_{p,\theta}}$ in *c)* is similar and simpler. □

4.4.2. Simplest Model Examples

Proof of Corollary 1.1. Thanks to either Theorem 4.4 for $Y_p \in \{l_p, L_p\}$, or Theorem 4.5 for $Y_p \in \{S_p, L_p(\mathcal{M}, \tau_p)\}$, X_p has the Markov type $\min(p, 2)$ that becomes Markov type 2 with the constant $C_{MT} = 2\sqrt{13/6}\sqrt{p-1}$ if $2 \in [q, p]$. Similarly, due to either Theorem 4.9 for $Y_q \in \{l_q, L_q\}$, or Theorem 4.10 for $Y_q \in \{S_q, L_q(\mathcal{M}, \tau_q)\}$, X_q has the Markov cotype $r \in [\max(q, 2), \infty)$ that becomes $r \in [2, \infty)$ with the constant $C_{MC} = 2^{2/r'}(q-1)^{-1/r}$. Applying, for instance, Theorem 4.11, we obtain the inclusion $S_b(X_p, X_q) \supset (0, \min(p, 2)/\max(q, 2)]$ meaning, for $2 \in [q, p]$, both the identity $S_b(X_p, X_q) = (0, 1]$ and that the pair (X_p, X_q) possesses the (d_r, α_r)-extension property with

$$\alpha_r = 2/r \in (0, 1] \text{ and } d_r = 4(39/2)^{1/r}\left(\frac{p-1}{q-1}\right)^{1/r},$$

thus, finishing the proof of *a)* and almost all *b)*. The rest of *b)* (the first identity) is provided by Theorem 4.15 assisted by Lemma 2.3 and its classical counterpart for the Lebesgue spaces.

In the case of Part c), every metric space E possesses the Markov type 1 with the constant $C_{MT} = 1$ according to Part b) of Remark 4.1. Similarly to b), due to either Theorem 4.9 for $Y_\sigma \in \{l_\sigma, L_\sigma\}$, or Theorem 4.10 for $Y_\sigma \in \{S_\sigma, L_\sigma(\mathcal{M}, \tau_\sigma)\}$, X_σ has the Markov cotype $t \in [\max(\sigma, 2), \infty)$ with the constant $C_{MC} = 2^{2/t'}(\max(\sigma, 2) - 1)^{-1/t}$. Applying Theorem 4.11, we observe that the pair (E, X_σ) possesses the (d_t, α_t)–extension property for

$$\alpha_t = 2/t \in (0, 1/\max(\sigma, 2)] \text{ and } d_t = 4(3/2)^{2/t}(\max(\sigma, 2) - 1)^{-1/t},$$

finishing the proof of c) (even implying the inclusion $S_b(X_1, X_\sigma) \supset (0, 1/\max(\sigma, 2)])$. Part d) is directly implied by Theorem 4.15 combined with Lemma 2.3 as in b). Indeed, one finds isometric 1–complemented copies of l_1 in Y_1 and l_σ in Y_σ using Lemma 2.3 (when applicable). $\square$

Let us consider another model case of the application of our algorithm based on the geometrically friendly norms on function spaces introduced in [42, Section 3.1]. Let us recall that, for a vector $v \in \mathbb{R}^k$, $v_{\min}$ ($v_{\max}$) is its minimal (maximal) component $\min_{i \in I_k} v_i$ ($\max_{i \in I_k} v_i$).

Theorem 4.18. *For $i \in \{0, 1\}$ and $Y_i \in \bigcup_{j \in I_5} \Gamma_j$, let X_i be either a subspace or a quotient, or a Banach space finitely represented in Y_i. Assume also that either $\alpha_i = \min(p_{i\min}, q_i, 2)$ and $\beta_i = \max(p_{i\max}, q_i, 2)$ if $Y_i \in \left\{Y_{p_i, q_i}^{s_i}(G_i), Y_{p'_i, q'_i}^{s_i}(G_i)^*\right\} \subset \Gamma_1 \cup \Gamma_3 \cup \Gamma_4$, or $\alpha_i = \min(p_{i\min}, 2)$ and $\beta_i = \max(p_{i\max}, 2)$ if $Y_i \in \left\{Y_{p_i}^{s_i}(G_i), Y_{p'_i}^{s_i}(G_i)^*\right\} \subset \Gamma_2$, or $\alpha_i = \min(p_i, 2)$ and $\beta_i = \max(p_i, 2)$ if $Y_i = Y_{p_i} \in \Gamma_5$. Then we have*
a) $S_b(X_0, X_1) \supset (0, \alpha_0/\beta_1]$;
b) $S_b(Y_0, Y_1] = (0, \alpha_0/\beta_1]$ if $\dim(X_0) = \dim(X_1) = \infty$ and $Y_i \notin \Gamma_4$ for $i \in \{0, 1\}$.

Remark 4.3. *a) The explicit uniform upper bounds for the extension operators can be easily traced as in Corollary 1.1.*
b) To treat the case of X_i being, for $r \in (1, \infty)$ an l_r–sum of the spaces under consideration with the supremum and infimum of all parameters in $(1, \infty)$, we need to consider infimum and supremum of r, 2 and all other parameters of the spaces constituting the l_r-sum as α_i and β_i correspondingly. Indeed, the identity $d_{BM}(l_\sigma(I_m), l_\gamma(I_m)) = m^{|1/\sigma - 1/\gamma|}$ with $\gamma \in \{\alpha_0, \beta_1\}$ shows that the conditions of the existence of uniformly isomorphic and complemented copies of $l_\gamma(I_m)$ is satisfied because Lemmas 2.1 – 2.3 and Theorems 2.1 – 2.3 provide the existence of such copies of $l_{\sigma_k}(I_m)$ with $\lim_{k \to \infty} \sigma_k = \gamma$, and the quantitative asymmetric uniform convexity and smoothness of the l_r–sums is covered by the approach in [23].

Proof of Theorem 4.18. Since the Markov type and cotype exponents and constants of X_i are provided in Theorems 4.5, 4.10 and 4.16, we employ Theorem 4.11 to obtain Part a) and complement these theorems with the proper combination of Lemmas 2.1 – 2.3, Theorems 2.1 – 2.3 and Remark 2.5 to verify the presence of uniformly isomorphic copies of $\{l_{\alpha_0}(I_m)\}_{m \in \mathbb{N}}$ in Y_0 and uniformly isomorphic and complemented copies of $\{l_{\beta_1}\}_{m \in \mathbb{N}}$ in Y_1 to verify the conditions of Theorem 4.15, thus, implying Part b). $\square$

ACKNOWLEDGMENTS

I am grateful to Professor Oleg V. Besov for the most essential part of my knowledge related to Besov, Lizorkin–Triebel and other function spaces and appreciates his encouragement and support over the years.

At different times and under different circumstances, the work related to this chapter had been partially supported by the the Australian Research Council (project numbers DP0881037, DP120100097 and DP150102758), the Russian Fund for Basic Research (project numbers 08–01–00443, 11–01–00444–a and 14–01–00684) and the EIS Research Grant held by James McCoy at the Institute for Mathematics and its Applications at the University of Wollongong.

It is also my pleasure to thank the President and Scientific and Technical Editor(s) of Nova Science Publishers, Inc. for the opportunity and related collaboration.

REFERENCES

[1] Bergh, J. and Löfström, J. 1976. *Interpolation spaces. An introduction.* (In *Die Grundlehren der mathematischen Wissenschaften, B. 223.*) Berlin–Heidelberg–New York–Tokyo: Springer–Verlag.

[2] Besov, Oleg Vladimirovich, Il'in, Valentin Petrovich, and Nikol'skiy, Sergey Mikhailovich. 1996. *Integral representations of functions and embedding theorems.* Moscow: Nauka–Fizmatgiz. (1st Ed. Moscow: Nauka. 1975.)

[3] Burenkov, Viktor Ivanovich. 1998. *Sobolev Spaces on Domains.* (In *Teubner Texts in Mathematics, V. 137*), Stuttgart: B. G. Teubner Verlagsgesellschaft GmbH.

[4] Triebel, Hans. 2006. *Theory of Function Spaces III.* (In *Monographs in Mathematics, V. 100.*) Basel–Boston–Berlin: Birkhäuser Verlag. P. 1–426+i–xii.

[5] Odell, Elton and Schlumprecht, Thomas. 1994. "The distortion problem." *Acta Math.* 173: 259–281.

[6] Milman, Vitaliy Davidovich. 1969. "Geometric theory of Banach spaces. II. Geometry of the unit ball." *Uspehi. Mat. Nauk* 26(6(162)): 73–149.

[7] Tsar'kov, I. G. 1992. "Theorems on global existence of an implicit function, and their applications." *Russian Acad. Sci. Dokl. Math.* 45: 638--640.

[8] Tsar'kov, I. G. 1993. "On global existence of an implicit function." *Russian Acad. Sci. Sbornik Math.* 184: 79--116. Engl. transl. 1994. 79: 287--313.

[9] Tsar'kov, I. G. 1993. "Smoothing of uniformly continuous mappings in L_p spaces." *Mat. Zametki* 54: 123–140. Engl. transl. Math. Notes. 1993. V. 54. P. 957–967.

[10] Ajiev, S. S. 2014. "Hölder analysis and geometry on Banach spaces: homogeneous homeomorphisms and commutative group structures, approximation and Tzar'kov's phenomenon. Part I." *Eurasian Math. J.* 5(1): 7–60.

[11] Ajiev, S. S. 2014. "Hölder analysis and geometry on Banach spaces: homogeneous homeomorphisms and commutative group structures, approximation and Tzar'kov's phenomenon. Part II." *Eurasian Math. J.* 5(2): 6–50.

[12] Minty, G. 1970. "On the extension of Lipschitz, Lipschitz-Hölder and monotone functions." *Bull. Amer. Math. Soc.* 76: 334–339.

[13] Williams, L. R., Wells, J. H., and Hayden, T. 1971. "On the extension of Lipschitz-Hölder maps on L_p spaces." *Studia. Math.* 39: 29–38.

[14] Wells, J. H. and Williams, L. R. 1975. *Embeddings and Extensions in Analysis.* (In *Ergebnisse der Mathematik und ihrer Grenzgebiete, b. 84.*) Berlin–Heidelberg–New York: Springer–Verlag. P. 1–108+i–vii.

[15] Williams, L. R. and Wells, J. H. 1978. "L_p-inequalities." *J. Math. Anal. Appl.* 64: 518–529.

[16] Ajiev, S. S. 2013. "Quantitative Hahn–Banach theorems and isometric extensions for wavelet and other Banach spaces." *Axioms* 2: 224–270. (Open access: doi:10.3390/axioms2020224.)

[17] Ball, K. M. 1992. "Markov chains, Riesz transforms and Lipschitz maps." *Geom. Funct. Anal.* 2(2): 137–172.

[18] Benyamini, Y. and Lindenstrauss, J. 2000. *Geometric nonlinear functional analysis. Volume 1.* (In *Colloquium Publications, V. 48.*) Providence, Rhode Island: American Mathematical Society. P. 1–489+i–xii.

[19] Konyagin, Sergey Vladimirovich and Tsar'kov, I. G. 1988. "On smoothing of maps in normed spaces." *Russian Math. Surveys* 43(262): 205–206.

[20] Naor, Assaf. 2001. "A phase transition phenomenon between the isometric and isomorphic extension problems for Hölder functions between L_p spaces." *Mathematika* 48(1–2): 253–271.

[21] Naor, Assaf, Peres, Yuval, Schramm, Oded, and Sheffield, Scott. 2006. "Markov chains in smooth Banach spaces and Gromov–hyperbolic metric spaces." *Duke Math. J.* 134: 165–197.

[22] Ostrovskii, Mikhail I. 2013. *Metric embeddings : bi-Lipschitz and coarse embeddings into Banach spaces.* (In *De Gruyter studies in mathematics, V. 49.*) Berlin-Boston: Walter de Gruyter GmbH.

[23] Ajiev, S. S. 2019. "Asymmetric convexity and smoothness: quantitative Lyapunov theorem and Lozanovskii factorisation, Walsh chaos and convex duality." In *Understanding Banach Spaces*, edited by Daniel González. Long Island: Nova Science Publishers, Inc.

[24] Ajiev, S. S. 2009. "On Chebyshev centres, retractions, metric projections and homogeneous inverses for Besov, Lizorkin–Triebel Sobolev and other Banach spaces." *East J. Approx.* 15(3): 375-428.

[25] Ajiev, S. S. 2019. "Copies of sequence spaces and basic properties of function spaces." In *Understanding Banach Spaces*, edited by Daniel González. Long Island: Nova Science Publishers, Inc.

[26] Ajiev, S. S. "Nonlocal geometry of function and other Banach spaces: constants, moduli and compact and Kato–Pełczyński operators." *Manuscript* 1–51.

[27] Besov, Oleg Vladimirovich. 2005. "Interpolation, embedding, and extension of spaces of functions of variable smoothness." *Tr. Matem. Inst. Steklova.* 248: 52–63. Engl. transl. *Proc. Steklov Math. Inst.* 248(1): 47–58.

[28] Sobolev, Sergey L'vovich. 1950. *Some applications of functional analysis in mathematical physics.* Leningrad: Izd. Leningr. Gos. Univ. 1950. P. 1–255. Third edition. Moscow: "Nauka". 1988. P. 1–334. English transl. by H. H. McFaden. (In *Translations of Mathematical Monographs, V. 90.*) Providence, Rhode Island: American Mathematical Society. 1991. P. 1–286+i–viii.

[29] Nikol'skii, Sergey Mikhailovich. 1969. *Approximation of function of several variables and embedding theorems.* Moscow: Nauka. Engl. transl. Berlin–Heidelberg–New York: Springer–Verlag. 1975.

[30] Triebel, Hans. 1995. *Interpolation Theory, Function Spaces, Differential Operators.* Amsterdam–Oxford.: North–Holland Publishing Company. P. 1–528.

[31] Maleev, R. P. and Trojanski, S. L. 1971/2. "Moduli of convexity and of smoothness for the spaces L_{pq}." *Annuaire Univ. Sofia Fac. Math.* 66: 331–339.

[32] Ajiev, S. S. "Optimal non–dual local non–stationary multiresolution analysis for anisotropic function spaces." *Manuscript* 1–51.

[33] Berkolaiko, M. Z. and Novikov, I. Ya. 1994. "Unconditional bases in spaces of functions of anisotropic smoothness." *Proc. Steklov Math. Inst.* 204(4): 27–41.

[34] Cohen, A., Daubechies, Ingrid, and Feauveau, J.–C. 1992. "Biorthogonal bases of compactly supported wavelets." *Comm. Pure Appl. Math.* 45: 485–560.

[35] Daubechies, Ingrid. 1992. *Ten lectures on wavelets.* Philadelphia: Soc. Industr. Appl. Math. P. 1–357+i–xix.

[36] Diestel, Joseph, Jarchow, Hans, and Tonge, Andrew. 1995. *Absolutely summing operators.* Cambridge: Cambridge University Press. P. 1–474+i–xv.

[37] Lindenstrauss, Joram and Tzafriri, Lior. 1979. *Classical banach spaces. Volume II.* (In *Ergebnisse, v. 97.*) Berlin-Heidelberg-New York: Springer Verlag. P. 1–243+i–x.

[38] Ajiev, S. S. 1996. "Characterization of the function spaces $B_{p,q}^s(G), L_{p,q}^s(G)$ and $W_p^s(G)$, and embeddings into BMO(G)." *Proc. Steklov Inst. Math.* 214(3): 1–18.

[39] Levy, Mireille. 1980. "L'espace d'interpolation réel $(A_0, A_1)_{\theta,p}$ contient l_p." In *Initiation Seminar on Analysis, 19th year: 1979/80. Publ. Math. Univ. Pierre et Marie Curie* 41(3): 9.

[40] Mastylo, M. 1991. "Banach spaces via sublinear operators." *Math. Japonica* 36(1): 85–92.

[41] Pisier, Gilles and Xu, Q. 2001. "Non–Commutative L_p–Spaces." In *Handbook of the geometry of Banach spaces, V. II*, 1459–1517. Amsterdam–Tokyo: Elsevier.

[42] Ajiev, S. S. 1999. "Characterizations of $B_{p,q}^s(G), L_{p,q}^s(G), W_p^s(G)$ and other function spaces. Applications." *Proc. Steklov Math. Inst.* 227(4): 1–36.

[43] Ajiev, S. S. 2003. "On the boundedness of singular integral operators from certain classes I." *Proc. Steklov Math. Inst.* 243: 11–38.

[44] Ajiev, S. S. 2006. "On the boundedness of singular integral operators from certain classes II." *Analysis Mathematica* 32(2): 81–112.

[45] Peetre, Jaak. 1969. "On the theory of $L_{p,\lambda}$ spaces." *J. Funct. Anal.* 4: 71–87.

[46] Il'in, Valentin Petrovich. 1972. "Function spaces $\mathcal{L}_{r,p,\theta}^{\lambda,a;b_s}(G)$." *Zap. Nauchn. Sem. LOMI. Acad. Sci. USSR* 23: 33–40.

[47] DeVore, Ronald A. and Sharpley, Robert C. 1984. "Maximal functions measuring smoothness." *Mem. Amer. Math. Soc.* 293.

[48] Calderón, A.P. and Torchinski, A. 1975 and 1977. "Parabolic maximal functions associated with distributions I and II." *Adv. Math.* 16: 1–64 and 24: 101–171.

[49] Ajiev, S. S. 2001. "Phillips–type theorems for Nikol'skii and certain other spaces." *Proc. Steklov Math. Inst.* 232(1): 27–38.

[50] Ajiev, S. S. 2010. "On concentration, deviation and Dvoretzky's theorem for Besov, Lizorkin–Triebel and other spaces." *Compl. Var. Ellipt. Eq.* (Special issue dedicated to Gérard Bourdaud.) 55(8–10): 693–726.

[51] Ajiev, S. S. "Amir–Franchetti problems, quantitative K–convexity test and Jacobi convexity hierarchy." *Manuscript* 1–35.

[52] Prus, B. and Smarzewski, R. 1987. "Strongly unique best approximations and centres in uniformly convex spaces." *J. Math. Anal. Appl.* 121: 10–21.

[53] Dixmier, J. 1953. "Formes linéaires sur un anneau d'opérateurs." *Bull. Soc. Math. France* 81: 222–245.

[54] Simon, B. 1979. *Trace ideals and their applications*. Cambridge: Cambridge University Press.

[55] Ball, K., Carlen, E. A., and E. H. Lieb. 1994. "Sharp uniform convexity and smoothness inequalities for trace norms." *Invent. Math.* V. 115. No. 3. 1994. P. 463–482.

[56] Pisier, Gilles. 1975. "Martingales with values in uniformly convex spaces." *Israel J. Math.* 20(3–4): 326-350.

[57] Pisier, Gilles. 1986. "Probabilistic Methods in the Geometry of Banach Spaces." (In *Probability and Analysis, Varenna, 1985*.) *Lect. Notes Math.* V. *1206*, 167–241.

[58] Lyons, T. J. and Zhang, T. S. 1994. "Decomposition of Dirichlet processses and its applications." *Ann. Probab.* 22: 494–524.

[59] Dunford, Nelson. 1943. "Spectral theory." *Bull. Amer. Math. Soc.* 49: 637–651.

In: Understanding Banach Spaces
Editor: Daniel González Sánchez
ISBN: 978-1-53616-745-0
© 2020 Nova Science Publishers, Inc.

Chapter 24

DIFFERENTIAL EQUATIONS WITH A SMALL PARAMETER IN A BANACH SPACE

Vasiliy Kachalov[*]
National Research University "MPEI", Moscow, Russia

Abstract

Many studies have been dedicated to the study of differential equations with a small parameter. In this case, regardless of whether a small parameter in the equation is introduced in a regular or singular way, as a rule asymptotic solutions are constructed. The aim of this chapter is to find the conditions for the usual convergence of series in the powers of a small parameter, representing solutions of both linear and nonlinear equations in Banach spaces. In the case of singularly perturbed problems, the author of the regularization method S.A. Lomov called such solutions pseudo–analytical (pseudo–holomorphic). When the problem is regularly perturbed, in some cases (for example, when an ordinary differential equation or system is investigated) the analytical dependence of the solution on the parameter is guaranteed by the Poincare decomposition theorem.

Keywords: differential equations with a small parameter, vector spaces of exponential type, analyticity of the solution with respect to a parameter, scale of Banach spaces, Navier–Stokes type equation

AMS Subject Classification: 34E05, 34K26

1. INTRODUCTION

The small parameter method allows you to build solutions of equations in the form of series in powers of the parameter included in the studied problem. After the series is constructed, the question about the nature of its convergence is usually arises. In the theory of singular perturbations, such convergence is usually asymptotic (therefore, this theory is often called the theory of asymptotic integration) [1, 2]. In fact, introducing the concept of an asymptotic series, A. Poincare probably meant that with some restrictions on the given problems,

---
[*]Corresponding Author's E-mail: vikachalov@rambler.ru.

the asymptotic series may also converge in the usual sense, i.e. as a power law, with coefficients singularly dependent on a small parameter. For the first time, such solutions were obtained in the basic concepts of S. A. Lomov's regularization method [2, 3] and they were called pseudanalytical (pseudoholomorphic). At the same time, it was proposed to use the theory of exponential type vector spaces to study ordinary convergence [4, 5]. However, in this way, it was possible to construct solutions of *linear* equations in Banach spaces. As for the *nonlinear* problems, here, in the finite–dimensional case, the existence of analytic with respect to a small parameter *integrals* was proved and sufficient conditions for the pseudo–holomorphism of solutions of such problems were given [6]. The situation in the regular perturbation theory is fundamentally different: while solutions of the singularly perturbed problems in the general case are not analytic in parameter ("Dyson's argument" [7]), solutions of normal systems of differential equations whose right-hand parts analytically depend on, the parameter are analytic in accordance with the Poincare decomposition theorem [8]. But even here, when entering Banach space, the analyticity of solutions is not always obtained, but only when special conditions are fulfilled such as the subordination of one operator to another by the operator [9].

In this chapter, firstly we investigate linear differential equations with a small parameter at the derivative, and then equations of the Boltzmann or Navier–Stokes type containing a bilinear operator.

2. SPACES OF VECTORS OF EXPONENTIAL TYPE

2.1. Basic Concepts

Let E be a Banach space, and $\mathcal{L}(E)$ be the algebra of bounded operators defined on E. Denote by $D_{z_0}^{r_0} = \{z : |z - z_0| < r_0\}$ a circle with center z_0 and radius r_0 on the complex plane $\mathbb{C}$. Other notation: $\mathcal{A}(\overline{D}_{z_0}^{r_0}; E)$ is the Banach space of functions analytic in the circle $D_{z_0}^{r_0}$ and continuous on its closure $\overline{D}_{z_0}^{r_0}$, with the norm $\|f(z)\|_\mathcal{A} = \max_{z \in \overline{D}_{z_0}^{r_0}} \|f(z)\|_E$; $\mathcal{A}(\overline{D}_{z_0}^{r_0}; \mathcal{L}(E))$ is a Banach algebra analytic in a specified circle of operators which are continuous on its closure with values from $\mathcal{L}(E)$.

Suppose that $A(z), A^{-1}(z) \in \mathcal{A}(\overline{D}_{z_0}^{r_0}; \mathcal{L}(E))$, and denote by $F = A^{-1}(z)\partial_z$: $\mathcal{A}(\overline{D}_{z_0}^{r_0}; E) \to \mathcal{A}(\overline{D}_{z_0}^{r_0}; E)$ is a closed unbounded operator with the domain $D(F) = \{f(z) \in \mathcal{A}_{z_0}^{r_0}(\overline{D}_{z_0}^{r_0}; E) : \partial_z f \in \mathcal{A}_{z_0}^{r_0}(\overline{D}_{z_0}^{r_0}; E)\}$. If $u_n(z) = F^n[u(z)]$, then, obviously,

$$\int_{z_0}^{z} A(z_1) u_n(z_1) dz_1 = u_{n-1}(z) - u_{n-1}(z_0), \quad n \in \mathbb{N}.$$

Using this equality, one can prove the general Taylor formula

$$u(z) = u(z_0) + \int_{z_0}^{z} A(z_1) dz_1 \cdot F[u(z)]\big|_{z=z_0} + \cdots$$

$$+ \int_{z_0}^{z} A(z_1) dz_1 \cdots \int_{z_0}^{z_{n-1}} A(z_n) dz_n \cdot F^n[u(z)]\big|_{z=z_0} \quad (1)$$

$$+ \int_{z_0}^{z} A(z_1) dz_1 \int_{z_0}^{z_1} A(z_2) dz_2 \cdots \int_{z_0}^{z_{n-1}} A(z_n) dz_n \int_{z_0}^{z_n} A(z_{n+1}) F^{n+1}[u(z_{n+1})] dz_{n+1},$$

with the residual term in integral form. In particular, if $A(z)$ coincides with the identity operator I, then the equality (1) turns into the usual Taylor formula.

Let us proceed to the construction and study of spaces of vectors of exponential type. Let us define $c > 0$ and denote by $\mathcal{U}^c$ the set of such elements $u(z) \in \mathcal{A}(\overline{D}_{z_0}^{r_0}; E)$ that $F^k[u(z)] \in \mathcal{A}(\overline{D}_{z_0}^{r_0}; E)$ for all $k = 0, 1, 2, \ldots$, and there exists a number $m(u, c) > 0$ such that

$$\|F^k[u(z)]\|_{\mathcal{A}} \leqslant m(u, c) c^k, \quad k = 0, 1, 2, \ldots.$$

Lemma 1. *The set $\mathcal{U}^c$ with norm*

$$\|u\|_{\mathcal{U}^c} = \sup_k \frac{\|F^k[u(z)]\|_{\mathcal{A}}}{c^k}$$

is Banach.

Proof. The fact that $\mathcal{U}^c$ is a linear normed space is obvious. Establish its completeness.

Let $\{u_n\}_{n=1}^{\infty} \subset \mathcal{U}^c$ be the fundamental sequence, that is, $\forall \varepsilon > 0 \; \exists N \colon \forall m, p > N$

$$\|u_m - u_p\|_{\mathcal{U}^c} = \sup_k \frac{\|F^k[u_n(z) - u_p(z)]\|_{\mathcal{A}}}{c^k} < \varepsilon.$$

It follows that for each $k = 0, 1, 2, \ldots$ sequence $\{F^k[u_n(z)]\}_{n=1}^{\infty}$ is fundamental in the space $\mathcal{A}(\overline{D}_{z_0}^{r_0}; E)$ and, therefore, due to its completeness, has a limit. Denote it by u^k. We have, $F^{k-1}[u_n(z)] \to u^{k-1}(z)$, $F^k[u_n(z)] \to u^k(z)$ with $n \to \infty$ and each $k \in \mathbb{N}$, and since the operator $F: \mathcal{A}(\overline{D}_{z_0}^{r_0}; E) \to \mathcal{A}(\overline{D}_{z_0}^{r_0}; E)$ is closed, then $u^{k-1} \in D(F)$ and $Fu^{k-1} = u^k$, $k = 1, 2, \ldots$. Therefore, if $u_n \to u^0$ with $n \to \infty$ in $\mathcal{A}(\overline{D}_{z_0}^{r_0}; E)$, then $F^k u_n \to F^k u^0$ in the same space and, therefore, $F^k u^0 \in D(F)$ for every natural k.

We now prove that $\exists m(u^0, c) > 0: \|F^k u^0\|_{\mathcal{A}} \leq m(u^0, c) c^k$. To do this, note that $\forall \varepsilon > 0 \; k = 0, 1, \ldots \; \exists N : \forall n > N \; \|F^k u^0\|_{\mathcal{A}} \leq \|F^k u_n\|_{\mathcal{A}} + \varepsilon \leq Mc^k + \varepsilon$ where $M > 0$ is such that $\|u_n\|_{\mathcal{U}^c} < M$ (such M exists because $\{u_n\}_{n=1}^{\infty}$ is fundamental to $\mathcal{U}^c$ sequence). Further, in view of the arbitrariness of $\varepsilon > 0$, we get that $\|F^k[u^0(z)]\|_{\mathcal{A}} \leqslant Mc^k$ $\forall k = 0, 1, \ldots$. Therefore, for $m(u^0, c) = M$, the required estimate is fulfilled and, thus, Lemma 1 is proved. $\square$

The space $\mathcal{U}^c$ is called *a space of vectors of exponential type $\leqslant c$* and as follows from Lemma 1, the operator F is bounded in it, and $\|F\| \leqslant c$ (in fact, $\|F\| = c$, which is easy to notice by considering the equation $\partial_z w = cA(z)w$).

Let us introduce the vector space $\exp_F E = \bigcup_{c=1}^{\infty} \mathcal{U}^c$ and let us define the inductive limit topology in it, i. e.
$$\exp_F E = \lim \operatorname{ind}_{c \to \infty} \mathcal{U}^c,$$
and we will call this space *–space of vectors of exponential type*.

Remark 2. *If $A(z) = I$, then the constructed space coincides with the space of entire functions of exponential type.*

2.2. Representation of Elements of the Space of Vectors of Exponential Type

In the case of entire exponential type functions, the coefficients of their expansions in the power series depend on the number in a special way [10]. As the following statement shows, the same is fulfilled in the general case defined by the formula (1).

Theorem 3. *Let the conditions for the operators $A(z)$ and $A^{-1}(z)$ be fulfilled. Then the vector $u(z) \in \mathcal{U}^c$ admits the representation*

$$u(z) = u_0 + \int_{z_0}^{z} A(z_1)dz_1 \cdot u_1 + \cdots + \int_{z_0}^{z} A(z_1)dz_1 \cdots \int_{z_0}^{z_{n-1}} A(z_n)dz_n \cdot u_n + \cdots, \quad (2)$$

where $u_n = F^n[u(z)]\big|_{z=z_0}$ and the series $\sum_{n=0}^{\infty} \|u_n\|_E \varepsilon^n$ converges in the circle $|\varepsilon| < 1/c$.

Proof. Let us write the formula 1 and the let us use the inequality $\|u_n\| \leqslant m(u,c)c^n$ to estimate the residual term $R_n(z)$:

$$\|R_n(z)\|_E \leqslant a^{n+1}\|F^{n+1}u\|_{\mathcal{A}} \left| \int_{z_0}^{z} dz_1 \int_{z_0}^{z_1} dz_2 \cdots \int_{z_0}^{z_n} dz_{n+1} \right|$$

$$\leqslant \|F^{n+1}u\|_{\mathcal{A}} \frac{|z-z_0|^{n+1}a^{n+1}}{(n+1)!} \leqslant \frac{r_0^{n+1}a^{n+1}c^{n+1}m(u,c)}{(n+1)!},$$

where $a = \max_{z \in \overline{D}_{z_0}^{r_0}} \|A(z)\|_{\mathcal{L}(E)}$.

Hence we get that for $n \to \infty$ $\|R\|_{\mathcal{A}} \to 0$ and, therefore, the representation (2) with coefficients $u_n = F^n[u(z)]\big|_{z=z_0}$ such that the series $\sum_{n=0}^{\infty} \|u_n\|\varepsilon^n$ converges for $|\varepsilon| < 1/c$, is proved. $\square$

2.3. Isomorphism of $\exp_F E$ to the Space of Analytic Functions with Values in E

Let us denote the Banach space of functions with values in E analytic in the circle $D^{\rho_n} = \{\varepsilon : |\varepsilon| < \rho_n\}$ and continuous on its closure $\overline{D}^{\rho_n}$ by $\beta(\rho_n; E)$, with the norm $\|\varphi\|_n =$

$\max\limits_{|\varepsilon|=\rho_n} \|\varphi(\varepsilon)\|_E$, and by $\alpha(\rho; E)$ we denote the space of functions analytic in the circle $D^\rho = \{\varepsilon : |\varepsilon| < \rho\}$, which are the projective limit of the sequence of Banach spaces $\{\beta(\rho_n; E)\}_{n=1}^\infty$ with $\rho_n \to \rho$.

The space $\alpha(\rho; E)$ is complete and metrizable [11]; its topology is defined using the seminorms $\|\varphi\|_n$.

Next, let
$$\widetilde{\mathcal{U}}^c = \bigcap_{s>c} \mathcal{U}^s,$$
moreover, the topology of the projective limit, the sequence of $\mathcal{U}^s$ Banach spaces is given on $\widetilde{\mathcal{U}}^c$, because $\mathcal{U}^{s_1} \supset \mathcal{U}^{s_2}$ if $s_1 > s_2 > c$. As well as $\alpha(\rho; E)$, the space $\widetilde{\mathcal{U}}^c$ is complete, metrizable locally convex with a system of seminorms, dictated by the norms of $\mathcal{U}^s$.

Theorem 4. *The topological spaces $\widetilde{\mathcal{U}}^c$ and $\alpha(\rho; E)$ are isomorphic for $\rho = c^{-1}$, and the operator performing this isomorphism has the following form:*

$$\Psi[\varphi(\varepsilon)] = \frac{1}{2\pi i} \oint_l \frac{U(z, \varepsilon)}{\varepsilon} \varphi(\varepsilon) d\varepsilon, \tag{3}$$

where $\varphi(\varepsilon) \in \alpha(\rho; E)$; l is the contour belonging to D^ρ and covering the center of the circle; $U(z, \varepsilon)$ is the evolution operator of the equation $\varepsilon \partial_z u = A(z) u$.

Proof. Firstly we prove that $\widetilde{\mathcal{U}}^c$ and $\alpha(c^{-1}; E)$ are isomorphic as linear spaces. Indeed, in accordance with Theorem 1, each $u(z) \in \widetilde{\mathcal{U}}^c$ can be associated with an element

$$\varphi(\varepsilon) = \sum_{k=0}^\infty \varepsilon^n F^n[\varphi(z)]\big|_{z=0} \in \alpha(c^{-1}; E).$$

Since $\varphi(\varepsilon) \in \alpha(s^{-1}; E)$ for all $s > c$. On the contrary, let

$$\varphi(\varepsilon) = \sum_{k=0}^\infty \varepsilon^n v_n \in \alpha(c^{-1}; E). \tag{4}$$

Let us investigate the series

$$u(z) = v_0 + \int_{z_0}^z A(z_1) dz_1 \cdot v_1 + \cdots + \int_{z_0}^z A(z_1) dz_1 \cdots \int_{z_0}^{z_{n-1}} A(z_n) dz_n \cdot v_n + \cdots$$

and prove that $u(z) \in \widetilde{\mathcal{U}}^c$, i.e. $\forall s > c\ u(z) \in \mathcal{U}^s$. Since the inclusion (4) takes place, then according to the Cauchy inequalities $\forall s > c\ \forall n = 0, 1, \ldots\ \exists M > 0: \|v_n\|_E \leqslant M s^n$, and, therefore, for each $z \in \overline{D}_{z_0}^{r_0}$

$$\|F^k[u(z)]\|_E = \|(A^{-1}(z)\partial_z)^k[u(z)]\|_E =$$
$$= \left\|v_k + \int_{z_0}^z A(z_1) dz_1 \cdot v_{k+1} + \cdots + \int_{z_0}^z A(z_1) dz_1 \cdots \int_{z_0}^{z_{n-1}} A(z_n) dz_n \cdot v_{n+k} + \cdots \right\|_E \leqslant$$
$$\leqslant \|v_k\| + a\|v_{k+1}\|_E \left|\int_{z_0}^z dz_1\right| + \cdots + a^n \|v_{n+k}\|_E \left|\int_{z_0}^z dz_1 \int_{z_0}^{z_1} dz_2 \cdots \int_{z_0}^{z_{n-1}} dz_n\right| + \cdots \leqslant$$
$$\leqslant M s^k + a M s^{k+1} |z - z_0| + \cdots + a^n M s^{k+n} \frac{|z-z_0|^n}{n!} + \cdots \leqslant M s^k e^{a r_0 s}.$$

This shows that $\forall k = 0, 1, \ldots$

$$\|F^k u\|_{\mathcal{A}} = \max_{z \in \overline{D}_{z_0}^{r_0}} \|F^k[u(z)]\|_E \leqslant m(u,s) c^k$$

for $m(u,s) = M e^{ar_0 s}$. Therefore $u(z) \in \mathcal{U}^s \ \forall s > c$.

Thus, the isomorphism of the linear spaces $\widetilde{\mathcal{U}}^c$ and $\alpha(c^{-1}; E)$ is proved. In order to prove their isomorphism as locally convex spaces, it is necessary to prove that $\widetilde{\mathcal{U}}^c \subset \alpha(c^{-1}; E)$ and $\alpha(c^{-1}; E) \subset \widetilde{\mathcal{U}}^c$, moreover, both inclusions are made by the continuous operators Ψ^{-1} and Ψ.

The formula (3) (which generalizes the Borel formula) can be proved using the residue theorem, taking into account that

$$U(z,\varepsilon) = I + \varepsilon^{-1} \int_{z_0}^{z} A(z_1) dz_1 + \cdots + \varepsilon^{-n} \int_{z_0}^{z} A(z_1) dz_1 \cdots \int_{z_0}^{z_{n-1}} A(z_n) dz_n + \cdots. \quad (5)$$

We prove that the operator Ψ implementing the embedding $\alpha(c^{-1}; E) \subset \widetilde{\mathcal{U}}^c$ is continuous, that is, $\forall s > c\ \exists n \in \mathbb{N}$: $\forall \varphi(\varepsilon) \in \alpha(c^{-1}; E)$ the inequality $\|\Psi_\varphi\|_{\widetilde{\mathcal{U}}^c} \leq \mathcal{K} \|\varphi\|_n$ for some $\mathcal{K} > 0$. We fix $s > c$, then, by the definition of norms in $\widetilde{\mathcal{U}}^s$ and $\beta(\rho_n; E)$, this inequality takes the following form:

$$\sup_k \frac{\|F^k[\Psi_\varphi]\|_{\mathcal{A}}}{s^k} \leqslant \mathcal{K} \max_{|\varepsilon|=\rho_n} \|\varphi(\varepsilon)\|_E. \quad (6)$$

Let us assume $\rho_n = s^{-1}$ in this inequality and let us prove its validity. Using the definition of the operator Ψ, the inequality (6) is rewritten as follows:

$$\sup_k \left(s^{-k} \left\| \oint_{|\varepsilon|=\rho_n} \frac{U(z,\varepsilon)}{2\pi i \varepsilon^{k+1}} \varphi(\varepsilon) d\varepsilon \right\|_{\mathcal{A}} \right) \leqslant \mathcal{K} \|\varphi\|_n. \quad (7)$$

From the formula (5) it follows that

$$\max_{\substack{|\varepsilon|=\rho_n, \\ z \in \overline{D}_{z_0}^{r_0}}} \|U(z,\varepsilon)\| \leqslant e^{ar_0 \rho_n^{-1}} = e^{ar_0 s},$$

i.e.,

$$\left\| \frac{1}{2\pi i} \oint_{|\varepsilon|=\rho_n} \frac{U(z,\varepsilon)}{\varepsilon^{k+1}} \varphi(\varepsilon) d\varepsilon \right\|_{\mathcal{A}} \leqslant \frac{e^{ar_0 s}}{2\pi \rho_n^k} \|\varphi\|_n,$$

which means the inequality (6). So, $\alpha(c^{-1}; E) \subset \widetilde{\mathcal{U}}^c$.

Let us establish the continuity of the inverse embedding $\widetilde{\mathcal{U}}^c \subset \alpha(c^{-1}; E)$, i.e., show that $\forall n \in \mathbb{N}\ \exists s > c$: $\forall u \in \mathcal{U}^s\ \|\varphi\|_n \leqslant \widetilde{\mathcal{K}} \|u\|_{\mathcal{U}^s}$ for some $\widetilde{\mathcal{K}} > 0$. In fact, since

$$\varphi(\varepsilon) = \sum_{k=0}^{\infty} \varepsilon^k F^k[u(z)]\Big|_{z=z_0},$$

that
$$\|\varphi\|_n \leq \sum_{k=0}^{\infty} \rho_n^k \|u\|_{\mathcal{U}^s} s^k \leq \widetilde{\mathcal{K}} \|u\|_{\mathcal{U}^s},$$

if $(\rho_n + (c^{-1} - \rho_n)/2)$ is taken as s, and $\widetilde{\mathcal{K}}$ is set to $(1 - \rho_n s)^{-1}$.

Thus, $\widetilde{\mathcal{U}}^c \Leftrightarrow \alpha(\rho^{-1}; E)$ and the theorem is proved. □

Corollary 5. *Insofar as*
$$\exp_F E = \bigcup_{c>0}^{\infty} \widetilde{\mathcal{U}}^c,$$
then there is an isomorphism $\exp_F E \Leftrightarrow \alpha_0(E)$, *where* $\alpha_0(E)$ *is the linear space of functions analytic at* $\varepsilon = 0$ *with values in the Banach space* E.

Corollary 6. *If E is finite–dimensional, then, as is well known, $\alpha(\rho; E)$ is the Grothendieck nuclear space* [11]. *We show that* $\exp_F E$ *is a nuclear space.*

Indeed, $\alpha(c^{-1}; E) \Leftrightarrow \widetilde{\mathcal{U}}^c$, and, therefore, $\widetilde{\mathcal{U}}^c$ is also nuclear Grothendieck space. If a
$$u \in \widetilde{\mathcal{U}}^{c_1} = \bigcap_{c>c_1} \mathcal{U}^c,$$
then $u \in \mathcal{U}^c \ \forall c > c_1$. Therefore, knowingly
$$u \in \widetilde{\mathcal{U}}^{c_2} = \bigcap_{c>c_2} \mathcal{U}^c$$
with $c_1 < c_2$. Therefore, $\widetilde{\mathcal{U}}^{c_1} \subset \widetilde{\mathcal{U}}^{c_2}$ with $c_1 < c_2$.

To prove the continuity of this embedding, we fix $p > c_2$ and find $q > c_1$ such that $\forall u \in \widetilde{\mathcal{U}}^{c_1} \ \|u\|_{\mathcal{U}^p} \leq \|u\|_{\mathcal{U}^q}$. The definition of the norm in $\mathcal{U}^s$ shows that this inequality holds for all p such that $c_1 < q < p$, since
$$\sup_k \frac{\|F^k u\|_{\mathcal{A}}}{p^k} \leq \sup_k \frac{\|F^k u\|_{\mathcal{A}}}{q^k}.$$

As stated in Corollary 1,
$$\exp_F E = \bigcap_{c>0} \widetilde{\mathcal{U}}^c,$$
means
$$\exp_F E = \lim_{c \to \infty} \widetilde{\mathcal{U}}^c.$$

It is known [11] that the inductive limit of nuclear spaces is also a nuclear space. Thus, it is proved that $\exp_F E$ is the Grothendieck nuclear space.

3. THE SMOOTHNESS OF SOLUTIONS OF LINEAR DIFFERENTIAL EQUATIONS IN A SINGULAR PARAMETER

3.1. The Case of a Limited Limit Operator

The linear theory of singular perturbations considers the equations of the form

$$\varepsilon \partial_z w - A(z)w = f(z), \quad z \in D_{z_0}^{r_0} \tag{8}$$

and studies the behavior of their solutions with $\varepsilon \to 0$.

Since the parameter ε enters this equation in an analytical (even whole) way, the natural question arises: can the solution $w(z, \varepsilon)$ also analytically depend on ε? The nontriviality of such a question is primarily due to the fact that for $\varepsilon = 0$, the equation (8) ceases to be differential, and therefore this parameter value is special for this equation. Since the answer turned out to be positive, this made it possible to construct a so-called analytic theory of singular perturbations [3, 8]. If we set the initial condition for the equation (8) and apply S. A. Lomov's regularization method, then under certain conditions [4–6], one can get a solution of the problem that converges in the usual sense (and not asymptotically) Cauchy.

Theorem 7. *In order for the equation (8) to have an analytical solution at $\varepsilon = 0$, it is necessary and sufficient that*

$$\int_{z_0}^{z} h(\xi) d\xi \in \exp_F E. \tag{9}$$

In this case, the resulting solution will be the only one.

Proof. *Necessity.* Let

$$w(z, \varepsilon) = w_0(z) + \varepsilon w_1(z) + \cdots + \varepsilon^n w_n(z) + \cdots \tag{10}$$

analytic at $\varepsilon = 0$ the solution of the equation (8) and the series representing it converges uniformly on $\overline{D}_{z_0}^{r_0}$. In accordance with the method of uncertain coefficients, we obtain that

$$w_0(z) = -A^{-1}(z)h(z), \quad w_n(z) = F^n[w_{n-1}(z)], \quad n = 1, 2, \ldots, \tag{11}$$

because of the uniform convergence in $z \in \overline{D}_{z_0}^{r_0}$ of the series (10) for $|\varepsilon| < c^{-1}$, where c is a positive number, means together with Cauchy inequalities for the coefficients of analytic functions that $w_0(z) \in \mathcal{U}^c$. According to Theorem 1, from the membership of $A^{-1}(z)h(z)$ to the space of vectors of exponential type the inclusion of (9) follows.

Sufficiency. We will seek a solution to the equation (8) as the series (9). Then, by substituting this series into the equation and using the method of indefinite coefficients, we get the formulas (11) and the convergence of the series (9) will be obvious due to the fact that $w_0(z) \in \mathcal{U}^c$ for some $c > 0$.

The uniqueness of the solution $w(z, \varepsilon)$ comes from the method of its construction. The theorem is proved. $\square$

Differential Equations with a Small Parameter in a Banach Space 423

All of the above easily applies to higher order equations

$$\varepsilon \partial_z^m w - A(z)w = h(z), \quad z \in \overline{D}_{z_0}^{r_0}, \quad m \geqslant 2, \tag{12}$$

with initial conditions $w^{(k)}(z_0, \varepsilon) = 0$ with $k = \overline{1, m-1}$.

In this case, we introduce the operator $F_m = A^{-1}(z)\partial_z^m$ with the domain $D(F_m) = \{f \in \mathcal{A}(\overline{D}_{z_0}^{r_0}; E) : \partial_z^m f \in \mathcal{A}(\overline{D}_{z_0}^{r_0}), f^{(k)} = 0, k = \overline{1, m-1}\}$, and the space $\mathcal{U}_m^c$ of vectors of exponential type $\leqslant c$ constructed from it will consist of elements kind of

$$u(z) = u(z_0) + \int_{z_0}^z d\tau_{m-1} \int_{z_0}^{\tau_{m-1}} d\tau_{m-2} \cdots \int_{z_0}^{\tau_1} A(z_1)dz_1 \cdot F_m[u(z)]\big|_{z=z_0}$$

$$+ \int_{z_0}^z d\tau_{m-1} \int_{z_0}^{\tau_{m-1}} d\tau_{m-2} \cdots \int_{z_0}^{\tau_1} A(z_1)dz_1 \int_{z_0}^{z_1} d\tau_{m-1} \int_{z_0}^{\tau_{m-1}} d\tau_{m-2} \cdots$$

$$\cdots \int_{z_0}^{\tau_1} A(z_2)dz_2 \cdot F_m^2[u(z)]\big|_{z=z_0} + \cdots$$

$$+ \int_{z_0}^z d\tau_{m-1} \int_{z_0}^{\tau_{m-1}} d\tau_{m-2} \cdots \int_{z_0}^{\tau_1} A(z_1)dz_1 \cdots \int_{z_0}^{z_{n-1}} d\tau_{m-1} \int_{z_0}^{\tau_{m-1}} d\tau_{m-2} \cdots$$

$$\cdots \int_{z_0}^{\tau_1} A(z_n)dz_n \cdot F_m^n[u(z)]\big|_{z=z_0} + \cdots.$$

Accordingly,

$$\exp_{F_m} E = \lim_{c \to \infty} \widetilde{\mathcal{U}}_m^c.$$

The statement is similar to the previous one.

Theorem 8. *In order for the (12) problem to have a unique solution at $\varepsilon = 0$, it is necessary and sufficient that $A^{-1}(z)h(z) \in \exp_{F_m} E$.*

3.2. Analyticity with Respect to the Parameter of Solutions of Singular Type Equations in the Case of an Unbounded Limit Operator

Here in the real domain we will investigate the equation of a form

$$\varepsilon \partial_t u - A(t)u = h(t), \tag{13}$$

when $t \in [0, T]$ and the operator $A(t) : E \to E$ is unboundedly closed, with the domain $D(A(t))$, which independent of t dense and everywhere in E. Let us denote by $C([0, T]; E)$ the space of functions continuous on $[0, T]$ with values in E and norm

$$\|v(t)\|_C = \max_{t \in [0,T]} \|v(t)\|_E,$$

in which it is banach. Let $C([0, T]; \mathcal{L}(E))$ be the algebra of bounded operator-valued functions continuous on the segment $[0, T]$.

Subsequent verbal proof will be based on the concept of a scale of Banach spaces [12]. First of all, this is necessary in order to prove a representation similar to the (2) representation.

Let s be a parameter varying on the segment $[0, l]$, and each value of s corresponds to a Banach space E_s and $E_s \subset E_{s'}$ if $s > s'$. The norm of the element $u \in E_s$ will be denoted by $\|u\|_s$, while let $\|u\|_{s'} \leqslant \|u\|_s$, when $u \in E_s$ ($s > s'$). Denote by L_γ the space of linear operators A defined in each space E_s with $0 < s \leqslant 1$, mapping E_s into $E_{s'}$, $0 \leqslant s' < s$, and such that for $u \in E_s$

$$\|Au\|_{s'} \leqslant \frac{\gamma}{s-s'}\|u\|_s.$$

If we put

$$\|A\|_{L_\gamma} = \sup_{0 \leqslant s' < s \leqslant 1} \sup_u \{(s-s')\|Au\|_{s'}\|u\|_s^{-1}\},$$

then the space L_γ will be Banach.

Consider the operator $F = A^{-1}(t)\partial_t$ with the domain $D(F) = \{v(t) : Fv \in C([0;T];E)\}$ in $C([0,T];E)$. Next, we take the elements $v(t) \in C([0;T];E)$ such that $F^k v \in D(F)$ and there is $m(v,c) > 0$ such that $\|F^k v\|_c \leq m(v,c)c^k$, $k = 0, 1, \ldots$ for some $c > 0$. Many such elements with the norm

$$\|u\|_{\mathcal{U}^c} = \sup_k \frac{\|F^k v\|_c}{c^k}$$

will be a Banach space. It is called the space of vectors of exponential type $\leq c$ and the operator F is bounded in it. Clearly, $\mathcal{U}^c \subset \mathcal{U}^{c'}$ as $c < c'$, therefore, if on the space

$$\exp_F E = \bigcup_{c>0} \mathcal{U}^c$$

set the topology of the inductive limit, then it will be a locally convex topological space.

Theorem 9. *Let $v(t) \in \mathcal{U}^c$, with $E = E_1 \subset \{E_s\}$ ($0 \leq s \leq 1$), operator $A(t) : E \to E$ is closed unbounded with a domain independent of t, and $A^{-1}(t) \in C([0;T];\mathcal{L}(E))$. Then, if $A(t)$ takes values in L_γ and is continuous on the segment $[0;T]$, then for each $s \in [0;1)$ there is $T_s > 0$, such that for any $t \in [0;T_s)$ the representation is valid*

$$v(t) = v(0) + \int_0^t A(t_1)dt_1 \cdot Fv\Big|_{t=0} + \cdots \qquad (14)$$

$$+ \int_0^t A(t_1)dt_1 \cdots \int_0^{t_{n-1}} A(t_n)dt_n \cdot F^n v\Big|_{t=0} + \cdots.$$

Proof. As in the case of a bounded statement, the representation

$$v(t) = v(0) + \int_0^t A(t_1)dt_1 \cdot Fv\Big|_{t=0} + \cdots$$

$$+ \int_0^t A(t_1)dt_1 \cdots \int_0^{t_{n-2}} A(t_{n-1})dt_{n-1} \cdot F^{n-1}v\Big|_{t=0} + R_n(t),$$

where
$$R_n(t) = \int_0^t A(t_1)dt_1 \cdots \int_0^{t_{n-2}} A(t_{n-1})dt_{n-1} \int_0^{t_{n-1}} A(t_n)F^n[v(t_n)]dt_n$$

— residual term in integral form.

Fix $s \in [0; 1)$, and let $1 > s_1 > s_2 > \cdots > s_{n-1} > s$ be an arbitrary partition of the interval $(s, 1)$. We have:

$$\|R_n(t)\|_s \leqslant m(v, c) \frac{\gamma^n t^n c^n}{n!(1-s_1)(s_1-s_2)\cdots(s_{n-1}-s)},$$

since $\|F^n v\|_s \leqslant m(v,c)c^n \ \forall n \in \mathbb{N} \ \forall s \in (0; 1]$.

Take a uniform partition, i.e.,
$$1 - s_1 = s_1 - s_2 = \cdots = s_{n-1} - s = (1-s)/n,$$

then
$$\|R_n(t)\|_s \leqslant m(v,c) \frac{\gamma^n t^n c^n n^n}{n!(1-s)^n} \longrightarrow 0, \quad n \to \infty,$$

uniformly in t from each compactum belonging to $[0; T_s)$, where $T_s = \min\{T; (1-s)(\gamma c e)^{-1}\}$. Theorem 5 is proved. □

Now we can formulate sufficient conditions for the existence of an analytic solution at the point $\varepsilon = 0$ of the equation (13).

Theorem 10. *Let the conditions of Theorem 5 be fulfilled and the right–hand part has a form: side $h(t) = A(t)v(t)$, where $v(t) \in \exp_F E$. Then the equation (13) has a unique solution with respect to ε over the entire interval $[0; T]$.*

Example 11. *Consider the equation for $t \geqslant 0$*
$$\varepsilon \partial_t u - (t+1)\partial_x u = tx^2, \quad x \in [0; R_0], \tag{15}$$

in which the domain of definition $D((t+1)\partial_x)$ of the limit operator consists of functions $g(x)$ allowing analytic continuation to the circle $\Omega_0 = \{z \in \mathbb{C} : |z| < R_0\}$, and such that $g(0) = 0$. We define $d > 0$ and denote by Ω_s a circle of radius $R_0 + sd$ with center at the point $z = 0$. If E_s $(0 \leqslant s \leqslant 1)$ is the space of functions analytic in the circle Ω_s and continuous on its closure $\overline{\Omega}_s$, with the norm of uniform convergence, then $\{E_s\}$ is the desired scale of Banach spaces. In this case, as follows from the Cauchy integral formula of,

$$\|\partial_z g(z)\|_{s'} \leqslant \frac{1}{(s-s')d} \|g(z)\|_s,$$

when $0 \leqslant s' < s$.

The solution of the equation (15), analytical at the point $\varepsilon = 0$, can be represented as the series
$$u(t, x, \varepsilon) = -\frac{tx^3}{3(t+1)} + 2 \sum_{n=1}^{\infty} \varepsilon^n \frac{(-1)^n (2n-1)!! x^{n+3}}{(n+3)!(t+1)^{2n+1}},$$

which converges at $|\varepsilon| < 1/2R_0$, uniformly in the domain $G = \{(t, x) : t \geq 0, \ 0 \leq x \leq R_0\}$.

4. NONLINEAR OPERATOR EQUATIONS IN A BANACH SPACE

Using the estimates made in the proof of Theorem 5, we can establish that the operator

$$U(t,\tau) = I + \int_\tau^t A(t_1)dt_1 + \cdots + \int_\tau^t A(t_1)dt_1 \cdots \int_\tau^{t_{n-1}} A(t_n)dt_n \cdots \quad (16)$$

is resolving for the equation $\partial_t u = A(t)u$, and the series (16) converges on the interval $[0; T_s]$. In the Banach space E we consider the Cauchy problem

$$\begin{aligned} \partial_t u - A(t)u &= B(u,u), \ t \in [0;T], \\ u(0) &= \nu w_0(\nu), \end{aligned} \quad (17)$$

where $A(t)$ is a closed unbounded operator satisfying the conditions stated earlier; $B(u,v)$ is a bilinear operator, with the domain $D(A(t)) \times D(A(t))$; $w_0(\nu)$ is a vector from E, analytically depending on viscosity ν.

It is easy to see that in the case of boundedness of the operator B, the equation (17) is a Boltzmann type equation, and if B is bounded in the first variable and closed in the second, then this Navier–Stokes equation (when A does not depend on t) [13].

Let us make the replacement $u = \nu w$ and move on to the following initial problem:

$$\begin{aligned} \partial_t w - A(t)w &= \nu B(w,w), \\ w(0,\nu) &= w_0(\nu). \end{aligned} \quad (18)$$

Here, the *nonlinear* member of the equation is subordinate and, therefore, the solution can be searched as a series in viscosity degrees, solving the *linear* problem at each step.

So let

$$w(t,\nu) = W_0(t,\nu) + \nu W_1(t,\nu) + \cdots + \nu^n W_n(t,\nu) + \cdots. \quad (19)$$

Then, in accordance with the method of uncertain coefficients, we obtain a series of problems:

$$\begin{aligned} \partial_t W_0 - A(t)W_0 &= 0, \ W_0(0,\nu) = w_0(\nu), \\ \partial_t W_1 - A(t)W_1 &= B(W_0, W_0), \ W_1(0,\nu) = 0, \\ &\vdots \\ \partial_t W_n - A(t)W_n &= \sum_{k=0}^{n-1} B(W_k, W_{n-k-1}), \ W_n(0,\nu) = 0, \\ &\vdots \end{aligned} \quad (20)$$

from which we successively determine the coefficients of the series.

Further, let D_0 be a linear manifold from $D(A(t))$ such that the following conditions are satisfied:

1°. The countable system of norms $\|\cdot\|_k$ is introduced in D_0 so that $\forall v \in D_0$ $\|v\|_1 \leqslant \|v\|_2 \leqslant \cdots$ and convergence for each of them implies convergence in the norm of E.

Differential Equations with a Small Parameter in a Banach Space

$2°$. There is a representation
$$D_0 = \bigcup_{c>0} Y^c,$$
where Y^c is the set of elements such that $\forall v \in Y^c \; \forall k \in \mathbb{N} \; \|v\|_k \leqslant e^{kc}$.

$3°$. If $u \in Y^{c_1}$, $v \in Y^{c_2}$, then $\|B(u,v)\|_k \leqslant c_2 e^{k(c_1+c_2)}$.

$4°$. The operator $U(t,\tau)$ uniformly in (t,τ) is bounded in a countably valued space D_0, i.e., $\exists q > 0$: $\forall v \in D_0 \; \forall k \in \mathbb{N} \; \|U(t,\tau)v\|_k \leqslant q\|v\|_k$.

Theorem 12. *Let the conditions $1°$–$4°$ be fulfilled. Then, if $w_0(\nu) \in D_0$, then the solution $w(t,\nu)$ of the initial problem (18) will be analytic at the point $\nu = 0$.*

Proof. Without loss of generality, we assume that $q = 1$ in the condition $4°$. Note that the inequality in the condition $3°$ in terms of exponentials is as follows:
$$\|B(u,v)\|_k \leqslant e^{c_1 k} \frac{d}{dx}(e^{c_2 x})\big|_{x=k}. \tag{21}$$

Using the operator $U(t,\tau)$, we solve the problems of the series (20):
$$W_0(t,\nu) = U(t,0)w_0(\nu), \tag{22}$$
$$W_1(t,\nu) = \int_0^t U(t,\tau)B(W_0(\tau,\nu), W_0(\tau,\nu))d\tau,$$
$$\vdots$$
$$W_n(t,\nu) = \int_0^t U(t,\tau)\left[\sum_{k=0}^{n-1} B(W_k(\tau,\nu), W_{n-k}(\tau,\nu))\right]d\tau,$$
$$\vdots$$

Using the formula (21) and the method of mathematical induction, we can prove that the norms $\|W_n(t,\nu)\|_k$ are majorized using functions $P_n(k,t)$
$$P_0(k,t) = e^{ck}, \tag{23}$$
$$P_1(k,t) = cte^{2ck},$$
$$P_2(k,t) = \frac{3}{2}c^2 t^2 e^{3ck},$$
$$P_3(k,t) = \frac{8}{3}c^3 t^3 e^{4ck},$$
$$\vdots$$
$$P_n(k,t) = \frac{(n+1)^{n-1}}{n!} c^n t^n e^{(n+1)ck},$$
$$\vdots$$

The analyticity of $W_n(t,\nu)$ follows from the equality (22), the condition $4°$ and the analyticity of $w_0(\nu)$; the convergence of the series (19) follows from the formulas (23). Theorem 7 is proved. □

Remark 13. *If not all coefficients of $W_n(t, \nu)$ depend regularly on ν (this is possible, for example, when $A(t)$ depends on ν) and the series (19) converges in some neighborhood of the value $\nu = 0$, then the solution $w(t, \nu)$ is called pseudo–holomorphic* [6].

REFERENCES

[1] Butuzov, V. F. and Vasil'eva, A. B. 1973. *Asymptotic Expansions of Solutions to Singularly Perturbed Equations*. Moscow: Nauka, in Russian.

[2] Lomov, S. A. 1992. *An Introduction to the General Theory of Singular Perturbations*. Amer. Math. Soc. Providence, RI.

[3] Lomov, S. A. and Lomov, I. S. 2011. *Fundamentals of the Mathematical Theory of Boundary Layers*. Moscow: Moscow State University, in Russian.

[4] Kachalov, V. I. and Lomov, S. A. 1988. "Smoothness of Solutions of Differential Equations with Respect to a Singular Parameter". *Soviet Math. Dokl.* 37(2): 465–467.

[5] Kachalov, V. I. and Lomov, S. A. 1989. "On the Analytic Properties of Solutions of Differential Equations with Singular Points". *Soviet Math. Dokl.* 39(1): 12–14.

[6] Kachalov, V. I. 2017. "On the Holomorphic Regularization of Singularly Perturbed Systems of Differential Equations". *Comput. Math. and Math. Phys.*, 57(4): 653–660.

[7] Krivoruchenko, M. I., Nadyozhin, D. K. and Yudin, A. V. 2018. "Hydrostatic Equilibrium of Stars without Electroneutrally constant". *Phys. Rev. D.* 97(15): id 083016.

[8] Coddington, E. A. and Levinson, N. 1955. *Theory of Ordinary Differential Equations*. New York: McGraw–Hill Inc.

[9] Kato, T. 1966. *Perturbation Theory for Linear Operators*. Berlin: Springer.

[10] Boas, R. P. 1944. Function of Exponential Type III. *Duke Math. J.* 11: 507–511.

[11] Yosida, K. 1965. *Functional Analysis*. Berlin: Springer.

[12] Ovsyannikov, L. V. 1965. "Singular Operator in Banach Spaces Scale". *Soviet Math. Dokl.* 163(4): 819–822, in Russian.

[13] Richtmyer, R. D. 1978. *Principles of Advanced Mathematical Physics*. Vol. II. New York: Springer–Verlag.

In: Understanding Banach Spaces
Editor: Daniel González Sánchez

ISBN: 978-1-53616-745-0
© 2020 Nova Science Publishers, Inc.

Chapter 25

ROLE OF HANSON–ANTCZAK–TYPE V–INVEX FUNCTIONS IN SUFFICIENT EFFICIENCY CONDITIONS FOR SEMIINFINITE MULTIOBJECTIVE FRACTIONAL PROGRAMMING

Ram U. Verma[*]
International Publications US, World Leading Publishers
Mathematical Sciences Division, Denton, TX, US

Abstract

In this chapter, we discuss numerous sets of global parametric sufficient efficiency conditions under various Hanson–Antczak–type generalized $(\alpha, \beta, \gamma, \xi, \eta, \omega, \rho, \theta, m)$–$V$–invexity assumptions for a semiinfinite multiobjective fractional programming problem.

A semiinfinite multiobjective fractional programming problem with a finite number of variables and infinitely many constraints is called a *semiinfinite programming problem*. Problems of this nature have been utilized for the modeling and analysis of a wide range of theoretical as well as concrete, real–world, practical problems. More specifically, semiinfinite programming concepts and techniques have found relevance and applications in approximation theory, statistics, game theory, engineering design (especially, design of structures, control systems, digital filters, electronic circuits, etc.), boundary value problems, defect minimization for operator equations, geometry, random graphs, graphs related to Newton flows, wavelet analysis, reliability testing, environmental protection planning, decision making under uncertainty, semidefinite programming, geometric programming, disjunctive programming, optimal control problems, robotics (still in testing stages, robots as bartenders at large casinos, robots in automobile assembly plants, and robots as workers stocking shelves at Walmart) and continuum mechanics, among other business and industry.

Furthermore, utilizing two partitioning schemes, we establish several sets of generalized parametric sufficient efficiency results each of which is in fact a family of such results whose members can easily be identified by appropriate choices of certain sets and functions in this chapter, which can be applied to various aspects of

---
[*]Corresponding Author's E-mail: verma99@msn.com.

semiinfinite programming, including, but not limited to, optimality conditions, duality relations, and numerical algorithms, anticipatory systems and gene–environment networks in [1–5], the analysis and implementation of Vasicek–type interest rate models in [6], an interesting gemstone cutting problem in [7]. Plus, we observe the obtained results can be applied as a major research based on the work of Verma [8] on general approximation–solvability of nonlinear equations involving A–regular operators under the inner- approximation- scheme by Zeidler [9].

Keywords: semiinfinite programming, multiobjective fractional programming, Hanson–Antczak–type generalized $(\alpha, \beta, \gamma, \xi, \eta, \omega, \rho, \theta, m)$-$V$-invex functions

AMS Subject Classification: 90C29, 90C30, 90C32, 90C34, 90C46

1. INTRODUCTION

In this chapter, we first introduce some multiparameter generalizations of the class of V–r–invex functions defined recently by Antczak [10], and then, utilizing the new functions, we state and prove a number of parametric sufficient efficiency results under various Hanson–Antczak–type generalized $(\alpha, \beta, \gamma, \xi, \eta, \omega, \rho, \theta, m)$-$V$-invexity assumptions for the following semiinfinite multiobjective fractional programming problem:

$$(P) \quad \text{Minimize } \varphi(x) = \big(\varphi_1(x), \ldots, \varphi_p(x)\big) = \left(\frac{f_1(x)}{g_1(x)}, \ldots, \frac{f_p(x)}{g_p(x)}\right)$$

subject to

$$G_j(x, t) \leqq 0 \text{ for all } t \in T_j, \ j \in \underline{q},$$
$$H_k(x, s) = 0 \text{ for all } s \in S_k, \ k \in \underline{r},$$
$$x \in X,$$

where p, q, and r are positive integers, X is a nonempty open convex subset of $\mathbb{R}^n$ (n–dimensional Euclidean space), for each $j \in \underline{q} \equiv \{1, 2, \ldots, q\}$ and $k \in \underline{r}$, T_j and S_k are compact subsets of complete metric spaces, for each $i \in \underline{p}$, f_i and g_i are real–valued functions defined on X, for each $j \in \underline{q}$, $G_j(\cdot, t)$ is a real–valued function defined on X for all $t \in T_j$, for each $k \in \underline{r}$, $H_k(\cdot, s)$ is a real–valued function defined on X for all $s \in S_k$, for each $j \in \underline{q}$ and $k \in \underline{r}$, $G_j(x, \cdot)$ and $H_k(x, \cdot)$ are continuous real–valued functions defined, respectively, on T_j and S_k for all $x \in X$, and for each $i \in \underline{p}$, $g_i(x) > 0$ for all x satisfying the constraints of (P).

Multiobjective programming problems like (P) but with a finite number of constraints, that is, when the functions G_j are independent of t, and the functions H_k are independent of s, have been widely under subject of numerous investigations in the past three decades. Several classes of static and dynamic optimization problems with multiple fractional objective functions have been studied and, consequently, a number of sufficient efficiency and duality results are currently available for these problems in the related literature. More details on various aspects of multiobjective fractional programming are available in [11–13]. For more information about the vast general area of multiobjective programming, the reader may consult [14–17].

We observe that despite a phenomenal proliferation of publications in several areas of multiobjective programming, there exists tremendous opportunity for *semiinfinite nonlinear multiobjective fractional programming problems*. It seems that so far no sufficient efficiency results based on generalized HA$(\alpha, \beta, \gamma, \xi, \eta, \omega, \rho, \theta, m)$–$V$–invexity concepts have been published in the related literature for any kind of semiinfinite multiobjective fractional programming problems. Therefore, to the best of our knowledge, all the efficiency results presented in this chapter are new in the area of semiinfinite programming. In the present chapter, we shall formulate a number of parametric sufficient efficiency results for our semiinfinite nonlinear multiobjective fractional programming problem (P) under various generalized HA$(\alpha, \beta, \gamma, \xi, \eta, \omega, \rho, \theta, m)$–$V$–invexity assumptions, which are associated with various parametric duality relations for (P). As a matter of fact, a mathematical programming problem with a finite number of variables and infinitely many constraints is called a *semiinfinite programming problem*, while problems of this class have been utilized for the modeling and analysis of a wide range of theoretical as well as concrete, real–world, practical problems. More specifically, semiinfinite programming concepts and techniques have found relevance and applications, including, but not limited to, approximation theory, statistics, game theory, engineering design (design of structures, control systems, digital filters, electronic circuits), boundary value problems, defect minimization for operator equations, geometry, random graphs, graphs related to Newton flows, wavelet analysis, reliability testing, environmental protection planning, decision making under uncertainty, semidefinite programming, geometric programming, disjunctive programming, optimal control problems, robotics (still in testing stages, use of robots in automobile industry, and bartenders in casinos), and continuum mechanics, among others. For more details on various aspects of semiinfinite programming, including areas of applications, optimality conditions, duality relations, and numerical algorithms, the reader is referred to [8, 18–31]. Some relatively more recent applications of semiinfinite programming to data envelopment are discussed in [32, 33], to anticipatory systems and gene-environment networks in [1–5], to the analysis and implementation of Vasicek–type interest rate models in [6], to infinite kernel learning in [34, 35], and to basket option pricing in [36].

The rest of this chapter is organized as follows. In Section 2, we present a number of definitions and auxiliary results which will be needed in the sequel. In Section 3, we begin our discussion of sufficient efficiency conditions where we formulate and prove several sets of sufficiency criteria under a variety of generalized $HA(\alpha, \beta, \gamma, \xi, \eta, \omega, \rho, \theta, m)$–$V$–invexity assumptions that are placed on certain vector–valued functions whose entries consist of the individual as well as some combinations of the problem functions. Utilizing two partitioning schemes, in Section 4 we establish several sets of generalized parametric sufficient efficiency results each of which is in fact a family of such results whose members can easily be identified by appropriate choices of certain sets and functions. Finally, in Section 5 we summarize our main results and also point out some further research opportunities arising from certain modifications of the principal problem model considered in this chapter.

Evidently, all the parametric sufficient efficiency results established in this chapter can easily be modified and restated for each one of the following seven classes of nonlinear programming problems, which are special cases of (P):

$(P1)$ $\quad$ $\displaystyle\operatorname*{Minimize}_{x\in\mathbb{F}}\ \bigl(f_1(x),\ldots,f_p(x)\bigr);$

$(P2)$ $\quad$ $\displaystyle\operatorname*{Minimize}_{x\in\mathbb{F}}\ \frac{f_1(x)}{g_1(x)};$

$(P3)$ $\quad$ $\displaystyle\operatorname*{Minimize}_{x\in\mathbb{F}}\ f_1(x),$

where $\mathbb{F}$ (assumed to be nonempty) is the feasible set of (P), that is,

$$\mathbb{F}=\{x\in X : G_j(x,t)\leqq 0 \text{ for all } t\in T_j,\ j\in \underline{q},\ H_k(x,s)=0 \text{ for all } s\in S_k,\ k\in \underline{r}\};$$

$(P4)$ $\quad$ $\displaystyle\operatorname{Minimize}\ \left(\frac{f_1(x)}{g_1(x)},\ldots,\frac{f_p(x)}{g_p(x)}\right)$

subject to
$$\tilde{G}_j(x)\leqq 0,\ j\in \underline{q},\ \tilde{H}_k(x)=0,\ k\in \underline{r},\ x\in X,$$

where f_i and g_i, $i\in \underline{p}$, are as defined in the description of (P), $\tilde{G}_j$, $j\in \underline{q}$, and $\tilde{H}_k$, $k\in \underline{r}$, are real–valued functions defined on X;

$(P5)$ $\quad$ $\displaystyle\operatorname*{Minimize}_{x\in\mathbb{G}}\ \bigl(f_1(x),\ldots,f_p(x)\bigr);$

$(P6)$ $\quad$ $\displaystyle\operatorname*{Minimize}_{x\in\mathbb{G}}\ \frac{f_1(x)}{g_1(x)};$

$(P7)$ $\quad$ $\displaystyle\operatorname*{Minimize}_{x\in\mathbb{G}}\ f_1(x),$

where $\mathbb{G}$ is the feasible set of $(P4)$, that is,

$$\mathbb{G}=\{x\in X: \tilde{G}_j(x)\leqq 0,\ j\in \underline{q},\ \tilde{H}_k(x)=0,\ k\in \underline{r}\}.$$

2. Preliminaries

In this section we recall, for convenience of reference, the definitions of certain classes of generalized convex functions which will be needed in the sequel. We begin by defining an invex function which has been instrumental in creating a vast array of interesting and important classes of generalized convex functions.

Definition 1. *Let f be a differentiable real–valued function defined on $\mathbb{R}^n$. Then f is said to be η–invex (invex with respect to η) at y if there exists a function $\eta: \mathbb{R}^n\times\mathbb{R}^n\to\mathbb{R}^n$ such that for each $x\in\mathbb{R}^n$,*

$$f(x)-f(y)\geqq \langle\nabla f(y),\eta(x,y)\rangle,$$

where $\nabla f(y)=(\partial f(y)/\partial y_1,\partial f(y)/\partial y_2,\ldots,\partial f(y)/\partial y_n)$ is the gradient of f at y, and $\langle a,b\rangle$ denotes the inner product of the vectors a and b; f is said to be η–invex on $\mathbb{R}^n$ if the above inequality holds for all $x,y\in\mathbb{R}^n$.

From this definition it is clear that every differentiable real–valued convex function is invex with $\eta(x, y) = x - y$. This generalization of the concept of convexity was originally proposed by Hanson [37] who showed that for a nonlinear programming problem of the form

Minimize $f(x)$ subject to $g_i(x) \leqq 0$, $i \in \underline{m}$, $x \in \mathbb{R}^n$,

where the differentiable functions $f, g_i : \mathbb{R}^n \to \mathbb{R}$, $i \in \underline{m}$, are invex with respect to the same function $\eta : \mathbb{R}^n \times \mathbb{R}^n \to \mathbb{R}^n$, the Karush–Kuhn–Tucker necessary optimality conditions are also sufficient. The term *invex* (for *in*variant con*vex*) was coined by Craven [38] to signify the fact that the invexity property, unlike convexity, remains invariant under bijective coordinate transformations.

In a similar manner, one can readily define η–pseudoinvex and η–quasiinvex functions as generalizations of differentiable pseudoconvex and quasiconvex functions.

Let the function $F = (F_1, F_2, \ldots, F_N) : \mathbb{R}^n \to \mathbb{R}^N$ be differentiable at x^*. The following generalizations of the notions of invexity, pseudoinvexity, and quasiinvexity for vector–valued functions were originally proposed in [39].

Definition 2. *The function F is said to be (α, η)–V–invex at x^* if there exist functions $\alpha_i : \mathbb{R}^n \times \mathbb{R}^n \to \mathbb{R}_+ \setminus \{0\} \equiv (0, \infty)$, $i \in \underline{N}$, and $\eta : \mathbb{R}^n \times \mathbb{R}^n \to \mathbb{R}^n$ such that for each $x \in \mathbb{R}^n$ and $i \in \underline{N}$,*

$$F_i(x) - F_i(x^*) \geqq \langle \alpha_i(x, x^*) \nabla F_i(x^*), \eta(x, x^*) \rangle.$$

Definition 3. *The function F is said to be (β, η)–V–pseudoinvex at x^* if there exist functions $\beta_i : \mathbb{R}^n \times \mathbb{R}^n \to \mathbb{R}_+ \setminus \{0\}$, $i \in \underline{N}$, and $\eta : \mathbb{R}^n \times \mathbb{R}^n \to \mathbb{R}^n$ such that for each $x \in \mathbb{R}^n$,*

$$\left\langle \sum_{i=1}^N \nabla F_i(x^*), \eta(x, x^*) \right\rangle \geqq 0 \Rightarrow \sum_{i=1}^N \beta_i(x, x^*) F_i(x) \geqq \sum_{i=1}^N \beta_i(x, x^*) F_i(x^*).$$

Definition 4. *The function F is said to be (γ, η)–V–quasiinvex at x^* if there exist functions $\gamma_i : \mathbb{R}^n \times \mathbb{R}^n \to \mathbb{R}_+ \setminus \{0\}$, $i \in \underline{N}$, and $\eta : \mathbb{R}^n \times \mathbb{R}^n \to \mathbb{R}^n$ such that for each $x \in \mathbb{R}^n$,*

$$\sum_{i=1}^N \gamma_i(x, x^*) F_i(x) \leqq \sum_{i=1}^N \gamma_i(x, x^*) F_i(x^*) \Rightarrow \left\langle \sum_{i=1}^N \nabla F_i(x^*), \eta(x, x^*) \right\rangle \leqq 0.$$

The concept of η–invexity has been extended in many ways, and various types of generalized invex functions have been utilized for establishing a wide range of sufficient optimality criteria and duality relations for several classes of nonlinear programming problems. For more information about invex functions, the reader may consult [40–50], and for recent surveys of these and related functions, the reader is referred to [51, 52].

Recently, Antczak [10] introduced the following variant of the class of V–invex functions.

Definition 5. *A differentiable function $f : X \to \mathbb{R}^k$ is called (strictly) $\zeta_i - \tilde{r}$–invex with respect to η at $u \in X$ if there exist functions $\eta : X \times X \to \mathbb{R}^n$ and $\zeta_i : X \times X \to \mathbb{R}_+ \backslash \{0\}$, $i \in \underline{k}$, such for each $x \in X$,*

$$\frac{1}{\tilde{r}} e^{\tilde{r} f_i(x)} (>) \geqq \frac{1}{\tilde{r}} e^{\tilde{r} f_i(u)} [1 + \tilde{r} \zeta_i(x, u) \langle \nabla f_i(u), \eta(x, u) \rangle] \text{ for } \tilde{r} \neq 0,$$

$$f_i(x) - f_i(u) \geqq \zeta_i(x, u) \langle \nabla f_i(u), \eta(x, u) \rangle \text{ for } \tilde{r} = 0.$$

This class of functions was used in [10] for establishing some sufficiency and duality results for a nonlinear programming problem with differentiable functions. Their nonsmooth analogues were discussed in [53].

In this chapter, we shall utilize the following slightly modified and more general versions of the V–r–invex functions defined in [10].

Let the function $F = (F_1, F_2, \ldots, F_p) : X \to \mathbb{R}^p$ be differentiable at x^*.

Definition 6. *The function F is said to be (strictly) $HA(\alpha, \beta, \gamma, \xi, \eta, \omega, \rho, \theta, m)$–$V$–invex at $x^* \in X$ if there exist functions $\alpha : X \times X \to \mathbb{R}$, $\beta : X \times X \to \mathbb{R}$, $\gamma_i : X \times X \to \mathbb{R}_+$, $\xi_i : X \times X \to \mathbb{R}_+ \backslash \{0\}$, $i \in \underline{p}$, $\eta, \omega : X \times X \to \mathbb{R}^n$, $\rho_i : X \times X \to \mathbb{R}$, $i \in \underline{p}$, and $\theta : X \times X \to \mathbb{R}^n$ such that for all $x \in X (x \neq x^*)$ and $i \in \underline{p}$,*

$$\frac{1}{\alpha(x, x^*)} \gamma_i(x, x^*) \left(e^{\alpha(x, x^*)[F_i(x) - F_i(x^*)]} - 1 \right)$$

$$(>) \geqq \frac{1}{\beta(x, x^*)} \langle \frac{1}{2} \xi_i(x, x^*) \nabla F_i(x^*), e^{\beta(x, x^*)\eta(x, x^*)} - \mathbf{1} \rangle$$

$$+ \frac{1}{\beta(x, x^*)} \langle \frac{1}{2} \xi_i(x, x^*) \nabla F_i(x^*), e^{\beta(x, x^*)\omega(x, x^*)} - \mathbf{1} \rangle$$

$$+ \rho_i(x, x^*) \|\theta(x, x^*)\|^m \quad \text{if } \alpha(x, x^*) \neq 0 \text{ and } \beta(x, x^*) \neq 0 \text{ for all } x \in X,$$

$$\frac{1}{\alpha(x, x^*)} \gamma_i(x, x^*) \left(e^{\alpha(x, x^*)[F_i(x) - F_i(x^*)]} - 1 \right) (>) \geqq \langle \frac{1}{2} \xi_i(x, x^*) \nabla F_i(x^*), \eta(x, x^*) \rangle$$

$$+ \langle \frac{1}{2} \xi_i(x, x^*) \nabla F_i(x^*), \omega(x, x^*) \rangle$$

$$+ \rho_i(x, x^*) \|\theta(x, x^*)\|^m \quad \text{if } \alpha(x, x^*) \neq 0 \text{ and } \beta(x, x^*) = 0 \text{ for all } x \in X,$$

$$\gamma_i(x, x^*)[F_i(x) - F_i(x^*)](>) \geqq \frac{1}{\beta(x, x^*)} \langle \frac{1}{2} \xi_i(x, x^*) \nabla F_i(x^*), e^{\beta(x, x^*)\eta(x, x^*)} - \mathbf{1} \rangle$$

$$+ \frac{1}{\beta(x, x^*)} \langle \frac{1}{2} \xi_i(x, x^*) \nabla F_i(x^*), e^{\beta(x, x^*)\omega(x, x^*)} - \mathbf{1} \rangle$$

$$+ \rho_i(x, x^*) \|\theta(x, x^*)\|^m \quad \text{if } \alpha(x, x^*) = 0 \text{ and } \beta(x, x^*) \neq 0 \text{ for all } x \in X,$$

$$\gamma_i(x, x^*)[F_i(x) - F_i(x^*)](>) \geqq \langle \frac{1}{2} \xi_i(x, x^*) \nabla F_i(x^*), \eta(x, x^*) \rangle$$

$$\langle \xi_i(x, x^*) \nabla F_i(x^*), \omega(x, x^*) \rangle + \rho_i(x, x^*) \|\theta(x, x^*)\|^m$$

$$\text{if } \alpha(x, x^*) = 0 \text{ and } \beta(x, x^*) = 0 \text{ for all } x \in X,$$

where $\|\cdot\|$ is a norm on $\mathbb{R}^n$ and

$$\left(e^{\beta(x,x^*)\eta(x,x^*)} - \mathbf{1}\right) \equiv \left(e^{\beta(x,x^*)\eta_1(x,x^*)} - 1, \ldots, e^{\beta(x,x^*)\eta_n(x,x^*)} - 1\right).$$

The function F is said to be (strictly) HA(α, β, γ, ξ, η, ω, ρ, θ, m)-V-invex on X if it is (strictly) HA(α, β, γ, ξ, η, ω, ρ, θ, m)–V–invex at each point $x^* \in X$, where m is a positive integer.

Definition 7. The function F is said to be (strictly) $HA(\alpha, \beta, \gamma, \xi, \eta, \omega, \rho, \theta, m)$–V–pseudoinvex at $x^* \in X$ if there exist functions $\alpha : X \times X \to \mathbb{R}$, $\beta : X \times X \to \mathbb{R}$, $\gamma : X \times X \to \mathbb{R}_+$, $\xi_i : X \times X \to \mathbb{R}_+\setminus\{0\}$, $i \in p$, $\eta, \omega : X \times X \to \mathbb{R}^n$, $\rho : X \times X \to \mathbb{R}$, and $\theta : X \times X \to \mathbb{R}^n$ such that for all $x \in X (x \neq x^*)$,

$$\frac{1}{\beta(x,x^*)}\left\langle \frac{1}{2}\sum_{i=1}^{p}\xi_i(x,x^*)\nabla F_i(x^*), e^{\beta(x,x^*)\eta(x,x^*)} - \mathbf{1}\right\rangle$$
$$+ \frac{1}{\beta(x,x^*)}\left\langle \frac{1}{2}\sum_{i=1}^{p}\xi_i(x,x^*)\nabla F_i(x^*), e^{\beta(x,x^*)\omega(x,x^*)} - \mathbf{1}\right\rangle$$
$$\geqq -\rho(x,x^*)\|\theta(x,x^*)\|^m$$
$$\Rightarrow \frac{1}{\alpha(x,x^*)}\gamma(x,x^*)\left(e^{\alpha(x,x^*)\sum_{i=1}^{p}[F_i(x)-F_i(x^*)]} - 1\right)(>) \geqq 0$$
$$\text{if } \alpha(x,x^*) \neq 0 \text{ and } \beta(x,x^*) \neq 0 \text{ for all } x \in X,$$

$$\left\langle \frac{1}{2}\sum_{i=1}^{p}\xi_i(x,x^*)\nabla F_i(x^*), \eta(x,x^*)\right\rangle + \left\langle \frac{1}{2}\sum_{i=1}^{p}\xi_i(x,x^*)\nabla F_i(x^*), \omega(x,x^*)\right\rangle$$
$$\geqq -\rho(x,x^*)\|\theta(x,x^*)\|^m$$
$$\Rightarrow \frac{1}{\alpha(x,x^*)}\gamma(x,x^*)\left(e^{\alpha(x,x^*)\sum_{i=1}^{p}[F_i(x)-F_i(x^*)]} - 1\right)(>) \geqq 0$$
$$\text{if } \alpha(x,x^*) \neq 0 \text{ and } \beta(x,x^*) = 0 \text{ for all } x \in X,$$

$$\frac{1}{\beta(x,x^*)}\left\langle \frac{1}{2}\sum_{i=1}^{p}\xi_i(x,x^*)\nabla F_i(x^*), e^{\beta(x,x^*)\eta(x,x^*)}\right\rangle$$
$$+ \frac{1}{\beta(x,x^*)}\left\langle \frac{1}{2}\sum_{i=1}^{p}\xi_i(x,x^*)\nabla F_i(x^*), e^{\beta(x,x^*)\omega(x,x^*)} - \mathbf{1}\right\rangle$$
$$\geqq -\rho(x,x^*)\|\theta(x,x^*)\|^m$$
$$\Rightarrow \gamma(x,x^*)\sum_{i=1}^{p}[F_i(x) - F_i(x^*)](>) \geqq 0$$
$$\text{if } \alpha(x,x^*) = 0 \text{ and } \beta(x,x^*) \neq 0 \text{ for all } x \in X,$$

$$\left\langle \frac{1}{2} \sum_{i=1}^{p} \xi_i(x, x^*) \nabla F_i(x^*), \eta(x, x^*) \right\rangle + \left\langle \frac{1}{2} \sum_{i=1}^{p} \xi_i(x, x^*) \nabla F_i(x^*), \omega(x, x^*) \right\rangle$$

$$\geqq -\rho(x, x^*) \|\theta(x, x^*)\|^m$$

$$\Rightarrow \gamma(x, x^*) \sum_{i=1}^{p} [F_i(x) - F_i(x^*)](>) \geqq 0$$

if $\alpha(x, x^*) = 0$ and $\beta(x, x^*) = 0$ for all $x \in X$.

The function F is said to be (strictly) $HA(\alpha, \beta, \gamma, \xi, \eta, \omega, \rho, \theta, m)$–V–pseudoinvex on X if it is (strictly) $HA(\alpha, \beta, \gamma, \xi, \eta, \omega, \rho, \theta, m)$–V–pseudoinvex at each point $x^* \in X$.

Definition 8. *The function F is said to be (prestrictly) $HA(\alpha, \beta, \gamma, \xi, \eta, \omega, \rho, \theta, m)$–V–quasiinvex at $x^* \in X$ if there exist functions $\alpha : X \times X \to \mathbb{R}$, $\beta : X \times X \to \mathbb{R}$, $\gamma : X \times X \to \mathbb{R}_+$, $\xi_i : X \times X \to \mathbb{R}_+ \setminus \{0\}$, $i \in \underline{p}$, $\eta, \omega : X \times X \to \mathbb{R}^n$, $\rho : X \times X \to \mathbb{R}$, and $\theta : X \times X \to \mathbb{R}^n$ such that for all $x \in X$,*

$$\frac{1}{\alpha(x, x^*)} \gamma(x, x^*) \left(e^{\alpha(x, x^*) \sum_{i=1}^{p} [F_i(x) - F_i(x^*)]} - 1 \right)(<) \leqq 0$$

$$\Rightarrow \frac{1}{\beta(x, x^*)} \left\langle \frac{1}{2} \sum_{i=1}^{p} \xi_i(x, x^*) \nabla F_i(x^*), e^{\beta(x, x^*) \eta(x, x^*)} - \mathbf{1} \right\rangle$$

$$+ \frac{1}{\beta(x, x^*)} \left\langle \frac{1}{2} \sum_{i=1}^{p} \xi_i(x, x^*) \nabla F_i(x^*), e^{\beta(x, x^*) \omega(x, x^*)} - \mathbf{1} \right\rangle$$

$$\leqq -\rho(x, x^*) \|\theta(x, x^*)\|^m$$

if $\alpha(x, x^*) \neq 0$ and $\beta(x, x^*) \neq 0$ for all $x \in X$,

$$\frac{1}{\alpha(x, x^*)} \gamma(x, x^*) \left(e^{\alpha(x, x^*) \sum_{i=1}^{p} [F_i(x) - F_i(x^*)]} - 1 \right)(<) \leqq 0$$

$$\Rightarrow \left\langle \frac{1}{2} \sum_{i=1}^{p} \xi_i(x, x^*) \nabla F_i(x^*), \eta(x, x^*) \right\rangle$$

$$+ \left\langle \frac{1}{2} \sum_{i=1}^{p} \xi_i(x, x^*) \nabla F_i(x^*), \omega(x, x^*) \right\rangle$$

$$\leqq -\rho(x, x^*) \|\theta(x, x^*)\|^m$$

if $\alpha(x, x^*) \neq 0$ and $\beta(x, x^*) = 0$ for all $x \in X$,

$$\gamma(x,x^*)\sum_{i=1}^{p}[F_i(x)-F_i(x^*)](<)\leqq 0$$

$$\Rightarrow \frac{1}{\beta(x,x^*)}\left\langle \frac{1}{2}\sum_{i=1}^{p}\xi_i(x,x^*)\nabla F_i(x^*), e^{\beta(x,x^*)\eta(x,x^*)}-\mathbf{1}\right\rangle$$

$$+\frac{1}{\beta(x,x^*)}\left\langle \frac{1}{2}\sum_{i=1}^{p}\xi_i(x,x^*)\nabla F_i(x^*), e^{\beta(x,x^*)\omega(x,x^*)}-\mathbf{1}\right\rangle$$

$$\leqq -\rho(x,x^*)\|\theta(x,x^*)\|^m$$

if $\alpha(x,x^*)=0$ and $\beta(x,x^*)\neq 0$ for all $x\in X$,

$$\gamma(x,x^*)\sum_{i=1}^{p}[F_i(x)-F_i(x^*)](<)\leqq 0 \Rightarrow$$

$$\left\langle \frac{1}{2}\sum_{i=1}^{p}\xi_i(x,x^*)\nabla F_i(x^*), \eta(x,x^*)\right\rangle$$

$$+\left\langle \frac{1}{2}\sum_{i=1}^{p}\xi_i(x,x^*)\nabla F_i(x^*), \omega(x,x^*)\right\rangle$$

$$\leqq -\rho(x,x^*)\|\theta(x,x^*)\|^m$$

if $\alpha(x,x^*)=0$ and $\beta(x,x^*)=0$ for all $x\in X$.

The function F is said to be (prestrictly) $HA(\alpha,\beta,\gamma,\xi,\eta,\omega,\rho,\theta,m)$–V–quasiinvex on X if it is (prestrictly) $HA(\alpha,\beta,\gamma,\xi,\eta,\omega,\rho,\theta,m)$–V–quasiinvex at each point $x^*\in X$.

In the proofs of the sufficient efficiency theorems, sometimes it may be more convenient to use certain alternative but equivalent forms of the above definitions. These are obtained by considering the contrapositive statements. For example, $HA(\alpha,\beta,\gamma,\xi,\eta,\omega,\rho,\theta,m)$–V–pseudoinvexity (when $\alpha(x,x^*)\neq 0$ and $\beta(x,x^*)\neq 0$ for all $x\in X$) can be defined in the following equivalent way:

The function F is said to be $HA(\alpha,\beta,\gamma,\xi,\eta,\omega,\rho,\theta,m)$–V–pseudoinvex at $x^*\in X$ if there exist functions $\alpha: X\times X\to\mathbb{R}$, $\beta: X\times X\to\mathbb{R}$, $\gamma: X\times X\to\mathbb{R}_+$, $\xi_i: X\times X\to\mathbb{R}_+\setminus\{0\}$, $i\in\underline{p}$, $\eta,\omega: X\times X\to\mathbb{R}^n$, $\rho: X\times X\to\mathbb{R}$, and $\theta: X\times X\to\mathbb{R}^n$ such that for all $x\in X$,

$$\frac{1}{\alpha(x,x^*)}\gamma(x,x^*)\left(e^{\alpha(x,x^*)\sum_{i=1}^{p}[F_i(x)-F_i(x^*)]}-1\right)<0$$

$$\Rightarrow \frac{1}{\beta(x,x^*)}\left\langle \frac{1}{2}\sum_{i=1}^{p}\xi_i(x,x^*)\nabla F_i(x^*), e^{\beta(x,x^*)\eta(x,x^*)}-\mathbf{1}\right\rangle$$

$$+\frac{1}{\beta(x,x^*)}\left\langle \frac{1}{2}\sum_{i=1}^{p}\xi_i(x,x^*)\nabla F_i(x^*), e^{\beta(x,x^*)\omega(x,x^*)}-\mathbf{1}\right\rangle$$

$$< -\rho(x,x^*)\|\theta(x,x^*)\|^m.$$

In the sequel, we shall also need a consistent notation for vector inequalities. For $a, b \in \mathbb{R}^m$, the following order notation will be used: $a \geqq b$ if and only if $a_i \geq b_i$ for all $i \in \underline{m}$; $a \geq b$ if and only if $a_i \geq b_i$ for all $i \in \underline{m}$, but $a \neq b$; $a > b$ if and only if $a_i > b_i$ for all $i \in \underline{m}$; and $a \not\geq b$ is the negation of $a \geq b$.

Consider the multiobjective problem

$$(P^*) \qquad \underset{x \in \mathbb{F}}{\text{Minimize}} \ F(x) = (F_1(x), \ldots, F_p(x)),$$

where F_i, $i \in \underline{p}$, are real–valued functions defined on $\mathbb{R}^n$.

An element $x^\circ \in \mathbb{F}$ is said to be an *efficient (Pareto optimal, nondominated, noninferior)* solution of (P^*) if there exists no $x \in \mathbb{F}$ such that $F(x) \leq F(x^\circ)$. In the area of multiobjective programming, there exist several versions of the notion of efficiency most of which are discussed in [14–17]. However, throughout this chapter, we shall deal exclusively with the efficient solutions of (P) in the sense defined above.

Next we recall the following result for a set of necessary efficiency conditions for (P).

Theorem 9. *[54] Let $x^* \in \mathbb{F}$, let $\lambda^* = \varphi(x^*)$, for each $i \in \underline{p}$, let f_i and g_i be continuously differentiable at x^*, for each $j \in \underline{q}$, let the function $G_j(\cdot, t)$ be continuously differentiable at x^* for all $t \in T_j$, and for each $k \in \underline{r}$, let the function $H_k(\cdot, s)$ be continuously differentiable at x^* for all $s \in S_k$. If x^* is an efficient solution of (P), if the generalized Guignard constraint qualification holds at x^*, and if for each $i_0 \in \underline{p}$, the set $\text{cone}(\{\nabla G_j(x^*, t) : t \in \hat{T}_j(x^*), j \in \underline{q}\} \cup \{\nabla f_i(x^*) - \lambda_i^* \nabla g_i(x^*) : i \in \underline{p}, i \neq i_0\}) + \text{span}(\{\nabla H_k(x^*, s) : s \in S_k, k \in \underline{r}\})$ is closed, then there exist $u^* \in U$ and integers ν_0^* and ν^*, with $0 \leqq \nu_0^* \leqq \nu^* \leqq n + 1$, such that there exist ν_0^* indices j_m, with $1 \leq j_m \leq q$, together with ν_0^* points $t^m \in \hat{T}_{j_m}(x^*)$, $m \in \underline{\nu_0^*}$, $\nu^* - \nu_0^*$ indices k_m, with $1 \leq k_m \leq r$, together with $\nu^* - \nu_0^*$ points $s^m \in S_{k_m}$ for $m \in \underline{\nu^*} \backslash \underline{\nu_0^*}$, and ν^* real numbers v_m^*, with $v_m^* > 0$ for $m \in \underline{\nu_0^*}$, with the property that*

$$\sum_{i=1}^{p} u_i^*[\nabla f_i(x^*) - \lambda_i^* \nabla g_i(x^*)] + \sum_{m=1}^{\nu_0^*} v_m^* \nabla G_{j_m}(x^*, t^m) + \sum_{m=\nu_0^*+1}^{\nu^*} v_m^* \nabla H_{k_m}(x^*, s^m) = 0,$$

where $\text{cone}(V)$ is the conic hull of the set $V \subset \mathbb{R}^n$ (i.e., the smallest convex cone containing V), $\text{span}(V)$ is the linear hull of V (i.e., the smallest subspace containing V), $\hat{T}_j(x^*) = \{t \in T_j : G_j(x^*, t) = 0\}$, $U = \{u \in \mathbb{R}^p : u > 0, \sum_{i=1}^{p} u_i = 1\}$, and $\underline{\nu^*} \backslash \underline{\nu_0^*}$ is the complement of the set $\underline{\nu_0^*}$ relative to the set $\underline{\nu^*}$.

With regard to the choice of the type of generalized V–invex functions, specified in Definitions 5 – 7, to be used in the statements and proofs of our sufficient efficiency theorems, we shall consistently use the cases in which the functions α and β are nonzero for all $(x, y) \in X \times X$. All the efficiency results established in this chapter can be modified, and proved for the other cases in a similar manner.

3. Sufficient Efficiency Conditions

In this section, we present several sets of sufficiency results in which various generalized HA($\alpha, \beta, \gamma, \eta, \omega, \xi, \rho, \theta,$ m) –V–invexity assumptions are imposed on certain vector functions whose components are the individual as well as some combinations of the problem functions.

Let the function $\mathcal{E}_i(\cdot, \lambda, u) : X \to \mathbb{R}$ be defined, for fixed λ and u, on X by

$$\mathcal{E}_i(z, \lambda, u) = u_i[f_i(z) - \lambda_i g_i(z)], \ i \in \underline{p}.$$

Theorem 10. *Let $x^* \in \mathbb{F}$, let $\lambda^* = \varphi(x^*)$, let the functions f_i, g_i, $i \in \underline{p}$, $G_j(\cdot, t)$, and $H_k(\cdot, s)$ be differentiable at x^* for all $t \in T_j$ and $s \in S_k$, $j \in \underline{q}$, $k \in \underline{r}$, and assume that there exist $u^* \in U$ and integers ν_0 and ν, with $0 \leqq \nu_0 \leqq \nu \leqq n+1$, such that there exist ν_0 indices j_m, with $1 \leqq j_m \leqq q$, together with ν_0 points $t^m \in \hat{T}_{j_m}(x^*)$, $m \in \underline{\nu_0}$, $\nu - \nu_0$ indices k_m, with $1 \leqq k_m \leqq r$, together with $\nu - \nu_0$ points $s^m \in S_{k_m}$, $m \in \underline{\nu}\backslash\underline{\nu_0}$, and ν real numbers v_m^*, with $v_m^* > 0$ for $m \in \underline{\nu_0}$, with the property that*

$$\sum_{i=1}^{p} u_i^*[\nabla f_i(x^*) - \lambda_i^* \nabla g_i(x^*)] + \sum_{m=1}^{\nu_0} v_m^* \nabla G_{j_m}(x^*, t^m) + \sum_{m=\nu_0+1}^{\nu} v_m^* \nabla H_{k_m}(x^*, s^m) = 0. \quad (1)$$

Assume, furthermore, that either one of the following two sets of conditions holds:

(a)   (i) $(f_1 - \lambda_1 g_1, \ldots, f_p - \lambda_p g_p)$ is $HA(\alpha, \beta, \bar{\gamma}, \xi, \eta, \omega, \bar{\rho}, \theta, m)$–$V$–invex at x^* and $\bar{\gamma}(x, x^*) > 0$ for all $x \in \mathbb{F}$;

  (ii) $(v_1^* G_{j_1}(\cdot, t^1), \ldots, v_{\nu_0}^* G_{j_{\nu_0}}(\cdot, t^{\nu_0}))$ is $HA(\alpha, \beta, \hat{\gamma}, \pi, \eta, \omega, \hat{\rho}, \theta, m)$–$V$–invex at x^*;

  (iii) $(v_{\nu_0+1}^* H_{k_{\nu_0+1}}(\cdot, s^{\nu_0+1}), \ldots, v_\nu^* H_{k_\nu}(\cdot, s^\nu))$ is $HA(\alpha, \beta, \check{\gamma}, \delta, \eta, \omega, \check{\rho}, \theta, m)$–$V$–invex at x^*;

  (iv) $\xi_i = \pi_k = \delta_l = \sigma$ for all $i \in \underline{p}$, $k \in \underline{\nu_0}$, and $l \in \underline{\nu}\backslash\underline{\nu_0}$;

  (v) $\sum_{i=1}^{p} u_i^* \bar{\rho}_i(x, x^*) + \sum_{m=1}^{\nu_0} \hat{\rho}_m(x, x^*) + \sum_{m=\nu_0+1}^{\nu} \check{\rho}_m(x, x^*) \geqq 0$ for all $x \in \mathbb{F}$;

(b) *the function* $(L_1(\cdot, u^*, v^*, \lambda^*, \bar{t}, \bar{s}), \ldots, L_p(\cdot, u^*, v^*, \lambda^*, \bar{t}, \bar{s}))$ *is* $HA(\alpha, \beta, \gamma, \xi, \eta, \omega, 0, \theta, m)$–$V$–*pseudoinvex at* x^* *and* $\gamma(x, x^*) > 0$ *for all* $x \in \mathbb{F}$, *where*

$$L_i(z, u^*, v^*, \lambda^*, \bar{t}, \bar{s}) = u_i^*\Bigg[f_i(z) - \lambda_i^* g_i(z) + \sum_{m=1}^{\nu_0} v_m^* G_{j_m}(z, t^m) + \sum_{m=\nu_0+1}^{\nu} v_m^* H_{k_m}(z, s^m)\Bigg], \ i \in \underline{p}.$$

Then x^* is an efficient solution of (P).

Proof. (a): Suppose to the contrary that x^* is not an efficient solution of (P), and hence for some $x \in \mathbb{F}$, $\varphi(x) \leqslant \varphi(x^*) = \lambda^*$. This implies that

$$f_i(x) - \lambda_i^* g_i(x) \leqq 0 \ \text{ for each } i \in \underline{p} \quad (2)$$

and

$$f_\ell(x) - \lambda_\ell^* g_\ell(x) < 0 \ \text{ for some } \ell \in \underline{p}. \quad (3)$$

Since the exponential function $z \to e^z$ is monotonically increasing, $u^* > 0$, $\alpha(x, x^*) \neq 0$, and $\bar{\gamma}_i(x, x^*) > 0$, $i \in \underline{p}$, for all $x \in \mathbb{F}$, it follows from (2) and (3) that the following strict inequality holds:

$$\frac{1}{\alpha(x, x^*)} \sum_{i=1}^{p} u_i^* \bar{\gamma}_i(x, x^*) \left(e^{\alpha(x,x^*)[f_i(x) - \lambda_i^* g_i(x)]} - 1 \right) < 0. \tag{4}$$

In view of our assumptions in (i) – (iv), we have

$$\frac{1}{\alpha(x, x^*)} \bar{\gamma}_i(x, x^*) \left(e^{\alpha(x,x^*)\{f_i(x) - \lambda_i^* g_i(x) - [f_i(x^*) - \lambda_i^* g_i(x^*)]\}} - 1 \right)$$

$$\geq \frac{1}{\beta(x, x^*)} \frac{1}{2} \left\langle \sigma(x, x^*)[\nabla f_i(x^*) - \lambda_i^* \nabla g_i(x^*)], e^{\beta(x,x^*)\eta(x,x^*)} - \mathbf{1} \right\rangle$$

$$+ \frac{1}{\beta(x, x^*)} \frac{1}{2} \left\langle \sigma(x, x^*)[\nabla f_i(x^*) - \lambda_i^* \nabla g_i(x^*)], e^{\beta(x,x^*)\omega(x,x^*)} - \mathbf{1} \right\rangle$$

$$+ \bar{\rho}_i(x, x^*) \|\theta(x, x^*)\|^m, \ i \in \underline{p}, \tag{5}$$

$$\frac{1}{\alpha(x, x^*)} \hat{\gamma}_m(x, x^*) \left(e^{\alpha(x,x^*)[v_m^* G_{j_m}(x, t^m) - v_m^* G_{j_m}(x^*, t^m)]} - 1 \right)$$

$$\geq \frac{1}{\beta(x, x^*)} \frac{1}{2} \left\langle \sigma(x, x^*) v_m^* \nabla G_{j_m}(x^*, t^m), e^{\beta(x,x^*)\eta(x,x^*)} - \mathbf{1} \right\rangle$$

$$+ \frac{1}{\beta(x, x^*)} \frac{1}{2} \left\langle \sigma(x, x^*) v_m^* \nabla G_{j_m}(x^*, t^m), e^{\beta(x,x^*)\omega(x,x^*)} - \mathbf{1} \right\rangle$$

$$+ \hat{\rho}_m(x, x^*) \|\theta(x, x^*)\|^m, \ m \in \underline{\nu_0}, \tag{6}$$

$$\frac{1}{\alpha(x, x^*)} \breve{\gamma}_m(x, x^*) \left(e^{\alpha(x,x^*)[v_m^* H_{k_m}(x, s^m) - v_m^* H_{k_m}(x^*, s^m)]} - 1 \right)$$

$$\geq \frac{1}{\beta(x, x^*)} \frac{1}{2} \left\langle \sigma(x, x^*) v_m^* \nabla H_{k_m}(x^*, s^m), e^{\beta(x,x^*)\eta(x,x^*)} - \mathbf{1} \right\rangle$$

$$+ \frac{1}{\beta(x, x^*)} \frac{1}{2} \left\langle \sigma(x, x^*) v_m^* \nabla H_{k_m}(x^*, s^m), e^{\beta(x,x^*)\omega(x,x^*)} - \mathbf{1} \right\rangle$$

$$+ \breve{\rho}_m(x, x^*) \|\theta(x, x^*)\|^m, \ m \in \underline{\nu} \setminus \underline{\nu_0}. \tag{7}$$

Multiplying (5) by u_i^* and then summing over $i \in \underline{p}$, summing (6) over $m \in \underline{\nu_0}$, and summing (7) over $m \in \underline{\nu}\setminus\underline{\nu_0}$, and finally adding the resulting inequalities, we get

$$\frac{1}{\alpha(x,x^*)}\Big\{\sum_{i=1}^{p} u_i^* \bar{\gamma}_i(x,x^*)\Big(e^{\alpha(x,x^*)\{f_i(x)-\lambda_i^* g_i(x)-[f_i(x^*)-\lambda_i^* g_i(x^*)]\}} - 1\Big)$$

$$+ \sum_{m=1}^{\nu_0} \hat{\gamma}_m(x,x^*)\Big(e^{\alpha(x,x^*)[v_m^* G_{j_m}(x,t^m)-v_m^* G_{j_m}(x^*,t^m)]} - 1\Big)$$

$$+ \sum_{m=\nu_0+1}^{\nu} \check{\gamma}_m(x,x^*)\Big(e^{\alpha(x,x^*)[v_m^* H_{k_m}(x,s^m)-v_m^* H_{k_m}(x^*,s^m)]} - 1\Big)\Big\}$$

$$\geqq \frac{1}{\beta(x,x^*)}\sigma(x,x^*)\frac{1}{2}\Big\langle \sum_{i=1}^{p} u_i^*[\nabla f_i(x^*) - \lambda_i^* \nabla g_i(x^*)] + \sum_{m=1}^{\nu_0} v_m^* \nabla G_{j_m}(x^*,t^m)$$

$$+ \sum_{m=\nu_0+1}^{\nu} v_m^* \nabla H_{k_m}(x^*,s^m), e^{\beta(x,x^*)\eta(x,x^*)} - 1\Big\rangle$$

$$+ \frac{1}{\beta(x,x^*)}\sigma(x,x^*)\frac{1}{2}\Big\langle \sum_{i=1}^{p} u_i^*[\nabla f_i(x^*) - \lambda_i^* \nabla g_i(x^*)] + \sum_{m=1}^{\nu_0} v_m^* \nabla G_{j_m}(x^*,t^m)$$

$$+ \sum_{m=\nu_0+1}^{\nu} v_m^* \nabla H_{k_m}(x^*,s^m), e^{\beta(x,x^*)\omega(x,x^*)} - 1\Big\rangle$$

$$+$$

$$\Big[\sum_{i=1}^{p} u_i^* \bar{\rho}_i(x,x^*) + \sum_{m=1}^{\nu_0} \hat{\rho}_m(x,x^*) + \sum_{m=\nu_0+1}^{\nu} \check{\rho}_m(x,x^*)\Big]\|\theta(x,x^*)\|^m.$$

Now using (1) and (v), and noticing that $\sigma(x,x^*) > 0$, $\varphi(x^*) = \lambda^*$; $x, x^* \in \mathbb{F}$, and $G_{j_m}(x^*,t^m) = 0$ for all $m \in \underline{\nu_0}$, the above inequality reduces to

$$\frac{1}{\alpha(x,x^*)}\sum_{i=1}^{p} u_i^* \bar{\gamma}_i(x,x^*)\Big(e^{\alpha(x,x^*)[f_i(x)-\lambda_i^* g_i(x)]} - 1\Big) \geqq 0,$$

which contradicts (4). Therefore, we conclude that x^* is an efficient solution of (P).

(b) : Let x be an arbitrary feasible solution of (P). From (1) we observe that

$$\frac{1}{\beta(x,x^*)}\Big\langle \sum_{i=1}^{p} u_i^*[\nabla f_i(x^*) - \lambda_i^* \nabla g_i(x^*)] + \sum_{m=1}^{\nu_0} v_m^* \nabla G_{j_m}(x^*,t^m)$$

$$+ \sum_{m=\nu_0+1}^{\nu} v_m^* \nabla H_{k_m}(x^*,s^m), e^{\beta(x,x^*)\eta(x,x^*)} - 1\Big\rangle$$

$$+ \frac{1}{\beta(x,x^*)}\Big\langle \sum_{i=1}^{p} u_i^*[\nabla f_i(x^*) - \lambda_i^* \nabla g_i(x^*)] + \sum_{m=1}^{\nu_0} v_m^* \nabla G_{j_m}(x^*,t^m)$$

$$+ \sum_{m=\nu_0+1}^{\nu} v_m^* \nabla H_{k_m}(x^*,s^m), e^{\beta(x,x^*)\omega(x,x^*)} - 1\Big\rangle = 0,$$

which in view of our $HA(\alpha, \beta, \gamma, \xi, \eta, \omega, 0, \theta, m)$–$V$–pseudoinvexity assumption implies that

$$\frac{1}{\alpha(x,x^*)}\gamma(x,x^*)\left(e^{\alpha(x,x^*)\sum_{i=1}^{p}[L_i(x,u^*,v^*,\lambda^*,\bar{t},\bar{s})-L_i(x^*,u^*,v^*,\lambda^*,\bar{t},\bar{s})]} - 1\right) \geqq 0.$$

We need to consider two cases: $\alpha(x, x^*) > 0$ and $\alpha(x, x^*) < 0$. If we assume that $\alpha(x, x^*) > 0$ and recall that $\gamma(x, x^*) > 0$, then the above inequality becomes

$$e^{\alpha(x,x^*)\sum_{i=1}^{p}[L_i(x,u^*,v^*,\lambda^*,\bar{t},\bar{s})-L_i(x^*,u^*,v^*,\lambda^*,\bar{t},\bar{s})]} \geqq 1,$$

which implies that

$$\sum_{i=1}^{p} L_i(x, u^*, v^*, \lambda^*, \bar{t}, \bar{s}) \geqq \sum_{i=1}^{p} \xi_i(x, x^*) L_i(x^*, u^*, v^*, \lambda^*, \bar{t}, \bar{s}).$$

Because $x^* \in \mathbb{F}$, $t^m \in \hat{T}_{j_m}(x^*)$, $m \in \underline{\nu_0}$, and $\lambda_i^* = \varphi_i(x^*)$, $i \in \underline{p}$, the right–hand side of the above inequality is equal to zero, and hence we have $L(x, u^*, v^*, \lambda^*, \bar{t}, \bar{s}) \geqq 0$. Inasmuch as $x \in \mathbb{F}$, and $v_m^* > 0$, $m \in \underline{\nu_0}$, this inequality simplifies to

$$\sum_{i=1}^{q} u_i^* \xi_i(x, x^*)[f_i(x) - \lambda_i^* g_i(x)] \geqq 0. \tag{8}$$

The above inequality implies that

$$\big(f_1(x) - \lambda_1^* g_1(x), \ldots, f_p(x) - \lambda_p^* g_p(x)\big) \not< (0, \ldots, 0),$$

which in turn implies that

$$\varphi(x) = \left(\frac{f_1(x)}{g_1(x)}, \ldots, \frac{f_p(x)}{g_p(x)}\right) \not< (\lambda_1^*, \ldots, \lambda_p^*) = \varphi(x^*).$$

Since $x \in \mathbb{F}$ was arbitrary, we conclude from this inequality that x^* is an efficient solution of (P).

If we assume that $\alpha(x, x^*) < 0$, we arrive at the same conclusion. $\square$

Theorem 11. *Let $x^* \in \mathbb{F}$, let $\lambda^* = \varphi(x^*)$, let the functions f_i, g_i, $i \in \underline{p}$, $G_j(\cdot, t)$, and $H_k(\cdot, s)$ be differentiable at x^* for all $t \in T_j$ and $s \in S_k$, $j \in \underline{q}$, $k \in \underline{r}$, and assume that there exist $u^* \in U$ and integers ν_0 and ν, with $0 \leqq \nu_0 \leqq \nu \leqq n+1$, such that there exist ν_0 indices j_m, with $1 \leqq j_m \leqq q$, together with ν_0 points $t^m \in \hat{T}_{j_m}(x^*)$, $m \in \underline{\nu_0}$, $\nu - \nu_0$ indices k_m, with $1 \leqq k_m \leqq r$, together with $\nu - \nu_0$ points $s^m \in S_{k_m}$, $m \in \underline{\nu}\setminus\underline{\nu_0}$, and ν real numbers v_m^*, with $v_m^* > 0$ for $m \in \underline{\nu_0}$, such that (1) holds. Assume, furthermore, that any one of the following four sets of hypotheses is satisfied:*

(a)   (i) $\big(\mathcal{E}_1(\cdot, \lambda^*, u^*), \ldots, \mathcal{E}_p(\cdot, \lambda^*, u^*)\big)$ is $HA(\alpha, \beta, \bar{\gamma}, \xi, \eta, \omega, \bar{\rho}, \theta, m)$–$V$–pseudoinvex at x^* and $\bar{\gamma}(x, x^*) > 0$ for all $x \in \mathbb{F}$;

     (ii) $\big(v_1^* G_{j_1}(\cdot, t^1), \ldots, v_{\nu_0}^* G_{j_{\nu_0}}(\cdot, t^{\nu_0})\big)$ is $HA(\alpha, \beta, \hat{\gamma}, \pi, \eta, \omega, \hat{\rho}, \theta, m)$–$V$–quasiinvex at x^*;

(iii) $(v^*_{\nu_0+1}H_{k_{\nu_0+1}}(\cdot,s^{\nu_0+1}),\ldots,v^*_\nu H_{k_\nu}(\cdot,s^\nu))$ is $HA(\alpha,\beta,\check{\gamma},\delta,\eta,\omega,\check{\rho},\theta,m)$–$V$–quasiinvex at x^*;

(iv) $\bar{\rho}(x,x^*)+\hat{\rho}(x,x^*)+\check{\rho}(x,x^*)\geqq 0$ for all $x\in\mathbb{F}$;

(b) (i) $(\mathcal{E}_1(\cdot,\lambda^*,u^*),\ldots,\mathcal{E}_p(\cdot,\lambda^*,u^*))$ is prestrictly $HA(\alpha,\beta,\bar{\gamma},\xi,\eta,\omega,\bar{\rho},\theta,m)$–$V$–quasiinvex at x^* and $\bar{\gamma}(x,x^*)>0$ for all $x\in\mathbb{F}$;

(ii) $(v^*_1 G_{j_1}(\cdot,t^1),\ldots,v^*_{\nu_0}G_{j_{\nu_0}}(\cdot,t^{\nu_0}))$ is $HA(\alpha,\beta,\hat{\gamma},\pi,\eta,\omega,\hat{\rho},\theta,m)$–$V$–quasiinvex at x^*;

(iii) $(v^*_{\nu_0+1}H_{k_{\nu_0+1}}(\cdot,s^{\nu_0+1}),\ldots,v^*_\nu H_{k_\nu}(\cdot,s^\nu))$ is $HA(\alpha,\beta,\check{\gamma},\delta,\eta,\omega,\check{\rho},\theta,m)$–$V$–quasiinvex at x^*;

(iv) $\bar{\rho}(x,x^*)+\hat{\rho}(x,x^*)+\check{\rho}(x,x^*)>0$ for all $x\in\mathbb{F}$;

(c) (i) $(\mathcal{E}_1(\cdot,\lambda^*,u^*),\ldots,\mathcal{E}_p(\cdot,\lambda^*,u^*))$ is prestrictly $HA(\alpha,\beta,\bar{\gamma},\xi,\eta,\omega,\bar{\rho},\theta,m)$–$V$–quasiinvex at x^* and $\bar{\gamma}(x,x^*)>0$ for all $x\in\mathbb{F}$;

(ii) $(v^*_1 G_{j_1}(\cdot,t^1),\ldots,v^*_{\nu_0}G_{j_{\nu_0}}(\cdot,t^{\nu_0}))$ is strictly $HA(\alpha,\beta,\hat{\gamma},\pi,\eta,\omega,\hat{\rho},\theta,m)$–$V$–pseudoinvex at x^*;

(iii) $(v^*_{\nu_0+1}H_{k_{\nu_0+1}}(\cdot,s^{\nu_0+1}),\ldots,v^*_\nu H_{k_\nu}(\cdot,s^\nu))$ is $HA(\alpha,\beta,\check{\gamma},\delta,\eta,\omega,\check{\rho},\theta,m)$–$V$–quasiinvex at x^*;

(iv) $\bar{\rho}(x,x^*)+\hat{\rho}(x,x^*)+\check{\rho}(x,x^*)\geqq 0$ for all $x\in\mathbb{F}$;

(d) (i) $(\mathcal{E}_1(\cdot,\lambda^*,u^*),\ldots,\mathcal{E}_p(\cdot,\lambda^*,u^*))$ is prestrictly $HA(\alpha,\beta,\bar{\gamma},\xi,\eta,\omega,\bar{\rho},\theta,m)$–$V$–quasiinvex at x^* and $\bar{\gamma}(x,x^*)>0$ for all $x\in\mathbb{F}$;

(ii) $(v^*_1 G_{j_1}(\cdot,t^1),\ldots,v^*_{\nu_0}G_{j_{\nu_0}}(\cdot,t^{\nu_0}))$ is $HA(\alpha,\beta,\hat{\gamma},\pi,\eta,\omega,\hat{\rho},\theta,m)$–$V$–quasiinvex at x^*;

(iii) $(v^*_{\nu_0+1}H_{k_{\nu_0+1}}(\cdot,s^{\nu_0+1}),\ldots,v^*_\nu H_{k_\nu}(\cdot,s^\nu))$ is strictly $HA(\alpha,\beta,\check{\gamma},\delta,\eta,\omega,\check{\rho},\theta,m)$–$V$–pseudoinvex at x^*;

(iv) $\bar{\rho}(x,x^*)+\hat{\rho}(x,x^*)+\check{\rho}(x,x^*)\geqq 0$ for all $x\in\mathbb{F}$.

Then x^* is an efficient solution of (P).

Proof. (a): Let x be an arbitrary feasible solution of (P). Since $G_{j_m}(x,t^m)\leqq 0 = G_{j_m}(x^*,t^m)$, it follows that

$$\sum_{m=1}^{\nu_0} v^*_m \pi_m(x,x^*)G_{j_m}(x,t^m) \leqq \sum_{m=1}^{\nu_0} v^*_m \pi_m(x,x^*)G_{j_m}(x^*,t^m),$$

and hence

$$\frac{1}{\alpha(x,x^*)}\hat{\gamma}(x,x^*)\left(e^{\alpha(x,x^*)\sum_{m=1}^{\nu_0}\pi_m(x,x^*)[v^*_m G_{j_m}(x,t^m)-v^*_m G_{j_m}(x^*,t^m)]}-1\right)\leqq 0$$

because $\alpha(x, x^*) \neq 0$ and $\hat{\gamma}(x, x^*) \geqq 0$. In view of (ii), this inequality implies that

$$\frac{1}{\beta(x, x^*)} \frac{1}{2} \Big\langle \sum_{m=1}^{\nu_0} v_m^* \nabla G_{j_m}(x^*, t^m), e^{\beta(x,x^*)\eta(x,x^*)} - \mathbf{1} \Big\rangle$$
$$+ \frac{1}{\beta(x, x^*)} \frac{1}{2} \Big\langle \sum_{m=1}^{\nu_0} v_m^* \nabla G_{j_m}(x^*, t^m), e^{\beta(x,x^*)\omega(x,x^*)} - \mathbf{1} \Big\rangle$$
$$\leqq -\hat{\rho}(x, x^*) \|\theta(x, x^*)\|^m. \tag{9}$$

In a similar manner, our assumptions in (iii) lead to the following inequality:

$$\frac{1}{\beta(x, x^*)} \frac{1}{2} \Big\langle \sum_{m=\nu_0+1}^{\nu} v_m^* \nabla H_{k_m}(x^*, s^m), e^{\beta(x,x^*)\eta(x,x^*)} - \mathbf{1} \Big\rangle$$
$$+ \frac{1}{\beta(x, x^*)} \frac{1}{2} \Big\langle \sum_{m=\nu_0+1}^{\nu} v_m^* \nabla H_{k_m}(x^*, s^m), e^{\beta(x,x^*)\omega(x,x^*)} - \mathbf{1} \Big\rangle$$
$$\leqq -\breve{\rho}(x, x^*) \|\theta(x, x^*)\|^m. \tag{10}$$

Now combining (1), (9) and (10), and using (iv), we find

$$\frac{1}{\beta(x, x^*)} \frac{1}{2} \Big\langle \sum_{i=1}^{p} u_i^* [\nabla f_i(x^*) - \lambda_i^* \nabla g_i(x^*)], e^{\beta(x,x^*)\eta(x,x^*)} - \mathbf{1} \Big\rangle$$
$$+ \frac{1}{\beta(x, x^*)} \frac{1}{2} \Big\langle \sum_{i=1}^{p} u_i^* [\nabla f_i(x^*) - \lambda_i^* \nabla g_i(x^*)], e^{\beta(x,x^*)\omega(x,x^*)} - \mathbf{1} \Big\rangle$$
$$\geqq -\bar{\rho}(x, x^*) \|\theta(x, x^*)\|^m.$$

This in view of (i) implies that

$$\frac{1}{\alpha(x, x^*)} \bar{\gamma}(x, x^*) \Big(e^{\alpha(x,x^*) \sum_{i=1}^{p} u_i^* \{f_i(x) - \lambda_i^* g_i(x) - [f_i(x^*) - \lambda_i^* g_i(x^*)]\}} - 1 \Big) \geqq 0.$$

Since $\bar{\gamma}(x, x^*) > 0$ and $\varphi(x^*) = \lambda^*$, this inequality results in

$$\sum_{i=1}^{p} u_i^* [f_i(x) - \lambda_i^* g_i(x)] \geqq 0.$$

Based on the proof of Theorem 10, this inequality leads to the conclusion that x^* is an efficient solution of (P).

(b) – (e) : The proofs are similar to that of part (a). $\square$

In the remainder of this section, we briefly discuss certain modifications of Theorems 10 and 11 obtained by replacing (1) with an inequality.

Theorem 12. Let $x^* \in \mathbb{F}$, let $\lambda^* = \varphi(x^*)$, let the functions f_i, g_i, $i \in \underline{p}$, $G_j(\cdot, t)$, and $H_k(\cdot, s)$ be differentiable at x^* for all $t \in T_j$ and $s \in S_k$, $j \in \underline{q}$, $k \in \underline{r}$, and assume that there exist $u^* \in U$ and integers ν_0 and ν, with $0 \leq \nu_0 \leq \nu \leq n+1$, such that there exist ν_0 indices j_m, with $1 \leq j_m \leq q$, together with ν_0 points $t^m \in \hat{T}_{j_m}(x^*)$, $m \in \underline{\nu_0}$, $\nu - \nu_0$ indices k_m, with $1 \leq k_m \leq r$, together with $\nu - \nu_0$ points $s^m \in S_{k_m}$, $m \in \underline{\nu} \backslash \underline{\nu_0}$, and ν real numbers v_m^*, with $v_m^* > 0$ for $m \in \underline{\nu_0}$, such that the following inequality holds:

$$\frac{1}{\beta(x, x^*)} \frac{1}{2} \Big\langle \sum_{i=1}^{p} u_i^* [\nabla f_i(x^*) - \lambda_i^* \nabla g_i(x^*)] + \sum_{m=1}^{\nu_0} v_m^* \nabla G_{j_m}(x^*, t^m)$$
$$+ \sum_{m=\nu_0+1}^{\nu} v_m^* \nabla H_{k_m}(x^*, s^m), e^{\beta(x, x^*)\eta(x, x^*)} - 1 \Big\rangle$$
$$+ \frac{1}{\beta(x, x^*)} \frac{1}{2} \Big\langle \sum_{i=1}^{p} u_i^* [\nabla f_i(x^*) - \lambda_i^* \nabla g_i(x^*)] + \sum_{m=1}^{\nu_0} v_m^* \nabla G_{j_m}(x^*, t^m)$$
$$+ \sum_{m=\nu_0+1}^{\nu} v_m^* \nabla H_{k_m}(x^*, s^m), e^{\beta(x, x^*)\omega(x, x^*)} - 1 \Big\rangle$$
$$\geqq 0 \quad \text{for all } x \in \mathbb{F}, \quad (11)$$

where $\beta : X \times X \to \mathbb{R}$ and $\eta, \omega : X \times X \to \mathbb{R}^n$ are given functions. Furthermore, assume that either one of the two sets of conditions specified in Theorem 10 is satisfied. Then x^* is an efficient solution of (P).

Although the proofs of Theorems 10 and 12 are essentially the same, their contents are somewhat different. This can easily be seen by comparing (1) with (11). We observe that any solution of (1) is also a solution of (11), but the converse is not necessarily true. Moreover, (1) is a system of n equations, whereas (11) is a single inequality. Evidently, from a computational point of view, (1) is preferable to (11) because of the dependence of the latter on the feasible set of (P).

The modified version of Theorem 11 can be stated in a similar manner.

4. Generalized Sufficiency Criteria

In this section, we discuss several families of sufficient efficiency results under various generalized HA(α, β, γ, ξ, η, ω, ρ, θ, m)–V–invexity hypotheses imposed on certain vector functions whose components are formed by considering different combinations of the problem functions. This is accomplished by employing a certain partitioning scheme which was originally proposed in [55] for the purpose of constructing generalized dual problems for nonlinear programming problems. For this we need some additional notation.

Let ν_0 and ν be integers, with $1 \leq \nu_0 \leq \nu \leq n+1$, and let $\{J_0, J_1, \ldots, J_M\}$ and $\{K_0, K_1, \ldots, K_M\}$ be partitions of the sets $\underline{\nu_0}$ and $\underline{\nu} \backslash \underline{\nu_0}$, respectively; thus, $J_i \subseteq \underline{\nu_0}$ for each $i \in \underline{M} \cup \{0\}$, $J_i \cap J_j = \emptyset$ for each $i, j \in \underline{M} \cup \{0\}$ with $i \neq j$, and $\cup_{i=0}^{M} J_i = \underline{\nu_0}$. Obviously, similar properties hold for $\{K_0, K_1, \ldots, K_M\}$. Moreover, if m_1 and m_2 are the numbers of the partitioning sets of $\underline{\nu_0}$ and $\underline{\nu}\setminus\underline{\nu_0}$, respectively, then $M = \max\{m_1, m_2\}$ and $J_i = \emptyset$ or $K_i = \emptyset$ for $i > \min\{m_1, m_2\}$.

In addition, we use the real–valued functions $\Phi_i(\cdot, u, v, \lambda, \bar{t}, \bar{s})$ and $\Lambda_\tau(\cdot, v, \bar{t}, \bar{s})$, $\tau \in \underline{M}$, defined, for fixed $u, v, \lambda, \bar{t} \equiv (t^1, t^2, \ldots, t^{\nu_0})$, and $\bar{s} \equiv (s^{\nu_0+1}, s^{\nu_0+2}, \ldots, s^{\nu})$, on X as follows:

$$\Phi_i(z, u, v, \lambda, \bar{t}, \bar{s}) = u_i \Big[f_i(z) - \lambda_i g_i(z) + \sum_{m \in J_0} v_m G_{j_m}(z, t^m) + \sum_{m \in K_0} v_m H_{k_m}(z, s^m) \Big], \quad i \in \underline{p},$$

$$\Lambda_\tau(z, v, \bar{t}, \bar{s}) = \sum_{m \in J_\tau} v_m G_{j_m}(z, t^m) + \sum_{m \in K_\tau} v_m H_{k_m}(z, s^m), \quad \tau \in \underline{M}.$$

Making use of the sets and functions defined above, we can now formulate our first collection of generalized sufficiency results for (P) as follows.

Theorem 13. *Let $x^* \in \mathbb{F}$, let $\lambda^* = \varphi(x^*)$, let the functions f_i, g_i, $i \in \underline{p}$, $G_j(\cdot, t)$, and $H_k(\cdot, s)$ be differentiable at x^* for all $t \in T_j$ and $s \in S_k$, $j \in \underline{q}$, $k \in \underline{r}$, and assume that there exist $u^* \in U$ and integers ν_0 and ν, with $0 \leq \nu_0 \leq \nu \leq n + 1$, such that there exist ν_0 indices j_m, with $1 \leq j_m \leq q$, together with ν_0 points $t^m \in \hat{T}_{j_m}(x^*)$, $m \in \underline{\nu_0}$, $\nu - \nu_0$ indices k_m, with $1 \leq k_m \leq r$, together with $\nu - \nu_0$ points $s^m \in S_{k_m}$, $m \in \underline{\nu}\setminus\underline{\nu_0}$, and ν real numbers v_m^*, with $v_m^* > 0$ for $m \in \underline{\nu_0}$, such that (1) holds. Assume, furthermore, that any one of the following three sets of hypotheses is satisfied:*

(a) (i) $(\Phi_1(\cdot, u^*, v^*, \lambda^*, \bar{t}, \bar{s}), \ldots, \Phi_p(\cdot, u^*, v^*, \lambda^*, \bar{t}, \bar{s}))$ is $HA(\alpha, \beta, \bar{\gamma}, \xi, \eta, \omega, \bar{\rho}, \theta, m)$–V–pseudoinvex at x^* and $\bar{\gamma}(x, x^*) > 0$ for all $x \in \mathbb{F}$;

   (ii) $(\Lambda_1(\cdot, v^*, \bar{t}, \bar{s}), \ldots, \Lambda_M(\cdot, v^*, \bar{t}, \bar{s}))$ is $HA(\alpha, \beta, \hat{\gamma}, \pi, \eta, \omega, \hat{\rho}, \theta, m)$–V–quasiinvex at x^*;

   (iii) $\bar{\rho}(x, x^*) + \hat{\rho}(x, x^*) \geqq 0$;

(b) (i) $(\Phi_1(\cdot, u^*, v^*, \lambda^*, \bar{t}, \bar{s}), \ldots, \Phi_p(\cdot, u^*, v^*, \lambda^*, \bar{t}, \bar{s}))$ is prestrictly $HA(\alpha, \beta, \bar{\gamma}, \xi, \eta, \omega, \bar{\rho}, \theta, m)$–V–quasiinvex at x^* and $\bar{\gamma}(x, x^*) > 0$ for all $x \in \mathbb{F}$;

   (ii) $(\Lambda_1(\cdot, v^*, \bar{t}, \bar{s}), \ldots, \Lambda_M(\cdot, v^*, \bar{t}, \bar{s}))$ is $HA(\alpha, \beta, \hat{\gamma}, \pi, \eta, \omega, \hat{\rho}, \theta, m)$–V–quasiinvex at x^*;

   (iii) $\bar{\rho}(x, x^*) + \hat{\rho}(x, x^*) > 0$;

(c) (i) $(\Phi_1(\cdot, u^*, v^*, \lambda^*, \bar{t}, \bar{s}), \ldots, \Phi_p(\cdot, u^*, v^*, \lambda^*, \bar{t}, \bar{s}))$ is prestrictly $HA(\alpha, \beta, \bar{\gamma}, \xi, \eta, \omega, \bar{\rho}, \theta, m)$–V–quasiinvex at x^* and $\bar{\gamma}(x, x^*) > 0$ for all $x \in \mathbb{F}$;

   (ii) $(\Lambda_1(\cdot, v^*, \bar{t}, \bar{s}), \ldots, \Lambda_M(\cdot, v^*, \bar{t}, \bar{s}))$ is strictly $HA(\alpha, \beta, \hat{\gamma}, \pi, \eta, \omega, \hat{\rho}, \theta, m)$–V–pseudoinvex at x^*;

   (iii) $\bar{\rho}(x, x^*) + \hat{\rho}(x, x^*) \geqq 0$.

Then x^* **is an efficient solution of** (P).

Proof. Let x be an arbitrary feasible solution of (P).

(a): It is clear that (1) can be expressed as follows:

$$\sum_{i=1}^{p} u_i^*[\nabla f_i(x^*) - \lambda_i^* \nabla g_i(x^*)] + \sum_{m \in J_0} v_m^* \nabla G_{j_m}(x^*, t^m) + \sum_{m \in K_0} v_m^* \nabla H_{k_m}(x^*, s^m)$$

$$+ \sum_{\tau=1}^{M} \Big[\sum_{m \in J_\tau} v_m^* \nabla G_{j_m}(x^*, t^m) + \sum_{m \in K_\tau} v_m^* \nabla H_{k_m}(x^*, s^m) \Big] = 0. \quad (12)$$

Since $x, x^* \in \mathbb{F}$, $v_m^* > 0$, and $t^m \in \hat{T}_{j_m}(x^*)$, $m \in \underline{\nu_0}$, it follows that

$$\sum_{\tau=1}^{M} \pi_\tau(x, x^*) \Lambda_\tau(x, v^*, \bar{t}, \bar{s}) = \sum_{\tau=1}^{M} \pi_\tau(x, x^*) \Big[\sum_{m \in J_\tau} v_m^* G_{j_m}(x, t^m) + \sum_{m \in K_\tau} v_m^* H_{k_m}(x, s^m) \Big]$$

$$\leqq 0$$

$$= \sum_{\tau=1}^{M} \pi_\tau(x, x^*) \Big[\sum_{m \in J_\tau} v_m^* G_{j_m}(x^*, t^m) + \sum_{m \in K_\tau} v_m^* H_{k_m}(x^*, s^m) \Big]$$

$$= \sum_{\tau=1}^{M} \pi_\tau(x, x^*) \Lambda_\tau(x^*, v^*, \bar{t}, \bar{s}),$$

and hence

$$\frac{1}{\alpha(x, x^*)} \hat{\gamma}(x, x^*) \Big(e^{\alpha(x, x^*) \sum_{\tau=1}^{M} \pi_\tau(x, x^*)[\Lambda_\tau(x, v^*, \bar{t}, \bar{s}) - \Lambda_\tau(x^*, v^*, \bar{t}, \bar{s})]} - 1 \Big) \leqq 0,$$

which because of (ii) implies that

$$\frac{1}{\beta(x, x^*)} \frac{1}{2} \Big\langle \sum_{\tau=1}^{M} \Big[\sum_{m \in J_\tau} v_m^* \nabla G_{j_m}(x^*, t^m) + \sum_{m \in K_\tau} v_m^* \nabla H_{k_m}(x^*, s^m) \Big], e^{\beta(x, x^*) \eta(x, x^*)} - \mathbf{1} \Big\rangle$$

$$+ \frac{1}{\beta(x, x^*)} \frac{1}{2} \Big\langle \sum_{\tau=1}^{M} \Big[\sum_{m \in J_\tau} v_m^* \nabla G_{j_m}(x^*, t^m) + \sum_{m \in K_\tau} v_m^* \nabla H_{k_m}(x^*, s^m) \Big], e^{\beta(x, x^*) \omega(x, x^*)} - \mathbf{1} \Big\rangle$$

$$\leqq -\hat{\rho}(x, x^*) \|\theta(x, x^*)\|^m. \quad (13)$$

Combining (12) and (13), and using (iii) we get

$$\frac{1}{\beta(x, x^*)} \frac{1}{2} \Big\langle \sum_{i=1}^{p} u_i^*[\nabla f_i(x^*) - \lambda_i^* \nabla g_i(x^*)] + \sum_{m \in J_0} v_m^* \nabla G_{j_m}(x^*, t^m)$$

$$+ \sum_{m \in K_0} v_m^* \nabla H_{k_m}(x^*, s^m), e^{\beta(x, x^*) \eta(x, x^*)} - \mathbf{1} \Big\rangle$$

$$+ \frac{1}{\beta(x, x^*)} \frac{1}{2} \Big\langle \sum_{i=1}^{p} u_i^*[\nabla f_i(x^*) - \lambda_i^* \nabla g_i(x^*)] + \sum_{m \in J_0} v_m^* \nabla G_{j_m}(x^*, t^m)$$

$$+ \sum_{m \in K_0} v_m^* \nabla H_{k_m}(x^*, s^m), e^{\beta(x, x^*) \omega(x, x^*)} - \mathbf{1} \Big\rangle$$

$$\geqq \hat{\rho}(x, x^*) \|\theta(x, x^*)\|^2 \geqq -\bar{\rho}(x, x^*) \|\theta(x, x^*)\|^m,$$

which by virtue of (i) implies that

$$\frac{1}{\alpha(x,x^*)}\bar{\gamma}(x,x^*)\left(e^{\alpha(x,x^*)\sum_{i=1}^{p}[\Phi_i(x,u^*,v^*,\lambda^*,\bar{t},\bar{s})-\Phi_i(x^*,u^*,v^*,\lambda^*,\bar{t},\bar{s})]}-1\right)\geqq 0.$$

Since $\bar{\gamma}(x,x^*)>0$, this inequality implies that

$$\sum_{i=1}^{p}\Phi_i(x,u^*,v^*,\lambda^*,\bar{t},\bar{s})\geqq\sum_{i=1}^{p}\Phi_i(x^*,u^*,v^*,\lambda^*,\bar{t},\bar{s})]=0,$$

where the equality follows from the fact that $\lambda_i^*=\varphi_i(x^*)$, $i\in p$, $t^m\in\hat{T}_{j_m}(x^*)$, and $x^*\in\mathbb{F}$. Because $x\in\mathbb{F}$ and $v_m^*>0$ for each $m\in\nu_0$, this inequality further reduces to

$$\sum_{i=1}^{p}u_i^*[f_i(x)-\lambda_i^*g_i(x)]\geqq 0.$$

In the proof of Theorem 10 it was shown that this inequality leads to the conclusion that x^* is an efficient solution of (P).

(b): Proceeding as in the proof of part (a), we see that (ii) leads to the following inequality:

$$\frac{1}{\beta(x,x^*)}\frac{1}{2}\Big\langle\sum_{\tau=1}^{M}\Big[\sum_{m\in J_\tau}v_m^*\nabla G_{j_m}(x^*,t^m)+\sum_{m\in K_\tau}v_m^*\nabla H_{k_m}(x^*,s^m)\Big],e^{\beta(x,x^*)\eta(x,x^*)}-\mathbf{1}\Big\rangle$$
$$+\frac{1}{\beta(x,x^*)}\frac{1}{2}\Big\langle\sum_{\tau=1}^{M}\Big[\sum_{m\in J_\tau}v_m^*\nabla G_{j_m}(x^*,t^m)+\sum_{m\in K_\tau}v_m^*\nabla H_{k_m}(x^*,s^m)\Big],e^{\beta(x,x^*)\omega(x,x^*)}-\mathbf{1}\Big\rangle$$
$$\leqq-\hat{\rho}(x,x^*)\|\theta(x,x^*)\|^m.$$

Now combining this inequality with (12) and using (iii), we obtain

$$\frac{1}{\beta(x,x^*)}\frac{1}{2}\Big\langle\sum_{i=1}^{p}u_i^*[\nabla f_i(x^*)-\lambda_i^*\nabla g_i(x^*)]+\sum_{m\in J_0}v_m^*\nabla G_{j_m}(x^*,t^m)$$
$$+\sum_{m\in K_0}v_m^*\nabla H_{k_m}(x^*,s^m),e^{\beta(x,x^*)\eta(x,x^*)}-\mathbf{1}\Big\rangle$$
$$+\frac{1}{\beta(x,x^*)}\frac{1}{2}\Big\langle\sum_{i=1}^{p}u_i^*[\nabla f_i(x^*)-\lambda_i^*\nabla g_i(x^*)]+\sum_{m\in J_0}v_m^*\nabla G_{j_m}(x^*,t^m)$$
$$+\sum_{m\in K_0}v_m^*\nabla H_{k_m}(x^*,s^m),e^{\beta(x,x^*)\omega(x,x^*)}-\mathbf{1}\Big\rangle$$
$$\geqq\hat{\rho}(x,x^*)\|\theta(x,x^*)\|^m>-\bar{\rho}(x,x^*)\|\theta(x,x^*)\|^m,$$

which by virtue of (i) implies that

$$\frac{1}{\alpha(x,x^*)}\bar{\gamma}(x,x^*)\left(e^{\alpha(x,x^*)\sum_{i=1}^{p}[\Phi_i(x,u^*,v^*,\lambda^*,\bar{t},\bar{s})-\Phi_i(x^*,u^*,v^*,\lambda^*,\bar{t},\bar{s})]}-1\right)\geqq 0.$$

The rest of the proof is identical to that of part (a).

(c): The proof is similar to those of parts (a) and (b). □

Each one of the six sets of conditions given in Theorem 13 and its modified version obtained by replacing (1) with (11) can be viewed as a family of sufficient efficiency conditions whose members can easily be identified by appropriate choices of the partitioning sets J_μ and K_μ, $\mu \in \underline{M} \cup \{0\}$.

In the remainder of this section we present another collection of sufficiency results which are somewhat different from those stated in Theorem 13 These results are formulated by using a partition of $\underline{p}$ in addition to those of $\underline{\nu_0}$ and $\underline{\nu}\backslash\underline{\nu_0}$, and by placing appropriate generalized $HA(\alpha, \beta, \bar{\gamma}, \xi, \eta, \omega, \rho, \theta, m)$-V-invexity requirements on certain vector functions involving $\mathcal{E}_i(\cdot, \lambda, u)$, $i \in \underline{p}$, $G_j(\cdot, t)$, $j \in \underline{q}$, and $H_k(\cdot, s)$, $k \in \underline{r}$.

Let $\{I_0, I_1, \ldots, I_d\}$, $\{J_0, J_1, \ldots, J_e\}$, and $\{K_0, K_1, \ldots, K_e\}$ be partitions of $\underline{p}$, $\underline{\nu_0}$, and $\underline{\nu}\backslash\underline{\nu_0}$, respectively, such that $D = \{0, 1, 2, \ldots, d\} \subseteq E = \{0, 1, \ldots, e\}$, and let the function $\Pi_\tau(\cdot, u, v, \lambda, \bar{t}, \bar{s}) : X \to \mathbb{R}$ be defined, for fixed $u, v, \lambda, \bar{t},$ and $\bar{s}$, by

$$\Pi_\tau(z, u, v, \lambda, \bar{t}, \bar{s}) = \sum_{i \in I_\tau} u_i[f_i(z) - \lambda_i g_i(z)] + \sum_{m \in J_\tau} v_m G_{j_m}(z, t^m)$$
$$+ \sum_{m \in K_\tau} v_m H_{k_m}(z, s^m), \quad \tau \in D.$$

Theorem 14. *Let $x^* \in \mathbb{F}$, let $\lambda^* = \varphi(x^*)$, let the functions f_i, g_i, $i \in \underline{p}$, $G_j(\cdot, t)$, and $H_k(\cdot, s)$ be differentiable at x^* for all $t \in T_j$ and $s \in S_k$, $j \in \underline{q}$, $k \in \underline{r}$, and assume that there exist $u^* \in U$, and integers ν_0 and ν, with $0 \leq \nu_0 \leq \nu \leq n + 1$, such that there exist ν_0 indices j_m, with $1 \leq j_m \leq q$, together with ν_0 points $t^m \in \hat{T}_{j_m}(x^*)$, $m \in \underline{\nu_0}$, $\nu - \nu_0$ indices k_m, with $1 \leq k_m \leq r$, together with $\nu - \nu_0$ points $s^m \in S_{k_m}$, $m \in \underline{\nu}\backslash\underline{\nu_0}$, and ν real numbers v_m^*, with $v_m^* > 0$ for $m \in \underline{\nu_0}$, such that (1) holds. Assume, furthermore, that any one of the following three sets of hypotheses is satisfied:*

(a) (i) $\left(\Pi_0(\cdot, u^*, v^*, \lambda^*, \bar{t}, \bar{s}), \ldots, \Pi_d(\cdot, u^*, v^*, \lambda^*, \bar{t}, \bar{s})\right)$ is $HA(\alpha, \beta, \bar{\gamma}, \xi, \eta, \omega, \bar{\rho}, \theta, m)$–V–pseudoinvex at x^*;

(ii) $\left(\Lambda_{d+1}(\cdot, v^*, \bar{t}, \bar{s}), \ldots, \Lambda_e(\cdot, v^*, \bar{t}, \bar{s})\right)$ is $HA(\alpha, \beta, \hat{\gamma}, \pi, \eta, \omega, \hat{\rho}, \theta, m)$–V–quasiinvex at x^*;

(iii) $\bar{\rho}(x, x^*) + \hat{\rho}(x, x^*) \geqq 0$ for all $x \in \mathbb{F}$;

(b) (i) $\left(\Pi_0(\cdot, u^*, v^*, \lambda^*, \bar{t}, \bar{s}), \ldots, \Pi_d(\cdot, u^*, v^*, \lambda^*, \bar{t}, \bar{s})\right)$ is prestrictly $HA(\alpha, \beta, \bar{\gamma}, \xi, \eta, \omega, \bar{\rho}, \theta, m)$–V–quasiinvex at x^*;

(ii) $\left(\Lambda_{d+1}(\cdot, v^*, \bar{t}, \bar{s}), \ldots, \Lambda_e(\cdot, v^*, \bar{t}, \bar{s})\right)$ is strictly $HA(\alpha, \beta, \hat{\gamma}, \pi, \eta, \omega, \hat{\rho}, \theta, m)$–V–pseudoinvex at x^*;

(iii) $\bar{\rho}(x, x^*) + \hat{\rho}(x, x^*) \geqq 0$ for all $x \in \mathbb{F}$;

(c) (i) $\left(\Pi_0(\cdot, u^*, v^*, \lambda^*, \bar{t}, \bar{s}), \ldots, \Pi_d(\cdot, u^*, v^*, \lambda^*, \bar{t}, \bar{s})\right)$ is prestrictly $HA(\alpha, \beta, \bar{\gamma}, \xi, \eta, \omega, \bar{\rho}, \theta, m)$–V–quasiinvex at x^*;

(ii) $\left(\Lambda_{d+1}(\cdot, v^*, \bar{t}, \bar{s}), \ldots, \Lambda_e(\cdot, v^*, \bar{t}, \bar{s})\right)$ is $HA(\alpha, \beta, \hat{\gamma}, \pi, \eta, \omega, \hat{\rho}, \theta, m)$–V–quasiinvex at x^*;

(iii) $\bar{\rho}(x, x^*) + \hat{\rho}(x, x^*) > 0$ for all $x \in \mathbb{F}$.

Then x^* is an efficient solution of (P).

Proof. (a): Suppose to the contrary that x^* is not an efficient solution of (P). Then there is $\bar{x} \in \mathbb{F}$ such that $\varphi(\bar{x}) \leqslant \varphi(x^*)$, and so it follows that

$$f_i(\bar{x}) - \lambda_i^* g_i(\bar{x}) \leq 0, \quad i \in \underline{p},$$

with strict inequality holding for at least one index $\ell \in \underline{p}$. Since $u^* > 0$, we see that for each $\tau \in D$,

$$\sum_{i \in I_\tau} u_i^*[f_i(\bar{x}) - \lambda_i^* g_i(\bar{x})] \leq 0, \tag{14}$$

with strict inequality holding for at least one index $\tau \in D$. Now using this inequality, we see that

$$\Pi_\tau(\bar{x}, u^*, v^*, \lambda^*, \bar{t}, \bar{s})$$
$$= \sum_{i \in I_\tau} u_i^*[f_i(\bar{x}) - \lambda_i^* g_i(\bar{x})] + \sum_{m \in J_\tau} v_m^* G_{j_m}(\bar{x}, t^m) + \sum_{m \in K_\tau} v_m^* H_{k_m}(\bar{x}, s^m)$$
$$\leq \sum_{i \in I_\tau} u_i^*[f_i(\bar{x}) - \lambda_i^* g_i(\bar{x})] \quad \text{(by the feasibility of } \bar{x} \text{ and positivity of } v_m^*, \ m \in \underline{\nu_0})$$
$$\leq 0 \quad \text{(by (14))}$$
$$= \sum_{i \in I_\tau} u_i^*[f_i(x^*) - \lambda_i^* g_i(x^*)] + \sum_{m \in J_\tau} v_m^* G_{j_m}(x^*, t^m) + \sum_{m \in K_\tau} v_m^* H_{k_m}(x^*, s^m)$$
$$\text{(since } \lambda^* = \varphi(x^*), \ x^* \in \mathbb{F}, \text{ and } t^m \in \hat{T}_{j_m}(x^*), \ m \in \underline{\nu_0})$$
$$= \Pi_\tau(x^*, u^*, v^*, \lambda^*, \bar{t}, \bar{s}),$$

with strict inequality holding for at least one index $\tau \in D$. Inasmuch as $\xi_\tau(\bar{x}, x^*) > 0$ for each $\tau \in D$, it follows that

$$\sum_{\tau \in D} \xi_\tau(\bar{x}, x^*) \Pi_\tau(\bar{x}, u^*, v^*, \lambda^*, \bar{t}, \bar{s}) < \sum_{\tau \in D} \xi_\tau(\bar{x}, x^*) \Pi_\tau(x^*, u^*, v^*, \lambda^*, \bar{t}, \bar{s}),$$

and thus

$$\frac{1}{\alpha(\bar{x}, x^*)} \bar{\gamma}(\bar{x}, x^*) \left(e^{\alpha(\bar{x}, x^*) \sum_{\tau \in D} \xi_\tau(\bar{x}, x^*)[\Pi_\tau(x, u^*, v^*, \lambda^*, \bar{t}, \bar{s}) - \Pi_\tau(x^*, u^*, v^*, \lambda^*, \bar{t}, \bar{s})]} - 1 \right) < 0,$$

which in view of (i) implies that

$$\frac{1}{\beta(\bar{x}, x^*)} \frac{1}{2} \Big\langle \sum_{i=1}^p u_i^*[\nabla f_i(x^*) - \lambda_i^* \nabla g_i(x^*)] + \sum_{\tau \in D} \Big[\sum_{m \in J_\tau} v_m^* \nabla G_{j_m}(x^*, t^m)$$
$$+ \sum_{m \in K_\tau} v_m^* \nabla H_{k_m}(x^*, s^m) \Big], e^{\beta(\bar{x}, x^*) \eta(\bar{x}, x^*)} - 1 \Big\rangle$$
$$+ \frac{1}{\beta(\bar{x}, x^*)} \frac{1}{2} \Big\langle \sum_{i=1}^p u_i^*[\nabla f_i(x^*) - \lambda_i^* \nabla g_i(x^*)] + \sum_{\tau \in D} \Big[\sum_{m \in J_\tau} v_m^* \nabla G_{j_m}(x^*, t^m)$$
$$+ \sum_{m \in K_\tau} v_m^* \nabla H_{k_m}(x^*, s^m) \Big], e^{\beta(\bar{x}, x^*) \omega(\bar{x}, x^*)} - 1 \Big\rangle$$
$$< -\bar{\rho}(\bar{x}, x^*) \|\theta(\bar{x}, x^*)\|^m. \tag{15}$$

As shown in the proof of Theorem 13, for each $\tau \in E\backslash D$, $\Lambda_\tau(\bar{x}, v^*, \bar{t}, \bar{s}) \leqq \Lambda_\tau(x^*, v^*, \bar{t}, \bar{s})$, and hence

$$\sum_{\tau \in E\backslash D} \pi_\tau(\bar{x}, x^*)\Lambda_\tau(\bar{x}, v^*, \bar{t}, \bar{s}) \leqq \sum_{\tau \in E\backslash D} \pi_\tau(\bar{x}, x^*)\Lambda_\tau(x^*, v^*, \bar{t}, \bar{s}).$$

Since $\hat{\gamma}(\bar{x}, x^*) \geqq 0$, this inequality implies that

$$\frac{1}{\alpha(\bar{x}, x^*)}\hat{\gamma}(\bar{x}, x^*)\left(e^{\alpha(\bar{x},x^*)\sum_{\tau \in E\backslash D}\pi_\tau(\bar{x},x^*)[\Lambda_\tau(\bar{x},v^*,\bar{t},\bar{s})-\Lambda_\tau(x^*,v^*,\bar{t},\bar{s})]} - 1\right) \leqq 0,$$

which in view of (ii) implies that

$$\frac{1}{\beta(\bar{x},x^*)}\frac{1}{2}\Big\langle \sum_{\tau \in E\backslash D}\Big[\sum_{m \in J_\tau}v_m^*\nabla G_{j_m}(x^*, t^m) + \sum_{m \in K_\tau}v_m^*\nabla H_{k_m}(x^*, s^m)\Big], e^{\beta(\bar{x},x^*)\eta(\bar{x},x^*)} - 1$$

$$+ \frac{1}{\beta(\bar{x},x^*)}\frac{1}{2}\Big\langle \sum_{\tau \in E\backslash D}\Big[\sum_{m \in J_\tau}v_m^*\nabla G_{j_m}(x^*, t^m) + \sum_{m \in K_\tau}v_m^*\nabla H_{k_m}(x^*, s^m)\Big], e^{\beta(\bar{x},x^*)\omega(\bar{x},x^*)} - 1\Big\rangle$$

$$\leqq -\hat{\rho}(\bar{x}, x^*)\|\theta(\bar{x}, x^*)\|^m. \quad (16)$$

Now combining (15) and (16) and using (iii), we see that

$$\frac{1}{\beta(\bar{x},x^*)}\frac{1}{2}\Big\langle \sum_{i=1}^{p}u_i^*[\nabla f_i(x^*) - \lambda_i^*\nabla g_i(x^*)]$$

$$+ \sum_{m=1}^{\nu_0}v_m^*\nabla G_{j_m}(x^*, t^m)$$

$$+ \sum_{m=\nu_0+1}^{\nu}v_m^*\nabla H_{k_m}(x^*, s^m), e^{\beta(\bar{x},x^*)\eta(\bar{x},x^*)} - 1\Big\rangle$$

$$+ \frac{1}{\beta(\bar{x},x^*)}\frac{1}{2}\Big\langle \sum_{i=1}^{p}u_i^*[\nabla f_i(x^*) - \lambda_i^*\nabla g_i(x^*)]$$

$$+ \sum_{m=1}^{\nu_0}v_m^*\nabla G_{j_m}(x^*, t^m)$$

$$+ \sum_{m=\nu_0+1}^{\nu}v_m^*\nabla H_{k_m}(x^*, s^m), e^{\beta(\bar{x},x^*)\omega(\bar{x},x^*)} - 1\Big\rangle$$

$$< -[\bar{\rho}(\bar{x}, x^*) + \hat{\rho}(\bar{x}, x^*)]\|\theta(\bar{x}, x^*)\|^m \leqq 0,$$

which contradicts (1). Therefore, x^* is an efficient solution of (P).

(b) and (c): The proofs are similar to that of part (a). □

Corollary 15. *Let $x^* \in \mathbb{F}$, let $\lambda^* = \varphi(x^*)$, let the functions f_i, g_i, $i \in \underline{p}$, $G_j(\cdot, t)$, and $H_k(\cdot, s)$ be differentiable at x^* for all $t \in T_j$ and $s \in S_k$, $j \in \underline{q}$, $k \in \underline{r}$, and assume that there exist $u^* \in U$, and integers ν_0 and ν, with $0 \leqq \nu_0 \leqq \nu \leqq n+1$, such that there exist ν_0 indices j_m, with $1 \leqq j_m \leqq q$, together with ν_0 points $t^m \in \hat{T}_{j_m}(x^*)$, $m \in \underline{\nu_0}$, $\nu - \nu_0$*

indices k_m, with $1 \leqq k_m \leqq r$, together with $\nu - \nu_0$ points $s^m \in S_{k_m}$, $m \in \underline{\nu} \backslash \underline{\nu_0}$, and ν real numbers v_m^*, with $v_m^* > 0$ for $m \in \underline{\nu_0}$, such that (1) holds. Assume, furthermore, that any one of the following three sets of hypotheses is satisfied:

(a) (i) $\big(\Pi_0(\cdot, u^*, v^*, \lambda^*, \bar{t}, \bar{s}), \ldots, \Pi_d(\cdot, u^*, v^*, \lambda^*, \bar{t}, \bar{s})\big)$ is $HA(\alpha, \beta, \bar{\gamma}, \xi, \eta, \bar{\rho}, \theta, m)$–V–pseudoinvex at x^*;

  (ii) $\big(\Lambda_{d+1}(\cdot, v^*, \bar{t}, \bar{s}), \ldots, \Lambda_e(\cdot, v^*, \bar{t}, \bar{s})\big)$ is $HA(\alpha, \beta, \hat{\gamma}, \pi, \eta, \hat{\rho}, \theta, m)$-V-quasiinvex at x^*;

  (iii) $\bar{\rho}(x, x^*) + \hat{\rho}(x, x^*) \geqq 0$ for all $x \in \mathbb{F}$;

(b) (i) $\big(\Pi_0(\cdot, u^*, v^*, \lambda^*, \bar{t}, \bar{s}), \ldots, \Pi_d(\cdot, u^*, v^*, \lambda^*, \bar{t}, \bar{s})\big)$ is prestrictly $HA(\alpha, \beta, \bar{\gamma}, \xi, \eta, \bar{\rho}, \theta, m)$–V–quasiinvex at x^*;

  (ii) $\big(\Lambda_{d+1}(\cdot, v^*, \bar{t}, \bar{s}), \ldots, \Lambda_e(\cdot, v^*, \bar{t}, \bar{s})\big)$ is strictly $HA(\alpha, \beta, \hat{\gamma}, \pi, \eta, \hat{\rho}, \theta, m)$–V–pseudoinvex at x^*;

  (iii) $\bar{\rho}(x, x^*) + \hat{\rho}(x, x^*) \geqq 0$ for all $x \in \mathbb{F}$;

(c) (i) $\big(\Pi_0(\cdot, u^*, v^*, \lambda^*, \bar{t}, \bar{s}), \ldots, \Pi_d(\cdot, u^*, v^*, \lambda^*, \bar{t}, \bar{s})\big)$ is prestrictly $HA(\alpha, \beta, \bar{\gamma}, \xi, \eta, \bar{\rho}, \theta, m)$–V–quasiinvex at x^*;

  (ii) $\big(\Lambda_{d+1}(\cdot, v^*, \bar{t}, \bar{s}), \ldots, \Lambda_e(\cdot, v^*, \bar{t}, \bar{s})\big)$ is $HA(\alpha, \beta, \hat{\gamma}, \pi, \eta, \hat{\rho}, \theta, m)$–V–quasiinvex at x^*;

  (iii) $\bar{\rho}(x, x^*) + \hat{\rho}(x, x^*) > 0$ for all $x \in \mathbb{F}$.

Then x^* is an efficient solution of (P).

We observe that one can readily identify numerous special cases of the six families of sufficiency results stated in Theorem 14 and its modified version obtained by replacing (1) with (11), by appropriate choices of the partitioning sets I_μ, $\mu \in D$, and J_τ and K_τ, $\tau \in E$.

REFERENCES

[1] Ergenç, T., Pickl, S. W., Radde, N. and Weber, G.–W. 2004. "Generalized semi–infinite optimization and anticipatory systems". *Int. J. Comput. Anticipatory Syst.* 15: 3–30.

[2] Weber, G.–W., Alparslan–Gök, S. Z. and Söyler, A. 2009. "A new mathematical approach in environmental and life sciences: gene–environment networks and their dynamics". *Environ. Model. Assess.* 14: 267–288.

[3] Weber, G.–W., Taylan, P., Alparslan–Gök, Z., Özögür–Akyüz, S. and Akteke–Öztürk, B. 2008. "Optimization of gene–environment networks in the presence of errors and uncertainty with Chebyshev approximation". *TOP* 16: 284–318.

[4] Weber, G.-W. and Tezel, A. 2007. "On generalized semi–infinite optimization of genetic networks". *TOP* 15: 65–77.

[5] Weber, G.-W., Tezel, A., Taylan, P., Söyler, A. and Çetin, M. 2008. "Mathematical contributions to dynamics and optimization of gene–environment networks". *Optimization* 57: 353–377.

[6] Chen, H. and Hu, C. F. 2009. "On the resolution of the Vasicek–type interest rate model". *Optimization* 58: 809–822.

[7] Winterfeld, A. 2008. "Application of general semi–infinite programming to lapidary cutting problems". *European J. Oper. Res.* 191: 838–854.

[8] Verma, R. U. 1994. "General approximation–solvability of nonlinear equations involving A–regular operators". *Zeitschrift fur Analysis und ihre Anwendungen* 13: 89–96.

[9] Zeidler, E. 1990. *Nonlinear Functional Analysis and its Applications II/B*. New York: Springer–Verlag.

[10] Antczak, T. 2005. "The notion of $V-r$–invexity in differentiable multiobjective programming". *J. Appl. Anal.* 11: 63–79.

[11] Zalmai, G. J. 1996. "Proper efficiency conditions and duality models for nonsmooth multiobjective fractional programming problems with operator constraints, part I: Theory". *Utilitas Math.* 50: 163–201.

[12] Zalmai, G. J. 1997. "Proper efficiency conditions and duality models for nonsmooth multiobjective fractional programming problems with operator constraints, part II: Applications". *Utilitas Math.* 51: 193–237.

[13] Zalmai, G. J. 1998. "Proper efficiency principles and duality models for a class of continuous–time multiobjective fractional programming problems with operator constraints". *J. Stat. Manag. Syst.* 1: 11–59.

[14] Miettinen, K. M. 1999. *Nonlinear Multiobjective Optimization*. Boston: Kluwer Academic Publishers.

[15] Sawaragi, Y., Nakayama, H. and Tanino, T. 1986. *Theory of Multiobjective Optimization*. New York: Academic Press.

[16] White, D. J. 1982. *Optimality and Efficiency*. New York: Wiley.

[17] Yu, P. L. 1985. *Multiple–Criteria Decision Making: Concepts, Techniques, and Extensions*. New York: Plenum Press.

[18] Brosowski, B. 1982. *Parametric Semiinfinite Optimization*. Frankfurt A. M.: Peter Lang.

[19] Fiacco, A. V. and Kortanek, K. O. (eds.) 1983. *Semi–infinite Programming and Applications. Lecture Notes in Economics and Mathematical Systems, Vol. 215*. Berlin: Springer.

[20] Glashoff, K. and Gustafson, S. A. 1983. *Linear Optimization and Approximation*. Berlin: Springer.

[21] Goberna, M. A. and López, M. A. 1998. *Linear Semi–Infinite Optimization*. New York: Wiley.

[22] Goberna, M. A. and López, M. A. (eds.) 2001. *Semi–infinite Programming – Recent Advances*. Kluwer, Dordrecht.

[23] Gribik, P. R. 1979. "Selected applications of semi–infinite programming", in *Constructive Approaches to Mathematical Models* (C. V. Coffman G. J. Fix, eds.) New York: Academic Press, 171–187.

[24] Gustafson, S. A. and Kortanek, K. O. 1983. "Semi–infinite programming and applications", in *Mathematical Programming: The State of the Art* (A. Bachem, *et al.*, eds.) Berlin: Springer, 132–157.

[25] Henn, R. and Kischka, P. 1976. "Über einige Anwendungen der semi–infiniten Optimierung" ["For some applications of semi - infinite optimization"], *Zeitschrift Oper. Res.* 20: 39–58.

[26] Hettich, R. (ed.) 1976. *Semi–infinite Programming*. Lecture Notes in Control and Information Sciences, Vol. 7. Berlin: Springer.

[27] Hettich, R. and Kortanek, K. O. 1993. "Semi–infinite programming: theory, methods, and applications". *SIAM Review* 35: 380–429.

[28] Hettich, R. and Zencke, P. 1982. *Numerische Methoden der Approximation und semi–infinite Optimierung*[*Numerical methods of approximation and semi - infinite optimization*]. Stuttgart: Teubner.

[29] López, M. and Still, G. 2007. "Semi–infinite programming". *European J. Oper. Res.* 180: 491–518.

[30] Reemtsen, R. and Rückmann, J. J. (eds.) 1998. *Semi–Infinite Programming*. Boston: Kluwer.

[31] Weber, G.–W. 2002. "Generalized semi–infinite optimization: theory and applications in optimal control and discrete optimization". *J. Stat. Manag. Syst.* 5: 359–388.

[32] Jess, A., Jongen, H. Th., Neralić, L. and Stein, O. 2001. "A semi–infinite programming model in data envelopment analysis". *Optimization* 49: 369–385.

[33] Neralić, L. and Stein, O. 2004. "On regular and parametric data envelopment analysis". *Math. Methods Oper. Res.* 60: 15–28.

[34] Özögür–Akyüz, S. and Weber, G.-W. 2010. "Infinite kernel learning via infinite and semi–infinite programming". *Optim. Methods Softw.* 25: 937–970.

[35] Özögür–Akyüz, S. and Weber, G.-W. 2010. "On numerical optimization theory of infinite kernel learning". *J. Global Optim.* 48: 215–239.

[36] Daum, S. and Werner, R. 2011. "A novel feasible discretization method for linear semi–infinite programming applied to basket option pricing. *Optimization* 60: 1379–1398.

[37] Hanson, M. A. 1981. "On sufficiency of the Kuhn–Tucker conditions", *J. Math. Anal. Appl.* 80: 545–550.

[38] Craven, B. D. 1981. "Invex functions and constrained local minima". *Bull. Austral. Math. Soc.* 24: 357–366.

[39] Jeyakumar, V. and Mond, B. 1992. "On generalised convex mathematical programming". *Austral. Math. Soc. Ser. B.* 34: 43–53.

[40] Ben–Israel, A. and Mond, B. 1986. "What is invexity?" *J. Austral. Math. Soc. Ser. B.* 28: 1–9.

[41] Giorgi, G. and Guerraggio, A. 1996. "Various types of nonsmooth invex functions". *J. Inform. Optim. Sci.* 17: 137–150.

[42] Giorgi, G. and Mititelu, Şt. 1993. "Convexités généralisées et propriétés". *Rev. Roumaine Math. Pures Appl.* 38: 125–172.

[43] Hanson, M. A. and Mond, B. 1982. "Further generalizations of convexity in mathematical programming". *J. Inform. Optim. Sci.* 3: 25–32.

[44] Martin, D. H. 1985. "The essence of invexity". *J. Optim. Theory Appl.* 47: 65–76.

[45] Mititelu, Şt. 2004. "Invex functions". *Rev. Roumaine Math. Pures Appl.* 49: 529–544.

[46] Mititelu, Şt. 2007. "Invex sets and nonsmooth invex functions". *Rev. Roumaine Math. Pures Appl.* 52: 665–672.

[47] Mititelu, Şt. and Postolachi, M. 2011. "Nonsmooth invex functions via upper directional derivative of Dini". *J. Adv. Math. Stud.* 4: 57–76.

[48] Mititelu, Şt. and Stancu–Minasian, I. M. 1993. "Invexity at a point: Generalizations and classification". *Bull. Austral. Math. Soc.* 48: 117–126.

[49] Pini, R. 1991. "Invexity and generalized convexity". *Optimization* 22: 513–525.

[50] Reiland, T. W. 1990. "Nonsmooth invexity". *Bull. Austral. Math. Soc.* 42: 437–446.

[51] Kanniappan, P. and Pandian, P. 1996. "On generalized convex functions in optimization theory – A survey". *Opsearch.* 33: 174–185.

[52] Pini, R. and Singh, C. 1997. "A survey of recent [1985 – 1995] advances in generalized convexity with applications to duality theory and optimality conditions". *Optimization* 39: 311–360.

[53] Antczak, T. 2009. "Optimality and duality for nonsmooth multiobjective programming problems with $V-r$–invexity". *J. Global Optim.* 45: 319–334.

[54] Zalmai, G. J. and Zhang, Q. 2008. "Global semiparametric sufficient efficiency conditions for semiinfinite multiobjective fractional programming problems involving generalized (α, η, ρ)–V–invex functions. *Southeast Asian Bull. Math.* 32: 573–599.

[55] Mond, B. and Weir, T. 1981. Generalized concavity and duality, in *Generalized Concavity in Optimization and Economics* (S. Schaible and W. T. Ziemba, eds.) New York: Academic Press.

ABOUT THE EDITOR

Daniel González Sánchez, PhD
Professor
Universidad de Las Américas, Quito, Ecuador
E-mail: daniel.gonzalez.sanchez@udla.edu.ec

Daniel González Sánchez has a degree in Mathematics and he is an Engineer in Computer Sciences. He earned a doctorate in Applied Mathematics studying initial value problems for the Newton method in Banach spaces. All his remarks were obtained at the University of La Rioja in Spain. González is a prolific author and researcher on top of the journals in the field of Numerical Analysis, focused principally on nonlinear, integral and differential equations using iterative processes. In addition, Daniel serves as an active reviewer in these journals. He is also interested in pedagogical processes to teach mathematics in higher education and currently heads a research group and several international projects.

INDEX

A

analytic function, 3, 195, 228, 243
analytic solution, 72, 425
anisotropic function space, 249, 308, 315, 330, 365, 374, 379, 412
anisotropic Sobolev space, 314
approximate computational order of convergence, 58, 63, 78, 79, 89, 90, 121, 145, 160, 173, 184
approximated zero, 78, 79, 89, 90, 194
associated space, x, 209, 210, 211, 222, 231, 238
asymmetric Fourier transform, 273
asymptotic error constant, 78, 79, 89, 90
atomic Lyapunov constant, 246, 302
atomic measure, 253, 267, 344, 378
autonomous differential equation, 63, 121

B

Babylonian/Pythagorean theorem, 246
ball convergence, 155, 167, 179
Banach algebra, 416
Banach lemma, 119, 130, 206
Banach space, ix, x, xi, 4, 5, 6, 7, 11, 15, 16, 17, 21, 22, 25, 26, 27, 30, 34, 35, 37, 43, 47, 57, 69, 71, 73, 83, 84, 95, 96, 115, 123, 125, 137, 138, 153, 154, 165, 166, 177, 178, 189, 191, 199, 200, 201, 209, 210, 219, 220, 228, 232, 243, 245, 246, 247, 248, 249, 250, 253, 254, 255, 258, 259, 264, 267, 268, 269, 271, 272, 274, 276, 277, 278, 279, 280, 281, 282, 283, 284, 288, 292, 293, 295, 296, 297, 298, 299, 300, 301, 302, 303, 304, 307, 308, 310, 314, 317, 318, 319, 320, 325, 326, 327, 335, 336, 337, 340, 342, 344, 345, 346, 350, 351, 352, 354, 364, 366, 367, 369, 370, 371, 372, 373, 374, 375, 376, 381, 385, 390, 391, 392, 393, 394, 395, 396, 397, 398, 399, 400, 401, 402, 403, 405, 406, 407, 409, 410, 411, 412, 413, 415, 416, 418, 419, 421, 424, 425, 426
Banach–Mazur distance, 250, 326, 335, 375
basin of attraction, 191
Besov, x, xi, 246, 249, 250, 251, 256, 257, 280, 294, 299, 304, 306, 307, 309, 313, 315, 319, 321, 322, 323, 324, 325, 331, 335, 342, 343, 344, 347, 348, 350, 351, 352, 353, 363, 364, 365, 367, 369, 370, 371, 372, 374, 377, 379, 380, 381, 382, 383, 387, 388, 392, 394, 399, 407, 408, 410, 412, 413
Birkhoff–Fortet orthogonality, 246, 248, 297, 298
Bochner, 5, 15, 16, 17, 21, 25, 26, 27, 34, 249, 264, 288, 318, 345, 346, 374
Bochner–measurable function, 249, 318, 374
Borel subset, 5
boundary condition, 18, 20
boundary value problem, xi, 90, 111, 204, 379, 429, 431
bounded extension, x, 245, 248, 307, 351, 364, 369, 371, 380, 391, 392, 402, 407
bounded linear operator, 57, 98, 201, 248, 250, 281
bounded outer inverse, 48, 51
Bynum–Drew–Gross inequality, 246, 288

C

Cauchy integral formula, 210, 232, 425
Cauchy sequence, 103, 105, 205
Cauchy–Pompeiu integral representation, 210
Cauchy–Riemann operator, 211, 212, 217, 232, 236, 242
Cauchy–Riemann system, 223
center Lipschitz condition, 37
Chebyshev, 36, 44, 81, 94, 95, 122, 123, 126, 145, 164, 175, 187, 200, 201, 202, 203, 206, 247, 264, 267, 285, 307, 364, 380, 412, 452
Chebyshev–like method, 36, 206
classical moduli, 259, 281, 282
Clifford algebra, x, 209, 210, 211, 214, 215, 217, 228, 229, 230, 231, 232, 235, 236, 238, 239, 242, 243, 244
Clifford analysis, 222, 228, 229, 231, 244
Clifford structure, 210, 232
Clifford valued function, 216, 219, 235, 236, 237
Clifford–type algebra, 209
closed ball, 74, 86, 98, 116, 141, 156, 168, 180
complemented subspace, 249, 250, 253, 288, 313, 314, 319, 354, 369, 374, 375

computational cost, ix, 39, 81, 94, 115, 123, 164, 176, 188
computational order of convergence, 58, 78, 79, 89, 90, 105, 121, 138, 145, 160, 172, 184
continuous operator, 63, 77, 88, 121, 137, 138, 161, 173, 184, 420
continuously differentiable function, 37, 222
convergence, ix, xi, 35, 37, 38, 41, 43, 44, 45, 46, 47, 48, 54, 55, 57, 58, 63, 67, 68, 69, 71, 72, 73, 77, 78, 82, 83, 84, 85, 88, 89, 94, 95, 97, 99, 100, 101, 105, 109, 112, 113, 115, 116, 121, 123, 125, 126, 127, 129, 135, 137, 138, 145, 147, 148, 149, 151, 153, 154, 155, 160, 163, 164, 165, 166, 167, 175, 176, 177, 178, 179, 180, 183, 187, 188, 189, 190, 191, 193, 194, 195, 196, 199, 200, 203, 415, 416, 422, 425, 426, 427
convergence ball, ix, 73, 77, 84, 88, 121
convergence criteria, 48, 57, 67, 125
convergence domain, 37, 72, 78, 83, 84, 85, 89, 154, 166, 178
convergence region, 47, 48
convex domain, 3, 11, 199, 258
convex duality, 364, 411
cosine theorem, 246, 298

D

derivative free method, 83
differentiable mapping, 47
differential equation, xi, 77, 88, 161, 162, 173, 174, 184, 186, 206, 210, 228, 232, 415, 416
Dirac operator, 210, 211, 217, 218, 219, 228, 236, 237, 244
divided difference, 68, 73, 74, 84, 86, 127, 128, 134, 138, 141, 147, 259
domain, 2, 14, 15, 24, 47, 55, 68, 89, 201, 219, 222, 226, 238, 251, 253, 257, 260, 287, 316, 319, 335, 351, 367, 370, 377, 379, 381, 383, 416, 423, 424, 425, 426
double moduli, x, 245, 246, 258
duality mapping, 247, 248, 303, 304

E

error bound, ix, 39, 50, 53, 57, 71, 73, 78, 85, 89, 100, 105, 184
Euclidean Dirac operator, 211
Euclidean space, x, 137, 211, 222, 226, 238, 255, 256, 299, 313, 314, 315, 316, 370, 372, 430
evolution operator, 419

F

Fefferman–Stein inequality, 253, 319
Fenchel–Young transform, 259

Fenchel–Young–Lindenstrauss duality, 246, 247, 258, 385
Fibonacci sequence, 130
finite mixed quasi–norm, 248, 318
fixed point iterative method, 83
fractional derivative, 5, 8, 16, 17, 26, 27
fractional inequality, 4, 6, 7, 9, 10, 19, 30
Fredholm integral equation, 198, 205, 206

G

gamma function, 16, 25, 26
Gauss–Newton method, 48, 113
generalized conjugate method, 63, 121
generalized convex function, 432, 455
generalized fractional derivative, 13
generalized inverse, 47, 48
generalized Lipschitz condition, 59, 155, 167, 179
generalized minimum residual method, 63, 121
gradient, 432
Green kernel, 111
Grothendieck nuclear space, 421

H

Halley, 44, 45, 94, 95, 122, 123, 126, 164, 176, 188, 200, 201, 207
Hanner convexity, 246, 283, 289
Hanner inequality, 285
Hausdorff measure, 5
high convergence order method, 116
high order characterisation, 245, 247
Hilbert space, 97, 98, 254, 258, 293, 297, 299, 300, 301, 394, 397, 398, 400
Hilbert-Pachpatte, ix
holomorphic, 2, 14, 24, 210, 211, 257, 315, 324, 372
holomorphic initial data, 210
hybrid method, 42

I

identity operator, 28, 190, 196, 200
IG–space, 246, 252, 255, 281, 290, 294, 295, 297, 369, 385, 389
ill–posed problems, 48, 112, 113, 114
initial guess, ix, 71, 73, 78, 79, 80, 84, 85, 109, 155, 167, 178, 179
initial point, 37, 42, 48, 72, 84, 116, 138, 148, 149, 154, 166, 178
initial value problem, 209, 210, 219, 222, 228, 230, 231, 232, 238, 243, 244
integral equation, 68, 110, 111, 161, 173, 185, 189, 195, 196, 197, 199, 206, 207
integral operator, 16, 17, 25, 366, 413
integro–differential operator, 210

interior estimate, 209, 210, 220, 221, 222, 232, 238, 242
inverse operator, 190, 199, 200, 201, 202, 203, 204, 278
invex function, 430, 432, 433, 434, 438, 455, 456
iterate, 113
iteration index, 78, 79, 80, 90
iterative method, 36, 37, 38, 43, 44, 45, 48, 55, 58, 59, 67, 68, 69, 72, 73, 78, 81, 84, 93, 94, 95, 97, 98, 112, 122, 123, 135, 137, 147, 151, 154, 155, 164, 166, 167, 175, 178, 187, 189, 191, 193, 194, 197, 199, 200, 201, 207
iterative regularization method, 98, 112, 113

K

Kadets constant, 267, 291, 299, 300, 301
Kantorovich, 37, 43, 45, 48, 55, 135, 190, 193, 197, 247, 267, 315, 366
Kantorovich hypothesis, 43
Krein–Milman property, 346, 367
Kronecker delta, 214

L

Lavrentiev regularization, 97, 112, 113, 114
Lebesgue measure, 5, 248, 317, 373
linear spline, 108
Lipschitz, x, 37, 39, 47, 48, 52, 63, 71, 73, 85, 97, 100, 109, 111, 125, 148, 153, 155, 160, 162, 165, 167, 172, 174, 177, 179, 183, 185, 190, 220, 303, 307, 351, 364, 369, 370, 371, 373, 375, 377, 379, 381, 383, 385, 387, 389, 390, 391, 392, 393, 395, 397, 399, 401, 402, 403, 405, 406, 407, 409, 411, 413
Lipschitz condition, 97, 109, 148, 220
Lipschitz constant, 52, 63, 71, 73, 85, 153, 160, 165, 172, 177, 183
Lipschitz parameter, 47, 48, 125, 155, 167, 179
Lipschitz type condition, 37
Littlewood–Paley decomposition, 257, 315, 324, 330, 383
Lizorkin–Triebel, 246, 250, 251, 256, 257, 294, 299, 304, 306, 307, 313, 315, 316, 321, 322, 323, 324, 325, 335, 342, 343, 344, 347, 348, 350, 351, 352, 363, 364, 367, 369, 370, 371, 374, 377, 379, 380, 381, 382, 383, 387, 388, 392, 399, 408, 410, 412, 413
local convergence, ix, 35, 38, 43, 47, 48, 50, 51, 53, 57, 58, 59, 65, 71, 72, 73, 74, 83, 84, 85, 86, 115, 116, 118, 125, 137, 138, 140, 141, 142, 147, 148, 153, 154, 155, 156, 165, 166, 167, 168, 177, 178, 179, 180, 189, 190
locally unique solution, 35, 71, 72, 83, 153, 154, 165, 177, 178
Logarithmic method, 36

Lozanovskii factorisation, 245, 246, 247, 248, 304, 364, 411
Lyapunov theorem, 246, 247, 267, 299, 307

M

majorizing sequence, 101, 104, 128, 129
Markov chain, 369, 371, 372, 380, 385, 389, 393, 394, 395, 396, 397, 398, 400, 401, 411
matrix representation, x, 211, 214, 215, 217, 219, 222, 229, 230, 231, 232, 233, 234, 235, 236, 237, 239, 243
maximum principle, 313
measurable function, 248, 273, 314, 317, 318, 321, 322, 337, 373, 380, 381, 382
memory, 71, 83, 84
metric space, x, 5, 303, 369, 370, 371, 372, 390, 392, 393, 394, 402, 403, 405, 406, 409, 411, 430
modified Newton's method, 113
monogenic function, 209, 210, 213, 221, 222, 228, 229, 232, 238
monogenic right–hand side, 210
multi–homogeneous inequality, 246, 247, 279, 286
multiobjective fractional programming, xi, 429, 430, 431, 453, 456

N

Newton, ix, xi, 35, 37, 39, 41, 43, 44, 45, 46, 47, 48, 49, 51, 52, 53, 55, 58, 59, 61, 63, 65, 67, 68, 69, 77, 81, 88, 93, 94, 95, 96, 113, 116, 121, 122, 123, 126, 135, 145, 148, 151, 160, 163, 164, 172, 175, 176, 183, 187, 188, 189, 190, 191, 193, 194, 195, 196, 197, 198, 199, 200, 201, 202, 203, 429, 431
Newton/finite projection method, 63, 121
nondominated, 437
nonlinear equation, ix, xi, 55, 69, 71, 73, 78, 79, 81, 82, 83, 84, 89, 90, 93, 94, 95, 96, 115, 122, 123, 135, 147, 151, 153, 155, 163, 164, 165, 167, 175, 176, 177, 178, 187, 188, 189, 190, 195, 196, 199, 415, 430, 453
norm, 2, 14, 24, 53, 67, 111, 162, 174, 186, 195, 206, 220, 221, 222, 228, 229, 243, 247, 251, 253, 254, 255, 256, 257, 258, 279, 281, 298, 299, 305, 314, 315, 317, 318, 320, 321, 322, 325, 329, 330, 335, 340, 341, 343, 345, 347, 348, 349, 352, 353, 371, 374, 375, 376, 378, 379, 380, 381, 382, 383, 384, 385, 388, 404, 405, 416, 418, 421, 423, 424, 425, 426
nuclear space, 421

O

open ball, 57
operational cost, 199, 200, 201
Opial, ix, 1, 4, 7, 10, 24, 34

order of convergence, 35, 39, 40, 43, 63, 68, 71, 78, 81, 89, 94, 105, 121, 122, 145, 148, 150, 154, 160, 163, 164, 172, 175, 184, 187, 199, 200, 201, 203
orthonormal basis, 211, 397, 398
Ostrowski, ix, 1, 3, 4, 6, 9, 11, 13, 15, 18, 20, 23, 25, 27, 29, 30, 31, 32, 33, 34, 81, 94

P

parametrization, 1, 13, 23
Pareto optimal, 437
parts formula, 2, 14, 24
Perverzev, 97, 112, 113
piecewise smooth, 1, 13, 23
Pisier inequality, 372
Poincare, ix, xi, 1, 415, 416
Polya, ix, 13, 15, 18, 19, 20
priori estimates, 39, 40
pseudo–holomorphic, 415, 428
pseudoinvex, 433, 435, 436, 437, 439, 442, 443, 446, 449, 452

Q

quadrangle condition, 315, 316, 335, 350
quasi–Banach space, 249, 250, 252, 309, 320, 337, 338, 351, 374, 375, 391
quasi–continuous, 39
quasiinvex, 433, 436, 437, 442, 443, 446, 449, 452
quasi–norm, 248, 279, 317, 318, 320, 373, 380
quaternion, 228, 243
quotient map, 348, 351, 376

R

Rademacher cotype, 295, 296, 314, 336, 341, 376, 388, 389, 400, 407
Rademacher function, 295, 336, 375
Rademacher type, x, 246, 247, 295, 296, 297, 300, 313, 315, 316, 326, 335, 336, 339, 341, 343, 375, 376, 380, 389, 396, 397, 399, 400, 401, 402, 407, 408
radii of convergence, ix, 57, 71, 73, 85, 116
radius of convergence, 52, 63, 72, 75, 84, 85, 132, 139, 147, 148, 149, 153, 154, 156, 157, 160, 165, 166, 168, 169, 172, 177, 178, 180, 183
range of convergence domain, 78, 89
real interpolation, x, 248, 369, 371, 390, 392, 403, 405
regularization method, 113, 415, 416, 422
residual error, 78, 79, 80, 89, 90, 91, 92, 93
restricted convergence region, 37, 125
reverse triangle inequality, 247, 259, 260, 261
right Caputo fractional derivative, 1

S

Schatten–von Neumann class, 246, 253, 254, 369, 378, 386
scheme, xi, 71, 78, 83, 85, 97, 154, 155, 166, 167, 179, 294, 369, 371, 372, 373, 391, 392, 400, 406, 430, 445
Schock, 97, 112
semi-embedding, x, 313
semiinfinite programming, xi, 429, 431
semi–local convergence, 35, 37, 38, 43, 47, 63, 72, 154, 166
separability, x, 309, 313, 316, 348, 349, 350
sequence space, x, 250, 251, 253, 305, 308, 313, 315, 319, 326, 335, 344, 351, 361, 374, 377, 378, 397, 400, 412
smooth, 1, 2, 3, 13, 14, 15, 23, 24, 258, 259, 261, 264, 267, 272, 273, 274, 276, 277, 280, 282, 283, 284, 293, 294, 295, 296, 298, 299, 300, 301, 302, 303, 304, 313, 315, 324, 372, 381, 383, 385, 387, 388, 390, 411
smoothness, x, xi, 245, 246, 247, 255, 256, 257, 258, 259, 261, 267, 275, 276, 277, 278, 279, 281, 282, 283, 284, 289, 290, 292, 293, 296, 300, 301, 303, 306, 307, 308, 309, 314, 321, 336, 343, 354, 364, 365, 366, 369, 370, 371, 372, 379, 385, 386, 387, 388, 389, 390, 395, 396, 407, 409, 411, 412, 413, 414
Sobolev, ix, x, 1, 4, 6, 10, 246, 251, 255, 257, 294, 304, 307, 308, 309, 313, 314, 315, 318, 321, 325, 344, 351, 364, 365, 367, 369, 370, 371, 377, 379, 383, 387, 392, 394, 399, 407, 410, 412
Sobolev generalised derivative, 257
Sobolev space, x, 246, 257, 304, 313, 314, 321, 325, 344, 379, 383, 392
space of vectors, 418, 422, 424
Steffensen–type method, 137
subgradient mapping, 245, 246
superreflexivity, x, 313, 316, 348, 380

T

test function, 79, 90
Traub method, 148
triangle inequality, 2, 14, 24, 246, 272, 279, 298, 306, 358, 393, 404

U

UMD, x, 253, 255, 257, 313, 316, 345, 346, 347, 348, 349, 380
unbounded, 45, 71, 72, 85, 116, 386, 416, 424, 426
uniform convexity, x, xi, 245, 246, 247, 255, 257, 258, 259, 263, 276, 277, 278, 281, 282, 284, 291, 292, 293, 296, 297, 309, 314, 336, 369, 371, 372,

379, 385, 386, 387, 388, 389, 390, 392, 407, 409, 414
uniqueness, 39, 48, 50, 52, 53, 59, 77, 85, 88, 120, 145, 149, 159, 171, 182, 210, 422
upper error estimate, 58

V

von Neumann algebra, 253, 254, 289, 378

W

Walsh chaos, 245, 246, 247, 364, 380, 385, 411
Walsh system, 246, 301, 302
wavelet, x, xi, 249, 250, 307, 313, 315, 319, 321, 351, 352, 353, 354, 355, 364, 372, 374, 411, 429, 431
w–conditioned, 39
Wirtinger, 14, 21

Related Nova Publications

MATHEMATICAL MODELING OF REAL WORLD PROBLEMS: INTERDISCIPLINARY STUDIES IN APPLIED MATHEMATICS

EDITORS: Zafer Aslan, Funda Dökmen, Enrico Feoli, and Abul H. Siddiqi

SERIES: Mathematics Research Developments

BOOK DESCRIPTION: Data mining provides avenues for proper understanding of real world problems. For researchers interested in data mining and new applications, this book is a multidisciplinary 'handbook' in data processes, engineering and medical applications.

HARDCOVER ISBN: 978-1-53616-267-7
RETAIL PRICE: $230

MATHEMATICAL MODELING FOR THE SOLUTION OF EQUATIONS AND SYSTEMS OF EQUATIONS WITH APPLICATIONS. VOLUME III

AUTHORS: Ioannis K. Argyros and Santhosh George

SERIES: Mathematics Research Developments

BOOK DESCRIPTION: These books contain a plethora of updated bibliography and provide comparison between various investigations made in recent years in the field of computational mathematics in the wide sense.

HARDCOVER ISBN: 978-1-53615-942-4
RETAIL PRICE: $270

To see a complete list of Nova publications, please visit our website at www.novapublishers.com

Related Nova Publications

NEXT GENERATION NEWTON-TYPE METHODS

AUTHOR: Ram U. Verma

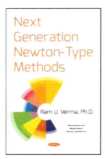

SERIES: Mathematics Research Developments

BOOK DESCRIPTION: This monograph is aimed at presenting "Next Generation Newton-Type Methods," which outperform most of the iterative methods and offer great research potential for new advanced research on iterative computational methods.

HARDCOVER ISBN: 978-1-53615-456-6
RETAIL PRICE: $160

NEW TRENDS IN FRACTIONAL PROGRAMMING

AUTHOR: Ram U. Verma

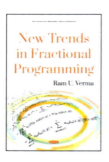

SERIES: Mathematics Research Developments

BOOK DESCRIPTION: This monograph presents smooth, unified, and generalized fractional programming problems, particularly advanced duality models for discrete min-max fractional programming.

HARDCOVER ISBN: 978-1-53615-371-2
RETAIL PRICE: $230

To see a complete list of Nova publications, please visit our website at www.novapublishers.com